The Seismogenic Zone of Subduction Thrust Faults

MARGINS Theoretical and Experimental Earth Science Series

MARGINS Theoretical and Experimental Earth Science Series

Continental margins are the Earth's principle loci for producing hydrocarbon and metal resources for earthquake, landslide, volcanic, and climatic hazards, and for the greatest population density. Despite the societal and economic importance of margins, many of the mechanical, fluid, chemical, and biological processes that shape them are poorly understood. Progress is hindered by the sheer scope of the problems and by the spatial-temporal scale and complexities of the processes.

The MARGINS Program (a research initiative supported by the U.S. National Science Foundation) seeks to understand the complex interplay of processes that govern continental margin evolution. The objective is to develop a self-consistent understanding of the processes that are fundamental to margin formation and evolution. The books in the MARGINS series investigate aspects of these active systems as a whole, viewing a margin not so much as a geological entity of divergent, translational, or convergent type, but more in terms of a complex physical, chemical, and biological system subject to a variety of influences.

The Seismogenic Zone of Subduction Thrust Faults

EDITED BY
Timothy H. Dixon and J. Casey Moore

Columbia University Press / New York

Columbia University Press and the authors wish to express their appreciation for assistance given by the National Science Foundation and the Margins office at Washington University in St. Louis in the publication of this book.

Columbia University Press
Publishers Since 1893
New York Chichester, West Sussex

Library of Congress Cataloging-in-Publication Data

The seismogenic zone of subduction thrust faults / edited by Timothy H. Dixon and J. Casey Moore.
p. cm. — (Margins theoretical and experimental earth science series)
Includes index.
ISBN 10: 0-231-13866-0 (cloth : alk. paper) / ISBN 13: 978-0-231-13866-6
ISBN 10: 0-231-51201-5 (e-book) / ISBN-13: 978-0-231-51201-5 (e-book)
1. Subduction zones. 2. Thrust faults (Geology). 3. Earthquake prediction. I. Dixon, Timothy H. II. Moore, J. Casey, 1945– III. Title.

QE511.46.S46 2007
551.22—dc22

2006101332

♾

Columbia University Press books are printed on permanent and durable acid-free paper.
Printed in China
c 10 9 8 7 6 5 4 3 2 1

Contents

Preface

The Seismogenic Zone of Subduction Thrust Faults originated from a MARGINS Theoretical Institute on "The Seismogenic Zone Revisited," held at Snowbird, Utah, March 16 to 21, 2003. More than 100 scientists from North, South, and Central America, Asia, and Europe attended this meeting. Theoretical Institutes strive to assemble active workers and young scientists, including graduate students, to review the subject matter, and foster interaction between the active fields of a discipline. A particular goal is to communicate the state of the art to a broader community, especially developing young professionals. The meeting itself involved spirited interaction and communication among participants. This volume attempts to capture some of that excitement and to also provide a broader perspective on the "Seismogenic Zone of Subduction Thrusts" to a larger audience. The volume evolved from manuscripts submitted by interested scientists after the meeting. Although not an exact image of the meeting, the papers provide representative snapshots and an excellent overview of the subject. We appreciate the efforts of the authors and reviewers.

Subduction zones, with their devastating earthquakes and tsunamis, are a principal type of plate boundary and commonly occur along continental margins. Continental margins are the focus of human activity, replete with natural resources and geologic hazards. The MARGINS Program, sponsored by the U.S. National Science Foundation (NSF), was developed to coordinate research activity along continental margins. The Seismogenic Zone Experiment or "SEIZE" is one initiative of the MARGINS Program and is responsible for organizing the Theoretical Institute that led to this volume. The MARGINS program sponsored the participation of U.S. scientists in the Theoretical Institute, while similar programs in other nations provided travel funds for non-U.S. scientists. We particularly thank Bilal Haq, the NSF MARGINS Program Director, for his support. The MARGINS Office (Julie Morris, Meredith Berwick, and Paul Wyler) tirelessly facilitated the assembly of this volume.

We thank the MARGINS Publications Series (Neal Driscoll, Gary Karner, Julie Morris, and Eli Silver) for their support. We are also indebted to the editorial staff at Columbia University Press for their work in final production of *The Seismogenic Zone of Subduction Thrust Faults*.

This volume is being published on the cusp of Integrated Ocean Drilling Program penetrations into active subduction zones in Japan and Central America. The drilling activity involves diverse related scientific investigations and will culminate in holes into the seismogenic zone of subduction thrusts, the first direct observation of these important features. We hope this volume will provide a partial foundation for this work, and look forward to significant advances in the study of subduction zones during the next decade.

PART I

Introduction

1

The Seismogenic Zone of Subduction Thrust Faults

INTRODUCTION

Timothy H. Dixon and J. Casey Moore

Subduction zones such as the circum-Pacific "ring of fire" are characterized by a dipping, relatively narrow (~100–200 km width) and shallow (<50 km depth) surface capable of generating large ($M > 7$) and great ($M > 8$) earthquakes. This surface represents the interface between subducting and overriding plates and is often termed the seismogenic zone (fig. 1.1). For reasons that are fundamental to the subduction process, including depression of isotherms and consequent broadening of that part of the plate interface where seismic processes can occur, these zones generate Earth's largest earthquakes and most destructive tsunamis. The earthquakes and subsequent tsunamis often impact densely populated coastal areas, causing large numbers of fatalities, as tragically demonstrated by the 26 December 2004 Sumatra earthquake and tsunami, believed to have killed more than 200,000 people.

Although we have a general understanding of the processes that occur in the seismogenic zone, many critical details remain obscure. For example, we know that earthquakes reflect the rapid release of strain associated with prior locking of the shallow plate interface and strain accumulation during interseismic periods lasting tens to hundreds of years. However, we do not understand why some interplate subduction earthquakes are relatively modest in size (M 7.5 or less), rupturing relatively small areas with limited along strike (trench-parallel) rupture length (<100 km), while others, such as the great ($M > 9$) 1960 Chile [*Plafker*, 1972], 1964 Alaska, and 2004 Sumatra events [*Lay et al.*, 2005; *Ammon et al.*, 2005; *Stein and Okal*, 2005], can rupture ~1200 km or more along strike. The effective rupture area for such events is huge, ~10^5 km^2. For comparison, the longest known rupture on the San Andreas fault, the 1906 San Francisco earthquake (400 km), had a rupture area of ~6×10^3 km^2.

Critical details of the strain accumulation process are also obscure. Why are some subduction zones fully locked, accumulating elastic strain at essentially the full plate convergence rate, while others appear to be only partially cou-

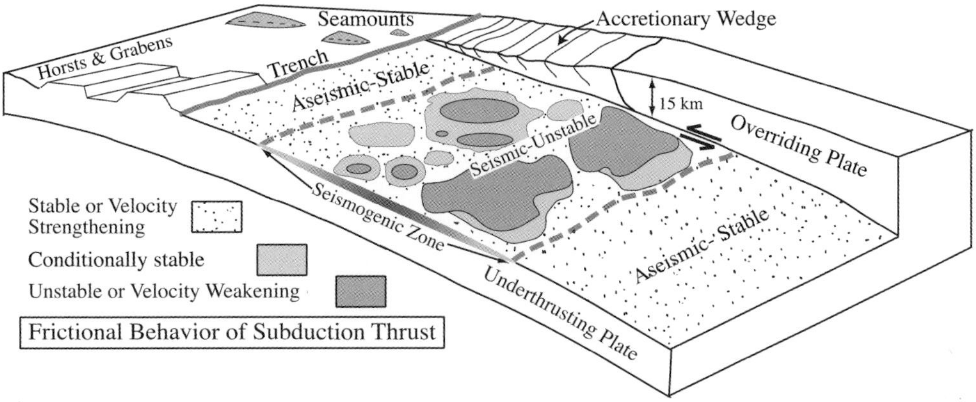

Figure 1.1 Perspective view of the seismogenic zone. The seismogenic zone represents the failure surface of large and great earthquakes and may be represented by aftershock distribution. The seismogenic zone is characterized mechanically by large areas or asperities that fail in a frictionally unstable or velocity-weakening manner. Failure by accelerating slip is the essence of an earthquake. Conversely, aseismic areas both updip and downdip of the seismogenic zone or within it may fail in a frictionally stable or velocity-strengthening manner. The absence of runaway slip in these areas precludes earthquakes or limits their size. Modified from *Bilek and Lay* [2002], *Lay and Schwartz* [2004], and *Lay and Bilek* (this volume).

pled or even freely slipping, accumulating little or no elastic strain? Are they locked all the way to the surface, or mainly locked at depth, perhaps in patches [*Wang and Dixon*, 2004; *Lay and Schwartz*, 2004]? Reasoning by analogy with the San Andreas fault in California, complete locking from seismogenic depths for this fault (~15 ± 5 km) all the way to the surface is common, but surface creep is also observed in places such as the Parkfield segment, implying that more complex patterns of locking and strain accumulation can occasionally occur. However, as pointed out by *Underwood* (this volume), sediments deposited on the downgoing plate of a subduction zone evolve as they are exposed to increasing pressure and temperature during the subduction process, leading to profound downdip changes in porosity, permeability, and pore-fluid pressure on and near the seismogenic plate interface. This means that the physical environment of subduction thrust faults is fundamentally different from transform and normal faults, perhaps limiting the utility of the San Andreas comparison and suggesting the possibility of complex spatial locking patterns (fig. 1.1).

Temporal variation is another important and poorly understood factor. Are patterns of strain accumulation on the seismogenic plate interface constant during a given earthquake cycle? If so, do these patterns translate into patterns of strain release during the next earthquake? During subsequent earthquakes? If patterns change, what are the controlling factors? Are variations in the frictional properties of subducted sediments critical or are variations in pore-fluid pressure and corresponding changes in effective normal stress more important?

What role do diagenetic and metamorphic reactions play in generating and modulating fluid flow and permeability along subduction thrust faults?

The MARGINS program of the U.S. National Science Foundation began in 1998 and includes the Seismogenic Zone Experiment (SEIZE) to study such questions. In March 2003, MARGINS hosted a 5-day workshop in Snowbird, Utah, to evaluate progress to date. The papers in this volume are based on talks presented at the workshop and subsequent discussion; they provide a comprehensive review of key seismogenic zone processes.

Defining the Seismogenic Zone

The seismogenic zone is that portion of the plate interface that produces earthquakes via stick-slip sliding [*Brace and Byerlee*, 1966]. For simplicity, in this volume we focus on interplate subduction fault zones capable of producing large ($M > 7$) and great ($M > 8$) earthquakes. Other regions, including intraplate fault zones such as the outer rise of subduction zones or upper crustal earthquakes within the overriding plate, are also interesting but generally are associated with smaller events.

In a simple "stick-slip" frictional mechanics model, stress and strain accumulate during the interseismic period ("stick"), subsequently failing in a catastrophic manner ("slip"). More sophisticated dynamical friction models are probably more relevant: that part of a fault zone undergoing stick-slip behavior is thought to be dominated by unstable or conditionally stable frictional slip [*Scholz*, 1998] (fig. 1.1). Frictionally unstable regions fail with velocity-weakening behavior. Once a small rupture initiates, slip acceleration occurs because frictional strength decreases as slip velocity increases (e.g., *Marone and Saffer*, this volume). In contrast, aseismic portions of the fault, usually considered to be shallower or deeper than the seismic portion, are dominated by stable frictional failure (aseismic creep). These regions are considered to be velocity strengthening: frictional strength increases as slip velocity increases. They are therefore incapable of runaway slip, and earthquakes cannot nucleate. New results show that aseismic portions of the fault can also occur embedded within portions of the fault generally considered seismic [e.g., *Norabuena et al.*, 2004].

The seismogenic zone as defined above corresponds roughly to a temperature range of ~100°–150°C at the updip end to ~350°C–450°C at the downdip end. The temperature limits of the seismogenic zone are established by comparing thermal models to the depth extent of earthquake ruptures and their aftershocks [*Currie et al.*, 2002; *Hyndman and Wang*, 1993; *Hyndman et al.*, 1995, 1997; *Oleskevich et al.*, 1999; *Wang et al.*, 1995] and hence have uncertainties associated with model limitations. This temperature range corresponds roughly to a depth range of 5–10 km (updip) to 40–50 km (downdip), but it is important to realize that significant inhomogeneities occur within this region in terms of frictional and seismic behavior.

The zone of seismic slip for a given major earthquake is often assumed to be approximately equivalent to the location of aftershocks (e.g., *Schwartz and DeShon*, this volume). The extent of earthquake dislocation on the fault surface may be estimated from seismic or tsunami waveform data and geodetic data measuring coseismic displacement [e.g., *Sataki*, 1993]. The location of the seismogenic zone may also be inferred from geodetic measurements during the interseismic period because the locked plate interface causes measurable strain accumulation at the surface [e.g., *Savage*, 1983; *Thatcher and Rundle*, 1984; *Dixon*, 1993; *Lundgren et al.*, 1999; *Norabuena et al.*, 1998, 2004]. It is usually assumed that the fault area that accumulates strain during the interseismic period is the same area that will subsequently rupture in the next large or great earthquake, although this remains to be fully tested. Microseismicity during the interseismic interval may only partially represent the extent of subsequent rupture, in part because this type of seismicity may actually represent slippage and partial or full strain release [*Norabuena et al.*, 2004; *Schwartz and DeShon*, this volume].

While the seismogenic zone is often referred to as being locked through most of the seismic cycle, locking may actually be both spatially and temporally heterogeneous (fig. 1.1). In terms of spatial complexity, seismic data have long suggested, and geodetic data have recently confirmed, that seismogenic zones may not be fully locked and slip at some fraction of the full plate rate during interseismic intervals [e.g., *McNally and Minster*, 1981; *Norabuena et al.*, 1998]. The mechanical basis for this behavior is not clear, but one model involves patches of frictionally unstable fault surface surrounded by conditionally stable and stable surfaces [*Bilek and Lay*, 2002; *Lay and Schwartz*, 2004]. The frictionally stable portions of the fault are subject to creep, while the frictionally unstable portions remain locked through most of the seismic cycle. Surface geodetic monitoring with low spatial density might "see" this fault surface as being "partially locked." Recent geodetic observations in Costa Rica, where the Nicoya peninsula enables surface observations quite close to the trench, have confirmed that strain accumulation in the seismogenic zone during the interseismic period can indeed be patchy [*Norabuena et al.*, 2004]. The locked, frictionally unstable portions of the fault surface presumably fail during the seismic intervals, and their displacement catches up with the surrounding creeping, frictionally stable portions. The frictionally unstable locked patches would thus constitute "asperities."

However, even this spatially complex picture of the seismogenic zone may be oversimplified. Observations of transient slip events on or near the seismogenic zone imply that locking and frictional properties can also vary in time throughout the seismic cycle. Such events have long been suspected on the basis of seismic and classical (terrestrial) geodetic data [e.g., *Sacks et al.*, 1978; *Linde and Silver*, 1989], but the development of space geodetic techniques such as GPS has provided a much better observational framework. Such events are now well known in Japan [e.g., *Kawasaki et al.*, 1995; *Heki et al.*, 1997;

Ozawa et al., 2002], Cascadia [*Dragert et al.*, 2001; *Miller et al.*, 2002], and Mexico [*Iglesias et al.*, 2004; *Larson et al.*, 2004] and are likely common in most subduction zones. As they become better studied, they will likely have important implications for our evolving picture of frictional conditions on the seismogenic plate interface.

Overview of Papers in This Volume

The papers in this volume document significant progress in understanding the issues discussed above since the initiation of the SEIZE program, but many important problems remain. *Hyndman* (this volume), who pioneered many of the topics discussed in the volume, sets the stage for our review by summarizing current knowledge and outstanding issues. His comprehensive review includes references to some of the key early work on seismogenic zone processes, providing historical context. The remainder of the volume is organized in five sections:

- The Incoming Plate
- Convergent Margin Structure, Fluids, and Subduction Thrust Evolution
- Laboratory Studies
- Seismic and Geodetic Studies
- Regional-Scale Deformation

The Incoming Plate

Knowledge of the nature of materials entering the subduction thrust fault is one of the advantages of studying seismogenesis at subduction zones. By looking at features of the incoming plate just outboard of the trench and assuming recently subducted lithosphere has the same or similar characteristics, it may be possible to draw conclusions about the seismogenic process and material properties in the fault zone through comparison of adjacent segments of subduction zones with contrasting characteristics. Three papers in this section discuss the nature of incoming sediments, the thermal state of the incoming lithosphere, and the effects of subducting topography on seismogenesis.

Most subducting seafloor is covered by a layer of sediment. *Underwood* (this volume) emphasizes that this sediment is essential in defining major hydrogeologic and geotechnical-mechanical aspects of the subduction system. Therefore it is essential to understand its lithologic architecture in order to predict its influence in the seismogenic zone. Underwood further points out that the simple model of sediments that coarsen and thicken upward as the plate approaches the trench is violated at most well-studied subduction zones. The specific documentation of the nature of incoming sediment will be essential at any locality for which deep drilling is planned to evaluate the seismogenic zone.

Hutnak et al. (this volume) discuss existing and new heat flow data and corresponding thermal models for the incoming Cocos plate offshore northern Costa Rica. They focus on the well-known but poorly understood contrast between relatively high heat flow characteristic of lithosphere created at the Cocos-Nazca spreading center (CNS) and anomalously low heat flow in the immediately adjacent East Pacific Rise (EPR) lithosphere. The strong heat flow contrast in oceanic crust off northern Costa Rica and the possible role of hydrothermal cooling were first identified by *Langseth and Silver* [1996]. New bathymetric, magnetic, seismic reflection, and heat flow data allow considerable refinement of this picture. These data illustrate the role of seamounts and other topographic features in guiding large fluxes of hydrothermal fluids. The new data support a model whereby EPR lithosphere is regionally cool because of efficient hydrothermal circulation in the uppermost basement.

The contrast in thermal state of the incoming lithosphere offshore Costa Rica provides a rare opportunity for a variety of comparative studies, because the northern half of the peninsula subducts cool EPR-generated lithosphere, while the southern half subducts warm CNS lithosphere. For example, *Newman et al.* [2002] note a change in the depth extent of interplate microseismicity, while *Norabuena et al.* [2004] and *Schwartz and DeShon* (this volume) point out correlated changes in the updip extent of locking on the plate interface as inferred from inversion of GPS geodetic data.

It has been known for some time that large subduction zone earthquakes exhibit considerable heterogeneity in slip and moment release across the fault plane, which may relate to strength contrasts (asperities) on the fault plane. *Bilek* (this volume) reviews the history of the asperity model and examines the role of bathymetric features such as seamounts in controlling regions of high versus low moment release in earthquakes.

Convergent Margin Structure, Fluids, and Subduction Thrust Evolution

Marine geophysical studies, drilling and associated laboratory results, and investigations of ancient accretionary prisms all provide a view of the nature of the subduction thrust, the overlying and underlying materials, and inferences on what controls seismogenesis. The papers noted below address these issues from the frontal part of the margin to greater depth as imaged by seismic reflection or revealed by exhumed rocks.

Bekins and Screaton (this volume) review the state of hydrogeologic knowledge for the Barbados accretionary complex, an intensively studied subduction-related sediment accretion zone. Data acquired from multiple drill holes and two long-term borehole observatories ("CORKs") allow observations of fluid flow and transient events. These data show that pore pressures in the upper few kilometers of the accretionary complex tend to be high, especially below major faults, and can exceed lithostatic values. Also, fluid flow tends to be channelized. Fluid budgets and models suggest that transient high-flow events are relatively rare and are spatially restricted to major faults and other channels.

Fluid pressure is a significant factor controlling fault strength and sliding behavior and hence is critical to understanding seismogenic zone behavior. *Saffer* (this volume) reviews fluid-pressure distribution and magnitude within underthrust sediments at the well-studied Nankai and Costa Rican margins. Techniques used to ascertain fluid pressure within underthrust sediments at modern convergent margins are discussed. A compilation of existing data is also presented. Comparison of the fluid-pressure observations to end member models affirms a system in which the décollement acts as a major fluid drain. Saffer shows that spatial patterns of underconsolidation suggest upward drainage to the décollement and outward toward the trench. Documentation of elevated fluid pressures within underthrust sediments indicates that the rate of geological forcing by burial exceeds the ability of sediments to drain, limited by low permeability and drainage path lengths of several hundred meters. The magnitude of elevated pore pressures is consistent with results from hydrologic modeling studies and suggests that sediment permeability is the primary control on fluid pressure. This implies that permeability ultimately exerts a strong control on the strength of subduction systems by modulating effective stress.

Morgan et al. (this volume) provide an extensive range of evidence showing that sediments incoming to the Nankai subduction zone along the Muroto transect undergo diagenetic changes that increase their strength. These authors argue that some of the porosity anomalies that might be interpreted as a result of overpressuring (*Saffer*, this volume) could also be explained by diagenetic cementation. They further suggest that the cementation and strengthening of the underthrust sediments protect it from deformation as it continues to be underthrust beneath the accretionary prism.

McIntosh et al. (this volume) present new seismic reflection data for the Nicaragua margin. They image the plate interface down to a depth of 40 km and are able to observe a number of subducted seamounts. They suggest that such seamounts played a critical role in promoting shallow rupture during the 1992 Nicaragua earthquake. This event was a tsunami earthquake, defined as an earthquake that generates a tsunami larger than expected given its seismic moment [*Kanamori*, 1972].

Moore et al. (this volume) summarize features of exhumed accretionary prisms that may explain their seismogenic behavior. These authors note that, in the presumed temperature ranges of seismic activity (*Hyndman*, this volume), quartz and carbonate veining localized along faults tend to introduce materials to fault surfaces that can increase their cohesion or alter their frictional properties. These changes may induce stick-slip behavior and fundamentally control the position of the seismogenic zone.

Laboratory Studies

As the lithologic composition and pressure-temperature conditions of subduction thrusts become better known, experimental studies of the frictional properties of key materials become important. In the past, much experimental work has focused on materials and conditions typifying continental plate boundaries, such as the San Andreas fault. Papers in this volume provide an overview of recent experimental efforts to understand the behavior of materials at a range of P-T-H_2O conditions relevant to subduction thrusts.

Moore and Lockner (this volume) review laboratory data on the smectite clay montmorillonite, a key mineral in many subduction thrust faults. This clay is stable to temperatures of ~150°C, whereupon transition to illite occurs. The authors point out that published literature on the frictional strength of montmorillonite gives coefficients of friction from 0.06 to 0.78, i.e., more than an order of magnitude variation, and that both velocity-weakening as well as velocity-strengthening behavior have been reported. They emphasize that some early experiments may not have controlled the water-saturation conditions very accurately and that variable water saturation probably accounts for most of the variation in frictional behavior.

Marone and Saffer (this volume) review some recent experiments on illite, smectite, and quartz. They demonstrate that the smectite-illite phase transition is unlikely to control the upper aseismic to seismic transition in subduction zones because illite is velocity strengthening and will not foster stick-slip behavior (Saffer and Marone, 2003). They note the tendency of unconsolidated materials including gouge to be velocity strengthening whereas lithified, especially quartz-rich rocks, tend to be velocity weakening, promoting stick-slip behavior. Overall they favor increasing consolidation/lithification, associated increase in effective stress, and quartz cementation as key processes controlling the upper aseismic-seismic transition.

Beeler (this volume) reviews laboratory results on faulting in weak materials, focusing on frictional strength, seismic coupling, dilatancy, and pore-fluid pressure. Over the last 4 decades, tectonophysicists have shown that, in general, it is not the strength of a fault that is the key to seismic slip but rather its stability or rate of change of strength with slip rate [e.g., *Scholz*, 1998]. Beeler carefully examines the basis of rate-dependent frictional parameters. He concludes that, in the absence of chemical weakening, the nominally stronger and more dilatant materials have a greater tendency toward velocity weakening and therefore unstable seismic slip compared to weaker materials. The reduction of contact area across asperities during dilation is the key to reducing friction. Accordingly, the instability that fosters accelerating slip is rooted in fault strength. Subduction thrusts initially develop in intrinsically weak, overpressured sediments (*Saffer*, this volume). Something must change at depth to spawn large and great earthquakes along these subduction faults.

Seismic and Geodetic Studies

Hasegawa et al. (this volume) examine rupture patterns for interplate earthquakes in northeast Japan, focusing on regions where successive earthquakes have occurred in the same location. They compare areas of maximum moment release (maximum slip) to the locked and slipping zones of the plate boundary during the corresponding interseismic period, as measured by inversion of geodetic data. They show that regions of maximum seismic slip in a given event tend to repeat in the next earthquake and also correspond to regions of maximum locking during the interseismic period. Thus interseismic geodetic data have significant value in hazard assessment, in the sense that regions of maximum locking will likely correspond to regions of subsequent high moment release.

Lay and Bilek (this volume) review seismic and tsunami data on moderate and large subduction zone earthquakes in the upper portion of the seismogenic zone. Numerous events with source depths <10 km tend to have anomalous faulting properties, typically involving either low rupture velocities or low stress drops. Tsunami earthquakes involve very large events that rupture the shallowest part of the seismogenic zone, with displacements extending all the way to the trench. Direct measurements for the 1992 Nicaragua tsunami event demonstrate that the rupture velocity was low, possibly because of low effective rigidity around the shallow fault zone. The variability of ruptures at shallow depths leads Lay and Bilek to infer that multiscale frictional heterogeneity, with patches of stable, conditionally stable and unstable friction, is a general attribute of shallow subduction zones.

Heki (this volume) summarizes key crustal deformation results from the dense array of continuous GPS stations in Japan, currently the best-instrumented subduction zone in the world. These results include an improved kinematic description of the major rigid blocks and bounding plates, secular subsidence in part of the fore-arc region, which may reflect subduction erosion of part of the upper plate, detailed coseismic rupture patterns on some major faults, and a variety of transient events. These include slow postseismic fault slip, which may be aseismic, slow-slip events independent of major earthquakes that may be aseismic or accompanied by microseismicity, and seasonal movements related to surface loading from groundwater and snow. Separating true transient deep crustal motion from station noise (e.g., from mismodeled orbits) and surface loading effects is important not only from a "signal-versus-noise" perspective but also because, in some cases, the transient surface loading or unloading may actually stimulate deeper transient processes.

Wang (this volume) reviews rheological models that may be combined with geodetic data to study subduction earthquakes and the earthquake cycle. He describes common elastic half-space and viscoelastic coupling models, the latter with an elastic layer overlying and coupled to one or more underlying viscoelastic layers, often implemented via finite element techniques. Wang points out that the way such models are implemented, including their kinematic boundary conditions, may influence results; hence caution is warranted. A case study of the Cascadia subduction zone is presented, illustrating some common trade-offs among model parameters.

Schwartz and Deshon (this volume) review seismic and geodetic data for the northern Costa Rica margin. The combination of unique geographic setting (the Nicoya peninsula is located quite close to the trench, essentially "perched" over the seismogenic zone) and a wealth of data from MARGINS-funded studies is leading to significant advances in our understanding of the behavior of seismogenic zones that subduct young oceanic lithosphere. The authors demonstrate that ongoing interseismic microseismicity is not spatially synonymous with the area of the seismogenic zone suggested by geodetic studies or failure in large earthquakes. They compare the thermal state of the incoming plate to the mechanical transitions inferred from seismic and geodetic data to investigate the role of sediment diagenesis and low-grade metamorphic reactions in controlling seismogenic behavior.

Regional-Scale Deformation

Most of the papers in this volume address the set of conditions and processes directly related to the earthquake cycle, a relatively short-term ($<10^3$ year) process. However, if the plate interface is highly coupled ("locked") during the interseismic part of the cycle, relatively high stresses may develop and propagate through the upper plate, which can act as a stress guide. Assuming this condition repeats through many earthquake cycles, geologically significant permanent strain and block rotation and translation may result. For example, oblique subduction may drive trench-parallel motion of fore-arc "sliver blocks" [*Fitch*, 1972; *Jarrard*, 1986]. In parts of Central America, coastal terrains migrate northwest at rates between 5 and 14 mm/yr [*DeMets*, 2001; *McCaffrey*, 2002; *LaFemina et al.*, 2002; *Norabuena et al.*, 2004]. Trench-normal compressive stresses related to a locked plate interface in southern Costa Rica may be responsible for out-of-sequence (nonplate boundary) thrust faulting and folding in both the fore-arc and back-arc regions of southern Costa Rica [*Norabuena et al.*, 2004]. If continued over millions of years, this can result in significant crustal shortening and thickening, i.e., Andean-style mountain building.

Seno (this volume) contrasts the subduction and collision process, and the connections between processes on the main subduction thrust fault and deformation in the back arc and hinterland. He emphasizes the importance of slab dehydration in controlling rheology and hence the subsequent response to compressive stresses, in particular in determining whether delamination of the lower crust or lithospheric mantle may occur.

Kley and Vietor (this volume) investigate the Cenozoic growth of the Andes. They focus on the interplay between factors controlling the direction and magnitude of horizontal compressive stress on the plate interface (e.g., global plate kinematics, slab dip, degree of mechanical coupling) and factors controlling the rheology of the upper plate.

Acknowledgments

We thank the authors of this volume for contributing comprehensive and insightful papers on the seismogenic zone and the reviewers for thoughtful and timely comments. We also thank NSF's MARGINS program, in particular Bilal Haq, for supporting the workshop that led to this volume. THD and JCM thank Thorne Lay and Susan Schwartz for comments on this paper. Finally, we thank the MARGINS Office (Julie Morris, Meredith Berwick, and Paul Wyer) and series editors (Neal Driscoll, Gary Karner, Julie Morris, and Eli Silver) as well as the staff of Columbia University Press for assistance with the publication of this volume.

References

Ammon, C. J., et al. (2005), Rupture process of the 2004 Sumatra-Andaman earthquake, *Science, 308*(5725), 1133–1139.

Beeler, N. M. (2007), Laboratory-observed faulting in intrinsically and apparently weak materials: Strength, seismic coupling, dilatancy and pore fluid pressure, this volume.

Bekins, B. A., and E. J. Screaton (2007), Pore pressure and fluid flow in the northern Barbados accretionary complex: A synthesis, this volume.

Bilek, S. L. (2007), Influence of subducting topography on earthquake rupture, this volume.

Bilek, S. L., and T. Lay (2002), Tsunami earthquakes possibly widespread manifestations of frictional conditional stability, *Geophys. Res. Lett., 29*(14), 1673, doi:10.1029/2002GL015215.

Brace, W., and J. D. Byerlee (1966), Stick-slip as a mechanism for earthquakes, *Science, 153*(3739), 990–992.

Currie, C. A., R. D. Hyndman, K. Wang, and V. Kostoglodov (2002), Thermal models of the Mexico subduction zone: Implications for the megathrust seismogenic zone, *J. Geophys. Res., 107*(B12), 2370, doi:10.1029/2001JB000886.

DeMets, C. (2001), A new estimate for present-day Cocos-Caribbean plate motion: Implications for slip along the Central American volcanic arc, *Geophys. Res. Lett., 28*, 4043–4046.

Dixon, T. (1993), GPS measurements of relative motion of the Cocos and Caribbean plates and strain accumulation across the Middle America Trench, *Geophys. Res. Lett., 20*, 2167–2170.

Dragert, H., K. Wang, and T. S. James (2001), A silent slip event on the deeper Cascadia subduction interface, *Science, 292*(5521), 1525–1528.

Fitch, T. J. (1972), Plate convergence, transcurrent faults, and internal deformation adjacent to Southeast Asia and western Pacific, *J. Geophys. Res., 23*, 4432–4460.

Hasegawa, A., N. Uchida., T. Igarashi., T. Matsuzawa, T. Okada, S. Miura, and Y. Suwa (2007), Asperities and quasi-static slip on the subducting plate boundary east off Tohoko, northeast Japan, this volume.

Heki, K. (2007), Secular, transient and seasonal crustal movements in Japan from a dense GPS array: Implication for plate dynamics in convergent boundaries, this volume.

Heki, K., S. Miyazaki, and H. Tsuji (1997), Silent fault slip following an interplate thrust earthquake at the Japan Trench, *Nature, 386*, 595–598.

Hutnak, M., A. T. Fisher, C. A. Stein, R. Harris, K. Wang, E. Silver, G. Spinelli, M. Pfender, H. Villinger, R. MacKnight, P. Costa Pisani, H. DeShon, and C. Diamente (2007), The thermal state of 18-24 Ma upper lithosphere subducting below the Nicoya Peninsula, northern Costa Rica margin, this volume.
Hyndman, R. D. (2007), The seismogenic zone of subduction thrust faults: What we know and what we don't know, this volume
Hyndman, R. D., and K. Wang (1993), Thermal constraints on the zone of major thrust earthquake failure: The Cascadia subduction zone, *J. Geophys. Res., 98*, 2039–2060.
Hyndman, R. D., K. Wang, and M. Yamano (1995), Thermal constraints on the seismogenic portion of the southwestern Japan subduction thrust, *J. Geophys. Res., 100*, 15,373–15,392.
Hyndman, R. D., M. Yamano, and D.A. Oleskevich (1997), Seismogenic zone of subduction thrust faults, *Island Arc, 6*, 244–260.
Iglesias, A., S. K. Singh, A. R. Lowry, M. Santoyo, V. Kostoglodov, K. M. Larson, and S. I. Franco-Sanchez (2004), The silent earthquake of 2002 in the Guerrero seismic gap, Mexico (*Mw*=7.6): Inversion of slip on the plate interface and some implications, *Geofis. Int. 43*(3), 1–9.
Jarrard, R. D. (1986), Terrane motion by strike-slip faulting of fore-arc slivers, *Geology, 14*(9), 780–783.
Kanamori, H. (1972), Tectonic implications of the 1944 Tonankai and 1946 Nankaido earthquakes, *Phys. Earth Planet. Inter., 5*, 129–139.
Kawasaki, I., Y. Asai, T. Tamura, N. Sagiya, Y. Mikami, M. Okada, M. Sakata, and M. Kasahara (1995), The 1992 Sanriku-Oki, Japan, ultra-slow earthquake, *J. Phys. Earth, 43*, 105–116.
Kley, J. and T. Vietor (2007), Subduction and mountain building in the Central Andes, this volume.
LaFemina, P., T. H. Dixon, and W. Strauch (2002), Bookshelf faulting in Nicaragua, *Geology, 30*(8), 751–754.
Langseth, M. G., and E. A. Silver (1996), The Nicoya convergent margin- A region of exceptionally low heat flow, *Geophys. Res. Lett., 23*, 891–894.
Larson, K. M., A. Lowry, V. Kostolodov, W. Hutton, O. Sanchez, K. Hudnut, and G. Suarez (2004), Crustal deformation measurements in Guerrero, Mexico, *J. Geophys. Res., 109*, B04409, doi:10.1029/2003JB002843
Lay, T., and S. Bilek (2007), Anomalous earthquake ruptures at shallow depths on subduction zone megathrusts, this volume.
Lay, T., and S. Y. Schwartz (2004), Comment on "Coupling semantics and science in earthquake research," *Eos Trans. AGU, 85*(36), 339–340.
Lay, T., et al. (2005), The great Sumatra-Andaman earthquake of 26 December 2004, *Science, 308*(5725), 1127–1133.
Linde, A., and P. Silver (1989), Elevation changes and the great 1960 Chilean earthquake: Support for aseismic slip, *Geophys. Res. Lett., 16*, 1305–1308.
Lundgren, P., M. Protti, A. Donnellan, M. Heflin, E. Hernandez, and D. Jefferson (1999), Seismic cycle and plate margin deformation in Costa Rica: GPS observations from 1994 to 1997, *J. Geophys. Res., 104*, 28,915–28,926.
Marone, C., and D. Saffer (2007), Fault friction and the upper transition from seismic to aseismic faulting, this volume.
McCaffrey, R. (2002), Crustal block rotations and plate coupling, in *Plate Boundary Zones, Geodyn. Ser.*, vol. 20, edited by S. Stein and J. Freymueller, pp. 101–122, AGU, Washington, D. C.
McIntosh, K. D., E. A. Silver, I. Ahmed, A. Berhorst, C. R. Ranero, R. K. Kelly, and E. R. Flueh (2007), The Nicaragua convergent margin: Seismic reflection imaging of the source of a tsunami earthquake, this volume.
McNally, K. C., and J. B. Minster (1981), Nonuniform seismic slip rates along the Middle America trench, *J. Geophys. Res., 86*, 4949–4959.

Miller, M., T. Melbourne, D. Johnson, and W. Sumner (2002), Periodic slow earthquakes from the Cascadia subduction zone, *Science, 295*(5564), 2423.

Moore, D. E., and D. A. Lockner (2007), Friction of the smectite clay montmorillonite: A review and interpretation of data, this volume.

Moore, J. C., C. Rowe, and F. Meneghini (2007), How accretionary prisms elucidate seismogenesis in subduction zones, this volume.

Morgan, J. K., E. B. Sunderland, and M. V. S. Ask (2007), Deformation and diagenesis at the Nankai subduction zone: Implications for sediment mechanics, décollement initiation and propagation, this volume.

Newman, A. V., S. Y. Schwartz, V. Gonzales, H. R. DeShon, J. M. Protti, and L. Dorman (2002), Along-strike variability in the seismogenic zone below Nicoya Peninsula, Costa Rica, *Geophys. Res. Lett., 29*(20), 1977, doi:10.1029/2002GL015409.

Norabuena, E., L. Leffler-Griffin, A. Mao, T. H. Dixon, S. Stein, I. S. Sacks, L. Ocola, and M. Ellis (1998), Space geodetic observations of Nazca-South America convergence across the central Andes, *Science, 279*(5349), 358–362.

Norabuena, E., et al. (2004), Geodetic and seismic constraints on some seismogenic zone processes in Costa Rica, *J. Geophys. Res., 109*, B11403, doi:10.1029/2003JB002931.

Oleskevich, D. A., R. D. Hyndman, and K. Wang (1999), The updip and downdip limits to great subduction earthquakes; thermal and structural models of Cascadia, south Alaska, SW Japan, and Chile, *J. Geophys. Res., 104*, 14,965–14,991.

Ozawa, S., M. Murakami, M. Kaidzu, T. Tada, T. Sagiya, Y. Hatanaka, H. Yarai, and T. Nishimura (2002), Detection and monitoring of ongoing aseismic slip in the Tokai region, central Japan, *Science, 298*(5595), 1009–1011.

Plafker, G. (1972), Alaskan earthquake of 1964 and Chilean earthquake of 1960: Implications for arc tectonics, *J. Geophys. Res., 77*, 901–925.

Sacks, I. S., S. Suyehiro, A. T. Linde, and J. A. Snoke (1978), Slow earthquakes and stress redistribution, *Nature, 275*(5681), 599–602.

Saffer, D. (2007), Pore pressure within underthrust sediments in subduction zones, this volume.

Saffer, D., and C. Marone (2003), Comparisons of smectite and illite-rich gouge, frictional properties: Applications to the updip limit of the seismogenic zone of subduction thrusts, *Earth Planet. Sci. Lett., 215*, 219–235.

Satake, K. (1993), Depth distribution of cosesmic slip along the Nankai trough, Japan, from joint inversion of geodetic and tsunami data, *J. Geophys. Res., 98*, 4553–4563.

Savage, J. C. (1983), A dislocation model of strain accumulation and release at a subduction zone, *J. Geophys. Res., 88*, 4984–4996.

Scholz, C. H. (1998), Earthquakes and friction laws, *Nature, 391*, 37–42.

Schwartz, S. Y., and H. R. DeShon (2007), Distinct up-dip limits to geodetic locking and microseismicity at the Northern Costa Rica seismogenic zone: Evidence for two mechanical transitions, this volume.

Seno, T. (2007), Collision versus subduction: The importance of slab dehydration, this volume.

Stein, S., and E. Okal (2005), Speed and size of the Sumatra earthquake, *Nature, 434*, 581–582.

Thatcher, W., and J. B. Rundle (1984), A viscoelastic coupling model for the cyclic deformation due to periodically repeated earthquakes at subduction zones, *J. Geophys. Res., 89*, 7631–7640.

Underwood, M. B. (2007), Sediment inputs to subduction zones: Why lithostratigraphy and clay mineralogy matter, this volume.

Wang, K. (2007), Elastic and viscoelastic models of crustal deformation in subduction earthquake cycles, this volume.

Wang, K., and T. H. Dixon (2004), Coupling semantics and science in earthquake research, *Eos Trans. AGU, 85*(18), 180.

Wang, K. H., R. D. Hyndman, and M. Yamano (1995), Thermal regime of the southwest Japan subduction zone, *Tectonophysics 248*, 53–69.

2 The Seismogenic Zone of Subduction Thrust Faults

WHAT WE KNOW AND DON'T KNOW

R. D. Hyndman

Abstract

There have been great advances recently in characterizing and understanding earthquakes on subduction thrust faults; this paper discusses some of the many questions that remain. Important seismic characteristics of subduction thrust faults and their physical associations include the following: (1) The maximum thrust earthquake magnitude, *Mx*, is highly variable among subduction zones; *Mx* may be related to the downdip seismogenic width, i.e., updip and downdip rupture limits, or to the physical characteristics and stress on the fault. (2) The term "seismic coupling," i.e., fraction of relative motion that is accommodated seismically, needs careful definition. Meaningful use of the term requires specification of the downdip seismogenic width. Some subduction zones appear to be completely locked, with no aseismic slip between megathrust events; others have mostly aseismic slip. (3) The term "seismic asperity" also needs careful definition; it commonly describes fault regions that had especially large slip in a great earthquake. However, inferences that such areas always have larger earthquake displacement and that they are associated with fault physical characteristics are not yet firmly established. (4) Subduction thrust faults are concluded to be weak. The commonly favored explanation is regionally elevated fluid pressures, but weak fault zone materials and dynamic rupture processes also have been proposed. (5) Most subduction thrusts have consistent updip and downdip seismogenic limits, i.e., an updip aseismic zone tens of kilometers wide commonly limited by a temperature of 100°–150°C. There is not yet agreement on the mechanism responsible. The downdip limit is frequently the intersection of the thrust with the fore-arc Moho, i.e., ~40 km for continent subduction, less for island arcs. However, deeper thrust events have been observed in some regions. For very hot subduction zones a critical seismogenic temperature limit of ~350°C is reached at a shallower depth.

(6) The reflection character of subduction thrust faults appears to change from a usually strong negative reflection in the updip aseismic zone, to a thin, sharp but weaker interface in the seismic portion, to a broad shear zone for the deeper aseismic zone. (7) Displacements on subduction thrust faults occur over a range of speeds, from earthquake rupture (seconds), to rates that generate tsunamis (minutes), to slower slip seen only in geodetic data. The speed controls are still unclear. (8) Immediately downdip of the seismogenic zone, slip on the aseismic zone in some areas occurs in slow slip events lasting a few weeks to months with intervals of a year to a few years. There are associated seismic tremors with no clear onset.

Introduction

Most of the world's great earthquakes ($M \geq 8$), many intermediate magnitude events, and most large tsunamis are generated by rupture on the "seismogenic zone" of subduction thrust faults (fig. 2.1). In this discussion, I outline some of important seismic characteristics of subduction thrust faults and their physical associations; what I think we know and what we don't know. Most of what we know has come from remote observation. There are no boreholes as yet into the seismogenic portion of subduction thrusts, although there has been drilling through the updip aseismic portion by the Ocean Drilling Program [e.g., *Moore et al.*, 2001, and reference therein]. There also has been limited drilling through active land faults, and there is important information from exhumed ancient subduction thrust faults [e.g., *Heermance et al.*, 2003; *Hashimoto et al.*, 2002, and references therein]. In the next few years we hope to have much

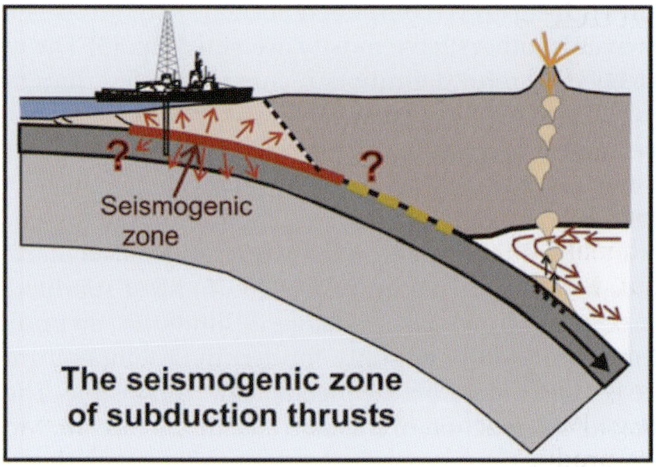

Figure 2.1 The seismogenic zone of subduction thrust faults. These generate most $M > 8$ earthquakes and large tsunamis globally.

new data from deep drilling by the Japanese research ship *CHIKYU*. Seafloor precision geodetic systems now being developed also should make a very important contribution. What we know about the behavior of subduction thrust seismogenic zones, and associations of that behavior with the physical composition and state of the thrust, comes mainly from (1) great and smaller thrust earthquakes, (2) land geodetic data, (3) ocean drilling and exhumed subduction thrusts, (4) seismic reflection and wide angle data, (5) thermal data and models, (6) electrical sounding and other geophysical data, and (7) stress indicators near the plate boundary.

The seismic behavior of subduction thrust faults is highly variable, both regionally and locally. Some subduction zones generate events with a maximum magnitude of *M* ~ 7; others have great earthquakes of magnitude over *M* 9. Some subduction thrusts are largely aseismic between infrequent great earthquakes and are inferred to be fully locked. Some have very frequent but only small to intermediate magnitude earthquakes. Some subduction zones have very large seismic moment release; others have very little. Below, I discuss some of the subduction thrust observations and their physical explanations, along with some of the more important questions about the seismic behavior of subduction faults. I do not discuss the seismological characteristics of great earthquakes in any detail; there are a number of excellent reviews of that subject [cf. *Kanamori,* 1986, 1983; *Nishenko,* 1991; *Pacheco and Sykes,* 1992; *Ruff,* 1996] and the nature of the great earthquake cycle and associated elastic and viscoelastic deformation [*Wang,* this volume]. I focus on the physical associations on the thrust with the characteristics and variations of great and smaller thrust earthquakes.

Subduction Thrust Earthquakes: Maximum Magnitude and "Seismic Coupling"

Some subduction zones generate thrust earthquakes of magnitude greater than *M* 9, whereas others have maximum earthquake magnitudes of *M* ~ 7 (fig. 2.2). The largest historical earthquakes, *M* ~ 9, have occurred in the subduction zones of southern Chile, Cascadia, southern Alaska, the Kuril trench, and north Sumatra [e.g., *Plafker,* 1969, 1972; *Abe and Kanamori,* 1980]. In contrast, only earthquakes of magnitude less than 7 or 7.5 have been recorded for most of the southwest Pacific island arcs; the smallest maximum, *Mx*, appears to be for most of the Mariana-Izu-Bonin subduction zones [e.g., *Pacheco and Sykes,* 1992; *Pacheco et al.,* 1993].

Since the magnitude of earthquakes increases systematically with the fault rupture area [e.g., *Wells and Coppersmith,* 1994], this difference means that subduction thrusts producing *Mx* < 7.5 earthquakes probably have seismic behavior only in small patches, at most a few tens of kilometers across, whereas *M* 9 earthquakes have seismic rupture that may be over areas of ~100 km

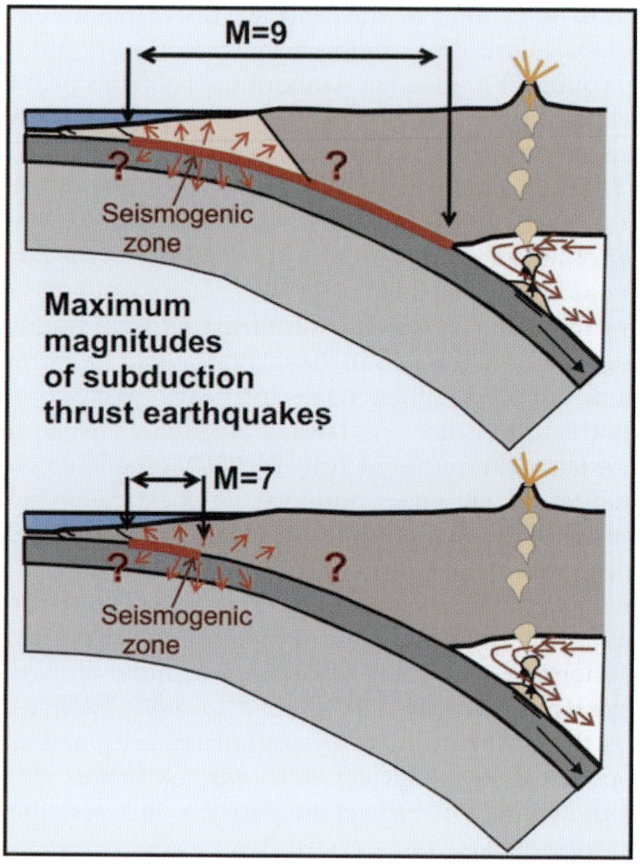

Figure 2.2 Some subduction thrusts produce *M* 9 earthquakes, others only less than *M* 7; what are the controls: maximum fault area; seismic coupling; seismic versus aseismic slip?

downdip and ~1000 km along strike, such as Alaska 1964 [e.g., *Plafker*, 1969, 1971], Chile 1960 [e.g., *Plafker and Savage*, 1970], and northern Sumatra 2004 events. The conclusion that subduction thrusts with only *Mx* ~ 7 earthquakes have only small patches that are seismic is supported by their generally small seismic moment release rate; their seismic efficiency ("seismic coupling," see below) is very small [e.g., *Pacheco et al.*, 1993]. Most of the plate convergence is inferred to be accommodated aseismically. Although they may be very infrequent, the large events represent a much greater seismic moment release rate than the many *M* < 7 earthquakes. The subduction zones that have exhibited great *M* ~ 9 earthquakes are found to have seismic efficiency close to 1 [e.g., *Pacheco et al.*, 1993]. Plate convergence is accommodated mainly seismically over a defined downdip width. For most of these great earthquake regions, land geodetic data also require almost complete thrust locking between earthquake events; Cascadia, southwest Japan, and Chile are examples [e.g., *Hyndman and Wang*, 1995; *Hyndman et al.*, 1995; *Brooks et al.*, 2003].

"Seismic coupling" is an expression that needs careful definition [*Wang and Dixon*, 2004]. It usually refers to the fraction of the plate convergence rate at a subduction zone that is accommodated in thrust earthquakes [e.g., *Ruff and Kanamori*, 1983; *Peterson and Seno*, 1984; *Jarrard*, 1986; *Pacheco et al.*, 1993; *McCaffrey*, 1997]. The remainder is inferred to be accommodated by some form of aseismic slip. In a kinematic description, a locked or fully coupled fault has no or low slip between great earthquakes. It is important to recognize that this definition does not involve any inferences of stress condition or fault properties [*Wang and Dixon*, 2004]. Partial coupling usually refers to the fraction of plate convergence that is accommodated seismically. The aseismic motion may occur in post seismic transient slip, in steady creep motion, or in slow slip events (discussed below). In estimating the seismic component of plate convergence, it is important to recognize that some component of plate convergence may be accommodated by crustal shortening in the back arc and possibly the fore arc [e.g., *Hindle et al.*, 2002; *Mazzotti and Hyndman*, 2002; *Norabuena et al.*, 2004]. Also, transient slip must be considered, including tsunami earthquakes and post seismic slip of great events that are not included in the seismic moment [*Wang*, this volume].

A critical parameter in the calculation of seismic coupling that often is not emphasized is the downdip seismogenic width. The calculated coupling is inversely proportional to the assumed seismogenic width. A less ambiguous expression may be "seismic efficiency," the fraction of relative convergence motion that is accommodated seismically over a defined downdip fault width. However, in the section below I will follow the common use of "seismic coupling." It probably is preferable to use the less ambiguous quantity, the average seismic moment release rate per unit length of subduction zone, because the downdip seismogenic width is often poorly known. It is also important to recognize that subduction zones with only infrequent great earthquakes have poor statistical sampling. There may have been no such events in the historical record.

To calculate the seismic coupling, the seismic slip from the average seismic moment release rate is compared to the convergence rate to give a seismic coupling efficiency "α" [e.g., *Pacheco et al.*, 1993, and references therein]. The coupling $\alpha = 1$ if all convergence is accommodated in earthquakes over the defined downdip seismogenic width. However, it is important to recognize that the calculation of α involves a number of poorly know parameters, especially the critical variable, the downdip width of the seismogenic zone "W." The computed seismic coupling is inversely proportional to the choice of W. The width W has often been fixed, taking the updip limit near the trench and the downdip limit at depth 40–50 km (i.e., width of ~100 km) based on a common maximum rupture depth of great earthquakes [e.g., *Pacheco et al.*, 1992; *Tichelaar and Ruff*, 1993]. In the discussion below I will use the expression "apparent seismic coupling" for use where the downdip seismogenic width is unconstrained and is assumed.

Three explanations have been proposed for the apparent variation in seismic coupling, one related to seismogenic fault area, one to stress state, and one to frictional conditions. The first is that the downdip width of the seismogenic or coupled zone is wide where there are great earthquakes and very narrow where there are only smaller thrust earthquakes [e.g., *Hyndman et al.*, 1997]. Within the defined downdip width *W*, the thrust then is fully or almost fully seismically coupled in both cases (seismic efficiency of 1). The rupture length of great earthquakes along the margin is commonly 2 to 3 times the downdip width [e.g., *Tichelaar and Ruff*, 1993] (there are important exceptions with very long ruptures along strike, such as Cascadia, southern Chile, and Sumatra) so the maximum magnitude decreases strongly with decreasing downdip width. The maximum downdip depth may be the most important variable (see below). This explanation is supported by a maximum earthquake depth of ~10 km for the subduction zones such as Mariana with $Mx \sim 7$, compared to 40–50 km for the subduction zones with great earthquakes such as southern Chile and Alaska, $Mx \sim 9$ [e.g., *Pacheco et al.*, 1993; *Zhang and Schwartz*, 1992; *Tichelaar and Ruff*, 1993]. If, for subduction zones like Mariana, the updip and downdip seismogenic limits approximately coincide, only a few patches will be seismic. This explains, at least in part, the very small seismic moment release rate and is one limiting case explanation. The other limiting case is that thrust earthquakes are distributed over a broad width to a depth limit of ~40 km but that only a few patches are seismic, producing only intermediate maximum magnitude earthquakes (see discussion of *Wang*, this volume). The updip and downdip seismogenic limits generally are nearly constant along strike for each subduction zone segment where there are no significant changes in plate age, plate convergence rate, and plate dip. The limited local variation in these limits suggests that they are related to thrust physical conditions or state common to significant subduction zone lengths along strike. These limits are discussed in the next section.

The second explanation for variations in apparent seismic coupling and for subduction zone maximum earthquake magnitude focuses on differences in the stress regimes, for example between southwest Pacific arcs and east Pacific continental subduction zones [e.g., *Scholz and Campos*, 1995]. The former are inferred to be in extensional regimes with the arcs receding from the trenches and subducting slabs, and the latter in compression with the continents overriding the subducting slab [*Hyndman*, 1972; *Uyeda and Kanamori*, 1979]. *Scholz and Campos* [1995] argue that extensional regimes have only small earthquakes; in contrast, compressional regimes have large megathrust earthquakes. However, the west Pacific subduction zones of Kuril, northeast Japan, and southwest Japan all have had great earthquakes. Also, the apparent seismic coupling factor is high in the southwest Japan subduction zone and lower in the northeastern Japan subduction zone [*Astiz et al.*, 1988], whereas the force interaction as reflected by upper plate stress is opposite [*Wang and Suyehiro*,

1999; *Wang and Dixon*, 2004]. For the areas with large earthquakes, subduction occurs beneath thick continental crust. A more consistent association seems to be that small maximum magnitudes are associated mainly with island arcs whereas great earthquakes are associated with subduction beneath continental crust (i.e., "Andean" type subduction). The latter association is at least partly explained above; the fore-arc crust in island arcs is thin so the fore-arc Moho and aseismic fore-arc mantle behavior are reached at a shallow depth.

The third explanation for variations in apparent seismic coupling is variations in fault properties, i.e., frictional state, or force required for rupture. An association that has not been well quantified is that very smooth incoming oceanic plates have infrequent but very large events (up to *M* 9), whereas rough incoming plates with seamounts, fracture zones, etc. that cause stress concentrations usually have smaller more frequent thrust earthquakes ($M < 8$) [e.g., *Bilek et al.*, 2003]. Thrust irregularities such as seamount chains and aseismic ridges may especially limit the along-strike lengths of subduction thrust earthquakes and their character [e.g., *Kodaira et al.*, 2002]. A related association is that great earthquakes commonly occur where there are large accretionary sedimentary prisms [e.g., *Ruff*, 1989]; the latter may smooth the subduction thrust such that there are few stress concentrations and strain can build up to very large ruptures. An example is the larger apparent coupling for the southwest Japan, Nankai trough, compared to the northeast Japan trench [e.g., *Astiz et al.*, 1988; *Pacheco et al.*, 1993]. Other examples of subduction zones with great thrust earthquakes and large accretionary prisms are southeastern Alaska, Cascadia, and southern Chile [e.g., *Oleskevich et al.*, 1999]. However, large accretionary prisms only occur for continental subduction zones where there is a large source of sediment, so the main association of large earthquakes may be with subduction beneath continents rather than with accretionary prisms.

Subduction Thrust Asperities

"Seismic asperity" is another expression that is used in a number of different ways and needs careful definition (see also *Wang* [this volume]). The simple and unambiguous use is for patches or regions within the overall area of rupture in a specific great earthquake that has especially large displacement [e.g., *Lay and Kanamori*, 1981; *Lay et al.*, 1982]. The earthquake slip distribution may be mapped through seismic waveform modeling, and the updip region in a few cases by tsunami modeling [e.g., *Johnson*, 1999; *Toshitaka et al.*, 2002]. Confusion comes with the extension of this definition to associations with spatial variations in the physical properties of the thrust interface (fig. 2.3). Asperities are often inferred to be "stronger" than the surrounding region of the thrust, and therefore they accommodate most of the plate convergence seismically, whereas adjacent areas have more aseismic slip (cf. reviews by *Scholz* [1990] and *Ruff* [1992]). Although a number of clear associations have been suggested between the patterns of subduction thrust seismic behavior and physical

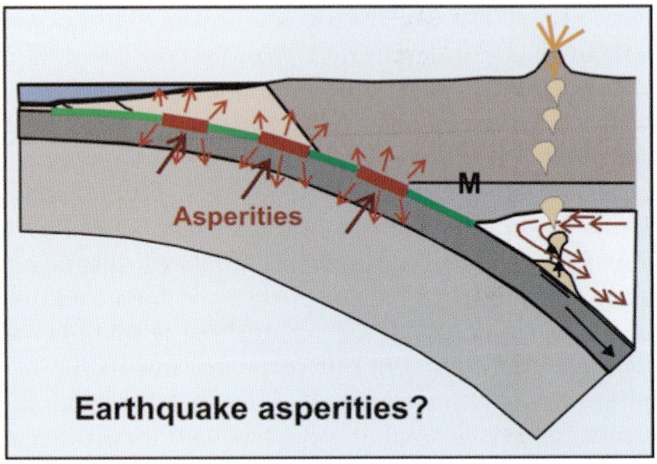

Figure 2.3 What is the nature of asperities that are inferred to be "stronger" patches on the thrust?

features on the incoming plate such as seamounts, aseismic ridges, and fracture zones [e.g., *Bilek et al.*, 2003; *Kodaira et al.*, 2002], only few associations of local variations in rupture displacement in great earthquakes with localized physical features on thrusts have been conclusively established [e.g., *Igarashi et al.*, 2003; *Zweck et al.*, 2002].

If there is a physical connection with areas of large rupture, we expect that there will be more slip in these asperity regions for all great earthquake ruptures that include them and that adjacent areas will always have more aseismic slip. This behavior is opposite to the concept of "seismic gaps" in which areas of little or no recent rupture displacement are expected to have greater rupture in future events [*McCann et al.*, 1979; *Wyss and Wiemer,* 1999, and references therein]. In the latter case, the long-term average earthquake slip may be nearly constant along a subduction zone. In one model, there may be fully locked patches surrounded by regions that are freely slipping. A number of repetitions of great earthquakes in the same region are required to establish which view is more correct. There have been only a few reported repeat ruptures of the same area by historical great earthquakes [e.g., *Wyss and Wiemer,* 1999; *Schwartz*, 1999], and the first of the two repeated events is usually too old for the high-quality seismic data required to model the rupture distribution. New geodetic data may allow connection between areas on the thrust that are locked and large earthquake rupture asperities.

Weak Subduction Thrust Faults

It has been recognized for many years that shallow angle thrust faults must be weak to accommodate large subhorizontal displacements [*Hubbert and Rubey,*

1959; *Raleigh and Evernden*, 1981; *Davis et al.*, 1983]. Other arguments based on regional stress and force balance [e.g., *Wang and He*, 1999], and on the lack of thermal anomalies due to frictional heating, have been developed subsequently [e.g., *Wang et al.*, 1995, and references therein] (fig. 2.4). On some margins the maximum principal stress is margin parallel rather than in the direction of convergence [e.g., *Wang et al.*, 1995]. Margin-normal stress on several margins is determined to be less than or equal to the vertical stress [e.g., *Wang and Suyehiro*, 1999; *Wang et al.*, 1995]. The conclusion of weak large faults also applies to continental strike-slip faults [e.g., *Zoback et al.*, 1987; *Hickman*, 1991; *Lachenbruch and McGarr*, 1990]. The arguments for weak strike-slip faults comes first from the maximum horizontal principal stress approaching orthogonal (65°–85°) to the fault as the fault is approached, suggesting that the fault is moving at very low levels of shear stress. Second, weak strike-slip faults are concluded from the lack of the thermal anomaly expected for motion on strong faults. The inferred strength of large faults is <20 MPa, similar to the stress relieved in large earthquakes (stress drop). In contrast, the overall strength of the brittle upper lithosphere is estimated to be much greater, >50–100 MPa, as predicted by friction laws [e.g., *Hickman*, 1991; *Zoback and Townend*, 2002].

The reasons for the weakness of large faults have been much discussed. The three main possibilities are (1) elevated fluid pressures [e.g., *Magee and Zoback*, 1993]; (2) low coefficients of friction due to the fault zone material, i.e., clay, serpentinite, silica gel [e.g., *Vrolijk*, 1990; *Toro et al.*, 2004]; and (3) dynamic weakening, i.e., slip shear heating, propagation of dilation waves along the fault, and fluidization of fault-zone material [e.g., *Brune*, 1993; *Shi et al.*, 1998; *Brodsky and Kanamori*, 2001; *Ma et al.*, 2003; *Melosh*, 1996; *Lachenbruch*, 1980]. The elevated fluid-pressure explanation appears favored by most authors [see *Davis et al.*,

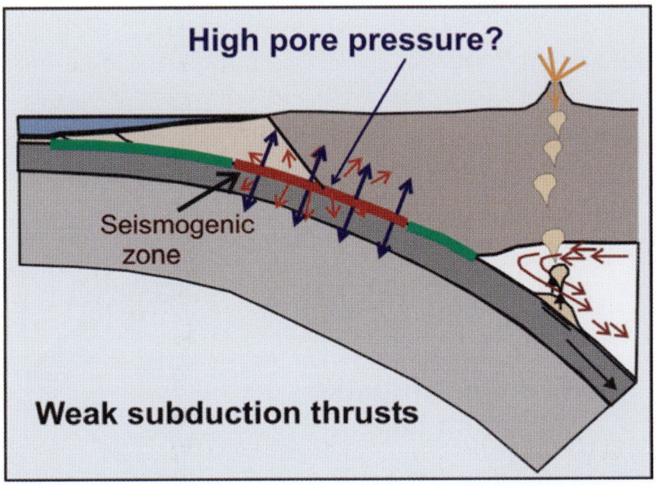

Figure 2.4 Subduction thrust faults are very weak based on regional stress, earthquake, and thermal data.

1983], especially for subduction thrust faults, following the well-established acceptance of this explanation for the low angle of thrust faults in sedimentary fold and thrust belts [*Hubbert and Rubey*, 1959]. For subduction thrusts, there is an ample supply of fluid from the consolidation of underthrust material and from dehydration reactions in the downgoing slab. The fluid supply for continental transcurrent faults is less obvious, but some suggestions have been made [e.g., *Kirby et al.*, 2002]. Secure confirmation awaits drilling of the seismogenic portion of subduction thrust faults by the International Ocean Drilling Program (IODP) drill ship *CHIKYU* and of active continental faults such as the San Andreas fault by the San Andreas Fault Observatory at Depth (SAFOD) project.

Updip Limit of Subduction Thrust Earthquakes

Both great subduction earthquakes and smaller thrust events usually do not extend to the trench (fig. 2.5); there is an updip aseismic zone commonly tens of kilometers wide (fig. 2.4) [e.g., *Byrne et al.*, 1988; *Byrne and Fisher*, 1990]. This updip limit, or upper stability transition depth, is defined by (see summary by *Oleskevich et al.* [1999]) (1) the updip rupture limit in great earthquakes, as determined by waveform modeling; (2) the updip limit from modeling of the tsunamis generated by great earthquakes; (3) the updip limit of great earthquake aftershocks; and (4) the updip limit of small thrust earthquakes on the subduc-

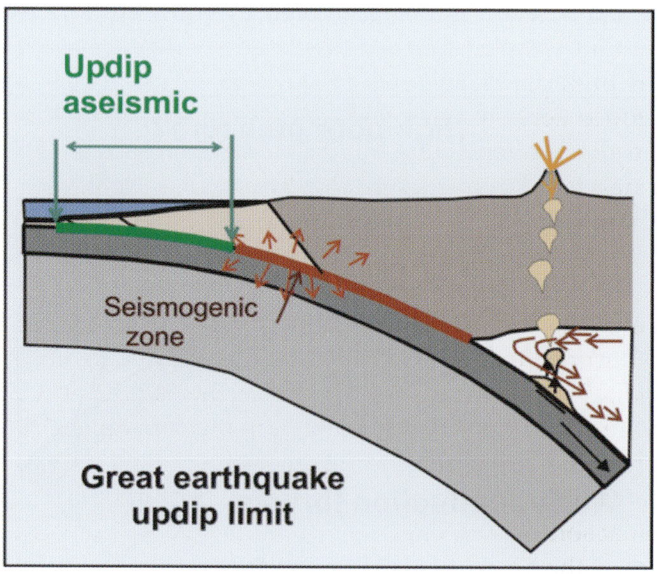

Figure 2.5 Great earthquakes usually do not rupture to the trench; there is an updip aseismic zone. This limit is important for tsunami generation.

tion thrust between great events. In several subduction zones the coast is close enough for land geodetic data to provide some constraint to the updip limit of the locked zones, although the resolution is low [e.g., *Lundgren et al.*, 1999; *Norabuena et al.*, 2004]. Seafloor geodetic measurements that are in progress by United States and Japanese groups should soon give additional information [e.g., *Spiess et al.*, 1998]. All of these definitions are based on different measures and sometimes give different locations [e.g., *Norabuena et al.*, 2004], but usually they are consistent within the common horizontal resolution of 10–20 km [e.g., *Oleskevich et al.*, 1999]. Correlation of seismic properties with composition and state changes is very dependent on how well this updip rupture or seismogenic limit can be defined. The accuracy varies greatly for different subduction zones, with horizontal definition relative to the trench ranging from about ±5 km for well-studied subduction zones with recent great earthquakes to only very generally (±30 km) for others. Where the limit has been determined, it ranges from ~30 km from the trench for the Nankai, southwest Japan subduction zone [e.g., *Ando*, 1975; *Toshitaka et al.*, 2002; *Obana et al.*, 2001], which is young and hot, to ~100 km for the cool shallow dipping southeast Alaska subduction zone [e.g., *Algermissen et al.*, 1969; *Johnson et al.*, 1996].

Because tsunami wave analyses show that the updip aseismic zone usually does not contribute significantly to tsunami generation, the motion in this region must be slower than over the few tens of minutes of the tsunami period. There is inadequate seafloor geodetic data as yet to determine if the updip aseismic zone moves mainly as a transient in the hours to days following great events or there is some continuous motion between great events. *Wang* (this volume) has pointed out that if the adjacent deeper part of the thrust is locked, there is no force to drive motion on the updip portion. The nature and timing of motion on the updip portion of the subduction thrust is an important application of high-precision seafloor geodetic data for the future. In a few events, this updip zone appears to rupture independently of the deeper seismic zone. The slip is over a time interval of a few minutes to tens of minutes such as to generate a large tsunami but to radiate little seismic energy, i.e., a tsunami earthquake. No clear special physical characteristics have as yet been identified for subduction zones having tsunami earthquakes, but they are too infrequent to as yet allow ready correlations [e.g., *Okal and Newman*, 2001].

The first explanation for the updip limit that was suggested is the reasonable one: the part of the thrust in contact with accreted sediments is aseismic [e.g., *Byrne et al.*, 1988]. Seismic behavior should then start landward of the thrust contact with the crystalline crust of the overlying fore arc, i.e., the "backstop." At least for smaller prisms, the sediments are generally quite unconsolidated and should have little strength to support elastic strain buildup. However, it is now clear that at some continental subduction zones, great earthquake rupture occurs at or near the base of large accretionary sedimentary prisms. At a few subduction zones that have great earthquakes, most if not all of the seismic rupture zone underlies accreted sedimentary material, i.e., Cascadia, southern

Alaska, and southwest Japan [e.g., *Oleskevich et al.*, 1999]. For other subduction zones where there also are great earthquakes, there is little accreted sediment and substantial undercutting erosion of the fore-arc crust is inferred, i.e., most of South America and northeast Japan [e.g., *von-Huene and Scholl*, 1991]. Thus the updip limit does not appear to be constrained primarily by downdip changes in the bulk physical composition of the overlying fore arc. Variations in the material within the fault zone décollement, however, may be important [e.g., *Cloos and Shreve*, 1996].

Where there are large accretionary prisms, there is the complication that part of the plate convergence is taken up by shortening of the seaward portion of the prism rather than by slip on the main subduction thrust. The shortening may be on splay thrusts ("out of sequence thrusts") in folding and in bulk shortening. Some earthquake ruptures may rise to the seafloor on faults landward of the trench, giving an updip limit in earthquake and tsunami modeling of that event (e.g., *Park et al.* [2000] for southwest Japan and *Plafker* [1972] for southern Alaska). However, most of the long-term thrust motion, including aseismic slip, must extend to near the trench because of the geometric constraint that most of the sediment thickening must occur near the toe of the sedimentary prism.

Two other associations of the updip limit that have been suggested are depth, i.e., pressure-controlled consolidation and temperature. Depth and temperature are roughly correlated for many subduction zones, but as discussed below, there are substantial variations in vertical temperature gradients due to incoming plate age, thickness of insulating sediments on the incoming plate, plate convergence rate, etc. [e.g., *Oleskevich et al.*, 1999]. Thermal models for subduction thrust temperatures are discussed below. For most subduction zones the updip limit for great earthquake rupture appears to correlate best with thrust temperature, with a critical temperature of 100°–150°C [e.g., *Vrolijk*, 1990; *Hyndman et al.*, 1997] rather than pressure (depth). This conclusion is supported by a recent study to define the updip seismogenic limit and along-strike variations in temperature for the Costa Rica subduction zone by *Harris and Wang* [2002], *Spinelli and Saffer* [2004], and *Norabuena et al.* [2004]. They found a good correlation between variations in the thermal state of the incoming oceanic plate and the depth to the updip seismogenic limit defined by microearthquakes. There is a sharp boundary in the incoming oceanic plate heat flow along the margin that is well correlated with a change in the distance from the trench and the depth of the microearthquake updip limit. Along strike across the boundary, there is a significant change in the seismicity updip limit, but the estimated temperature at the limit on both sides of the boundary is 100°–150°C.

The physical explanation for a temperature-controlled updip limit is still much debated [e.g., *Moore and Saffer*, 2001; *Saffer and Marone*, 2003]. The earlier favored explanation was temperature-dependent clay dehydration [*Vrolijk*, 1990; *Hyndman and Wang*, 1993] following a proposal that this mechanism was responsible for the lack of shallow earthquakes in land faults [e.g., *Marone and Scholz*, 1988]. Laboratory data show that the transition from smectite to illite/chlorite occurs in the appropriate temperature range of 100°–150°C [e.g., *Moore and Vrolijk*, 1992; *Chamley*, 1989]. Also, laboratory data show smectite clays to be very weak [e.g., *Wang and Mao*, 1979]. The slip behavior of smectite and illite/chorite appear to be complex, but recent laboratory studies of their sliding behavior do not confirm the transition from stable-sliding velocity strengthening to stick-slip velocity weakening behavior [*Saffer and Marone*, 2003]. Other temperature-dependent mechanisms are therefore being examined. This limit is important for tsunami generation and thus the hazard at inland cities. Possible mechanisms include silica and carbonate diagenesis and consolidation changes of permeability that control pore-fluid pressure.

Downdip Limit of Subduction Thrust Earthquakes

The downdip limit, or lower stability transition depth, on subduction thrust faults (fig. 2.6) is important for the closest approach of the seismic source and thus the hazard at inland cities. This limit is defined by (1) the downdip rupture limit of great earthquakes as determined by waveform modeling, (2) the downdip limit of great earthquake aftershocks, (3) small thrust earthquakes occurring between great events, (4) land geodetic data that define the rupture downdip limit, and (5) geodetic data that provide constraint to the downdip limit of

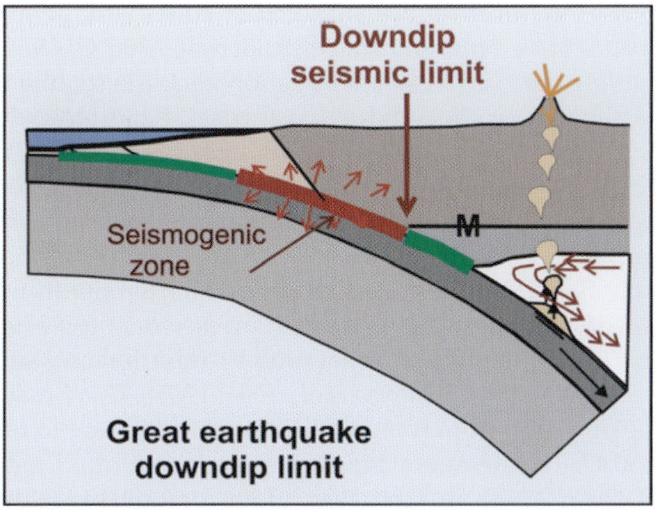

Figure 2.6 Great earthquakes have a variable maximum depth of rupture ~10–50 km, with deeper aseismic motion. This limit is important for closest approach and thus seismic hazard at inland cities.

the interseismic locked zone. Correlation of earthquake characteristics with downdip composition and state changes is very dependent on how well this limit can be defined. Moreover, there appears to be a transition zone of substantial width between fully seismic and fully aseismic zones [e.g., *Wang et al.*, 2003]. The accuracy of the downdip limit varies greatly for different subduction zones, depths resolved to approximately ±5 km (10–20 km in horizontal distance) for well-studied margins, to ±10 or 20 km (~50 km horizontal) for others (see summary by *Oleskevich et al.* [1999]). The downdip limits range from ~10 km for the Mariana subduction zone to ~50 km for some continental subduction zones. The consistent downdip depth limit along the length of a number of subduction zones gives encouragement that the limit is controlled by some physical composition or state change downdip that varies regionally rather than random variations in stress or state. A good example is the South America subduction zone where the maximum depth of rupture in great earthquakes and of the interseismic locked zone is consistently between 40 and 50 km [*Tichelaar and Ruff*, 1991; *Oleskevich et al.*, 1999]. Two downdip state and composition changes have been suggested for the limit. The downdip seismic limit for most, but not all, subduction zones appears to agree with either a maximum temperature of 350°C or the thrust intersection with the fore-arc serpentinized mantle [e.g., *Hyndman et al.*, 1997]. Further study is needed to determine if the apparent exceptions are due to inadequate knowledge, for example, of the depth of the fore-arc mantle intersection, or if other processes control the limit in these cases [e.g., *Seno*, 2005].

Temperature Limit

The downdip rupture limit for hot subduction zones subducting young oceanic crust appears to be temperature controlled at ~350°C. Detailed numerical thermal models specific to each subduction zone are required for accurate temperature estimates on the subduction thrusts. Examples where the temperature limit appears to apply are Cascadia [*Savage et al.*, 1991; *Hyndman and Wang*, 1993, 1995], Mexico [*Currie et al.*, 2002], southern Chile [*Tichelaar and Ruff*, 1993; *Oleskevich et al.*, 1999], and southwest Japan [*Hyndman et al.*, 1995]. The maximum temperature of 350°C agrees with the seismic temperature limits estimated for continental strike-slip fault zones. This temperature corresponds well to the transition from velocity-weakening (seismic) to velocity-strengthening (aseismic) behavior based on laboratory data for crustal composition rocks [e.g., *Tse and Rice*, 1986; *Blanpied et al.*, 1991, 1995]. There may be a tapered transition zone downdip from full great earthquake rupture to no motion that extends to ~450°C. That temperature corresponds to a rapid increase in fault instantaneous shear stress in laboratory data for quartzo-feldspathic rocks [*Tse and Rice*, 1986]. At higher temperatures the shear stress increases rapidly with increasing shear velocity, so rapid seismic slip should not occur. It appears that rupture does not initiate in but may extend into the transition zone. The temperature of 450°C approximately corresponds to the "brittle-ductile" transition; at higher temperatures, shear may no longer be sharply localized. For subduction zones with great earthquakes, during the interseismic period, there may be a transition from no motion on the thrust to nearly steady creep over the depth range from 350° to 450°C.

Hydrated Fore-Arc Mantle Limit

In many areas the maximum seismogenic depth from the thermal limit is at great depths, much greater than those observed. In these cases the maximum depth usually appears to be close to the intersection of the subduction thrust with the fore-arc Moho [e.g., *Ruff and Tichelaar*, 1996; *Hyndman et al.*, 1997; *Oleskevich et al.*, 1999; *ANCORP Working Group*, 2003]. Accurate verification of the fore-arc mantle limit is subject to the problem of defining the depth of the intersection of the thrust with the fore-arc Moho. First, the fore-arc Moho itself is usually difficult to define because the Moho velocity contrast is small. This is interpreted to be because the *P* wave velocity of the fore-arc mantle is reduced by serpentinization to velocities similar to the crust [e.g., *Christensen*, 2004; *Hyndman and Peacock*, 2003, and references therein]. Second, there is evidence in a few places of the fore-arc Moho turning upward or turning downward close to its intersection with the subduction thrust fault. Therefore, even a well-determined regional depth to the fore-arc Moho does not provide assurance of defining the intersection accurately. Receiver function studies giving *S*-wave structure are providing much improved constraints across the thrust-Moho intersection [e.g., *Bostock et al.*, 2002], and these structure questions should soon be better resolved. Another problem is the potential of sediment or of fore-arc crustal material being carried downward in a channel along the décollement. This process could result in the thrust contact being within felsic crustal composition material (i.e., seismic behavior) in places well below the fore-arc Moho contact.

An explanation for the fore-arc Moho seismic limit may be that the fore-arc mantle is serpentinized, providing an aseismic thrust plane [*Hyndman et al.*, 1997; *Peacock and Hyndman*, 1999]. There is considerable data supporting the hypothesis that the fore-arc mantle contains substantial serpentinite [e.g., *Hyndman and Peacock*, 2003, and references therein]. The laboratory data, however, do not yet give a clear story of its seismic/aseismic behavior [*Moore et al.*, 1997; *Reinen et al.*, 1991], but the weak layered structure of serpentinite is such that seismic behavior is not expected. An additional factor that may make the thrust below the fore-arc Moho aseismic is that it may contain significant amounts of talc owing to rising silica-rich fluids from the dehydrating underlying slab [*Peacock and Hyndman*, 1999]. Talc has a very weak layered mineral structure, and it is unlikely to behave seismically.

Seismic and Aseismic Reflection Character of the Subduction Thrust

The subduction thrust exhibits substantial variations in multichannel seismic reflection character downdip that may allow mapping the updip and downdip seismogenic limits. Good-quality deep multichannel seismic reflection is available for only a few subduction zones, and the change from seismogenic to deeper aseismic behavior often occurs near the coast so reflection data must be combined. However, there is evidence for changes in thrust reflection character at both the updip and downdip seismogenic limits (fig. 2.7). Ocean Drilling Program (ODP) has shown that the thrust in the updip aseismic region is quite thin and, at least in sedimentary prisms, has high fluid pressure [e.g., *Brown et al.*, 2003]. The multichannel reflection character in this updip region is variable from inferred positive to negative impedance contrasts that may relate to lateral pore-pressure variations [e.g., *Shipley et al.*, 1992, 1994]. A clear change in décollement reflection amplitude has been observed for the southwest Japan subduction thrust from strong negative polarity to very weak at 30–40 km from the deformation front; this is close to the updip limit from other constraints [*Bangs et al.*, 2004]. This is also the position of an increase in accretionary wedge taper, indicating an increase in fault frictional strength.

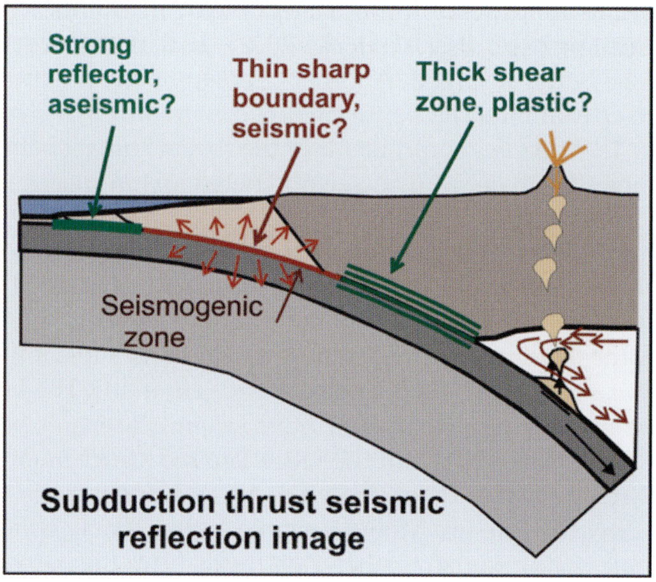

Figure 2.7 The subduction thrust seismic reflection image appears to be sharp where seismic and a broad band deeper where aseismic.

Bangs et al. [2004] interpret these changes as due to loss of fluids and loss of excess fluid-pressure downdip.

Within the locked seismogenic zone, the multichannel seismic reflection images generally show a thin although weak seismic reflector and inferred sharp thrust contact. At greater depth beneath the downdip seismogenic limit, where resolved, the multichannel reflection images in several areas show a ~5 km thick reflective band that may be a shear zone [*Nedimovic et al.*, 2003; *ANCORP Working Group*, 2003] above the downgoing plate. This layer may have high Poisson's ratio indicative of high pore pressure [*Cassidy and Ellis*, 1991]. There is support for this shear zone thickening with depth in exhumed faults [e.g., *Sibson*, 1992]. The thrust reflection thickness change thus may allow mapping the downdip seismic-aseismic transition [*Nedimovic et al.*, 2003].

Speed of Slip on Subduction Thrust Faults

Displacements on subduction thrust faults occur over a range of speeds, from earthquake rupture (seconds), to rates that generate tsunamis (minutes), to slower slip seen only in geodetic data. The speed controls are still unclear. Some slip events are too slow to generate much seismic energy (fig. 2.8) but fast enough to generate tsunamis ("tsunami earthquakes"), i.e., slip duration of minutes [e.g., *Kanamori*, 1972; *Pelayo and Wiens*, 1992; *Okal and Newman*, 2001; *Bilek and Lay*, 2002]. Tsunami events are primarily defined for the shallower updip portions of subduction thrusts where slip can generate tsunamis. A few slip events are still slower and are seen only in geodetic data [e.g., *Sato et al.*, 2004]. In this case, slower slip events are defined mainly on the deeper

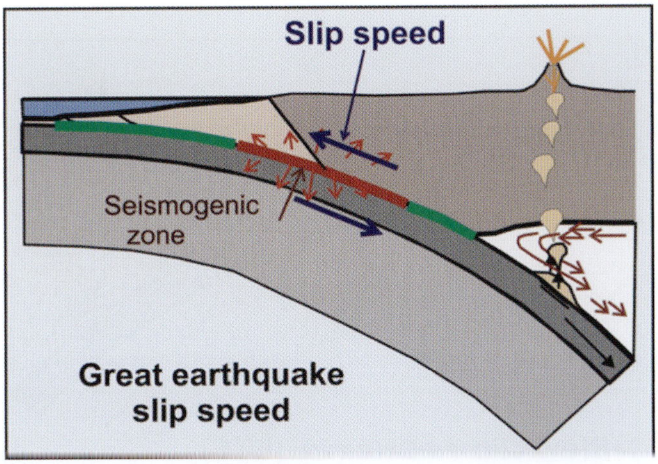

Figure 2.8 Some thrust events are fast (seismic), some are intermediate (tsunami earthquakes), and some are slow (seen only by geodetic measurements).

part of subduction thrusts that affect land geodetic data. Models for fault slip behavior must accommodate this range of event slip rates. Such models also must accommodate the evidence noted above that subduction thrust faults are very weak, from thermal data and models, from regional stress estimates, and from earthquake data. High fluid pressures that reduce the effective fault normal stress are suggested, as discussed above.

Downdip Slow Slip Events and Seismic Tremor

Recent GPS geodetic data have indicated slow slip events downdip of the thrust seismogenic zone, not associated with normal abrupt onset earthquakes (fig. 2.9) [*Dragert et al.*, 2001; *Lowrey et al.*, 2001; *Miller et al.*, 2002; *Ozawa et al.*, 2002]. The subduction thrust just downdip of the seismogenic zone also has been found to exhibit seismic tremor [e.g., *Obara*, 2002]. On one subduction zone, the seismic tremor has now been associated with the slow slip events, i.e., Episodic Tremor and Slip (ETS) [*Rogers and Dragert*, 2003]. The mechanism of slip is beginning to be examined [*Shibazaki and Yosihisa*, 2003; *Yoshida and Naoyuki*, 2003]. The location of the slip and tremor events may provide a new method for defining the downdip end of the locked seismogenic zone. Of special significance is that each of these slow slip events loads the locked part of the thrust farther updip. Therefore the time interval of their occurrence represents a period of increased probability of great earthquake rupture. This time-dependent risk is now being examined [*Mazzotti and Adams*, 2004].

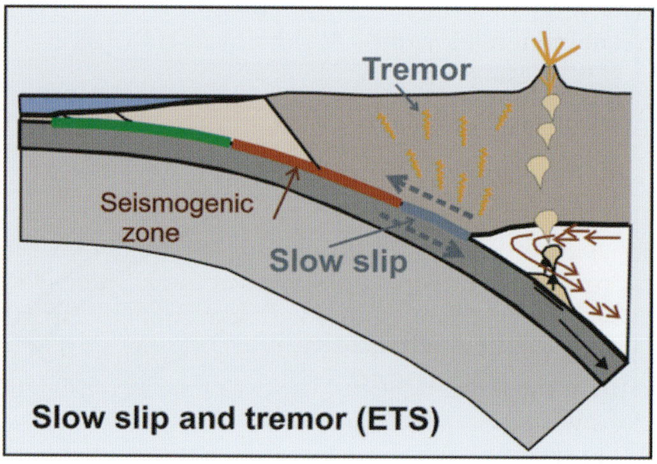

Figure 2.9 Slow slip and seismic tremor on the thrust downdip of the seismogenic zone.

Conclusions

Until recently, most of what was known about the seismic behavior of subduction thrust faults came from teleseismic studies of the earthquakes that they generate. However, there have been great advances recently in characterizing and understanding earthquakes on subduction thrust faults based on seismic reflection and wide-angle structure studies, ocean bottom seismograph recording of microearthquakes, geodetic data, thermal data, stress indicators, drilling the updip portion of the thrust, exhumed ancient subduction faults, and a wide variety of modeling work. Important seismic characteristics of subduction thrust faults and their physical associations include the following:

1. The subduction thrust maximum earthquake magnitude *Mx* is highly variable among subduction zones; *Mx* is at least partly related to the downdip seismogenic width, i.e., updip and downdip rupture limits, because these limits are a principal control on rupture area. Variations in the fault physical characteristics and stress on the fault also may be important. Continental subduction zones tend to have large maximum thrust earthquake magnitudes and island arcs usually have *Mx* less than ~7.5.

2. The term "seismic coupling," i.e., fraction of relative motion that is accommodated seismically compared to by aseismic slip, needs careful definition. Meaningful use of the term requires specification of the downdip seismogenic width. Average moment release rate is a less ambiguous quantity. Some subduction zones appear to be completely locked with almost no aseismic slip between great megathrust events (after post earthquake transients). These faults tend to have very few thrust earthquakes between great events. Other subduction zones have very small seismic moment release rates and are interpreted to have mostly aseismic slip. They usually have small maximum magnitudes and many small thrust earthquakes. Continental subduction zones with large maximum magnitudes usually have large average moment release rates. Island arcs with small maximum magnitudes tend to have only small rates of seismic moment release; the small moment rates may be due in part to very small downdip seismogenic widths.

3. The term "seismic asperity" also needs careful definition. There is no confusion if this term is used to describe fault areas with especially large slip in particular great earthquakes. However, inferences that such areas always have larger displacement in successive great earthquakes and that they are associated with special physical characteristics on the fault are not yet firmly established. There also have been associations of the locations and boundaries of great earthquakes and rupture limits with a number of fault physical characteristics such as seamounts, aseismic ridges, fracture zones etc., but again, in most areas, there have not been enough repeated great events to establish these associations with assurance.

4. Subduction thrust faults are concluded to be weak, i.e., the maximum shear strength is low, on the basis of stress arguments and the lack of a measur-

able effect in observed heat flow. This low strength is consistent with the small stress drop of large thrust earthquakes. The favored explanation is regionally elevated fluid pressures, but several other processes have been proposed, including weak fault materials and dynamic weakening in the rupture process.

5. Most continental subduction thrusts have consistent updip and downdip seismogenic limits. There usually is an updip aseismic zone tens of kilometers wide. This limit commonly corresponds to a temperature of 100°–150°C. There is not yet agreement on the mechanism responsible; the dehydration of stable-sliding smectite clays to illite/chlorite occurs at about that temperature, but recent laboratory data have not supported this mechanism. Other chemical alterations that increase the permeability and allow lower pore pressure and seismic behavior at greater depth also have been suggested. The downdip limit is frequently close to the intersection of the thrust with the fore-arc Moho, i.e., ~40 km for continental subduction, less for island arcs since their fore-arc crusts are usually much thinner. For very hot subduction zones the downdip limit appears to be shallower and has been found to correspond to the critical seismogenic temperature limit of ~350°C found in laboratory studies of crustal rocks.

6. From a few observations the reflection character of subduction thrust faults appears to change downdip from the updip aseismic zone, to the seismogenic zone, to the downdip aseismic zone. A strong reflection, commonly negative, has been observed in the updip aseismic zone that has been associated with high pore pressure. In the seismogenic zone, there usually is a thin sharp but weaker interface. For the deeper aseismic zone a broad shear zone has been observed by deep reflection measurements in a few areas. Such a broad shear zone is observed in field studies of the deeper parts of exhumed faults.

7. Displacements on subduction thrust faults occur over a range of speeds, from earthquake rupture (seconds), to rates that generate tsunamis (minutes), to slower slip seen only in geodetic data. The speed controls are still unclear; variations in fault materials and pore pressure have been suggested.

8. Immediate downdip of the seismogenic zone, the slip on the aseismic zone in some areas, occurs in slow slip events lasting a few weeks with intervals of a year to a few years. There are associated seismic tremors with no clear onset. The mechanism for these episodic tremor and slip events is not yet well understood, but the times of slow slip may represent a period of higher probability for great thrust earthquakes.

Acknowledgments

I acknowledge the many discussions with colleagues and associates on the subject of the seismogenic zone of subduction thrust faults, including a series of meetings and advisory groups involved in the planning for drilling of subduction thrust faults by ODP and IODP. The MARGINS workshop on the Seismogenic Zone of Subduction Faults was especially helpful. Thorough reviews of the manuscript by C. Moore and T. Dixon were much appreciated. Publication of the Geological Survey of Canada 1996202.

References

Abe, K., and H. Kanamori (1980), Magnitudes of great shallow earthquakes from 1953 to 1977, *Tectonophysics, 62,* 191–203.

Algermissen, S. T., W. A. Rinehart, R. W. Sherburne, and W. Dillinger Jr. (1969), Preshocks and aftershocks of the Prince William Sound earthquake of March 28, 1964, *U.S. Coast Geod. Surv., 211,* 23–43.

ANCORP Working Group (2003), Seismic imaging of a convergent continental margin and plateau in the central Andes (Andean Continental Research Project 1996 (ANCORP96)), *J. Geophys. Res., 108*(B7), 2328, doi:10.1029/2002JB001771.

Ando, M. (1975), Source mechanisms and tectonic significance of historical earthquakes along the Nankai Trough, Japan, *Tectonophysics, 27,* 119–140.

Astiz, L., T. Lay, and H. Kanamori (1988), Large intermediate-depth earthquakes and the subduction process, *Phys. Earth Planet. Inter., 53,* 80–166.

Bangs, N. L., T. H. Shipley, S. P. S. Gulick, G. F. Moore, S. Kuromoto, and Y. Nakamura (2004), Evolution of the Nankai Trough décollement from the trench into the seismogenic zone: Inferences from three-dimensional seismic reflection imaging, *Geology, 32*(4), 273–276.

Bilek, S. L., and T. Lay (2002), Tsunami earthquakes possibly widespread manifestations of frictional conditional instability, *Geophys. Res. Lett., 29*(14), 1673, doi:10.1029/2002GL015215.

Bilek, S. L, S. Y. Schwartz, and H. R. DeShon (2003), Control of seafloor roughness on earthquake rupture behavior, *Geology, 31*(5), 455–458.

Blanpied, M. L., D. A. Lockner, and J. D. Byerlee (1991), Fault stability inferred from granite sliding experiments at hydrothermal conditions, *Geophys. Res. Lett., 18,* 609–612.

Blanpied, M. L., D. A. Lockner, and J. D. Byerlee (1995), Frictional slip of granite at hydrothermal conditions, *J. Geophys. Res., 100,* 13,045–13,064.

Bostock, M. G., R. D. Hyndman, S. Rondenay, and S. M. Peacock (2002), An inverted Moho and serpentinization of the fore arc mantle, *Nature, 417,* 536–538.

Brodsky, E. E., and H. Kanamori (2001), Elastohydrodynamic lubrication of faults, *J. Geophys. Res., 106,* 16,357–16,374.

Brooks, B. A., M. Bevis, R. Smalley, E. Kendrick, R. Manceda, E. Lauria, R. Maturana, and M. Araujo (2003), Crustal motion in the southern Andes (26–36 S): Do the Andes behave like a microplate, *Geochem. Geophys., Geosyst., 4*(10), 1085, doi:10.1029/2003GC000505.

Brown, K., A. Kopf, M. Underwood, and L. Weinberger (2003), Composition and fluid pressure controls on the state of stress on the Nankai subduction thrust: A weak plate boundary, *Earth Planet. Sci. Lett., 214,* 589–603.

Brune, J. N. (1993), Elastohydrodynamic lubrication of faults, *J. Geophys. Res., 106,* 16, 357–16,374.

Byrne, T., and D. M. Fisher (1990), Evidence for a weak and overpressured decollement beneath sediment-dominated accretionary prisms, *J. Geophys. Res., 95,* 9081–9097.

Byrne, D. E., D. M. Davis, and L. R. Sykes (1988), Loci and maximum size of thrust earthquakes and the mechanics of the shallow region of subduction zones, *Tectonics, 7,* 833–857.

Cassidy, J. F., and R. M. Ellis (1991), Shear wave constraints on deep crustal reflective zone beneath Vancouver Island, *J. Geophys. Res., 96,* 19,843–19,851.

Chamley, H. (1989), *Clay Sedimentology,* Springer, New York.

Christensen, N. I. (2004), Serpentinites, peridotites, and seismology, *Int. Geol. Rev., 46,* 795–816.

Cloos, M., and R. L. Shreve (1996), Shear-zone thickness and the seismicity of Chilean- and Marianas-type subduction zones, *Geology, 24*(2), 107–110.
Currie, C. A., R. D. Hyndman, K. Wang, and V. Kostoglodov (2002), Thermal models of the Mexico subduction zone: Implications for the megathrust seismogenic zone, *J. Geophys. Res., 107*(B12), 2370, doi:10.1029/2001JB000886.
Davis, D. M., J. Suppe, and F. A. Dahlen (1983), Mechanics of fold-and-thrust belts and accretionary wedges, *J. Geophys. Res., 88*, 1153–1172.
Dragert, H., K. Wang, and T. S. James (2001), A silent slip event on the deeper Cascadia subduction interface, *Science, 292*(5521), 1525–1528.
Harris, R. N., and K. Wang (2002), Thermal models of the Middle America trench at the Nicoya Peninsula, Costa Rica, *Geophys. Res. Lett., 29*(21), 2010, doi:10.1029/2002GL015406.
Hashimoto, Y., M, Enjoji, A. Sakaguchi, and G. Kimura (2002), P-T conditions of cataclastic deformation associated with underplating: An example from the Cretaceous Shimanto complex, Kii Peninsula, SW Japan, *Earth Planets Space, 54*, 1133–1138.
Heermance, R., Z. K. Shipton, and J. P. Evans (2003), Fault structure control on fault slip and ground motion during the 1999 rupture of the Chelungpu Fault, Taiwan, *Bull. Seismol. Soc. Am., 93*, 1034–1050.
Hickman, S. (1991), Stress in the lithosphere and the strength of active faults, *U.S. Natl. Rep. Int. Union Geod. Geophys. 1987–1990, Rev. Geophys., 29*, 759–775.
Hindle, D., J. Kley, E. Klosko, S. Stein, T. Dixon, and E. Norabuena (2002), Consistency of geologic and geodetic displacements during Andean orogenesis, *Geophys. Res. Lett., 29*(8), 1188, 10.1029/2001GL013757.
Hubbert, M. K., and W. W. Rubey (1959), Role of fluid pressure in the mechanics of overthrust faulting I: Mechanics of fluid-filled porous solids and its application to overthrust faulting, *Geol. Soc. Am. Bull., 70*, 115–166.
Hyndman, R. D. (1972), Plate motions relative to the deep mantle and the development of subduction zones, *Nature, 238*, 263–265.
Hyndman, R. D., and S. M. Peacock (2003), Serpentinization of the forearc mantle, *Earth Planet. Sci. Lett., 212*, 417–432.
Hyndman, R. D., and K. Wang (1993), Thermal constraints on the zone of major thrust earthquake failure: The Cascadia subduction zone, *J. Geophys. Res., 98*, 2039–2060.
Hyndman, R. D., and K. Wang (1995), The rupture area of Cascadia great earthquakes from current deformation and the thermal regime, *J. Geophys. Res., 100*, 22,133–22,154.
Hyndman, R. D., K. Wang, and M. Yamano (1995), Thermal constraints on the seismogenic portion of the southwestern Japan subduction thrust, *J. Geophys. Res., 100*, 15,373–15,392.
Hyndman, R. D., M. Yamano, and D. A. Oleskevich (1997), The seismogenic zone of subduction thrust faults, *Island Arc, 6*, 244–260.
Igarashi, T., T. Matsuzawa, and A. Hasegawa (2003), Repeating earthquakes and interplate aseismic slip in the northeastern Japan subduction zone, *J. Geophys. Res., 108*(B5), 2249, doi:10.1029/2002JB001920.
Jarrard, R. D. (1986), Relations among subduction parameters, *Rev. Geophys., 24*, 217–284.
Johnson, J. M. (1999), Heterogeneous coupling along Alaska-Aleutians as inferred from tsunami, seismic, and geodetic inversions, in *Tsunamigenic Earthquakes and Their Consequences*, edited by R. Dmowska-Renata and B. Saltzman, *Adv. Geophys., 39*, 1–116.
Johnson, M., K. Satake, R. Holdahl, and J. Sauber (1996), The 1964 Prince William Sound earthquake: Joint inversion of tsunami and geodetic data, *J. Geophys. Res., 101*, 523–532.
Kanamori, H. (1972), Mechanism of tsunami earthquakes, *Phys. Earth Planet. Inter., 6*, 346–359.
Kanamori, H. (1983), Global seismicity, in *Earthquakes: Observation, Theory, and Interpretation*, edited by H. Kanamori and E. Boschi, 696–608, Elsevier, New York.
Kanamori, H. (1986), Rupture processes of subduction zone earthquakes, *Annu. Rev. Earth Planet. Sci., 14*, 293–322.

Kirby, S., K. Wang, and T. Brocher (2002), A possible deep, long-term source for water in the northern San Andreas Faults system: A ghost of Cascadia subduction past?, *Eos Trans. AGU, 83*(47) Fall Meet. Suppl., Abstract S22B-1038.

Kodaira, S., E. Kurashimo, J. O. Park, N. Takahashi, A. Nakanishi, S. Miura, T. Iwasaki, N. Hirata, K. Ito, and Y. Kaneda (2002), Structural factors controlling the rupture process of a megathrust earthquake at the Nankai Trough seismogenic zone, *Geophys. J. Int., 149*, 815–835.

Lachenbruch, A. H. (1980), Frictional heating, fluid pressure, and the resistance to fault motion, *J. Geophys. Res., 85*, 6097–6112.

Lachenbruch, A. H., and A. McGarr (1990), Stress and heat flow, in The San Andreas Fault System, California, edited by R. E. Wallace, *U.S. Geol. Surv. Prof. Pap., 1515*, 261–277.

Lay, T. and H. Kanamori (1981), An asperity model of great earthquake sequences, in *Earthquake Prediction; An International Review, Maurice Ewing Ser.*, vol. 4, edited by D. W. Simpson and P. G. Richards, pp. 579–592, AGU, Washington, D. C. .

Lay, T., H. Kanamori, and L. Ruff (1982), The asperity model and the nature of large subduction zone earthquakes, *Earthquake Predict. Res., 1*, 3–71.

Lowrey, A. R., K. M. Larson, V. Kostoglodov, and R. Bilham (2001), Transient fault slip in Guerrero, southern Mexico, *Geophys. Res. Lett., 28*, 3753–3756.

Lundgren, P., M. Protti, A. Donnellan, M. Heflin, E. Hernandez, and D. Jefferson (1999), Seismic cycle and plate margin deformation in Costa Rica: GPS observations from 1994 to 1997, *J. Geophys. Res., 104*, 28,915–28,926.

Ma, K. F., E. Brodsky, J. Mori, J. Chen, T. Song, and H. Kanamori (2003), Evidence for fault lubrication during the 1999 Chi-Chi, Taiwan, earthquake (Mw7.6), *Geophys. Res. Lett., 30*(5), 1244, doi:10.1029/2002GL015380.

Magee, M. E., and M. D. Zoback (1993), Evidence for a weak interplate thrust fault along the northern Japan subduction zone and implications for the mechanics of thrust faulting and fluid expulsion, *Geology, 21*(9), 809–812.

Marone, C., and C. Scholz (1988), The depth of seismic faulting and the upper transition from stable sliding to unstable slip regimes, *Geophys. Res. Lett., 15*, 621–624.

Mazzotti, S., and J. Adams (2004), Time variability of near-term probabilities for a great earthquake on the Cascadia subduction zone, *Bull. Seismol. Soc. Am., 95*, 1954–1959.

Mazzotti, S., and R. D. Hyndman (2002), Yakutat collision and strain transfer across the northern Canadian Cordillera, *Geology, 30*(6), 495–498.

McCaffrey, R. (1997), Statistical significance of the seismic coupling coefficient, *Bull. Seismol. Soc. Am., 87*, 1069–1073.

McCann, W. E., S. P. Nishenko, L. R. Sykes, and J. Krause (1979), Seismic gaps and plate tectonics: Seismic potential for major plate boundaries, *Pure Appl. Geophys., 117*, 1082–1147.

Melosh, H. J. (1996), Dynamical weakening of faults by acoustic fluidization, *Nature, 379*, 601–606.

Miller, M. M., T. Melbourne, D. J. Johnson, and W. Q. Sumner (2002), Periodic slow earthquakes from the Cascadia Subduction Zone, *Science, 295*(5564), 2423.

Moore, J. C., and D. Saffer (2001), Updip limit of the seismogenic zone beneath the accretionary prism of southwest Japan: An effect of diagenetic to low-grade metamorphic processes and increasing effective stress, *Geology, 29*(2), 183–186.

Moore, J. C., and P. Vrolijk (1992), Fluids in accretionary prisms, *Rev. Geophys., 30*, 113–135.

Moore, D. E., D. A. Lockner, M. Shengli, R. Summers, and J. D. Byerlee (1997), Strengths of serpentinite gouges at elevated temperatures, *J. Geophys. Res., 102*, 14,787–14,801.

Moore, G. F., et al. (2001), Deformation and fluid flow processes in the Nankai Trough accretionary prism, *Proc. Ocean Drill. Program Initial Rep.* [CD-ROM], *190*. (Available at http://www-odp.tamu.edu/publications/190_IR/190ir.htm).

Nedimovic, M. R., R. D. Hyndman, K. Ramachandran, and G. D. Spence (2003), Reflection signature of seismic and aseismic slip on the northern Cascadia subduction interface, *Nature, 424*, 416–420.

Nishenko, S. P. (1991), Circum-Pacific seismic potential: 1989–1999, *Pure Appl. Geophys., 135*, 169–259.

Norabuena, E., et al. (2004), Geodetic and seismic constraints on seismogenic zone processes in Costa Rica, *J. Geophys. Res., 109*, B11403, doi:10.1029/2003JB002931.

Obana, K., S. Kodaira, K. Mochizuki, and M. Shinohara (2001), Micro-seismicity around the seaward limit of the 1946 Nankai earthquake dislocation area, *Geophys. Res. Lett., 28*, 2333–2336.

Obara, K. (2002), Nonvolcanic deep tremor associated with subduction in southwest Japan, *Science, 296*(5573), 1679–1681.

Okal, E. A., and A. Newman (2001), Tsunami earthquakes; the quest for a regional signal, *Phys. Earth Planet. Inter., 124*, 45–70.

Oleskevich, D. A., R. D. Hyndman, and K. Wang (1999), The updip and downdip limits to great subduction earthquakes: Thermal and structural models of Cascadia, south Alaska, SW Japan, and Chile, *J. Geophys. Res., 104*, 14,965–14,991.

Ozawa, S., M. Murakami, M. Kaidzu, T. Tada, T. Sagiya, Y. Hatanaka, H. Yarai, and T. Nishimura (2002), Detection and monitoring of ongoing aseismic slip in the Tokai region, central Japan, *Science, 298*(5595), 1009–1012.

Pacheco, J. F., and L. R. Sykes (1992), Seismic moment catalogue of large shallow earthquakes, 1900 to 1989, *Bull. Seismol., Soc. Am., 82*, 1306–1349.

Pacheco, J. F., L. R. Sykes, and C. H. Scholz (1993), Nature of seismic coupling along simple plate boundaries of the subduction type, *J. Geophys. Res., 98*, 14,133–14,159.

Park, J. O., T. Tsuru, S. Kodaira, A. Nakanishi, S. Miura, Y. Kaneda, and Y. Kono (2000), Out-of-sequence thrust faults developed in the coseismic slip zone of the 1946 Nankai earthquake (Mw=8.2) off Shikoku, southwest Japan, *Geophys. Res. Lett., 27*, 1033–1036.

Peacock, S. M., and R. D. Hyndman (1999), Hydrous minerals in the mantle wedge and the maximum depth of subduction thrust earthquakes, *Geophys. Res. Lett., 26*, 2517–2520.

Pelayo, A. M., and D. A. Wiens (1992), Tsunami earthquakes; slow thrust-faulting events in the accretionary wedge, *J. Geophys. Res., 97*, 15,321–15,337.

Peterson, E. T., and T. Seno (1984), Factors affecting seismic moment release rates in subduction zones, *J. Geophys. Res., 89*, 10,233–10,248.

Plafker, G. (1969), Tectonics of the March 27, 1964, Alaska earthquake, *U.S. Geol. Surv. Prof. Pap., 543–I.*

Plafker, G. (1971), Tectonics, in *The Great Alaska Earthquake of 1964*, pp. 47–122, Natl. Acad. of Sci., Washington, D. C.

Plafker, G. (1972), Alaskan earthquake of 1964 and Chilean earthquake of 1960: Implications for arc tectonics, *J. Geophys. Res., 77*, 901–925.

Plafker, G., and J. C. Savage (1970), Mechanism of the Chilean earthquakes of May 21 and 22, 1960, *Geol. Soc. Am. Bull., 81*, 1001–1030.

Raleigh, B., and J. F. Evernden (1981), Case for low deviatoric stress in the lithosphere, in *Mechanical Behaviour of Crustal Rocks; The Handin Volume, Geophys. Monogr. Ser.*, vol. 24, edited by N. L. Carter et al., pp. 173–186, AGU, Washington, D. C.

Reinen, L. A., J. D. Weeks, and T. E. Tullis (1991), The frictional behavior of serpentinite: Implications for aseismic creep on shallow crustal faults, *Geophys. Res. Lett., 18*, 1921–1924.

Rogers, G. C., and H. Dragert (2003), Episodic tremor and slip on the Cascadia subduction zone: The chatter of silent slip, *Science, 300*(5627), 1942–1943.

Ruff, L. J. (1989), Do trench sediments affect great earthquake occurrence in subduction zones?, *Pure Appl. Geophys., 129*, 263–282.

Ruff, L. J. (1992), Asperity distributions and large earthquake occurrence in subduction zones, *Tectonophysics, 211*, 61–83.

Ruff, L. J. (1996), Large earthquakes in subduction zones: Segment interactions and recurrence times, in *Subduction: Top to Bottom*, Geophys. Monogr. Ser., vol. 96, edited by G. E. Bebout et al., pp. 91–104, AGU, Washington, D. C.

Ruff, L., and H. Kanamori (1983), Seismic coupling and uncoupling at subduction zones, *Tectonophysics, 99*, 99–117.

Ruff, L. J., and B. W. Tichelaar (1996), What controls the seismogenic plate interface in subduction zones, in *Subduction: Top to Bottom*, Geophys. Monogr. Ser., vol. 96, edited by G. E. Bebout et al., pp. 105–111, AGU, Washington, D. C.

Saffer, D. M., and C. Marone (2003), Comparison of smectite- and illite-rich gouge frictional properties: Application to the updip limit of the seismogenic zone along subduction megathrusts, *Earth Planet. Sci. Lett., 215*, 219–235.

Sato, T., K. Imanishi, N. Kato, and T. Sagiya (2004), Detection of a slow slip event from small signal in GPS data, *Geophys. Res. Lett., 31*, 1–4.

Savage, J. C., M. Lisowski, and W. H. Prescott (1991), Geodetic strain measurements in Washington, *J. Geophys. Res., 96*, 14,493–14,507.

Scholz, C. H. (1990), *The Mechanics of Earthquakes and Faulting*, Cambridge Univ. Press, New York.

Scholz, C. H., and J. Campos (1995), On the mechanism of seismic decoupling and back-arc spreading at subduction zones, *J. Geophys. Res., 100*, 22,103–22,115.

Schwartz, S. Y. (1999), Non-characteristic behaviour and complex recurrence of large subduction zone earthquakes, *J. Geophys. Res., 104*, 23,111–23,125.

Seno, T. (2005), Variation of downdip limit of the seismogenic zone near the Japanese islands: Implications for the serpentinization mechanism of the forearc mantle wedge, *Earth Planet. Sci. Lett., 231*(3–4), 249–262, doi:10.1016/j.epsl.2004.12.027.

Shi, B., A. Anooshehpoor, J. N. Brune, and Y. Zeng (1998), Dynamics of thrust faulting; 2D lattice model, *Bull. Seism. Soc. Am., 88*, 1484–1494.

Shibazaki, B., and I. Yosihisa (2003), On the physical mechanism of silent slip events along the deeper part of the seismogenic zone, *Geophys. Res. Lett., 30*(9), 1489, doi:10.1029/2003GL017047.

Shipley, T. H., K. D. McIntosh, E. A. Silver, and P. L. Stoffa (1992), 3–dimensional seismic imaging of the Costa-Rica accretionary prism: Structural diversity in a small volume of the lower slope, *J. Geophys. Res., 97*, 4439–4459.

Shipley, T. H., G. F. Moore, N. L. Bangs, J. C. Moore, and P. L. Stoffa (1994), Seismically inferred dilatancy distribution, northern Barbados Ridge décollement: Implications for fluid migration and fault strength, *Geology, 22*(5), 411–414.

Sibson, R. H. (1992), Implications of fault valve behavior for rupture nucleation and recurrence, *Tectonophysics, 211*, 283–293.

Spiess, F. N., C. D. Chadwell, J. A. Hildebrand, L. E. Young, G. H. Purcell Jr., and H. Dragert (1998), Precise GPS/ acoustic positioning of seafloor reference points for tectonic studies, *Phys. Earth Planet. Inter., 108*, 101–112.

Spinelli, G. A., and D. M. Saffer (2004), Along-strike variations in underthrust sediment dewatering on the Nicoya margin, Costa Rica related to the updip limit of seismicity, *Geophys. Res. Lett., 31*, L04613, doi:10.1029/2003GL018863.

Tichelaar, B. W., and L. J. Ruff (1991), Seismic coupling along the Chilean subduction zone, *J. Geophys. Res., 96*, 11,997–12,022.

Tichelaar, B. W., and L. J. Ruff (1993), Depth of seismic coupling along subduction zones, *J. Geophys. Res., 98*, 2017–2037.

Toro, G. D., D. L. Goldsby, and T. E. Tullis (2004), Friction falls towards zero in quartz rock as slip velocity approaches seismic rates, *Nature, 427*, 436–439.

Toshitaka, B., Y. Tanioka, P. R. Cummins, and K. Uhira (2002), The slip distribution of the 1946 Nankai earthquake estimated from tsunami inversion using a new plate model, in *Subduction Zone Structure and Megathrust Earthquakes, Phys. Earth Planet. Inter, 132*, 59–73.

Tse, S. T., and J. R. Rice (1986), Crustal earthquake instability in relation to the depth variation of frictional slip properties, *J. Geophys. Res., 91*, 9452–9472.
Uyeda, S., and H. Kanamori (1979), Back-arc opening and the mode of subduction, *J. Geophys. Res., 84*, 1049–1061.
von-Huene, R., and D. W. Scholl (1991), Observations at convergent margins concerning sediment subduction, subduction erosion, and the growth of continental crust, *Rev. Geophys., 29*, 279–316.
Vrolijk. P. (1990), On the mechanical role of smectite in subduction zones, *Geology, 18*(8), 703–707.
Wang, C. Y., and N. H. Mao (1979), Shearing of rock joints with saturated clays at high confining pressures, *Geophys. Res. Lett., 6*, 825–828.
Wang, K., Elastic and viscoelastic models of crustal deformation in subduction earthquake cycles, this volume.
Wang, K., and K. Suyehiro (1999), How does plate coupling affect crustal stresses in Northeast and Southwest Japan?, *Geophys. Res. Lett. 26*, 2307–2310.
Wang, K., and T. Dixon (2004), "Coupling" semantics and science in earthquake research, *Eos Trans. AGU, 85*, 180.
Wang, K., and J. He (1999), Mechanics of low-stress forearcs: Nankai and Cascadia, *J. Geophys. Res., 104*, 15,191–15,205.
Wang, K., T. Mulder, G. C. Rogers, and R. D. Hyndman (1995), Case for very low coupling stress on the Cascadia subduction thrust fault, *J. Geophys. Res., 100*, 12, 907–12,918.
Wang, K., R. Wells, S. Mazzotti, R. D. Hyndman, and T. Sagiya (2003), A revised dislocation model of interseismic deformation of the Cascadia subduction zone, *J. Geophys. Res., 108*(B1), 2009, doi:10.1029/2001JB001227.
Wells, D. L., and K. J. Coppersmith (1994), New empirical relationships among magnitude, rupture length, rupture width, rupture area, and surface displacement, *Bull. Seismol. Soc. Am., 84*, 974–1002.
Wyss, M., and S. Wiemer (1999), How can one test the seismic gap hypothesis? The case of repeated ruptures in the Aleutians, *Pure Appl. Geophys., 155*, 259–278.
Yoshida, S., and K. Naoyuki (2003), Episodic aseismic slip in a two-degree-of-freedom block-spring model, *Geophys. Res. Lett.*, 30(13), 1681, doi:10.1029/2003GL017439.
Zhang, Z., and S. Y. Schwartz (1992), Depth distribution of moment release in underthrusting earthquakes at subduction zones, *J. Geophys. Res., 97*, 537–544.
Zoback, M. D., and J. Townend (2002), Steady-state failure equilibrium and deformation of intraplate lithosphere, *Int. Geol. Rev., 44*, 383–401.
Zoback, M. D., et al. (1987), New evidence on the state of stress of the San Andreas fault system, *Science, 238*, 1105–1111.
Zweck, C., J. T. Freymueller, and S. C. Cohen (2002), Three-dimensional elastic dislocation modeling of the postseismic response to the 1964 Alaska earthquake, *J. Geophys. Res., 107*(B4), 2064, doi:10.1029/2001JB000409.

PART II

The Incoming Plate

3 Sediment Inputs to Subduction Zones

Why Lithostratigraphy and Clay Mineralogy Matter

Michael B. Underwood

Abstract

Lithostratigraphic architecture exerts a first-order influence over the material properties and tectonic behavior of subduction zones. The amount and types of clay-size particles affect any sediment's coefficient of friction and permeability. Some minerals (e.g., smectite and opal) significantly increase fluid production during diagenesis. Compartments of excess pore-fluid pressure are, in many depositional systems, set up initially by confined sand-body geometries. Generic facies models for trenches show upward thickening and coarsening trends (from basalt at the base through pelagic ooze, hemipelagic mud, silty turbidites, and sandy turbidites), but there are many exceptions to this paradigm that can affect fault-zone behavior. Inputs to Nankai Trough change in three dimensions because irregular basement topography deflected turbidity currents during early stages of abyssal-floor sedimentation; sandy turbidites are the norm, rather than the exception, beneath the frontal décollement. The content of smectite changes along strike in Nankai because of local perturbations of heat flow and presubduction diagenesis. Distal deposits in Cascadia Basin are also unusually sandy, and concentrations of smectite are erratic because plumes from diverse point sources (e.g., Columbia River discharge) mix and interfinger inconsistently. The effect of Cascadia lithostratigraphy on décollement behavior remains uncertain. The Barbados subduction zone also includes anomalously coarse-grained turbidites below the décollement. The fault zone itself contains unusually high concentrations of smectite, which probably contributes to the décollement's intrinsic weakness and downdip fluid production. The Costa Rica segment of Middle America Trench is, in comparison, starved of terrigenous sediment, so basement there is overlain successively by open-ocean calcareous ooze and diatomaceous smectite-rich hemipelagic mud. Spatial variations in the thermal structure of the Cocos plate modulate diagenetic reaction progress, fluid production, pore pressure, and effective stress, both within and beneath the plate boundary. As the Seismogenic Zone Experiment progresses, we will need to quantify the thermal and sediment inputs to these (and other) systems in three dimensions before pinpointing how and why geotechnical and hydrogeologic properties change near the updip limit of seismicity.

Introduction

One of the more ambitious goals of the Seismogenic Zone Experiment (SEIZE) is to separate the effects of intrinsic frictional strength (i.e., gouge versus wall rock) from the effects of gradual or transient increases in pore pressure as they collectively modulate the downdip evolution of plate-boundary faults. As subduction proceeds, sedimentary strata on the downgoing plate change progressively in the downdip direction. These transformations create a fundamental difference between subduction megathrusts and transform boundaries (e.g., the San Andreas fault). The basal décollement marks the structural contact between domains of offscraped and underthrust strata [*Moore,* 1989]. Presumably, the faults nucleate within, or shift their positions preferentially, to take advantage of weaker stratigraphic intervals. Several factors might be responsible for zones of weakness, however, including the first-order differences in sediment texture, grain fabric, and mineral composition [*Lupini et al.*, 1981; *Mitchell*, 1993]. An equally important parameter is pore-fluid pressure, which modulates effective normal stress within many types of sedimentary basins and fault zones [*Hubbert and Rubey*, 1959; *Byerlee*, 1990; *Osborne and Swarbrick*, 1997; *Faulkner and Rutter*, 2001; *Flemings et al.*, 2002]. Stratigraphic architecture, especially sand-body geometry, imparts first-order controls over shear strength, permeability, and fluid flow both prior to and during subduction. It is vital therefore to characterize the initial texture, composition, and hydrogeologic properties of sediment inputs as a first step in predicting how geotechnical behavior changes toward the updip limit of a seismogenic zone. A second question to answer is: Through which stratigraphic interval does structural partitioning occur across the décollement? The fault's initial location is important because it determines which portions of the input stratigraphy get offscraped and which continue downdip toward the seismogenic front.

Given the diversity of subduction-zone stratigraphy worldwide [*Underwood and Moore*, 1995], and the prospects for substantial along-strike heterogeneity in sediment composition and thermal structure across any individual system, the SEIZE community must characterize subduction inputs on a case-by-case basis. With this community goal in mind, the objectives of this paper are to highlight some of the basic principles of subduction-zone sedimentology

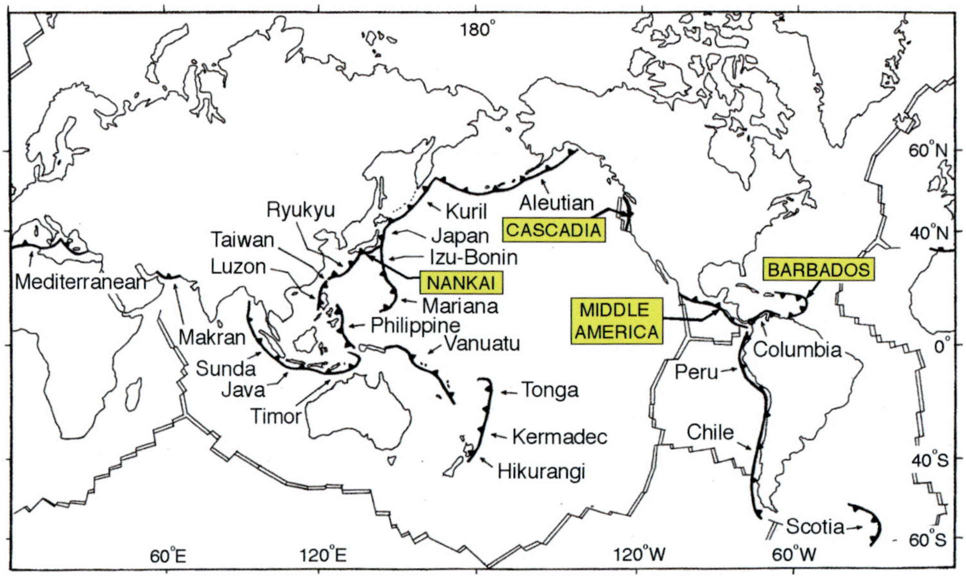

Figure 3.1 Global distribution of active subduction zones.

and to summarize what is already known about key input parameters at four well-documented localities: Nankai Trough (offshore Japan), Cascadia Basin (offshore Vancouver Island, Washington, and Oregon), Barbados Ridge (east of the Lesser Antilles island arc), and the Middle America Trench (offshore Costa Rica) (fig. 3.1). The discussion places emphasis on lithostratigraphic architecture, spatial variations in the content of hydrous minerals (e.g., smectite), and the relation between three-dimensional (3-D) thermal structure and diagenetic fluid production.

Stratigraphic Paradigm

Most marine geologists probably expect subduction-margin stratigraphic successions to gradually thicken and coarsen upward (from pelagic ooze to hemipelagic mud to terrigenous turbidites) as an oceanic plate migrates toward a given trench [*Piper et al.*, 1973; *von Huene*, 1974]. Examples that conform to this facies pattern include the eastern Aleutian Trench [*Piper et al.*, 1973], the southern and central Chile Trench [*Schweller and Kulm*, 1978], and the northern Middle America Trench off Mexico [*Moore and Shipley*, 1988]. This style of stratigraphy should favor a deformation response in which the frontal décollement occupies a position between the base of the sand-rich turbidite facies and the top of the hemipelagic facies. The generic facies model, however, does not fit all subduction zones [*Underwood and Moore*, 1995]. In fact, a global view of subduction-zone sedimentation and lithostratigraphy provides ample motivation for SEIZE to evaluate the inputs to each system.

Unusually thick abyssal-plain deposits, or submarine fans that accumulate seaward of a trench, will change the accretionary dynamics of subduction. Some of the more prominent abyssal-plain systems that fall into this category are the Indus Fan [*Kolla and Coumes*, 1984; *Clift et al.*, 2001], the Bengal-Nicobar Fan [*Emmel and Curray*, 1984; *Einsele et al.*, 1996; *Weber et al.*, 1997], the Surveyor and Chirikof channel systems in the Gulf of Alaska [*Carlson et al.*, 1996], and the Zodiac Fan outboard of the central Aleutian Trench [*Stevenson et al.*, 1983]. With highly thickened input sections, frontal décollements occur within, rather than below, the sandy sediment, as illustrated by the central Aleutian Trench [*McCarthy and Scholl*, 1985], the eastern Aleutian Trench [*von Huene et al.*, 1998], the Sunda Trench [*Moore et al.*, 1982; *Kopp and Kukowski*, 2003], and the Makran accretionary prism [*Kukowski et al.*, 2001; *Schluter et al.*, 2002]. The connection between stratigraphy and the position of the frontal décollement, in these types of systems, may be cryptic unless deep drilling can be used to ground-truth seismic-reflection data.

At the other extreme of sedimentary inputs, some ocean-continent subduction margins lack coarse-grained turbidites altogether because arid climate dampens terrigenous influx (e.g., the Peru Trench) [*Schweller and Kulm*, 1978; *von Huene et al.*, 1985]. With minimal sediment to accrete, the frontal décollement propagates through fine-grained hemipelagic mud, pelagic ooze, or the basalt-sediment interface [*Shipley and Moore*, 1986; *Shipley et al.*, 1990]. This situation is exacerbated along sediment-starved intra-oceanic arc-trench systems such as the Mariana Trench [*Fryer*, 1996] and the Tonga-Kermadec Trench [*Ballance*, 1991; *Clift et al.*, 1998; *Ballance et al.*, 1999]. Most sediment-starved fore arcs are sites of subduction erosion rather than accretion, especially along corridors where seamounts, aseismic ridges, and horst-graben relief punctuate the downgoing plate [*von Huene and Culotta*, 1989; *von Huene and Scholl*, 1991; *Sosson et al.*, 1994; *Clift and MacLeod*, 1999; *von Huene et al.*, 1999; *Ranero and von Huene*, 2000; *Clift et al.*, 2003; *Vannucchi et al.*, 2003, 2004]. Under these circumstances, fault-zone materials at seismogenic depths may be derived largely from the base of the overriding plate [*von Huene et al.*, 2004].

Geotechnical Paradigm

Another compelling reason for SEIZE to document sediment inputs to subduction zones is to pinpoint their impact on geotechnical properties. Empirical and experimental studies demonstrate that clay-rich fault gouges are weaker than surrounding wall rocks, typically because of their constituent particle size and mineralogy [*Wang*, 1980; *Shimamoto and Logan*, 1981; *Chester and Logan*, 1986; *Morrow et al.*, 1982]. Clay-size particles, especially expandable minerals of the smectite group (e.g., montmorillonite), significantly affect strata's coefficient of internal friction, compressibility, and permeability [*Mesri and Olson*, 1971; *Olson*, 1974; *Lupini et al.*, 1981; *Moore et al.*, 1982; *Morrow et al.*, 1984, 1992; *Logan and Rauenzahn*, 1987; *Freed and Peacor*, 1989a; *Robinson and Allam*, 1998]. Typical coefficients of friction for pure kaolinite and illite are 0.18–0.21, as compared to the value for quartz (0.70); pure smectite usually yields a frictional coefficient of 0.07–0.17 [*Mitchell*, 1993; *Moore and Lockner*, this volume]. Values for multicomponent clay-mineral mixtures range from 0.20 to 0.40, depending on the proportion of smectite [*Wang*, 1980; *Shimamoto and Logan*, 1981; *Logan and Rauenzahn*, 1987]. The coefficient for typical illite-quartz shale is 0.41–0.63 [*Saffer and Marone*, 2003]. Simple two-component mixtures (e.g., smectite plus quartz) display linearly proportional changes in the frictional coefficient if the volumetric contribution of the weaker mineral is between 25 and 70% [*Lupini et al.*, 1981; *Logan and Rauenzahn*, 1987; *Brown et al.*, 2003].

Laboratory-to-laboratory comparisons of direct-shear and ring-shear test results can be confusing because of differences in testing parameters, including strain rate, normal stress load, moisture content, temperature, and unaccounted for mineral impurities [*Lupini et al.*, 1981; *Brown et al.*, 2003; *Kopf and Brown*, 2003; *Saffer and Marone*, 2003; *Marone and Saffer*, this volume; *Moore and Lockner*, this volume]. These difficulties notwithstanding, the bulk content of total clay minerals appears to be a more important variable to consider than the abundance of any specific clay type. As stressed by *Vrolijk* [1990], smectite-rich claystones certainly deserve special attention because of their extreme mechanical weakness, their ability to swell, and their tendency to sustain higher bulk water contents and porosity by retaining water in both interlayer and intergranular sites [*Bird*, 1984; *Colten-Bradley*, 1987; *Chatterji and Morgenstern*, 1990; *Mesri et al.*, 1994; *Fitts and Brown*, 1999]. The body of geotechnical evidence cited above has led to what is sometimes referred to as the "smectite-to-illite hypothesis," which suggests that the downdip transition from aseismic (stable sliding) to seismic (stick-slip) behavior on subduction megathrusts is controlled by the diagenetic replacement of mechanically weak smectite by the stronger clay minerals illite and chlorite [*Byrne et al.*, 1988; *Vrolijk*, 1990]. The completion temperature of the smectite-to-illite reaction ranges from ~90° to 140°C [*Perry and Hower*, 1970; *Freed and Peacor*, 1989b]. The smectite-illite reaction window roughly coincides with the temperature of the updip limit of seismicity [*Hyndman and Wang*, 1995; *Hyndman et al.*, 1995, 1997]. It remains to be proven whether this apparent match is a coincidence or cause and effect [*Marone and Saffer*, this volume] and whether its significance is local or global.

Hydrogeology Paradigm

In addition to sediment's cohesion and coefficient of internal friction, the shear strength of any given stratigraphic interval depends on fluid pressure and total normal stress. Thus it is also essential for SEIZE to determine how three-dimensional facies architecture affects permeability and fluid flow within the subduction inputs and to track the three-dimensional distribution of sediment-derived fluid sources along their downdip reaction paths. The release of fluids from mudstones during diagenesis (including volatiles generated from organic matter) is modulated by a combination of mechanical compaction, tectonic consolidation, and mineral reactions [*Powers*, 1967; *Skempton*, 1970; *Hunt*, 1990; *Audet and McConnell*, 1992; *Hart et al.*, 1995]. When unconsolidated sediment first enters a subduction zone, most of the water is stored in pore spaces, but the pores collapse near the deformation front owing to rapid vertical loading and tectonic consolidation [*Bekins and Dreiss*, 1992; *Moore and Vrolijk*, 1992]. Consolidation then decreases exponentially as effective stress increases along the subduction path [*Bray and Karig*, 1985]. At the same time, fluid production from dehydration reactions becomes increasingly important as the burial temperature increases and the pore volume is reduced by compaction. To complicate matters further, the thermal structure of subducting lithosphere varies with crustal age, basement relief, and patterns of hydrothermal circulation [*Lister*, 1977; *Parsons and Sclater*, 1977; *Stein and Stein*, 1994; *Fisher et al.*, 2003a, 2003b; *Harris et al.*, 2004; *Hutnak et al.*, this volume]. The zones of maximum fluid production therefore will shift spatially in response to variations in two independent input parameters: hydrous mineral abundance and heat flow.

In general, a larger proportion of uncemented sand in the upper part of an approaching stratigraphic section should favor higher friction coefficients, higher net permeability, better drainage, and fluid pressures closer to hydrostatic, as compared to subduction inputs of clay-rich hemipelagic mudstone or pelagic ooze [*Lambe and Whiteman*, 1969; *Mitchell*, 1993; *Neuzil*, 1994; *Giambalvo et al.*, 2000]. One of the ways in which accretionary prisms adjust to these hydrogeologic variables is the taper angle, which steepens in well-drained systems and flattens in poorly drained systems [*Saffer and Bekins*, 2002]. Pathways of focused fluid migration, moreover, will track the architecture of high-permeability sand bodies inherited from the depositional environment (i.e., isolated shoestring channel sands, stacked channel-levee complexes, unconfined sand sheets, depositional lobes). Conversely, if subduction inputs consist of low-permeability mud throughout a system, then diffuse drainage will be retarded and focused flow will follow structural conduits of fracture-enhanced permeability [*Moore and Vrolijk*, 1992; *Saffer and Bekins*, 1999].

As compaction progresses, the water stored within the structures of such minerals as opal and smectite contributes more and more to the total fluid budget. Transformations of opal to quartz and smectite to illite are controlled by reaction kinetics [*Earnst and Calvert*, 1969; *Huang et al.*, 1993]. Liberation of mineral-hosted water occurs in discrete temperature-dependent steps [*Bird*, 1984; *Bruce*, 1984; *Keller and Isaacs*, 1985; *Colten-Bradley*, 1987]. Secondary factors (e.g., solubility, bulk rock composition) can affect the timing of opal-to-quartz diagenesis [*Pisciotto*, 1981; *Isaacs*, 1982; *Williams et al.*, 1985], but that reaction usually reaches completion between 45°C and 85°C [*Murata et al.*, 1977; *Behl and Garrison*, 1994]. Higher contents of detrital silt and clay also drive reaction temperatures down [*Keller and Isaacs*, 1985]. During smectite diagenesis, two pulses of dehydration usually occur at ~85°–95°C and ~120°–130°C [*Perry and Hower*, 1972]. Thus we should see evidence of fluid production within several temperature-dependent loci that shift through subduction systems in time and space. If strata within the loci of dehydration are clay-rich and poorly drained, water liberated from the mineral structures into the pore space can drive pore pressure well above hydrostatic values [*Freed and Peacor*, 1989a; *Moore and Vrolijk*, 1992].

Merging the Paradigms

In its elementary form, the "smectite-to-illite hypothesis" provides scientists with a tractable idea to test, but that concept is oversimplified because it does not take into account three-dimensional variations in lithostratigraphy, either within or among individual subduction margins. Similarly, the hypothesis fails to account adequately for local and global variability in total clay abundance, the absolute amount of smectite, grain fabric, pore pressure, heat flow, and reaction progress prior to subduction. A more holistic view stressed by *Moore and Saffer* [2001] is that several interrelated factors and processes (e.g., precipitation of carbonate and silicate cements, progressive growth of phyllosilicates, clay-mineral dehydration, changes in frictional properties of clay) act in consort to modulate the strength and mechanical behavior of strata on opposite sides of the updip limit of the seismogenic zone. The recipe for blending each factor and process almost certainly changes from one subduction margin to another, within each individual margin, and through time. This variability creates a challenge for SEIZE; lessons from one system cannot necessarily be imported into another.

Many scholars of subduction have emphasized the ramifications of abnormally high pore pressure [*von Huene and Lee*, 1982; *Brown and Westbrook*, 1987; *Moore*, 1989; *Moore et al.*, 1991; *Moore and Vrolijk*, 1992; *Brown et al.*, 1994; *Tobin et al.*, 1994; *Saffer and Bekins*, 1999; *Saffer*, this volume]. Unbalanced loading, compaction disequilibrium, stratigraphic isolation of sand bodies by low-permeability mudstones, and thermal expansion of pore fluids can all enhance lateral fluid flow and help create overpressured compartments even without subduction-induced effects [*Barker*, 1972; *Bradley*, 1975; *Shi and Wang*, 1986; *Bredehoeft et al.*, 1988; *Hart et al.*, 1995; *Neuzil*, 1995; *Osborne and Swarbrick*, 1997; *Flemings et al.*, 2002]. Episodic or cyclic buildup of fluid overpressure helps account for the global phenomenon of weak faults [*Hubbert and Rubey*, 1959; *Sibson*, 1981; *Davis et al.*, 1983; *Byerlee*, 1990, 1993; *Magee and Zoback*, 1993; *Hickman et al.*, 1995; *Matthai and Fisher*, 1996]. One of the goals of SEIZE therefore is to track strata as they approach the seismogenic zone of a subduction megathrust and tease apart the respective contributions of changing frictional strength versus pore pressure and effective normal stress.

Nankai Trough

The Nankai accretionary prism is located where the Philippine Sea plate (Shikoku Basin) is subducting beneath the Eurasian plate at a rate of ~4 cm/yr [*Seno et al.*, 1993; *DeMets et al.*, 1994; *Shipboard Scientific Party*, 2001a; *Sella et al.*, 2002]. Two transects have been drilled over 3 decades: Legs 31 and 87 of the Deep Sea Drilling Project (DSDP) and Legs 131, 190, and 196 of the Ocean Drilling Program (ODP). The Muroto (central) transect (fig. 3.2) sits above a subducting basement high that formed by seafloor spreading and seamount volcanism (Kinan seamounts) on the back-arc side of the Izu-Bonin subduction zone [*Moore et al.*, 2001]. The Ashizuri (western) transect is located on the distal western flank of the fossil spreading center where the crust is older and relatively smooth. Some of the Kinan seamounts are as young as 7 Ma, and heat flow increases sharply in proximity to the fossil volcanic centers [*Yamano et al.*, 2003]. Two reference sites (1173 and 1177) provide the baseline data for characterizing along-strike changes in sediment inputs (fig. 3.2).

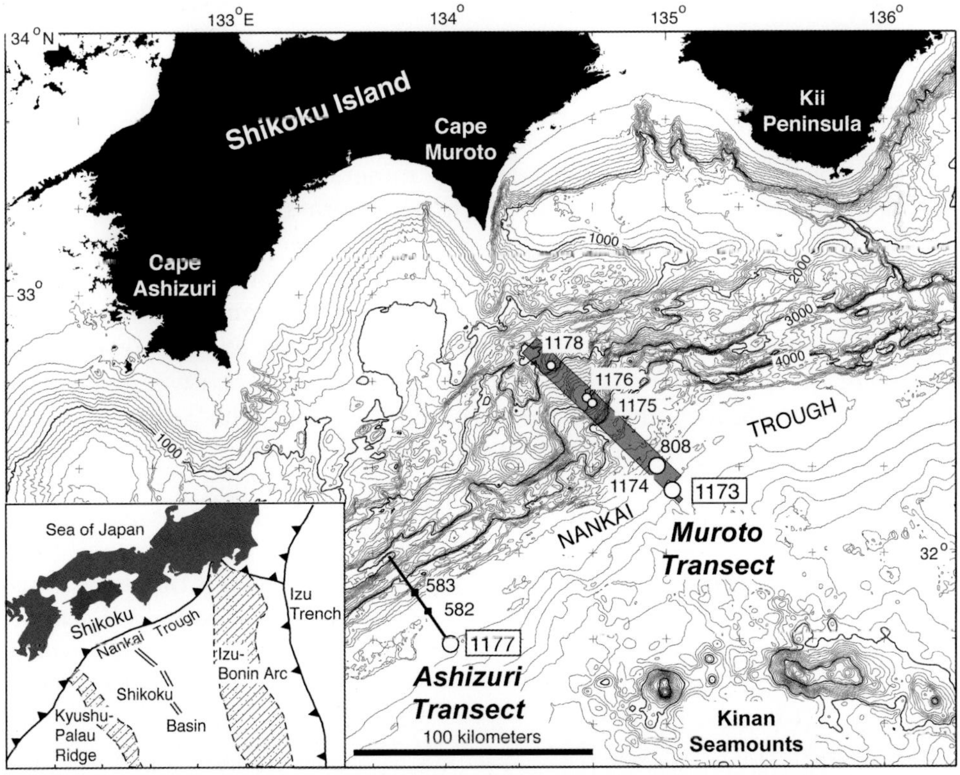

Figure 3.2 Index map of Nankai Trough and Shikoku Basin. Numbers refer to Deep Sea Drilling Project (DSDP) and Ocean Drilling Program (ODP) drill sites.

Lithostratigraphy

As demonstrated during ODP Leg 131 at ODP Site 808 [*Taira et al.*, 1992; *Pickering et al.*, 1993; *Taira and Ashi*, 1993], the stratigraphic section entering central Nankai Trough (Muroto transect) is more or less consistent with generic facies models [*Piper et al.*, 1973]. At Site 1173, the trench-wedge facies is 102 m thick, and the turbidites thicken and coarsen upward (fig. 3.3). The upper Shikoku Basin facies extends from 102 to 446 m below seafloor (mbsf) and includes interbeds of hemipelagic mud and volcanic ash, whereas the lower Shikoku Basin facies (446–790 mbsf) is a monotonous unit of bioturbated hemipelagic mudstone. At the Muroto deformation front (Site 1174), the décollement passes through homogeneous mudstone of the lower Shikoku Basin [*Moore et al.*, 2001; *Bangs et al.*, 2004].

The Ashizuri transect (fig. 3.2) differs fundamentally from Muroto because sandy turbidites are widespread within the lower Shikoku Basin. At Site 1177 the upper Shikoku Basin facies (300–402 mbsf) consists of hemipelagic mud and volcanic ash, similar to what was found at Site 1173. The lower Shikoku Basin contains a hemipelagic facies (402–449 mbsf) and a thick Miocene turbidite facies (449–748 mbsf) with uncemented layers of siliciclastic sand, gravel, and mudstone-clast conglomerate. Similar quartz-rich turbidites of Miocene-Pliocene age were recovered during DSDP Leg 31 at nearby Site 297 [*Shipboard Scientific Party*, 1975].

Sediment Composition

X-ray diffraction (XRD) analyses of clay minerals within the Quaternary trench-wedge facies of Nankai show a preponderance of detrital illite and chlorite [*Chamley*, 1980; *Chamley et al.*, 1986; *Underwood et al.*, 1993a, 1993b; *Steurer and Underwood*, 2003]. When clay-provenance indicators are combined with data from sand petrography, seafloor bathymetry, and paleoflow [*Taira and Niitsuma*, 1986; *Shimamura*, 1989; *Marsaglia et al.*, 1992; *Pickering et al.*, 1992; *Fergusson*, 2003], it becomes clear that the Quaternary trench wedge sustains itself mostly through southwest-directed axial transport, with primary sources located in the Izu-Honshu collision zone of central Japan.

Figure 3.3 Stratigraphic columns for ODP Sites 1177 and 1173, Nankai Trough. Also shown are calculated values of total clay minerals (from bulk-powder X-ray diffraction) and estimated percentages of smectite and illite in bulk mudstone. Relative percentages were calculated using singular value deconvolution (SVD) weighting factors for bulk powders [*Shipboard Scientific Party*, 2001b] and clay-size oriented aggregates [*Underwood et al.*, 2003]. Bulk percentages for smectite are products of %smectite relative to illite + chlorite (<2 μm) times total clay minerals relative to quartz + feldspar + calcite. See *Steurer and Underwood* [2003] for data. ▶

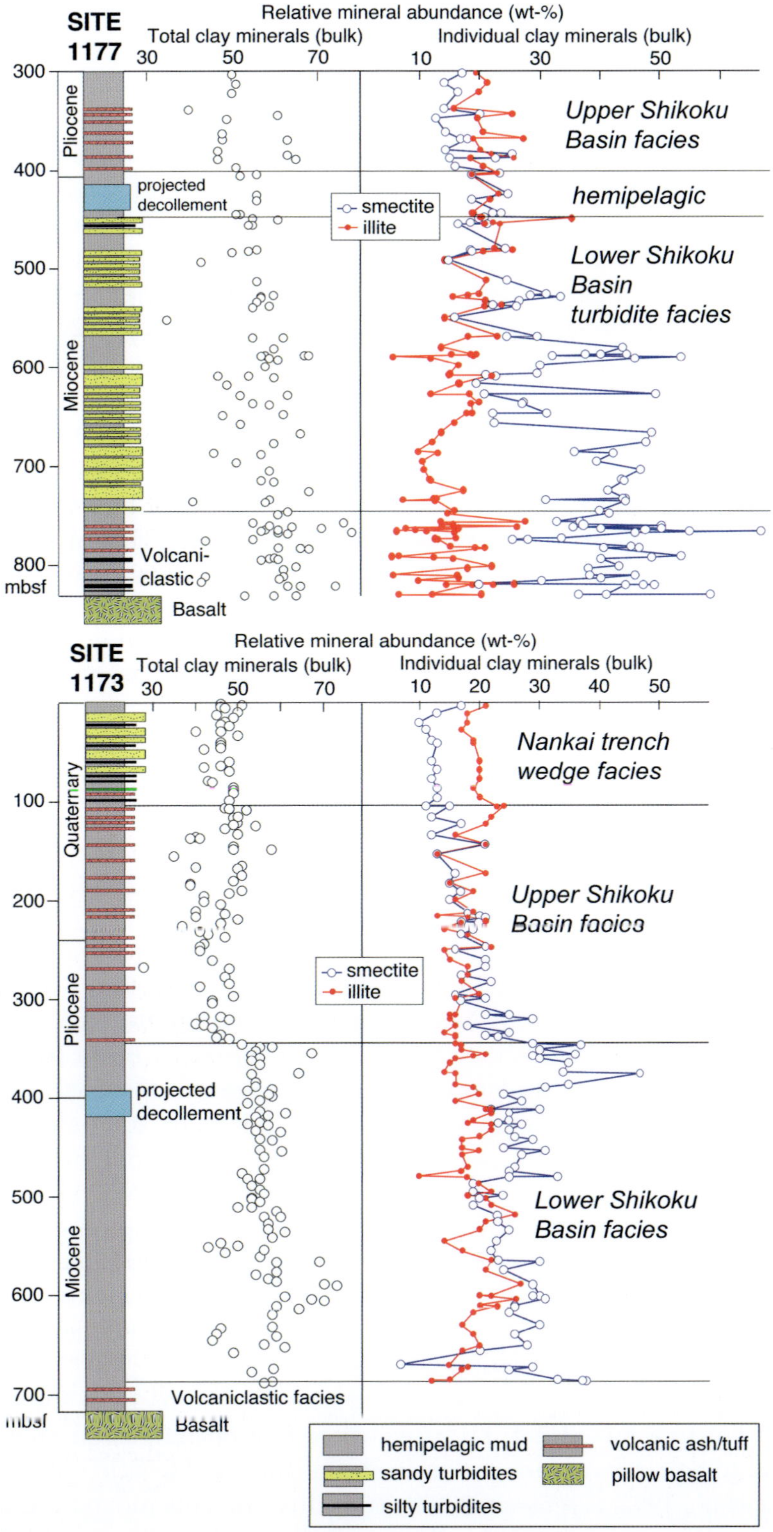
SITE 1177
Relative mineral abundance (wt-%)
Total clay minerals (bulk)
30 50 70
Individual clay minerals (bulk)
10 30 50
300
400
500
600
700
800
mbsf
Pliocene
Miocene
projected decollement
Volcani-clastic
Basalt
smectite
illite
Upper Shikoku Basin facies
hemipelagic
Lower Shikoku Basin turbidite facies
SITE 1173
Relative mineral abundance (wt-%)
Total clay minerals (bulk)
30 50 70
Individual clay minerals (bulk)
10 20 30 40 50
100
200
300
400
500
600
700
mbsf
Quaternary
Pliocene
Miocene
projected decollement
Volcaniclastic facies
Basalt
smectite
illite
Nankai trench wedge facies
Upper Shikoku Basin facies
Lower Shikoku Basin facies
hemipelagic mud
sandy turbidites
silty turbidites
volcanic ash/tuff
pillow basalt

Unlike the trench wedge, strata within the distal parts of Shikoku Basin (DSDP Sites 297, 442, 443, and 444) display evidence of progressive smectite depletion with decreasing age due to changes in detrital sources [*Chamley*, 1980]. Smectite also increases monotonically within the upper Shikoku Basin facies at Site 808, but *Underwood and Pickering* [1996] were unable to separate the effects of shifting provenance from in situ ash-to-smectite and smectite-to-illite diagenesis. Smectite at Site 1173 increases down section from ~10 bulk wt % at the top of the trench-wedge facies to >35 bulk wt % near the top of the lower Shikoku Basin facies; from there, even though total clay-mineral content increases, smectite decreases gradually down section to ~25 bulk wt % before reaching erratic proportions just above basement (fig. 3.3). Site 1177 exhibits a much different trend (fig. 3.3). The lower Shikoku Basin turbidite facies yields some compositional scatter due to its lithologic heterogeneity (interbedded turbidite-hemipelagite), but the average content of smectite in mudstone increases dramatically from ~20 to 50 bulk wt %. Miocene mudstones from DSDP Site 297 (Ashizuri transect) are also enriched in smectite and total clay minerals [*Underwood et al.*, 2003]. The higher content of smectite is a consistent trait of Miocene strata throughout the region, including accreted mudstones at Site 1178 [*Underwood and Steurer*, 2003] and coeval deposits in the Sea of Japan [*Fagel et al.*, 1992]. The temporal change in clay-mineral suites has been attributed to gradual intensification of the proto-Kuroshio Current around 3 Ma, dampening of weathered-volcanic influx from the Izu-Bonin arc, and recycling more clay minerals from sedimentary rocks that were uplifted and exposed in the Outer Zone of Japan [*Underwood and Steurer*, 2003].

Implications

The décollement off Muroto occurs within a monotonous interval of hemipelagic mudstone, along an initially nonreflective interface [*Bangs et al.*, 2004]. Seismic-reflection records from Ashizuri indicate that the frontal décollement in that corridor passes just above the top of the Shikoku Basin turbidite facies [*Morgan et al.*, 1995]. Mudstones with high contents of smectite are common below the stratigraphic position of the décollement. Ring-shear tests of samples from DSDP Site 297 prove that the high smectite contents reduce the mudstone's coefficient of friction to unusually low values (<0.25) [*Brown et al.*, 2003]. This frictional effect, however, has little or no bearing on geotechnical properties at the position of the frontal décollement. Nowhere does the fault's stratigraphic position coincide with any anomalous increases in the amount of total clay or percent smectite (fig. 3.3). Factors other than clay mineralogy must be considered to explain the fault's location.

Petrographic analyses of Pliocene and Miocene sands from the Ashizuri transect point to a mixed detrital provenance that encompassed recycling of the Shimanto Belt, coeval fore-arc-basin strata, and local felsitic-volcanic rocks [*Marsaglia et al.*, 1992; *Fergusson*, 2003]. Turbidity currents moved the sand toward the southwest into Shikoku Basin, mostly as sheet flows, with the distances of transport extending perhaps 600 km beyond the base of slope. Turbidites are absent from the Miocene-Pliocene section at Site 1173, however, so the fossil spreading ridge and Kinan seamounts must have retained enough seafloor relief during that time to deflect turbidity currents and inhibit transport of sand onto basement highs. Thus, when viewed in three dimensions, the Ashizuri sand layers pinch out along strike against a coeval Muroto mudstone facies (fig. 3.4). Vertical drainage above the turbidite facies should be retarded by 47 m of low-permeability mudstone (fig. 3.4). As noted by *Bradley* [1975], overpressured compartments in sedimentary basins must be viewed as independently isolated with hydrologic seals on all sides. In theory, the three-dimensional configuration of highly permeable, uncemented sand bodies in Shikoku Basin should set up analogous overpresssured compartments at the updip pinch-outs. Compaction disequilibrium, moreover, should magnify the effect once strata pass beneath the trench wedge and deformation front (fig. 3.4). These regional-scale hydrogeologic effects, rather than lithology-dependent frictional effects, probably dictate where the plate-boundary fault is positioned within the subducting sedimentary section.

Prior to subduction, several diagenetic reactions are superimposed on temporal shifts in the flux of detrital clay into Shikoku Basin. These changes in composition are important because they affect fluid budgets, pore-water chemistry, and physical properties. Because of fluctuating heat flow in Shikoku

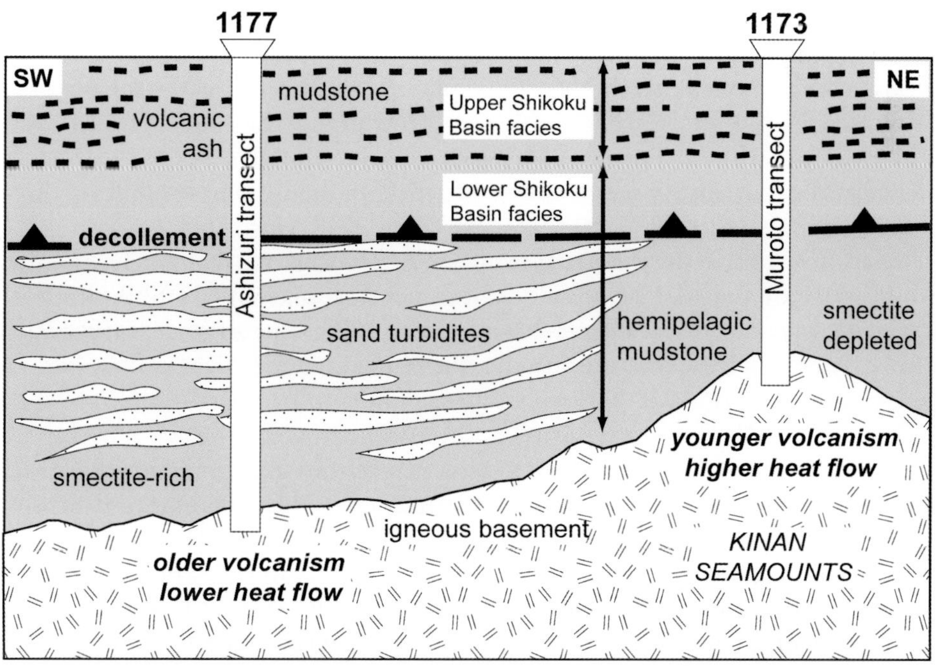

Figure 3.4 Schematic cross section through ODP Sites 1177 and 1173, Nankai Trough (see fig. 3.2 for location). The plane of the section is oriented parallel to the strike of the margin and shows pinch-out of Miocene to Pliocene turbidites against the seafloor relief of the Kinan seamounts.

Basin [*Kinoshita and Yamano*, 1986; *Wang et al.*, 1995; *Yamano et al.*, 2003], reaction progress varies considerably along strike. Heat flow within the Muroto transect is unusually high (typically 180–240 mW/m^2 seaward of the deformation front and 130–200 mW/m^2 at the prism toe). Glass shards in discrete volcanic-ash layers at Site 808 have altered to dioctahedral smectite (Al-rich beidellite) and K-rich clinoptilolite [*Masuda et al.*, 1996]. "Spikes" in smectite content (up to 91% of the clay-size fraction) occur locally; these cryptic bentonites are products of in situ alteration of volcanic-ash layers [*Underwood and Pickering*, 1996]. Replacement of disseminated glass shards first enhances the smectite content of Shikoku Basin mudstones [*Steurer and Underwood*, 2003], but higher burial temperatures (projected to 140°C at the base of Hole 1174) have pushed smectite-to-illite diagenesis nearly to completion in the lowermost strata [*Underwood et al.*, 1993b; *Underwood and Pickering*, 1996; *Masuda et al.*, 2001]. Clay-mineral alteration actually begins seaward of the trench, and reaction progress advances in a landward direction. At Site 1173, the proportion of illite in illite/smectite (I/S) mixed-layer clays increases from ~15% at 390 mbsf to a maximum of 60% near the bottom of the hole, but there is considerable scatter in the values and no I/S ordering. Reaction progress is more advanced at Site 1174 (up to 84% illite in I/S), and the reaction profile there is tighter [*Steurer and Underwood*, 2003]. At Site 808, the reaction reaches 89% illite in I/S with $R = 1$ ordering [*Underwood and Pickering*, 1996]. What this means, in essence, is that inputs to the Muroto corridor of the Nankai subduction zone are "pre-cooked." Accordingly, the locus of smectite-to-illite diagenesis remains unusually close to the prism toe. Reaction progress continues toward completion beneath the décollement, but the hypothetical downdip consequences (i.e., increased coefficient of friction, exhaustion of clay dehydration, dissipation of fluid overpressure, higher effective stress) will be less intense and occur earlier along the subduction path, as compared to corridors where heat flow is lower.

Heat flow along the Ashizuri transect, in contrast to Muroto, is relatively low because of thermal decay along the distal ridge flank of Shikoku Basin [*Kinoshita and Yamano*, 1986; *Ashi and Taira*, 1993; *Shipboard Scientific Party*, 2001a]. The projected bottom-hole temperature at Site 1177 is only ~45°C, so the I/S mixed-layer clays do not change systematically with depth [*Steurer and Underwood*, 2003]. For most samples, the proportion of illite in I/S clays remains <25%, with random ordering, as one might expect for a mostly detrital clay assemblage. This nonexistence of smectite-to-illite diagenesis is actually representative of Nankai inputs overall; the thermal state of the Muroto transect is aberrant. Once strata move deeper beneath the Ashizuri accretionary prism (i.e., 30–40 km landward of the trench), release of interlayer water from the smectite and I/S must contribute to changes in hydrogeology and geotechnical properties below and within the deeper plate boundary. Drainage of that pore water into the subducted Miocene sand beds probably promotes fluid flow in the updip direction, but flow focusing is within stratigraphic conduits beneath the décollement.

Cascadia Basin

Subduction of the Juan de Fuca plate beneath western North America occurs at a rate of 4.2 cm/yr [*DeMets et al.*, 1990, 1994]. Cascadia Basin extends westward from the base of slope to where sediment laps onto juvenile basement of the Juan de Fuca Ridge (fig. 3.5). Sediment supply is unusually large throughout

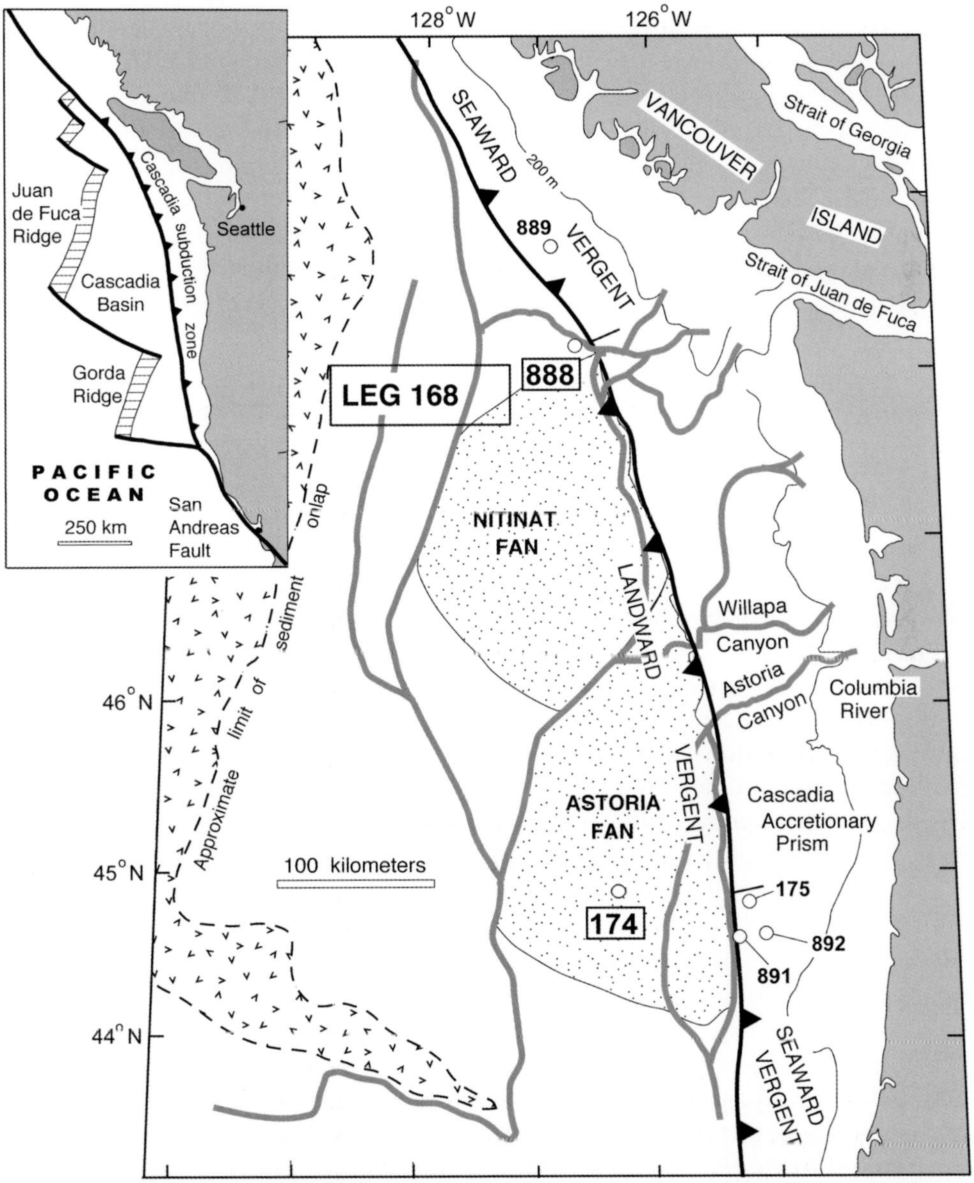

Figure 3.5 Index map of Cascadia Basin with locations of ODP Leg 168 transect, ODP Site 888 (Leg 146), and DSDP Site 174 (Leg 18). Note that the Fraser River discharges into the Strait of Georgia just beyond the limit of the map.

this region because of highly active submarine canyons and voluminous fluvial watersheds [*Carlson and Nelson*, 1969; *Barnard*, 1978; *Carson et al.*, 1986; *Davis and Hyndman*, 1989]. This delivery network drives total sediment thickness to a maximum of >3000 m [*Carlson and Nelson*, 1987; *MacKay*, 1995]. The basin floor contains several individual channel-levee complexes and two submarine fans [*Griggs and Kulm*, 1970; *Hampton et al.*, 1989]. ODP Site 888 is located on the northern fringes of Nitinat Fan (fig. 3.5), where total sediment thickness is ~2500 m [*Shipboard Scientific Party*, 1994]. Astoria Fan radiates asymmetrically from the mouths of Astoria and Willapa Canyons [*Carlson and Nelson*, 1969; *Nelson et al.*, 1970; *Nelson*, 1976]. The distal edge of that fan was cored at DSDP Site 174 (fig. 3.5), where total sediment thickness is ~910 m. Beneath the fan sediments, the distal abyssal-plain facies changes texture and thickness in response to basement relief and basement age, reaching a maximum thickness of ~1500 m [*Davis and Hyndman*, 1989; *MacKay*, 1995]. The basal décollement of the accretionary prism propagates into the abyssal-plain facies, well beneath the lowermost submarine-fan deposits [*Goldfinger et al.*, 1996].

Lithostratigraphy

ODP Leg 168 cored nine sites along the distal edge of northwest Cascadia Basin, where the basement age is 0.86–3.59 Ma (fig. 3.6). In general, strata within that area include three lithostratigraphic units that coarsen and thicken upward, from basalt at the base to hemipelagic mud, silt turbidites, and sand turbidites. The thickness of each unit changes considerably among sites (fig. 3.6), and most unit boundaries are time transgressive [*Su et al.*, 2000]. Basement relief is typically 300–500 m, and some basement highs pierce the sediment cover [*Davis et al.*, 1992; *Wheat and Mottl*, 1994; *Mottl et al.*, 1998]. ODP Sites 1030 and 1031, for example, are situated above a basement ridge that rises to within 41 m of the seafloor; the overlying sediments are mostly hemipelagic mud and carbonate-rich mud. Collectively, results from Leg 168 show how heterogeneous the facies architecture and sand-to-mud ratios can be, even within the distal extremities of Cascadia Basin [*Underwood et al.*, 2005]. This intricate stratigraphic variability is important because it probably influences décollement behavior once strata reach the deformation front.

ODP Site 888 (fig. 3.5) was cored during Leg 146 as the oceanic reference site for Vancouver subduction-accretion [*Shipboard Scientific Party*, 1994]. Magnetic anomalies indicate a basement age of ~6 Ma beneath the drill site [*Riddihough*, 1977], but the age of the recovered section is late Pleistocene (<0.6 Ma). Drilling penetrated only 567 m into sand and silt turbidites, pebbly sand, and hemipelagic mud associated with Nitinat Fan (fig. 3.7) and stopped well short of the décollement horizon, which rests almost 2500 m below seafloor at the prism toe [*Westbrook*, 1994]. The sandiest interval recovered at Site 888 (channelized middle fan facies) extends from ~190 to 452 mbsf (fig. 3.7).

DSDP Site 174 is located on the southwest fringes of Astoria Fan (fig. 3.5). Acoustic basement is 911 m below sea floor [*von Huene and Kulm*, 1973], and the

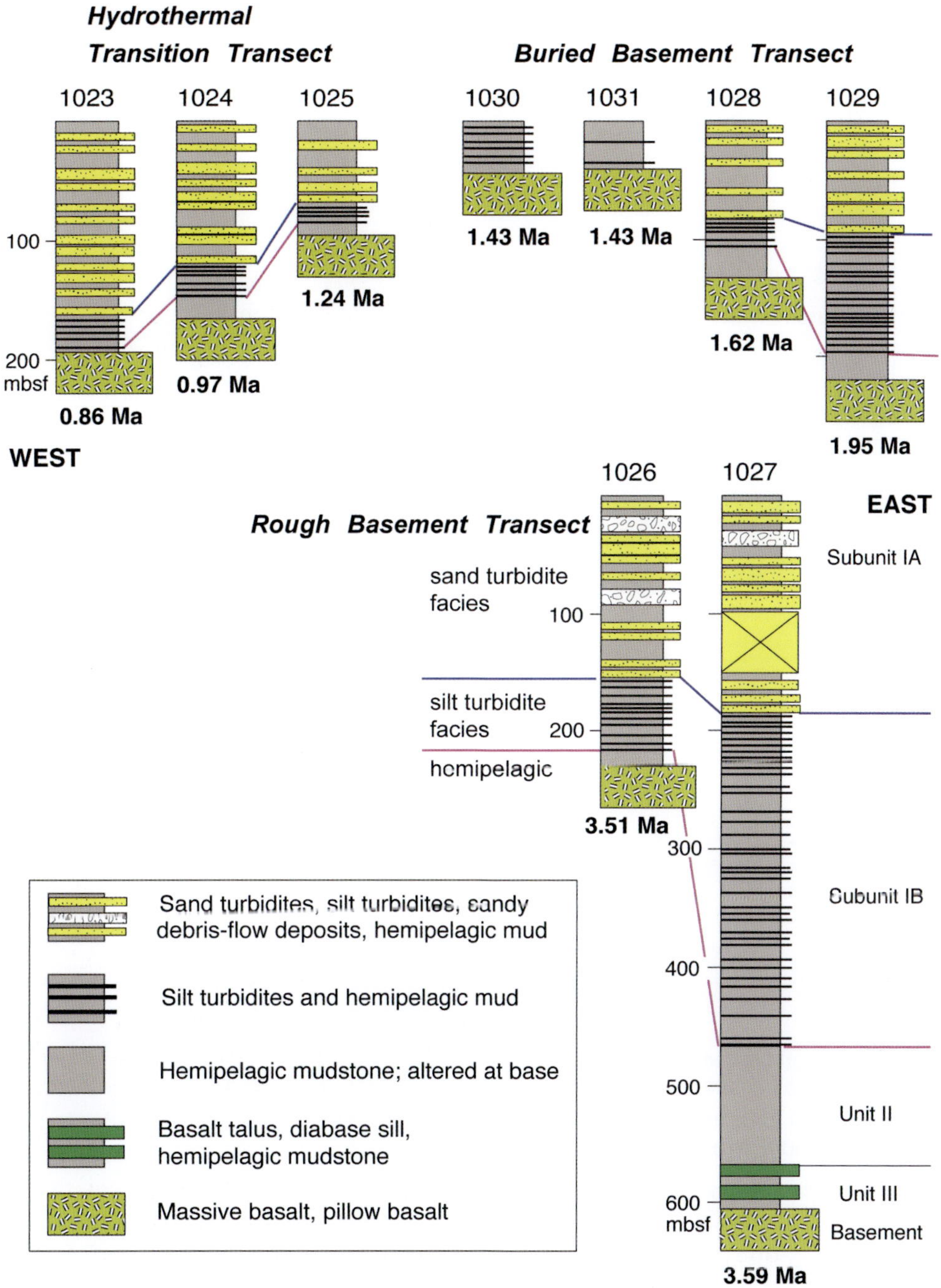

Figure 3.6 Stratigraphic columns for ODP sites cored in Cascadia Basin during Leg 168. Numbers at the base refer to ages of igneous basement (Ma).

igneous age beneath Site 174 is ~8 Ma [*Heirtzler et al.*, 1968]. This fan is >1000 m thick near its apex but thins to only 284 m at Site 174 [*Nelson*, 1984; *Goldfinger et al.*, 1996]. The distal-fan facies contains substantial amounts of sand through-

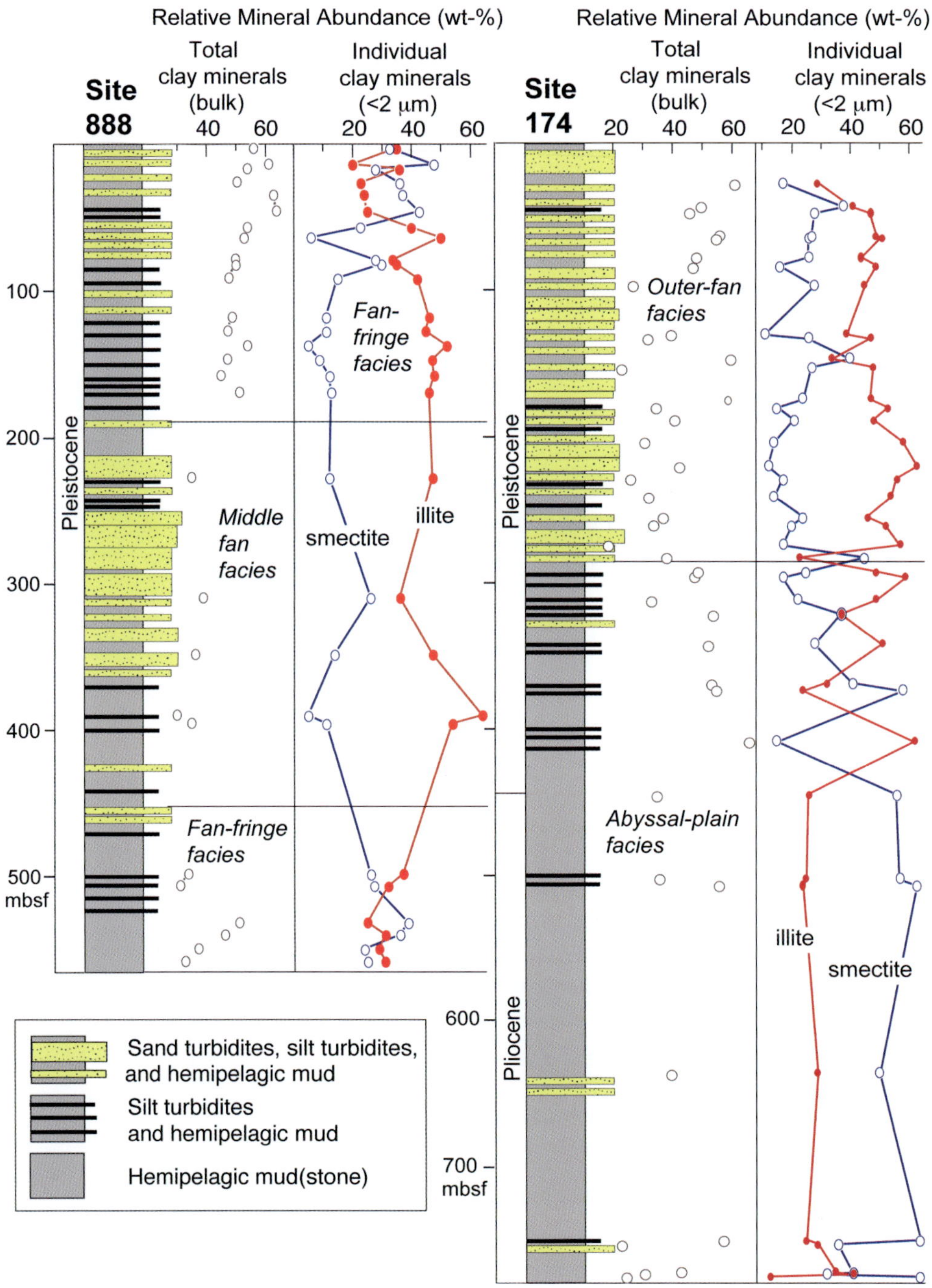

Figure 3.7 Stratigraphic columns for ODP Site 888 and DSDP Site 174, Cascadia Basin. Also shown are calculated values of total clay minerals (from bulk-powder X-ray diffraction; relative to quartz + feldspar + calcite) and relative percentages of smectite and illite in clay-size fraction (<2 μm). Relative percentages were calculated using SVD weighting factors for bulk powders [*Underwood et al.*, 2003] and Biscaye weighting factors clay-size oriented aggregates [*Biscaye*, 1965].

out, whereas the underlying abyssal-plain facies consists of thinner sand-to-silt turbidites and hemipelagic mudstone (fig. 3.7). The oldest fan sediments at Site 174 are ~760 ka [*Goldfinger et al.*, 1996], but the facies boundary is time-transgressive because of subduction-accretion. The onset of fan sedimentation system-wide has been dated at ~1.3–1.4 Ma [*McNeil et al.*, 2000].

Sediment Composition

Delivery of sediment to the floor of the Cascadia Basin is currently dominated by two large fluvial systems: the Columbia and Fraser Rivers (fig. 3.5). Physical breakdown and chemical alteration of volcanic rocks within the Columbia River watershed combine to create a fluvial clay-mineral suite that averages 53% smectite [*Knebel et al.*, 1968]. The Fraser River watershed, which discharges in the Strait of Georgia, is more heterolithic with fewer volcanic sources of weathered clay; its higher latitude and cooler climate promote a higher ratio of mechanical to chemical weathering and more illite plus chlorite. Relative smectite contents in near-surface mud on the continental margin change in response to the fluvial discharge and ocean currents. Smectite exceeds 50% of the clay minerals near the Columbia River mouth [*Karlin*, 1980], but values decrease toward the north. Percentages of smectite also decrease southward to 20% or less because of a shift in provenance to the Klamath Mountains and California Coast Ranges (see also *Griggs and Hein* [1980]). An added complication is that clay dispersal on the floor of Cascadia Basin funnels through channels. Near-surface clays from the thalweg and levees of Cascadia Channel, for example, contain an average of 52% smectite from the Columbia River point source [*Duncan et al.*, 1970], but relative percentages on the surrounding abyssal floor are considerably lower.

Along the Leg 168 transect (fig. 3.6), the most abundant clay mineral is chlorite; its contribution to the <2 μm size fraction (relative to smectite + illite) varies from 29 to 78% with an average value of 52% [*Underwood and Hoke*, 2000]. Illite content is 18–59%, with a mean of 40%. Relative percentages of smectite range from 0 to 25%, with an average of only 8%. When these data are combined with values of percent total clay, it suggests that bulk percentages of smectite are probably 5 wt % or less. The likely sources for a chlorite-rich, smectite-poor clay assemblage include Vancouver Island (fig. 3.5), the Fraser River watershed via the Straits of Juan de Fuca, and perhaps the Olympic Peninsula [*Underwood and Hoke*, 2000].

Compared to the Leg 168 transect area, samples from the upper 90 m of Nitinat Fan (ODP Site 888) contain significantly higher percentages of smectite, ranging from 20 to 50% relative to chlorite + illite (fig. 3.7). This clay-mineral assemblage probably incorporated some of its suspended sediment from the Columbia River plume, which drifts well to the north of the river mouth during the winter [*Karlin*, 1980]. Compositional scatter also reflects variable amounts of mixing or interfingering of mud plumes from different sources. Below 90 mbsf (meters below seafloor), smectite decreases to between 5 and

39% of the clay-mineral budget; this shift toward more illite and chlorite is probably a response to greater amounts of discharge through the Strait of Juan de Fuca. The content of total clay minerals (relative to quartz, feldspar, and calcite) decreases steadily down section; values within the sand-rich middle-fan facies are <40 wt %. This shift in bulk composition is probably a grain-size effect (i.e., more silt-size quartz and feldspar). Merging data from bulk-powder and clay-size X-ray diffraction indicates that smectite in Nitinat Fan reaches a maximum of ~30 bulk wt % but averages only 10 bulk wt %.

Data from DSDP Site 174 share several similarities with the results from Site 888, but there are also some significant differences (fig. 3.7). Astoria Fan also shows a gradual down-section decrease in total clay minerals (relative to quartz, feldspar, and calcite), probably due to grain-size effects. The abundance of smectite (relative to illite + chlorite) within the turbidite-rich fan facies ranges from 11 to 38%, with a mean of 22%. The abyssal-plain deposits beneath Astoria Fan show an increase in total clay minerals, and there is a substantial increase in smectite in the clay-size fraction, with a maximum value of 65% and a mean of 42% (fig. 3.7). The highest estimate of smectite in bulk is 38 wt %, and the average bulk value for abyssal-plain deposits is 18 wt %.

Implications

Because the subducting Juan de Fuca plate is young (<9 Ma), heat flow values near the Cascadia prism toe are high (up to 160 mW/m^2) and calculated temperatures at the top of the igneous crust reach ~225° to 275 °C [*Hyndman and Wang*, 1993, 1995]. Guided by these thermal constraints, plus the unusual thickness of strata in Cascadia Basin, one might conclude that the smectite-to-illite reaction advances to completion seaward of the deformation front and that detrital smectite content, regardless of absolute abundance, has little or no bearing on downdip changes in material properties near the top of basement. According to *Hyndman and Wang* [1993, 1995], the temperature of transition from aseismic to seismic behavior is close to 150°C. If that suggestion is correct, then presubduction diagenesis helps explain why the entire megathrust seems to be locked and potentially seismogenic at relatively shallow depths [*Rogers et al.*, 1996; *Hyndman et al.*, 1997]. The exact depth where the smectite-to-illite reaction reaches completion probably shifts along strike, however, and those variations in reaction products could influence basal friction along the frontal décollement.

One of the enigmatic aspects of Cascadia accretion is the segmentation of thrust vergence [*Seely*, 1977; *MacKay et al.*, 1992; *MacKay*, 1995; *Fisher et al.*, 1999; *Gutscher et al.*, 2001]. Landward-vergent faults occur along a central corridor from northern Oregon to northern Washington (fig. 3.5). That part of Cascadia Basin seems to be dominated by smectite-rich sediment in the near-surface abyssal-plain facies [*Underwood*, 2002]. Faults flip abruptly to seaward vergent south of latitude 45°N, where illite and chlorite increase in abundance (fig. 3.5). The Vancouver margin (with relatively low contents of smectite) also contains seaward-vergent structures [*Davis and Hyndman*, 1989]. In theory, changes in thrust vergence relate to how large the contrast in shear strength happens to be across the base of the accretionary prism [*Seely*, 1977], but a definitive explanation for the exact cause of the effect has proven elusive.

Looking at the prism toe, *MacKay* [1995] showed how depth to the décollement jumps up and down section in ensemble with fault vergence (i.e., ~1200 m above igneous basement in the zone of seaward vergence, but only ~400 m above basement in the zone of landward vergence). The reasons for the fault's 800-m adjustment in stratigraphic position are unknown. Regardless of the cause, variations in temperature within and below the fault zone at these contrasting depths should foster coincident shifts in smectite-to-illite reaction progress. Within the southern corridor of seaward vergence, the fault's propagating tip (at 2400 mbsf) coincides with an estimated temperature of ~120°–145°C (using near-surface heat-flow data of *Hyndman and Wang* [1995] and the thermal conductivity data of *Davis et al.* [1997]). Illitic I/S mixed-layer clay should survive at those temperatures, especially if the rate of heating is fast [*Freed and Peacor*, 1989b]. In theory, this set of circumstances should push the final stage of smectite dehydration farther landward and deeper beneath the prism. At the toe, where frontal thrusts develop, the incomplete I/S reaction within smectite-poor mudstones should maintain values of shear strength that do not vary significantly across the base of the wedge. Conversely, within the zone of landward vergence, a temperature estimate for the frontal décollement is 135°–160°C (at 3000 mbsf). Warming the subducted smectite-rich hemipelagic mud by an additional 15°–20°C probably expels the last of three layers of water from the clay's interlayer site [*Colten-Bradley*, 1987] soon after subduction. Once that water is liberated, pore fluids should migrate updip and drain into the décollement. This sets up a condition in which migrating fluids encounter less-reacted smectite-rich mudstones with lower permeabilities (relative to illite-chlorite-rich mudstones along strike), thereby enhancing the likelihood of fluid overpressure. Thus earlier (shallower) clay dehydration within the zone of landward vergence may be responsible for increases of pore pressure, reductions of effective stress, and lowering of basal shear strength closer to the deformation front, as compared to later (deeper) dehydration of smectite-poor strata along strike within the zones of seaward vergence [*Underwood*, 2002].

Barbados Ridge

Barbados Ridge marks the subduction boundary between the North American plate and the Caribbean plate (fig. 3.8). The convergence rate resolved perpendicular to the trench is ~2.8 cm/yr [*Dixon et al.*, 1998]. The accretionary prism narrows significantly toward the north because supplies of sediment from South America decrease in that direction [*Westbrook*, 1982]. Ocean drilling (DSDP Leg 78A, ODP Legs 110, 156, and 171A) has been restricted to a single anomalous corridor where Tiburon Rise impinges on the subduction front and total sedi-

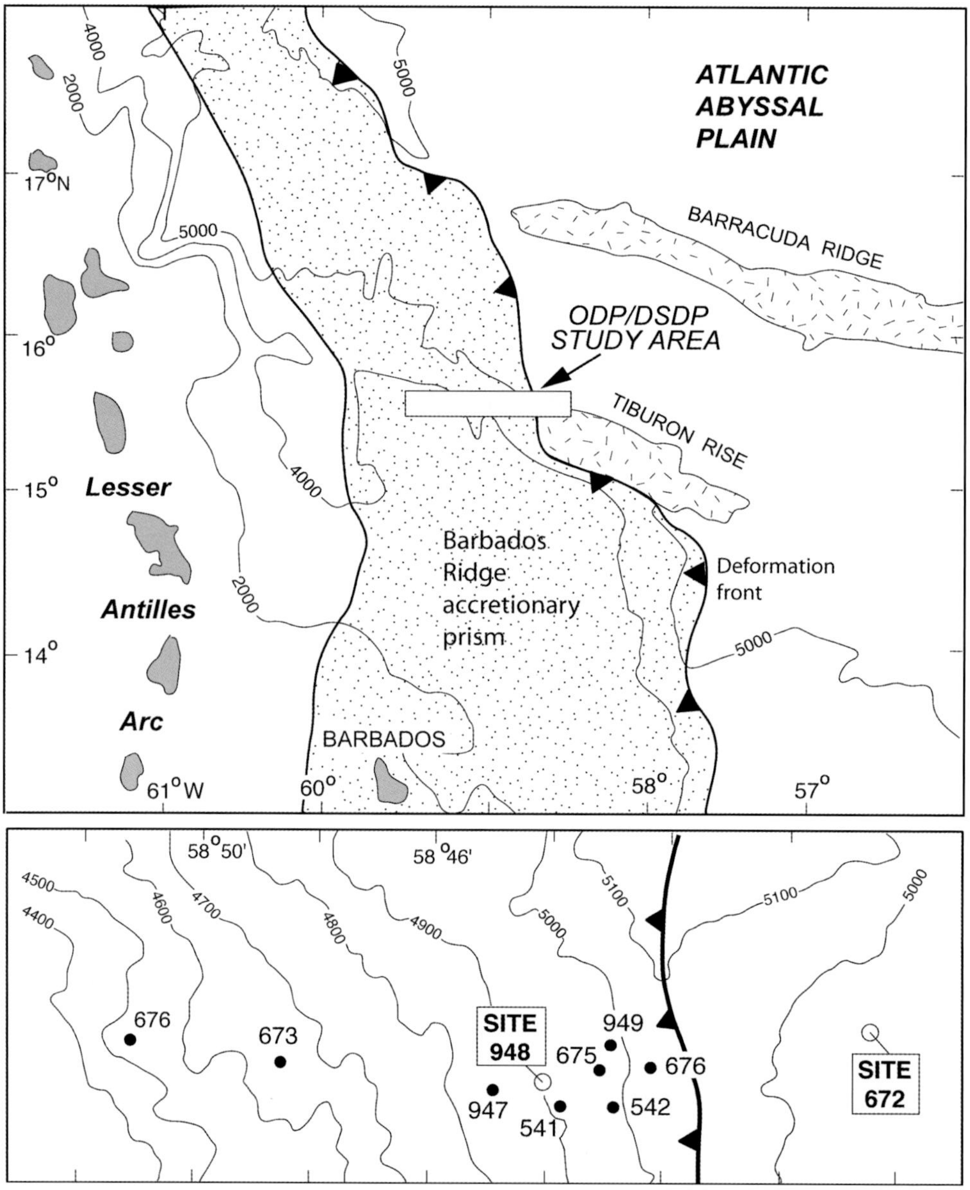

Figure 3.8 Index map of northern Barbados Ridge with location of DSDP/ODP drilling transect. Inset map below shows positions of individual drilling sites.

ment is thin (fig. 3.8). West of the deformation front, imbricate thrust faults characterize the structural domain of frontal offscraping [*Moore et al.*, 1982, 1988]. The décollement is a long, potentially dilatant zone that typically displays high amplitude and negative polarity on seismic reflection profiles [*Westbrook and Smith*, 1983; *Westbrook et al.*, 1988; *Bangs et al.*, 1996; *Shipley et al.*, 1994].

Lithostratigraphy

Strata entering the northern Barbados subduction zone include five fundamental subdivisions (fig. 3.9): (1) lower Pleistocene to Pliocene calcareous clay, mud, marl, and volcanic ash; (2) upper to middle Miocene mudstone and calcareous mudstone with sparse ash; (3) lower Miocene radiolarian-bearing claystone; (4) Eocene-Oligocene interbeds of claystone, calcareous claystone, chalk

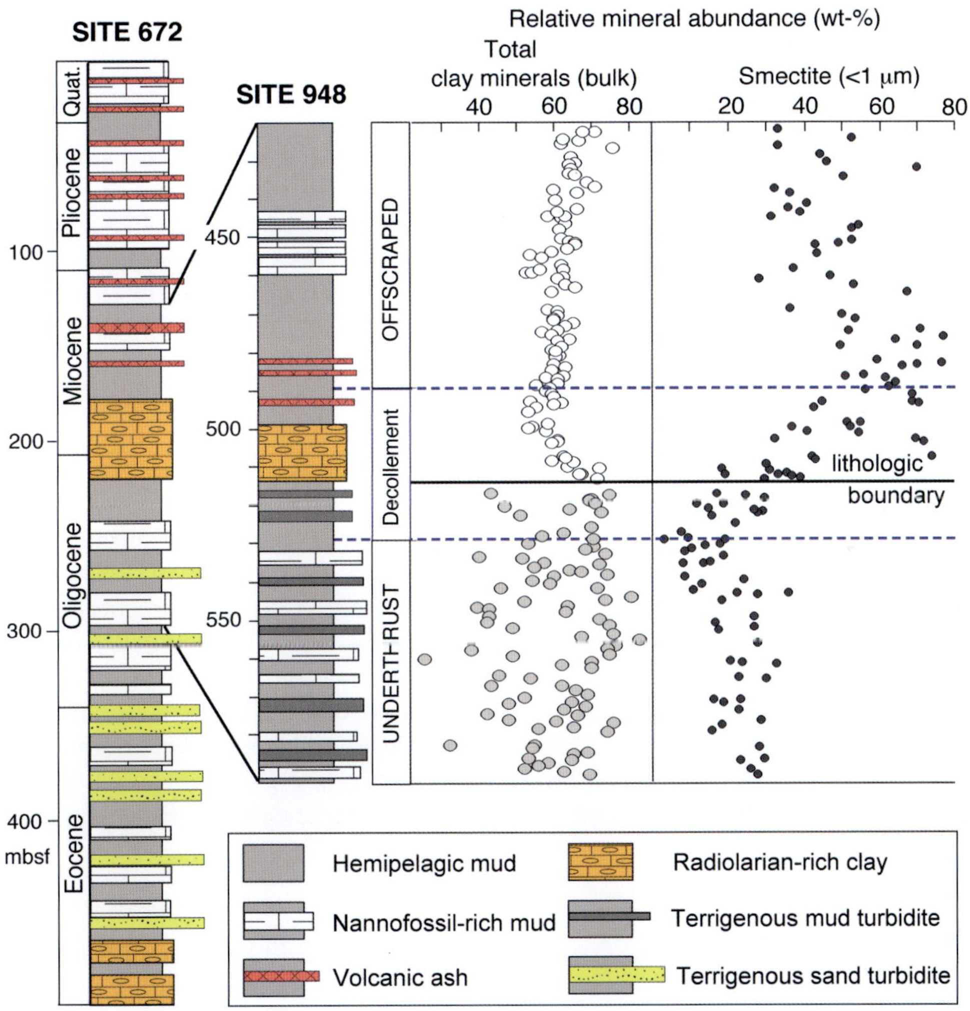

Figure 3.9 Stratigraphic columns for ODP Sites 672 and 948, Barbados Ridge. Note expansion of the depth scale for Site 948. Also shown are calculated values of total clay minerals (from bulk-powder X-ray diffraction, relative to quartz + feldspar + calcite) and relative percentages of smectite in clay fraction (<1 μm). Relative percentages were calculated using SVD weighting factors for bulk powders [*Fisher and Underwood*, 1995] and Biscaye weighting factors for clay-size oriented aggregates [*Biscaye*, 1965]. See *Underwood and Deng* [1997] for data tables.

turbidites, silt turbidites, and quartz-sand turbidites; and (5) calcareous claystone overlying pillow basalt [*Moore et al.*, 1982, 1988, 1998]. The anomalously coarse-grained Eocene turbidites evidently were transported from continental sources on South America upslope onto the basement high of Tiburon Rise [*Dolan et al.*, 1989]. The frontal décollement was analyzed very thoroughly at Site 948, where a first-order lithologic boundary separates the uppermost turbidites from radiolarian claystone (fig. 3.9). The base of the décollement zone is at 530 mbsf, ~15 m beneath the lithologic boundary, whereas the top is at ~490 mbsf, 10 m above the radiolarian claystone [*Housen et al.*, 1996].

Sediment Composition

The clay mineral suite in the Barbados subduction system includes smectite, kaolinite, and illite [*Latouche and Maillet*, 1984; *Pudsey*, 1984]. Smectite content at Site 672 (oceanic reference site) increases to 85–90% of the clay-size fraction (<2 μm) near the proto-décollement; lower Miocene radiolarian claystones at Site 671 (prism toe) also tend to be smectite-rich [*Capet et al.*, 1990; *Tribble*, 1990]. The smectite's origin has been debated. One school of thought favors an authigenic source [*Pudsey*, 1984; *Tribble*, 1990; *Tribble and Yeh*, 1994], whereas others [*Latouche and Maillet*, 1984; *Capet et al.*, 1990] have argued for temporal changes in the detrital balance between continental transport from South America (richer in illite and chlorite) and volcanogenic input from the Lesser Antilles arc (richer in smectite and kaolinite).

At Site 948, the contents of total clay minerals and smectite both increase near the top of the décollement. The clay-size fraction of radiolarian claystone is 32–71% smectite, but even higher relative percentages (70–80% smectite) occur within the upper part of the décollement zone (fig. 3.9). The elevated percentages of total clay minerals and smectite convert to bulk smectite estimates of 35–55 wt %. The monotonic reduction of smectite down the décollement zone is particularly noteworthy (fig. 3.9). Below the lithologic boundary, interbeds of hemipelagic mudstone and mud turbidites contain ~20% smectite, on average, and the lowest value (3%) occurs immediately below the décollement zone.

Implications

Heat flow is erratic near the Barbados deformation front because of transient fluid flow, but the calculated values (92–192 mW/m^2) are much higher than expected for the age of subducting lithosphere [*Fisher and Hounslow*, 1990]. Using samples from Site 671, *Tribble* [1990] showed that some of the smectite is diagenetically altered to I/S mixed-layer clay (with ~30–65% illite) at depths below 500 mbsf. This discovery was both curious and unexpected given the projected burial temperature of ~25°C at 500 mbsf, but *Buatier et al.* [1992] provided verification of the X-ray diffraction interpretation using transmission electron microscopy (TEM) and geochemical evidence. One explanation for incipient illitization at such low temperatures is enhanced fluid flow, which would replenish the supply of essential solutes such as K^+. *Underwood and Deng* [1997] showed that a stepwise shift in apparent reaction progress occurs precisely at the lithologic boundary between fine-grained turbidites and radiolarian claystone; this correlation to the stratigraphy means that all or part of the I/S reaction effect may be due to temporal changes in detrital clay provenance. In addition, there is no evidence for anomalous illitization within the stratigraphic intervals most likely to experience enhanced flow. The top candidate, in this regard, would be near the top of the frontal décollement (~495 mbsf), where pore-water concentrations of chloride (i.e., independent evidence of enhanced fluid flow) are diluted significantly [*Kastner et al.*, 1997]. In conclusion, the full extent and cause of presubduction clay diagenesis remain inconclusive.

Tribble [1990] also demonstrated that lower Oligocene and Eocene strata (with abundant sandy turbidites) beneath the décollement horizon contain substantial amounts of smectite (32–67% in the clay fraction). The sand beds probably provide high-permeability stratigraphic conduits for enhanced drainage below the décollement. Dehydration of the mudstone interbeds, moreover, probably contributes to anomalies in both physical and chemical hydrogeology. There is compelling geophysical and geotechnical evidence for transient fluid flow along the décollement, fracture-enhanced permeability, high pore pressure, and dilation [*Moore et al.*, 1995; *Fisher and Hounslow*, 1990; *Fisher et al.*, 1996; *Fisher and Zwart*, 1997; *Moore and Tobin*, 1997; *Screaton et al.*, 1997; *Bekins and Screaton*, this volume]. Sand layers below the décollement doubtlessly furnish preferred pathways for fluid flow until the permeability is occluded by cementation, but once that occurs, drainage probably focuses along the fault. In support of this notion, the chloride content of pore fluids decreases dramatically within and immediately above the fault zone [*Gieskes et al.*, 1990; *Kastner et al.*, 1991; *Vrolijk et al.*, 1991]. Most workers have attributed those geochemical gradients to updip fluid migration [*Bekins et al.*, 1994, 1995; *Kastner et al.*, 1997], although some of the freshening may be a laboratory artifact of squeezing interlayer water from smectite-rich clays [*Fitts and Brown*, 1999].

Kopf and Brown [2003] showed that the majority of clays and claystones they tested from Barbados have unusually low coefficients of friction (below 0.25); the minimum value (0.13) coincides with the décollement zone. Several interrelated factors merge together to create a zone of unusual weakness near the frontal décollement [*Deng and Underwood*, 2001; *DiLeonardo et al.*, 2002; *Wallace et al.*, 2003]. Some of the factors are inherited from presubduction lithostratigraphy. The interval of core-scale deformation stretches to nearly 40 m in thickness at Site 948 and extends across several diverse sediment types (fig. 3.9). A sharp minimum in percent smectite and a maximum in percent illite mark the base of the décollement, but the top of the décollement passes through a diffuse interval of high smectite (fig. 3. 9). Thus the geotechnical properties of just one lithology (e.g., radiolarian claystone) cannot be solely responsible for the décollement's location. First, individual slip events are more likely to occur where smectite lowers frictional strength [*Deng and Underwood*, 2001]. A second factor is the local abundance of radiolarians, which contribute to higher than normal initial porosities followed by abnormal porosity collapse [*Wallace et al.*, 2003]. A third factor is that the shear strength must decrease wherever pore pressure within the fault zone rises above hydrostatic [*Moore et al.*, 1995]. Updip fluid advection is probably the principal cause of the excess pore pressure, but fluid flow then triggers another potential feedback: clay-swelling pressure. The "imported" fluid is less saline (relative to normal ocean bottom water), presumably because it migrated from zones of deeper-seated dehydration reactions. Decreases in pore-fluid salinity and porosity collapse should increase the physicochemical stress generated by swelling clay minerals [*Chatterji and Morgenstern*, 1990; *Mesri et al.*, 1994]. If fresher pore water does migrate to the propagating tip of the décollement, its arrival should increase smectite swelling and reduce the mudstone's frictional strength even more [*Deng and Underwood*, 2001].

Costa Rica

Most of the recent research on sediment inputs to the Middle America Trench has concentrated on the Nicoya Peninsula region offshore Costa Rica (fig. 3.10). The Cocos plate in this region is subducting beneath the Caribbean plate at a rate of ~85 mm/yr [*DeMets*, 2001]. Terrigenous sedimentation is subdued (relative to Nankai, Cascadia, and Barbados) because of the comparatively small watersheds of Central America. Seaward of the subduction front, magnetic anomalies can be interpreted to delineate between older lithosphere (20–25 Ma) that formed at the East Pacific Rise (EPR) versus lithosphere (15–20 Ma in age) that originated at the Cocos-Nazca spreading center (CNS) [*Hey*, 1977; *Protti et al.*, 1995; *Barckhausen et al.*, 1998]. Basement roughness also changes across the boundary; seamount chains (e.g., Fisher Seamount) and large aseismic ridges (e.g., Cocos Ridge) are more prevalent on the CNS crust [*Hey*, 1977; *von Huene et al.*, 2000]. However, within the TicoFlux study area (fig. 3.10), the CNS crust is smooth and the EPR crust displays greater basement roughness.

Lithostratigraphy

One reference site was cored on the floor of Middle America Trench during ODP Leg 170 (Site 1039), within the area of EPR-crust subduction (fig. 3.10). Two sites at the margin toe (Sites 1040 and 1043) revealed the same basic stratigraphic succession, with nearly all of the incoming sediment thrust beneath the frontal décollement [*Kimura et al.*, 1997; *Saito and Goldberg*, 2001]. The reference section at Site 1039 starts at the base with gabbro and contains ~225 m of pelagic sediment, which is composed of mostly of calcareous nannofossils with variable amounts of diatoms (fig. 3.11). A transitional unit of mixed sediment separates the chalk from overlying deposits of siliceous hemipelagic mud. The hemipelagic facies is ~150 m thick.

Sediment is thin or absent above the seamounts; as a result, there is a substantial amount of variability in total sediment thickness to the northwest of

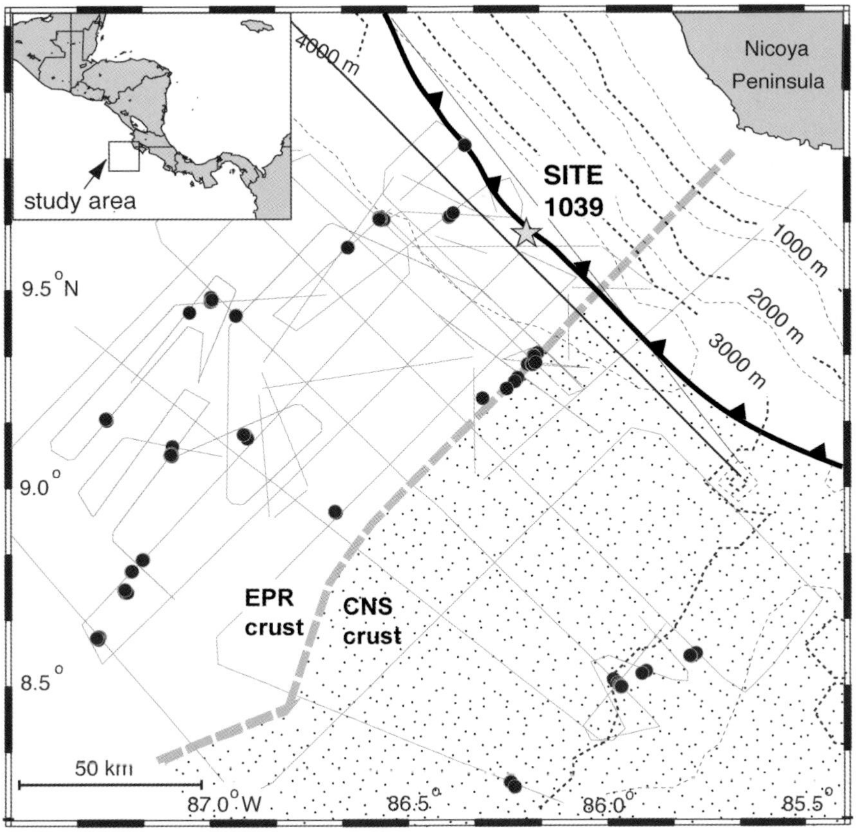

Figure 3.10 Index map of TicoFlux study area, Costa Rica segment of Middle America Trench. Dots represent locations of stations for gravity and piston cores. Star shows location of ODP Site 1039 on trench floor. Heavy dashed lines separates crust formed along the East Pacific Rise (EPR) and crust formed by Cocos-Nazca spreading (CNS). Thin lines represent seismic-reflection tracklines used for isopach mapping of subduction inputs [*Spinelli and Underwood*, 2004].

the EPR-CNS boundary [*Spinelli and Underwood*, 2004]. Differences in acoustic character [*Shipley et al.*, 1990] allow isopach mapping of an upper hemipelagic interval (relatively transparent) and a lower pelagic interval (higher amplitude, more continuous reflections). The hemipelagic interval reaches comparable thicknesses over both the EPR and CNS crust, ranging up to 207 m, with an average of 130 m [*Spinelli and Underwood*, 2004]. The pelagic interval, in contrast, is consistently thicker on the older EPR crust (average = 221 m) as compared to the CNS crust (average = 183 m).

Dozens of new piston and gravity cores were collected seaward of the trench during the TicoFlux project [*Fisher et al.*, 2003a]. These samples now permit mapping of lithologic variability on the Cocos Plate with greater 3-D clarity (fig. 3.10). Dark olive gray hemipelagic mud with abundant diatoms is,

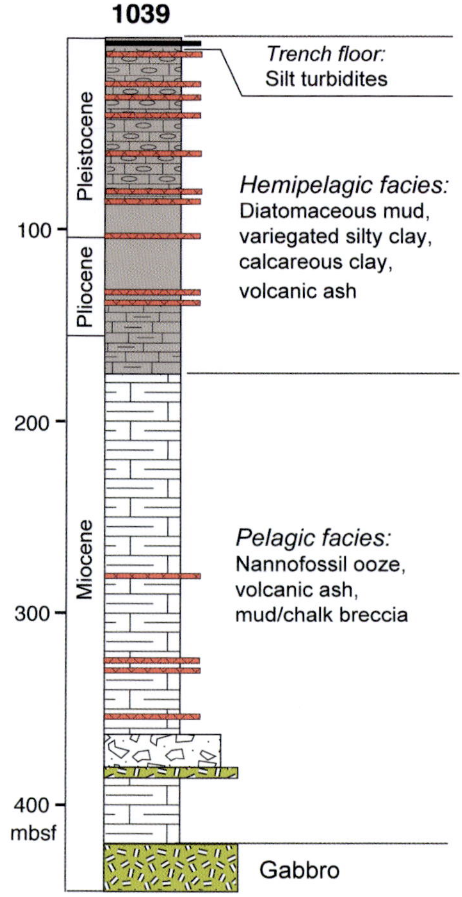

Figure 3.11 Stratigraphic column for ODP Site 1039, Costa Rica subduction zone. See fig. 3.10 for location.

by far, the most common variety near-surface sediment. Locally, layers of nannofossil chalk grade upward through mixed sediment (with variable carbonate content) into siliceous hemipelagic sediment. Seafloor exposures of nannofossil chalk are restricted to basement highs (e.g., seamounts and smaller basalt outcrops) where total sediment thickness is at a minimum. Winnowing by bottom currents probably thwarted deposition of hemipelagic mud at those localities. Overall, the most significant variable to affect the lithostratigraphic inputs is basement relief.

Sediment Composition

Compositional analyses of the near-surface sediments show three principle sediment types with minimal intralithologic variability (fig. 3.12). The nannofossil chalk is nearly pure, with an average opal content of 2 wt % [*Spinelli and Underwood*, 2004]. Hemipelagic mud has the highest content of biogenic opal (mean = 10 wt %), and variegated clay (mixed clay-carbonate sediment)

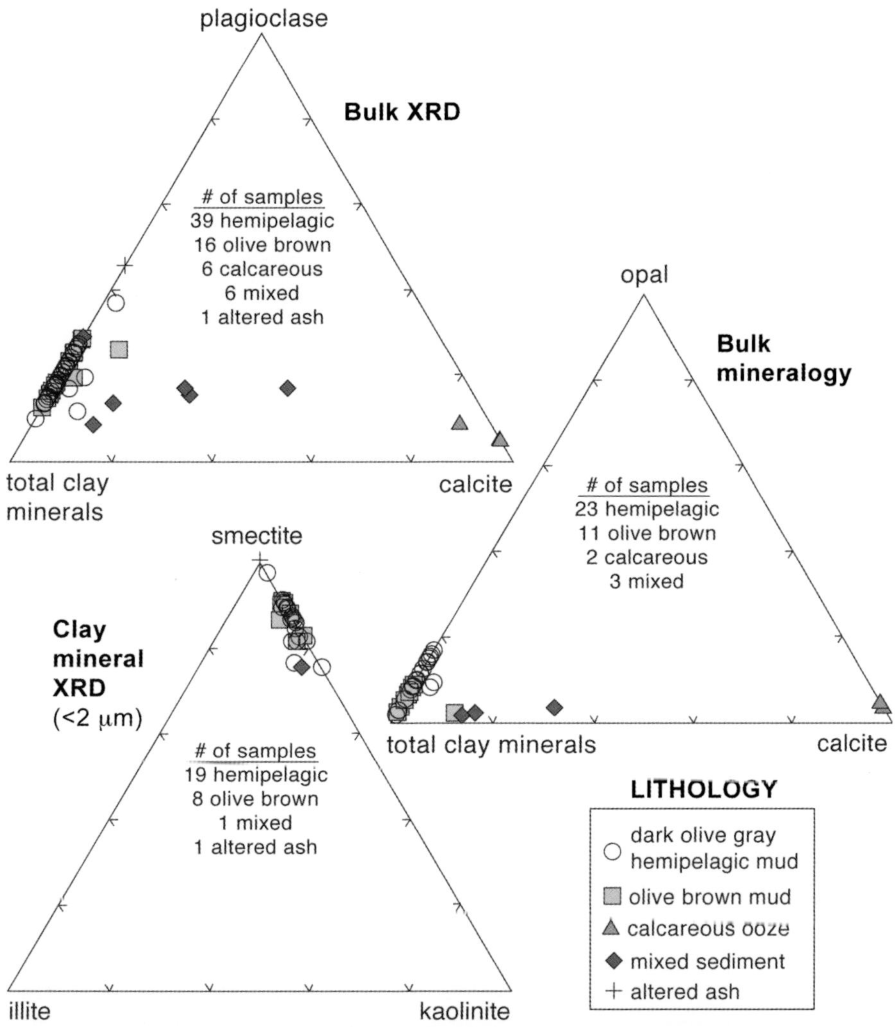

Figure 3.12 Ternary diagrams showing compositional modes for near-surface sediments seaward of the Costa Rica subduction front. See fig. 3.10 for core locations. Data are based on bulk-powder X-ray diffraction, alkaline leaching for opal, and clay-fraction XRD (<2 μm). Relative percentages were calculated using singular value deconvolution (SVD) weighting factors for bulk powders [*Spinelli and Underwood*, 2004] and Biscaye weighting factors for clay-size oriented aggregate [*Biscaye*, 1965]. See *Spinelli and Underwood* [2004] for data tables.

is intermediate in opal content (mean = 5 wt %). Bulk percentages of total clay minerals in the hemipelagic mud range from 59 to 88 wt % (fig. 3.12). Percentages of smectite in the clay-size fraction (relative to illite and kaolinite) average 86% and range from 75 to 97% (fig. 3.12). The smectite is dioctahedral (montmorillonite or beidellite), with ~60% expandability, which suggests a detrital origin (i.e., tropical weathering of the Central America volcanic arc). The average content of kaolinite is 13%, and many samples show traces of zeolite (clinoptilolite or heulandite) in the clay-size fraction.

Implications

On average, the inputs of diatomaceous hemipelagic mud to the Costa Rica subduction zone contain ~60 bulk wt % smectite [*Spinelli and Underwood*, 2004], which is consistently higher than the three other systems described herein. Judging from geotechnical tests of similar smectite-rich clay mixtures [*Lupini et al.*, 1981; *Brown et al.*, 2003], it seems reasonable to suggest that the frontal décollement is weak everywhere because of the incoming sediment's unusually low coefficient of friction ($\mu < 0.25$). Several additional factors, however, need to be considered. First, the deforming upper plate forms a steeply tapered wedge [*Christeson et al.*, 1999], which implies that the coefficient of basal friction is relatively high [*Dahlen*, 1990]. Second, if patches of anomalous permeability or frictional strength exist anywhere within the Costa Rica system, they should be concentrated in zones where basement topography is elevated and accumulations of hemipelagic mud are thin to nonexistent. Subduction of bare seamounts to seismogenic depths may, in fact, influence the earthquake rupture process [*Bilek et al.*, 2003; *Bilek*, this volume]. These "asperities" may develop, not just because the basement is elevated, but also because of the pelagic sediment's higher coefficient of friction (no smectite) and enhanced capacity to drain pore fluid (less clay).

In addition, the frontal décollement within the ODP transect area (Legs 170 and 205) localizes at the boundary between a deformed slope apron and the incoming sediment [*Kimura et al.*, 1997; *Shipboard Scientific Party*, 2003]. Laboratory consolidation tests indicate that the underthrust sediments are underconsolidated beneath the prism toe, and this anomaly is probably caused by retarded drainage and build up of excess pore pressure [*Saffer et al.*, 2000]. Because the Costa Rica margin is in a state of tectonic erosion, the plate boundary migrates up section into the overriding plate as one moves toward the east [*Ranero and von Huene*, 2000; *Vannucchi et al.*, 2003]. Consequently, with increasing depth along the décollement, the frictional properties of the incoming sediment probably play a smaller and smaller role in fault-zone behavior. Alternatively, dehydration of smectite and opal within strata that pass beneath the décollement enhances the likelihood of fluid overpressure [*Spinelli and Saffer*, 2004]. The quantitative details of this pressure effect are not well understood because at the same time the clay-mineral solids shrink in volume, water from the mineral structure enters the pore space and expands due to a change in density. Some of the expelled water probably drains updip along the décollement, contributing to geochemical anomalies. According to the conceptual model of *von Huene et al.* [2004], overpressured fluids may also migrate across the fault into the upper plate, where they trigger hydrofracturing and acceleration of basal erosion.

Pore fluids from the frontal décollement zone [*Kimura et al.*, 1997] provide several geochemical indications of reactions that occurred at temperatures warmer than 80°–120°C but cooler than 150°C [*Silver et al.*, 2000; *Chan and Kastner*, 2001]. The in situ temperature at ODP Site 1040 (margin toe) is <5°C, however, so transient updip fluid migration along the décollement may be to blame for the geochemical anomalies. Another explanation for the sharp pore-water chemistry profiles is continuous flow along the décollement combined with sustained upward flow from compacting underthrust sediments, with strata above the fault acting as a hydrologic seal [*Saffer and Screaton*, 2003]. In either scenario, when the mud's high smectite content (60 bulk wt %) combines with 10 wt % biogenic silica, the circumstances are favorable for substantial volumes of fluid production. Once the burial temperatures warm up enough to trigger opal and smectite dehydration [*Perry and Hower*, 1972; *Keller and Isaacs*, 1984], the hemipelagic inputs to Costa Rica subduction become an increasingly important factor to consider [*Spinelli and Saffer*, 2004].

Sediment composition is relatively uniform within the input section, but thermal structure is not. The temperature gradient at ODP Site 1040 is only 7.2°C/km [*Kimura et al.*, 1997], so the requisite thermal conditions for dehydration reactions may not kick in until after the subducting sediment reaches depths of 7–12 km. Heat flow surveys have demonstrated that the thermal state of the Cocos plate changes northwest the Leg 170 transect [*Fisher et al.*, 2003a; *Hutnak et al.*, this volume]. The CNS crust is warmer, overall, than the older EPR crust, but bare seamounts also cause local anomalies due to hydrothermal inflow and discharge. Whether warming of basement at the margin toe is local or regional, it should push the diagenetic fronts for opal and smectite dehydration farther updip and closer to the trench [*Spinelli and Saffer*, 2004]. This phenomenon is germane to SEIZE because the inferred along-strike responses of diagenesis and hydrogeology seem to match spatial shifts in the depths of microseismicity and strain accumulation [*Newman et al.*, 2002; *Spinelli and Underwood*, 2004; *Spinelli and Saffer*, 2004; *Norabuena et al.*, 2004].

Conclusions

Quantitative three-dimensional characterization of sediment inputs to subduction zones is an essential component of the SEIZE science plan. Sediment texture, mineral composition, and thermal structure within the incoming sedimentary section act together to influence physical hydrogeology, fluid budgets, and geotechnical properties above, within, and below plate-boundary faults. This paper provides a sedimentologist's overview of four well-documented subduction margins (Nankai, Cascadia, Barbados, and

Costa Rica). The overarching message is: Subduction inputs are highly variable in both time and space.

It would be misguided to assume that all stratigraphic sections coarsen and thicken upward from pelagic ooze (distal abyssal plain) to hemipelagic mud (proximal abyssal plain) to silty and sandy turbidites (trench wedge). As exceptions to that norm, substantial thicknesses of highly permeable sand (abyssal-plain deposits) continue along the subduction path beneath the frontal décollement in Nankai (Ashizuri transect), Barbados, and Cascadia. Until those sands are lithified by cementation, their high net permeability will control three-dimensional patterns of fluid circulation. The hydrology of stratigraphic conduits will mimic, in three dimensions, their original sand-body architecture (shoestring channel sands, stacked channel-levee complexes, unconfined sheet sands, convex-up depositional lobes). Local buildup of fluid overpressure creates compartments of low effective stress and low shear strength. These conditions should be anticipated wherever irregular sand facies terminate laterally against comparatively impermeable mudstones (e.g., the pinch-outs on basement highs in Shikoku Basin). The third dimension of a hydrologic seal will be created where sandy intervals are capped by thick mudstone (e.g., top of the lower Shikoku Basin facies), thereby retarding vertical drainage. These conditions can develop prior to subduction, but rapid loading beneath the margin wedge will exacerbate the situation.

Most subduction margins display significant amounts of compositional heterogeneity along strike. Spatial variability in sediment composition may be the result of mud plumes from diverse sources that overlap and mix to various degrees (e.g., abyssal-plain facies of Cascadia Basin). Delivery of detrital clay minerals usually changes through time, as well, because of gradual or punctuated shifts in climate, continental weathering, fluvial drainage, volcanic activity, and ocean currents (e.g., Shikoku Basin, Barbados). In some cases (Barbados), a relatively thin interval of smectite-rich clay seems to bracket the frontal décollement. In other cases (Costa Rica), sections of smectite-rich diatomaceous clay reach 100–200 m in thickness. Laboratory tests demonstrate that such high contents of smectite (>50 bulk wt %) lead to unusually low coefficients of friction. Some predisposed zones of weakness are, no doubt, exploited by propagating faults, but it would be incorrect to push these observations into a universal paradigm of cause and effect. The frontal Nankai décollement, for example, does not pass through smectite-rich sediments along either the Muroto or Ashizuri transect. Without deep drilling, the stratigraphic location of the Cascadia décollement continues to defy explanation.

As a final consideration, variability in the thermal state of subducting lithosphere can lead to substantial differences in how quickly diagenetic reactions progress, both outboard of a subduction front and downdip along the subduction path. This effect is particularly significant in Nankai Trough, where heat flow varies with distance from the fossil spreading ridge of Shikoku Basin. By the time strata pass beneath the toe of the Muroto accretionary prism, the smectite-to-illite reaction advances nearly to completion at the sediment-basalt interface. In contrast, coeval strata within the nearby Ashizuri transect remain virtually unaltered. Accurate predictions of where (in the downdip direction) and how much fluid volume might be released through dehydration reactions demand accurate knowledge of the starting materials and thermal structure.

The objective of quantifying subduction inputs is vital because of the links between sediment type, geotechnical properties, mineral dehydration, fluid flow, pore pressure, effective stress, and seismicity. The amount of total clay is probably more important than the amount of any particular type of clay (e.g., smectite) as a control on the sediment's coefficient of friction and permeability. It is also important, however, to show how geotechnical and hydrologic properties change with diagenesis. To attribute the updip limit of a seismogenic zone to just one type of reaction (e.g., smectite to illite) is almost certainly oversimplified. True interdisciplinary integration will be needed to construct a better model. In fact, long-term monitoring and spatial mapping of inputs from multiple transects of boreholes will be required to achieve the goals of SEIZE. Otherwise, we will fail to capture four-dimensional heterogeneities in parameters that likely govern fault-zone behavior.

Acknowledgments

This summary would have been impossible without the valuable contributions of many former students at the University of Missouri, the dozens of fellow shipboard scientists on ODP Legs 131, 156, 168, and 190, and participants in the TicoFlux expeditions. Funding was provided by NSF-MARGINS grants OCE-0304946, OCE-02106921, and OCE-0001768. This research also used samples provided by the Ocean Drilling Program, sponsored by the U.S. National Science Foundation and participating countries under management of Joint Oceanographic Institutions, Inc. Comments by Demian Saffer and an anonymous reviewer significantly improved the form and substance of the paper.

References

Ashi, J., and A. Taira (1993), Thermal structure of the Nankai accretionary prism as inferred from the distribution of gas hydrate BSRs, in *Thermal Evolution of the Tertiary Shimanto Belt, Southwest Japan: An Example of Ridge-Trench Interaction*, edited by M. B. Underwood, *Spec. Pap. Geol. Soc. Am., 273*, 137–150.

Audet, D. M., and J. D. C. McConnell (1992), Forward modeling of porosity and pore pressure evolution in sedimentary basins, *Basin Res., 4*, 147–162.

Ballance, P. F. (1991), Gravity flows and rock recycling on the Tonga landward trench slope: Relation to the trench-slope tectonic process, *J. Geol., 99*, 817–828.

Ballance, P. F., A. G. Ablaev, I. K. Pushchin, S. P. Pletnev, M. G. Birylina, T. Itaya, H. A. Follas, and G. W. Gibson (1999), Morphology and history of the Kermadec trench-arc-backarc basin-remnant arc system at 30 to 32°S: Geophysical profile, microfossil and K-Ar data, *Mar. Geol., 159*, 35–62.

Bangs, N. L., T. H. Shipley, and G. F. Moore (1996), Elevated fluid pressure and fault zone dilation inferred from seismic models of the northern Barbados Ridge décollement, *J. Geophys. Res., 101*, 627–642.

Bangs, N. L., T. H. Shipley, S. P. S. Gulick, G. F. Moore, S. Kuromoto, and Y. Nakamura (2004), Evolution of the Nankai Trough décollement from the trench into the seismogenic zone: Inferences from three-dimensional seismic reflection imaging, *Geology, 32*, 273–276.

Barckhausen, U., H. A. Roeser, and R. von Huene (1998), Magnetic signature of upper plate structures and subducting seamounts at the convergent margin off Costa Rica, *J. Geophys. Res., 103*, 7079–7093.

Barker, C. (1972), Aquathermal pressuring: Role of temperature in development of abnormal-pressure zones, *Am. Assoc. Pet. Geol. Bull., 56*, 2068–2071.

Barnard, W. D. (1978), The Washington continental slope: Quaternary tectonics and sedimentation, *Mar. Geol., 27*, 79–114.

Behl, R. J., and R. E. Garrison (1994), The origin of chert in the Monterey Formation of California (USA), in *Siliceous, Phosphatic and Glauconitic Sediments of the Tertiary and Mesozoic: Proceedings of the 29th International Geological Congress, Part C*, edited by A. Iijima, A. Abed, and R. Garrison, pp.101–132, VSP, Utrecht, Netherlands.

Bekins, B. A., and S. J. Dreiss (1992), A simplified analysis of parameters controlling dewatering in accretionary prisms, *Earth Planet. Sci. Lett., 109*, 275–287.

Bekins, B. A., and E. J. Screaton (2007), Pore pressure and fluid flow in the northern Barbados accretionary complex: A synthesis, this volume.

Bekins, B. A., A. M. McCaffrey, and S. J. Dreiss (1994), Influence of kinetics on the smectite to illite transition in the Barbados accretionary prism, *J. Geophys. Res., 99*, 18,147–18,158.

Bekins, B. A., A. M. McCaffrey, and S. J. Dreiss (1995), Episodic and constant flow models for the origin of low-chloride waters in modern accretionary complex, *Water Resour. Res., 31*, 3205–3215.

Bilek, S. L. (2007), Influence of subducting topography on earthquake rupture, this volume.

Bilek, S. L., S. Y. Schwartz, and H. R. DeShon (2003), Control of seafloor roughness on earthquake rupture behavior, *Geology, 31*(5), 455–458.

Bird, P. (1984), Hydration phase diagrams and friction of montmorillonite under laboratory and geological conditions, with implication for shale compaction, slope stability, and strength of fault gouge, *Tectonophysics, 107*, 235–260.

Biscaye, P. E. (1965), Mineralogy and sedimentation of recent deep-sea clays in the Atlantic Ocean and adjacent seas and oceans, *Geol. Soc. Am. Bull., 76*, 803–832.

Bradley, J. S. (1975), Abnormal formation pressure, *Am. Assoc. Pet. Geol. Bull., 59*, 957–973.

Bray, C. J., and D. E. Karig (1985), Porosity of sediments in accretionary prisms and some implications for dewatering processes, *J. Geophys. Res., 90*, 768–778.

Bredehoeft, J. D., R. D. Djevanshir, and K. Belitz (1988), Lateral fluid flow in a compacting sand-shale sequence: South Caspian Basin, *Am. Assoc. Pet. Geol. Bull., 72*, 416–424.

Brown, K. M. and G. K. Westbrook (1987), Mud diapirism and subcretion in the Barbados Ridge Complex, *Tectonics, 7*, 613–640.

Brown, K. M., B. A. Bekins, C. Clennell, G. Dewhurst, and G. Westbrook (1994), Heterogeneous hydrofracture development and accretionary fault dynamics, *Geology, 22*(3), 259–262.

Brown, K. M., A. Kopf, M. B. Underwood, and J. L. Weinberger (2003), Compositional and fluid pressure controls on the state of stress on the Nankai subduction thrust, *Earth Planet. Sci. Lett., 21*, 589–603.

Bruce, C. H. (1984), Smectite dehydration - its relations to structural development and hydrocarbon accumulation in northern Gulf of Mexico basin, *Am. Assoc. Pet. Geol. Bull., 68*, 673–683.

Buatier, M. D., D. R. Peacor, and J. R. O'Neil (1992), Smectite-illite transition in Barbados accretionary wedge sediments: TEM and AEM evidence for dissolution/crystallization at low temperature, *Clays Clay Miner., 40*, 65–80.

Byerlee, J. (1990), Friction, overpressure and fault normal compression, *Geophys. Res. Lett., 17*, 2109–2112.

Byerlee, J. (1993), Model for episodic flow of high-pressure water in fault zones before earthquakes, *Geology, 21*(4), 303–306.

Byrne, D. E., D. M. Davis, and L. R. Sykes (1988), Loci and maximum size of thrust earthquakes and the mechanics of the shallow region of subduction zones, *Tectonics, 7*, 833–857.

Capet, X., H. Chamley, C. Beck, and T. Holtzapffel (1990), Clay mineralogy of sites 671 and 672, Barbados Ridge accretionary complex and Atlantic Abyssal Plain: Paleoenvironmental and diagenetic implications, *Proc. Ocean Drill. Program Sci. Results, 110*, 85–96.

Carlson, P. R., and C. H. Nelson (1969), Sediments and sedimentary structures of the Astoria submarine canyon-fan system, northeast Pacific, *J. Sediment. Petrol., 39*, 1269–1282.

Carlson, P. R., and C. H. Nelson (1987), Marine geology and resource potential of Cascadia Basin, in *Geology and Resource Potential of the Continental Margin of Western North America and Adjacent Ocean Basins - Beaufort Sea to Baja California*, edited by D. W. Scholl, A. Grantz, and J. Vedder, pp. 523–536, Circum-Pac. Counc. for Energy and Miner. Resour., Houston, Tex.

Carlson, P. R., A. J. Stevenson, T. R. Bruns, D. M. Mann, and Q. Huggett (1996), Sediment pathways in Gulf of Alaska from beach to abyssal plain, in *Geology of the United States Seafloor, The View From GLORIA*, edited by J. V. Gardner, M. E. Field, and D. C. Twichell, pp. 255–277, Cambridge Univ. Press, New York.

Carson, B., E. T. Baker, B. M. Hickey, C. A. Nittrouer, D. J. DeMaster, K. W. Thorbjarnarson, and G. W. Snyder (1986), Modern sediment dispersal and accumulation in Quinault submarine canyon: A summary, *Mar. Geol., 71*, 1–13.

Chamley, H. (1980), Clay sedimentation and paleoenvironment in the Shikoku Basin since the middle Miocene (Deep Sea Drilling Project Leg 58, North Philippine Sea), *Proc. Ocean Drill. Program Initial Rep., 58*, 669–681.

Chamley, H., J-P. Cadet, and J. Charvet (1986), Nankai Trough and Japan Trench late Cenozoic paleoenvironments deduced from clay mineralogic data, *Proc. Ocean Drill. Program Initial Rep., 87*, 633–642.

Chan, L. H., and M. Kastner (2001), Lithium isotopic compositions of pore fluids and sediments in the Costa Rica subduction zone: Implications for fluid processes and sediment contribution to the arc volcanoes, *Earth Planet. Sci. Lett., 183*, 275–290.

Chatterji, P. K. and N. R. Morgenstern (1990), A modified shear strength formulation for swelling clay soils, in *Physico-Chemical Aspects of Soils and Related Materials*, ASTM STP 1095, edited by K. B. Hoddinott and R. O. Lamb, pp. 118–135, Am. Soc. for Test. and Mater., Philadelphia, Pa.

Chester, F. H. and J. M. Logan (1986), Implications for mechanical properties of brittle faults from observations of the Punchbowl fault zone, Califormia, *Pure App. Geophys., 124*, 79–106.

Christeson, G. I., K. D. McIntosh, T. H. Shipley, E. R. Flueh, and H. Goedde (1999), Structure of the Costa Rica convergent margin offshore Nicoya peninsula, *J. Geophys. Res., 104*, 443–468.

Clift, P. D., and C. J. MacLeod (1999), Slow rates of tectonic erosion estimated from the subsidence and tilting of the Tonga forearc, *Geology, 27*, 411–414.

Clift, P. D., C. J. MacLeod, D. R. Tappin, D. J. Wright, and S. H. Bloomer (1998), Tectonic controls on sedimentation and diagenesis in the Tonga Trench and forearc, southwest Pacific, *Geol. Soc. Am. Bull., 110*, 483–496.

Clift, P. D., N. Shimizu, G. D. Layne, J. S. Blusztajn, C. Gaedicke, H. -U. Schluter, M. K. Clark, and S. Amjad (2001), Development of the Indus Fan and its significance for the erosional history of the western Himalaya and Karakoram, *Geol. Soc. Am. Bull., 113*, 1039–1051.

Clift, P. D., I. Percher, N. Kukowski, and A. Hampel (2003), Tectonic erosion of the Peruvian forearc, Lima Basin, by subduction and Nazca Ridge collision, *Tectonics, 22*(3), 1023, doi:10.1029/2002TC001386.

Colten-Bradley, V. A. (1987), Role of pressure in smectite dehydration; Effects on geopressure and smectite-to-illite transformation, *Am. Assoc. Pet. Geol. Bull.*, *71*, 1414–1427.

Dahlen, F. A. (1990), Critical taper model of fold-and-thrust belts and accretionary wedges, *Annu. Rev. Earth Planet. Sci.*, *18*, 55–99.

Davis, E. E., and R. D. Hyndman (1989), Accretion and recent deformation of sediments along the northern Cascadia subduction zone, *Geol. Soc. Am. Bull.*, *101*, 1465–1480.

Davis, E. E., et al. (1992), FlankFlux: An experiment to study the nature of hydrothermal circulation in young oceanic crust, *Can. J. Earth Sci.*, *29*, 925–952.

Davis, E. E., A. T. Fisher, and J. V. Firth (Eds.) (1997), *Proceedings of the Ocean Drilling Program, Initial Results*, vol. 168, Ocean Drill. Program, College Station, Tex.

Davis, D., J. Suppe, and F. A. Dahlen (1983), The mechanics of fold-and-thrust belts, *J. Geophys. Res.*, *95*, 1153–1172.

DeMets, C., R. G. Gordon, D. F. Argus, and S. Stein (1990), Current plate motions, *Geophys. J. Int.*, *101*, 425–478.

DeMets, C., R. G. Gordon, D. F. Argus, and S. Stein (1994), Effect of recent revisions to the geomagnetic reversal time scale on estimates of current plate motions, *Geophys. Res. Lett.*, *21*, 2191–2194.

DeMets, D. C. (2001), A new estimate for present-day Cocos-Caribbean plate motion: Implications for slip along the Central American volcanic arc, *Geophys. Res. Lett.*, *28*, 4043–4046.

Deng, X., and M. B. Underwood (2001), Abundance of smectite and the location of a plate-boundary fault, Barbados accretionary prism, *Geol. Soc. Am. Bull.*, *113*, 495–507.

DiLeonardo, C. G., J. C. Moore, S. Nissen, and N. Bangs (2002), Control of internal structure and fluid-migration pathways within the Barbados Ridge décollement zone by strike-slip faulting: Evidence from coherence and three-dimensional seismic amplitude imaging, *Geol. Soc. Am. Bull.*, *114*, 51–62.

Dixon, T. H., F. Farina, C. DeMets, P. Jansma, P. Mann, and E. Calais (1998), Relative motion between the Caribbean and North American plates and related boundary zone deformation from a decade of GPS observations, *J. Geophys. Res.*, *103*, 15,157–15,182.

Dolan, J. F., C. Beck, and Y. Ogawa (1989), Upslope deposition of extremely distal turbidites: An example from the Tiburon Rise, west-central Atlantic, *Geology*, *17*(11), 990–994.

Duncan, J. R., L. D. Kulm, and G. B. Griggs (1970), Clay mineral composition of Late Pleistocene and Holocene sediments of Cascadia Basin, northeastern Pacific Ocean, *J. Geol.*, *78*, 213–221.

Earnst, W. G., and S. E. Calvert (1969), An experimental study of the recrystallization of porcelanite and its bearing on the origin of some bedded cherts, *Am. J. Sci.*, *267-A*, 114–133.

Einsele, G., L. Ratschbacher, and A. Wetzel (1996), The Himalaya-Bengal Fan denudation-accumulation system during the past 20 Ma, *J. Geol.*, *104*, 163–184.

Emmel, F. J., and J. R. Curray (1984), The Bengal submarine fan, northeastern Indian Ocean, *Geo Mar. Lett.*, *3*, 119–124.

Fagel, N., L. Andre, H. Chamley, P. Debrabant, and L. Jolivet (1992), Clay sedimentation in the Japan Sea since the Early Miocene: Influence of source-rock and hydrothermal activity, *Sediment. Geol.*, *80*, 27–40.

Faulkner, D. R., and E. H. Rutter (2001), Can the maintenance of overpressured fluids in large strike-slip fault zones explain their apparent weakness?, *Geology*, *29*(6), 503–506.

Fergusson, C. L. (2003), Provenance of Miocene-Pleistocene turbidite sands and sandstones, Nankai Trough, Ocean Drilling Program Leg 190 [online], *Proc. Ocean Drill. Program Sci. Results*, *190/196*. (Available at http://www-odp.tamu.edu/publications/190196SR)

Fisher, A. T., and M. W. Hounslow (1990), Transient fluid flow through the toe of the Barbados accretionary complex: Constraints from Ocean Drilling Program Leg 110 heat flow studies and simple models, *J. Geophys. Res.*, *95*, 8845–8858.

Fisher, A. T., and M. B. Underwood (1995), Calibration of an X-ray diffraction method to determine relative mineral abundances in bulk powders using matrix singular value

deconvolution: A test from the Barbados accretionary complex, *Proc. Ocean Drill. Program Initial Results, 156,* 29–38.
Fisher, A., and G. Zwart (1997), Packer experiments along the decollement of the Barbados accretionary complex: Measurements of in situ permeability, *Proc. Ocean Drill. Program Sci. Results, 156,* 199–218.
Fisher, A., G. Zwart, and the Leg 156 Scientific Party (1996), Relation between permeability and effective stress along a plate-boundary fault, Barbados accretionary complex, *Geology, 24*(4), 307–310.
Fisher, A. T., C. A. Stein, R. N. Harris, K. Wang, E. A. Silver, M. Pfender, M. Hutnak, A. Cherkaoui, R. Bodzin, and H. Villinger (2003a), Abrupt thermal transition reveals hydrothermal boundary and role of seamounts within the Cocos Plate, *Geophys. Res. Lett., 30*(11), 1550, doi:10.1029/2002GL016766.
Fisher, A. T., et al. (2003b), Hydrothermal recharge and discharge across 50 km guided by seamounts on a young ridge flank, *Nature, 421,* 618–621.
Fisher, M. A., E. R. Flueh, D. W. Scholl, T. Parsons, R. E. Wells, A. Trehu, U. ten Brink, and C. S. Weaver (1999), Geologic processes of accretion in the Cascadia subduction zone west of Washington state, *J. Geodyn., 27,* 277–288.
Fitts, T. G., and K. M. Brown (1999), Stress induced smectite dehydration: Ramifications for patterns of freshening and fluid expulsion in the N. Barbados accretionary wedge, *Earth Planet. Sci. Lett., 172,* 179–197.
Flemings, P. B., B. B. Stump, T. Finkbeiner, and M. Zoback (2002), Flow focusing in overpressured sandstones: Theory, observations, and applications, *Am. J. Sci., 302,* 827–855.
Freed, R. L., and D. R. Peacor (1989a), Geopressured shale and sealing effect of smectite to illite transition, *Am. Assoc. Pet. Geol. Bull., 73,* 1223--1232.
Freed, R. L., and D. R. Peacor (1989b), Variability in temperature of the smectite/illite reaction in Gulf Coast sediments, *Clays Clay Miner., 24,* 171–180.
Fryer, P. (1996), Evolution of the Mariana convergent plate margin system, *Rev. Geophys., 34,* 89–125.
Giambalvo, E. R., A. T. Fisher, J. T. Martin, L. Darty, and R. P. Lowell (2000), Origin of elevated sediment permeability in a hydrothermal seepage zone, eastern flank of the Juande Fuca Ridge, and implications for transport of fluid and heat, *J. Geophys. Res., 105,* 913–928.
Gieskes, J. M., P. Vrolijk, and G. Blanc (1990), Hydrogeochemistry of the northern Barbados accretionary complex transact: Ocean Drilling Program Leg 110, *J. Geophys. Res., 95,* 8809–8818.
Goldfinger, C., L. D. Kulm, R. S. Yeats, B. Appelgate, M. E. MacKay, and G. R. Cochrane (1996), Active strike-slip faulting and folding of the Cascadia subduction-zone plate boundary and forearc in central and northern Oregon, in *Assessing Earthquake Hazards and Reducing Risk in the Pacific Northwest, U.S. Geol. Surv. Prof. Pap., 1560,* 223–256.
Griggs, G. B., and J. R. Hein (1980), Sources, dispersal, and clay mineral composition of fine-grained sediment off central and northern California, *J. Geol., 88,* 541–566.
Griggs, G. B., and L. D. Kulm (1970), Sedimentation in Cascadia deep-sea channel, *Geol. Soc. Am. Bull., 81,* 1361–1384.
Gutscher, M-A., D. Klaeschen, E. Flueh, and J. Malavieille (2001), Non-Coulomb wedges, wrong-way thrusting, and natural hazards in Cascadia, *Geology, 29*(5), 379–382.
Hampton, M. A., H. A. Karl, and N. H. Kenyon (1989), Sea-floor drainage features of Cascadia Basin and the adjacent continental slope, northeast Pacific Ocean, *Mar. Geol., 87,* 249–272.
Harris, R. N., A. T. Fisher, and D. S. Chapman (2004), Fluid flow through seamounts and implications for global mass fluxes, *Geology, 32*(8), 725–728.
Hart, B. S., P. B. Flemings, and A. Deshpande (1995), Porosity and pressure: Role of compaction disequilibrium in the development of geopressures in a Gulf Coast Pleistocene basin, *Geology, 23*(1), 45–48.

Heirtzler, J. R., G. O. Dickson, E. M. Herron, W. C. Pitman, and X. Le Pichon (1968), Marine magnetic anomalies, geomagnetic field reversals, and motions of the ocean floor and continents, *J. Geophys. Res.*, *73*, 2119–2136.

Hey, R. N. (1977), A new class of pseudofaults and their bearing on plate tectonics: A propagating rift model, *Earth Planet. Sci. Lett.*, *37*, 321–325.

Hickman, S., R. Sibson, and R. Bruhn (1995), Introduction to special section: Mechanical involvement of fluids in faulting, *J. Geophys. Res.*, *100*, 12,831–12,840.

Housen, B. A., H. J. Tobin, P. Labaume, E. C. Leitch, A. J. Maltman, and ODP Leg 156 Shipboard Science Party (1996), Strain decoupling across the décollement of the Barbados accretionary prism, *Geology*, *24*(2), 127–130.

Huang, W. L., J. M. Longo, and D. R. Pevear (1993), An experimentally derived kinetic model for smectite-to-illite conversion and its use as a geothermometer, *Clays Clay Miner.*, *41*, 162–177.

Hubbert, M. K., and W. W. Rubey (1959), Role of fluid pressure in the mechanics of overthrust faulting, I: Mechanisms of fluid-filled porous solids and its application to overthrust faulting, *Geol. Soc. Am. Bull.*, *70*, 115–166.

Hunt, J. M. (1990), Generation and migration of petroleum from abnormally pressured fluid compartments, *Am. Assoc. Pet. Geol. Bull.*, *74*, 1–12.

Hutnak, M., et al. (2007), The thermal state of 18–24 Ma upper lithosphere subducting below the Nicoya Peninsula, northern Costa Rica margin, this volume.

Hyndman, R. D., and K. Wang (1993), Thermal constraints on the zone of major thrust earthquake failure: The Cascadia subduction zone, *J. Geophys. Res.*, *98*, 2039–2060.

Hyndman, R. D., and K. Wang (1995), The rupture zone of Cascadia great earthquakes from current deformation and the thermal regime, *J. Geophys. Res.*, *100*, 22,133–22,154.

Hyndman, R. D., K. Wang, and M. Yamano (1995), Thermal constraints on the seismogenic portion of the southwestern Japan subduction thrust, *J. Geophys. Res.*, *100*, 15,373–15,392.

Hyndman, R. D., M. Yamano, and D. A. Oleskevich (1997), The seismogenic zone of subduction thrust faults, *Island Arc*, *6*, 244–260.

Isaacs, C. M. (1982), Influence of rock composition on kinetics of silica phase changes in the Monterey Formation, Santa Barbara area, California, *Geology*, *10*(6), 304–308.

Karlin, R. (1980), Sediment sources and clay mineral distributions off the Oregon Coast, *J. Sediment. Petrol.*, *50*, 543–560.

Kastner, M., H. Elderfield, and J. B. Martin (1991), Fluids in convergent margins: What do we know about their composition, origin, role in diagenesis and importance for oceanic chemical fluxes?, *Philos. Trans. R. Soc. London, Ser. A*, *335*, 243–259.

Kastner, M., Y. Zheng, T. Laier, W. Jenkins, and T. Ito (1997), Geochemistry of fluids and flow regime in the decollement zone at the northern Barbados Ridge, *Proc. Ocean Drill. Program Sci. Results*, *156*, 311–319.

Keller, M. A., and C. M. Isaacs (1985), An evaluation of temperature scales for silica diagenesis in diatomaceous sequences including a new approach based on the Miocene Monterey Formation, California, *Geo –Mar. Lett.*, *5*, 31–35.

Kimura, G., et al. (1997), *Proceedings of the Ocean Drilling Program, Initial Results*, vol. 170, Ocean Drill. Program, College Station, Tex.

Kinoshita, M., and M. Yamano (1986), The heat flow anomaly in the Nankai Trough area, *Proc. Ocean Drill. Program Initial Rep.*, *87*, 737–743.

Knebel, H. J., J. C. Kelly, and J. T. Whetten (1968), Clay minerals of the Columbia River: A qualitative, quantitative and statistical evaluation, *J. Sediment. Petrol.*, *38*, 600–611.

Kolla, V., and F. Coumes (1984), Morpho-acoustic and sedimentologic characteristics of the Indus Fan, *Geo Mar. Lett.*, *3*, 133–140.

Kopf, A., and K. M. Brown (2003), Friction experiments on saturated sediments and their implications for the stress state of the Nankai and Barbados subduction thrusts, *Mar. Geol.*, *202*, 193–210.

Kopp, H., and N. Kukowski (2003), Backstop geometry and accretionary mechanics of the Sunda margin, *Tectonics, 22*(6), 1072, doi:10.1029/2002TC001420.
Kukowski, N., T. Schillhorn, K. Huhn, U. von Rad, S. Husen, and E. R. Flueh (2001), Morphotectonics and mechanics of the central Makran accretionary wedge off Pakistan, *Mar. Geol., 173*, 1–19.
Lambe, T. W., and R. V. Whiteman (1969), *Soil Mechanics*, John Wiley, Hoboken, N. J.
Latouche, C., and N. Maillet (1984), Evolution of Cenozoic clay assemblages in the Barbados Ridge (Deep Sea Drilling Project Sites 541, 542, 543), *Proc. Ocean Drill. Program Initial Rep., 78A*, 343–356.
Lister, C. R. B. (1977), Estimators for heat flow and deep rock properties based on boundary layer theory, *Tectonophysics, 41*, 157–171.
Logan, J. M., and K. A. Rauenzahn (1987), Frictional dependence of gouge mixtures of quartz and montmorillonite on velocity, composition and fabric, *Tectonophysics, 144*, 87–108.
Lupini, J. F., A. E. Skinner, and P. R. Vaughan (1981), The drained residual strength of cohesive soils, *Geotechnique, 31*, 181–213.
MacKay, M. E. (1995), Structural variation and landward vergence at the toe of the Oregon accretionary prism, *Tectonics, 14*, 1309–1320.
MacKay, M. E., G. F. Moore, G. R. Cochrane, J. C. Moore, and L. D. Kulm (1992), Landward vergence and oblique structural trends in the Oregon margin accretionary prism: Implications and effect on fluid flow, *Earth Planet. Sci. Lett., 109*, 477–491.
Magee, M. E., and M. D. Zoback (1993), Evidence for a weak interplate thrust fault along the northern Japan subduction zone and implications for the mechanics of thrust faulting and fluid expulsion, *Geology, 21*(9), 809–812.
Marone, C., and D. Saffer (2007), Fault friction and the upper transition from seismic to aseismic faulting, this volume.
Marsaglia, K. M., R. V. Ingersoll, and B. Packer (1992), Tectonic evolution of the Japanese Islands as reflected in modal compositions of Cenozoic forearc and backarc sand and sandstone, *Tectonics, 11*, 1028–1043.
Masuda, H., J. R. O'Neil, W-T. Jiang, and D. R. Peacor (1996), Relation between interlayer composition of authigenic smectite, mineral assemblages, I/S reaction rate and fluid composition in silicic ash of the Nankai Trough, *Clays Clay Miner., 44*, 443–459.
Masuda, H., D. R. Peacor, and H. Dong (2001), Transmission electron microscopy study of conversion of smectite to illite in mudstones of the Nankai trough: Contrast with coeval bentonites, *Clays Clay Miner., 49*, 109–118.
Matthai, S. K., and G. Fisher (1996), Quantitative modeling of fault-fluid-discharge and fault-dilation-induced fluid-pressure variations in the seismogenic zone, *Geology, 24*(2), 183–186.
McCarthy, J., and D. W. Scholl (1985), Mechanism of subduction accretion along the central Aleutian trench, *Geol. Soc. Am. Bull., 96*, 691–701.
McNeil, L. C., C. Goldfinger, L. D. Kulm, and R. S. Yeats (2000), Tectonics of the Neogene Cascadia forearc basin: Investigations of a deformed late Miocene unconformity, *Geol. Soc. Am. Bull., 112*, 1209–1224.
Mesri, G., and R. E. Olson (1971), Consolidation characteristics of montmorillonite, *Geotechnique, 21*, 341–352.
Mesri, G., M. C. Pakbaz, and A. F. Cepeda-Diaz (1994), Meaning, measurement and field application of swelling pressure of clay shales, *Geotechnique, 44*, 129–145.
Mitchell, J. K. (1993), *Fundamentals of Soil Behavior*, 2nd ed., John Wiley, Hoboken, N. J.
Moore, D. E., and D. A. Lockner (2007), Friction of the smectite clay montmorillonite: A review and interpretation of data, this volume.
Moore, D. E., C. A. Morrow, and J. D. Byerlee (1982), Use of swelling clays to reduce permeability and its potential application to nuclear waste repository sealing, *Geophys. Res. Lett., 9*, 1009–1012.

Moore, G. F., and T. H. Shipley (1988), Mechanisms of sediment accretion in the Middle America Trench off Mexico, *J. Geophys. Res., 93*, 8911–8927.

Moore, G. F., J. R. Curray, and F. J. Emmel (1982), Sedimentation in the Sunda Trench and forearc region, in *Trench-Forearc Geology*, edited by J. K. Leggett, *Geol. Soc. Spec. Publ., 10*, 245–258.

Moore, G. F., et al. (2001), New insights into deformation and fluid flow processes in the Nankai Trough accretionary prism; results of Ocean Drilling Program Leg 190, *Geochem. Geophys. Geosyst., 2*(10), doi:10.1029/2001GC000166.

Moore, J. C. (1989), Tectonics and hydrogeology of accretionary prisms: Role of the décollement zone, *J. Struct. Geol., 11*, 95–106.

Moore, J. C., and D. Saffer (2001), Updip limit of the seismogenic zone beneath the accretionary prism of southwest Japan: An effect of diagenetic to low-grade metamorphic processes and increasing effective stress, *Geology, 29*, 183–186.

Moore, J. C., and H. Tobin (1997), Estimated fluid pressures of the Barbados accretionary prism and adjacent sediments, *Proc. Ocean Drill. Program Sci. Results, 156*, 229–238.

Moore, J. C., and P. Vrolijk (1992), Fluids in accretionary prisms, *Rev. Geophys., 30*, 113–135.

Moore, J. C., et al. (1982), Offscraping and underthrusting of sediment at the deformation front of the Barbados Ridge: Deep Sea Drilling Project Leg 78A, *Geol. Soc. Am. Bull., 93*, 1065–1077.

Moore J. C., et al. (1988), Tectonic and hydrology of the northern Barbados Ridge: Results from Ocean Drilling Program Leg 110, *Geol. Soc. Am. Bull., 100*, 1578–1593.

Moore, J. C., K. M. Brown, F. Horath, G. Cochrane, M. MacKay, and G. Moore (1991), Plumbing accretionary prisms: Effects of permeability variations, *Philos. Trans. R. Soc. London, Ser. A, 335*, 275–288.

Moore, J. C., et al. (1995), Abnormal fluid pressures and fault-zone dilation in the Barbados accretionary prism: Evidence from logging while drilling, *Geology, 23*(7), 605–608.

Moore, J. C., et al. (1998), Consolidation patterns during initiation and evolution of a plate boundary decollement zone: Northern Barbados accretionary prism, *Geology, 26*(9), 811–814.

Morgan, J. K., D. E. Karig, and A. Maniatty (1995), The estimation of diffuse strains in the toe of the western Nankai accretionary prism: A kinematic solution, *J. Geophys. Res., 99*, 7019–7032.

Morrow, C. A., L. Q. Shi, and J. D. Byerlee (1982), Strain hardening and strength of clay-rich fault gouges, *J. Geophys. Res., 87*, 6771–6780.

Morrow, C. A., L. Q. Shi, and J. D. Byerlee (1984), Permeability of fault gouge under confining pressure and shear stress, *J. Geophys. Res., 89*, 3193–3200.

Morrow, C., B. Radney, and J. Byerlee (1992), Frictional strength and the effective pressure law of montmorillonite and illite clays, in *Fault Mechanics and Transport Properties of Rocks*, edited by B. Evans and T. F. Wong, pp. 69–88, Elsevier, New York.

Mottl, M. J., et al. (1998), Warm springs discovered on 3.5 Ma oceanic crust, eastern flank of the Juan de Fuca Ridge, *Geology, 26*(1), 51–54.

Murata, K. J., I. Friedman, and J. D. Gleason (1977), Oxygen isotope relations between diagenetic silica minerals in Monterey Shale, Temblor Range, California, *Am. J. Sci., 277*, 259–272.

Nelson, C. H. (1976), Late Pleistocene and Holocene depositional trends, processes and history of Astoria deep-sea fan, northeast Pacific, *Mar. Geol., 20*, 129–173.

Nelson, C. H. (1984), The Astoria Fan: An elongate type fan, *Geo Mar. Lett., 3*, 65–70.

Nelson, C. H., P. R. Carlson, J. V. Byrne, and T. R. Alpha (1970), Development of the Astoria canyon-fan physiography and comparison with similar systems, *Mar. Geol., 8*, 259–291.

Neuzil, C. E. (1994), How permeable are clays and shales?, *Water Resour. Res., 30*, 145–150.

Neuzil, C. E. (1995), Abnormal pressures as hydrodynamic phenomena, *Am. J. Sci., 295*, 742–786.

Newman, A. V., S. Y. Schwartz, V. Gonzalez, H. R. Deshon, J. M. Protti, and L. M. Dorman (2002), Along-strike variability in the seismogenic zone below Nicoya Peninsula, Costa Rica, *Geophys. Res. Lett., 29,* 1977–1981.

Norobuena, E., T. H. Dixon, S. Schwartz, H. DeShon, A. Newman, M. Protti, V. Gonzalez, et al., (2004), Geodetic and seismic constraints on some seismogenic zone processes in Costa Rica, *J. Geophys. Res., 109*(B11), 403, doi:10.1029/2003JB002931.

Olson, R. E. (1974), Shearing strength of kaolinite, illite, and montmorillonite, *J. Soil Mech. Found. Div. Am. Soc. Civ. Eng., 100,* 1215–1229.

Osborne, M. J., and R. E. Swarbrick (1997), Mechanisms for generating overpressure in sedimentary basins: A reevaluation, *Am. Assoc. Pet. Geol. Bull., 81,* 1023–1041.

Parsons, B., and J. G. Sclater (1977), An analysis of the variation of ocean floor bathymetry and heat flow with age, *J. Geophys. Res., 82,* 803–827.

Perry, E. A., and J. Hower (1970), Burial diagenesis in Gulf Coast pelitic sediments, *Clays Clay Miner., 18,* 165–177.

Perry, E. A., and J. Hower (1972), Late stage dehydration in deeply buried peletic sediments, *Am. Assoc. Pet. Geol. Bull., 56,* 2013–2021.

Pickering, K. T., M. B. Underwood, and A. Taira (1992), Open-ocean to trench turbidity-current flow in the Nankai Trough: Flow collapse and reflection, *Geology, 20*(12), 1099–1102.

Pickering, K. T., M. B. Underwood, and A. Taira (1993), Stratigraphic synthesis of the DSDP-ODP sites in the Shikoku Basin, Nankai Trough, and accretionary prism, *Proc. Ocean Drill. Program Sci. Results, 131,* 313–330.

Piper, D. J. W., R. von Huene, and J. R. Duncan (1973), Late Quaternary sedimentation in the active eastern Aleutian Trench, *Geology, 1*(1), 19--22.

Pisciotto, K. A. (1981), Diagenetic trends in the siliceous facies of the Monterey Shale in the Santa Maria region, California, *Sedimentology, 28,* 547–571.

Powers, M. C. (1967), Fluid-release mechanisms in compacting marine mudrocks and their importance in oil exploration, *Am. Assoc. Pet. Geol. Bull., 51,* 1240–1254.

Protti, M., F. Gundel, and K. McNally (1995), Correlation between age of the subducting Cocos plate and geometry of the Wadati-Benioff zone under Nicaragua and Costa Rica, in *Geologic and Tectonic Development of the Caribbean Plate Boundary in Southern Central America,* edited by P. Mann, *Spec. Pap. Geol. Soc. Am., 295,* 309–326.

Pudsey, C. J. (1984), X-ray mineralogy of Miocene and older sediments from Deep Sea Drilling Project Leg 78A, *Proc. Ocean Drill. Program Initial Rep., 78A,* 325–342.

Ranero, C. R., and R. von Huene (2000), Subduction erosion along the Middle America convergent margin, *Nature, 404,* 748–752.

Riddihough, R. P. (1977), Recent movements of the Juan de Fuca plate system, *J. Geophys. Res., 89,* 6980–6994.

Robinson, R. G., and M. M. Allam (1998), Effect of clay mineralogy on coefficient of consolidation, *Clays Clay Miner., 46,* 596–600.

Rogers, A. M., T. J. Walsh, W. J. Kockelman, and G. R. Priest (1996), Earthquake hazards in the Pacific Northwest - An overview, in *Assessing Earthquake Hazards and Reducing Risk in the Pacific Northwest,* U.S. Geol. Surv. Prof. Pap., 1560, 1–54.

Saito, S., and D. Goldberg (2001), Compaction and dewatering processes of the oceanic sediments in the Costa Rica and Barbados subduction zones: Estimates from in situ physical properties measurements, *Earth Planet. Sci. Lett., 191,* 283–293.

Saffer, D. (2007), Pore pressure within underthrust sediments in subduction zones, this volume.

Saffer, D. M., and B. A. Bekins (1999), Fluid budgets at convergent plate margins: Implications for the extent and duration of fault-zone dilation, *Geology, 27*(12), 1095–1098.

Saffer, D. M., and B. A. Bekins (2002), Hydrologic controls on the morphology and mechanics of accretionary wedges, *Geology, 30*(3), 271–274.

Saffer, D. M., and C. Marone (2003), Comparison of smectite- and illite-rich gouge frictional properties: Application to the updip limit of the seismogenic zone along subduction megathrusts, *Earth Planet. Sci. Lett., 215*, 219–235.
Saffer, D. M., and E. J. Screaton (2003), Fluid flow at the toe of convergent margins: Interpretation of sharp pore-water geochemical gradients, *Earth Planet. Sci. Lett., 213*, 261–270.
Saffer, D. M., E. A. Silver, A. T. Fisher, H. Tobin, and K. Moran (2000), Inferred pore pressures at the Costa Rica subduction zone: Implications for dewatering processes, *Earth Planet. Sci. Lett., 177*, 193–207.
Schluter, H. U., A. Prexl, Ch. Gaedicke, H. Roeser, Ch. Reichert, H. Meyer, and C. von Daniels (2002), The Makran accretionary wedge: Sediment thickness and ages and the origin of mud volcanoes, *Mar. Geol. 185*, 219–232.
Schweller, W. J., and L. D. Kulm (1978), Depositional patterns and channelized sedimentation in active eastern Pacific trenches, in *Sedimentation in Submarine Canyons, Fans, and Trenches*, edited by D. J. Stanley and G. Kelling, pp. 311–324, John Wiley, Hoboken, N. J.
Screaton, E. J., A. T. Fisher, B. Carson, and K. Becker (1997), Barbados Ridge hydrogeologic tests: Implications for fluid migration along an active décollement, *Geology, 25*(3), 239–242.
Seely, D. R. (1977), The significance of landward vergence and oblique structural trends on trench inner slopes, in *Island Arcs, Deep Sea Trenches, and Back-Arc Basins, Maurice Ewing Ser.*, vol. 1, edited by M. Talwani and W. C. Pitman, pp. 187–198, AGU, Washington, D.C.
Sella, G. F., T. H. Dixon, and A. Mao (2002), REVEL: A model for Recent plate velocities from space geodesy, *J. Geophys. Res., 107*(B4), 2081, doi:10.1029/2000JB000033.
Seno, T., S. Stein, and A. E. Gripp (1993), A model for the motion of the Philippine Sea plate consistent with NUVEL-1 and geological data, *J. Geophys. Res., 98*, 17,941–17,948.
Shi, Y., and C-Y. Wang (1986), Pore pressure generation in sedimentary basins: Overloading versus aquathermal, *J. Geophys. Res., 91*, 2153–2162.
Shimamoto, T., and J. M. Logan (1981), Effects of simulated clay gouges on the sliding behavior of Tennessee sandstone, *Tectonophysics, 75*, 243–255.
Shimamura, K. (1989), Topography and sedimentary facies of the Nankai Deep Sea Channel, in *Sedimentary Facies in the Active Plate Margin*, edited by A. Taira and F. Masuda, pp. 529–556, Terra Sci., Tokyo.
Shipboard Scientific Party (1975), Site 297, *Proc. Ocean Drill. Program Initial Rep., 31*, 275–316.
Shipboard Scientific Party (1994), Site 888, *Proc. Ocean Drill. Program Initial Rep., 146*, 55–125.
Shipboard Scientific Party (2001a), Leg 190 summary [online], *Proc. Ocean Drill. Program Initial Rep., 190*, 1–87. (Available at http://www-odp.tamu.edu/publications/190IR)
Shipboard Scientific Party (2001b), Explanatory notes [online], *Proc. Ocean Drill. Program Initial Rep., 190*. (Available at http://www-odp.tamu.edu/publications/190IR)
Shipboard Scientific Party (2003), Leg 205 summary notes [online], *Proc. Ocean Drill. Program Initial Rep., 205*. (Available at http://www-odp.tamu.edu/publications/205IR)
Shipley, T. H., and G. F. Moore (1986), Sediment accretion, subduction and dewatering at the base of the trench slope off Costa Rica: A seismic reflection view of the décollement, *J. Geophys. Res., 91*, 2019–2028.
Shipley, T. H., P. Soffa, and D. F. Dean (1990), Underthrust sediments, fluid migration paths, and mud volcanoes associated with the accretionary wedge off Costa Rica: Middle America Trench, *J. Geophys. Res., 95*, 8743–8752.
Shipley, T. H., G. F. Moore, N. Bangs, J. C. Moore, and P. L. Stoffa (1994), Seismically inferred dilatancy distribution, northern Barbados Ridge décollement: Implication for fluid migration and fault strength, *Geology, 22*(5), 411–414.
Sibson, R. H. (1981), Fluid flow accompanying faulting: Field evidence and models, in *Earthquake Prediction, An International Review, Maurice Ewing Ser.*, vol. 4, edited by D. W. Simpson and R. G. Richards, pp. 593–603, AGU, Washington, D. C.
Silver, E., M. Kastner, A. Fisher, J. Morris, K. McIntosh, and D. Saffer (2000), Fluid flow paths in the Middle America Trench and Costa Rica margin, *Geology, 28*(8), 679–682.

Skempton, A. W. (1970), The consolidation of clays by gravitational compaction, *Q. J. Geol. Soc. London*, *125*, 373–411.
Sosson, M., J. Bourgois, and B. Mercier de Lepinay (1994), SeaBeam and deep-sea submersible Nautile surveys in the Chiclayo canyon off Peru (7°S): Subsidence and subduction-erosion of an Andean-type convergent margin since Pliocene times, *Mar. Geol.*, *118*, 237–256.
Spinelli, G. A., and D. M. Saffer (2004), Along-strike variations in underthrust sediment dewatering on the Nicoya margin, Costa Rica related to the updip limit of seismicity, *Geophys. Res. Lett.*, *31*, L04613, doi:10.1029/2003GL018863.
Spinelli, G. A., and M. B. Underwood (2004), Character of sediments entering the Costa Rica subduction zone: Implications for partitioning of water along the plate interface, *Island Arc*, *13*, 432–451.
Stein, C., and S. Stein (1994), Constraints on hydrothermal heat flux through the oceanic lithosphere from global heat flow, *J. Geophys. Res.*, *99*, 3081–3095.
Steurer, J. F., and M. B. Underwood (2003), Clay mineralogy of mudstones from the Nankai Trough reference sites and frontal accretionary prism [online], *Proc. Ocean Drill. Program Sci. Results*, *190/196*, . (Available at http://www-odp.tamu.edu/publications/190196SR)
Stevenson, A. J., D. W. Scholl, and T. L. Vallier (1983), Tectonic and geologic implications of the Zodiac Fan, Aleutian Abyssal Plain, northeast Pacific, *Geol. Soc. Am. Bull.*, *94*, 259–273.
Su, X., K-H. Baumann, and J. Thiede (2000), Calcareous nannofossils from Leg 168: Biochronology and diagenesis, *Proc. Ocean Drill. Program, Sci. Results*, *168*, 39–50.
Taira, A., and J. Ashi (1993), Sedimentary facies evolution of the Nankai forearc and its implications for the growth of the Shimanto accretionary prism, *Proc. Ocean Drill. Program Sci. Results*, *131*, 331–341.
Taira , A., and N. Niitsuma (1986), Turbidite sedimentation in the Nankai Trough as interpreted from magnetic fabric, grain size, and detrital modal analyses, *Proc. Ocean Drill. Program Initial Rep.*, *87*, 611–632.
Taira, A. et al. (1992), Sediment deformation and hydrogeology of the Nankai Trough accretionary prism: Synthesis of shipboard results of ODP Leg 131, *Earth Planet. Sci. Lett.*, *109*, 431–450.
Tobin, H. J., J. C. Moore, and G. F. Moore (1994), Fluid pressure in the frontal thrust of the Oregon accretionary prism: Experimental constraints, *Geology*, *22*(11), 979–982.
Tribble, J. S. (1990), Clay diagenesis in the Barbados accretionary complex: Potential impact on hydrology and subduction dynamics, *Proc. Ocean Drill. Program Sci. Results*, *110*, 97–110.
Tribble, J. S., and H-W. Yeh (1994), Origin of smectite and illite-smectite in the Barbados accretionary complex: Oxygen isotopic evidence, *Geology*, *22*(3), 219–222.
Underwood, M. B. (2002), Strike-parallel variations in clay minerals and fault vergence in the Cascadia subduction zone, *Geology*, *30*(2), 155–158.
Underwood, M. B., and X. Deng (1997), Clay mineralogy and clay geochemistry in the vicinity of the decollement zone, northern Barbados Ridge, *Proc. Ocean Drill. Program Sci. Results*, *156*, 3–30.
Underwood, M. B., and K. D. Hoke (2000), Composition and provenance of turbidite sand and hemipelagic mud in northwestern Cascadia Basin, Leg 168 of the Ocean Drilling Program, *Proc. Ocean Drill. Program Sci. Results*, *168*, 51–65.
Underwood, M. B., and G. F. Moore (1995), Trenches and trench-slope basins, in *Tectonics of Sedimentary Basins*, edited by C. J. Busby and R. V. Ingersoll, pp. 179–219, Blackwell Sci., Malden, Mass.
Underwood, M. B., and K. T. Pickering (1996), Clay-mineral provenance, sediment dispersal patterns, and mudrock diagenesis in the Nankai accretionary prism, southwest Japan, *Clays Clay Miner.*, *44*, 339–356.
Underwood, M. B., and J. F. Steurer (2003), Composition and sources of clay from the trench slope and shallow accretionary prism of Nankai Trough [online], *Proc. Ocean Drill.*

Program Sci. Results, 190/196. (Available at http://www-odp.tamu.edu/publications/190196SR)

Underwood, M. B., R. Orr, K. Pickering, and A. Taira (1993a), Provenance and dispersal patterns of sediments in the turbidite wedge of Nankai Trough, *Proc. Ocean Drill. Program Sci. Results, 131*, 15–34.

Underwood, M., K. Pickering, J. M. Gieskes, M. Kastner, and R. Orr (1993b), Sediment geochemistry, clay mineralogy, and diagenesis: A synthesis of data from Leg 131, Nankai Trough, *Proc. Ocean Drill. Program Sci. Results, 131*, 343–363.

Underwood, M. B., N. Basu, J. Steurer, and S. Udas (2003), Data report: Normalization factors for semi-quantitative X-ray diffraction analysis, with application to DSDP Site 297 [online], Shikoku Basin, *Proc. Ocean Drill. Program Sci. Results, 190/196*. (Available at http://www-odp.tamu.edu/publications/190196SR)

Underwood, M. B., K. D. Hoke, A. T. Fisher, E. E. Davis, E. Giambalvo, L. Zuhlsdorff, and G. A. Spinelli (2005), Provenance, stratigraphic architecture, and hydrogeologic influence of turbidites on the mid-ocean ridge flank of northwestern Cascadia Basin, Pacific Ocean, *J. Sediment. Res., 75*, 149–164.

Vannucchi, P., C. R. Ranero, S. Galeotti, S. M. Straub, D. W. Scholl, and D. McDougall-Ried (2003), Fast rates of subduction erosion along the Costa Rica Pacific margin: Implications for nonsteady rates of crustal recycling at subduction zones, *J. Geophys. Res., 108*(B11), 2511, doi:10.1029/2002JB002207.

Vannucchi, P., S. Galeotti, P. D. Clift, C. R. Ranero, and R. von Huene (2004), Long-term subduction-erosion along the Guatemalan margin of the Middle America Trench, *Geology, 32*(7), 617–620.

von Huene, R. (1974), Modern trench sediments, in *The Geology of Continental Margins*, edited by C. A. Burk and C. L. Drake, pp. 207–211, Springer, New York.

von Huene, R., and R. C. Culotta (1989), Tectonic erosion at the front of the Japan Trench convergent margin, *Tectonophysics, 16*, 75–90.

von Huene, R., and L. D. Kulm (1973), Tectonic summary of Leg 18, *Proc. Ocean Drill. Program Sci. Results, 18*, 961–976.

von Huene, R., and H. Lee (1982), The possible significance of pore fluid pressures in subduction zones, *Am. Assoc. Pet. Geol. Mem., 34*, 781–791.

von Huene, R., and D. W. Scholl (1991), Observations at convergent margins concerning sediment subduction, subduction erosion, and the growth of continental crust, *Rev. Geophys., 29*, 279–316.

von Huene, R., L. D. Kulm, and J. Miller (1985), Structure of the frontal part of the Andean convergent margin, *J. Geophys. Res., 90*, 5429–5442.

von Huene, R., D. Klaeschen, M. Gutscher, and J. Fruehn (1998), Mass and fluid flux during accretion at the Alaskan margin, *Geol. Soc. Am. Bull., 110*, 468–482.

von Huene, R., W. Weinrebe, and F. Heeren (1999), Subduction erosion along the North Chile margin, *J. Geodyn., 27*, 345–358.

von Huene, R., C. R. Ranero, W. Weinrebe, and K. Hinz (2000), Quaternary convergent margin tectonics of Costa Rica, segmentation of the Cocos Plate, and Central American volcanism, *Tectonics, 19*, 314–334.

von Huene, R., C. R. Ranero, and P. Vannucchi (2004), Generic model of subduction erosion, *Geology, 32*(10), 913–916.

Vrolijk, P. (1990), On the mechanical role of smectite in subduction zones, *Geology, 18*(8), 703–707.

Vrolijk, P., A. Fisher, and J. Gieskes (1991), Geochemical and geothermal evidence for fluid migration in the Barbados accretionary prism (ODP Leg 110), *Geophys. Res. Lett., 18*, 947–950.

Wallace, G., J. C. Moore, and C. G. DiLeonardo (2003), Controls on localization and densification of a modern décollement: Northern Barbados accretionary prism, *Geol. Soc. Am. Bull., 115*, 288–297.
Wang, C-Y. (1980), Sediment subduction and frictional sliding in a subduction zone, *Geology, 8*(11), 530–533.
Wang, K., R. D. Hyndman, and M. Yamano (1995), Thermal regime of the Southwest Japan subduction zone: Effects of age history of the subducting plate, *Tectonophysics, 248*, 53–69.
Weber, M. E., M. H. Wiedicke, H. R. Kudrass, C. Hubscher, and H. Erlenkeuser (1997), Active growth of the Bengal Fan during sea-level rise and highstand, *Geology, 25*(4), 315–318.
Westbrook, G. K. (1982), The Barbados Ridge Complex: Tectonics of a mature forearc system, in *Trench-Forearc Geology*, edited by J. K. Leggett, *Spec. Publ. Geol. Soc. London, 10*, 275–290.
Westbrook, G. K. (1994), Growth of accretionary wedges off Vancouver Island and Oregon, *Proc. Ocean Drill. Program Sci. Results, 146*, 381–388.
Westbrook, G. K., and M. J. Smith (1983), Long decollements and mud volcanoes: Evidence from the Barbados Ridge complex for the role of high pore-fluid pressure in the development of an accretionary complex, *Geology, 11*(5), 279–283.
Westbrook, G. K., J. W. Ladd, P. Buhl, N. Bangs, and G. J. Tiley (1988), Cross section of an accretionary wedge: Barbados Ridge complex, *Geology, 16*(7), 631–635.
Wheat, C. G., and M. J. Mottl (1994), Hydrothermal circulation, Juan de Fuca Ridge eastern flank: Factors controlling basement water composition, *J. Geophys. Res., 99*, 3067–3080.
Williams, L. A., G. A. Parks, and D. A. Crerar (1985), Silica diagenesis, I. Solubility controls, *J. Sediment. Petrol., 55*, 301–311.
Yamano, M., M. Kinoshita, S. Goto, and O. Matsubayashi (2003), Extremely high heat flow anomaly in the middle part of the Nankai Trough, *Phys. Chem. Earth, 28*, 487–497.

4 The Thermal State of 18–24 Ma Upper Lithosphere Subducting Below the Nicoya Peninsula, Northern Costa Rica Margin

M. Hutnak, A. T. Fisher, C. A. Stein, R. Harris, K. Wang, E. Silver, G. Spinelli, M. Pfender, H. Villinger, R. MacKnight, P. Costa Pisani, H. DeShon, and C. Diamente

Abstract

Early studies of the thermal state of the Cocos plate subducting beneath the northern Costa Rica margin identified two adjacent areas of higher (>100 mW/m^2) and lower (<30 mW/m^2) heat flow. Measurements on crust formed at the Cocos-Nazca spreading center (CNS) were generally in agreement with later predictions from conductive lithospheric cooling models, whereas measurements on East Pacific Rise (EPR) crust were commonly lower by ~70%. However, the sparse distribution of early data precluded confident determination of the cause(s) for the heat-flow deficit on EPR-generated seafloor, nor did they allow an assessment of the nature of the transition between warmer and cooler regions. We present new thermal, swath map, and seismic data collected on the incoming plate offshore of Costa Rica, seaward of the Middle America Trench (MAT), in an area containing several tectonic boundaries, including the plate suture between EPR- and CNS-generated seafloor. We find that most transitions from warm to anomalously cool thermal conditions are not directly associated with any of the tectonic boundaries and that vigorous local hydrothermal circulation in uppermost basement is responsible for low heat-flow values on the cooler part of the plate. Thermal data in the vicinity of Ocean Drilling Program (ODP) Site 1039, located within the trench, indicate that the drill site is surrounded on three sides by regions of warmer seafloor. Heat-flow data across normal faults outboard of the MAT suggest that these faults do not allow vertical advective extraction of lithospheric heat in this location. Satellite and hydrosweep data show that seamounts are common features on EPR-generated seafloor, whereas CNS-generated seafloor immediately adjacent to the plate suture and the MAT is devoid of seamounts. It appears that most of the EPR-generated crust in this area is cooled on a regional basis by hydrothermal fluids recharging and discharging through seamounts and other basement outcrops. Analytical model calculations suggest that the vertical extent of hydrothermal cooling may be limited to the upper 100–600 m of basement, and effective upper-basement permeabilities are on the order of $10^{-10} - 10^{-8}$ m^2.

Introduction

The thermal state of subducting oceanic plates influences melting conditions in the overlying mantle wedge, the extent of sediment dewatering associated with mineralogical and rheological transitions, the occurrence of gas hydrates along the active margin, and the nature of tectonic and seismic processes at depth [e.g., *Fisher et al.*, 2003b; *Gaetani and Grove*, 1998; *Hyndman and Wang*, 1993; *Ruppel and Kinoshita*, 2000; *Saffer and Marone*, 2003; *Stein*, 2003; *Zhao et al.*, 1997]. Determining thermal conditions within subduction zones is difficult because drilling to necessary depths poses severe technical challenges, and seafloor thermal data collected on the margin are typically influenced by processes such as fluid flow and frictional heating along faults [e.g., *Henry and Wang*, 1991; *Iwamori*, 1998; *Peacock and Wang*, 1999]. In this paper we examine the thermal state of the upper lithosphere approaching the subduction zone along the Middle America Trench, offshore of the Nicoya Peninsula, Costa Rica. Much recent work in this region has focused on conditions within the upper plate and along the plate boundary within the subduction zone, but there has been relatively little emphasis on the nature of thermal conditions within the incoming plate prior to subduction. Understanding the thermal state of the incoming plate is important because it comprises both boundary and initial conditions for thermal models of subduction processes.

Any rigorous assessment of thermal conditions within oceanic lithosphere should include consideration of hydrothermal circulation. The upper oceanic crust is typically highly permeable, and hosts large-scale flows of water, energy, and solutes [e.g., *Schultz and Elderfield*, 1997; *Sclater et al.*, 1980]. These fluxes help to maintain the chemistry of the ocean, change the physical and geochemical state of the lithosphere, transport volatiles into the mantle, and support a vast subsurface biosphere [e.g., *Alt et al.*, 1996; *Cowen et al.*, 2003; *Elderfield and Schultz*, 1996; *Jacobson*, 1992; *Plank and Langmuir*, 1993; *Wheat et al.*, 2003].

Although much attention in the past has focused on high-temperature (>300°C) hydrothermal vents at mid-ocean ridge spreading centers, global heat flow data indicate that advective heat loss from ridge flanks is more than 3 times that at the ridge axis [*Mottl and Wheat*, 1994; *Stein and Stein*, 1994]. Because this off-axis flux occurs at much lower temperatures (<100°C), the corresponding mass flux of seawater through the crust is inferred to be at least 10–100 times greater than at the ridge axis [*Mottl and Wheat*, 1994; *Schultz and Elderfield*, 1997; *Stein and Stein*, 1994]. Ridge-flank fluxes of some solutes approach or even exceed those on the ridge crest and from riverine sources [*Elderfield and Schultz*, 1996; *Mottl*, 2003]. Advective heat loss from the ocean crust extends, on average, to a global mean age of ~65 Ma, but some regions lose heat advectively to greater ages, and fluid circulation within basement is capable of redistributing heat well beyond 100 Ma [*Noel and Hounslow*, 1988; *Stein and Stein*, 1994; *Von Herzen*, 2004].

In this synthesis, we briefly summarize the tectonic setting (discussed in greater detail elsewhere in this volume) and review early studies of thermal conditions within the lithosphere of the Cocos plate offshore northern Costa Rica. This includes a short discussion of results associated with Ocean Drilling Program (ODP) Leg 170, which examined the margin wedge and trench immediately offshore of the Nicoya Peninsula, and results from a predrilling survey cruise that focused on thermal conditions at planned drilling sites. The main focus of this paper concerns new results from two recent expeditions to this region (comprising the "TicoFlux" program), which sought to elucidate the extent, nature, and causes for differences in thermal conditions within adjacent parts of the Cocos plate prior to subduction.

Regional Setting

Costa Rica lies at the southern end of the Central American island arc system, where the Cocos plate subducts at ~85 mm/yr beneath the Caribbean plate [*DeMets*, 2001], forming the Middle America Trench (MAT; fig. 4.1). The 18–24 Ma seafloor offshore the Nicoya Peninsula, Costa Rica, has had a complex tectonic history and is commonly considered as comprising several distinct tectonic subregions [e.g., *Barckhausen et al.*, 2001; *Meschede et al.*, 1998; *Ranero and von Huene*, 2000; *von Huene et al.*, 2000].

Toward the middle of the Nicoya Peninsula, offshore magnetic lineations abruptly switch orientation by roughly 90°, distinguishing seafloor formed at the East Pacific Rise (EPR, trench-parallel isochrons) from that formed at the Cocos-Nazca Spreading Center (CNS, trench-perpendicular isochrons). The boundary between EPR- and CNS-generated seafloor is a combination of a triple junction trace and a fracture zone trace (referred to by *Fisher et al.* [2003b] as a "plate suture"). During earlier studies on a regional scale, the plate suture was called the "rough-smooth boundary (RSB)," a reference to the Cocos Ridge and associated rough topography related to the passing of the Cocos plate over the Galapagos hot spot [*Hey*, 1977]. Seafloor seaward of the northern segment of the Nicoya Peninsula was noted to be relatively smooth in comparison to that off the southern portion [e.g., *von Huene et al.*, 2000], but as discussed below, previously unidentified seamounts are important thermal (and hydrogeologic) features to the north of the plate suture.

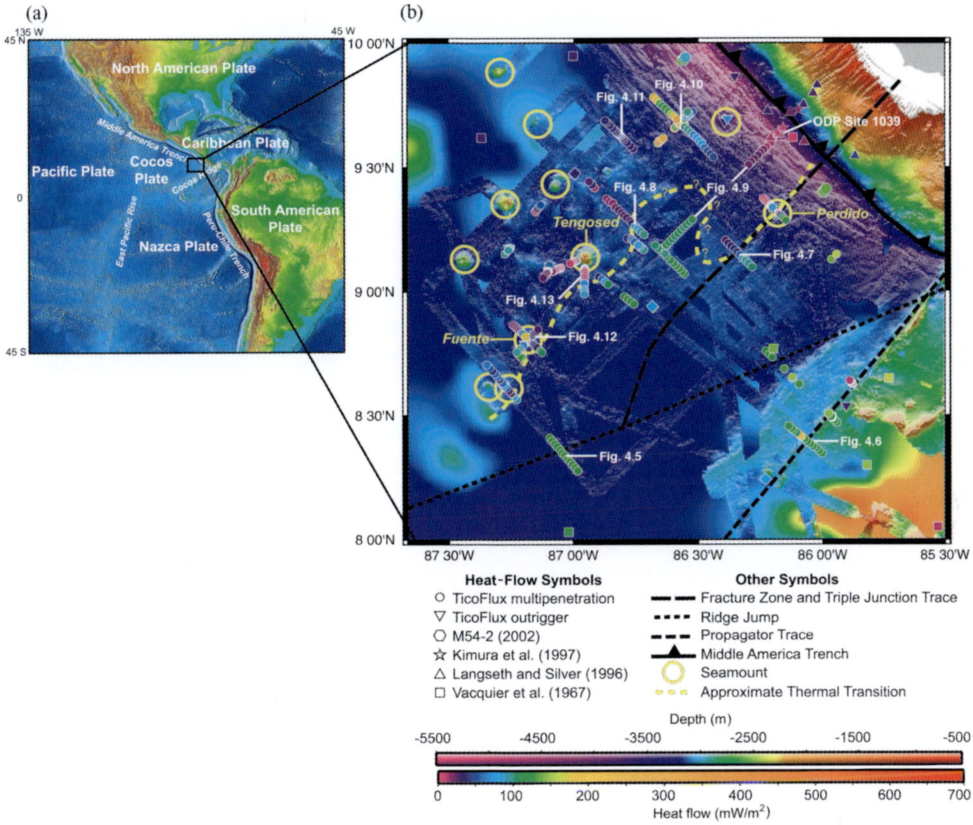

Figure 4.1 (a) Regional map showing tectonic plates, spreading centers, and regional bathymetry. Dark box indicates the TicoFlux study region in figure 4.1b. (Image courtesy NOAA-NGDC: http://www.ngdc.noaa.gov). (b) Seafloor bathymetry based on satellite gravimetry and swath mapping [*Barckhausen et al.*, 2001; *Fisher et al.*, 2003b; *von Huene et al.*, 2000], overlain by tectonic boundaries derived from bathymetry, gravity, and magnetics [*Barckhausen et al.*, 2001; *Hey*, 1977; *Meschede et al.*, 1998]. Color-coded symbols indicate heat-flow values: circles correspond to TicoFlux studies, while other sources are as labeled (M54-2 data courtesy of I. Grevemeyer (personal communication, 2002). ODP Leg 170 drilled a series of holes across the trench and onto the margin wedge northwest of the triple-junction trace [*Kimura et al.*, 1997]. Dashed yellow line indicates approximate thermal transition along seismic transects from anomalously low heat flow to areas with elevated heat flow. EPR-generated crust to the north of the triple junction and fracture zone traces has heat flow suppressed by roughly 70%, whereas CNS-generated crust to the south generally has heat flow that matches conductive lithospheric cooling models. Of note are the prevalence seamounts (gold circles) on EPR-generated crust, which allow recharging and discharging fluids to bypass the 300–500 m thick sediments and allow for the efficient extraction of large amounts of lithospheric heat.

Two additional tectonic boundaries within this part of the Cocos plate are important for the discussion of lithospheric thermal state: traces of a ridge jump and a propagating ridge [*Barckhausen et al.*, 2001; *Meschede et al.*, 1998] (fig. 4.1). The ridge jump helps to define the southern edge of a triangular block of CNS-generated lithosphere adjacent to the plate suture and the MAT and forms part of the boundary between CNS-generated and EPR-generated seafloor. The propagator trace cuts entirely across CNS-generated seafloor and offsets the magnetic anomalies by several tens of kilometers.

The Costa Rica margin is seismically very active, with at least four earthquakes of magnitude >7 having occurred in the last 200 years. A significant change of ~5–10 km in the depth of the updip limit of background microseismicity along the Nicoya Peninsula loosely coincides with the plate suture, suggesting that the thermal state of the crust (colder EPR versus warmer CNS) influences the seismogenic updip limit along the Nicoya portion of the MAT [*Newman et al.*, 2002].

ODP Leg 170 [*Kimura et al.*, 1997] drilled several sites on the subduction wedge offshore of the Nicoya Peninsula, and Site 1039 in the MAT, which has become a de facto reference site for regional studies of stratigraphy and sedimentation history on the incoming plate (fig. 4.1). Drilling at Site 1039 penetrated 180 m of diatom-rich hemipelagic claystone overlying 200 m of nannofossil-rich carbonate ooze, with the oldest recovered sediment dated at 16.5 Ma. The base of the sediment section consisted of brown claystone intruded by gabbroic sills. Frontal accretion of the stratigraphic section does not occur as the plate passes beneath the toe of the margin wedge [*Kimura et al.*, 1997; *Silver et al.*, 2001], although there is seismic evidence for underplating at deeper levels [*McIntosh and Sen*, 2000].

Thermal Studies Along the Costa Rica Margin

Studies Prior to TicoFlux

The earliest heat-flow data from seaward of the MAT were collected at a time when researchers were trying to understand the most fundamental aspects of heat loss from oceanic lithosphere [e.g., *Bullard et al.*, 1956; *Langseth et al.*, 1965; *Vacquier et al.*, 1967; *Von Herzen and Uyeda*, 1963] and before the precise age of the seafloor in this area was known [*Anderson and Hobart*, 1976]. It was thought that relatively high heat-flow values found near the EPR and other mid-ocean ridges might result from convection or heat sources in the mantle, and there was speculation as to the possible occurrence of hydrothermal circulation in these areas of generally high, but scattered, values [e.g., *Elder*, 1965; *Langseth et al.*, 1966; *Lee and Uyeda*, 1965]. Because heat-flow technology at the time made each measurement difficult and time consuming, heat-flow stations on early expeditions tended to be widely separated, with between-measurement spacing commonly being 100 km or more. In addition, the lack of continuous bathymetric maps and the absence of high-quality seismic data prevented a thorough assessment of the geological context of many early measurements, including the influence of buried basement relief and seafloor refraction on heat-flow measurements [e.g., *Sclater et al.*, 1976; *Von Herzen and Uyeda*, 1963].

Early studies of the eastern Pacific Ocean basin were able to clearly identify areas of relatively high (>100 mW/m^2) and low (<30 mW/m^2) heat flow. Higher values were generally found close to the EPR, but there were also elevated values found near the Cocos Ridge [*Vacquier et al.*, 1967; *Von Herzen and Uyeda*, 1963], which were attributed to recent igneous intrusions. Although elevated heat flow associated with the EPR was generally related to the upwelling limb of a mantle convection cell, there was no corresponding decrease in heat flow with increasing proximity to the MAT, which seemed to preclude the occurrence of mantle downwelling in this region. Low heat-flow values were found in the Guatemala Basin to the north, another observation that could not be attributed to mantle convection, and it was suggested that there might be a relative deficiency in heat producing elements within the mantle in this region.

In fact, the data collected and compiled by *Vacquier et al.* [1967] suggested that there were considerable differences in oceanic heat flow between EPR- and CNS-generated seafloor offshore of the Nicoya Peninsula (fig. 4.1). Conductive lithospheric cooling models suggest that 18–24 Ma seafloor should have heat flow of 95–120 mW/m^2 [e.g., *Parsons and Sclater*, 1977; *Stein and Stein*, 1992]. Pre-1990s heat-flow data from offshore of the Nicoya Peninsula yielded mean heat flow values of 31 mW/m^2 on EPR-generated seafloor and 110 mW/m^2 on CNS-generated seafloor (heat-flow fractions of ~0.3 and ~1.0, respectively) [*Stein and Stein*, 1994]. Seafloor heat-flow values below the conductive prediction are generally attributed to advective heat loss resulting from hydrothermal circulation within permeable basement. The sparse distribution of early data offshore Costa Rica precluded confident determination of the cause(s) of the observed heat-flow patterns.

Additional heat-flow measurements were collected near the MAT offshore of Costa Rica in the mid-1990s during a site survey in preparation for ocean drilling. *Langseth and Silver* [1996] collected >100 new heat-flow values organized mainly along several transects on the margin wedge and close to the trench axis on EPR-generated crust. Heat flow in the deepest part of the trench averaged 17 mW/m^2, and values on a fault-bounded bench seaward of the trench averaged 12 mW/m^2. Mean heat flow was somewhat higher, generally 25–30 mW/m^2, on the margin wedge within 30 km of the trench. This survey was conducted on and above oceanic crust produced entirely along the EPR, confirming with modern equipment that the low heat-flow values detected during previous decades were characteristic of this part of the Cocos plate.

Heat-flow data were collected at greater subseafloor depths during ODP Leg 170 [*Kimura et al.*, 1997]. The measured value within the trench at Site 1039 was 9 mW/m^2, about half the value measured at this location using surface probes. Borehole values from higher up on the wedge were also considerably lower than those determined with shallow seafloor probes. The discrepancy between borehole and seafloor values can be reconciled if there is upward fluid seepage at rates of 5–20 mm/yr [*Ruppel and Kinoshita*, 2000]. This rate of flow is too low to cause curvature in shallow thermal gradients measured close to the seafloor but will result in a significant reduction in conductive heat flow with depth [e.g., *Bredehoeft and Papadopulos*, 1965; *Fisher*, 2004; *Langseth et al.*, 1988]. *Langseth and Silver* [1996] attributed the low heat flow on the incoming plate near Site 1039 to efficient hydrothermal cooling in basement, with the extent of cooling possibly being enhanced by reactivation of normal faults outboard of the MAT due to plate flexure. A similar hypothesis was applied to the incoming plate along the Peru Margin [*Yamano and Uyeda*, 1990]. Models of the thermal regime of the incoming plate on the Costa Rica margin that account for hydrothermal cooling are in general agreement with observations of a landward increase in surface heat flow across the margin wedge [*Harris and Wang*, 2002; *Langseth and Silver*, 1996; *Spinelli and Saffer*, 2004]. All three models are able to fit the observations reasonably well when hydrothermal circulation is assumed to penetrate 500–1000 m into basement.

As of the late 1990s, thermal studies had established that EPR-generated seafloor northwest of the plate suture offshore of the Nicoya Peninsula was cooler than the global mean for crust of this age, whereas CNS-generated seafloor southeast of the plate suture was warmer than the global mean, but close to the conductive prediction. However, the available data did not permit determination of the location of the thermal transition between areas of warm and anomalously cool seafloor. As a result, the association of this transition with one or more tectonic boundaries and the cause(s) of the difference in seafloor thermal state remained uncertain. In addition, it was unclear how typical the ODP drilling transect was with respect to regional thermal conditions or how well two-dimensional thermal models of subduction might be expected to simulate subseafloor conditions. If there is a difference in the efficiency of hydrothermal heat extraction within EPR-generated seafloor, why does this occur? Why is the region of anomalously cool seafloor located within a segment that had been previously characterized as being "smooth," whereas the region of anomalously warm seafloor is located in an area that had been characterized as being "rough" [*von Huene et al.*, 2000]? This correlation is the opposite to what one might expect if the efficiency of hydrothermal heat extraction were enhanced by the presence of basement outcrops, since rougher seafloor might be expected to have more areas of basement exposure. Is it important that the overall structural "fabric" (as determined from magnetic lineations) of EPR-generated seafloor is oriented parallel to the trench, whereas CNS-generated seafloor crustal structure trends perpendicular or oblique to the trench? Perhaps this difference in structural orientation influences the extent of fault reactivation during flexure outboard of the trench and thus leads to a difference in the development of enhanced permeability in basement. The TicoFlux program was designed to address these questions and to assess the nature of hydrothermal processes in this region.

The TicoFlux Program

The TicoFlux program included two expeditions to investigate the nature of hydrothermal activity and its influence on lithospheric and subduction processes on the Cocos plate. The first cruise (TicoFlux 1) focused on regional patterns and processes, and the second (TicoFlux 2) assessed the influence of seamounts, faults, and other local features in guiding hydrothermal fluid flow. Both programs focused on a 2° × 2° region of the Cocos plate seaward of the MAT (fig. 4.1), where fundamental transitions in the structural, hydrological, and thermal state of the plate were thought to occur. Each expedition collected seismic, hydrosweep, and heat-flow data, as well as cores for analysis of sediment properties and pore-fluid geochemistry. The present paper focuses on preliminary thermal and limited seismic results; other TicoFlux work will be published elsewhere.

Data Acquisition and Processing We collected 1800 km of 480-channel seismic data during TicoFlux 1 and an additional 1200 km of four-channel data during TicoFlux 2. During TicoFlux 1, shots were generated with a 10-gun array and spaced 37.5 m apart. The 6000-m streamer provided 6.25-m hydrophone spacing and 80-fold subsurface coverage. During TicoFlux 2, shots were fired with two guns at 33.3-m intervals. The receiver cable consisted of four 50-m active sections and yielded threefold coverage. Data were processed initially aboard ship using SIOSEIS and later processing onshore was accomplished using SeisWorks. Hydrosweep data were edited manually to remove bad pings and noise. Data collected during TicoFlux 1 were combined with those collected by numerous other workers in the region (most of which was from landward of the trench) to generate a single bathymetric grid with 200-m pixel resolution (C. Ranero, personal communication, 2002). These data were overlain on coarser bathymetric data derived from satellite gravimetry, having a pixel resolution of ~3 km [*Smith and Sandwell*, 1997]. Hydrosweep data from TicoFlux 2 were draped across the composite data set.

Multipenetration heat-flow data were collected during both expeditions using a 3.5-m, 11-thermistor, violin-bow heat flow probe with in situ thermal conductivity [*Hyndman et al.*, 1979]. Heat-flow stations were colocated along seismic reflection profiles in order to merge information on sediment thickness and basement structure with thermal data. An additional 57 estimates of heat flow were made using autonomous, outrigger temperature probes [*Pfender and Villinger*, 2002] attached to gravity and piston cores. Data were processed on the basis of techniques described by *Villinger and Davis* [1987]. We used a Monte Carlo analysis to assess uncertainties in heat-flow values resulting from ambiguity as to the thickness of sediment layers having differing thermal conductivities and used the "scatter" parameter to aid in assessing how many thermistors should be used for each heat-flow estimate [*Stein and Fisher*, 2001]. Outrigger data were analyzed using the same method, but we added an additional uncertainty in the Monte Carlo analysis to account for representation of thermal conductivity with a depth function rather than individual measurements associated with each core.

Extrapolation of Seafloor Heat-Flow Data to Depth We estimated thermal conditions within sediments and shallow basement across the field area using a method similar to that of *Davis et al.* [1999]. Heat-flow data were combined with seismic estimates of sediment thickness and thermal-conductivity data derived from drilling to extrapolate downward from the seafloor to uppermost basement. Sediment cores recovered from ODP Site 1039, on the incoming plate in the trench, were tested during Leg 170 to determine sediment thermal conductivity and seismic velocity at a core scale, and additional information on seismic interval velocities was available from synthetic seismograms and drilling data [*Kimura et al.*, 1997]. Uncertainties in estimated temperatures at depth are based on uncertainties in point measurements of heat flow, depth to basement as picked from seismic profiles, and thermal resistance versus travel time relations developed from drilling data.

A comparison between *P*-wave velocities measured on core samples and interval velocities determined from seismic and drilling data at ODP Site 1039. *Kimura et al.* [1997] demonstrated that the core measurements were essentially correct in the upper 180 m of hemipelagic sediment (mean velocity of 1570 m/s), but core measurements underpredicted seismic velocities by ~9% within the deeper, pelagic section (fig. 4.2a). We attribute this disparity to (1) damage to the core samples that commonly occurs during rotary drilling and is most pronounced in lithified materials that are readily cracked during recovery and handling and (2) systematic underrepresentation of coarse-grained (sandy) material in the recovered core.

We tested two options for correcting the seismic velocities: applying a constant offset of 9% and applying an offset whose magnitude increased linearly with depth within the pelagic section. Cumulative two-way travel time (t) was computed from individual shipboard measurements (v) weighted by the depth intervals (dz) between them

$$t(z) = \int_0^z \frac{dz}{v} \tag{4.1}$$

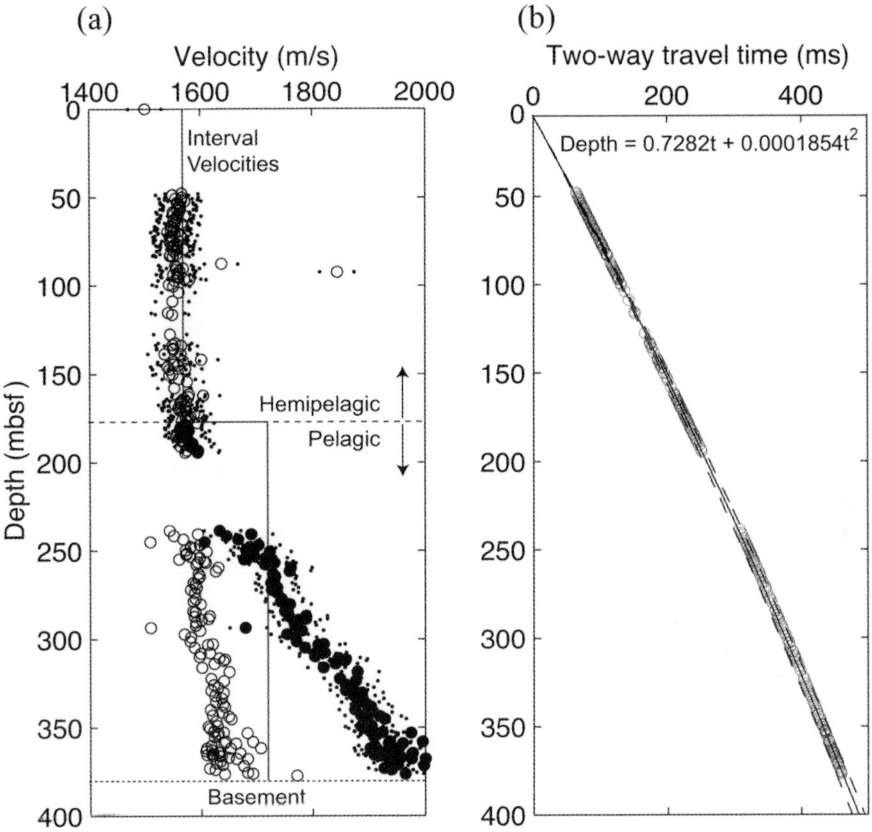

Figure 4.2 (a) Open circles show ODP shipboard seismic velocity measurements on cores recovered from ODP Leg 170 [*Kimura et al.*, 1997]. The pelagic/hemipelagic boundary occurs at a depth of 180 mbsf. Solid circles below the pelagic/hemipelagic boundary were adjusted linearly to be consistent with interval velocities obtained from synthetic seismograms. Single dots denote ±1 standard deviation applied to data from above and below the pelagic/hemipelagic boundary. Solid lines indicate average seismic velocities for each section. (b) Cumulative travel time computed from individual shipboard measurements weighted by the depth interval between them. Solid line shows the fit of a second-order polynomial regression. Dashed lines show uncertainty by regressing the data with ±1 standard deviation.

and regressed against depth below seafloor to derive the best-fitting, second-order polynomial function that allowed a transformation from two-way travel time to sediment thickness (fig. 4.2b). The function based on a depth-varying velocity correction resulted in a polynomial having a smaller second-order term, and we consider this to be preferable because it will help to minimize errors associated with extrapolation of the velocity-depth function beyond the sediment thickness at Site 1039 (~400 m). After applying the correction described above, we further estimate an uncertainty in this time-to-depth conversion by shifting the seismic velocities by ±1 standard deviation of the mean within each sediment interval.

The thermal conductivity of sediment cores recovered from Site 1039 was also determined on ODP Leg 170 [*Kimura et al.*, 1997] (fig. 4.3). Thermal-conductivity measurements are likely to be influenced by damage to the core much like seismic velocities, but we lack an independent estimate of upper-basement temperature that is necessary to apply rigorous corrections [e.g., *Davis et al.*, 1999]. Rather than attempting to correct thermal-conductivity values for core damage, we add an additional uncertainty to individual conductivity values [e.g., *Bevington and Robinson*, 1992] of ±1 standard deviation of the mean

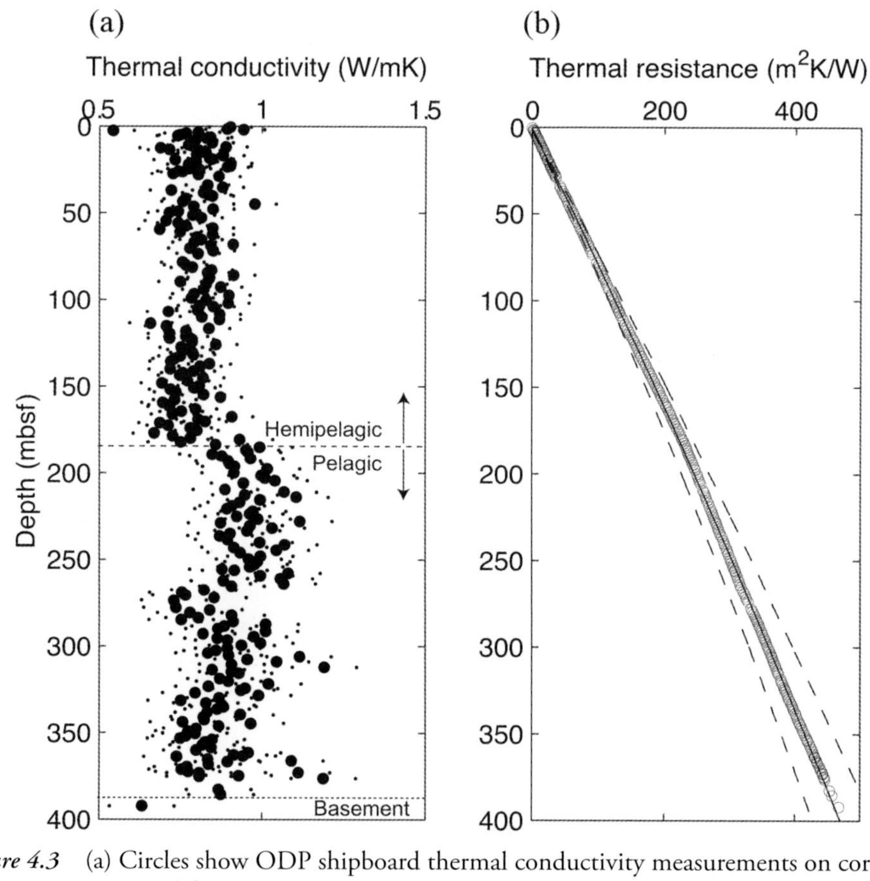

Figure 4.3 (a) Circles show ODP shipboard thermal conductivity measurements on cores recovered from ODP Leg 170 [*Kimura et al.*, 1997]. The pelagic/hemipelagic boundary occurs at a depth of 180 mbsf. Single dots denote ±1 standard deviation applied to data from above and below the pelagic/hemipelagic boundary. (b) Cumulative thermal resistance computed from individual shipboard thermal conductivity measurements weighted by the depth interval between them. Solid line shows the fit of a second order polynomial regression. Dashed lines show uncertainty by regressing the data with ±1 standard deviation.

thermal conductivity within each of the hemipelagic and pelagic intervals. Cumulative thermal resistance (R) was computed from individual shipboard measurements (λ) weighted by the depth intervals (dz) between them,

$$R(z) = \int_0^z \frac{dz}{\lambda} \tag{4.2}$$

and regressed against depth below seafloor to derive a function allowing for the transformation from sediment thickness to cumulative thermal resistance.

Uncertainties in seismic velocities and thermal conductivity described above generate combined uncertainties in thermal resistance versus two-way time (depth) of ±10%. There are additional uncertainties in picking the upper-basement reflector from seismic data (varies from site to site, particularly where the upper-basement surface is steep), in individual heat-flow values, in extrapolation of Site 1039 seismic velocity and thermal-conductivity data across the field area, and in the assumption that heat transport within the sediment is one-dimensional and vertical. Collectively, these factors may lead to uncertainties in the temperature at the sediment-basement interface and in the locations of isotherms within the sediment section, of 15–25% (fig. 4.4).

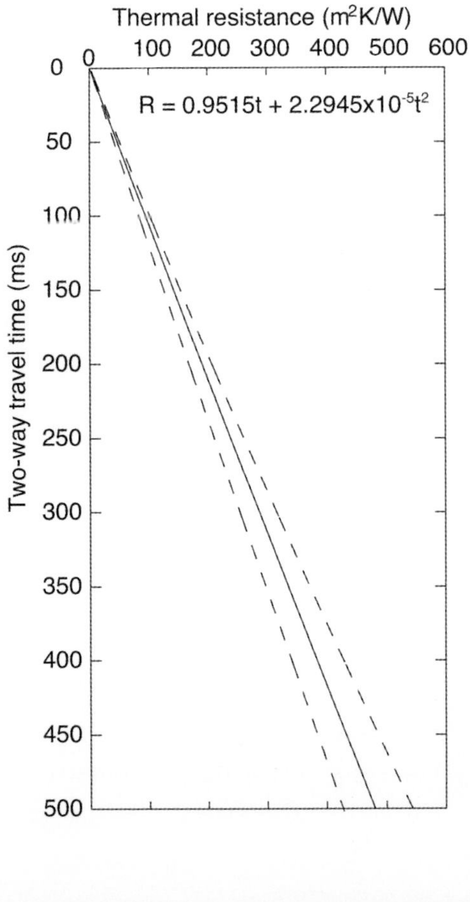

Figure 4.4 Relationships between cumulative thermal resistance and sediment two-way travel time resulting from regressions in figures 4.2 and 4.3. Solid line shows the fit of a second order polynomial regression, and dashed lines show uncertainty from regressions where ±1 standard deviation had been applied. As sediment two-way travel time increases, the fractional difference in the uncertainty in cumulative thermal resistance remains constant at roughly 8.5%. We apply a conservative 10% uncertainty in cumulative thermal resistance to all calculations of temperatures along the sediment-basement interface.

Thus it is risky to draw conclusions from individual temperature estimates. Instead, we focus on trends apparent in profiles of extrapolated temperatures and temperature contours, all of which are subject to the same assumptions and limitations.

In the absence of additional physical property data, the thermal structure within oceanic basement is even less well-constrained. For illustrative purposes, and in order to continue extrapolated temperature contours below the sediment-basement interface, we assume a thermal conductivity of upper-basement rocks of 1.6 W/m-K [e.g., *Becker et al.*, 1983; *Busch et al.*, 1992; *Karato*, 1983] and *P*-wave velocity of 4500 m/s [e.g., *Carlson*, 1998; *Jacobson*, 1992; *Rohr*, 1994]. However, vigorous local circulation may homogenize temperatures locally within upper basement [e.g., *Davis et al.*, 1989; *Fisher et al.*, 1990; *Langseth and Herman*, 1981], so temperature contours within upper basement should be viewed cautiously.

Thermal and Related Analyses

The focus of this section is on variations in thermal conditions within sediments and uppermost basement across the TicoFlux survey area. We begin with a brief assessment of seafloor and basement relief, as this topic is important for understanding the distribution of warmer and cooler crustal areas. Next, we show examples of heat flow and seismic transects across tectonic boundaries, thermal transitions, faults, and buried and exposed basement highs. Individual profiles are shown with thermal data and interpretations superimposed on seismic data, with line locations shown in figure 4.1. These examples are not exhaustive but illustrate characteristic patterns that help in delineating the thermal structure of the upper oceanic crust in this area. Finally, we conclude by discussing what these features and thermal patterns mean for local and regional hydrothermal conditions in basement, how the thermal state of the subducting crust could impact the seismogenic zone below the Nicoya Peninsula.

Seafloor and Basement Relief

On the basis of newly acquired hydrosweep coverage, seamounts are common features on EPR-generated seafloor (fig. 4.1) north of the plate suture. The largest of these features were apparent in bathymetric data derived from satellite gravimetry prior to the TicoFlux program [e.g., *Wessel*, 2001], consisting of massive edifices 3–5 km or more in diameter and rising 1–2 km above the surrounding seafloor. Swath data collected during the TicoFlux expeditions confirmed (or shifted) the locations of these features and identified numerous smaller basaltic outcrops that were not apparent from the satellite data. Because the spacing between seamounts in this area is typically several to tens of kilometers, a randomly positioned seismic line is unlikely to cross a basaltic outcrop. It is therefore not surprising that outcrops are relatively uncommon features on seismic data collected prior to swath mapping. High-resolution TicoFlux seismic data further revealed numerous buried basement highs that were outcrops before being covered by accumulating sediment. These local basement highs (buried or exposed at the seafloor) were found to be relatively common on EPR-generated seafloor in this area, whereas CNS-generated seafloor immediately adjacent to the plate suture and MAT appears to be entirely devoid of both exposed seamounts and large buried basement highs (fig. 4.1). Continuing to the south, across the ridge jump and propagator trace, seamounts and outcrops become much more prominent in association with the Cocos Ridge.

Heat-Flow Transects and Seismic Profiles

Thermal Transitions and Tectonic Boundaries Heat-flow station TF1-HF02 comprises 12 heat-flow measurements colocated along seismic line TF1-07, crossing the ridge jump near 87°03′N, 8°22′W (figs. 4.1 and 4.5). Basement is generally flat below 400 m of sediment, except for a 1000-m-wide, 160-m-tall buried basement high that corresponds to the ridge-jump trace.

Measured heat flow is 112–124 mW/m^2 across the transect, consistent with predictions from lithospheric cooling models for 19–21 Ma seafloor. Extrapolated temperatures along the sediment-basement interface remain generally uniform at 51°–59°C, except at the ridge jump where the estimated upper-basement temperature is just 35°C. This temperature estimate is consistent with purely conductive conditions; the lower value results from shallowing of basement coupled with slight thermal refraction, and there is no thermal evidence for hydrothermal processes in basement on either side of the ridge jump.

Conditions are markedly different along heat-flow station TF1-HF06. This station consisted of 13 heat-flow measurements colocated along seismic line TF1-14, crossing the propagator trace near 86°4′W, 8°23′N (figs. 4.1 and 4.6). As along line TF1-07 traversing the ridge jump to the west, basement is generally flat with the exception of a 2000-m-wide, 200-m-tall buried basement high associated with the propagator trace. Heat flow varies between 104 and 128 mW/m^2 along most of the transect but is elevated to >200 mW/m^2 above the propagator trace. Extrapolation of seafloor thermal conditions to depth indicates a relatively consistent temperature at the sediment-basement interface across this transect of 32°–41°C (essentially constant within estimated uncertainties), with only slightly lower apparent temperatures above the propagator trace. The consistency of temperatures across and adjacent to the propagator trace, despite a 200-m difference in elevation of the top of basement, requires that heat be redistributed efficiently within basement in this area, most likely by hydrothermal circulation. Similar processes have been documented in other

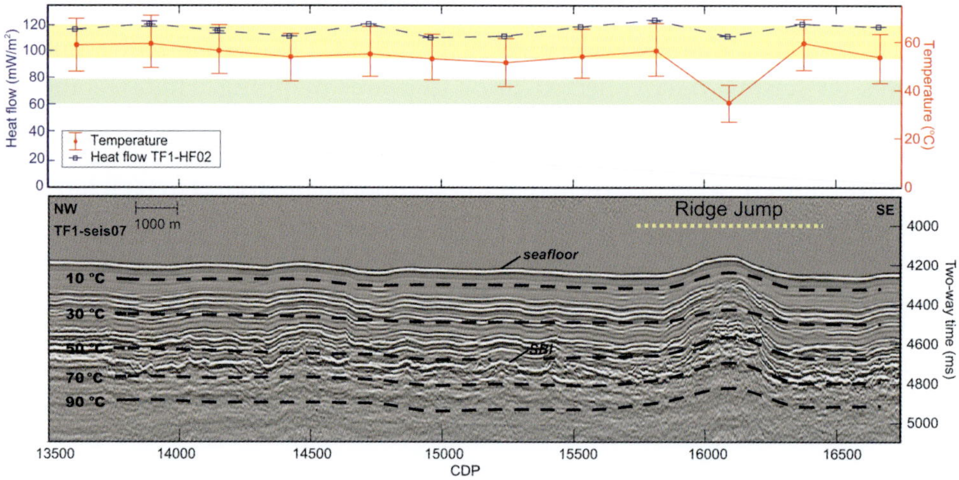

Figure 4.5 Seismic, heat flow, and calculated temperature data traversing the ridge jump along TF1-seis07. Top panel shows measured seafloor heat flow (blue squares) and calculated sediment-basement interface temperatures (red circles), with yellow and green bands indicating the conductive prediction and global mean, respectively, for heat flow through 18–24 Ma seafloor. Bottom panel shows seismic data overlain by isotherms calculated from heat flow and cumulative thermal resistance relationships shown in figure 4.4. The tectonic transition from *Barckhausen et al.* [2001] is centered along the buried basement high at CDP 16100. The seafloor and sediment-basement interface (SBI) are highlighted. Heat flow remains relatively constant across the transition while the sediment basement temperature decreases, indicative of vertical heat conduction and no hydrothermal fluid flow.

ridge-flank sites where vigorous circulation in upper basement homogenizes basement temperatures [e.g., *Davis et al.*, 1992; *Fisher et al.*, 1990]. There is no evidence for advective heat loss from the crust along heat-flow station TF2-HF06; all heat associated with lithospheric cooling is accounted for in the measured seafloor values. Instead, vigorous convection in basement moves heat around locally, elevating temperatures in upper basement along the propagator trace and increasing seafloor heat flow above this feature. This condition contrasts strikingly with that seen along heat-flow transect TF1-HF02 across the ridge jump (fig. 4.5), where thermal conditions are completely conductive.

In both of these examples, there are no thermal transitions associated with the major tectonic boundaries; both boundaries occur within "warm" parts of the plate, where heat flow is approximately that predicted by conductive lithospheric cooling models. Heat-flow transect TF1-HF02 crosses between seafloor created at the EPR and seafloor generated at the CNS, and conditions on both sides of the ridge jump are entirely conductive (fig. 4.5). Heat-flow transect TF2-HF06 crosses a propagator trace entirely within CNS-generated

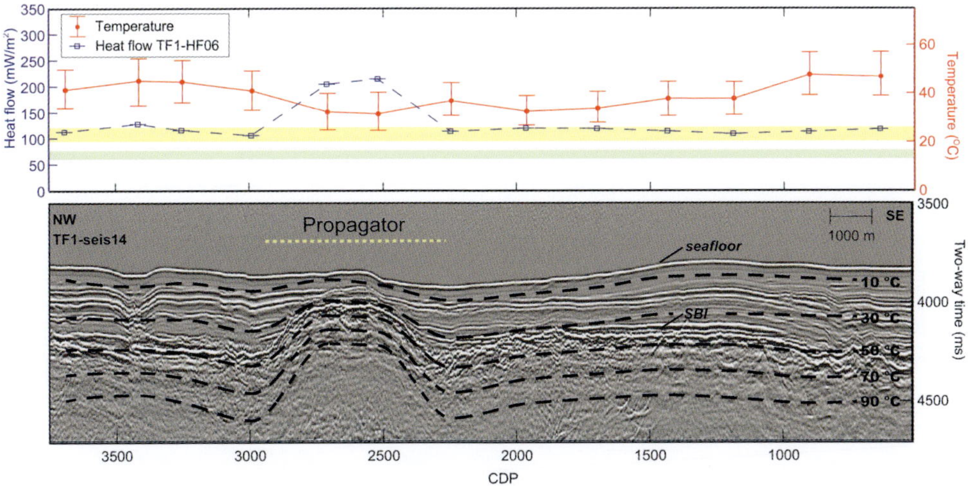

Figure 4.6 Seismic, heat flow, and calculated temperature data traversing the propagator trace along TF1-seis14. Symbols and nomenclature are the same as figure 4.5. Heat flow more than doubles across the buried basement high centered at CDP 2600, while sediment thins to roughly 100 m. Temperatures along the sediment-basement interface remain relatively uniform around 37°C across the basement high, indicating vigorous hydrothermal circulation within the feature.

seafloor, and hydrothermal circulation redistributes heat locally within this feature (fig. 4.6).

We found one thermal transition during the TicoFlux expeditions that corresponds to a tectonic boundary between different parts of the Cocos plate. Heat-flow station TF1-HF08 consisted of 11 heat-flow measurements colocated along seismic line TF1-11, crossing the triple-junction trace near 86°21′W, 9° 0′N (figs. 4.1 and 4.7).

The top of basement is ~100 m shallower on CNS-generated seafloor than on EPR-generated seafloor, with the change in depth occurring across a lateral distance of 1.5 km. Seafloor heat flow is ~30 mW/m^2 on the northwest (EPR) side of the transect and rises to ~100 mW/m^2 on the southeast (CNS) side. Temperatures along the sediment-basement interface on the EPR side of the boundary are 15°–30°C (rising with proximity to the boundary), whereas upper-basement temperatures on CNS side approach 45°C, even though the basement is shallower. Similar thermal transitions have been modeled by *Fisher et al.* [2003b], and as discussed below, the abruptness of this thermal transition (heat flow varies by a factor of 3 over a lateral distance of <3 km) suggests that it results from shallow hydrothermal circulation within the uppermost crust.

Other Thermal Transitions Most thermal transitions documented during the TicoFlux program were not associated with any tectonic boundary [*Fisher et al.*,

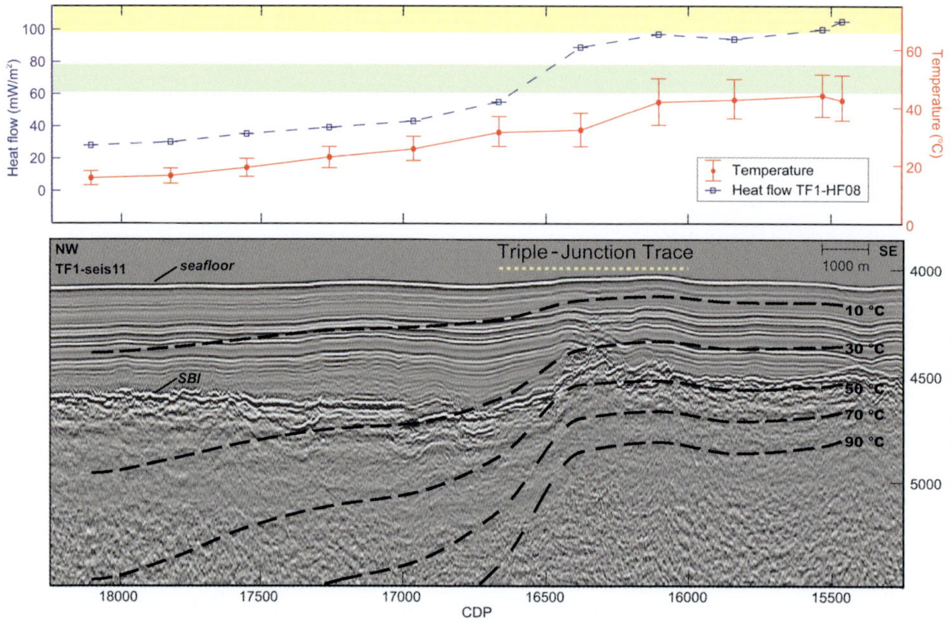

Figure 4.7 Seismic, heat flow, and calculated temperature data traversing the triple junction trace from EPR- to CNS-generated crust along TF1-seis11. Symbols and nomenclature are the same as fig. 4.5. Heat flow increases from suppressed values around 30 mW/m^2 on the EPR side to over 100 mW/m^2 on the CNS side. Isotherms are severely depressed on EPR-generated crust, indicative of efficient removal of large portions of lithospheric heat.

2003b] (fig. 4.1). As an example, heat-flow stations TF1-HF03,10,09 comprising 27 heat-flow measurements colocated along seismic line TF1-13, are entirely on EPR-generated seafloor and ~40 km from the nearest tectonic boundary (fig. 4.8).

Seafloor topography across the transect is generally flat, with bathymetric relief <50 m and sediment thicknesses of 350–450 m. The reflection character of basement in the northwestern part of the transect consists of a high-amplitude, generally smooth surface, which may be indicative of widely distributed sills near the base of the sediments [*Fisher et al.*, 2003b]. This area of smooth basement ends abruptly near CDP 9650, abutting a 4000-m-wide, 150 m-tall basement high. On the southeast side of this feature, the upper-basement reflector is much less distinct, and basement relief is considerably rougher.

The transition in basement appearance coincides with a rapid transition in seafloor heat flow, with values increasing from <30 mW/m^2 to the northwest to >100 mW/m^2 over a lateral distance of <2 km. Corresponding temperatures along the sediment-basement interface increase from <20°C to >40°C, from northwest to southeast. The abruptness and magnitude of this thermal transition indicates an extreme contrast in the nature and extent of heat loss from the upper oceanic crust in this area, with 70% of lithospheric heat being extracted

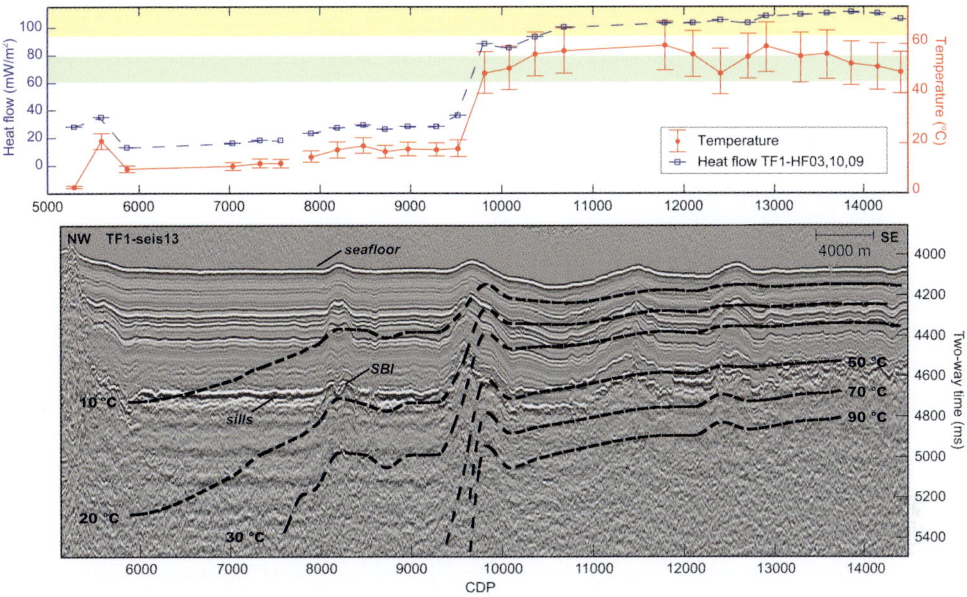

Figure 4.8 Seismic, heat flow, and calculated temperature data along TF1-seis13 on EPR-generated crust. Symbols and nomenclature are the same as figure 4.5. Heat flow transitions abruptly from <30 to >100 mW/m^2 over a lateral distance <2 km, while temperatures along the sediment-basement interface change concurrently. The abruptness and degree of this thermal transition indicate an extreme contrast in the nature and extent of hydrothermal circulation in shallow basement. The high-amplitude, smooth surface along the SBI is likely indicative of widely distributed sills.

from crust on the northwest side of this transect, most likely by hydrothermal circulation. In a later section, we discuss how local outcrops may facilitate rapid fluid exchange between the crust and ocean, cooling vast areas of EPR-generated seafloor in this region.

Heat-flow stations TF2-HF17 and TF1-HF05 comprise 22 heat-flow measurements colocated along seismic line TF1-03, with the two stations separated by an along-track distance of ~34 km (figs 4.1 and 4.9).

Basement is generally flat along transect TF2-HF17 and is covered with ~400 m of sediments (fig. 4.9a). Measured heat flow is 102–107 mW/m^2, consistent with predictions from conductive lithospheric cooling models for crust of this age. Extrapolated temperatures along the sediment-basement interface remain generally uniform at 48°–55°C. These thermal data are entirely consistent with purely conductive conditions.

Heat-flow and sediment-basement interface temperatures drop by ~80% 34 km northeast at station TF1-HF05 (fig. 4.9b), which crosses four fault-bounded benches immediately southwest of the ODP Leg 170 drilling transect (fig. 4.1). Station TF1-HF05 consisted of 12 heat-flow measurements extending roughly 15 km into the trench. Overall seafloor depth increases from ~3600 to over

4200 m, while seafloor relief across the faults ranges between 25 and 130 m. Basement relief varies between 50 and 140 m, with the largest offsets corresponding to faults closer to the trench axis. Sediment thickness across the transect is very uniform at 380–400 m. Heat-flow values range between 13 and 35 mW/m^2, with values elevated locally directly above basement faults. Extrapolated temperatures along the sediment-basement interface decrease from >12°C to <8°C from southwest to northeast along this transect.

Although the exact location and extent of the thermal transition along seismic line TF1-03 was not identified, the 80% decrease in lithospheric heat flow between stations TF2-HF17 and TF1-HF05 indicates an extreme contrast in the character and extent of heat extraction along a profile perpendicular to the MAT. Bend-related normal faults have been suggested to create permeable pathways allowing fluids to hydrate the lower crust and upper mantle in similar settings [*Peacock*, 2001; *Ranero et al.*, 2003]. The pervasiveness of bend-related faults along the trench slope outboard of the MAT on EPR-generated crust is hypothesized to provide pathways for deep penetration of seawater and an efficient mechanism for considerable lithospheric heat extraction. The influx of seawater is thought to promote serpentinization of chemically reactive lithospheric mantle and crust, leading to physical and chemical changes within the subducting slab. An important by-product of serpentinization is the generation of substantial amounts of heat [e.g., *Kelley et al.*, 2001; *Lowell and Rona*, 2002], and we would expect to see some evidence of this (either as direct fluid venting or as increased conductive heat flow) at the seafloor. If additional heat were being generated by serpentinization within the upper lithosphere and mantle in this region, then this additional heat must also be efficiently mined by hydrothermal circulation.

In order to investigate the origin of locally elevated heat-flow values measured above faults along heat-flow transect TF1-HF05 (fig. 4.9b), we constructed a two-dimensional finite element model using Finite Element Heat and Mass (FEHM) [*Zyvoloski et al.*, 1996]. Our model domain consisted of a 20-km grid, with eight sedimentary and six basalt units (seafloor and sediment-basement interface horizons were digitized from seismic line TF1-03), and a typical horizontal node spacing of 25 m. Physical properties from ODP Leg 170 Hole 1039 [*Kimura et al.*, 1997] were used as model constraints. Model output from a purely conductive simulation shows locally elevated heat flow of roughly 10–40% above the faults, with the largest increases corresponding to the greatest seafloor relief changes near CDP's 2050 and 2625, in the central portion of the transect (fig. 4.9b). We conclude that seafloor thermal refraction alone is sufficient to explain the locally elevated heat flow measured above the faults, and there is no evidence for either thermally significant subvertical fluid flow or serpentinization at depth.

Variations in Heat Flow Around Buried Basement Highs It is well known that basement relief enhances hydrothermal circulation and heat advection within

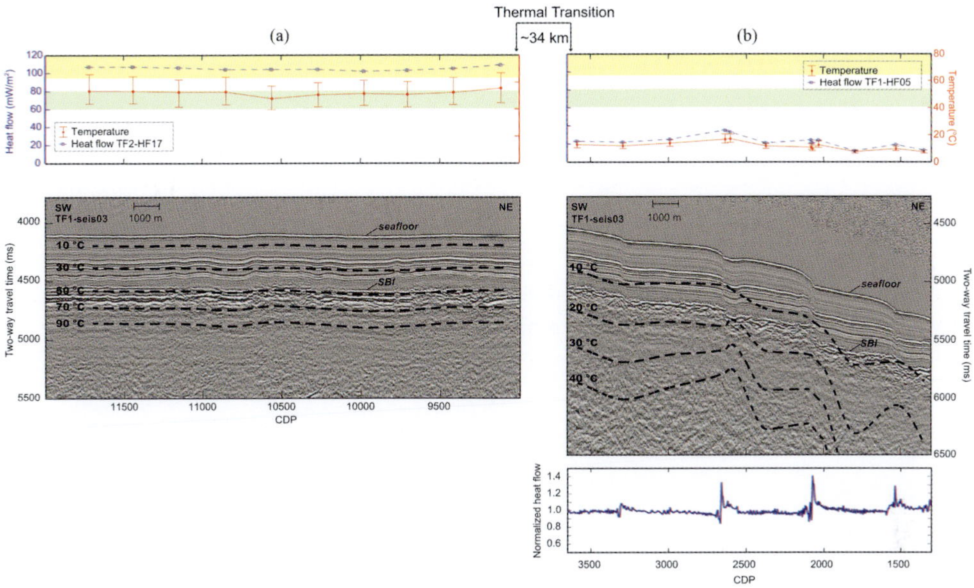

Figure 4.9 Seismic, heat flow, and calculated temperature data along TF1-seis03 (orientated perpendicular to the MAT) on EPR-generated crust. Symbols and nomenclature are the same as figure 4.5. (a) Station TF2-HF17 heat flow remains uniform across the transect at ~110 mW/m^2, and temperatures along the SBI remain constant at ~55°C. A significant thermal transition occurs along the 34 km separating stations TF2-HF17 and TF1-HF05, where heat flow drops by ~80%. (b) Station TF2-HF05 crosses four fault-bounded benches extending ~15 km seaward of the MAT. Heat flow varies between 12 and 35 mW/m^2, with locally elevated values corresponding to locations directly above the faults. Temperature along the SBI varies between 8° and 18°C. Bottom panel shows results from a two-dimensional conductive model simulation, indicating that the locally elevated heat flow values above the faults are the result of seafloor thermal refraction, and not due to vertical fluid flow.

upper oceanic crust [e.g., *Davis et al.*, 1989; *Fisher et al.*, 1990; *Johnson et al.*, 1993; *Lowell*, 1980] by focusing fluid flow toward basement highs. In the TicoFlux area we find areas on both EPR- and CNS-generated seafloor where vigorous hydrothermal circulation associated with buried basement highs homogenizes basement temperatures. We have already shown one example of this phenomenon on CNS-generated seafloor (fig. 4.6), where the mean heat flow is roughly equal to the conductive prediction, and in this section we show two additional examples from EPR-generated seafloor that demonstrate the variability and extent of this process.

The most dramatic example of thermal homogenization occurs on heat-flow stations TF1-HF04,11, which consisted of 17 heat-flow measurements co-located along seismic line TF1-05 (figs. 4.1 and 4.10).

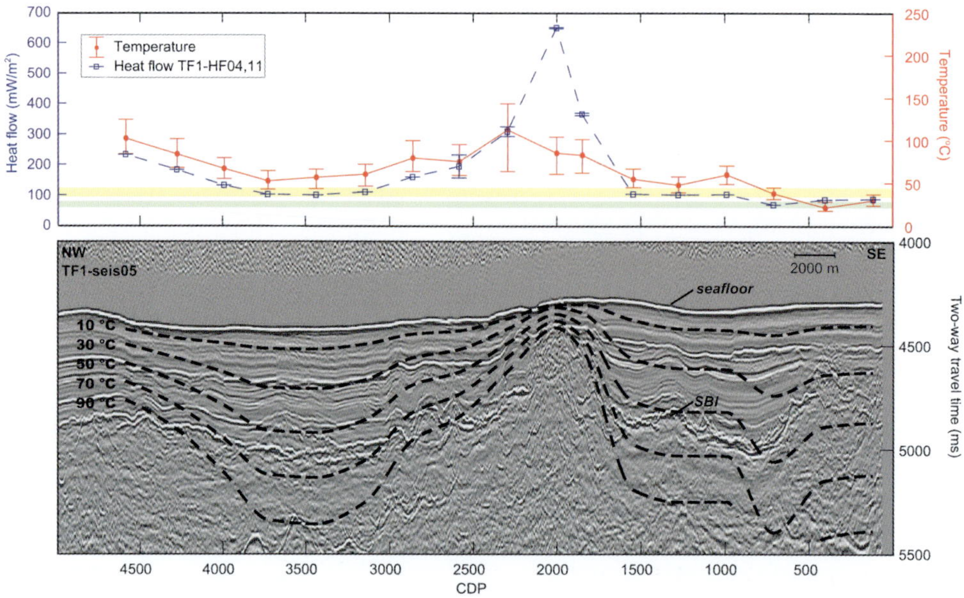

Figure 4.10 Seismic, heat flow, and calculated temperature data along TF1-seis05 on EPR-generated crust over an anomalously warm region. Symbols and nomenclature are the same as figure 4.5. Heat flow away from the basement high roughly matches conductive lithospheric cooling predictions for crust of this age, while over the basement high it rises by over sixfold as sediment thins to <100 m. Temperatures along the SBI remain remarkably uniform over the basement high despite the >500 m basement elevation change, indicative of vigorous hydrothermal circulation resulting in thermal homogenization within uppermost basement.

Seafloor heat flow is generally 30 mW/m^2 on EPR-generated seafloor in this region, but stations TF1-HF04,11 are in an isolated area where heat flow is considerably greater. Seafloor relief is <75 m along this transect, whereas a buried basement high rises by >550 m and sediment thins to <80 m above a 2.5-km-wide buried basement high. Heat flow rises along with basement elevation, from <100 to >650 mW/m^2, while estimated upper-basement temperatures remain remarkably uniform at ~80°C. The consistency of these temperatures requires vigorous fluid circulation. In the absence of hydrothermal circulation to homogenize upper-basement temperatures, we would expect to see a temperature difference of 50°–60°C between the shallowest and deepest basement along this heat-flow transect. The occurrence of such vigorous hydrothermal circulation requires considerable permeability in basement, in addition to heating from below, but it is not clear why this small area is also unusually warm relative to the surrounding seafloor. As discussed later, there is a large-scale flow system that extracts the majority of crustal heat from EPR-generated seafloor in this area on a regional basis, but it appears that the crust in the vicinity of this buried basement high is not hydrologically well-connected to the larger-scale flow system.

Another example of thermal homogenization around a buried basement high on EPR-generated seafloor occurs on heat-flow transect TF2-HF01, which consisted of 17 heat-flow measurements colocated along seismic line TF1-11 (figs. 4.1 and 4.11).

Seafloor relief is less than 25 m along this transect, whereas the top of basement rises 300 m and sediment thins to ~275 m above a 1000-m-wide basement high. Heat flow along most of the transect is suppressed by roughly 70% relative to conductive lithospheric cooling models, but heat flow is locally elevated immediately above the buried basement high, rising to >60 mW/m^2. Temperatures along the sediment-basement interface over the entire transect remain extremely uniform around 25°C, including the basement high, indicating vigorous circulation in uppermost basement. This profile is very similar to profile TF2-HF06 on CNS-generated seafloor (fig. 4.6) except that the "background" heat-flow value, away from the buried basement high, is lower. Along both profiles, heat flow doubles immediately above the basement high as a result of local circulation.

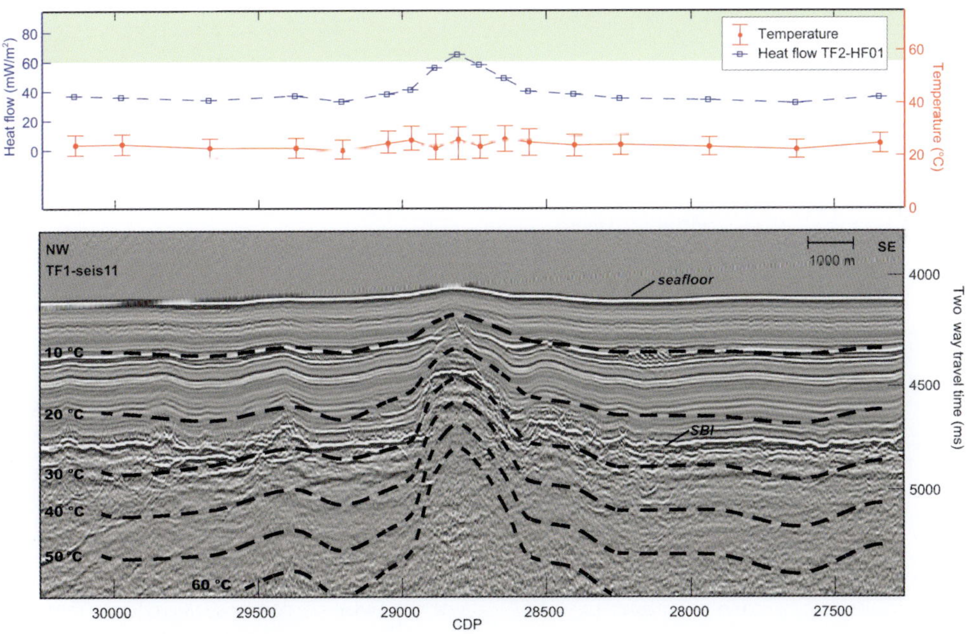

Figure 4.11 Seismic, heat flow, and calculated temperature data along TF1-seis11 on EPR-generated crust. Symbols and nomenclature are the same as figure 4.5. Heat flow doubles above the basement high centered at CDP 28800, while sediment thins to ~300 m. Hydrothermal circulation in uppermost basement is evidenced by temperatures along the SBI remaining extremely constant above the basement high, as is the case with TF01-HF(04,11) on seismic line TF1-05 (fig. 4.10).

Heat-Flow Measurements Near Discharging and Recharging Outcrops Seamounts and other basaltic basement outcrops are common on EPR-generated crust within the TicoFlux study area and are notably absent on CNS-generated seafloor southwest of the plate suture. These features appear to provide permeable pathways for the advective exchange of fluids and heat between the crust and overlying ocean, allowing hydrothermal fluids to bypass low-permeability sediments and efficiently extract large amounts of lithospheric heat [e.g., *Fisher et al.*, 2003a; *Villinger et al.*, 2002]. Modest pressure differences at the base of recharging and discharging columns of water provide the driving force for fluid exchange in this ridge-flank setting, and it is not possible to move fluids on a crustal scale at thermally significant rates through sediments thicker than a few meters [e.g., *Davis and Becker*, 2002; *Fisher and Becker*, 2000; *Giambalvo et al.*, 2000; *Mottl and Wheat*, 1994]. In this section, we present examples of heat-flow patterns around a discharging and a recharging seamount.

Fuente seamount is located on EPR-generated crust 56 km northwest of the intersection between the fracture zone and triple junction trace (figs. 4.1 and 4.12).

This conical feature has a radius of 2.5 km and rises 600 m above the surrounding seafloor. It was not apparent on bathymetric maps derived from

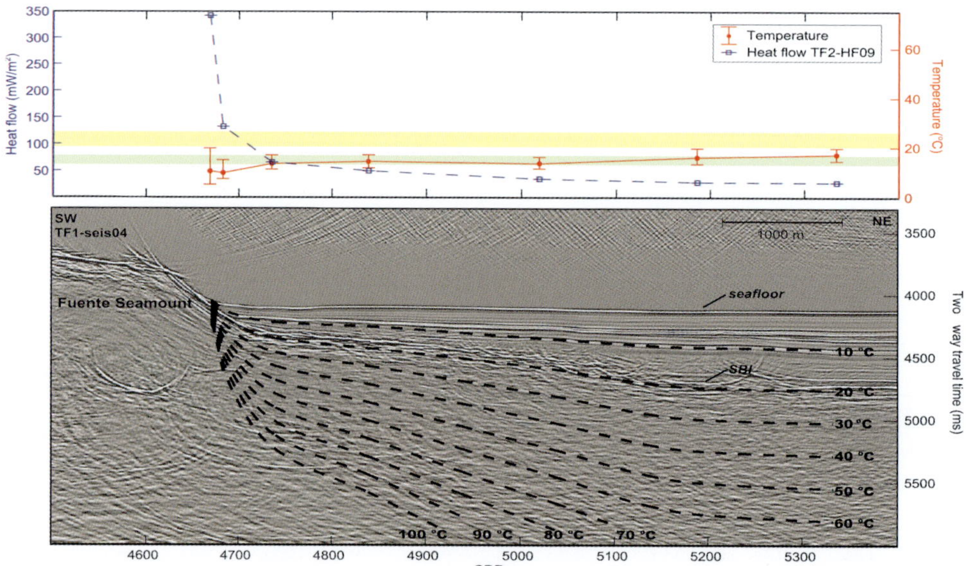

Figure 4.12 Seismic, heat flow, and calculated temperature data along TF1-seis04 crossing Fuente seamount. Symbols and nomenclature are the same as figure 4.5. Heat flow away from the outcrop is <30 mW/m^2, and increases by over 10-fold at the edge of basement exposure. Projected isotherms dip subparallel to the SBI where temperatures remain relatively constant, indicative of lateral advection of hydrothermal fluid in uppermost basement. A recharging outcrop is shown in figure 4.13.

satellite gravimetry data but was discovered during seismic and swath-map surveys during TicoFlux 1. Seafloor topography is extremely flat in the region surrounding Fuente seamount, whereas basement shoals by over 500 m before reaching the seafloor. Heat-flow station TF2-HF09 consisted of seven measurements colocated along seismic line TF1-04 and was intended to map the heat-flow distribution within several kilometers of the seamount (fig. 4.12). Heat flow ≥2.5 km from the limit of basement exposure is <30 mW/m^2, consistent with regional values on EPR-generated seafloor, but rises rapidly to 340 mW/m^2 (>10-fold) near the edge of the edifice. This increase in heat flow is several orders of magnitude greater than that which can be accounted for by conductive refraction alone. Temperatures along the sediment-basement interface remain remarkably constant at ~15°C along the entire length of the transect. If conditions in basement were conductive, we would expect to see a decrease in basement temperature as basement shallows upon approaching the outcrop. The data suggest instead that hydrothermal circulation locally homogenizes basement temperatures and are likely indicative of rapid lateral fluid flow toward (and out of) the seamount. Basement fluid temperatures are likely to be too low to generate a significant thermal anomaly in the water column around Fuente seamount, and venting fluids probably have a chemistry very close to bottom seawater because of the low temperature of reactions in basement [e.g., *Elderfield et al.*, 1999; *Kimura et al.*, 1997; *Mottl and Wheat*, 1994; *Wheat and Mottl*, 1994].

Tengosed seamount is located on EPR-generated crust 46 km northwest of the fracture zone trace and 44 km northeast of Fuente seamount (figs. 4.1 and 4.13).

This conical feature has a radius of 4.25 km and rises 1150 m above the surrounding seafloor. Unlike Fuente seamount, Tengosed is large enough to be readily apparent on bathymetric maps generated from satellite gravity data. Heat-flow station TF2-HF05 consisted of 21 stations colocated along seismic line TF2-23b, extending radially 12 km from the southern side of the seamount. In contrast to conditions around discharging Fuente seamount, heat flow near Tengosed decreases from >80 to <8 mW/m^2 over a lateral distance of roughly 7 km. In the absence of any significant variations in sediment thickness, temperatures along the sediment-basement interface are very low immediately adjacent to the outcrop, only slightly warmer than bottom water, and rise with increasing radial distance from the edifice. This pattern of suppressed heat flow and basement temperatures near the edge of the outcrop is indicative of substantial entrainment of cold seawater into upper basement, severely suppressing isotherms out to >4 km from the limit of basement exposure. This heat-flow profile is further complicated by the presence of a nontectonic thermal transition occurring roughly 6 km from the outcrop around CDP 860, where most of the change in heat flow occurs over a lateral distance of 2–3 km.

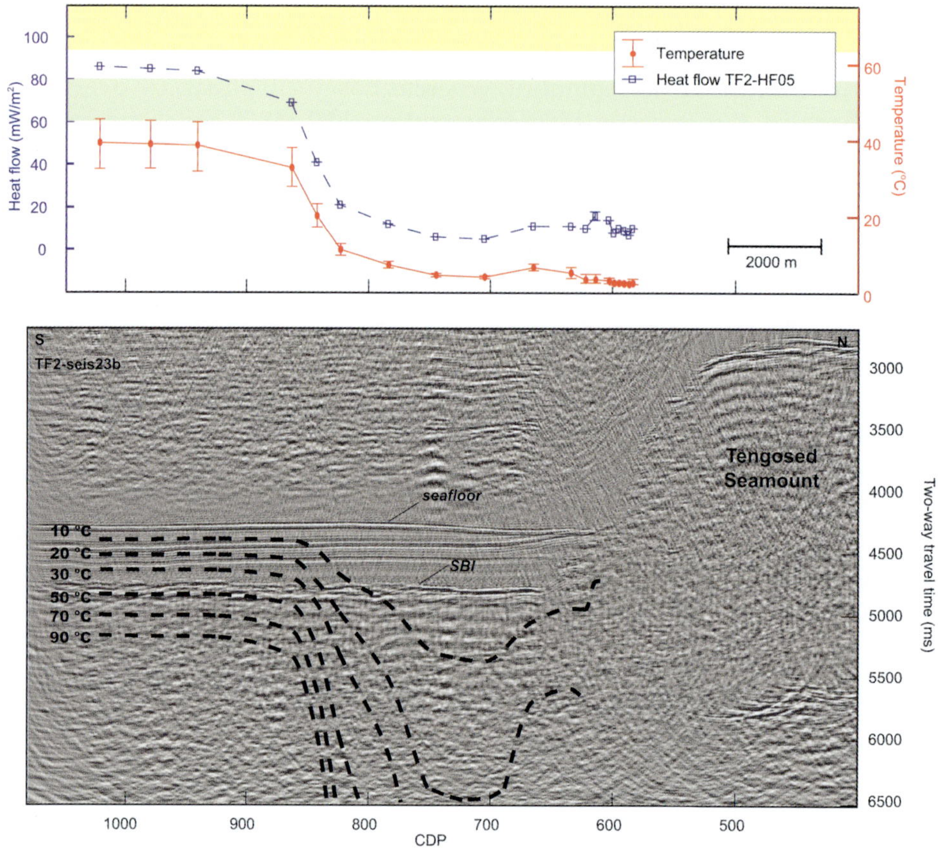

Figure 4.13 Seismic, heat flow, and calculated temperature data along TF2-seis23b crossing Tengosed seamount. Symbols and nomenclature are the same as figure 4.5. Heat flow in the vicinity of the outcrop drops by over 10-fold with temperatures along the SBI dropping concurrently. Isotherms in the vicinity of the outcrop are depressed out to 8 km from the limit of basement exposure, consistent with cold seawater being entrained and recharged into uppermost basement.

Hydrothermal Processes and Thermal Transitions

In order to evaluate the depth to the "source" of cooling within EPR-generated seafloor in the TicoFlux area, *Fisher et al.* [2003b] used numerical models of two-dimensional cross sections of the upper lithosphere within the Cocos plate. Heat was input into the base of the model domain, and heat sinks were distributed at various depths to simulate the extraction of heat by rapidly flowing hydrothermal fluids (fig. 4.14a).

Heat sinks were used to represent the advective extraction of heat without requiring specification of flow paths, three-dimensional modeling, or other

(a)

(b)

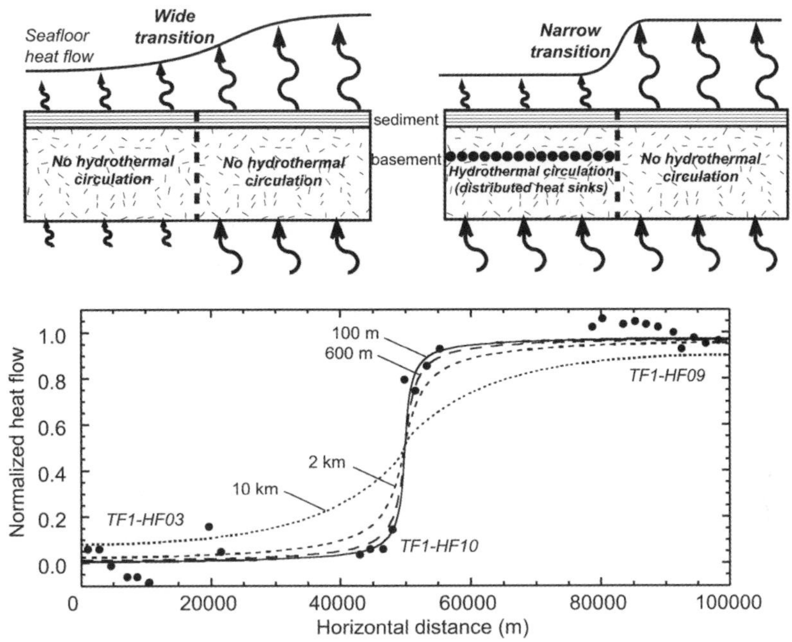

Figure 4.14 Conceptualization and results of models used to determine whether hydrothermal circulation was required within EPR-generated seafloor and to estimate the depth to the top of the hydrothermally active region [*Fisher et al.*, 2003b]. (a) Cartoons illustrating how the occurrence and depth to the top of the hydrothermal layer were tested. If the difference in seafloor heat flow detected on CNS- and EPR-generated seafloor resulted from differences in input at the base of the plate (or other deeply seated processes), we would expect to see a broad, gradual transition between warm and cool sides of the thermal transition. Alternately, if shallow hydrothermal circulation were responsible for the thermal contrast (as represented by point sinks for heat distributed on the left side of the transition), the transition should be narrow and abrupt. (b) Results of modeling of heat flow transect TF1-HF03,10,09 (fig. 4.8). Heat-flow values were normalized on a scale of 0–1, representing heat flow of 30–104 mW/m^2, for comparison to models. The abruptness and narrowness of the thermal transition requires relatively shallow hydrothermal circulation, a result consistent with modeling of other transitions in this area.

complexities. *Fisher et al.* [2003b] showed two seismic profiles and heat-flow transects across thermal transitions and compared observations to model predictions. We add results from another comparison between models and observations, based on heat-flow stations TF1-HF03,10,09 (figs. 4.8 and 4.14b). As with the other thermal transitions, the rapid increase in heat flow from ~30 to ~100 mW/m^2 occurs over a lateral distance of just a few kilometers, requiring the depth to the top of the hydrothermally active layer to be only a few hundred meters into basement.

It thus appears that most of the EPR-generated seafloor in the TicoFlux area is cooled by hydrothermal fluids and that basaltic outcrops facilitate the recharge and discharge of these fluids. *Langseth and Silver* [1996] argued that vigorous hydrothermal circulation was likely responsible for extreme cooling of the oceanic crust around ODP Site 1039, and *Silver et al.* [2001] used the observed extent of cooling to estimate a range of possible fluid-flow rates, driving forces, and crustal permeabilities. The basis for these calculations was the well-mixed aquifer model [*Langseth and Herman*, 1981], which is a one-dimensional representation of rapid lateral fluid flow superimposed on efficient local mixing in upper basement. We refine the results of earlier calculations [*Silver et al.*, 2001] on the basis of new data collected during the TicoFlux program.

Perdido outcrop is the closest seafloor basement exposure to ODP Site 1039 (fig. 4.1) and thus the most likely source of recharging hydrothermal fluid in this area. The upper oceanic crust is broken, uplifted, and exposed on the southwest side of the fracture zone [*Barckhausen et al.*, 2001], and heat-flow measurements made on the northwest side of this feature included the lowest values measured in the area (<4–9 mW/m^2). These low heat-flow values extend 3–10 km from the point of basement exposure, helping to place a strong constraint on the rate of lateral fluid flow within basement.

Assuming a typical sediment thickness of 500 m (as seen in seismic lines across the fracture zone), lateral fluid flow having a specific discharge (rate of discharge per unit area) of 3–30 m/yr within the upper 100–600 m of basement is required to suppress heat flow by 90–96% at a distance of 3–10 km from the recharge point (fig. 4.15a).

Thermal data from the cool part of the EPR-generated plate indicate upper-basement temperatures on the order of 20°C. As mentioned earlier, the primary driving force for ridge-flank hydrothermal circulation is the difference in pressure at the base of recharging and discharging water columns. We can calculate the greatest possible driving force for ridge-flank circulation by assuming that fluid recharge and discharge are adiabatic and that no energy is lost during fluid descent or ascent through the crust [e.g., *Fisher and Becker*, 2000]. The difference in pressures at the base of columns of recharging and discharging hydrothermal fluid is 20–50 kPa, assuming penetration to basement depths of 100–1000 m (fig. 4.15b). Finally, we can see from newly acquired swath-map data that the nearest exit points for hydrothermal fluids are outcrops located 75–110 km from Perdido seamount (fig. 4.1). As mentioned earlier, heat-flow data collected across normal faults on the seaward side of the trench give no evidence for fluid discharge (fig. 4.9) nor did an outcrop exposed near the trench in this area [*Fisher et al.*, 2003b].

Given a lateral specific discharge of 3–30 m/yr and a driving force of 20–50 kPa, upper-basement permeability in this area must be no less than 10^{-10} m^2 and could be as great as 10^{-8} m^2 (fig. 4.15b). These values at first seem surprisingly high for 18–24 Ma seafloor, but they are comparable to observations in other settings and earlier global predictions. Crustal permeabilities of 10^{-13} to 10^{-11} m^2 were estimated on the basis of single-hole packer and free-flow experi-

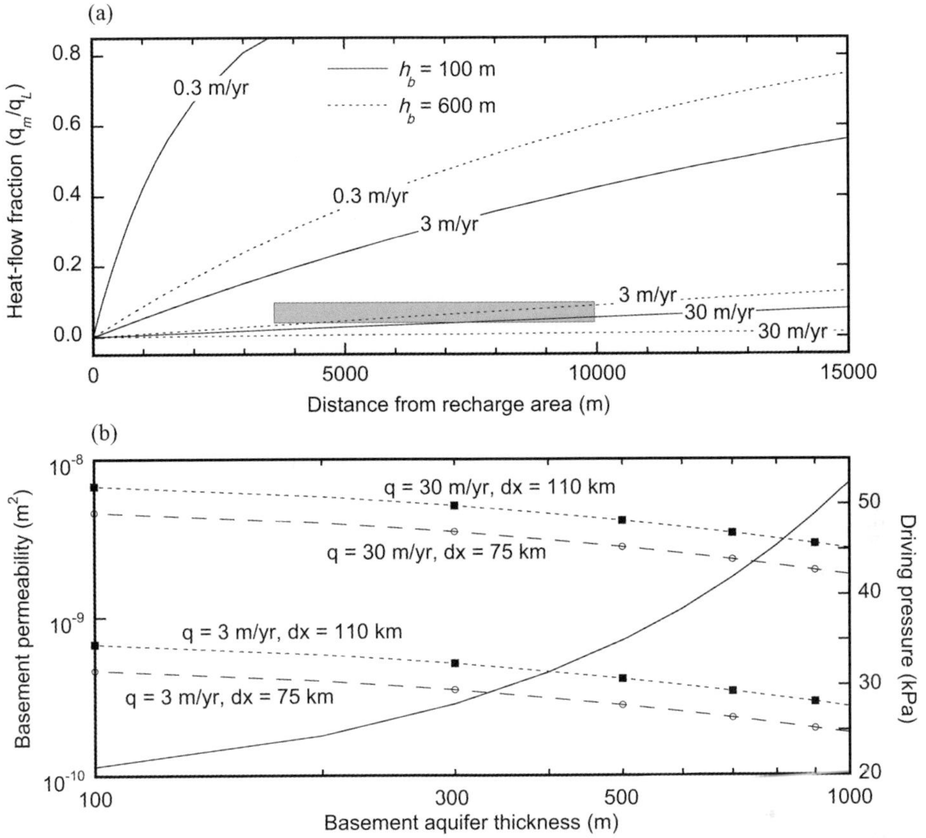

Figure 4.15 Calculations of fluid-flow rates, driving forces, and basement permeabilities required by TicoFlux data. (a) Results of well-mixed aquifer calculations [*Langseth and Herman*, 1981]. Specific discharge values are as indicated, h_b is the thickness of basement aquifer below 500 m of sediments. Heat-flow fraction is the ratio of measured to lithospheric heat-flow values. Gray box represents observations near Perdido outcrop, where measured heat flow is suppressed by 90–96% at a distance of 3–10 km from the outcrop. (b) Calculations of driving forces (solid line) and bulk basement permeabilities (dashed lines), based on consideration of fluid density differences between recharging and discharging outcrops (methods described in text and in greater detail by *Fisher and Becker* [2000] and *Fisher et al.* [2003a]). The temperature of fluids in upper basement was assumed to be 20°C, consistent with observations on the cool part of EPR-generated crust in this area. Outcrop spacing was constrained to be at least 75–110 km, based on available satellite-derived and swath maps. If the actual spacing between recharge and discharge sites is greater, bulk permeability in upper basement must be higher than shown.

ments within 3.5 Ma seafloor on the eastern flank of the Juan de Fuca Ridge, but higher permeabilities were estimated on the basis of crustal-scale flow and tidal analyses [e.g., *Becker and Davis*, 2003; *Becker and Fisher*, 2000; *Davis et al.*, 2000; *Fisher et al.*, 1997; *Fisher et al.*, 2003a]. In a study of global heat flow and permeability patterns, it was suggested that crustal-scale permeabilities of 10^{-8} to 10^{-10} m^2 are required by thermal observations on 20 Ma seafloor [*Fisher and Becker*, 2000]. Estimated driving forces and flow rates, and thus apparent basement permeabilities, scale roughly linearly with assumed depths of fluid penetration. Unfortunately, there are no observations from the TicoFlux surveys that help to constrain the depth extent of the large-scale lateral flow system responsible for very low heat-flow values common on EPR-generated seafloor in this area. However, data collected in other ridge-flank settings suggest that the uppermost few hundred meters are likely to be the most permeable [e.g., *Alt et al.*, 1986; *Bartetzko et al.*, 2001; *Fisher*, 1998].

The 3–30 m/yr specific discharge values calculated for the TicoFlux area are similar to flow rates inferred between widely spaced outcrops on the 3.5 Ma Juan de Fuca eastern flank [*Fisher et al.*, 2003a]. Unlike conditions within most EPR-generated seafloor in the TicoFlux area, there is no regional heat-flow deficit between the Juan de Fuca flank outcrops. However, rapid fluid flow in that area appears to be largely restricted to the region between the outcrops. In contrast, extraction of heat from most of the EPR-generated seafloor along the MAT in the TicoFlux area requires that much of the crust in this area host vigorous fluid circulation.

Within the region of generally cool, EPR-generated seafloor, there are several relatively small areas of elevated heat flow (figs. 4.1 and 4.10). These are likely to result from heterogeneities in the distribution of basement permeability; either "stagnant" zones in which no fluid circulation occurs or regions where vigorous local fluid circulation is hydrologically isolated from the regional flow system. It should not be surprising that such regions exist, because fractured hydrogeologic systems tend to be heterogeneous on a range of scales [e.g., *Fisher*, 1998; *Neuman*, 1994; *Neuman and Di Federico*, 2003]. In addition, we found small areas on CNS-generated seafloor with anomalously low heat flow, and none of these are located immediately southeast of the fracture zone and adjacent to the MAT, where there are no basement outcrops penetrating through the seafloor. Even closer to the Cocos Ridge, where we measured a few low heat-flow values, these are associated with localized circulation; there is no evidence for large-scale, lateral fluid flow systems on CNS-generated seafloor in this area.

There are several possible explanations for the difference in thermal state across the two regions of the Cocos plate surveyed during the TicoFlux expeditions. First, there may be a difference in overall crustal permeability between warm and cool sides of the plate. Second, there could be one or more hydrogeologic boundaries within crust that otherwise has similar properties. Finally, it may be that the density and proximity of seamounts and other basement outcrops limit the lateral extent of penetration of cool, rapidly flowing fluids.

If the first or second explanations were correct, we might expect that the thermal transition across the Cocos plate would coincide with one or more of the major tectonic boundaries. Instead, the thermal transition coincides with the plate suture (fracture zone trace) only where the plate is very close to the MAT (fig. 4.1). It is possible that there are one or more smaller-scale tectonic features associated with the part of the thermal transition that is offset from the plate suture but that such features are too small to be imaged clearly with the available seismic data. Subvertical fault offsets in basement of 20 m of less would likely be below the resolution of our instrumentation.

Another possible explanation for the lack of tectonic features associated with the thermal transitions is that seismic layer 2A (generally thought to be the most permeable part of oceanic basement on ridge flanks) is masked by sills, which are common in both the low and high heat-flow regions [*Silver et al.*, 2004]. The presence of sills could prevent seismic identification of faults in upper basement.

It is striking that, where the Cocos plate thermal transition is offset by 30–40 km from the plate suture, it runs close to a series of seamounts and other basement outcrops. It may be that there is no fundamental discontinuity in structure or hydrogeologic properties associated with the thermal transition in this area, but that the lack of outcrops on the warm side of the plate prevents significant advective heat extraction from basement. If this explanation is correct, there may be a critical combination of basement hydrogeologic properties and outcrop density that allows for the observed hydrothermal patterns. Moreover, there may be unusually high basement permeability within EPR-generated seafloor in this area, such that highly efficient advective heat extraction results from the imposition of closely spaced basaltic outcrops [e.g., *Fisher et al.*, 2003b]. We will explore these concepts in future studies using two- and three-dimensional numerical models.

Finally, we note that low heat flow on EPR-generated seafloor in this area occurs well seaward of the MAT, long before there is creation or reactivation of normal faults near the trench. The average heat flow of the incoming crust is the same as crust of its age far away from subduction zones [*Stein*, 2003]. There may be additional cooling in the trench associated with faulting and other subduction-related deformation [*Kimura et al.*, 1997; *Langseth and Silver*, 1996], but hydrothermal fluids extract an average of ~70% of lithospheric heat from most EPR-generated seafloor well before the plate reaches the trench. Thus vigorous fluid circulation must result from fundamental crustal construction and modification processes and is not directly related to subduction.

Implications of Hydrothermal Circulation and Thermal State of the Oceanic Crust on Subduction

One of the primary factors influencing the thermal regime of a subduction zone is the thermal state of the incoming oceanic plate as it enters the trench. The ability of hydrothermal circulation to redistribute heat and cool the incoming oceanic crust, as well as chemically bind additional water in the form of mineralogical and rheological transformations (i.e., peridotite to serpentinite), may have an important influence on the updip limit of the seismogenic zone. The extent to which hydrothermal cooling influences subduction processes depends, in part, on the depth extent of fluid circulation and heat extraction.

Unfortunately, new data collected during the TicoFlux expeditions do not provide a robust constraint on the depth extent of circulation. Numerical and analytical models require that the top of the hydrothermally active basement is no more than 100–600 m below the sediments and allow observational constraints to be satisfied with flow restricted to the upper 600 m of basement.

If high-angle normal faults associated with plate flexure penetrate to several kilometers, it is possible that circulating fluids cool the plate to similar or greater depths. Any additional heat generated by serpentinization due to the influx of seawater along bend-related normal faults outboard of the MAT must also be extracted advectively, since heat flow within the trench is only ~10% of lithospheric.

We note that the ODP Leg 170 drilling transect is surrounded on three sides by regions of elevated heat flow (fig. 4.1). Temperature of upper basement at Site 1039 on the incoming plate is ~6°C [*Kimura et al.*, 1997], whereas upper-basement temperatures ~40 km to the south near Perdido outcrop and ~50 km to the west at station TF1-HF04,11 (fig. 4.10) are ~50°C. Warm and cool subduction zones have been shown to exhibit differences in seismicity [*Peacock and Wang*, 1999], reflecting the differences in metamorphic reactions occurring within the subducted crust. The along-strike variability of ~50°C in upper-basement temperature surrounding the ODP Leg 170 transect will have an impact on the thermal structure of the subduction thrust. If hydrothermal circulation is restricted to uppermost basement, its thermal influence on subduction will be limited. Assuming efficient hydrothermal cooling to a depth of 1 km into basement, two-dimensional thermal models of subduction suggest that thermal rebound will occur by the time the plate has moved ~50 km landward of the trench axis (see fig. 4.3, *Harris and Wang* [2002] and fig. 4.2, *Spinelli and Saffer* [2004]), which may influence the updip limit of seismicity [*Hyndman et al.*, 1997; *Hyndman and Wang*, 1993; *Oleskevich et al.*, 1999]. Studies of the seismogenic zone in this region show a shallowing of the updip limit in seismicity toward the south [*Newman et al.*, 2002] consistent with warmer basement temperatures.

Conclusions

Early studies on the Costa Rica margin identified areas of higher (>100 mW/m^2) and lower (<30 mW/m^2) heat flow on CNS- and EPR-generated seafloor, respectively. However, the sparse distribution of early heat flow, seismic, and bathymetric data precluded confident determination of the causes of the heat-flow deficit on EPR-generated crust. In addition, there were no heat-flow measurements close to regional tectonic boundaries or basement exposures, making it impossible to determine if the thermal transition between anomalously cool and warm seafloor was associated with any of these features.

On the basis of newly acquired hydrosweep coverage, we find that seamounts with a characteristic spacing of 10–40 km are common features on EPR-generated seafloor north of the plate suture, whereas CNS-generated seafloor immediately adjacent to the plate suture and MAT appears to be entirely devoid of seamounts. These outcrops provide permeable pathways for the advective exchange of fluids and heat between the crust and overlying ocean, allowing hydrothermal fluids to bypass low-permeability sediments and efficiently extract large amounts of lithospheric heat. Most of the EPR-generated crust in the TicoFlux study region is cooled by rapidly flowing hydrothermal fluids, but we find no evidence for large-scale advective heat extraction in CNS-generated seafloor. Using observational data in the vicinity of Perdido outcrop as model constraints, lateral fluid flow having a specific discharge of 3–30 m/yr within the upper 100–600 m of basement can suppress heat flow by over 90% at a distance of 3–10 km from the recharge point. This requires that basement permeability in the area be on the order of 10^{-10} to 10^{-8} m^2. Although these values may seem high for this age of seafloor, they are comparable to estimates from other settings and are consistent with earlier global predictions.

We find purely conductive thermal conditions on one crossing of the ridge jump, and one crossing of the plate suture demonstrates vigorous hydrothermal circulation within uppermost basement. We find no thermal evidence for upward expulsion of fluids along bend-related normal faults outboard of the MAT near ODP Site 1039. There are areas on both EPR- and CNS-generated seafloor where rapid hydrothermal circulation associated with buried basement highs homogenizes upper-basement temperatures. Because the entire region of low heat flow on EPR-generated crust developed well seaward of the MAT, vigorous fluid circulation must result from fundamental crustal construction and modification processes and is not directly related to subduction.

The lack of a clear association between much of the interplate thermal transition with major tectonic boundaries suggests that other factors control fluid and heat flow in basement. If variations in crustal permeability or hydrogeologic boundaries were guiding or limiting flow, then we would expect the thermal transition between warm and cool parts of the Cocos plate to coincide with one or more tectonic boundaries (some of which may be seismically unresolvable). Where the thermal transition on EPR-generated seafloor is offset by 30–40 km from the plate suture, it runs alongside a series of seamounts and exposed basement outcrops that are actively recharging and discharging hydrothermal fluids between the crustal aquifer and the overlying ocean. This advective exchange of fluid and heat may be sufficient to create the observed thermal patterns, without the requirement for fundamental discontinuities in structural or hydrogeologic properties.

Acknowledgments

We thank the officers, crew, and technicians of the R/V *Maurice Ewing* and R/V *Melville* for their assistance and advice during the TicoFlux expeditions. Data from the R/V *Meteor* M54-2 cruise were generously provided by Ingo Grevemeyer. C. Ranero provided a compilation of earlier bathymetric data. SeisWorks and Promax software were provided by a grant from Landmark Graphics Corporation. This work was supported by NSF grants OCE-0001892 (A.F. and E.S.), OCE-0001944 (R.H.), and OCE-0001941 (C.S.), and IGPP/LANL grant 1317 (A.F.). This research used samples and/or data provided by the Ocean Drilling Program (ODP). The ODP is sponsored by the U.S. National Science Foundation (NSF) and participating countries under management of Joint Oceanographic Institutions (JOI), Inc. Funding for this research was provided by a Schlanger Ocean Drilling Fellowship, which is sponsored by the NSF-funded U.S. Science Support Program (USSSP).

References

Alt, J. C., J. Honnorez, C. Laverne, and R. Emmermann (1986), Hydrothermal alteration of a 1-km section through the upper oceanic crust, Deep Sea Drilling Project Hole 504B: Mineralogy, chemistry and evolution of seawater-basalt interactions, *J. Geophys. Res.*, *91*, 10,309–10,335.

Alt, J. C., D. A. H. Teagle, C. Laverne, D. A. Vanko, W. Bach, J. Honnorez, K. Becker, M. Ayadi, and P. A. Pezard (1996), Ridge-flank alteration of upper oceanic crust in the eastern Pacific: Synthesis of results for volcanic rocks of Holes 504B and 896A, *Proc. Ocean Drill. Program Sci. Results*, *148*, 435–450.

Anderson, R., and M. Hobart (1976), The relation between heat flow, sediment thickness, and age in the eastern Pacific, *J. Geophys. Res.*, *81*, 2968–2989.

Barckhausen, U., C. Renaro, R. von Huene, S. C. Cande, and H. A. Roeser (2001), Revised tectonic boundaries in the Cocos Plate off Costa Rica: Implications for the segmentation of the convergent margin and for plate tectonic models, *J. Geophys. Res.*, *106*, 19,207–19,220.

Bartetzko, A., P. Pezard, D. Goldberg, Y.-F. Sun, and K. Becker (2001), Volcanic stratigraphy of DSDP/ODP Hole 395A: An interpretation using well-logging data, *Mar. Geophys. Res.*, *22*, 111–127.

Becker, K., and E. Davis (2003), New evidence for age variation and scale effects of permeabilities of young oceanic crust from borehole thermal and pressure measuements, *Earth Planet. Sci. Lett.*, *210*, 499–508.

Becker, K., and A. Fisher (2000), Permeability of upper oceanic basement on the eastern flank of the Endeavor Ridge determined with drill-string packer experiments, *J. Geophys. Res.*, *105*, 897–912.

Becker, K., M. Langseth, R. P. Von Herzen, and R. Anderson (1983), Deep crustal geothermal measurements, Hole 504B, Costa Rica Rift, *J. Geophys. Res.*, *88*, 3447–3457.

Bevington, P. R., and D. K. Robinson (1992), *Data Reduction and Error Analysis for the Physical Sciences*, 328 pp., McGraw-Hill, New York.

Bredehoeft, J. D., and I. S. Papadopulos (1965), Rates of vertical groundwater movement estimated from the Earth's thermal profile, *Water Resour. Res., 1*, 325–328.

Bullard, E. C., A. E. Maxwell, and R. Revelle (1956), Heat flow through the deep sea floor, *Adv. Geophys., 3*, 153–181.

Busch, W. H., P. R. Castillo, P. A. Floyd, and G. Cameron (1992), Effects of alteration on physical properties of basalts from the Pigafetta and East Mariana basins, *Proc. Ocean Drill. Program Sci. Results 129*, 485–500.

Carlson, R. L. (1998), Seismic velocities in the uppermost oceanic crust: Age dependence and the fate of layer 2A, *J. Geophys. Res., 103*, 7069–7077.

Cowen, J. P., S. J. Giovannoni, F. Kenig, H. P. Johnson, D. Butterfield, M. S. Rappé, M. Hutnak, and P. Lam (2003), Fluids from aging ocean crust that support microbial life, *Science, 299*, 120–123.

Davis, E. E., and K. Becker (2002), Observations of natural-state fluid pressures and temperatures in young oceanic crust and inferrences regarding hydrothermal circulation, *Earth Planet. Sci. Lett., 204*, 231–248.

Davis, E. E., D. S. Chapman, C. Forster, and H. Villinger (1989), Heat-flow variations correlated with buried basement topography on the Juan de Fuca Ridge flank, *Nature, 342*, 533–537.

Davis, E. E., et al. (1992), FlankFlux: An experiment to study the nature of hydrothermal circulation in young oceanic crust, *Can. J. Earth Sci., 29*(5), 925–952.

Davis, E. E., D. S. Chapman, K. Wang, H. Villinger, A. T. Fisher, S. W. Robinson, J. Grigel, D. Pribnow, J. Stein, and K. Becker (1999), Regional heat-flow variations across the sedimented Juan de Fuca Ridge eastern flank: Constraints on lithospheric cooling and lateral hydrothermal heat transport, *J. Geophys. Res., 104*, 17,675–17,688.

Davis, E. E., K. Wang, K. Becker, and R. E. Thomson (2000), Formation-scale hydraulic and mechanical properties of oceanic crust inferred from pore-pressure response to periodic seafloor loading, *J. Geophys. Res., 105*, 13,423–13,435.

DeMets, C. (2001), A new estimate for present-day Cocos-Caribbean plate motion: Implications for slip along the Central American volcanic arc, *Geophys. Res. Lett., 28*, 4043–4046.

Elder, J. W. (1965), Physical processes in geothermal areas, in *Terrestrial Heat Flow, Geophys. Monogr. Ser.*, vol. 8, edited by W. H. K. Lee, pp. 211–239, AGU, Washington, D. C.

Elderfield, H., and A. Schultz (1996), Mid-ocean ridge hydrothermal fluxes and the chemical composition of the ocean, *Annu. Rev. Earth Planet. Sci., 24*, 191–224.

Elderfield, H., C. G. Wheat, M. J. Mottl, C. Monnin, and B. Spiro (1999), Fluid and geochemical transport through oceanic crust: A transect across the eastern flank of the Juan de Fuca Ridge, *Earth Planet. Sci. Lett., 172*, 151–165.

Fisher, A. (2004), Rates and patterns of fluid circulation, in *Hydrogeology of the Oceanic Lithosphere*, edited by E. E. Davis and H. Elderfield, pp. 339–377, Cambridge Univ. Press, New York.

Fisher, A. T. (1998), Permeability within basaltic oceanic crust, *Rev. Geophys., 36*, 143–182.

Fisher, A. T., and K. Becker (2000), Channelized fluid flow in oceanic crust reconciles heat-flow and permeability data, *Nature, 403*, 71–74.

Fisher, A. T., K. Becker, T. N. Narasimhan, M. G. Langseth, and M. J. Mottl (1990), Passive, off-axis convection on the southern flank of the Costa Rica Rift, *J. Geophys. Res., 95*, 9343–9370.

Fisher, A. T., K. Becker, and E. E. Davis (1997), The permeability of young oceanic crust east of Juan de Fuca Ridge determined using borehole thermal measurements, *Geophys. Res. Lett., 24*, 1311–1314.

Fisher, A. T., et al. (2003a), Hydrothermal recharge and discharge across 50 km guided by seamounts on a young ridge flank, *Nature, 421*, 618–621.

Fisher, A. T., C. A. Stein, R. N. Harris, K. Wang, E. A. Silver, M. Pfender, M. Hutnak, A. Cherkaoui, R. Bodzin, and H. Villinger (2003b), Abrupt thermal transition reveals hydrothermal boundary and role of seamounts within the Cocos Plate, *Geophys. Res. Lett., 30*(11), 1550, doi:10.1029/2002GL016766.

Gaetani, G. A., and T. L. Grove (1998), The influence of water on mantle melting, *Contrib. Mineral. Petrol., 131*, 323–346.

Giambalvo, E., A. T. Fisher, L. Darty, J. T. Martin, and R. P. Lowell (2000), Origin of elevated sediment permeability in a hydrothermal seepage zone, eastern flank of Juan de Fuca Ridge, and implications for transport of fluid and heat, *J. Geophys. Res., 105*, 913–928.

Harris, R. N., and K. Wang (2002), Thermal models of the Middle America Trench at the Nicoya Peninsula, Costa Rica, *Geophys. Res. Lett., 29*(21), 2010, doi:10.1029/2002GL015406.

Henry, P., and C.-Y. Wang (1991), Modeling of fluid flow and pore pressure at the tow of the Barbados and Oregon accretionary wedges, *J. Geophys. Res., 96*, 20,109–20,130.

Hey, R. N. (1977), Tectonic evolution of the Cocos-Nazca spreading center, *Geol. Soc. Am. Bull., 88*, 1404–1420.

Hyndman, R., M. Yamano, and D. Oleskevich (1997), The seismogenic zone of subduction thrust faults, *Island Arc, 6*(3), 244–260.

Hyndman, R. D., and K. Wang (1993), Thermal constraints on the seismogenic portion of the southwestern Japan subduction zone of subduction thrust faults, *J. Geophys. Res., 100*, 2039–2060.

Hyndman, R. D., E. E. Davis, and J. A. Wright (1979), The measurement of marine geothermal heat flow by a multipenetration probe with digital acoustic telemetry and in situ conductivity, *Mar. Geophys. Res., 4*, 181–205.

Iwamori, H. (1998), Transportation of H_2O and melting in subduction zones, *Earth Planet. Sci. Lett., 60*, 65–80.

Jacobson, R. S. (1992), Impact of crustal evolution on changes of the seismic properties of the uppermost oceanic crust, *Rev. Geophys., 30*, 23–42.

Johnson, H. P., K. Becker, and R. P. V. Herzen (1993), Near-axis heat flow measurements on the northern Juan de Fuca Ridge: Implications for fluid circulation in oceanic crust, *Geophys. Res. Lett., 20*, 1875–1878.

Karato, S. (1983), Physical properties of basalts from Deep Sea Drilling Project Hole 504B, *Initial Rep. Deep Sea Drill. Proj., 69*, 687–695.

Kelley, D. S., et al. (2001), An off-axis hydrothermal vent field near the Mid-Atlantic Ridge at 30°N, *Nature, 412*, 145–149.

Kimura, G., E. Silver, and P. Blum (1997), *Proceedings of the Ocean Drilling Program, Initial Reports*, 458 pp., Ocean Drill. Program, College Station, Tex.

Langseth, M. G., and B. Herman (1981), Heat transfer in the oceanic crust of the Brazil Basin, *J. Geophys. Res., 86*, 10,805–10,819.

Langseth, M. G., and E. A. Silver (1996), The Nicoya convergent margin - A region of exceptionally low heat flow, *Geophys. Res. Lett., 23*, 891–894.

Langseth, M. G., P. J. Grim, and M. Ewing (1965), Heat-flow measurements in the east Pacific Ocean, *J. Geophys. Res., 70*, 367–380.

Langseth, M. G., X. LePichon, and M. Ewing (1966), Crustal structure of the midocean ridges, 5, heat flow through the Atlantic Ocean floor and convection currents, *J. Geophys. Res., 71*, 5321–5355.

Langseth, M. G., M. J. Mottl, M. A. Hobart, and A. T. Fisher (1988), The distribution of geothermal and geochemical gradients near Site 501/504, implications for hydrothermal circulation in the oceanic crust, in *Proc. Ocean Drill. Program Initial Rep., 111*, 23–32.

Lee, W. H. K., and S. Uyeda (1965), Review of heat flow data, in *Terrestrial Heat Flow, Geophys. Monogr. Ser.*, vol. 8, edited by W. H. K. Lee, pp. 87–190, AGU, Washington, D. C.

Lowell, R. P. (1980), Topographically driven subcritical hydrothermal convection in the oceanic crust, *Earth Planet. Sci. Lett., 49*, 21–28.

Lowell, R. P., and P. A. Rona (2002), Seafloor hydrothemal systems driven by the serpentinization of peridotite, *Geophys. Res. Lett.*, *29*(11), 1531, doi:10.1029/2001GL014411.
McIntosh, K., and M. Sen (2000), Geophysical evidence for dewatering and deformation processes in the ODP Leg 170 area offshore Costa Rica, *Earth Planet. Sci. Lett.*, *178*, 125–138.
Meschede, M., U. Barckhausen, and H.-U. Worm (1998), Extinct spreading on the Cocos Ridge, *Terra Nova*, *10*, 211–216.
Mottl, M. J., and C. G. Wheat (1994), Hydrothermal circulation through mid-ocean ridge flanks: Fluxes of heat and magnesium, *Geochim. Cosmochim. Acta*, *58*, 2225–2237.
Neuman, S. P. (1994), Generalized scaling of permeabilities: Validation and effect of support scale, *Geophys. Res. Lett.*, *21*, 349–352.
Neuman, S. P., and V. Di Federico (2003), Multifaceted nature of hydrogeologic scaling and its interpretation, *Rev. Geophys.*, *41*(3), 1014, doi:10.1029/2003RG000130.
Newman, A. V., S. Y. Schwartz, H. R. DeShon, J. M. Protti, V. Gonzalez, and L. Dorman (2002), Along-strike variability in the seismogenic zone below Nicoya Peninsula, Costa Rica, *Geophys. Res. Lett.*, *29*(20), 1977, doi:10.1029/2002GL015409.
Noel, M., and M. W. Hounslow (1988), Heat flow evidence for hydrothermal convection in Cretaceous crust of the Madiera Abyssal Plain, *Earth Planet. Sci. Lett.*, *90*, 77–86.
Oleskevich, D., R. Hyndman, and K. Wang (1999), The updip and downdip limits to great subduction earthquakes: Thermal and structural models of Cascadia, south Alaska, SW Japan, and Chile, *J. Geophys. Res.*, *104*, 14,965–14,991.
Parsons, B., and J. G. Sclater (1977), An analysis of the variation of ocean floor bathymetry and heat flow with age, *J. Geophys. Res.*, *82*, 803–829.
Peacock, S. M. (2001), Are the lower planes of double seismic zones caused by serpentine dehydration in subducting oceanic mantle, *Geology*, *29*, 299–302.
Peacock, S. M., and K. Wang (1999), Seismic consequences of warm versus cool subduction metamorphism: Examples for southwest and northeast Japan, *Science*, *286*, 937–939.
Pfender, M., and H. Villinger (2002), Miniaturized data loggers for deep sea sediment temperature gradient measurements, *Mar. Geol.*, *186*(3–4), 557–570.
Plank, T., and C. H. Langmuir (1993), Tracing trace elements from sediment input to volcanic output at subduction zones, *Nature*, *362*, 739–742.
Ranero, C. R., and R. von Huene (2000), Subduction erosion along the Middle America convergent margin, *Nature*, *404*, 748–752.
Ranero, C. R., J. Phipps Morgan, and R. von Huene (2003), Bending-related faulting and mantle serpentinization at the Middle America Trench, *Nature*, *405*, 367–373.
Rohr, K. (1994), Increase of seismic velocities in upper oceanic crust and hydrothermal circulation in the Juan de Fuca plate, *Geophys. Res. Lett.*, *21*, 2163–2166.
Ruppel, C., and M. Kinoshita (2000), Fluid, methane, and energy flux in an active margin gas hydrate province, offshore Costa Rica, *Earth Planet. Sci. Lett.*, *179*, 153–165.
Saffer, D., and C. Marone (2003), Comparison of smectite- and illite-rich gouge frictional properties: Application to the updip limit of the seismogenic zone along subduction megathrusts, *Earth Planet. Sci. Lett.*, *215*, 219–235.
Schultz, A., and H. Elderfield (1997), Controls on the physics and chemistry of seafloor hydrothermal circulation, *Philos. Trans. R. Soc. London, Ser. A*, *355*(1723), 387–425.
Sclater, J. G., J. Crowe, and R. Anderson (1976), On the reliability of oceanic heat flow averages, *J. Geophys. Res.*, *81*, 2997–3006.
Sclater, J. G., C. Jaupart, and D. Galson (1980), The heat flow through oceanic and continental crust and the heat loss of the earth, *Rev. Geophys.*, *18*, 269–311.
Silver, E., M. Kastner, A. Fisher, J. Morris, K. McIntosh, and D. Saffer (2001), Fluid flow paths in the crust of the Middle America Trench, Costa Rica margin, *Geology*, *28*(8), 679–682.
Silver, E., P. C. Pisani, M. Hutnak, A. Fisher, H. R. DeShon, and B. Taylor (2004), An 8–10 Ma tectonic event on the Cocos Plate offshore Costa Rica: Result of Cocos Ridge collision?, *Geophys. Res. Lett.*, *31*, L18601, doi:10.1029/2004GL020272.

Smith, W. H. F., and D. T. Sandwell (1997), Global sea floor topography from satellite altymetry and ship depth soundings, *Science*, *277*, 1956–1962.

Spinelli, G. A., and D. Saffer (2004), Along-strike variations in underthrust sediment dewatering on the Nicoya margin, Costa Rica related to the updip limit of seismicity, *Geophys. Res. Lett.*, *31*, L04613, doi:10.1029/2003GL018863.

Stein, C. (2003), Heat flow and flexure at subduction zones, *Geophys. Res. Lett.*, *30*(23), 2197, doi:10.1029/2003GL018478.

Stein, C., and S. Stein (1992), A model for the global variation in oceanic depth and heat flow with lithospheric age, *Nature*, *359*, 123–129.

Stein, C., and S. Stein (1994), Constraints on hydrothermal heat flux through the oceanic lithosphere from global heat flow, *J. Geophys. Res.*, *99*, 3081–3095.

Stein, J. S., and A. T. Fisher (2001), Multiple scales of hydrothermal circulation in Middle Valley, northern Juan de Fuca Ridge: Physical constraints and geologic models, *J. Geophys. Res.*, *106*, 8563–8580.

Vacquier, V., J. G. Sclater, and C. E. Corry (1967), Studies in the thermal state of the earth, the 21st paper: Heat-flow, Eastern Pacific, *Bull. Earthquake Res. Inst.*, *45*, 375–393.

Villinger, H., and E. E. Davis (1987), HFRED: A program for reduction of marine heat flow data on a microcomputer, report, 77 pp., Geol. Surv. of Can., Sydney, B. C.

Villinger, H., I. Grevemeyer, N. Kaul, J. Hauschild, and M. Pfender (2002), Hydrothermal heat flux through aged oceanic crust: Where does the heat escape?, *Earth Planet. Sci. Lett.*, *202*(1), 159–170.

Von Herzen, R., and S. Uyeda (1963), Heat flow through the eastern Pacific floor, *J. Geophys. Res.*, *68*, 4219–4250.

Von Herzen, R. P. (2004), Geothermal evidence for continuing hydrothermal circulation in older (>60 Ma) ocean crust, in *Hydrogeology of the Oceanic Lithosphere*, edited by E. E. Davis and H. Elderfield, pp. 414–450, Cambridge Univ. Press, New York.

von Huene, R., C. R. Ranero, W. Wienrebe, and K. Hinz (2000), Quarternary convergent margin tectonics of Costa Rica, segmentation of the Cocos Plate, and Central American volcanism, *Tectonics*, *19*, 314–334.

Wessel, P. (2001), Global distribution of seamounts inferred from gridded Geosat/ERS-1 altimetry, *J. Geophys. Res.*, *106*, 19,431–19,442.

Wheat, C. G., and M. J. Mottl (1994), Hydrothermal circulation, Juan de Fuca Ridge eastern flank: Factors controlling basement water composition, *J. Geophys. Res.*, *99*, 3067–3080.

Wheat, C. G., J. McManus, M. Mottl, and E. G. Giambalvo (2003), Oceanic phosphorus imbalance: Magnitude of the mid-ocean ridge flank hydrothermal sink, *Geophys. Res. Lett.*, *30*(17), 1895, doi:10.1029/2003GL017318.

Yamano, M., and S. Uyeda (1990), Heat flow studies in the Peru Trench subduction zone, *Proc. Ocean Drill. Program Sci. Results*, *112*, 653–662.

Zhao, D., Y. Xu, D. A. Wiens, L. Dorman, J. Hildebrand, and S. Webb (1997), Depth extent of the Lau back arc spreading center and its relation to subduction processes, *Science*, *278*, 254–257.

Zyvoloski, G. A., B. A. Robinson, Z. D. Dash, and L. L. Trease (1996), *Users Manual for the FEHMN Application*, Los Alamos Natl. Lab., Los Alamos, N. M.

5

Influence of Subducting Topography on Earthquake Rupture

Susan L. Bilek

Abstract

Rupture models for large subduction zone earthquakes suggest significant heterogeneity in slip and moment release over the fault plane. Attempts to characterize the nature of the high-slip regions led to descriptions of these earthquakes in the context of the asperity model and allowed for subduction zones to be classified on the basis of likely earthquake asperity distributions. Understanding the physical features in the fault zone that produce these high-slip regions is still a challenge. Detailed bathymetric and seismic data collection in several subduction zones has led to correlations between large earthquake rupture patterns and subducting features such as seamounts and ridges. This is not a simple relationship, and it is not consistent in all subduction zones. Earthquakes at several subduction zones appear to concentrate slip on subducted features, but others have features that act as barriers to earthquake rupture. Additional questions remain about the temporal stability of high-slip regions. More research is needed to understand the connections between the upper plate, fault friction, subducting features, and overall subduction zone stresses and the patterns of large earthquakes in this complex environment.

Earthquake Slip and Asperities

Early definitions of asperities go back to Amonton, who proposed that static friction on a fault relates to the overall roughness of the fault surface. Increased roughness on the surface may be caused by increased occurrence of protrusions on the surface. He defined asperities as welded contacts along the fault. Early asperity models for earthquakes were developed using results from laboratory rock friction experiments [*Byerlee*, 1967; *Scholz and Engelder*, 1976].

These models suggest that faults are held together by high strength contacts, or Amonton's asperities. Stresses at the asperities were typically higher than at other portions of the fault. The physical causes of asperities along faults were related to variations in geometry, friction, or material properties along the fault.

Understanding the relationship between earthquakes and asperities advanced with analysis of higher-quality seismic data. Earthquake seismograms from a global distribution of seismic stations were used to examine the seismic moment release distribution and slip during large earthquakes. Many of these observations for continental and subduction zone earthquakes show high moment release, high slip, occurring in distinct patches on the fault plane rather than smoothly varying over the fault plane [e.g., *Wyss and Brune*, 1967; *Trifunac and Brune*, 1970; *Kanamori*, 1978; *Ruff and Kanamori*, 1983; *Schwartz and Ruff*, 1987; *Mendoza and Hartzell*, 1988; *Heaton*, 1990; *Wald et al.*, 1991; *Wald and Heaton*, 1994; *Ma et al.*, 2001; *Ji et al.*, 2002] (fig. 5.1). High-slip patches can be converted to relatively high stress drops, suggesting that these patches are areas of increased strength compared to the surrounding regions and can be considered asperities.

Small-magnitude earthquakes have also been used to describe asperity distribution on faults. *Nadeau and Johnson* [1998] suggest clustering of microearthquakes along the Parkfield segment of the San Andreas fault indicates asperity locations. They suggest that the earthquake clusters define high-strength patches able to break within weaker zones along the fault, although the high-stress drop values calculated for these earthquakes appear to be higher than

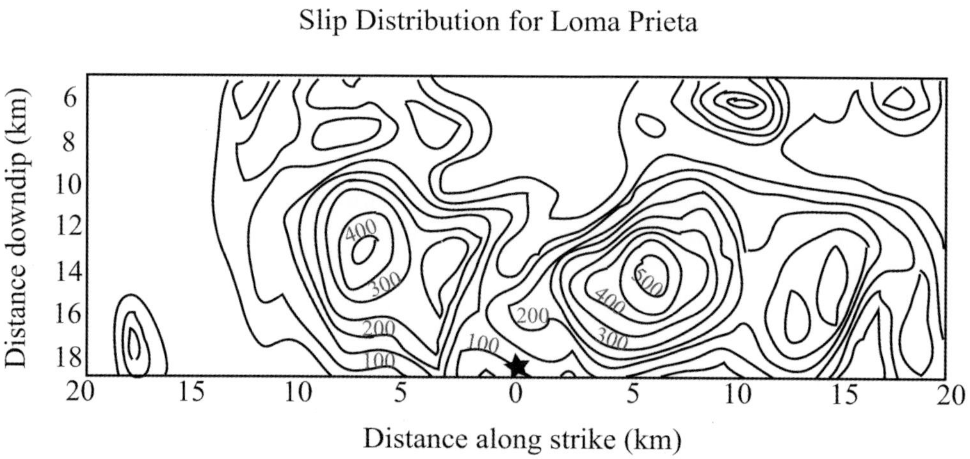

Figure 5.1 Slip distribution for the 1989 Loma Prieta earthquake, modified from *Beroza* [1991], with contours of slip in centimeters. A clear pattern of heterogeneous slip is observed for this continental earthquake, with two significant patches of high slip (asperities). Other models of strong motion and teleseismic data show consistent features [e.g., *Wald et al.*, 1991].

estimates from other small earthquakes [*Sammis and Rice*, 2001]. Detailed locations of small earthquakes in subduction zones have also yielded observations of clusters and suggestions of heterogeneous asperity distribution along these faults [e.g., *Hino et al.*, 1996; *Matsuzawa et al.*, 2002; *Igarashi et al.*, 2003; *Obana et al.*, 2003; *DeShon et al.*, 2003].

Seismic Observations of Heterogeneity

Several observations are used to suggest heterogeneity in earthquake rupture. Locations and recurrence patterns of large and small earthquakes as well as aftershock patterns provide some indication of heterogeneity levels on faults. Additionally, analysis of seismic waveforms yields source time functions that describe the time history of seismic moment release.

Earthquake Locations

Earthquake location and recurrence patterns provide estimates of heterogeneity. Along the Nankai margin, large events ($M \sim 8$) appear to break similar zones over time in fairly regular patterns [*Ando*, 1975], suggesting some level of segmentation or high-strength zones that persist over a few earthquake cycles. To the north in the Japan Trench, large earthquake locations are less regular. The occurrence of doublets, similar magnitude large earthquakes that occur in pairs with small spatial and temporal separation, also have implications for distribution of high-strength zones. They appear to be a failure at one asperity, triggering failure at an adjacent asperity [*Lay and Kanamori*, 1980; *Xu and Schwartz*, 1993]. Additionally, locations of small earthquakes and microseismicity have also been used to map out zones of strength heterogeneity along faults.

Source Time Functions

The source time function provides the seismic moment release history for the earthquake. An earthquake with a smooth rupture is best modeled with a trapezoid-like shape source time function, implying smooth average slip over the entire fault plane. However, source time functions from most earthquakes are not trapezoidal because slip does not occur smoothly over the fault area (e.g., fig. 5.1).

In order to obtain estimates of earthquake source time functions, seismologists model seismic waveforms, deconvolving estimates of Earth structure effects on wave propagation from the source time function [e.g., *Kikuchi and Kanamori*, 1986; *Ammon et al.*, 1993]. This modeling provides estimates for the locations of moment release on the fault plane as well as the time history of moment release. Because seismic moment is related to fault area and earthquake slip, the slip distribution along the fault can also be determined.

Models for Subduction Zone Earthquakes

Because most of the world's largest earthquakes occur in the shallow subduction zone, significant data exist about earthquake rupture patterns in this tectonic environment. Many observations of complex waveforms and source time functions for subduction zone earthquakes were used to formulate asperity models in the 1970s and 1980s specifically for subduction zone environments. *Kanamori* [1978] suggested that small subduction zone earthquakes result from rupture at a single asperity; complex multiple earthquakes or doublets can result when asperity failure induces large stresses on adjacent asperities to generate multiple failures. *Lay et al.* [1982] use a variety of observations to characterize the relative distribution of asperities in subduction zones and connect this distribution to great subduction zone earthquakes. By summarizing rupture process observations for many large earthquakes, such as location, aftershock distributions, and source time function shapes, they relate rupture patterns to features in trench morphology. Observations of time function variations and complex waveforms suggest that high-frequency seismic body waves are radiated from small parts of the rupture plane, the asperities or strong patches of the fault. Low-frequency seismic radiation and aftershocks originate from the rest of the rupture area described as a weak zone relative to the asperities. Within this model, more complicated ruptures suggest smaller and more heterogeneously distributed strong patches on the fault.

Using the foundations of the asperity model, *Lay et al.* [1982] divide subduction zones into several classes, roughly on the basis of the rupture length of large and great earthquakes and thus asperity distribution. Each class is used to characterize the expected regional stress conditions and seismic coupling (fig. 5.2). A Chile-type margin, characterized by great earthquakes with rupture lengths of over 500 km, describes the Chile, Alaska, and Kamchatka subduction zones. These great earthquakes tend to repeat at nearly identical locations. Large sections of the margin typically have few large subducting features such as seamounts or ridges, allowing for fairly smooth contact between the subducting and overriding plates. Representative earthquakes for the margins in this class (group I), such as the M_w = 9.5 1960 Chile and the M_w = 9.2 1964 Alaska, have very long duration and relatively simple source time functions, suggesting that these margins are strongly coupled with smooth plate contacts and almost uniform asperity distribution.

Subduction zones characterized by large earthquakes with rupture lengths of up to 500 km fall into the next class of margins (group II). These regions, such as the western Aleutians, Nankai, and Colombia, have variations in earthquake rupture patterns, with occasional earthquakes rupturing over several sections of the margin and others breaking only single segments. Source time functions for earthquakes at these margins typically have more complexity than those from the Chile-type margins. These regions have been described as having large asperities, although not as large as in the

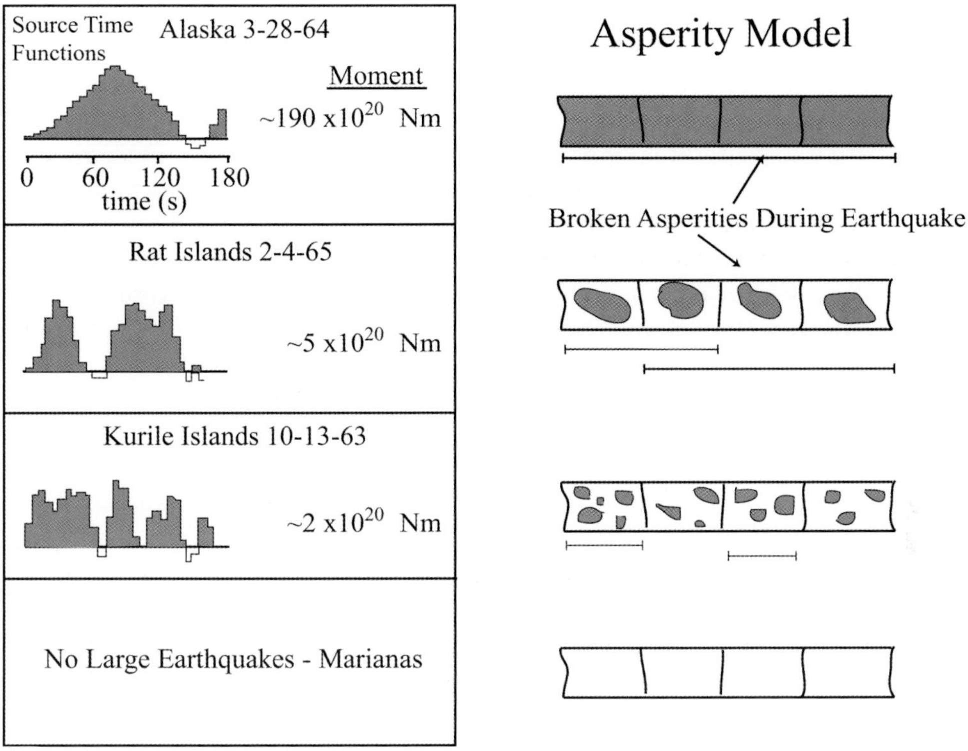

Figure 5.2 The asperity model for subduction zone earthquakes, modified from *Ruff and Kanamori* [1980] and *Lay et al.* [1982]. (left) An example source time functions (moment release during earthquake rupture), for representative earthquakes in each of the main asperity classes (right). The great 1964 Alaska earthquake had a fairly simple source time function, suggesting little heterogeneity in body-wave radiation and slip distribution along the fault plane. The corresponding asperity distribution for this earthquake (as well as the 1960 Chile earthquake) is a very large, mostly continuous strong fault area, shown in gray. Large earthquakes farther along the Aleutians also have large strong patches of fault area, but somewhat smaller than the 1960 Chile or 1964 Alaska case. Source time functions for earthquakes in the Aleutians (and regions such as Nankai and Colombia) have a few large pulses of moment release, reflecting the change in asperity distribution on the fault plane. Regions such as the Kuriles, Peru, and Central America make up the next class of margin, with a very heterogeneous asperity pattern on the fault plane. These earthquakes have significantly more complex source time functions, representing the rupture of multiple small asperities on the fault. The final class of margin includes the Marianas and Izu-Bonin arcs, regions which do not produce great earthquakes because of the lack of large strong patches on the subduction megathrust.

Chile-type margin. Rupture of one asperity may cause stress changes large enough to cause other large asperities to fail, as suggested by *Ruff* [1992], although this may depend on the relative size and separation of these large asperities.

The next group of margins (group III) shows increased complexity, as the Kurile Islands, Peru, central Chile, and Japan subduction zones typically have large earthquakes with rupture zones of ~100–300 km in length. These earthquakes, which tend to repeat in similar locations, typically do not grow into great earthquakes by breaking other segments. Earthquakes in these regions typically produce complex body waves, leading to more complex source time functions. Asperities for these regions tend to vary in size but are typically smaller than expected for the Aleutian (group II) type. Because of the smaller and more heterogeneous asperity size distribution, these earthquakes rarely rupture continuous sections of the margin as seen in the Chile-type margins (group I). An intermediate subgroup includes margins such as New Hebrides and Middle America, which have earthquakes with rupture lengths of 100–150 km and tend to produce fairly simple body waves and source time functions. These margins may have smaller asperities and larger separations between them, making it less likely that an earthquake could rupture more than one asperity.

Subduction zones such as Marianas, Izu-Bonin, and Ryukyu make up the final group (IV) in the asperity model. Great or very large earthquakes are rare in these subduction zones, suggesting that there is a lack of asperities large enough to produce these earthquakes. Small earthquakes and repeating clusters of small earthquakes may occur, but because the *Lay et al.* [1982] study focused only on large and great earthquakes, these margins were classified as relatively aseismic because of the lack of large and great earthquakes.

Physical Nature of Asperities

Using the asperity model framework, regional variations in subduction zone earthquake rupture patterns are ascribed to a distribution of variously sized asperities. Although an asperity is defined as a region of increased strength on the fault plane and location of high slip during an earthquake [e.g., *Byerlee*, 1967; *Lay et al.*, 1982; *Kanamori*, 1986], the question of what can make an asperity is more complicated. Reviewed below are several possibilities for increasing fault zone strength, including an influence from the upper plate, frictional variations within the fault zone, and specific features with high or variable relief on the subducting plate.

Upper Plate Strength

Variability in fore-arc conditions may impact strength on the subduction interface. *McCaffrey* [1993, 1994] suggests changes in fore-arc rheology controls the locations of great earthquakes. By classifying fore-arc strength using deviations between slip vectors and the plate convergence directions, *McCaffrey* [1993, 1994] finds that great megathrust earthquakes occur in regions where fore arcs are more rigid (i.e., only small deviation between slip and convergence directions).

Beck and Christensen [1991] tie heterogeneous slip during the 1965 $M_w = 8.7$ Rat Islands earthquake to segmentation in the Aleutian fore arc. The fore arc contains fault controlled canyons and basins that separate distinct blocks of undeformed material. High moment release occurred near the centers of the undeformed blocks, but little moment release was observed at the edges of the blocks. *Ryan and Scholl* [1993] support these conclusions with additional data from other large earthquakes and aftershock locations, suggesting that the strong fore-arc blocks dictate asperity locations on the deeper interface in the Aleutians.

Friction Variations

In addition to plate strength, variations in friction along the fault plane likely influence slip during earthquakes. Using numerical simulations of earthquake slip, *Rice* [1993] suggests that high-slip patches may be controlled by frictional characteristics on a fault plane. Within subduction zones, *Song and Simons* [2003] correlate strong negative trench parallel gravity anomalies with slip in large and great earthquakes. They relate the observed negative gravity variations to high-shear traction on the interplate thrust. The negative anomalies observed in the zones of great earthquake slip suggest a reduction in normal stress, which tends to reduce stick-slip behavior. *Song and Simons* [2003] suggest an increase in the effective coefficient of friction to offset the decrease in normal stress, producing high-shear stresses along the fault.

Subducting Plate Features

In their discussion of the asperity model, *Lay et al.* [1982] suggest seafloor features can dictate asperity size in subduction zones. Specifically, large segments of smooth plate, possibly in the presence of subducted sediment, can provide large strong regions on the thrust interface [*Hilde*, 1983; *Ruff*, 1989]. Additionally, there appears to be a positive correlation between the occurrence of great earthquakes and the amount of sediment near the trench [*Ruff*, 1989]. However, detailed slip patterns of large and great earthquakes have not been compared with detailed maps of sediment distribution along these margins.

Features that produce roughness or topography on the subducting plate are also suggested to influence earthquake slip distributions. Horst and graben structures provide a rough plate interface with alternating strong and weak contact zones with the overriding plate [*Tanioka et al.*, 1997]. Other subducted high-relief features, such as fracture zones, seamounts, and ridges, also make likely candidates to influence earthquake behavior, as these topographic highs could cause isolated strong coupling with the overriding plate [e.g., *Kelleher and McCann*, 1976, 1977; *Cloos*, 1992; *Cloos and Shreve*, 1996;

Scholz and Small, 1997]. There have been several correlations between subducting features and specific earthquakes within several subduction zones, as described below.

Examples of Subducting Features and Large Earthquake Occurrence

In recent years, there have been significant strides toward the identification of features on the subducting plate before and after it enters the trench using high-quality bathymetry and seismic data. In several subduction zones, large earthquakes have occurred that may be related to these subducting features, suggesting that bathymetric features such as seamounts and ridges may act as regions of high stress and high seismic coupling, providing a physical link to asperity distributions.

NE Japan

Seamounts have been imaged near the trench along the northeastern margin of Japan [Lallemand and Pichon, 1987; Lallemand et al., 1989]. In addition, variation in subducting plate roughness, from smooth segments to portions featuring several horst and graben structures, has been mapped using bathymetry data. These features may impact the size and types of earthquakes possible along the margin, as suggested by Tanioka et al. [1997]. In regions of smooth, sedimented seafloor to the northern part of Honshu, large underthrust earthquakes such as the 1968 M_w = 8.2 Tokachi-Oki and 1994 M_w = 7.7 Sanriku events are typical. In regions of rough plate subducting, smaller earthquakes occur, along with more unusual events such as the 1896 tsunami earthquake.

Java

Within the Java subduction zone, sonar data show seamounts on the incoming plate as well as entering the subduction zone at the Java trench [*Masson et al.*, 1990]. A M_w = 7.8 earthquake in 1994 occurred in this region, with principal slip located primarily in the region of a previously identified subducted seamount [*Abercrombie et al.*, 2001] (fig. 5.3). In general, this subduction zone does not produce many large underthrusting events, so *Abercrombie et al.* [2001] suggest that most of the Java subduction zone is weakly coupled except in the areas of subducted seamounts. These seamounts provide higher coupling with the overriding plate and are locations for the large earthquakes and high slip in this region.

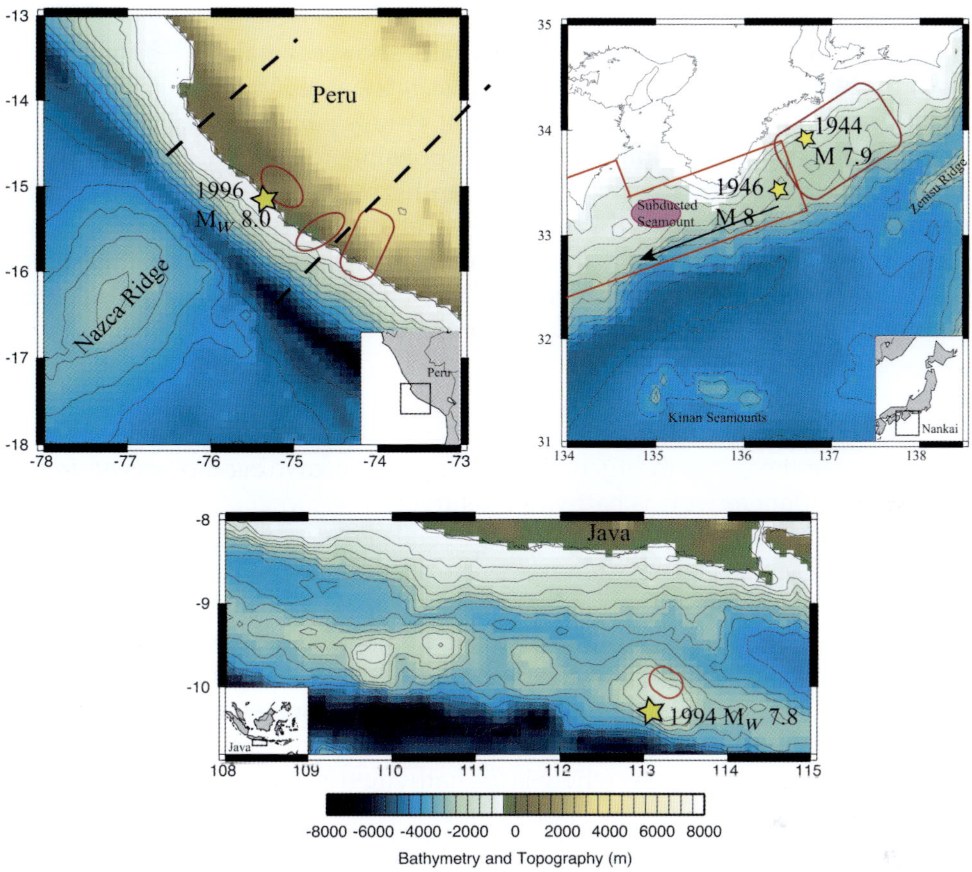

Figure 5.3 Examples of documented earthquake slip distributions impacted by subducting features. Bathymetry data are from the National Geophysical Data Center (http://www.ndgc.noaa.gov/mgg/bathymetry). The 1996 underthrust earthquake (star represents epicenter) in Peru had high slip (patches outlined in red) in the region of Nazca Ridge subduction as well as along the southern flank of the ridge [*Spence et al.*, 1999]. The dashed lines represent the likely location of previously subducted Nazca Ridge [*Spence et al.*, 1999]. Along the Nankai Trough, previously subducted seamounts (purple) and ridges have been imaged in the areas of the 1944 and 1946 great earthquakes. *Kodaira et al.* [2000] suggests that main seismogenic rupture for the 1946 propagated westward from the epicenter, stopping at the subducted seamount, with slow tsunamigenic slip southwest of the seamount. Using the pattern of the high slip for the 1994 earthquake, *Kodaira et al.* [2003] suggests a previously subducted Zenisu Ridge type feature prevented further NE directed rupture from the 1944 great earthquake. Along the Java subduction zone (bottom), *Abercrombie et al.* [2001] find high slip for the 1994 underthrust earthquake, one of the few large underthrust earthquakes in this region, collocated with a previously subducted seamount, as defined by the high spot in the bathymetry data.

Alaska-Aleutians

Bathymetry data from the eastern Aleutians show several seamount groups seaward of the Alaska-Aleutian Trench [*Estabrook et al.*, 1994]. Previously subducted seamounts, similar to the seamount groups currently observed such as the Gulf of Alaska seamounts, may be responsible for several earthquakes of M ~ 7 that repeat with almost identical depths, source radii, and focal mechanisms [*Estabrook et al.*, 1994]. Within the western Aleutians, earthquake catalogs do not contain many of these repeating-type earthquakes with similar source parameters, and corresponding bathymetry data do not show many seamounts on the subducting plate in this area.

Tonga

Seamounts in the Tonga-Kermedec subduction zone have been observed in satellite gravity data as well as bathymetry surveys [*Balance et al.*, 1989; *Chapel and Small*, 1996]. Within the Tonga Trench, several of these seamounts also correspond with locations of large underthrusting earthquakes. An M_w ~ 8.0 earthquake occurred in 1919 in the region of the Capricorn seamounts [*Pacheco and Sykes*, 1992]. In the region of the subducting Louisville Ridge, an M_w = 7.5 earthquake occurred in 1982 [*Christensen and Lay*, 1988]. *Scholz and Small* [1997] use these earthquakes and corresponding subducting features in the context of their model of seamount subduction, suggesting increases in normal stress and local seismic coupling in the areas where large bathymetric features are subducted.

Peru-Chile

Along the Peru-Chile margin, several large features, such as Nazca and Carnegie Ridges, are subducted. The 1996 M_w 8.0 Peru earthquake occurred in the region of the subducting Nazca Ridge. The earthquake source model produced by *Spence et al.* [1999] shows patches of high slip collocated with ridge subduction, as well as high slip to the southeastern side of the projected subducted ridge (fig. 5.3). *Spence et al.* [1999] suggest that the subducted ridge provides increased coupling in this region and that the 1942 M_w ~ 8 earthquake was also likely related to the subduction of the Nazca Ridge. *Gutscher et al.* [1999] uses a large historical earthquake database in conjunction with detailed bathymetry data and projections of continued ridge subduction to suggest the Carnegie Ridge significantly affects seismic coupling. They find unlike the Nazca Ridge, great earthquakes occur on the flanks of the subducted Carnegie ridge, not within the ridge-upper plate contact area.

Nankai

Using detailed bathymetric and seismic data, *Park et al.* [1999] and *Kodaira et al.* [2000, 2003] have imaged seamounts and ridges impacting the Nankai subduction zone (fig. 5.3). Observations of a subducting seamount in the rupture area of the great 1946 Nankaido earthquake have been related to seismic coupling variations. Analysis using seismic data suggests a smaller rupture area than geodetic data analysis, with the westernmost edge of the seismic area estimate abutting the subducted seamount [*Kodaira et al.*, 2000]. Geodetic estimates suggest additional rupture to the southwest of this seamount, leading to the idea that the subducted seamount increased local coupling enough to stop seismogenic rupture propagation to the southwest; however, slow slip was accommodated on an accretionary wedge splay fault southwest of the subducted seamount [*Kato and Ando*, 1997; *Kodaira et al.*, 2000]. To the east of the 1946 earthquake slip region, seismic reflection data show a feature at depth similar in size and likely origin to the present-day Zenisu Ridge. This subducted ridge is located in the area of rupture termination for the great 1944 Tonankai earthquake [*Kodaira et al.*, 2003]. Given the correlation between subducted feature locations and rupture termination for great earthquakes in this subduction zone, many have suggested that these features act as barriers rather than rupture asperities at least in these earthquakes [*Kodaira et al.*, 2000; 2003]. It is unknown whether their role will change over several earthquake cycles.

Costa Rica

Along the Middle America Trench offshore Costa Rica, detailed bathymetry data show considerable diversity in high-relief features on the subducting Cocos plate. Large earthquakes have occurred along this complex margin and show evidence for rupture diversity that has been linked to the diverse subducting features. Because of the quality of data and ability to show connections between several earthquakes and subducting features, the next few sections detail the structures and seismicity of this region.

Costa Rica Tectonic Structures The Cocos plate offshore Costa Rica can be divided into several domains on the basis of morphologic features, as seen in detailed bathymetry data from *Ranero et al.* [2003] (fig. 5.4). Near the Nicoya Peninsula, the oceanic crust is smooth with a 500 m thick sediment blanket covering the plate [*von Huene et al.*, 2000]. Offshore central Costa Rica, between the Nicoya and Osa peninsulas, the seafloor is disrupted by the Quepos Plateau and many seamounts. Approximately 40% of the oceanic crust offshore central Costa Rica is covered by seamounts that range in size between 1–2.5 km high and 10–20 km wide [*Ranero and von Huene*, 2000]. The ~200–300 km wide Cocos Ridge, an area of thickened oceanic crust and elevated seafloor [*Walther*, 2003; *Kolarsky et al.*, 1995], subducts to the southeast offshore the Osa Peninsula. This highly heterogeneous feature is likely a trace of the Galapagos hot spot as the Cocos plate passed over the hot spot during the Miocene [*Barckhausen et al.*, 2001]. These features create diversity in the overriding plate deformation with increased local relief in regions where these features are subducting [*Gardner*

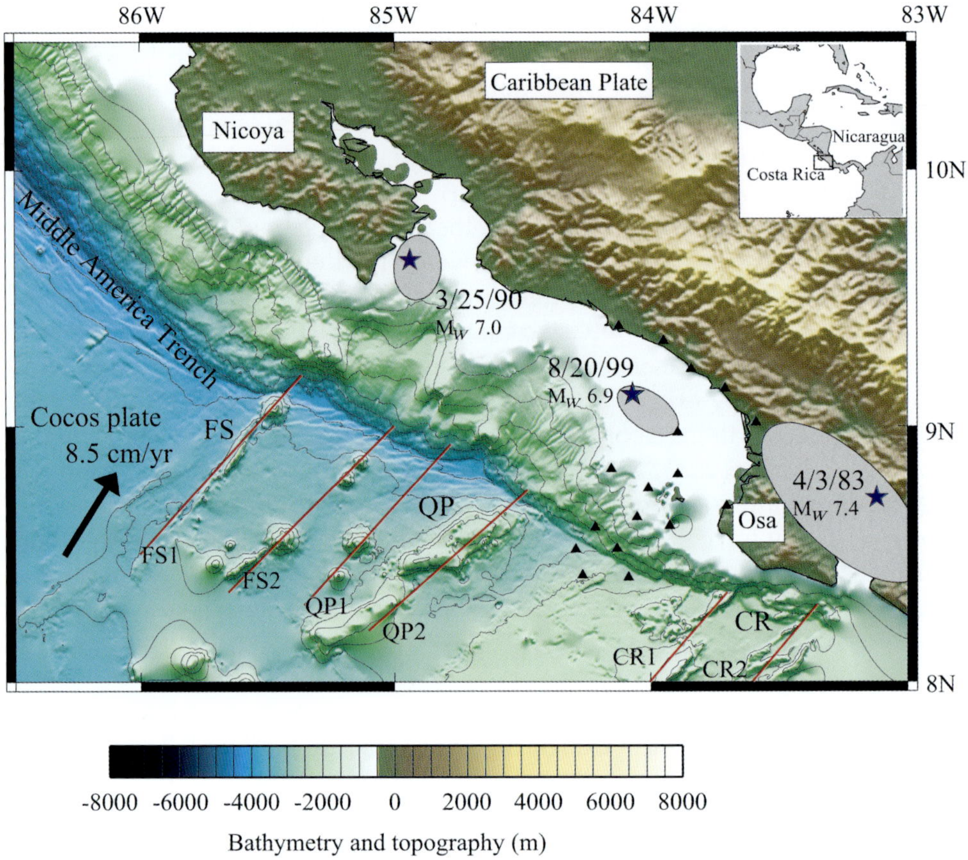

Figure 5.4 Bathymetric features and seismicity in the Middle America Trench offshore Costa Rica. Bathymetry data are from *Ranero et al.* [2003]. There is significant diversity in the subducting Cocos plate features, with the Fisher seamounts (FS), Quepos Plateau (QP), and Cocos Ridge (CR) subducting along this part of the margin. Large earthquakes in 1990, 1999, and 1983 (gray shaded areas show rupture areas for each earthquake) have occurred downdip of these subducting features. Figure 5.6 compares source time functions and rupture areas of these earthquakes with the size of subducting features, with feature size estimated in cross-section profiles (along the red lines).

et al., 1992; *von Huene et al.*, 1995; *Fisher et al.*, 1998; *Dominquez et al.*, 1998; *von Huene et al.*, 2000].

Relationship to Costa Rica Earthquakes These subducting features have been correlated with large historical earthquakes along the margin. *Protti et al.* [1995a] note that the largest underthrusting earthquakes with magnitude ≥ 7.5 have historically occurred only along the Nicoya Peninsula, where they infer the smooth subducted seafloor creates strong seismic coupling to act as a single large asperity. However, GPS data from *Norabuena et al.* [2004] show a more patchy distribution of locking. Other large thrust events (M_s up to 7.4) occur in southern Costa Rica where subduction of the buoyant Cocos Ridge creates moderate plate coupling. Earthquakes with maximum magnitudes of only $M < 7.0$ characterize the central Costa Rica region where isolated seamounts are subducted. *von Huene et al.* [2000] suggest that seamounts associated with the Fisher Seamount Chain near the southern tip of Nicoya are related to large earthquakes (M_s = 6.4–7.0) in the region. As there is no evidence for sheared off seamounts along the slope region, *von Huene et al.* [2000] propose that the seamounts enter the seismogenic zone and act as asperities for future earthquakes. Shallow, smaller-magnitude seismicity is more common in regions of seamount and disrupted crust subduction than in the smoother region subducting off northern Costa Rica.

During the 1980s and 1990s, three large ($M_w \geq 6.9$) earthquakes occurred within the different morphological domains in central Costa Rica (fig. 5.4). These earthquakes, well recorded with modern seismic instruments, provide an opportunity to go beyond correlating only the locations of large earthquakes and seafloor features to also examining possible influence of variable seafloor topography on earthquake rupture behavior. The M_w = 6.9 underthrusting earthquake in 1999 occurred northwest of the Osa Peninsula in the area of seamount and Quepos Plateau subduction. The other earthquakes occurred in the vicinity of the Fisher Seamount Chain (1990 M_w = 7.0) and near the Osa Peninsula and Cocos Ridge (1983 M_w = 7.4).

Source time functions for these earthquakes show variations in moment release patterns. For the 1999 Quepos earthquake, *Bilek et al.* [2003] use a variety of methods and data sets, producing consistent source models using both surface- and body-wave inversions. Most of the seismic moment was released in one main event (event 1), with minor moment release occurring in later, smaller subevents within the first 10 s, as shown in the source time function in figure 5.5. The time function overall is very simple, suggesting a simple rupture for this earthquake. Figure 5.5 also shows the locations of the subevents on the fault plane; the moment was concentrated spatially, with some rupture propagating to the southeast with a direction of ~100° (going from subevent 1 to subevent 2).

The 1990 M_w = 7.0 Nicoya Gulf earthquake had slightly more complexity in the rupture history compared to the 1999 Quepos earthquake. *Protti et al.* [1995b] describe two large subevents of moment release between 18 and 24 km depth (fig. 5.6). Aftershock patterns suggest an initial rupture area of ≤600 km^2 [*Protti et al.*, 1995b].

The 1983 Osa earthquake (M_w = 7.4) was one of the largest underthrusting earthquakes to occur in the Osa Peninsula region, with others occurring in 1904 (M_s = 7.6) and 1941 (M_s = 7.5) [*Tajima and Kikuchi*, 1995]. Body-wave analyses of the 1983 earthquake consistently show a more complicated rupture process than either the 1990 and 1999 events with longer source duration, more

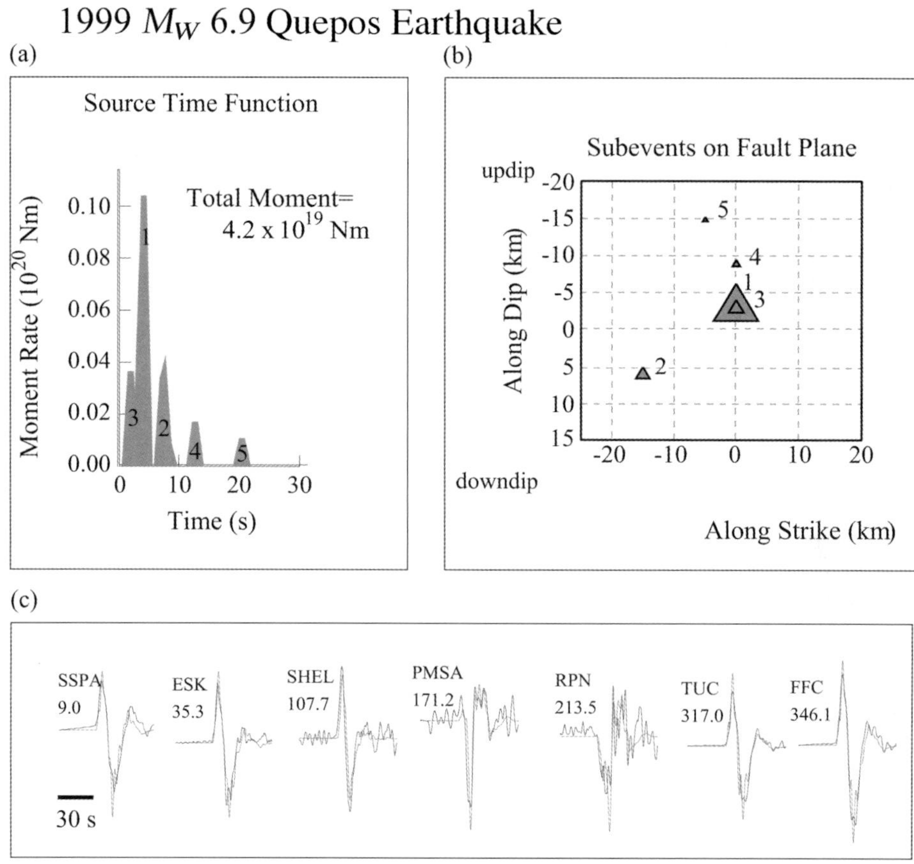

Figure 5.5 Determination of source time function and locations of main moment release for the 1999 Quepos earthquake, modified from *Bilek et al.* [2003]. This analysis involved inverting body-wave seismograms to find the best spatial and temporal distribution of moment release using the method of *Kikuchi and Kanamori* [1986]. (a) The source time function shows the majority of moment release occurred in the first large pulse, with minor moment release in pulses 2 and 3, during 10 s of rupture. (b) Pulses of moment release seen in the source time function are located on the two-dimensional (2-D) fault plane, oriented along strike (313°). The main pulses of moment release are fairly concentrated spatially, with minor rupture toward the southeast (going from 1 to 2). (c) Representative seismograms (black) used in the inversion, with station name and azimuth from the earthquake listed, as well as the predicted seismograms (red) for the best inversion model.

complexity in the source time function, and possible mechanism variations along strike [*Adamek et al.*, 1987; *Tajima and Kikuchi*, 1995].

Bathymetric data offshore Costa Rica allow for comparisons between bathymetric feature size with rupture areas of these large earthquakes (fig. 5.6). Cross sections through these features provide estimates of feature size that are com-

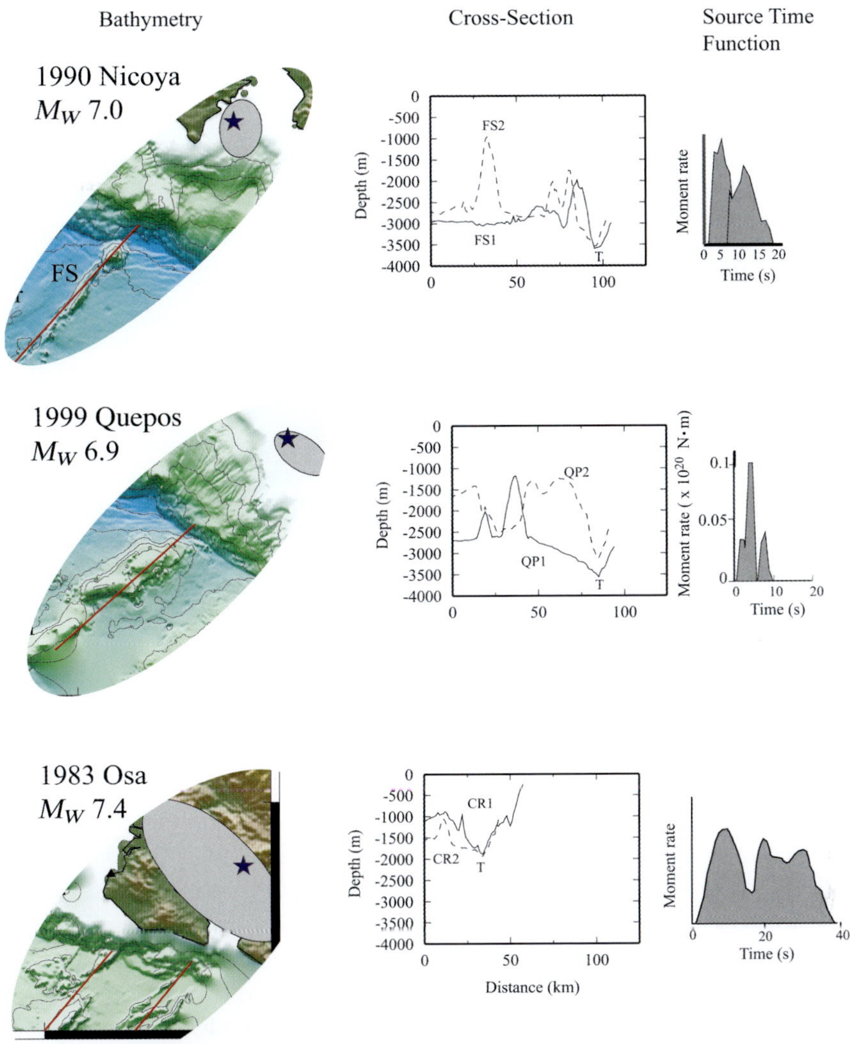

Figure 5.6 Comparison of Costa Rica earthquake rupture details with subducting features. For the 1990 Nicoya earthquake, two main pulses of moment release in the source time function suggest the rupture of 2 closely spaced asperities. Early aftershock patterns suggest a rupture area of ~600 km^2. Seamounts in the Fisher Seamount Group range in size between ~300 and 400 km^2. As the earthquake source details are comparable to the sizes of the subducting seamounts, *Protti et al.* [1995b] and *Bilek et al.* [2003] suggest that the 1990 earthquake was the rupture of seamount contacts with the upper plate. The 1999 Quepos earthquake had 1–2 main pulses of moment release in the source time function and an aftershock region of ~400 km^2, which can be explained by rupture of a portion of the Quepos Plateau [*Bilek et al.,* 2003]. The source time function for the 1983 earthquake contains more complexity than either the 1990 or 1999 events. On the basis of its location and rupture details from *Tajima and Kikuchi* [1995], this earthquake ruptured much of the along-strike extent of the Cocos Ridge, likely causing the complexity seen in the rupture process.

parable to the size of earthquake rupture area. Projecting the Quepos Plateau section of subducting crust along the plate motion vector intersects the location of the 1999 Quepos earthquake, suggesting that the earthquake slip may represent the rupture of specific plateau contact zones with the overriding plate. This scenario can explain the concentrated and simple nature of moment release for this earthquake. Figure 5.6 shows two cross sections through the Quepos Plateau area (QP1 and QP2 lines in fig. 5.4). Several of the subducting features in this region have dimensions of 12–25 km along the dip direction (as represented in the cross section across one of these features) and comparable dimensions in the strike dimension (150–400 km^2 in area), including several high patches of the Quepos Plateau and large seamounts to the northwest of the Quepos Plateau. The aftershock region (~400 km^2) does not extend beyond the along-strike width of the Quepos Plateau (fig. 5.7). *Bilek et al.* [2003] and *DeShon et al.* [2003] suggest that the Quepos Plateau served as the rupture asperity for the 1999 earthquake.

In the region of the 1990 Nicoya Gulf earthquake, the Fisher seamounts are clearly seen in the bathymetry data and are subducting at the southern end of the Nicoya Peninsula, seaward of the location of the 1990 earthquake. Individual seamounts within the chain can reach ~300–400 km^2, while small seamount pairs have composite area of ~200–250 km^2 [*von Huene et al.*, 2000]. On the basis of the bathymetry data and the rupture process determined for this event, *Protti et al.* [1995b] suggest that the 1990 earthquake represents the rupture of seamount contacts with the overriding plate. Rupture of more than one seamount within the seamount chain could produce the separate subevents of moment release determined for the 1990 Nicoya Gulf event.

Additional support for subducted seamounts at depth in the region comes from regional geology as well as seismic tomography images produced for this region. Seamounts at depth would likely produce some fore-arc uplift. *Marshall and Anderson* [1995] and *Gardner et al.* [2001] use radiocarbon dating of terraces and rotation in southeastern Nicoya to estimate ~400 km^2 of uplift at the tip of the peninsula. *Husen et al.* [2002] use data from over 3700 earthquakes to tomographically image a thick feature on the slab at approximately the same location and depth of the 1990 earthquake. They suggest the imaged feature is the subducted seamount that served as the asperity for the 1990 earthquake rupture.

The complex 1983 earthquake occurred to the southeast near the highly disturbed crust containing the Cocos Ridge. The broad arch of the ridge has topography of ~1 km high where it enters the Middle America Trench with variable sediment thickness along the ridge [*von Huene et al.*, 2000; *Walther*, 2003]. Comparing the locations of moment release obtained by *Tajima and Kikuchi* [1995] with regional bathymetry in the south suggests that this earthquake ruptured both of the elevated features (between CR1 and CR2) on the high area of the Cocos Ridge along with additional rupture propagation to the SE (fig. 5.4). The high subducted relief and sediment thickness in the area of

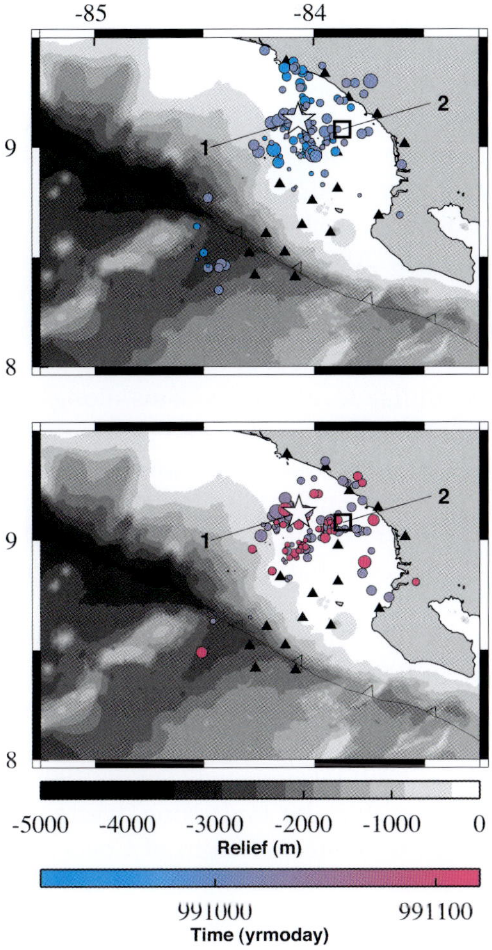

Figure 5.7 Aftershocks for the 1999 Quepos earthquake as recorded by a temporary onshore-offshore seismic network (triangles) [*DeShon et al.*, 2003]. The large star and box indicate the locations of the main two subevents of moment release for the 1999 event. Small circles are aftershock locations, colored by time after the main earthquake. The top panel shows that the earliest (first 2 months) aftershocks are spread between the areas of main moment release but do not extend much beyond the projected lateral extent of the Quepos Plateau. The bottom panel shows the final month of aftershock recording, when seismicity appears to be mainly located on what may be the flanks of the previously subducted Quepos Ridge.

the 1983 rupture led to the more complicated rupture behavior of this event (fig. 5.6) [*Bilek et al.*, 2003]. On the basis of these correlations between earthquake patterns and subducted features, it seems likely that earthquake magnitudes along the Costa Rica margin may be controlled by feature size, an important consideration for hazard assessments.

Discussion

Temporal Variability of Asperities

An important question connected with determining the physical or geologic nature of asperities is that of consistency of these features over several earthquake cycles. Some evidence suggests that these features are consistent over at least relatively short timescales. For instance, GPS data in a few subduction zones suggest that rupture zone patches of high slip during past large earthquakes currently accumulate more strain than adjacent regions that did not produce high slip. This appears to be the case for central and southern Japan [*Sagiya*, 1999; *Sagiya et al.*, 2000] as well as in the region of the 1964 great Alaska earthquake [*Freymueller et al.*, 2000, *Zweck et al.*, 2002]. In Costa Rica, part of the rupture area of the 1950 *M* 7.7 earthquake is also locked and accumulating strain at close to the plate rate [*Norabuena et al.*, 2004].

Seismic data collected offshore NE Japan also suggest temporal stability of asperities as defined by small earthquakes. Using 8 years of data, *Matsuzawa et al.* [2002] and *Igarashi et al.* [2003] locate several clusters of repeating small magnitude ($M \sim 4.8$) earthquakes with similar focal mechanisms along the plate boundary. They suggest that these earthquakes are due to repeating slips of small asperities (on the order of 0.1–1 km) surrounded by regions of stable sliding. They find that these small earthquakes are not collocated with regions of high moment release during large earthquakes but in areas surrounding the high moment release zones. This model suggests that several different sized strong regions exist in the region but that the smaller strong regions persist over at least short time periods. *Okada et al.* [2003] compare rupture areas of small earthquakes in NE Japan occurring in 1995 and 2001, finding significant overlap that supports a persistent asperity hypothesis. This would be consistent with repeating small earthquakes and persistent small asperities observed for strike slip faults such as the San Andreas fault [*Nadeau and Johnson*, 1998].

Using this idea of consistent asperities within the fault zone over some timescale, variations in earthquakes along the western margin of Costa Rica can be used to make assessments of the coupling and earthquake recurrence possibilities for this region. *Protti et al.* [1995a] and *Nishenko* [1991] segment the margin on the basis of magnitudes of previous earthquakes and inferred seismic coupling. In terms of the earthquakes described in the previous section, the 1999 earthquake is an example of the typical central segment event, whereas the more complicated 1983 earthquake is a typical Osa segment event. Size and location of earthquakes in the Quepos segment are likely limited by the number and size of incoming seamount and Quepos Plateau features, although heterogeneous coupling allows for the rupture of one or two closely spaced seamount features but does not continue beyond these features. The Nicoya and Osa segments experience more frequent large earthquakes because of smoother plate subduction (Nicoya) or continued subduction of the Cocos Ridge (Osa).

However, recurrence of large earthquakes in several subduction zones suggests that temporal persistence of asperities is complicated. *Schwartz* [1999] compares high-slip regions for recent subduction zone that occurred in regions of past large earthquakes for which slip information is available. In some regions, large earthquake slip for recent events is collocated with high slip in past events (i.e., Kuriles and Aleutian earthquakes). However, high slip for earthquakes in NE Japan and the Solomon Islands filled areas of low slip in previous earthquakes. This suggests that simple geometric or material variations in the subduction zones may not be solely responsible for controlling slip heterogeneity.

Subducted Features as Asperities and Barriers

Correlations between subducting features and high slip during earthquakes are complicated, as it appears that features can act as either asperities or barriers to rupture (fig. 5.8). Additionally, this behavior may not be consistent over many earthquake cycles, possibly related to variable strength of asperities. Observations from Nankai suggest these features, once subducted, act as barriers to seismic rupture at least through one earthquake cycle [*Kodaira et al.*, 2000]. In Costa Rica, Java, and other areas, it appears that these features are the locations of high earthquake slip. The roles of the subducted seamounts might depend on the degree of seismic coupling in each region such that weakly coupled margins might have seamounts that only act as asperities, while seamounts in strongly coupled margins could act as either. Variations in stresses or friction might allow for the role of seamounts to change over the seismic

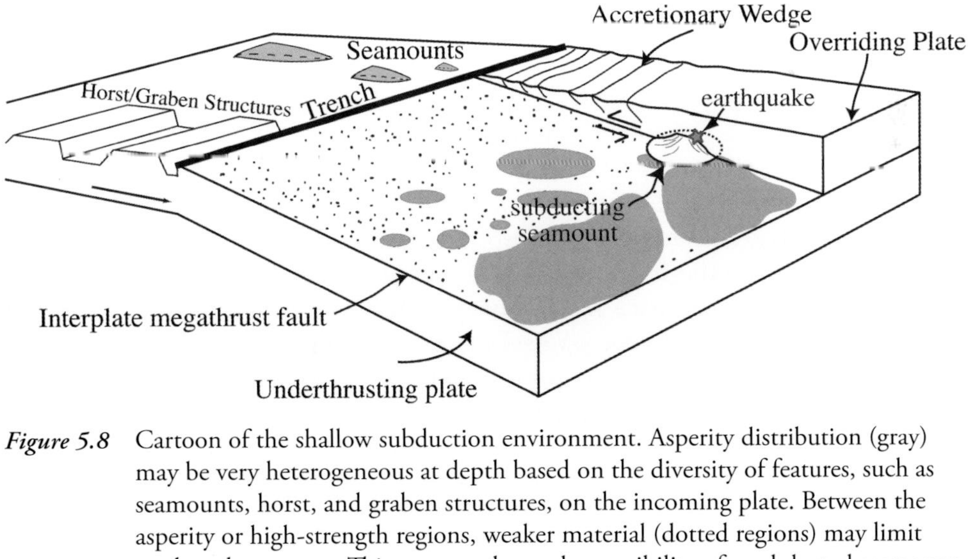

Figure 5.8 Cartoon of the shallow subduction environment. Asperity distribution (gray) may be very heterogeneous at depth based on the diversity of features, such as seamounts, horst, and graben structures, on the incoming plate. Between the asperity or high-strength regions, weaker material (dotted regions) may limit earthquake rupture. This cartoon shows the possibility of a subducted seamount being a location for earthquake rupture, as suggested for Costa Rica, Java, and Alaska. However, other factors may be important, such as the relative strength of the upper plate material and frictional variations on the plate interface. In addition, regions such as Nankai appear to have different behavior, with earthquakes beginning away from but terminating at previously subducted high-relief features.

cycle or several cycles and may impact regions such as Nankai, Costa Rica, and Java. Further identification of subducted features and determination of their roles in regional seismicity is needed to address this question.

Conclusions

Heterogeneous slip is found in earthquakes of all tectonic environments, allowing for these patterns to be broadly defined with models of heterogeneous asperity distribution along faults. As shallow subduction zones produce much of the global seismicity, understanding the nature of the strong patches and heterogeneous slip patterns for subduction-zone earthquakes is an important endeavor as it may provide insight into seismic hazards. The physical nature of the strong patches on subduction-zone megathrust faults may range from fault-friction variations, strength of adjacent fore-arc material, and various structural features on the subducting plate, as correlations have been made between earthquake behavior and all of these possibilities. However, more work is needed in several areas, such as increased mapping of subducting features in all subduction zones, better knowledge of frictional properties of subduction fault zone materials, and understanding of subduction-zone stress patterns, before a complete understanding of the interaction between subduction-zone features and large earthquakes is achieved.

Acknowledgments

Several ideas for this paper resulted from discussions during the 2003 MARGINS SEIZE Theoretical Institute meeting. H. Kanamori and A. Newman provided constructive reviews to improve the manuscript. Earthquake data were obtained from the IRIS Data Management Center. Bathymetry data used for several of the Costa Rica figures were kindly provided by C. Ranero and R. von Huene. GMT software [*Wessel and Smith,* 1991] was used for some figure preparation. This work was partially supported by a Turner postdoctoral fellowship from the University of Michigan.

References

Abercrombie, R. E., M. Antolik, K. Felzer, and G. Ekström (2001), The 1994 Java tsunami earthquake: Slip over a subducting seamount, *J. Geophys. Res., 106,* 6595–6607.

Adamek, S., F. Tajima, and D. A. Wiens (1987), Seismic rupture associated with subduction of the Cocos Ridge, *Tectonics, 6,* 757–774.

Ammon, C. J., A. A. Velasco, and T. Lay (1993), Rapid estimation of rupture directivity: Application to the 1992 Landers ($M_s = 7.4$) and Cape Mendecino ($M_s = 7.2$) California earthquakes, *Geophys. Res. Lett., 20,* 97–100.

Ando, M. (1975), Source mechanisms and tectonic significance of historical earthquakes along the Nankai Trough, Japan, *Tectonophysics, 27*, 119–140.

Balance, P. F., D. W. Scholl, T. L. Vallier, and R. H. Herzer (1989), Subduction of a late Cretaceous seamount of the Louisville Ridge at the Tonga Trench: A model of normal and accelerated tectonic erosion, *Tectonics, 8*, 853–962.

Barckhausen, U., C. R. Ranero, R. von Huene, S. C. Cande, and H. A. Roeser (2001), Revised tectonic boundaries in the Cocos Plate off Costa Rica: Implications for the segmentation of the convergent margin and for plate tectonic models, *J. Geophys. Res., 106*, 19, 207–19,220.

Beck, S. L., and D. H. Christensen (1991), Rupture process of the February 4, 1965 Rat Islands earthquake, *J. Geophys. Res., 96*, 2205–2221.

Beroza, G. C. (1991), Near-source modeling of the Loma Prieta earthquake: Evidence for heterogeneous slip and implications for earthquake hazard, *Bull. Seismol. Soc. Am., 81*, 1603–1621.

Bilek, S. L., S. Y. Schwartz, and H. R. DeShon (2003), Control of seafloor roughness on earthquake rupture behavior, *Geology, 31*(5), 455–458.

Byerlee, J. D. (1967), Theory of friction based on brittle fracture, *J. Appl. Phys., 38*, 2928–2934.

Chapel, D., and C. Small (1996), Distribution of large seamounts in the Pacific Ocean, *Eos Trans. AGU, 77*, Fall Meet. Suppl., F770.

Christensen, D., and T. Lay (1988), Large earthquakes in the Tonga region associated with subduction of the Louisville Ridge, *J. Geophys. Res., 93*, 13,367–13,389.

Cloos, M. (1992), Thrust-type subduction zone earthquakes and seamount asperities: A physical model for seismic rupture, *Geology, 20*(7), 601–604.

Cloos, M., and R. L. Shreve (1996), Shear-zone thickness and the seismicity of Chilean- and Marianas-type subduction zones, *Geology, 24*(2), 107–110.

DeShon, H. R., S. Y. Schwartz, S. L. Bilek, L. M. Dorman, V. Gonzalez, J. M. Protti, E. R. Flueh, and T. H. Dixon (2003), Seismogenic zone structure off the southern Middle America Trench, Costa Rica, *J. Geophys. Res., 108*(B10), 2491, doi:10:1029/2002JB002294.

Dominquez, S., S. E. Lallemand, J. Malavieille, and R. von Huene (1998), Upper plate deformation associated with seamount subduction, *Tectonophysics, 293*, 207–224.

Estabrook, C. H., K. H. Jacob, and L. R. Sykes (1994), Body wave and surface wave analysis of large and great earthquakes along the eastern Aleutian Arc, 1923–1993: Implications for future events, *J. Geophys. Res., 99*, 11,643–11,662.

Fisher, D. M., T. W. Gardner, J. S. Marshall, P. B. Sak, and M. Protti (1998), Effect of subducting sea-floor roughness on fore-arc kinematics, Pacific coast, Costa Rica, *Geology, 26*(5), 467–470.

Freymueller, J. T., S. C. Cohen, and H. J. Fletcher (2000), Spatial variations in present-day deformation, Kenai Peninsula, Alaska, and their implications, *J. Geophys. Res., 105*, 8079–8101.

Gardner, T. W., D. Verdonck, N. M. Pinter, R. Slingerland, K. P. Furlong, T. F. Bullard, and S. G. Wells (1992), Quaternary uplift astride the aseismic Cocos Ridge, Pacific coast, Costa Rica, *Geol. Soc. Am. Bull., 104*, 219–232.

Gardner, T. W., J. Marshall, D. Merritts, B. Bee, R. Burgette, E. Burton, J. Cooke, N. Kehrwald, M. Protti, D. Fisher, and P. Sak (2001), Holocene forearc block rotation in response to seamount subduction, southeastern Peninsula de Nicoya, Costa Rica, *Geology, 29*(2), 151–154.

Gutscher, M.-A., J. Malavieille, S. Lallemand, and J.-Y. Collot (1999), Tectonic segmentation of the North Andean margin: Impact of the Carnegie Ridge collision, *Earth Planet. Sci. Lett., 168*, 255–270.

Heaton, T. H. (1990), Evidence for and implications of self-healing pulses of slip in earthquake rupture, *Phys. Earth Planet. Inter., 64*, 1–20.

Hilde, T. W. C. (1983), Sediment subduction vs. accretion around the Pacific, *Tectonophysics, 99*, 381–397.

Hino, R., T. Kanazawa, and A. Hasegawa (1996), Interplate seismic activity near the northern Japan Trench deduced from ocean bottom and land-based seismic observations, *Phys. Earth Planet. Inter., 93*, 37–52.

Husen, S., E. Kissling, and R. Quintero (2002), Tomographic evidence for a subducted seamount beneath the Gulf of Nicoya, Costa Rica: The cause of the 1990 $M_w = 7.0$ Gulf of Nicoya earthquake, *Geophys. Res. Lett., 29*(8), 1238, doi:10.1029/2001GL014045.

Igarashi, T., T. Matsuzawa, and A. Hasegawa (2003), Repeating earthquakes and interplate aseismic slip in the northeastern Japan subduction zone, *J. Geophys. Res., 108*(B5), 2249, doi:10.1029/2002JB001920.

Ji, C., D. J. Wald, and D. V. Helmberger (2002), Source description of the 1999 Hector Mine, California, earthquake, part II: Complexity of slip history, *Bull. Seismol. Soc. Am., 92*, 1208–1226.

Kanamori, H. (1978), Use of seismic radiation to infer source parameters, *U.S. Geol. Surv. Open File Rep., 380*, 283–318.

Kanamori, H. (1986), Rupture process of subduction-zone earthquakes, *Annu. Rev. Earth Planet. Sci., 14*, 293–322.

Kato, T., and M. Ando (1997), Source mechanisms of the 1944 Tonankai and 1946 Nankaido earthquakes: Spatial heterogeneity of rise times, *Geophys. Res. Lett., 24*, 2055–2058.

Kelleher, J., and W. McCann (1976), Buoyant zones, great earthquakes, and unstable boundaries of subduction, *J. Geophys. Res., 81*, 4885–4896.

Kelleher, J., and W. McCann (1977), Bathymetric highs and the development of convergent plate boundaries, in *Island Arcs, Deep Sea Trenchs, and Back-Arc Basins, Maurice Ewing Ser.*, vol. 1, edited by M. Talwani and W. C. Pitman, pp. 115–122, AGU, Washington, D. C.

Kikuchi, M., and H. Kanamori (1986), Inversion of complex body waves–II, *Phys. Earth Planet. Int., 43*, 205–222.

Kodaira, S., N. Takahashi, A. Nakanishi, S. Miura, and Y. Kaneda (2000), Subducted seamount imaged in the rupture zone of the 1946 Nankaido earthquake, *Science, 289*, 104–106.

Kodaira, S., A. Nakanishi, J.-O. Park, A. Ito, T. Tsuru, and Y. Kaneda (2003), Cyclic ridge subduction at an inter-plate locked zone off central Japan, *Geophys. Res. Lett., 30*(6), 1339,doi:10.1029/2002GL016595.

Kolarsky, R. A., P. Mann, and W. Montero (1995), Island arc response to shallow subduction of the Cocos Ridge, Costa Rica, in *Geologic and Tectonic Development of the Caribbean Plate Boundary in Southern Central America,* edited by P. Mann, *Geol. Soc. Am. Spec. Pap., 295*, 235–262.

Lallemand, S., and X. Le Pinchon (1987), Coulomb wedge model applied to subduction of seamounts in the Japan Trench, *Geology, 15*, 1065–1069.

Lallemand, S., R Culotta, and R. von Huene (1989), Subduction of the Daiichi Kashima seamount in the Japan Trench, *Tectonophysics, 160*, 231–247.

Lay, T., and H. Kanamori (1980), Earthquake doublets in the Solomon Islands, *Phys. Earth Planet. Int., 21*, 283–304.

Lay, T., H. Kanamori, and L. Ruff (1982), The asperity model and the nature of large subduction zone earthquakes, *Earthquake Predict. Res., 1*, 3–71.

Ma, K.-F., J. Mori, S.-J. Lee, and S. B. Yu (2001), Spatial and temporal distribution of slip for the 1999 Chi-Chi, Taiwan, earthquake, *Bull. Seismol. Soc. Am., 91*, 1069–1087.

Marshall, J., and R. Anderson (1995), Quaternary uplift and seismic cycle deformation, Peninsula de Nicoya, Costa Rica, *Geol. Soc. Am. Bull., 107*, 463–473.

Masson, D. G., L. M. Parson, J. Milsom, G. Nichols, N. Sikumbang, B. Dwiyanto, and H. Kallagher (1990), Subduction of seamounts at the Java Trench: A view with long-range sidescan sonar, *Tectonophysics, 185*, 51–65.

Matsuzawa, T., T. Igarashi, and A. Hasegawa (2002), Characteristic small-earthquake sequence off Sanriku, northeastern Honshu, Japan, *Geophys. Res. Lett., 29*(11), 1543, doi:10.1029/2001GL014632.

McCaffrey, R. (1993), On the role of the upper plate in great subduction zone earthquakes, *J. Geophys. Res., 98,* 11,953–11,966.

McCaffrey, R. (1994), Global variability in subduction thrust fore-arc systems, *Pure Appl. Geophys., 142,* 173–224.

Mendoza, C., and S. H. Hartzell (1988), Inversion for slip distribution using GDSN P waves: North Palm Springs, Borah Peak, and Michoacan earthquakes, *Bull. Seismol. Soc. Am, 78,* 1092–1111.

Nadeau, R. M., and L. R. Johnson (1998), Seismological studies at Parkfield VI: Moment release rates and estimates of source parameters for small repeating earthquakes, *Bull. Seismol. Soc. Am, 88,* 790–814.

Nishenko, S. P. (1991), Circum-Pacific seismic potential: 1989–1999, *Pure Appl. Geophys., 135,* 169–259.

Norabuena, E., et al. (2004), Geodetic and seismic constraints on some seismogenic zone processes in Costa Rica, *J. Geophys. Res., 109,* B11403, doi:10.1029/2003JB002931.

Obana, K., S. Kodaira, Y. Kaneda, K. Mochizuki, M. Shinohara, and K. Suyehiro (2003), Microseismicity at the seaward updip limit of the western Nankai Trough seismogenic zone, *J. Geophys. Res., 108*(B10), 2459, doi:10.1029/2002JB002370.

Okada, T., T. Matsuzawa, and A. Hasegawa (2003), Comparison of source areas of M4.8 ± 0.1 repeating earthquakes off Kamaishi, NE Japan: Are asperities persistent features?, *Earth Plant. Sci. Lett., 213,* 361–374.

Pacheco, J. F., and L. R. Sykes (1992), Seismic moment catalog of large, shallow earthquakes, 1900–1989, *Bull. Seismol. Soc. Am., 82,* 1306–1349.

Park, J.-O.,T. Tsura, Y. Kaneda, Y. Kono, S. Kodaira, N. Takahashi, and H. Kinoshita (1999), A subducting seamount beneath the Nankai accretionary prism off Shikoku, southwestern Japan, *Geophys. Res. Lett., 26,* 931–934.

Protti, M., F. Güendel, and K. McNally (1995a), Correlation between the age of the subducting Cocos plate and the geometry of the Wadati-Benioff zone under Nicaragua and Costa Rica, in *Geologic and Tectonic Development of the Caribbean Plate Boundary in Southern Central America,* edited by P. Mann, *Spec. Pap. Geol. Soc. Am., 295,* 309–326.

Protti, M., et al. (1995b), The March 25, 1990(M_w = 7.0, M_L = 6.8) earthquake at the entrance of the Nicoya Gulf, Costa Rica: Its prior activity, foreshocks, aftershocks, and triggered seismicity, *J. Geophys. Res., 100,* 20,345–20,358.

Ranero, C. R., and R von Huene (2000), Subduction erosion along the Middle America convergent margin, *Nature, 404,* 748–752.

Ranero, C. R., J. Phipps Morgan, K. McIntosh, and C. Reichert (2003), Bending-related faulting and mantle serpentinization at the Middle America Trench, *Nature, 425,* 367–373.

Rice, J. R. (1993), Spatiotemporal complexity of slip on a fault, *J. Geophys. Res., 98,* 9885–9907.

Ruff, L. J. (1989), Do trench sediments affect great earthquake occurrence in subduction zones? *Pure Appl. Geophys., 129,* 263–282.

Ruff, L. J. (1992), Asperity distributions and large earthquake occurrence in subduction zones, *Tectonophysics, 211,* 61–83.

Ruff, L. J., and H. Kanamori (1980), Seismicity and the subduction process, *Phys. Earth Planet. Int., 23,* 240–252.

Ruff, L. J., and H. Kanamori (1983), The rupture process and asperity distribution of three great earthquakes from long period diffracted *P*-waves, *Phys. Earth Planet. Inter., 31,* 202–230.

Ryan, H. F., and D. W. Scholl (1993), Geologic implications of great interplate earthquakes along the Aleutian arc, *J. Geophys. Res., 98,* 22,135–22,146.

Sagiya, T. (1999), Interplate coupling in the Tokai District, central Japan, deduced by continuous GPS data, *Geophys. Res. Lett., 26,* 2315–2318.

Sagiya, T., S. Miyazaki, and T. Tada (2000), Continuous GPS array and present day crustal deformation of Japan, *Pure Appl. Geophys., 157,* 2303–2322.

Sammis, C. G., and J. R. Rice (2001), Repeating earthquakes as low-stress-drop events at a border between locked and creeping fault patches, *Bull. Seismol. Soc. Am., 91*, 532–537.
Scholz, C. H., and J. T. Engelder (1976), The role of asperity indentation and plough in rock friction 1. Asperity creep and stick slip, *Int. J. Rock Mech. Min. Sci. Geomech. Abstr., 13*, 149–154.
Scholz, C. H., and C. Small (1997), The effect of seamount subduction on seismic coupling, *Geology, 25*(6), 487–490.
Schwartz, S. Y. (1999), Noncharacteristic behavior and complex recurrence of large subduction zone earthquakes, *J. Geophys. Res., 104*, 23,111–23,125.
Schwartz, S. Y., and L. J. Ruff (1987), Asperity distribution and earthquake occurrence in the southern Kurile Islands arc, *Phys. Earth Planet. Inter., 49*, 54–77.
Song, T.-R., and M. Simons (2003), Large trench-parallel gravity variations predict seismogenic behavior in subduction zones, *Science, 301*, 630–633.
Spence, W., C. Mendoza, E. R. Engdahl, G. L. Choy, and E. Norabuena (1999), Seismic subduction of the Nazca Ridge as shown by the 1996–1997 Peru earthquake, *Pure Appl. Geophys. 154*, 753–776.
Tajima, F., and M. Kikuchi (1995), Tectonic implications of the seismic ruptures associated with the 1983 and 1991 Costa Rica earthquakes, in *Geologic and Tectonic Development of the Caribbean Plate Boundary in Southern Central America,* edited by P. Mann, *Spec. Pap. Geol. Soc. Am., 295*, 327–340.
Tanioka, Y., L. Ruff, and K. Satake (1997), What controls the lateral variation of large earthquake occurrence along the Japan Trench? *Island Arc, 6*, 261–266.
Trifunac, M. D., and J. N. Brune (1970), Complexity of energy release during the Imperial Valley, California earthquake of 1940, *Bull. Seismol. Soc. Am., 60*, 137–160.
von Huene, R., et al. (1995), Morphotectonics of the Pacific convergent margin of Costa Rica, in *Geologic and Tectonic Development of the Caribbean Plate Boundary in Southern Central America,* edited by P. Mann, *Spec. Pap. Geol. Soc. Am., 295*, 291–307.
von Huene, R., C. R. Ranero, and W. Weinrebe (2000), Quaternary convergent margin tectonics of Costa Rica, segmentation of the Cocos Plate, and Central American volcanism, *Tectonics, 19*, 314–334.
Wald, D. J., and T. H. Heaton (1994), Spatial and temporal distribution of slip for the 1992 Landers, California earthquake, *Bull. Seismol. Soc. Am., 84*, 668–691.
Wald, D. J., D. V. Helmberger, and T. H. Heaton (1991), Rupture model of the 1989 Loma Prieta earthquake from the inversion of strong motion and broadband teleseismic data, *Bull. Seismol. Soc. Am., 81*, 1540–1572.
Walther, C. H. E (2003), The crustal structure of Cocos Ridge off Costa Rica, *J. Geophys. Res., 108*(B3), 2136, doi:10.1029/2001JB000888.
Wessel, P., and W. H. F. Smith (1991), Free software helps map and display data, *Eos Trans. AGU, 72*, 441–445.
Wyss, M., and J. N. Brune (1967), The Alaska earthquake of 28 March 1964: A complex multiple source rupture, *Bull. Seismol. Soc. Am., 57*, 1017–1023.
Xu, Z. Y., and S. Y. Schwartz (1993), Large earthquake doublets and fault plane heterogeneity in the northern Solomon-Islands subduction zone, *Pure Appl. Geophys., 140*, 365–390.
Zweck, C., J. T. Freymueller, and S. C. Cohen (2002), Three-dimensional elastic dislocation model of the post-seismic response to the 1964 Alaska earthquake, *J. Geophys. Res., 107*(B4), 2064, doi:10.1029/2001JB000409.

PART III

Convergent Margin Structure, Fluids, and Subduction Thrust Evolution

6

Pore Pressure and Fluid Flow in the Northern Barbados Accretionary Complex

A Synthesis

Barbara A. Bekins
Elizabeth J. Screaton

Abstract

The northern Barbados sedimentary accretionary complex has been the focus of more hydrogeology-related studies than any other subduction zone. The results of these studies are beginning to provide a comprehensive understanding of the pore-pressure distribution and flow system in this active complex. As subduction begins, loading of the low-permeability incoming sediments leads to a rapid increase in pore pressure. Sediments drilled in the protodécollement ~6 km seaward of the deformation front are at hydrostatic pressure but have unusually high porosities of ~60%. Shear-enhanced consolidation of the décollement causes a rapid decrease in porosity to ~40% except for a high-porosity patch that is 1–2 km-wide along strike and extends at least 10 km down dip. Outside of the high-porosity patch, long-term monitoring in the décollement measured excess pressure ratios of $\lambda^* \sim 0.36$ at 1.8 km arcward, increasing to $\lambda^* \sim 0.5$ at 4 km arcward, which are consistent with results of steady-state models. Maintenance of the high-porosity patch requires estimated excess pressure ratios of $\lambda^* > 0.7$. Such high fluid pressures probably require transient flow from depth along the décollement. Low-chloride pore-water anomalies along the décollement also suggest transient flow. In contrast, elevated linear temperature gradients near the toe are more consistent with either steady-state flow along underthrust sand layers or elevated basement heat flow. Estimated rates of flow needed to produce the temperature anomalies are 5–20 times higher than the rate of fluid subduction. Two possible explanations for producing the required flow rates are discussed: (1) fluid draining from a large area is focused into a small flow channel and/or (2) fluid is stored during long periods of low flow and vented during short periods of high flow.

Introduction

At the northern Barbados accretionary complex, the Atlantic plate is thrust beneath the Caribbean plate (fig. 6.1) at ~2.8 cm yr^{-1} [*Dixon et al.*, 1998]. The shallow décollement dips from 180 m below seafloor (mbsf) at the deformation front to 10 km below the seafloor at the back of the accretionary complex, located 125 km arcward [*Westbrook et al.*, 1988]. Results from three Ocean

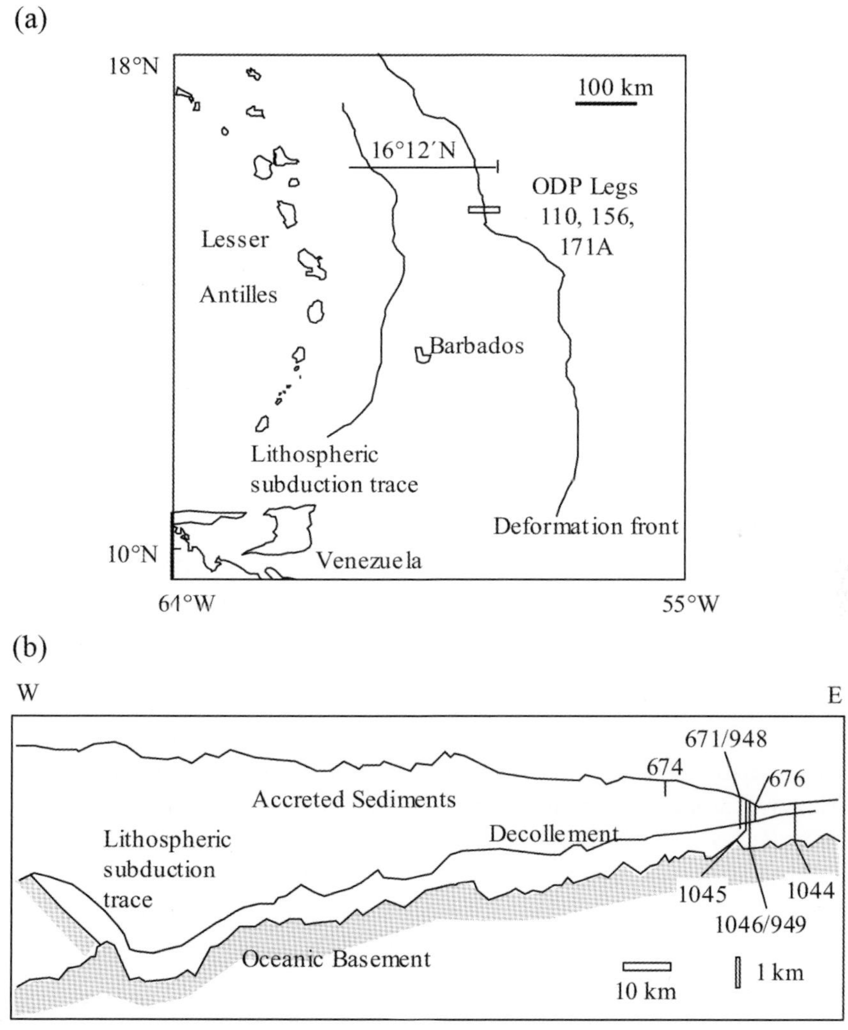

Figure 6.1 (a) Location map of the northern Barbados accretionary complex together with (b) interpreted seismic profile along 16°12′ [*Westbrook et al.*, 1988] showing boreholes from ODP Legs 110, 156, and 171A. Modified from *Bekins et al.* [1995].

Drilling Program (ODP) Legs, 110, 156, and 171A, and one Deep Sea Drilling Project (DSDP) Leg, 78A, indicate high fluid pressure and provide information about flow rates in the toe of the complex.

The Barbados complex represents a slow convergence rate, low taper angle, clay-rich end member for sedimentary accretionary prisms worldwide. Sediments of the northern Barbados Ridge accretionary prism consist primarily of fine-grained pelagic mudstones and hemipelagic clays. The underthrust sediments are composed of well-bedded hemipelagic clays with fine-grained turbiditic sand layers [*Shipboard Scientific Party*, 1988]. The northern segment at 15°30′ was chosen for detailed study because the décollement lies just 180 mbsf at the deformation front, making it easily accessible to drilling. In addition to multiple drilling legs, the area has also been the focus of three-dimensional (3-D) seismic imaging, heat-flow surveys, hydrologic tests, long-term pressure monitoring, and multiple generations of fluid-flow modeling. From these studies, there is broad agreement that pore pressures at the toe are a significant fraction of lithostatic, that fluid flow is focused along faults and sand layers, and that at least some flow events are transient. However, there are still a number of unresolved questions about pore pressures, flow rates, and duration of transient flow.

This paper provides a synthesis of current understanding of fluid-flow processes in northern Barbados. The first part of the paper summarizes fluid-pressure estimates and measurements near the toe. Comparison of the values yields insights into the expected spatial variations in pressure and the mechanisms that create overpressures. The second part of the paper summarizes the published estimates for fluid-flow rates. We compare the flow estimates to the rate that fluids enter the complex in the pore space and discuss the implications for the flow system.

Pore Pressures

A variety of approaches have been used to determine in situ pore pressures in the northern Barbados complex. These include long-term monitoring [*Foucher et al.*, 1997; *Becker et al.*, 1997], fluid flow modeling [*Shi and Wang*, 1988; *Screaton et al.*, 1990; *Bekins et al.*, 1995; *Henry*, 2000; *Cutillo et al.*, 2003], inferred consolidation state [*Moore et al.*, 1995], and seismic impedance analysis [*Shipley et al.*,

TABLE 6.1 **Pressures Measured at the Two CORK Sites**

Site	Distance Arcward, km	Water Depth, mbsl	Seafloor Pressure, MPa	Depth, mbsf
949C	1.8	5004.7	51.02	398.0
948D	4.0	4938.0	50.34	508.8

1994; *Bangs et al.*, 1996, 1999]. Attempts to reconcile the results using the different techniques have been limited. In the sections below, we compare model results to the pressures directly measured by long-term monitoring and inferred from consolidation state.

Two expressions are commonly used to relate pore pressure to the overburden stress in the sediments:

$$\lambda^* = (P_f - P_{\text{hydro}})/(P_{\text{litho}} - P_{\text{hydro}}) \quad (6.1)$$

$$\lambda = (P_f - P_{\text{seafloor}})/(P_{\text{litho}} - P_{\text{seafloor}}) \quad (6.2)$$

where P_f is the pore-fluid pressure and P_{litho} is the total overburden stress. The other terms are pressure at the seafloor, P_{seafloor}, and the weight of the water column at the location of interest, P_{hydro}. Both expressions are equal to 1 when the pore pressure is equal to the overburden stress. In the case where pressure is equal to hydrostatic, λ^* is equal to 0, while λ depends on the bulk density typically ranging from 0.7 at the seafloor to 0.45 at depth. Hydrogeologists more often use the λ^* because the value for hydrostatic pressure is fixed at 0. The quantity λ is used more often by structural geologists because the effective vertical stress σ'_v is obtained from total vertical stress σ_v, using $\sigma'_v = (1 - \lambda)\sigma_v$ [*Davis et al.*, 1983].

Long-Term Monitoring

Long-term monitoring was initiated at Holes 948D and 949C during ODP Leg 156 (fig. 6.1). Each hole was cased and screened along the décollement interval. Pressure and temperature sensors were then installed and the tops of the boreholes were sealed with Circulation Obviation Retrofit Kits (CORKs). The pressure and temperature records were retrieved a year and a half later using the submersible Nautile. Results of the pressure monitoring at the two sites differ substantially (table 6.1). Hole 949C, located 1.8 km from the deformation front, had a modest overpressure of 1 MPa [*Becker et al.*, 1997]. Hole 948D, located 4 km from the deformation front, recorded significantly greater overpressure of 2 MPa [*Foucher et al.*, 1997]. The greater value at Hole 948D led to speculation that the lack of bridge plug at the base of the casing, and the need to add mud to stabilize the hole, caused anomalously high pressures in the

Screened Interval, mbsf	Pore Pressure, MPa	Hydrostatic, MPa	Lithostatic, MPa	λ	λ^*	Reference
398–451	56.02	55.0	57.8	0.74	0.36	*Becker et al.* [1997]
509–551	57.50	55.5	59.4	0.79	0.50	*Foucher et al.* [1997]

borehole. However, the fact that the observed value was stable for the duration of the 512-day monitoring period suggests that the recorded pressure reflects in situ conditions.

Steady-State Fluid-Flow Modeling

The increase in overpressure with distance from the deformation front observed during the long-term monitoring is consistent with progressive loading of the underthrust sediments. *Screaton and Ge* [2000] showed that the very low sedimentation rates of 2–20 m Myr^{-1} result in hydrostatic pressures on the incoming plate. Progressive loading of the sediments near the toe occurs by imbricate thrusting in the prism and subduction of the underthrust sediments beneath the growing wedge. Although the load increase occurs by overthrusting of discrete sediment sections and progresses intermittently in time, it is not necessary to simulate the fate of individual sediment packets to understand the overall development of pore pressure in the complex. Models based on a prism deforming under pure shear strain with the reference frame pinned to the deformation front have been successfully used to understand the large-scale sediment consolidation and pore-pressure development of the décollement and underthrust sediments [*Screaton et al.*, 1990; *Le Pichon et al.*, 1990; *Bekins et al.*, 1995].

Using a hydrostatic seaward boundary, *Bekins et al.* [1995] simulated steady-state pore pressures in the complex using a permeability-depth relation for clay-rich sediments. The model results indicate that excess pressures extend ~5 km seaward of the deformation front and increase rapidly as the underthrust sediments are subducted beneath the toe of the prism (fig. 6.2a). Modeled values of λ^* along the décollement are ~0.3 at Site 949 and ~0.45 at Site 948. The estimated values are strikingly similar to those obtained during long-term monitoring (table 6.1) and the increase of 0.15 over a lateral distance of 2.2 km is nearly identical to the observed increase between the two sites. Simulations of 5 km of underthrusting and consolidation by *Stauffer and Bekins* [2001] also showed an increase in λ^* of 0.15 between sites 949 and 948.

Consolidation State

Excess fluid pressure influences sediment porosity by decreasing effective stress. Thus, in theory, the in situ effective stress can be estimated by comparing in situ porosity to values predicted from laboratory consolidation test results. Fluid pressure is then computed by subtracting the estimated in situ effective stress from the total vertical stress. *Moore et al.* [1995] used consolidation data from *Taylor and Leonard* [1990] together with logging while drilling (LWD) bulk density logs to infer the effective stresses and pore pressures at Site 948. Figure 6.2b shows a comparison between their estimated pressures and steady-state model results of *Bekins et al.* [1995].

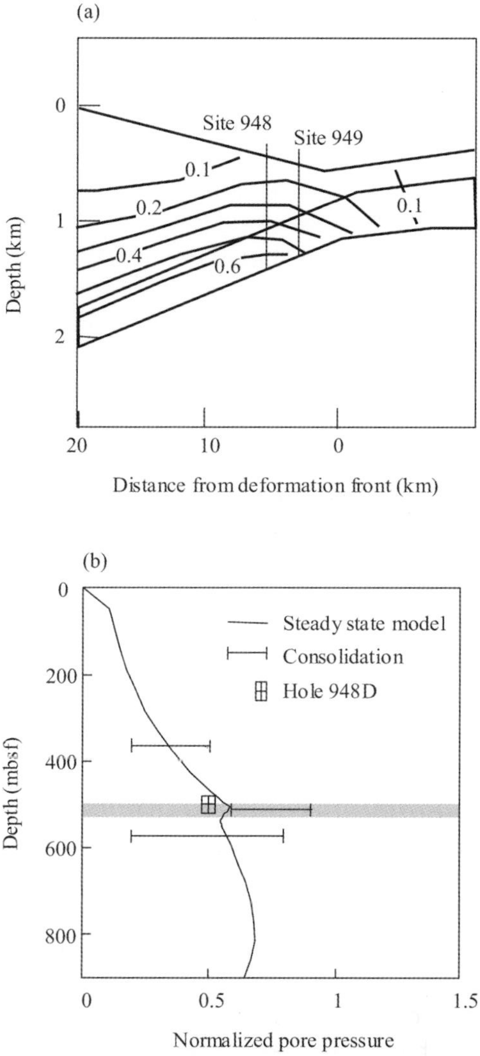

Figure 6.2 (a) Modeled overpressures at steady state along the décollement zone with locations of the two long-term monitoring sites marked. Model results are consistent with pressures observed during long-term monitoring. (b) Modeled overpressures versus depth at Site 948 compared to pressures inferred by *Moore et al.* [1995] using consolidation tests and LWD bulk densities. Estimated value of λ^* from long-term monitoring at Hole 948D is also marked as a rectangle covering the depth of the screened interval (480.7–522.6 mbsf) and spanning the range of uncertainty. The inferred décollement is the stippled interval.

Moore et al. [1995] found pore-pressure ratios, λ, ranging from 0.7 to 0.95 (λ^* ~ 0.35–0.8). Their estimated values varied widely over small scales (indicated by error bars in fig. 6.2b), but some trends were noted. The highest pore-pressure ratios in the prism were located below major thrust faults. At the base of the prism, a high-pressure, clay-rich interval with λ reaching 0.9 (λ^* ~ 0.8) extended downward from 380 mbsf to the center of the décollement. The overall trend of a rapid rise downward toward the décollement and more constant values in the underthrust sequence corresponds to the trend of the model profile from *Bekins et al.* [1995].

Seismic Impedance

Seismic images delineate a low-impedance area that creates a negative-polarity reflector in the plane of the décollement with a width of 1–2 km along strike (fig. 6.3). The negative-polarity reflector extends over 10 km downdip, from the deformation front to the arcward boundary of the surveyed area [*Shipley et al.*, 1994]. LWD bulk density logs from Site 1045 show that the décollement in this area is comprised of sediments with the same low bulk density (~1.4 g cm^{-1}) as the protodécollement [*Moore et al.*, 1998; *Bangs et al.*, 1999]. In contrast, Site 1046, located just 1 km away, has a normal polarity décollement reflection and LWD bulk density of ~1.6 g cm^{-1}. *Moore et al.* [1998] concluded that shear-enhanced consolidation caused the observed increase in bulk density between the protodécollement and Site 1046. They further postulated that the low bulk density retained at Site 1045 was due to arrested consolidation during ~3 km of underthrusting.

A number of observations from the negative-polarity area are consistent with greater pore pressures within that region compared to the area with a normal-polarity décollement reflection. The first attempt to drill through the negative-polarity area occurred at Site 947 during ODP Leg 156. The hole was abandoned ~60–80 m above the décollement (583.7 mbsf) owing to high formation pressures and hole collapse. The LWD data from Site 947 show gradually decreasing porosities with depth to ~50% at 500 mbsf. Below this depth, porosities increased dramatically to over 70% by the base of the hole [*Shipboard Scientific Party*, 1995a]. During ODP Leg 171A the negative-polarity area was successfully drilled to ~35 m below the base of the décollement at Site 1045. Significant operational difficulties were again encountered, requiring considerable hole conditioning, back reaming, and a 30-barrel sepiolite mud sweep to advance the drill string beyond the top of the décollement at 424 mbsf [*Shipboard Scientific Party*, 1998]. *Bangs et al.* [1999] noted that seismic images of thrusts

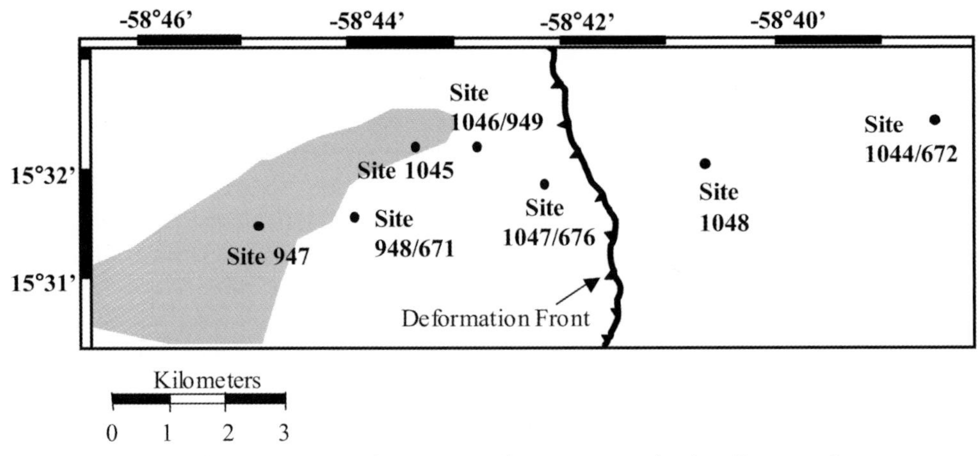

Figure 6.3 Map showing extent of negative-polarity area on the décollement after *Shipley et al.* [1994].

above the negative-polarity area have low densities and high porosities, suggesting that high pore pressures migrate upward from the décollement.

The direct and indirect evidence for greater excess pore pressures in the low-density, negative-polarity portion of the décollement indicate that the hydrologic regime there differs from the higher density, normal-polarity area. We examine three hypotheses for the contrasting décollement bulk densities at Site 1045, within the negative-polarity area, and Site 1046, within a normal polarity region. First, the low bulk densities may represent simple one-dimensional loading disequilibria. Second, arrested consolidation may result from continuous resupply of fluids along the décollement from deeper in the subduction complex, as suggested by *Bangs et al.* [1999]. Third, the low bulk densities may be due to dilation of sediments in the décollement during shear failure under decreasing effective stress (increasing pore pressure), as proposed by *Karig* [1990].

Consolidation and Flow Modeling

We first examine the scenario in which there is no fluid resupply from depth, and the low density in the décollement at 4 km arcward is due to one-dimensional loading disequilibria. If an additional ~340 m of sediment is added to the incoming sediment sequence such that the entire increased load is added to the pore pressure, the overpressure would reach $\lambda^* \sim 0.74$, similar to the value of 0.73 that *Stauffer and Bekins* [2001] estimated is required to maintain the low bulk densities during 4 km of subduction. However, it is also necessary to consider that excess pressures may have dissipated in the estimated 140 kyr. since subduction. Using an estimated specific storage of $\sim 1 \times 10^{-5}$ m^{-1} [*Screaton et al.*, 2000] and an analytical solution for one-dimensional transient flow [e.g., *Deming*, 2002], permeabilities of 10^{-21} m^2 or less would be required to maintain $\lambda^* \sim 0.74$ for 140 kyr Although one sample from ODP Leg 110 yielded a permeability of 10^{-20} m^2, the other 14 measured samples had values greater than 10^{-18} m^2 [*Taylor and Leonard*, 1990]; samples from ODP Leg 156 yielded permeabilities ranging from 3×10^{-18} m^2 to 8×10^{-16} m^2 [*Zwart et al.*, 1997]. Using a more reasonable lower bound for permeability of 10^{-19} m^2, excess pressures drop to $\lambda^* \sim 0.29$ after 140 kyr. Thus maintaining the low bulk densities by one-dimensional loading would require an unreasonably low permeability value over the entire negative-polarity area and throughout the entire overlying sediment column.

The second hypothesis, arrested consolidation caused by continuous resupply of fluids from depth was explored by *Stauffer and Bekins* [2001] using a coupled consolidation and flow model. They simulated loading, fluid flow, and consolidation of the prism base, décollement and underthrust for the first 5 km of subduction or 186 kyr. The goal of the modeling was to reproduce both the low décollement density in the negative-polarity area (~1.4 g cm^{-3} at Site 1045) and also the higher décollement density normal-polarity area (1.6–1.8 g cm^{-3} at Site 1046) (fig. 6.3). Two mechanisms to explain the difference in bulk densities were tested: (1) varying permeabilities at the base of the prism and

TABLE 6.2 **Summary of Specified Pressure Arcward Boundary Simulations**

Case	Prism Permeability, m^2		Specified Arcward Excess Pressure, kPa	Décollement Density, g cm^3	Décollement λ^*	
	k_x	k_z			Site 949	Site 948
F-A	10^{-17}	10^{-18}	1550	1.40–1.45	0.60	0.55
F-B	10^{-17}	10^{-18}	500	1.62–1.66	0.35	0.27
F-C	10^{-16}	10^{-17}	2300	1.45–1.48	0.55	0.73
F-D	10^{-16}	10^{-17}	800	1.70–1.75	0.28	0.36

(2) varying pore pressure at the arcward boundary of the model (table 6.2). Figure 6.4a shows décollement bulk density versus distance from the deformation front for four model scenarios. It was not possible to maintain the observed difference in bulk densities for the entire 5 km of subduction by varying only the permeability of the overlying prism sediments (cases F-A and F-B). Attempts to use even lower prism permeabilities with fluid resupply from depth caused pore pressures in the décollement to exceed lithostatic values. Varying the pore pressure at the arcward boundary best matched the observed décollement bulk densities, achieving both the low value at Site 1045 and the high value at Site 1046 (cases F-C and F-D).

In the third scenario, representing dilation, the negative-polarity area of the décollement can be envisioned as the seaward end of a pulse of high pressure that has migrated up the décollement [*Shipley et al.*, 1994]. On the basis of experimental observations, normally consolidated clays with an initially high void ratio contract when they are sheared in drained tests [*Wood*, 1990]. In contrast, clays that have an initially low void ratio expand when they are sheared under decreasing effective stress or increasing pore pressure. Once the clays reach critical state, they will continue to fail plastically with no change in void ratio. Thus, depending on the stress history and pore pressure during failure, shear strain along the clay-rich décollement is a mechanism that can produce both the increased density of the normal polarity sections and either decreased or constant density in the negative-polarity section. *Karig* [1990] presented evidence of fault-zone dilation during shear failure under decreasing effective stress conditions for the Barbados décollement. *Brown et al.* [1994] argued on the basis of ring-shear permeability measurements that propagation of hydrofracture or extensional shear along the décollement from depth best explains evidence for enhanced flow and high pore pressures observed in the negative-polarity anomaly. The key element in this scenario is that pore pressure must increase as the décollement sediments approach failure in the reverse-polarity area and remain the same or decrease in the normal-polarity area.

Bekins et al. [1995] simulated pore pressure changes that might occur when the décollement permeability increases from a steady-state value of 10^{-14} m^2 [*Screaton et al.*, 2000] to a dilated value of 10^{-12} m^2 [*Fisher and Zwart*, 1996]. Figure 6.4b shows simulated transient pressure profiles at Site 948 for 1000 and 10,000 years after the permeability increase. Transient simulated pressures

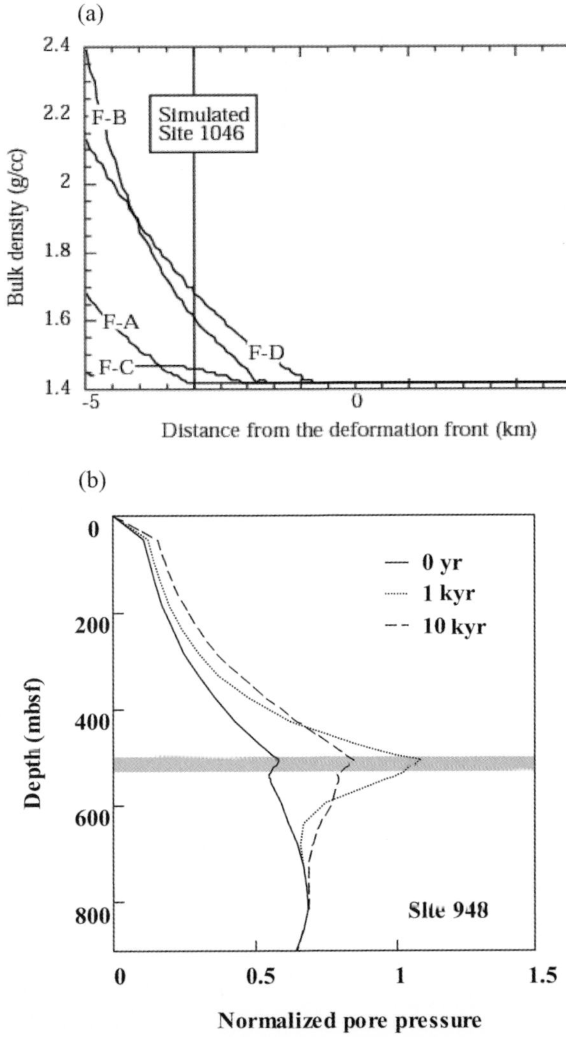

Figure 6.4 (a) Décollement consolidation versus distance from the deformation front for model scenarios given in table 6.2. Case F-C with the highest specified pressure at the arcward boundary best matched the low bulk density observed at Site 1045 located in the negative-polarity area. Modified from *Stauffer and Bekins* [2001]. (b) Modeled overpressures versus depth at Site 948 for a transient scenario with décollement permeability increased from 10^{-14} to 10^{-12} m^2 at the start of the model run. The inferred décollement is the stippled interval. Modified from *Bekins et al.* [1995].

exceed the value of $\lambda^* > 0.7$ needed to create the observed décollement low densities within 1000 years and remain high for more than 10,000 years. *Henry* [2000] calculated that for a basal friction of $\mu_b = 0.45$, shear failure would occur along the décollement between 1 and 4 km of the toe at $\lambda^* = 0.78–0.79$. He modeled the propagation of failure and associated permeability increase along the décollement and estimated a propagation rate of 15 m yr^{-1}, which is rapid compared to the amount of time that high pore pressures can be sustained. Thus the transient propagation of a pulse of pressure and associated shear failure from depth can reproduce the observed low densities and maintain them for 10,000 years. The transient model predicts that high pore pressures exist in the prism above the décollement. Drilling results from Site 947 and the seismic impedance analyses also suggest that pressures are elevated in the base of the prism over the negative-polarity area.

Spatial Variation of Pore Pressures

The measured pressures in the normal-polarity area are consistent with the results of steady-state modeling. For the negative-polarity area, however, it is difficult to maintain the required pressures in a steady-state model using the known rate of loading and measured permeabilities. Our simple one-dimensional analysis above suggests that loading disequilibria without fluid resupply is not feasible unless permeabilities are unrealistically low. Furthermore, the estimated pressure to maintain low density at 4 km arcward, $\lambda^* = 0.73$ (table 6.2) exceeds the maximum pressure simulated with steady state, two-dimensional modeling, $\lambda^* \sim 0.6$ (fig. 6.2a). Three-dimensional simulations [*Cutillo et al.*, 2003] show that additional fluid flow from the south, driven by consolidation in the thicker portion of the complex, cannot produce significantly higher steady-state pressures. Thus both the arrested consolidation through resupply and the transient dilation scenarios require more efficient propagation of pore pressures to shallow depths in the negative-polarity area than in the normal-polarity region.

Unfortunately, the low-density portion of the décollement has never been cored so there are no sediment samples to distinguish between the transient fault

TABLE 6.3 **Estimated Flow Rates and Percent of Influx Compiled from the Literature**

Method	Location	Permeability, m^2	Velocity, m yr^{-1}
Heat transport	prism (first 2 thrust packets)	10^{-13}	0.04
Heat transport	décollement Sites 676 and 672	10^{-13}	20
Heat transport	décollement Sites 676 and 672	10^{-12}	6.2
Heat transport	underthrust Sites 672, 948 and 949	10^{-12}	13
Heat transport	underthrust Sites 672, 948 and 949	3×10^{-13}	12
Heat transport	underthrust Sites 672, 948 and 949	10^{-12}	31
Heat and chloride	décollement Sites 948	10^{-13}	6.6
		10^{-12}	66
Soil mechanics	décollement Sites 1045 and 1046	10^{-14}	0.29
		10^{-13}	2.9

zone dilation and arrested consolidation explanations for the low density. Distinguishing between them is important because the two hypotheses have very different implications for fluid budgets. The scenario of arrested consolidation would require continuous fluid input for the estimated ~300 kyr since subduction at the back of the 3-D surveyed area (~10 km from the deformation front), whereas the dilation scenario allows for transient pressure pulses of shorter duration. In the next section, a quantitative analysis comparing flow rate estimates to fluid budgets shows that high-flow rates are difficult to sustain for 300 kyr.

Flow Rates

Estimates for the duration, location, and rate of flow have been based on both temperature [*Fisher and Hounslow*, 1990; *Henry*, 2000; *Cutillo et al.*, 2003] and chloride data [*Blanc et al.*, 1991; *Bekins et al.*, 1995; *Henry*, 2000]. Although *Fitts and Brown* [1999] questioned the interpretation of much of the Barbados chloride data, they concluded that a narrow low-chloride anomaly is present at the top of the décollement at Site 948. In general, narrow pore-water chemical anomalies along faults are recognized as indicative of transient flow [e.g., *Blanc et al.*, 1991]. In contrast, *Henry* [2000] postulated high steady-state flow velocities below the décollement to explain the linear temperature profiles with elevated thermal gradients measured at Sites 948 and 949, although elevated thermal gradients near the deformation front have also been attributed to crustal thinning [*Müller and Smith*, 1993]. Table 6.3 provides a summary of published flow velocity estimates for northern Barbados. The values range from 0.29 to 66 m yr^{-1} with corresponding permeabilities of 10^{-14} to 10^{-12} m^2.

Porosity	Darcy Flux, m yr^{-1}	Available Fluid, m^3 yr^{-1}	Input/ Output	Reference
0.50	0.02	4	20%	*Foucher et al.* [1990]
0.50	10	9–18	2–5%	*Foucher et al.* [1990]
0.56	3.5	1.53	4%	*Fisher and Hounslow* [1990]
0.30	4	3.12	8%	*Screaton and Ge* [1997]
0.54	6.3	3.12	5%	*Henry* [2000]
0.30	9.4	3.12	3%	*Cutillo et al.* [2003]
0.56	3.7	1.53	4%	*Saffer and Bekins* [1999]
0.56	36.7	1.53	0.4%	
0.49	0.14	1.53	110%	*Stauffer and Bekins* [2001]
0.49	1.4	1.53	11%	

The range of estimated rates is 10–100 times higher than steady-state modeled values of 0.02–0.07 m yr^{-1}. The implication is that flow rates must be enhanced either by transient cycles of storage and release or by focusing of fluids from large areas into spatially restricted conduits [*Saffer and Bekins*, 1999].

A number of questions have arisen during the course of determining these estimates. One question concerns the distribution of flow between the décollement and turbidite sand layers in the underthrust sequence. There is compelling evidence for flow in both zones, but the relative amount in each is still unclear. Another question is the difficulty in balancing outflow based on estimated flow durations and rates with the supply of fluid entering the complex. In the sections below, these questions will be described and possible ways to resolve them will be discussed.

Heat Flow

Seafloor and borehole measurements indicate highly variable heat flow with values averaging 40% above that expected from conductive thermal transport [e.g., *Langseth et al.*, 1990; *Ferguson et al.*, 1993]. A borehole profile through a shallow fault zone at Site 676 (location shown in fig. 6.3) suggests a transient perturbation [*Fisher and Hounslow*, 1990], requiring high fault-zone permeability and flow rates. *Foucher et al.* [1990] draw a distinction between shallow heat-flow perturbations associated with thrust faults in the prism and deeper, uniformly elevated heat flow observed at Sites 948 and 949 in the prism toe and Site 672 located 6 km seaward of the deformation front.

The uniformly elevated heat flow results in linear thermal profiles at Sites 948 and 949 (fig. 6.5). The linearity of the Site 949 temperature data across the décollement suggests that the source of the heat-flow perturbation must occur below this depth. *Henry* [2000] postulated that high fluid-flow rates transport heat in the turbidite sand units in the underthrust section. Elevated fluid flow in the underthrust sands is consistent with *Tokunaga* [2000], who showed that the turbidites act as relatively permeable drains that accelerate consolidation of interbedded clays. *Henry* [2000] also discusses the possibility of hydrothermal convection in the ~80 Ma basement along the major fracture zone bounding the northern edge of the Tiburon Rise.

Using advective heat transport models, a number of studies have estimated the rates of fluid flow required to produce the high thermal gradients at Sites 672, 676, 948, and 949. Table 6.3 lists the modeling studies, locations of the temperature data that constrain the models, and the resulting flow-rate estimates. These flow estimates significantly exceed values that can be produced with 2-D steady-state flow models. *Cutillo et al.* [2003] used a 3-D model of the Barbados complex to examine the along strike component of fluid flow from the thicker, southern portion of the complex but did not examine effects of laterally restricted conduits. They concluded that the small amount of additional fluid flow from the south could not produce the flow rates needed to explain

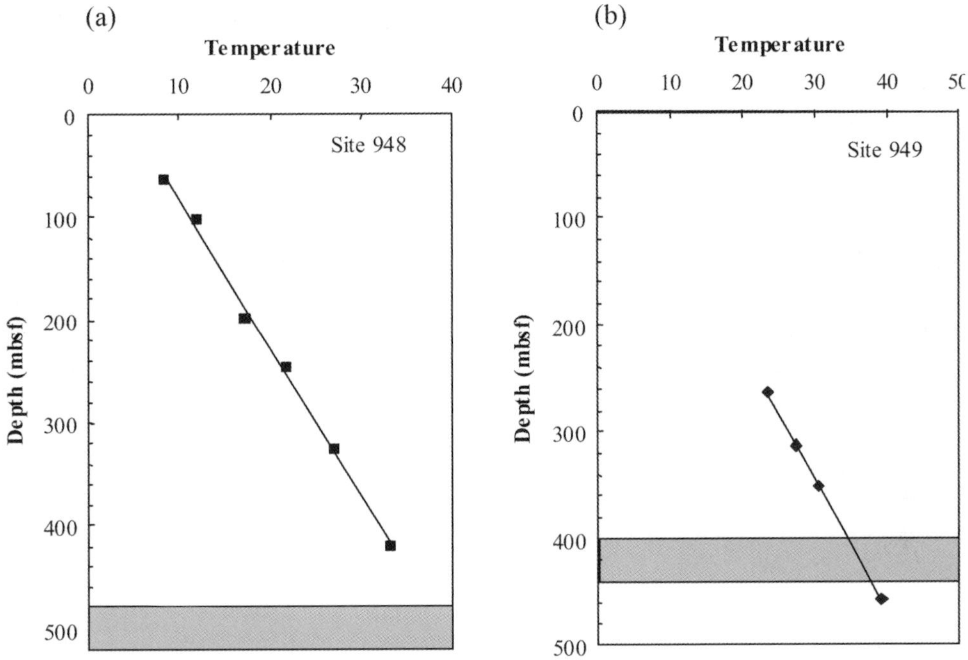

Figure 6.5 Temperature profiles for (a) Site 948 and (b) Site 949. Stippled rectangle shows inferred décollement depth from structural observations. Modified from *Henry* [2000].

the thermal gradients at the drill sites. Either fluid flow from large areas must be focused into spatially restricted conduits or some other explanation is required for the high heat flow.

Müller and Smith [1993] argue that the uniformly high background heat flow in the ODP drilling area could be due to thinning of the crust below the Tiburon Rise. Thinning of the crust probably occurred prior to 72 Ma during a period of extension along the boundary between the North and South American plates [*Müller and Smith*, 1993]. Subsequently, transpression along the plate boundary uplifted Tiburon Rise during the Tertiary [*Dixon and Mao*, 1997]. *Müller and Smith* [1993] present gravity data to show that the Moho beneath Tiburon Rise is also uplifted 2–4 km resulting in extreme thinning of the crust there. The large area of elevated heat flow with linear temperature profiles, and the excessively high fluid-flow rates required to produce the anomaly by fluid flow, support their hypothesis that the observed high heat flow in the vicinity of Tiburon Rise is due to crustal thinning rather than fluid flow.

Chloride

A pore-water chloride concentration anomaly of 20% less than seawater was found just above the décollement at Site 948 [*Shipboard Scientific Party*, 1995b].

The origin of this pore-water anomaly has been attributed to flow from deeper in the complex where temperatures are high enough to drive smectite dehydration [*Gieskes et al.*, 1989, 1990; *Vrolijk et al.*, 1991; *Blanc et al.*, 1991; *Shipboard Scientific Party*, 1995b]. Experimental work by *Fitts and Brown* [1999] indicated that much of the low-chloride pore water could be produced by water released from the smectite interlayers during squeezing for pore-water sampling. However, the two samples with the lowest chloride concentrations (fig. 6.6) could not be explained entirely by this sampling artifact. After accounting for the effect of the sample squeezing process, the low-chloride anomaly is still 12% fresher than seawater.

Other pore-water proxies for advective solute transport from higher-temperature regions deeper in the complex have been found in the Barbados décollement. *Kastner et al.* [1997] found a maximum in CH_4 at Site 948 at the same depth as the chloride minimum (~494 mbsf), close to the top of the décollement zone. Anomalous methane concentrations were also observed at nearby Site 671 [*Gieskes et al.*, 1990]. Anomalies in Br/Cl and I/Cl also occur at ~503 mbsf in Site 948, i.e., within the décollement zone [*Kastner et al.*, 1997].

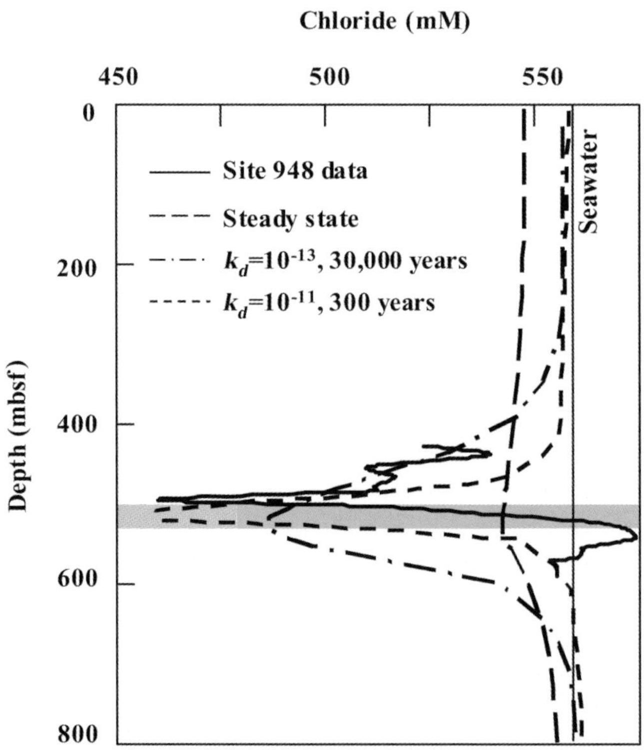

Figure 6.6 Transient chloride anomaly for two different décollement permeabilities 10^{-13} and 10^{-11} m^2, after 30,000 years and 300 years, respectively. Location of the inferred décollement between 498 and 529 mbsf is stippled and measured chloride concentrations are also plotted [*Shipboard Scientific Party*, 1995b].

You et al. [1993] report a potential maximum in boron below the base of the décollement of Site 671 at ~560 mbsf. These data may, in principle, indicate potential advection of fluids either in, or associated with, the décollement zone. The reported anomalies often do not appear to be located at the same depths, probably indicating periodicity of inputs and differing sources.

Kinetic modeling of clay dehydration in the Barbados complex predicts that the reaction takes place more than 50 km arcward of the deformation front, indicating that low-chloride fluids must migrate over 46 km laterally to reach Site 948 [*Bekins et al.*, 1994]. After accounting for potential low-chloride artifacts introduced during pore-water extraction, *Fitts and Brown* [1999] estimated that the vertical extent of the chloride anomaly is <10 m. The width and magnitude of this anomaly can be used to estimate the duration of flow and transport of the low-chloride water. The observed vertical interval of the anomaly can best be fit with transient flow lasting for ~300 years (fig. 6.6). This time limit on establishing the low-chloride pulse requires a décollement permeability of 10^{-11} m^2 and unreasonably large flow rates if the low-chloride water must travel updip for ~50 km.

Henry [2000] analyzed the trade-off between flow distance and time and proposed that the low-chloride pulse could instead have traveled only 500 m laterally, from the negative-polarity area. In this scenario, the low-chloride water would travel updip from the zone of smectite dehydration to the negative-polarity area, where the low-chloride anomaly would be greater than that observed at Site 948. It would then migrate laterally along a narrow flow zone to Site 948. The flow rate required for traveling 500 m in less than 500 years is ~1 m yr^{-1}, which is consistent with a reasonable décollement permeability value of 10^{-13} m^2. The extremely limited vertical extent of the low-chloride anomaly at Site 948 may help explain how this transient flow does not greatly impact the measured pore pressures at this site, which, as discussed above, are consistent with steady-state models.

Budget Considerations

A key aspect to understanding the fluid-flow system is to compare flow rates out of the complex to the rate that fluids enter in the pore space [*Le Pichon et al.*, 1990; *Saffer and Bekins*, 1999]. Modeling studies show that prism fluids do not contribute significantly to the décollement flow. Thus, at steady state the maximum flow that can exit the complex along the décollement or underthrust sediments is equal to the rate that fluids enter the complex in the underthrust sediments.

Integrating porosities over the average thickness of the underthrust section and multiplying by the plate convergence rate gives 5.7 m^3 yr^{-1} entering the complex in a vertical column of underthrust sediments that is 1 m wide along strike [*Saffer and Bekins*, 1999]. Although the thickness of the underthrust sediments varies by up to a factor of 2, the rate of subduction (2.8 cm yr^{-1})

and thus the rate that the entering column thickness changes is slow compared to the rate that fluids exit the complex (~1 m yr^{-1}). A significant fraction of the incoming fluid is expelled very near the toe of the complex. *Zhao and Moore* [1998] estimated that after 3.5 km of subduction ~1 m^3 had already been expelled from the upper two underthrust units (for a 1-m-wide slice along strike). Their estimate is very similar to other estimates of the fluid expelled from the underthrust section in the first 4 km. Converting their rate to cubic meters of fluid expelled per cubic meter of sediment per second gives 3×10^{-14} s^{-1}, which is comparable to the published range of $1–3 \times 10^{-14}$ s^{-1} [*Screaton et al.*, 1990; *Bekins et al.* 1995; *Saito and Goldberg*, 2001; *Stauffer and Bekins*, 2001].

Table 6.3 shows a comparison of the fluid influx with flux out for the various published flow rates. The budget for each domain is based on a hypothetical capture zone. For flow in the décollement we estimate that the top one third of the underthrust sediments drain to the décollement. These fluids will travel laterally within the décollement if its permeability is at least 10^{-14} m^2 [*Screaton et al.*, 1990; *Saffer and Bekins*, 1999]. The lower two thirds of the underthrust sediments are assumed to drain through the turbidite section [*Tokunaga*, 2000]. All budgets are figured for the fluids available arcward of Site 948, so the total incoming flux is decreased by 1 m^3 yr^{-1} [*Zhao and Moore*, 1998]. Two comparisons are given: one for outflow through a 1-m^2 area (Darcy flux) and another for total estimated outflow. For all but the estimate by *Foucher et al.* [1990] for flow out through the prism surface, total flow is computed for a 10-m-thick flow conduit with the given porosity. In general, for the range of published estimates for flow out the décollement and underthrust section, the available influx is a small fraction (0.4–11%) of the computed outflow.

To balance the fluid budget, either high-flow rates must last only a fraction of time or they must be spatially focused such that fluid draining from a large area flows out a small fraction of the toe [*Saffer and Bekins*, 1999]. Transient flow in the décollement may be related to fault movement either through the rupturing of permeability barriers or through dilation during shear failure as previously discussed [e.g., *Karig*, 1990]. In contrast, it is difficult to define a potential mechanism for transient permeability changes in the underthrust sand units, and therefore maintenance of high-flow rates in the underthrust sands would require to spatial focusing of steady-state flow.

Conceptual Models for Transient and Spatially Focused Flow

Figure 6.7 shows a series of conceptual models for balancing the fluid budget. Although table 6.3 indicates that outflow must be 20 times the inflow rate, for simplicity, the fig. 6.7 illustrates outflow rates that are 6 times the inflow rate. Two end-member methods for balancing the budget are depicted (figs. 6.7a and 6.7b). The first is spatially focused flow that drains fluid from a large area, and the second is transient flow with limited duration in time [*Saffer and Bekins*, 1999].

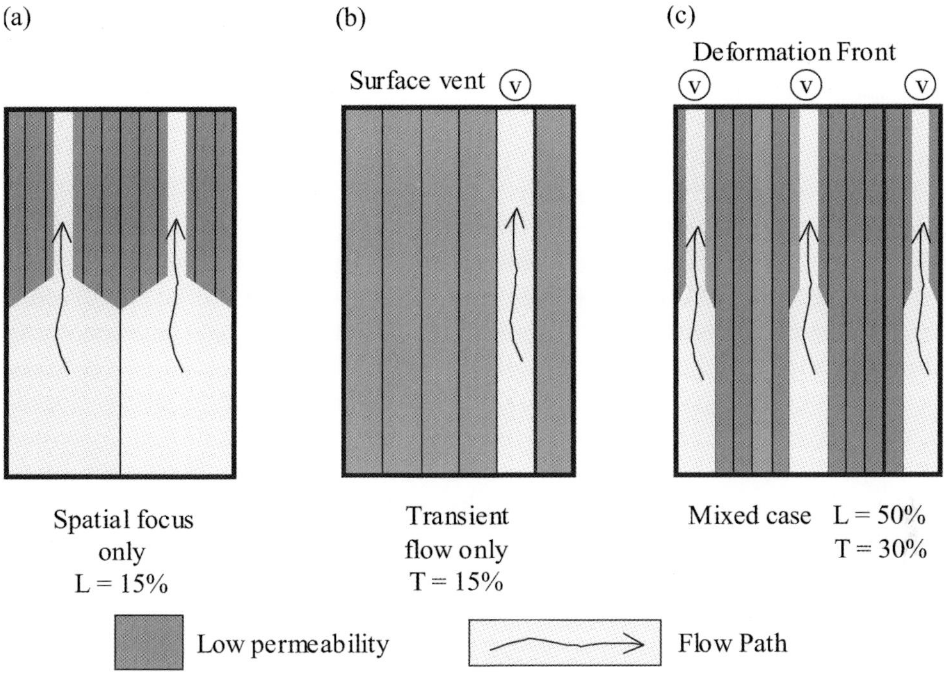

Figure 6.7 Map view cartoon of mechanisms for balancing the fluid budget for northern Barbados. (a) Steady-state flow from a large area at depth is focused into a narrow-flow conduit near the trench. This scenario is envisioned as the most likely way to produce high-flow rates in the underthrust turbidite sands. For focusing by a factor of 6, the percentage of distance along strike hosting high-flow rates, L, spans 16% of the trench. (b) Transient high-flow rates occur through an enhanced permeability, dilated portion of the fault, and the dilated area changes locations with time. To increase flow by a factor of 6 over steady-state values, the percent of time that the fault is dilated, T, is 16% (c) Mixed case where a flow rate equal to six times the steady-state rate is produced by focusing of flow out a conduit that narrows by 50% and is dilated 30% of the time. The location of the dilated portion migrates over time. This scenario is proposed for enhanced flow rates along the décollement in the negative-polarity area.

Spatial Focusing Figure 6.7a illustrates a scenario of spatial focusing of steady-state flow. This process is proposed as a way to achieve the high steady-state flow rates in the turbidite units that are postulated to explain the thermal data. The hypothesized mechanism for spatial focusing of flow is vertical or horizontal narrowing of high permeability units that host flow from depth in the complex. In this scenario, high pressures and flow rates are maintained continuously because flow paths out of the complex are constricted. Focusing in sands could occur due to disruption by faults [*DiLeonardo et al.*, 2002] or stratigraphic pinch-outs. For a 10-m-thick conduit with steady-state flow, table 6.3 indicates that lateral focusing must be at least a factor of 20. In other words, fluid from a 1-km-wide area at depth must be focused into a 50-m-wide area at the deformation front. Other possible mechanisms for increasing flow velocity without increasing the quantity of fluid include narrowing the vertical thickness of the conduit or reducing the porosity in the sand layers. Porosity reduction causes an increase in flow velocity because the same amount of fluid must pass through a smaller area of pore space.

Observations of the distribution of the turbidite unit in the underthrust sediments are consistent with both lateral and vertical focusing of flow. In the main drilling area, where there is elevated heat flow, the incoming sediment column is 672 m thick with ~100 m of turbidites (Site 1044). At the DSDP reference site, located 20 km north of the other drill sites (Site 543), the incoming sediment column is 411 m thick, with no turbidite layers, and elevated heat flow does not occur [*Henry*, 2000]. The thinner sediments at Site 543 are deposited on the west side of a basement high that runs parallel to the deformation front, extending ~50 northward from the Tiburon Rise [*Mauffret et al.*, 1984]. Pinch-out of turbidites at basement highs could lead to substantial lateral focusing of flow. Vertical focusing of flow is indicated by the 3-D seismic images, which show at least 50-m offset in the turbidites along normal faults extending upward from the basement [*Zhao and Moore*, 1998]. Whether these combined effects can result in focusing by a factor of 20 is not clear.

A number of authors have noted that the outflow from the underthrust seems excessive compared to the inflow [*Foucher et al.*, 1990; *Le Pichon et al.*, 1990; *Fisher and Hounslow*, 1990; *Saffer and Bekins*, 1999; *Cutillo et al.*, 2003]. The widespread agreement in the modeling estimates and the continued inability to account for the source of the high fluid-flow rates argues for revisiting alternate hypotheses for the excess heat flow observed at Sites 672, 948, and 949. One hypothesis that should be explored further is that thinning of the crust under Tiburon Rise causes the regionally high heat flow [*Müller and Smith*, 1993]. A comparison of the heat-flow data to the expected value and spatial extent of elevated conductive heat flow from this effect would be very useful.

Transient Focusing Figure 6.7b depicts a cartoon of a purely transient scenario to balance the budget. This cartoon applies to the décollement because the décollement is subject to shear-strain events that may cause transient increases in permeability resulting in increased flow rates. A narrow slice of the décollement, perpendicular to the deformation front, is depicted as dilated and flowing part of the time and closed for the remainder of the time. When the section is closed, fluids are stored and pressure builds in the underthrust section. When the section is open, the higher permeability results in higher flow rates for a short period of time. The simplistic budget of table 6.3 requires that each section can only be dilated 5% of the time. Although episodic flow is widely recognized in active tectonic systems [e.g., *Sibson*, 1981], the causes are still not well understood.

Combined Spatial and Transient Focusing The third case illustrated in fig. 6.7c represents a mixed scenario in which fluids are both spatially and temporally focused in the décollement. We feel that this scenario best matches the observations from the drilling and 3-D seismic survey of the northern Barbados décollement. Transient flow best explains the chloride data, and transient pressure propagation is needed to produce the very high inferred pressures in the negative-polarity area. In addition, spatial focusing is also indicated by the changing width of the negative-polarity area from ~1 km at the deformation front to ~2 km at the back of the surveyed area 10 km arcward (fig. 6.3). *DiLeonardo et al.* [2002] also showed that fluid flow is focused into channels along the décollement by cross-cutting strike-slip faults. Growth in the width of the negative-polarity area arcward is reasonable because the deeper sediments are more lithified and able to sustain open fractures [e.g., *Townend and Zoback*, 2000] compared to shallow depths where clay-rich sediments behave more plastically [e.g., *Chen and Baladi*, 1985]. Continuing this trend 50 km arcward to the smectite dehydration zone results in spatial focusing by a factor of 5. In a combined scenario of spatial focusing and transient flow, fault dilation for 25% of the time achieves a factor of 20 increase in flux.

Foucher et al. [1990] examined the flow rates required to produce the elevated heat flow observed in the first two thrust packets in the prism (table 6.3). They found that only a fivefold increase in outflow over inflow was required to explain these data. They speculated that both spatial and transient focusing of flow out thrust faults in the prism could produce this effect. *Brown et al.* [1994] proposed a scenario of both temporal and spatial variability of flow conduits in the décollement on the basis of permeability data. They noted that the distribution of dilated sections and associated high pore pressures must be spatially heterogeneous to maintain a critical wedge taper. Our summary based on pore-water data, pressure observations, modeling results, and budget computations similarly indicates that spatial and temporal heterogeneity are the best explanation for the pattern of flow along the northern Barbados décollement.

Summary

The low permeability of the incoming fine-grained sediment sequence at the northern Barbados accretionary complex results in rapid pore-pressure increases by 5 km arcward of the deformation front. Simulated steady-state fluid pressures are consistent with observed excess pressures increasing from ~1 MPa ($\lambda^* = 0.36$) measured just 1.8 km arcward of the deformation front to ~2 MPa ($\lambda^* = 0.50$) measured 4 km arcward. A 1–2 km-wide area that extends from the deformation front to at least 10 km arcward requires even higher excess pressures of $\lambda^* = 0.73$ at Site 1045 located 2.7 km arcward of the deformation front. Such high overpressures probably require transient flow from depth along the décollement. Low-chloride pore-water anomalies along the décollement also reflect transient flow. In contrast, observed temperature profiles near the toe are most consistent with steady-state flow of warm pore waters along sand layers in the underthrust sediments. Reconciling rates of flow necessary for the temperature and chloride anomalies with fluid budgets based on the rates pore-fluid subduction indicates spatial and temporal focusing of flow by a factor of at least 20. Whether this degree of focusing is reasonable requires greater understanding of the distribution of flow conduits and permeabilities in the décollement and underthrust.

Acknowledgments

This research used data provided by the Ocean Drilling Program (ODP). ODP is sponsored by the U.S. National Science Foundation (NSF) and participating countries under management of Joint Oceanographic Institutions (JOI), Inc. Screaton's Barbados research was funded by NSF OCE-9810959. We also wish to thank Nathan Bangs, Kevin Brown, Joris Gieskes, and Dorit Matmon for valuable reviews and discussions of the material in the manuscript. Figures 1, 4, and 5 were reproduced/modified by permission of American Geophysical Union.

References

Bangs, N. L., T. H. Shipley, and G. F. Moore (1996), Elevated fluid pressure and fault zone dilation inferred from seismic models of the northern Barbados Ridge décollement, *J. Geophys. Res., 101*, 627–642.

Bangs, N. L., T. H. Shipley, J. C. Moore, and G. F. Moore (1999), Fluid accumulation and channeling along the northern Barbados Ridge décollement thrust, *J. Geophys. Res., 104*, 20,399–20,414.

Becker, K., A. T. Fisher, and E. E. Davis (1997), The CORK experiment in Hole 949C: Long-term observations of pressure and temperature in the Barbados accretionary prism, *Proc. Ocean Drill. Program Sci. Results, 156*, 247–252.

Bekins, B. A., A. M. McCaffrey, and S. J. Dreiss (1994), The influence of kinetics on the smectite to illite transition in the Barbados accretionary prism, *J. Geophys. Res., 99*, 18,147–18,158.

Bekins, B. A., A. M. McCaffrey, and S. J. Dreiss (1995), Modeling the origin of low chloride pore waters in a modern accretionary complex, *Water Resour. Res., 31*, 3205–3215.

Blanc G., C. Doussan, C. Thomas, and J. Boulegue (1991), Non-steady state diffusion and advection model of transient concentration-depth profiles from the Barbados accretionary complex, *Oceanol. Acta, 16*, 363–372.

Brown, K. M., B. A. Bekins, B. Clennell, D. Dewhurst, and G. Westbrook (1994), Heterogeneous hydrofracture development and accretionary fault dynamics, *Geology, 22*, 259–262.

Chen, W. F., and G. Y. Baladi (1985), *Soil Plasticity*, 234 pp., Elsevier, New York.

Cutillo, P. A., E. J. Screaton, and S. Ge (2003), Three-dimensional numerical simulation of fluid flow and heat transport within the Barbados Ridge accretionary complex, *J. Geophys. Res., 108*(B12), 2555, doi:10.1029/2002JB002240.

Davis, D., J. Suppe, and F. A. Dahlen (1983), Mechanics of fold-and-thrust belts and accretionary wedges, *J. Geophys. Res., 88*, 1143–1172.

Deming, D. (2002), *Introduction to Hydrogeology*, 468 pp., McGraw-Hill, New York.

DiLeonardo, C. G., J. C. Moore, S. Nissen, and N. Bangs (2002), Control of internal structure and fluid migration pathways within the Barbados Ridge décollement zone by strike-slip faulting: Evidence from coherence and three dimensional seismic amplitude imaging, *Geol. Soc. Am. Bull.*, *114*, 51–63.

Dixon, T. H., and A. Mao (1997), A GPS estimate of relative motion between North and South America, *Geophys. Res. Lett.*, *24*, 535–538.

Dixon, T. H., F. Farina, C. DeMets, P. Jansma, P. Mann, and E. Calais (1998), Relative motion between the Caribbean and North American plates and related boundary zone deformation from a decade of GPS observations, *J. Geophys. Res.*, *103*, 15,157–15,182.

Ferguson, I. J., G. K. Westbrook, and M. G. Langseth (1993), Heat flow and thermal models of the Barbados Ridge accretionary complex, *J. Geophys. Res.*, *98*, 4121–4142.

Fisher, A. T., and M. W. Hounslow (1990), Heat flow through the toe of the Barbados accretionary complex: constraints from ODP Leg 110 and simple models, *J. Geophys. Res.*, *95*, 8845–8858.

Fisher, A. T., and G. Zwart (1996), Relationship between permeability and effective stress along a plate boundary fault, Barbados accretionary complex, *Geology*, *24*, 307–310.

Fitts, T. J., and K. M Brown (1999), Stress-induced smectite dehydration: ramifications for patterns of freshening and fluid expulsion in the N. Barbados accretionary wedge, *Earth Planet. Sci. Lett.*, *172*, 179–197.

Foucher, J. P., X. Le Pichon, S. Lallement, M. A. Hobart, P. Henry, M. Benedetti, G. K. Westbrook, and M. G. Langseth (1990), Heat flow, tectonics, and fluid circulation at the toe of the Barbados Ridge accretionary prism, *J. Geophys. Res.*, *95*, 8859–8867.

Foucher, J. P., P. Henry, and F. Harmegnies (1997), Long-term observations of pressure and temperature in ODP Hole 948D, Barbados accretionary prism, *Proc. Ocean Drilling Program Sci. Results*, *156*, 229–238.

Gieskes, J., et al. (1989), Hydrogeochemistry in the Barbados accretionary complex; Leg 110 ODP, *Palaeogeogr. Palaeoclimatol. Palaeoecol.*, *71*, 83–96.

Gieskes, J. M., P. Vrolijk, and G. Blanc (1990), Hydrogeochemistry of the Northern Barbados Accretionary Complex transect: Ocean Drilling Project Leg 110, *J. Geophys. Res.*, *95*, 8809–8818.

Henry, P. (2000), Fluid flow at the toe of the Barbados accretionary wedge constrained by thermal, chemical, and hydrogeologic observations and models, *J. Geophys. Res.*, *105*, 25,855–25,872.

Karig, D. E. (1990), Experimental and observational constraints on the mechanical behavior in the toes of accretionary prisms, in *Deformation Mechanisms, Rheology and Tectonics*, edited by R. J. Knipe and E. H. Rutter, *Geol. Soc. Spec. Publ. London*, *54*, 383–398.

Kastner, M., Y. Zheng, L. Troels, W. Jenkins, and T. Ito (1997), Geochemistry of fluids and flow regime in the décollement zone at the northern Barbados Ridge, in *Proc. Ocean Drill. Program Sci. Results*, *156*, 311–319.

Langseth, M. G., G. K. Westbrook, and M. Hobart (1990), Contrasting geothermal regimes of the Barbados Ridge accretionary complex, *J. Geophys. Res.*, *95*, 8829–8843.

Le Pichon, X., P. Henry, and S. Lallement (1990), Water flow in the Barbados accretionary complex, *J. Geophys. Res.*, *95*, 8945–8967.

Mauffret, A., G. K. Westbrook, M. Truchan, and J. Ladd (1984), The relief of the oceanic basement and the structure of the front of the accretionary complex in the region of Sites 541, 542, and 543, edited by B. Biju-Duval et al., *Initial Rep. Deep Sea Drill. Proj.*, *78A*, 49–62.

Moore, J. C., et al. (1995), Abnormal fluid pressures and fault-zone dilation in the Barbados accretionary prism: Evidence from logging while drilling, *Geology*, *23*(7), 605–608.

Moore, J. C., et al. (1998), Consolidation patterns during initiation and evolution of a plate-boundary décollement zone: Northern Barbados accretionary prism, *Geology*, *26*(9), 811–814.

Müller, R. D., and W. H. F. Smith (1993), Deformation of the oceanic crust between the North American and South American plates, *J. Geophys. Res.*, *98*, 8275–8291.

Saffer, D. M., and B. A. Bekins (1999), Fluid budgets at convergent plate margins: Implications for the extent and duration of fault-zone dilation, *Geology*, *27*(12), 1095–1098.

Saito, S., and D. Goldberg (2001), Compaction and dewatering processes of the oceanic sediments in the Costa Rica and Barbados subduction zones: Estimates from in situ physical property measurements, *Earth Planet. Sci. Lett., 191,* 283–293.

Screaton, E. J., and S. Ge (1997), An assessment of along-strike fluid and heat transport within the Barbados Ridge accretionary complex: Results of preliminary modeling, *Geophys. Res. Lett., 24,* 3085–3088.

Screaton, E. J., and S. Ge (2000), Anomalously high porosities in the proto-décollement zone of the Barbados accretionary complex: Do they indicate overpressures?, *Geophys. Res. Lett., 27,* 1993–1996.

Screaton, E. J., D. R. Wuthrich, and S. J. Dreiss (1990), Permeabilities, fluid pressures, and flow rates in the Barbados Ridge Complex, *J. Geophys. Res., 95,* 8997–9007.

Screaton, E. J., B. Carson, E. Davis, and K. Becker (2000), Permeability of a décollement zone: Results from a two-well experiment in the Barbados accretionary complex, *J. Geophys. Res., 105,* 21,403–21,410.

Sibson, R. H. (1981), Fluid flow accompanying faulting: Field evidence and models, in *Earthquake Prediction: An International Review, Maurice Ewing Ser.*, vol. 4, edited by D. W. Simpson and P. G. Richards, pp. 593–603, AGU, Washington D. C.

Shi, Y., and C-Y. Wang (1988), Generation of high pore pressures in accretionary prisms: Inferences from the Barbados subduction complex, *J. Geophys. Res., 93,* 8893–8910.

Shipboard Scientific Party (1988), Site 671, *Proc. Ocean Drill. Program Initial Rep., 110,* 67–204.

Shipboard Scientific Party (1995a), Site 947, *Proc. Ocean Drill. Program Initial Rep., 156,* 71–86.

Shipboard Scientific Party (1995b), Site 948, *Proc. Ocean Drill. Program Initial Rep., 156,* 87–192.

Shipboard Scientific Party (1998), Site 1045, *Proc. Ocean Drill. Program Initial Rep., 171A,* 39–54.

Shipley, T. H., G. F. Moore, N. Bangs, J. C. Moore, and P. L. Stoffa (1994), Seismically inferred dilatancy distribution, northern Barbados Ridge décollement: Implications for fluid migration and fault strength, *Geology, 22*(5), 411–414.

Stauffer, P., and B. A. Bekins (2001), Modeling consolidation and dewatering near the toe of the northern Barbados accretionary complex, *J. Geophys. Res., 106,* 6369–6383.

Taylor, E., and J. Leonard (1990), Sediment consolidation and permeability at the Barbados forearc, *Proc. Ocean Drill. Program Sci. Rep., 110,* 289–308.

Tokunaga, T. (2000), The role of turbidites on compaction and dewatering of underthrust sediments at the toe of the northern Barbados accretionary prism: New evidence from Logging While Drilling, ODP Leg 171A, *Earth Planet. Sci. Lett., 178,* 385–395.

Townend, J., and M. D. Zoback (2000), How faulting keeps the crust strong, *Geology, 28*(5), 399–402.

Vrolijk, P., A. Fisher, and J. Gieskes (1991), Geochemical and geothermal evidence for fluid migration in the Barbados accretionary prism, ODP Leg 110, *Geophys. Res. Lett., 18,* 947–950.

Westbrook, G. K., J. W. Ladd, P. Buhl, N. Bangs, and G. J. Tiley (1988), Cross section of an accretionary wedge: Barbados Ridge complex, *Geology, 16*(7), 631–635.

You, C. F., A. J. Spivack, J. H.Smith, and J. M. Gieskes (1993), Mobilization of boron in convergent margins; implications for the boron geochemical cycle, *Geology, 21*(3), 207–210.

Wood, D. M. (1990), *Soil Behaviour and Critical State Soil Mechanics*, 462 pp., Cambridge Univ. Press, New York.

Zhao, Z., and G. F. Moore (1998), Deformation and dewatering of the subducting plate beneath the lower slope of the northern Barbados accretionary prism, *J. Geophys. Res., 103,* 30,431–30,449.

Zwart, G., W. Brückmann, K. Moran, A. K. MacKillop, A. J. Maltman, A. Bolton, P. Vrolijk, T. Miller, M. J. Gooch, and A. Fisher (1997), Evaluation of hydrogeologic properties of the Barbados accretionary prism: A synthesis of Leg 156 results, *Proc. Ocean Drilling Program Sci. Results, 156,* 303–310.

7

Pore Pressure within Underthrust Sediment in Subduction Zones

Demian M. Saffer

Abstract

Fluid-pressure distribution within subduction zones profoundly affects mechanical processes and chemical transport; yet few direct measurements of this important quantity exist. In underthrust sediment that undergoes progressive uniaxial (vertical) consolidation, in situ pore pressure can be estimated indirectly by several methods, including laboratory consolidation tests on core samples, observed compaction trends in boreholes, and compaction state inferred from seismic velocity. Here, I review these methods, highlighting with examples from the Nankai and Costa Rica subduction zones. The different methods yield consistent results at each margin and indicate development of significant overpressures. Patterns of inferred pore pressure are most consistent with upward drainage to a permeable décollement. The tendency for vertical drainage increases with increased distance from the trench. Inferred pore pressures are generally consistent with simple models of consolidation, which demonstrate that excess fluid pressures result from a balance between loading rate and sediment permeability. These results hold potentially important implications for the role of sediment and fault permeability in affecting fault strength and the updip limit of the seismogenic zone via their control on pore pressure and effective stress. However, significant uncertainty exists in projecting shallow results landward; more detailed investigations are clearly needed to better understand the development of pore pressures and fault strength at greater depths.

Introduction

Fluid pressure exerts a fundamental control on a wide range of faulting processes at both subduction zones and in continental settings, including: fault strength [e.g., *Rice*, 1992; *Hubbert and Rubey*, 1959], rupture propagation [e.g., *Johnson and McEvilly*, 1995], strain localization [e.g., *Moore and Byrne*, 1987], and the taper angle of orogenic wedges [e.g., *Hubbert and Rubey*, 1959; *Davis et al.*, 1983]. In addition, pore pressure has been postulated to influence the updip limit of seismogenic faulting through its control on effective stress and consolidation state [e.g., *Moore and Saffer*, 2001; *Scholz*, 1998]. In this respect, quantifying fluid pressure and its distribution are central aspects of understanding subduction zone fault mechanics. The compaction state and fluid-flow patterns associated with subsurface pore pressure are also important for fluid and solute mass fluxes [e.g., *Bekins et al.*, 1995]. Fluids are important agents in the flux of dissolved chemical species through subduction zones; thus quantifying these fluxes in the upper portion of subduction zones is a fundamental step toward understanding volatile budgets at convergent margins.

At subduction zones, pore pressures significantly in excess of hydrostatic result from rapid burial and tectonic loading, combined with low permeabilities typical of marine sediment [e.g., *von Huene and Lee*, 1982; *Moore and Vrolijk*, 1992; *Saffer and Bekins*, 2002]. In general, direct and indirect constraints on pore-pressure magnitude in active subduction zones are rare. Where estimates of pore pressure exist, they are often limited to the shallow portions of the subduction system near the trench (<1.5 km below seafloor (bsf)) [e.g., *Saffer*, 2003]. In a few cases, direct measurements of pore pressure have been obtained from long-term monitoring within sealed intervals (Circulation Obviation Retrofit Kit (CORK) installations) [e.g., *Foucher et al.*, 1997; *Becker et al.*, 1997]. At a larger scale, critical taper theory also provides a means to indirectly estimate average pore pressure in active accretionary complexes but requires that the coefficient of friction for wedge and décollement materials be defined independently [e.g., *Davis et al.*, 1983; *Brown et al.*, 2003] and does not provide detailed constraints on the spatial distribution of pore pressure.

Commonly, in situ pore pressure is calculated from effective stress inferred from (1) index properties data observed in boreholes (porosity η, or void ratio [$e = \eta/(1 - \eta)$]) [e.g., *Rubey and Hubbert*, 1959; *Hart et al.*, 1995], (2) e or η inferred from compressional wave velocity [e.g., *Cochrane et al.*, 1994, 1996], or (3) laboratory consolidation experiments that define a maximum past effective stress (P'_c) [e.g., *Karig*, 1993; *Saffer et al.*, 2000]. These techniques are best suited to pore-pressure estimation in underthrust sediment, which typically undergoes uniaxial vertical strain in response to loading by subduction [e.g., *Housen et al.*, 1996; *Saffer*, 2003]. In an accreting sediment wedge, where strain and stress are generally non-uniaxial [e.g., *Henry et al.*, 2003], pressure estimation from compaction state [e.g., *Flemings*, 2002] or laboratory experiments [e.g., *Morgan and Ask*, 2004] is subject to considerable uncertainty because several important aspects of deformation are poorly constrained (see discussion in the work of *Morgan and Ask* (this volume). These include (1) the relative magnitude of lateral and vertical stresses, (2) the coefficient of lateral stress (K_0) in the sediment's initial state prior to incorporation into the wedge, (3) the critical state failure criterion for sediment, and (4) anisotropy of sediment compressibility.

From a mechanical viewpoint, the fluid content and physical properties of sediment underthrust at the trench are especially important because the main décollement localizes immediately above this sediment, and at many subduction zones steps downward into it [e.g., *Saffer*, 2003; *Moore, G. F. et al.*, 2001]. Fluid expulsion from underthrust sediment is also larger than that from sediment incorporated into the overriding wedge and thus has a primary impact on fluid budgets. This occurs because the underthrust sediment is transported arcward rapidly relative to compacting sediment in the overlying wedge [e.g., *Bekins and Dreiss*, 1992; *Saffer and Bekins*, 1998].

Studies using indirect methods have documented the development of elevated pore pressures within underthrust sediment at several subduction zones. For example, at the Nankai margin, *Screaton et al.* [2002] compared porosity-depth profiles at underthrust sites with a reference site to estimate depth-averaged pore pressure. At the Costa Rican and Nankai subduction zones, the evolution of excess pore pressures has been investigated by combining laboratory consolidation tests and borehole porosity data [e.g., *Saffer et al.*, 2000; *Saffer*, 2003] and by using seismic reflection data to constrain porosity and consolidation of sediment layers [*Bray and Karig*, 1988]. At the Cascadia subduction zone, *Cochrane et al.* [1996] inverted compressional wave velocity for porosity and then inferred pore pressure from porosity-depth profiles.

At both Costa Rica and Nankai, Ocean Drilling Program (ODP) drilling penetrated the incoming sediment at a reference site and through the décollement and underthrust sediment at two sites landward of the trench; core data and samples for laboratory consolidation tests were collected at both margins (figs. 7.1a and 7.1b). At both sites, CORKs were emplaced on recent ODP Legs; data from these instruments may ultimately provide ground-truthing for pore-pressure estimates discussed in this paper. Here, I review indirect methods for estimating pore pressure within underthrust sediment, highlighting with examples from the Costa Rican and eastern Nankai subduction zones. I then discuss (1) the hydrologic and mechanical implications of inferred overpressures at several convergent margins and (2) additional insights into factors controlling the generation and maintenance of excess pore pressures that can be gained from simple models.

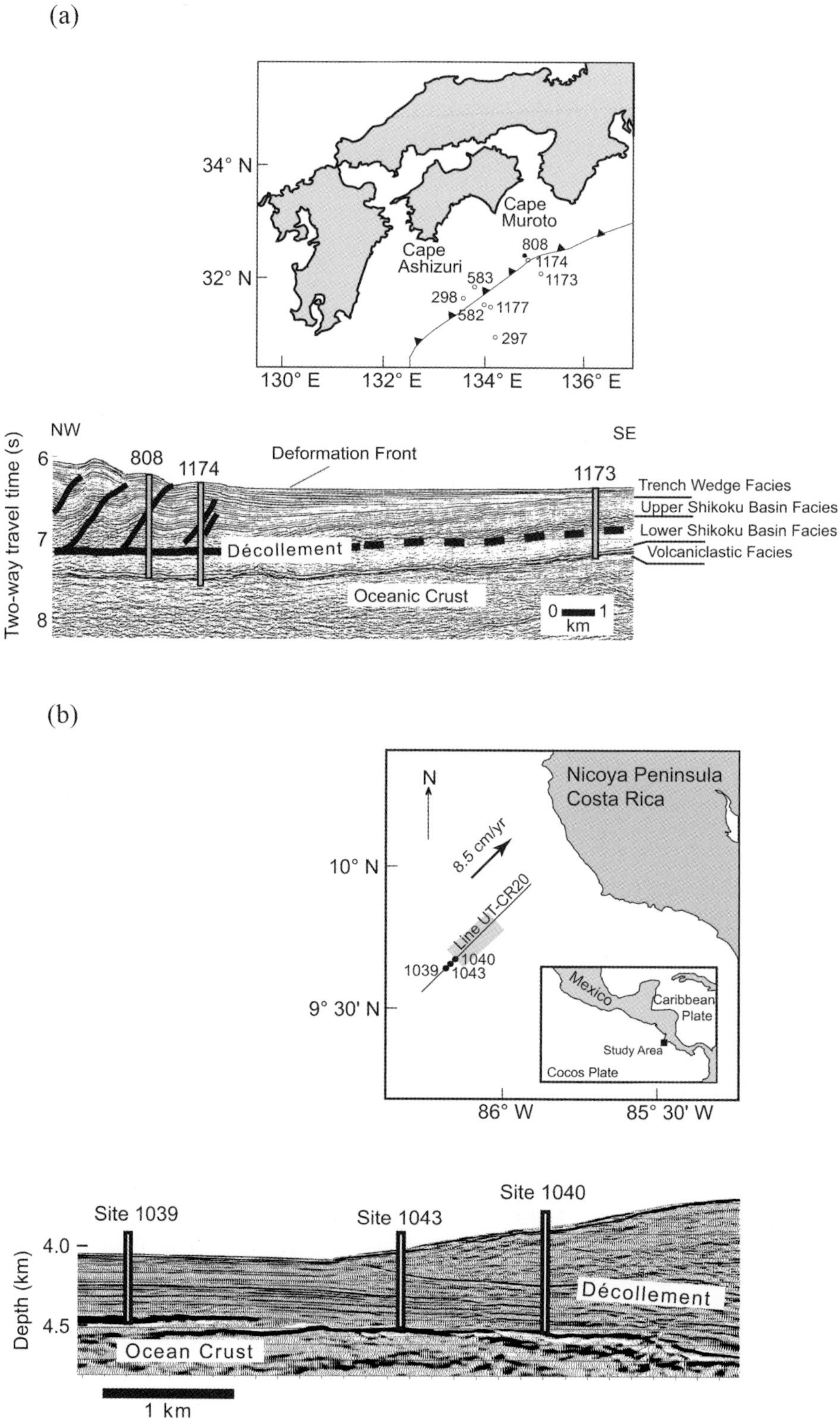
(a)
34° N
32° N
130° E
132° E
134° E
136° E
Cape Muroto
Cape Ashizuri
808
1174
1173
583
298
1177
582
297
NW
SE
Two-way travel time (s)
6
7
8
808
1174
Deformation Front
1173
Décollement
Oceanic Crust
0 1 km
Trench Wedge Facies
Upper Shikoku Basin Facies
Lower Shikoku Basin Facies
Volcaniclastic Facies
(b)
N
Nicoya Peninsula Costa Rica
8.5 cm/yr
10° N
9° 30' N
Line UT-CR20
1039
1040
1043
Mexico
Caribbean Plate
Study Area
Cocos Plate
86° W
85° 30' W
Site 1039
Site 1043
Site 1040
Depth (km)
4.0
4.5
Décollement
Ocean Crust
1 km

Figure 7.1 (a) Map of the Nankai Trench area, showing locations of ODP drill sites, and seismic line across the deformation front offshore Cape Muroto, showing general stratigraphy and drill site locations from ODP Legs 131 and 190. (b) Map of ODP Leg 170 and 205 study area offshore the Nicoya Peninsula, Costa Rica, and seismic line UT-CR20 across the trench. After Saffer [2003]. Note the marked thinning of incoming sediment in response to underthrusting.

Terminology for Reporting Pore Pressure and Consolidation Behavior

The symbols and nomenclature used throughout this paper are listed in table 7.1. Pore pressure, in particular, is reported in a variety of ways in existing literature, which can lead to some confusion when comparing estimates from different settings or different depths below seafloor. In order to remain consistent with previous work, and to facilitate comparison of pore pressures reported from different margins, I report excess pore pressures in four ways (tables 7.1 and 7.2): (1) excess fluid pressure, equal to the pressure above hydrostatic (in Pa); (2) pore-pressure ratio, λ (dimensionless); (3) modified pore-pressure ratio, λ^* (dimensionless); and (4) underconsolidation ratio, U (dimensionless). The pore pressure ratios are defined by

$$\lambda = \frac{P_f}{P_l} \tag{7.1a}$$

$$\lambda^* = \frac{(P_f - P_h)}{(P_l - P_h)} \tag{7.1b}$$

where P_f, P_l, and P_h are the fluid pressure, lithostatic pressure, and hydrostatic pressure (with all three values adjusted by subtracting the hydrostatic pressure associated with ocean depth), respectively (table 7.1) [e.g., *Davis et al.*, 1983; *Shi and Wang*, 1988].

Many mechanical formulations, such as critical taper theory, consider the geometry and stresses within subduction-accretion complexes in terms of the pore pressure ratio (λ). Although it is widely used in mechanical analyses, comparison of overpressures reported in terms of λ between sites is problematic because λ varies significantly as a function of sediment bulk density and thus as a function of depth. For example, at shallow depths, hydrostatic pore pressures can result in large values of λ (>0.6) simply because the bulk density of shallow sediment is low (1.2–1.4 kg m^{-3}). In this respect, the modified pore-pressure ratio (λ^*) is a more useful measure of excess pressure because it accounts for the pore pressure in excess of hydrostatic. A value for λ^* of zero indicates hydrostatic fluid pressure (zero excess fluid pressure), whereas

TABLE 7.1 **Notation**

Symbol	Description	Units
C_c	compression index ($\delta e/\log \delta \sigma'_v$)	no units
C_v	coefficient of consolidation	L^2/T
e	void ratio	dimensionless
e_i	initial void ratio	dimensionless
e_o	reference void ratio (at $\sigma'_v = 1$ kPa as used in this manuscript)	dimensionless
H	height of underthrusting sediment section	L
k	intrinsic permeability	L^2
K	hydraulic conductivity	L/T
m_v	volumetric compressibility	LT^2/M
P'_c	preconsolidation stress	M/LT^2
P_f	fluid pressure	M/LT^2
P_h	hydrostatic pressure	M/LT^2
P_l	lithostatic pressure	M/LT^2
t	time	T
U	underconsolidation ratio	dimensionless
v_p	plate velocity	L/T
x	distance arcward from trench	L
α	surface slope of margin wedge	radians
β	dip of décollement surface or subducting slab	radians
γ_w	unit weight of water	M/L^2T^2
η	fractional porosity	dimensionless
λ	pore pressure ratio (P_f/P_l)	dimensionless
λ^*	normalized pore pressure ratio ($[P_f - P_h]/[P_l - P_h]$)	dimensionless
σ_i	initial effective stress	M/LT^2
σ_v	total vertical stress	M/LT^2
σ'_v	effective vertical stress	M/LT^2
μ	dynamic viscosity	M/LT
g	gravitational acceleration	L/T^2
ρ_w	water density	M/L^3

a value for λ^* of unity indicates fluid pressures equal to the overburden and an effective normal stress of zero acting on horizontal planes.

Drainage in response to imposed loading can also be expressed as degree of underconsolidation (U), defined here as the ratio of excess pore pressure to the undrained excess pore pressure. This is sometimes also termed "percent underconsolidation" [e.g., *Lambe and Whitman*, 1969; *Holtz and Kovacs*, 1981; *von Huene and Lee*, 1982] and is equal to the mean excess pore pressure in a consolidating layer, divided by the increase in effective stress at its top caused by loading. Values of U range from zero for fully drained conditions (hydrostatic pressure) to unity for undrained loading. It is important to note that a value of $U = 1.0$ indicates undrained pore pressure increase in response to loading, but because the incoming sediment is characterized by a finite thickness prior to loading, this typically corresponds to values of λ and $\lambda^* < 1.0$.

Sediment Mechanics Approaches

In the following sections, I describe several methods grounded in soil mechanics that are commonly used to estimate in situ vertical effective stress (σ'_v). These include laboratory consolidation tests, analysis of borehole porosity-depth trends, and combined analyses of porosity inferred from seismic velocity with porosity-depth trends defined from borehole data. With downhole measurements of bulk density, which allow calculation of the total overburden stress (σ_v), estimates of σ'_v derived by the methods discussed below can be used to infer in situ pore fluid pressure (P_f) by

$$P_f = \sigma_v - \sigma'_v \tag{7.2}$$

These methods for estimating in situ effective stress, and the use of equation (7.2) to derive pore pressure from effective stress, are applicable where strain is one-dimensional (i.e., under conditions of uniaxial consolidation).

Laboratory Consolidation Tests

When reconsolidated in the laboratory, the stress-strain behavior of a sediment sample is controlled by its compressibility and stress history. When a volume of sediment or soil is unloaded (i.e., the effective stress acting on it is decreased), it experiences an increase in void ratio. The increase in void ratio, or rebound, as a function of stress decrease is described by a primarily elastic unloading curve [e.g., *Lambe and Whitman*, 1969]. Unloading typically occurs by removing material from its buried state but can also be caused in situ by an increase in fluid pressure not accompanied by an increase in total stress.

Upon reconsolidation in the laboratory a sample undergoes reduction in void ratio with increased effective stress closely following the rebound curve (also known as a reloading or reconsolidation curve) until reaching loads exceeding its maximum past effective stress, or preconsolidation stress (P'_c) [e.g., *Casagrande*, 1936; *Lambe and Whitman*, 1969]. At stresses beyond P'_c the sample deforms plastically along a virgin consolidation curve (fig. 7.2). The

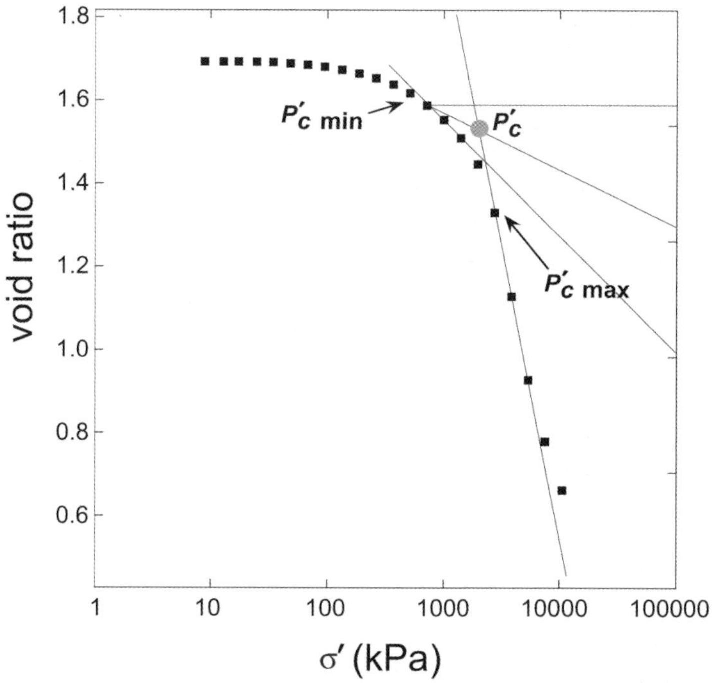

Figure 7.2 Example uniaxial consolidation test result for a sample from ODP Site 1173, showing graphical method for determination of P'_c [*Casagrande*, 1936]. Estimates of $P'_{c\ min}$ and $P'_{c\ max}$ reflect stresses at which laboratory consolidation behavior no longer plots on the reloading curve ($P'_{c\ min}$) or the virgin compression curve ($P'_{c\ max}$).

slope of the virgin consolidation curve describes the stress-strain behavior expected for undisturbed sediment (i.e., not unloaded by the coring process). The virgin consolidation curve is defined by

$$e = e_o - C_c \cdot \log(\sigma'_v) \tag{7.3}$$

where e is void ratio, e_o is a reference void ratio (defined at $\sigma'_v = 1$ kPa), C_c is the compression index, and σ'_v is vertical effective stress [e.g., *Lambe and Whitman*, 1969; *Holtz and Kovacs*, 1981]. The slopes of both the elastic (reloading or rebound) and plastic (virgin) portions of the stress-strain curve from laboratory tests provide important information about sediment compressibility.

The results of laboratory experiments can be used in two ways to obtain estimates of in situ effective stress within underthrust sediment. First, P'_c defined from the stress-strain behavior of an individual sample represents an estimate of maximum past effective stress, which can be interpreted as the actual in situ effective stress if the sample has been subjected to a monotonically increasing load in the field [e.g., *Moore*, 1989]. Second, using the slope of the virgin compression curve defined in experiments, observed changes in void ratio within correlated sediment packages can be used to estimate effective stress (fig. 7.3) [e.g., *Saffer*, 2003]. The latter technique requires that (1) data are available from a reference borehole where stress and consolidation state are known, (2) the incoming sediment section does not vary in thickness or mechanical properties laterally, (3) sediment packages can be correlated between the reference site and sites where the sediment has been buried and potentially dewatered, and (4) cementation does not cause porosity decrease or differences in mechanical properties between the reference site and buried sites.

It is important to recognize that interpretation of P'_c values from laboratory consolidation tests is subject to considerable uncertainty. These uncertainties include (1) a gradual transition between the reloading and virgin consolidation curves caused by even slight disturbance of sediment structure during the coring and sampling process, which leads to poor resolution of P'_c [e.g., *Mitchell*, 1993], (2) a potential component of non-uniaxial strain if lateral stresses are transferred across the décollement, and (3) cementation. Nonetheless, the results of consolidation tests are useful for identifying significant departures from hydrostatic pore pressure, for documenting spatial trends, and for estimating, to first order, in situ effective stress. If unloading has occurred in the field, as might be the case if pore pressure increased after consolidation was complete, void ratio will increase following a rebound, or unloading, curve. In

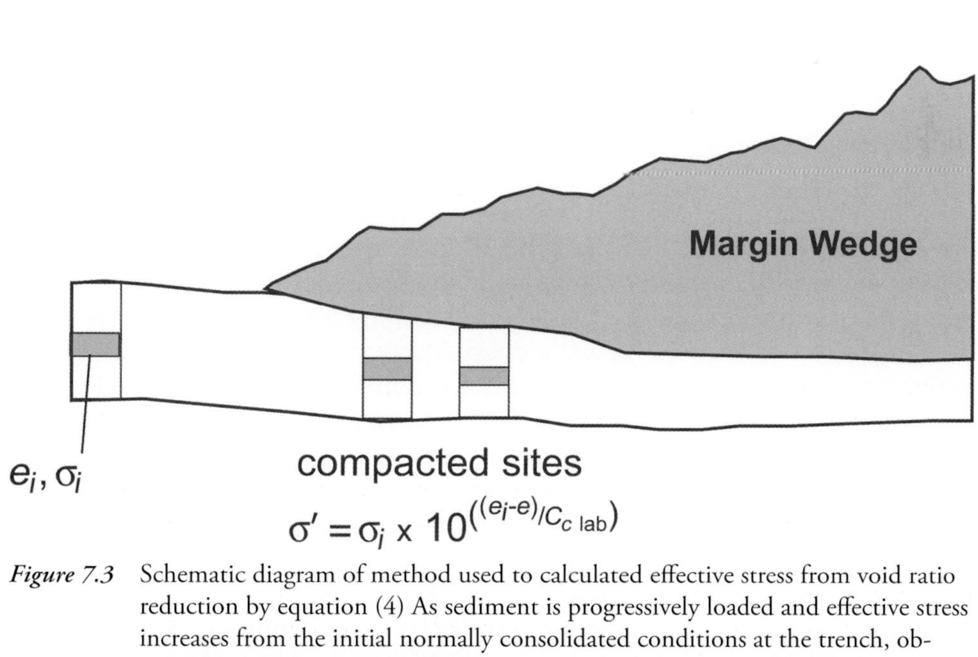

Figure 7.3 Schematic diagram of method used to calculated effective stress from void ratio reduction by equation (4) As sediment is progressively loaded and effective stress increases from the initial normally consolidated conditions at the trench, observed changes in void ratio within correlated sediments (gray shaded rectangle) can be combined with laboratory measurements of C_c to quantify the increase in effective stress.

this scenario, the unloading would not necessarily be evident from either the observed void ratio or the laboratory-derived estimate of P'_c, and thus effective stress estimated from sediment consolidation would not accurately reflect in situ conditions [e.g., *Bowers*, 2001]. Cementation typically causes an increase in yield strength leading to apparent overconsolidation (i.e., $P'_c > \sigma_v$) and a sharp transition between reloading and virgin consolidation curves [e.g., *Mitchell*, 1993]. Significant effects on consolidation behavior caused by cementation are evident and can be ruled out in some cases (for examples, see discussion by *Saffer et al.* [2000] and *Saffer* [2003]) but must be carefully evaluated.

Method Overview Uniaxial consolidation tests are typically conducted using a hydraulically or mechanically driven oedometer. A core sample is trimmed to a height of 10–20 mm and a diameter of 38–60 mm. In general, sample diameter is >2.5 times sample height [e.g., *Crawford*, 1986]. In most cases, the core sample is placed in a steel ring that maintains a condition of zero lateral strain. In other cases, confining fluid pressure is used to maintain a condition of zero lateral strain in a triaxial vessel. This method provides the advantage of eliminating the typically small sidewall friction present in fixed-ring tests [e.g., *Mitchell*, 1993], but it requires a considerably larger intact sample. A seating load of a few kPa is often applied to insure that the axial loading device (a diaphragm in the case of a hydraulic oedometer, and a piston or ram in the case of a mechanical oedometer) is docked against the sample. Usually, the sample is then saturated and backpressured for 24 hours using de-aired water. A backpressure of a few hundred kPa ensures sample saturation and drives any gases present into solution.

After backpressure is complete, an axial load is applied. Axial load is typically applied in one of three ways: (1) constant rate of strain (CRS), (2) constant rate of loading, or (3) stepped loading. In CRS tests, axial displacement is imposed by the ram, and the resulting stress and sample internal pore pressure are recorded [*Crawford*, 1986; *Lee*, 1985]. In constant rate of load tests, axial load is incremented at a constant rate (expressed as $\Delta\sigma_v/\Delta t$), and axial strain and pore pressure are recorded. In stepped loading tests the axial load is incremented as a step function, typically every 24 hours. For stepped loading tests, each successive axial load increment is a multiple of the previous axial load [e.g., *Crawford*, 1986]; thus, when axial load is plotted on a logarithmic axis, individual data points are evenly spaced.

In the examples below, I discuss laboratory consolidation test results for samples from Costa Rica, which were originally reported by *Saffer et al.* [2000]. Results are from consolidation tests on three samples from a reference borehole (Site 1039) and eight samples from a borehole inboard of the trench (Site 1040) (fig 7.1b). The samples represent the two main lithologies sampled by drilling on ODP Leg 170 [*Kimura et al.*, 1997]: a basal, ~200-m-thick pelagic carbonate section, and a ~180-m-thick hemipelagic clay-rich diatomaceous ooze. The samples were consolidated under axial loads up to 6 MPa using a high-precision hydraulic oedometer and following the incremental loading procedures described above [*Crawford*, 1986; *Lee*, 1985; *Saffer et al.*, 2000]. The samples were 6.25 cm in diameter, and initial height ranged from 1.5 to 1.6 cm. All samples were saturated with de-aired water and backpressured to 500 kPa for 24 hours before initiating consolidation tests.

Data Analyses and Estimation of σ'_v (Costa Rica Example) Values of P'_c are reported from laboratory tests following the Casagrande method, as shown in figure 7.2 [e.g., *Casagrande*, 1936; *Holtz and Kovacs*, 1981]. Error bounds on P'_c values are based on observed stress-strain behavior: minimum and maximum P'_c values reflect effective stresses at which the laboratory stress-strain curve no longer plots on the rebound curve or virgin consolidation curve, respectively (fig. 7.2). Bounds on in situ pore pressure are calculated by equation (7.2), with effective stress defined by P'_c values.

In addition, the second method described above can be used to obtain an independent estimate of effective stress, using the observed change in void ratio and the slope of the virgin consolidation curve defined in laboratory tests (fig. 7.3). This method also provides an estimate of effective stress at an intermediate borehole (Site 1043), where core samples were not collected, and thus no laboratory consolidation tests were conducted.

In order to project effective stress as shown in figure 7.3, the initial void ratio of the buried sediment must be known. In the regionally continuous sedimentary section drilled offshore Costa Rica [e.g., *Kimura et al.*, 1997], sediment packages can be correlated between the reference site (Site 1039) and subsequently buried sites; in this scenario the void ratio at the reference site can be treated as the initial void ratio of correlated sediment at the buried sites.

I correlate packets of sediment using the equivalent solid thickness within the underthrust sediment section, which is conserved under conditions of uniaxial strain (see *Saffer* [2003] for detailed discussion of this correlation). I project effective stress within correlated sediment packets by rearranging equation (7.3):

$$\sigma' = \sigma_i \cdot 10^{\left(\frac{(e_i - e)}{C_c}\right)} \tag{7.4}$$

where σ_i is initial effective stress at the reference site calculated from overburden stress and assuming hydrostatic pore pressure, e_i is initial void ratio, and e is observed void ratio at the buried site. Each correlated sediment packet is defined by 5 m of solid thickness, and I average values of e used in equation (7.4) over this interval.

Results for Costa Rica Example P'_c values for samples from Site 1040 indicate significant overpressures, even when accounting for uncertainties in P'_c as a measure of in situ effective stress as discussed above (fig. 7.4a, solid circles).

Pore pressures inferred from P_c' by equation (7.2) range from ~1.4 MPa above hydrostatic at the top of the section to ~3 MPa above hydrostatic at the base [*Saffer et al.*, 2000]. For comparison, the excess pore pressure expected for undrained conditions is ~3 MPa (figure 7.4a, gray shaded area). Thus the section appears partly drained at its top and nearly undrained at its base. At Site 1040, excess pore pressures determined from P_c' values reflect 43% underconsolidation (U = 0.43) at the top of the section, increasing to 90–100% underconsolidation (U = 0.90–1.0) at the base (figure 7.4b, solid circles). When normalized to the lithostatic load, these excess pore pressures reflect values of λ^* ranging from 0.4 at the top of the section to λ = 0.8–0.91 at its base (equivalent to values of λ = 0.74 and 0.87–0.97, respectively; for reference, a value of λ = 0.56 corresponds to hydrostatic pore pressure at site 1040) (fig. 7.4c, table 7.2).

Pore pressures estimated from observed changes in void ratio by equation (7.4) are remarkably consistent with those from P_c' values (figs. 7.4a and 7.4b, open squares). Again, excess pore pressures range from ~1.5 MPa above hydrostatic at the top of the section to ~3 MPa between 500 and 600 mbsf (fig. 7.4a), representing values of U ranging from 0.45 to 1.0 (fig. 7.4b, open squares) and λ^* from 0.43 to 0.79 (λ = 0.78 to 0.90). In contrast to those determined from P_c' values, below 600 mbsf, excess pore pressure inferred from changes in void ratio decreases to ~2 MPa (fig. 7.4a).

At Site 1043, no cores were collected below the décollement; as a result, pore-pressure estimates from P_c' values are not available. In situ effective stresses estimated from observed void ratio reduction (equation (7.4), see fig. 7.3) yield excess pore-pressure estimates that are relatively uniform throughout the section (figs. 7.4d and 7.4e). Estimated excess pore-pressure ranges from 850 kPa near the top of the section to >1.0 MPa at the base, corresponding to values of U ranging from 0.85 to ~1.0 and values of λ^* from 0.4 to 0.83 (λ from 0.78 to 0.95) (fig. 7.4f). The high values of U (near unity) indicate essentially undrained conditions. Interestingly, under nearly undrained conditions, both λ^* and λ decrease downsection because pore pressure (P_f) follows a hydrostatic gradient with depth, whereas lithostatic pressure (P_l) increases more rapidly with depth. A few correlated sediment packages between 250 and 310 mbsf yield significantly lower estimates of in situ excess pore pressure (50–750 kPa) and correspondingly smaller values of U (0.05–0.75) and λ (0.63–0.78) (figs. 7.4d and 7.4f; table 7.2). For reference, hydrostatic pore pressures at site 1043 correspond to λ = 0.61.

Sediment Porosity as an Indicator of Effective Stress

In many cases, samples are not collected for laboratory consolidation experiments, or sediment thickness and properties vary laterally. Under such conditions the methods described above and applied to the Costa Rican margin cannot be used reliably to obtain estimates of in situ effective stress or pore

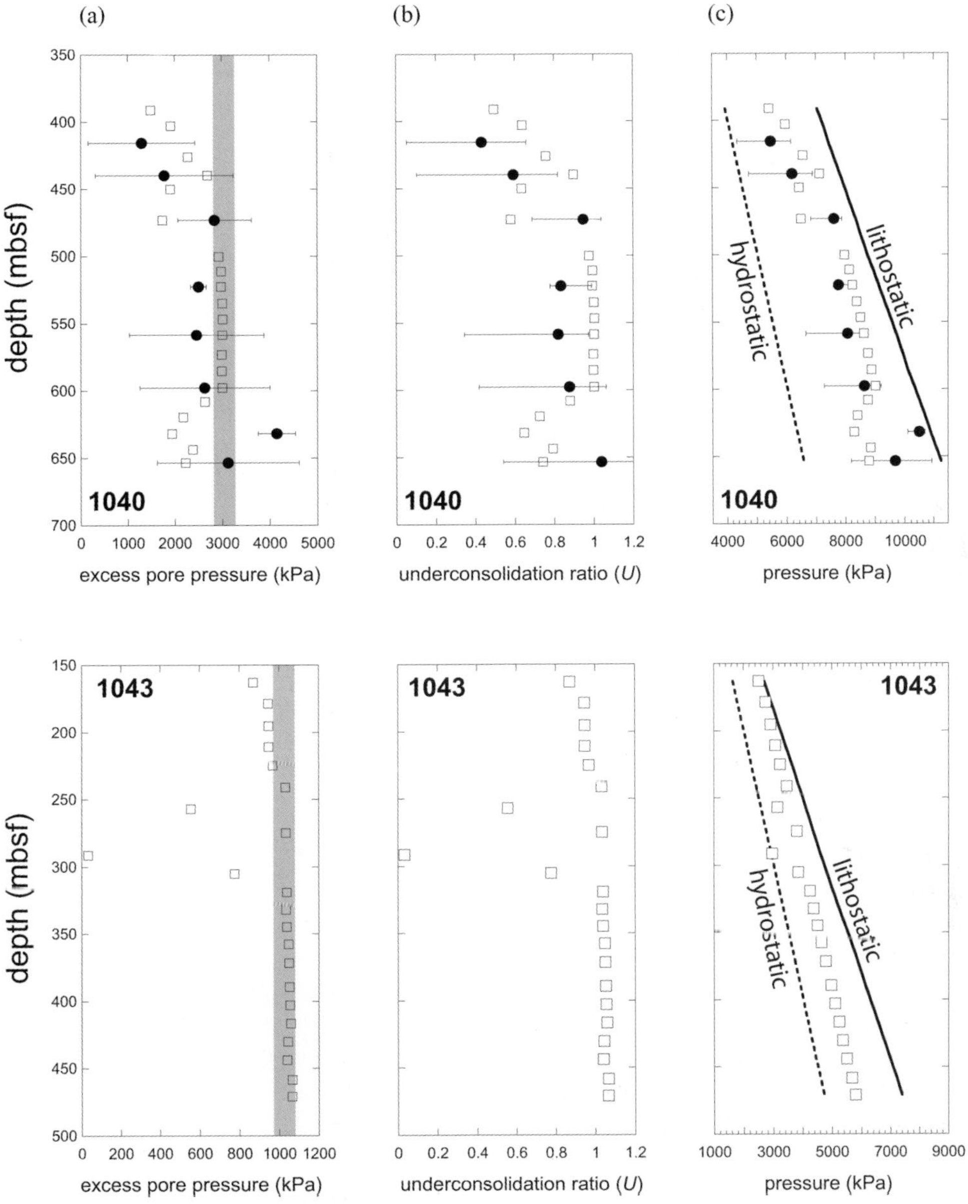

Figure 7.4 Excess pore pressure (a, d), underconsolidation ratio, U, (b, e), and pore pressure (c, f), for Sites 1043 and 1040. Values calculated by projecting changes in void ratio using equations (2) and (4) are shown by open squares. Values based on laboratory estimates of P'_c are shown as black circles (Site 1040 only). Error bars show range between $P'_{c\,min}$ and $P'_{c\,max}$ as described in text and in figure 7.2. In figures 7.4a and 7.4d, gray shaded region shows excess pore pressure expected for undrained loading beneath the margin wedge. In figures 7.4c and 7.4f, hydrostatic and lithostatic pressures are shown by the dashed and solid lines, respectively.

TABLE 7.2 **Estimates of Excess Pore Pressure Within Underthrust Sediment**

Location	Method	Distance From Trench	Pf Excess, MPa
Barbados			
Site 1047/676	equation (5)	0.85 km	0.35–1.2
Site 1046/949	equation (5)	2.2 km	0.70–2.1
	CORK[a]		1.02
Site 1045	equation (5)	3.2 km	1.1–2.7
Site 948	equation (5)	4.2 km	1.2–3.3
	CORK[a]		2.2
	P_c' (530 mbsf)		1.3
	equation (5)		2.0–2.5
16° 12′ N	seismic	0–60 km	—
14° 22′ N	seismic	0–35 km	—
East Nankai			
Site 1174	$\eta(z)$ offset	2 km	3.6
	equation (5)		2.5–4.6
Site 808	$\eta(z)$ offset	3 km	4.2
	equation (5)		4.0–5.5
	P_c' (2 samples)		4.9–6.3
Leg 190 Transect	seismic velocity	0–12.5 km	2–12
West Nankai			
Site 582 area	seismic velocity	outboard[c]	0.1–1.2
Cascadia			
N. Oregon	seismic velocity	proto-thrust zone[d]	15–28
Vancouver	seismic velocity	0–20 km	—
Costa Rica			
Site 1043	equation (4)	0.5 km	0.85–1.05
Site 1040	P_c' (8 samples)	1.6 km	1.4–3.0
Site 1040	equation (4)	1.6 km	1.5–3.0

[a] CORK measurement is from within décollement and top of underthrust section.
[b] Here λ and λ^* increase systematically from the trench landward; maximum values are 0.80–0.85 at 45–70 km from the trench along the transect at 14°22′ and 0.85–0.90 at 12–25 km from the trench along the transect at 16°12′.
[c] Estimates are from sediment beneath thick turbidite deposits, immediately outboard of trench.
[d] Proto-thrust zone is seaward of the frontal thrust, in a zone of incipient thrusting.
[e] Here λ and λ^* are highest at the top of the section at Site 1043 and decrease systematically downward.

λ*	λ	Reference
0.20–0.55	0.68–0.84	*Saffer* [2003]
0.30–0.57	0.67–0.83	*Saffer* [2003]
0.33–0.35	0.73	*Becker et al.* [1997]
0.40–0.76	0.73–0.90	*Saffer* [2003]
0.49–0.76	0.70–0.92	*Saffer* [2003]
~0.55	~0.8	*Foucher et al.* [1997]
0.39	0.77	*Vrolijk et al.* [1998]
~0.57–0.64	~0.85	*Moore and Tobin* [1997]
—	0.60–0.85[b]	*Westbrook* [1991]
—	0.50–0.80[b]	*Westbrook* [1991]
0.42	0.74	*Screaton et al.* [2002]
0.34–0.46	0.67–0.75	this paper, *Saffer* [2003]
0.47	0.76	*Screaton et al.* [2002]
0.42–0.49	0.69–0.76	this paper, *Saffer* [2003]
0.52–0.59	0.77–0.8	*Karig* [1993]
0.34–0.65[b]	0.69–0.82[b]	*Tobin and Saffer* [2003]
0.03–0.29	0.61–0.66	*Bray and Karig* [1988]
0.45–0.52	~0.5–0.74	*Cochrane et al.* [1994]
—	~0.5–0.8	*Wang* [1994]
0.4–0.83[e]	0.95–0.78[e]	this paper, *Saffer* [2003]
0.4–0.91	0.74–0.97	*Saffer et al.* [2000]
0.43–0.79	0.78–0.90	this paper *Saffer* [2003]

pressure. In particular, if stress state prior to underthrusting (σ_i in equation (7.4)) is not constrained because sediment thickness is variable away from the modern trench, projection of effective stress by equation (7.4) is prone to substantial uncertainty.

However, if sediments are lithologically and mechanically homogeneous downsection, a field-based porosity-depth or void ratio-effective stress relationship can be defined using borehole data from an undisturbed reference site [e.g., *Screaton et al.*, 2002; *Saffer*, 2003; *Dugan and Flemings*, 2000]. Such field-based relationships between compaction state and stress incorporate the effects of secondary consolidation; as a result, they often parallel the laboratory-derived virgin consolidation curve for the same sediment but intercept the strain axis at a lower void ratio (lower value of e_o) [e.g., *Karig and Ask*, 2003; *Holtz and Kovacs*, 1981].

One-Dimensional Method Overview Observations in a wide range of geologic settings indicate that sediment porosity decreases exponentially with depth [e.g., *Athy*, 1930; *Bray and Karig*, 1985]. Thus, if a porosity-depth trend can be defined at a reference location where pore pressure is hydrostatic (a state termed normal consolidation for the case of uniaxial vertical loading), any offset in this trend at other locations can be used to quantify the divergence from hydrostatic pore pressure [e.g., *Rubey and Hubbert*, 1959; *Hart et al.*, 1985; *Screaton et al.*, 2002].

Alternatively, porosity data from a reference borehole can be considered in the context of expected consolidation behavior described by equation (7.3) by plotting void ratio as a function of effective stress, where effective stress is calculated from equation (7.2) and assuming hydrostatic pore pressure. This approach defines a virgin consolidation curve derived directly from field data, which includes the effects of both primary and secondary consolidation (fig. 7.5), and allows inversion of observed void ratio values within buried sediment for in situ effective stress. Inversion of index properties data provides an estimate of effective stress for each porosity measurement, resulting in a depth-profile of estimated effective stress. Even in boreholes where samples are collected for laboratory consolidation experiments, this method is valuable because it provides a second constraint on in situ effective stress and at a higher spatial resolution than is typical of core sample collection for geotechnical testing.

Analyses and Estimation of σ'_v (Nankai Example) At Nankai, the entire underthrust section is composed of the Lower Shikoku Basin facies, a homogeneous clay-rich (~50–70 wt % clay fraction) hemipelagic mudstone with interbedded altered ash layers [*Moore, G. F. et al.*, 2001]. The clay minerals are dominantly illite and smectite, with minor amounts of chlorite. The section was penetrated and cored on ODP Leg 190 at a reference site ~10 km seaward of the trench (ODP Site 1173) and at sites 1174 (ODP Leg 190; 2 km landward of the trench)

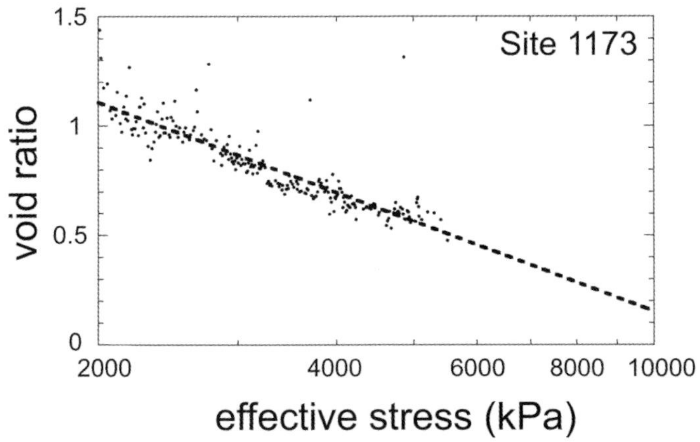

Figure 7.5 Void ratio versus effective stress for reference Site 1173. Values of void ratio were determined from shipboard measurement and corrected for rebound; values of effective stress were calculated from rebound-corrected bulk density measurements, assuming hydrostatic pore pressure. Dashed line shows best-fit virgin compression curve.

and 808 (ODP Leg 131; 3 km landward of the trench) (fig. 7.1a). At Site 1173, the Lower Shikoku exhibits a systematic decrease in porosity with depth as would be expected for mechanical compaction described by equation (7.3) (fig. 7.5).

I define the consolidation index (C_c) and void ratio intercept (e_0) of the sediment using observed void ratios at Site 1173, corrected for rebound by calibration to downhole logging data [*Saffer*, 2003; *Tobin and Saffer*, 2003]. In order to define effective stress downsection at site 1173, I assume that pore pressure there is hydrostatic [e.g., *Screaton et al.*, 2002]. This assumption is consistent with (1) consolidation test results that yield P_c' values in the upper 460 mbsf at site 1173 indicating normal consolidation (fig. 7.2; D. M. Saffer, unpublished data, 2002), (2) the large distance from the trench (~10 km), and (3) generally low sedimentation rates for most of the section (33.6–83 m Myr^{-1} in the lower 600 m, and 500 m Myr^{-1} in the upper 104 m at Nankai) [*Moore, G. F. et al.*, 2001]. On the basis of the sedimentation history at Site 1173, only the Quaternary deposition of distal trench turbidites within the upper 104 m could result in small overpressures; transient sedimentation models that incorporate laboratory measurements of permeability indicate that an overpressure of ~100–200 kPa throughout the section is possible [e.g., *Screaton et al.*, 2002].

At the two drill sites landward of the trench, sediment porosity is offset to higher values beneath the décollement than above it [e.g., *Screaton et al.*, 2002]. This has been interpreted to indicate a combination of increased mean stress above the décollement caused by tectonic compression and arrested consolidation of underthrust sediment [e.g., *Morgan and Karig*, 1993; *Screaton et al.*, 2002; *Flemings*, 2002]. The arrested consolidation of underthrust sediment is likely a result of burial-driven vertical stress increasing more rapidly than water can drain [e.g., *von Heune and Lee*, 1982; *Nuezil*, 1995].

Effective stress within underthrust sediment at the two buried sites (1174 and 808) can be determined from observed void ratios by rearranging equation (7.3):

$$\sigma' = 10^{\frac{(e_o - e)}{C_c}} \tag{7.5}$$

where e_o and C_c describe the intercept and slope of the consolidation curve defined by field data from the reference Site 1173 (fig. 7.5). For the data shown in figure 7.5, regression yields values of $e_o = 5.3$ and $C_c = 1.32$. If overpressures of 200 kPa were developed throughout the section by recent deposition of the upper 104 m of trench wedge turbidites, the values of σ' shown in figure 7.5 would be shifted to lower values by this amount; regression then yields values of $e_o = 5.01$ and $C_c = 1.23$, a decrease of ~6% for both parameters. Notably, small uncertainty in the stress state at the reference site causes only a small change in the inversion results because the effects of decreasing both e_o and C_c offset one-another in equation (7.5).

One-Dimensional Results On the basis of the offset of porosity-depth trends, *Screaton et al.* [2002] calculated an excess pore-pressure averaged over the underthrust section of 3.6 MPa at Site 1174 and 4.2 MPa at Site 808. These values correspond to values of $\lambda^* = {\sim}0.42$ at Site 1174 and $\lambda^* = {\sim}0.47$ at Site 808. *Screaton et al.* [2002] also note that a constant offset of the porosity-depth curve does not fit all of the underthrust porosity data equally; porosities in the upper portion of the section are lower than the shifted porosity-depth curve predicts. This misfit is interpreted to indicate lower excess pore pressures in the upper part of the underthrust section, possibly indicating upward drainage.

Inversion of observed void ratio for in situ effective stress at Sites 1174 and 808 yields similar results to the method of *Screaton et al.* [2002] but with higher spatial resolution downsection because each individual porosity value is inverted for effective stress, rather than estimating an average value for the entire section (figs. 7.6a–7.6f). At Site 1174, estimated excess pore pressure increases downsection, from ~2.5 MPa at 840 mbsf to a maximum of ~4.6 MPa at 1100 mbsf (fig. 7.6a). Under undrained conditions, excess pore pressures of ~4.9 MPa would be expected (fig. 7.6a, gray shaded area). Inferred excess pore pressures correspond to values of U ranging from 0.43 to 0.96 (fig. 7.6b) and values of λ^* ranging from 0.34 to 0.46 ($\lambda = 0.67$–0.75) (fig. 7.6c, table 7.2). Estimated excess pore pressures at Site 808 exhibit a similar pattern, increasing from 4.0 MPa at the top of the section to ~5.5 MPa at its base (fig. 7.6d), corresponding to values of U from 0.6 to 0.84 (fig. 7.6e) and λ^* from 0.42-0.49 ($\lambda = 0.69$-0.76) (fig. 7.6f, table 7.2). Undrained conditions would yield excess pore pressures of ~6.3 MPa (fig. 7.6d, gray shaded area). Average values of excess pore pressure from this inversion are 3.6 MPa at Site 1174 and 4.8 MPa

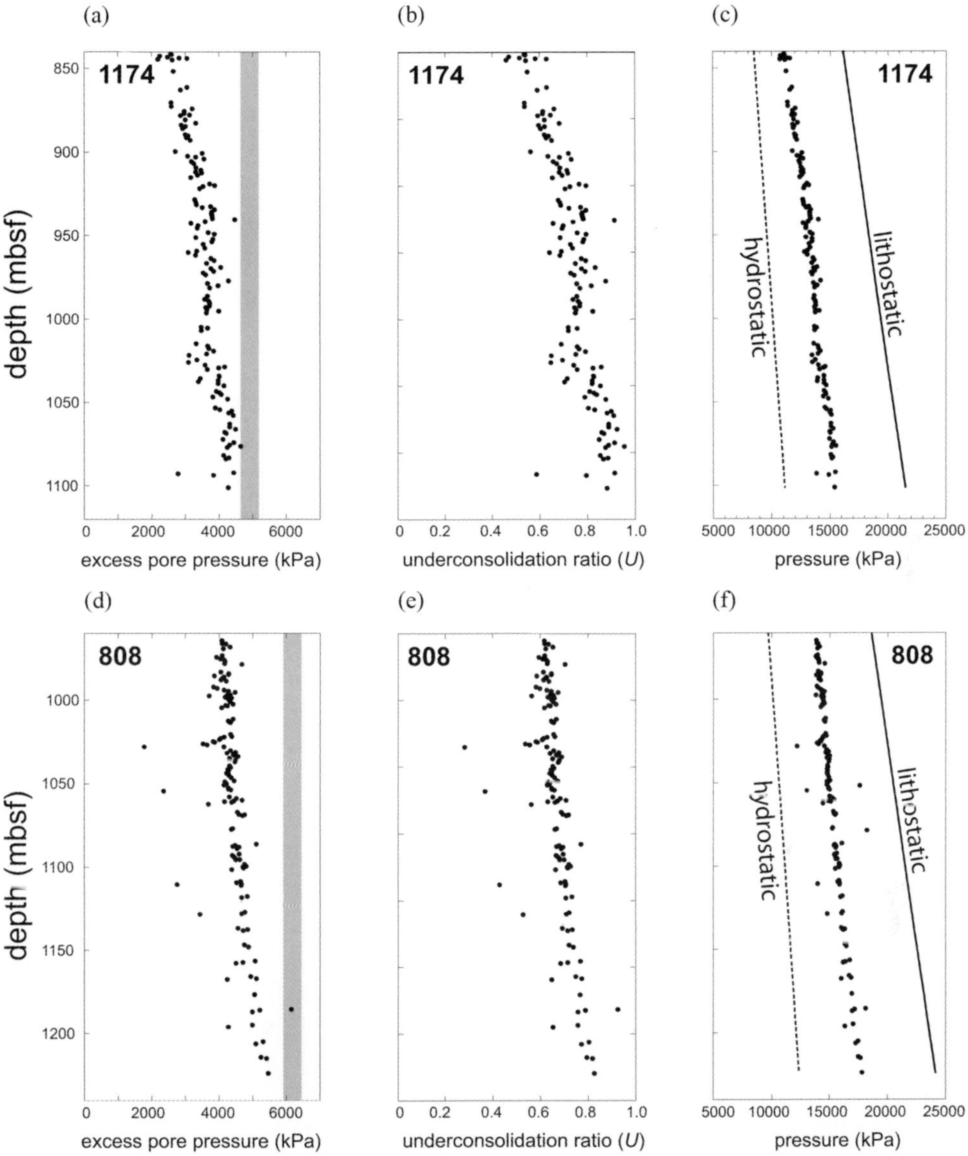

Figure 7.6 Excess pore pressure (a, d), underconsolidation ratio, *U*, (b, e), and pore pressure (c, f), for Sites 1174 and 808. In all cases, excess pore pressure is determined from the field-based virgin compression curve shown in figure 7.5, using equations (2) and (3). As with figure 7.4, gray shaded area in figures 7.6a and 7.6d show excess pore pressure expected for undrained loading, and hydrostatic and lithostatic pressures are shown by the dashed and solid lines, respectively, in figures 7.6c and 7.6f.

at Site 808, in good agreement with average values determined by *Screaton et al.* [2002] from offset of porosity profiles. Inversion using the slightly lower values of e_o and C_c determined for 200 kPa of overpressure at the reference Site 1173 result in estimated pore pressures that are consistently 255–280 kPa greater than those estimated assuming the reference site is not overpressured. This offset is consistent downsection and for both Sites 1174 and 808.

At both drill sites, values of excess pore pressure throughout the section are a significant fraction of the increase in overburden caused by underthrusting. As shown in figures 7.6b and 7.6d, the section remains 40–90% underconsolidated. This indicates that the sediment, like that at Costa Rica, is poorly drained in response to rapid tectonic burial [e.g., *von Heune and Lee*, 1982]. Furthermore, increased excess pore pressures with depth indicate that the upper portion if the sediment section is better drained than the base, consistent with the interpretation of *Screaton et al.* [2002].

Porosity From Seismic Velocity

Method Overview As discussed in the preceding sections, sediment physical properties data collected at boreholes, combined with laboratory consolidation studies on core samples, provide estimates of effective stress and pore pressure at isolated locations. However, sampling for physical properties and geotechnical experiments is limited to accessible areas near the trench, typically at depths less than ~1.5 km bsf. Thus pore pressures and consolidation state of underthrust sediment beyond a few kilometers from the trench and at greater depths are poorly known. At subduction zones where the incoming section is greater than 2–3 km thick, such as the Cascadia or the Eastern Aleutian Trench, even drilling at or near the trench can only access a fraction of the sediment column, and may not penetrate beneath the décollement. Furthermore, the applicability of these constraints to understanding pore-pressure generation, its spatial variability, and its ultimate effects on shear strength are limited by lack of spatial coverage.

Seismic reflection/refraction data, which provide seismic velocity and reflection amplitude, offer one means to constrain sediment properties over broad regions away from boreholes. In several studies, seismic velocities have been used to infer porosity and pore pressure within sediment [e.g., *Fowler et al.*, 1985; *Cochrane et al.*, 1994, 1996; *Bangs et al.*, 1990, 1999; *von Huene et al.*, 1998; *Westbrook*, 1991; *Wang*, 1994] (table 7.2). Although considerable uncertainties exist in inferences from seismic velocity, this technique is the only available means to extend spatial estimates of fluid content and pore pressure away from individual boreholes and to depths greater than ~1.5 km bsf. Reliable estimation of in situ pore pressure from seismic velocity requires (1) seismic data of sufficient quality to obtain interval velocities, (2) a laboratory- or field-derived transformation to obtain porosity from seismic velocity, and (3) a laboratory- or field-based transformation to obtain effective stress from porosity (cf. fig. 7.5).

Application to Nankai Here, velocities are taken from two-dimensional (2-D) prestack depth migration (PSDM) of a seismic line (line 215) perpendicular to

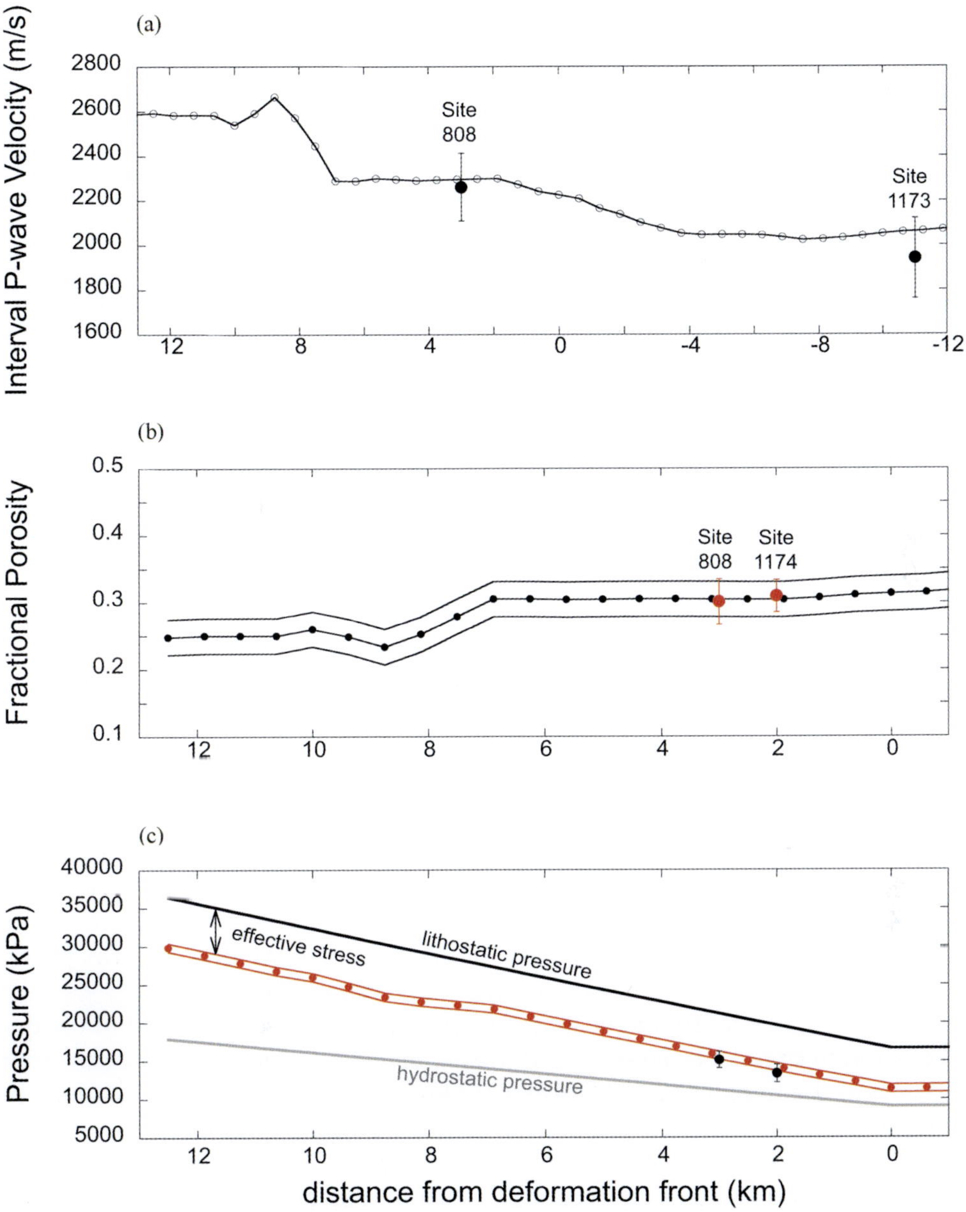

Figure 7.7 (a) Acoustic velocity within entire underthrust sediment interval determined from 2-D PSDM versus. distance from the trench (negative values are seaward of the trench). (b) Mean porosity within underthrust interval, determined from seismic velocities as described in text, vs. distance from the trench. (c) Mean pore pressure within underthrust section calculated by inversion of porosity values shown in figure 7.7b versus distance from the trench. Gray and black lines show hydrostatic and lithostatic pressures, respectively. Error bounds for porosity (figure 7.7b) and pore pressure (figure 7.7c) are based on estimated uncertainty in porosities derived from seismic velocities (see text for further explanation).

the trench, located within a 1999 3-D survey area (fig. 7.1a) [*Hills et al.*, 2002, also unpublished data; *Tobin and Saffer*, 2003]. The underthrust sediment column is seismically transparent; reliable interval velocities are therefore only available for the entire section between the décollement zone and the top of ocean crust. Interval velocities used in the following analysis represent average values for the entire ~320-m-thick underthrust sediment section. This provides an estimate of mean porosity, and thus mean excess pore pressure, within the underthrust section.

In order to invert seismic velocity for porosity, *Tobin and Saffer* [2003] developed a model for the modification of *P*-wave velocity and elastic moduli during burial and tectonic transport of the underthrust section specific to the transect defined by drilling at Nankai. Core data from the underthrust interval at Sites 1173, 1174, and 808 indicate that the approach of *Erickson and Jarrard* [1998] is most appropriate for mapping seismic velocity to porosity at this location, where diagenetic and consolidation processes likely lead to velocity enhancement below a "threshold" porosity [*Erickson and Jarrard*, 1998; *Tobin and Saffer*, 2003]. To invert mean porosities derived from seismic velocities for effective stress, the field-based consolidation relationship shown in figure 7.5 is used.

Results for Nankai Example Interval velocity within the underthrust section increases steadily with distance landward from the deformation front, from ~2050 m/s at the trench (average depth of ~900 mbsf) to ~2600 m/s 9–12 km arcward (depth of ~1500–1800 mbsf) (fig. 7.7a). Although the errors on preliminary PSDM-derived velocities are substantial (as large as ±5–10%) (G. Moore, personal communication, 2004) they are in good agreement with shipboard values from ODP Leg 190 (fig. 7.7a, solid circles); shipboard values probably slightly underestimate in situ velocity because measurements were taken at atmospheric pressure. A similar trend is apparent along a nearby parallel seismic line (line 284, not shown) [Tobin and Saffer, 2003].

The trend of slightly increasing velocity with distance landward of the deformation front reflects decreasing porosity. Estimated porosity decreases from 29–34% at the trench to 20–27% by 9–12 km landward (fig. 7.7b). Seismically derived porosity estimates are in excellent agreement with rebound-corrected shipboard porosity values at Sites 1174 and 808 (fig. 7.7b, solid circles). Uncertainty in porosity values derived from seismic velocity shown in figure 7.7b over the velocity range shown in figure 7.7a is estimated at ±2–3.1%, defined by the standard error of fitting the velocity-porosity transform to data from Sites 1173 and 808 [*Tobin and Saffer*, 2003].

As reported by *Tobin and Saffer* [2003], porosity values shown in figure 7.7b are inverted for effective stress using equation (7.5) and the void ratio-effective stress relationship defined in figure 7.5. Assuming an exponential decrease in porosity with depth in the overlying wedge to define the total overburden stress [e.g., *Screaton et al.*, 2002], mean pore pressure within the underthrust section is given by equation (7.2). Estimated pore pressure increases from near-hydrostatic values at the trench to a significant fraction of lithostatic values within a few kilometers from the trench (fig. 7.7c). Excess pore pressure increases from a ~1–2 MPa at the trench to ~12 MPa by 12 km landward, corresponding to values of λ^* increasing from 0.34 to 0.65 (λ = ~0.7 to 0.82). Over the first 6–7 km from the trench, average pore pressures nearly parallel the increase in lithostatic stress, indicating that the pore pressure increases with burial and tracks the increase in total overburden driven by underthrusting, whereas the effective stress remains essentially constant (fig. 7.7c). Inferred effective stresses and pore pressures are in excellent agreement with those derived from downhole data at Sites 1174 and 808 (fig. 7.7c, solid circles).

Discussion

Utility and Applicability of Sediment Mechanics Approaches

In some cases, pore pressures inferred from laboratory consolidation tests and porosity data are in good agreement with (1) each other (fig. 7.4), (2) directly measured values where data are available, and (3) constraints from mechanical formulations, such as critical taper theory [e.g., *Saffer*, 2003] (table 7.2). In addition, in examples from underthrust sediment at Barbados and Costa Rica, laboratory-derived values of C_c are in excellent agreement with values inferred from field consolidation curves [e.g., *Taylor and Leonard*, 1990; *Saffer*, 2003] despite the fact that laboratory tests do not account for the effects of secondary consolidation or thermal or diagenetic compaction processes. Furthermore, sediments from reference sites at both Barbados and Nankai exhibit a log-linear relationship between void ratio and effective stress, exactly as would be expected for mechanical consolidation and as observed in laboratory tests (fig. 7.5) [e.g., *Holtz and Kovacs*, 1981]. At Costa Rica this is not the case, mainly because the sediment lithology, and thus mechanical properties, varies significantly downsection [e.g., *Kimura et al.*, 1997].

Effective stresses and pore pressures projected using laboratory- and field-derived values of C_c are also remarkably consistent with constraints from available field data. At Costa Rica and Nankai, pore pressures determined by inverting porosity data reach a significant fraction of the load emplaced by underthrusting but generally do not exceed it (figs. 7.4, 7.6, and 7.7). Likewise, projected effective stresses do not exceed the burial load (cf. Figs. 7.4, 7.6, and 7.7), although there is nothing implicit in equation (7.4) that precludes it. This result is of interest because inferred effective stress is highly sensitive to the value of C_c used in inverting compaction state for effective stress; if the value of C_c is not applicable to the sediment, very large errors in estimated effective stress can result because the relationship between void ratio and stress is log-linear [e.g., *Moore and Tobin*, 1997].

In addition, numerical models of loading that incorporate observed loading rates and directly measured sediment permeabilities and compressibilities yield pore pressures in good agreement with inferred values [e.g., *Gamage and Screaton*, 2003; *Screaton and Ge*, 2000]. In these models, sediment compressibility is used to define both loading efficiency and porosity loss in response to simulated fluid expulsion; notably, simulated porosities are also in excellent agreement with observations [e.g., *Gamage and Screaton*, 2003; *Screaton and Saffer*, 2005].

Taken together, these results suggest that in some cases, laboratory consolidation test results (conducted over days to weeks) can be applied to sediment consolidating over much longer times (tens of kyr to several Myr). The compatibility of inferred effective stresses with (1) the known overburden and hydrostat, (2) P_c' estimates, and (3) modeling results further suggests that, at least in these examples, laboratory- and field-based consolidation indices provide a reasonable approximation of in situ properties. The applicability of laboratory results to in situ geologic consolidation depends on many factors, including (but not limited to) lithology and its variability within the sediment column, cementation, timescale of loading, and temperature-dependent fabric development [e.g., *Lambe and Whitman*, 1969; *Mitchell*, 1993]. More detailed work is clearly needed to evaluate the relationship between laboratory data and geologic consolidation and to carefully explore the factors that control it [e.g., *Karig and Hou*, 1992; *Karig and Ask*, 2003].

Implications for Fluid Escape Pathways

Effective stresses estimated using the methods described in the Sediment Mechanics Approaches section indicate that significantly elevated fluid pressures within underthrust sediment are common. Estimated pore pressures range from partly drained behavior at Site 1040 offshore Costa Rica to undrained behavior at drill sites offshore Barbados. A limited body of work also indicates elevated pore pressures within accreting wedges, although such analyses are less constrained than for underthrust sediment because the magnitude of horizontal stresses is not well known [e.g., *Flemings*, 2002; *Moore and Tobin*, 1997]. In all cases, elevated pore pressures and poorly drained behavior are attributed to rapid loading by subduction, combined with low permeability. As sediment is buried and loaded rapidly beneath the overriding wedge, fluid pressure builds more rapidly than it can be dissipated by diffusion [e.g., *Gibson*, 1958; *von Huene and Lee*, 1982].

At drill sites along the Costa Rican and Nankai margins, inferred underconsolidation ratios increase with depth, indicating that the uppermost sediments are better drained than those deeper in the section (figs. 7.4 and 7.6). This result suggests a component of upward drainage. On the basis of inferred pore pressure profiles (cf. figs. 7.4 and 7.6), the maximum pressure gradient is upward and toward the trench near the margin toe and becomes dominantly

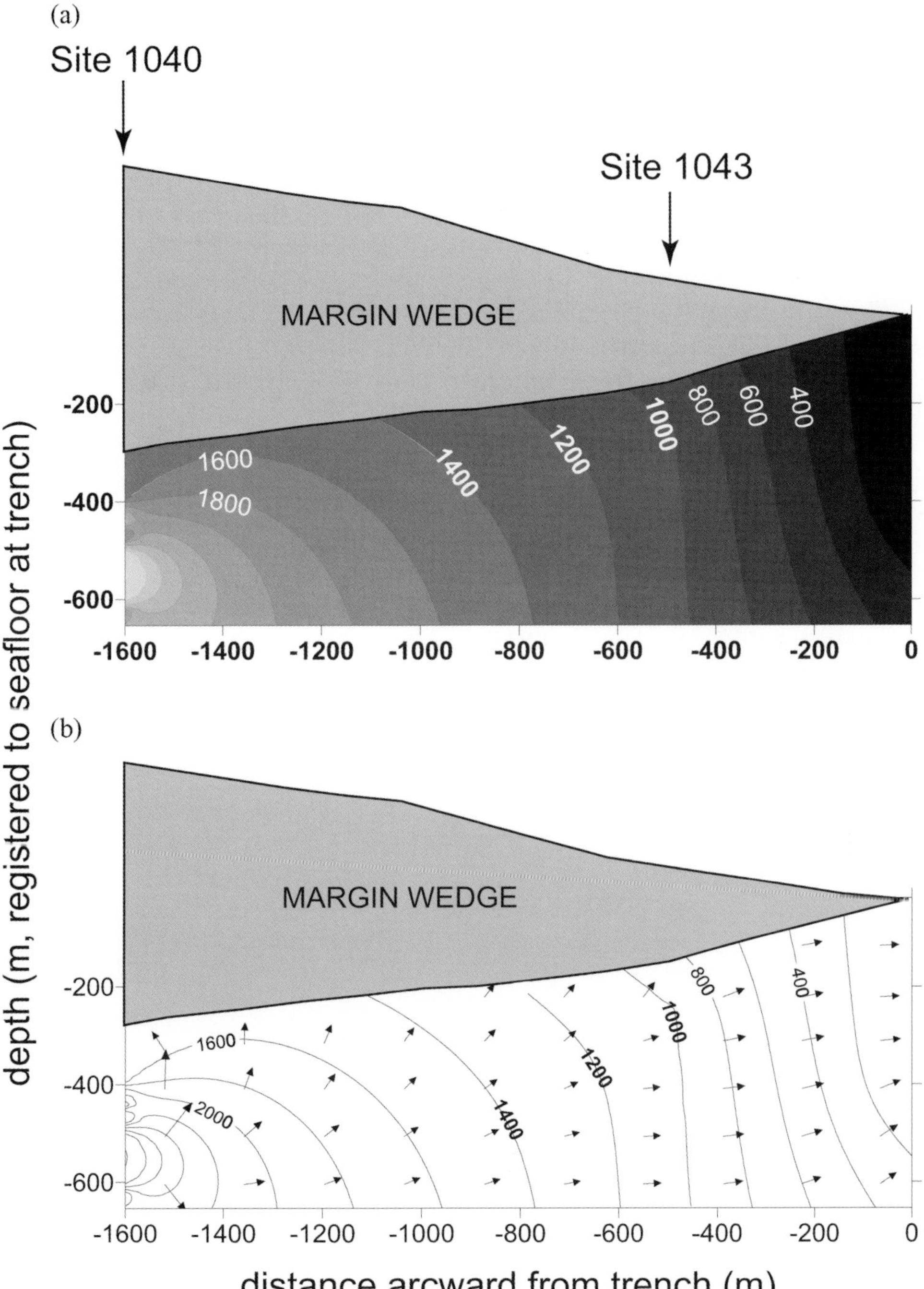

Figure 7.8 Cross section showing contours of excess pore pressure (in kPa) within (a) underthrust sediment at the toe of the Costa Rican margin, and (b) showing fluid flow vectors for isotropic hydraulic conductivity (i.e. direction of maximum excess pressure gradient). Contours were generated based on profiles of inferred pore pressure at Sites 1040 and 1043 shown in figures 7.4a and 7.4d and assuming hydrostatic pore pressures throughout the section at the trench.

vertical with distance landward from the trench (fig. 7.8a). If bulk hydraulic conductivity is near-isotropic and homogeneous, fluid flow will follow the maximum gradient (fig. 7.8b). If significant anisotropy exists, flow directions will be skewed in the direction of maximum hydraulic conductivity. Most typically in bedded sediment, layer-parallel hydraulic conductivity is larger than layer-normal conductivity. This scenario would result in a similar pattern of increased vertical flow with increased distance from the trench to that shown in figure 7.8b but would delay the decrease in horizontal flow to greater distances from the trench [e.g., *Saffer et al.*, 1997]. Two-dimensional hydrologic modeling studies also indicate that within a few kilometers from the trench, dewatering of underthrust sediment should be dominantly vertical, either to a permeable décollement zone or through the overriding wedge to the seafloor [e.g., *Saffer and Bekins*, 1998; *Screaton and Saffer*, 2005].

As noted by *von Huene and Lee* [1982], underconsolidation of underthrusting sediment, and associated overpressures, can be evaluated in the context of two end-member one-dimensional consolidation models. One end-member model, developed by *Gibson* [1958], describes consolidation in response to continuous sedimentation. This model applies to underthrust sediment if drainage occurs through both the consolidating underthrust section and the overriding wedge (figs. 7.9a and 7.9b). A second end-member model considers

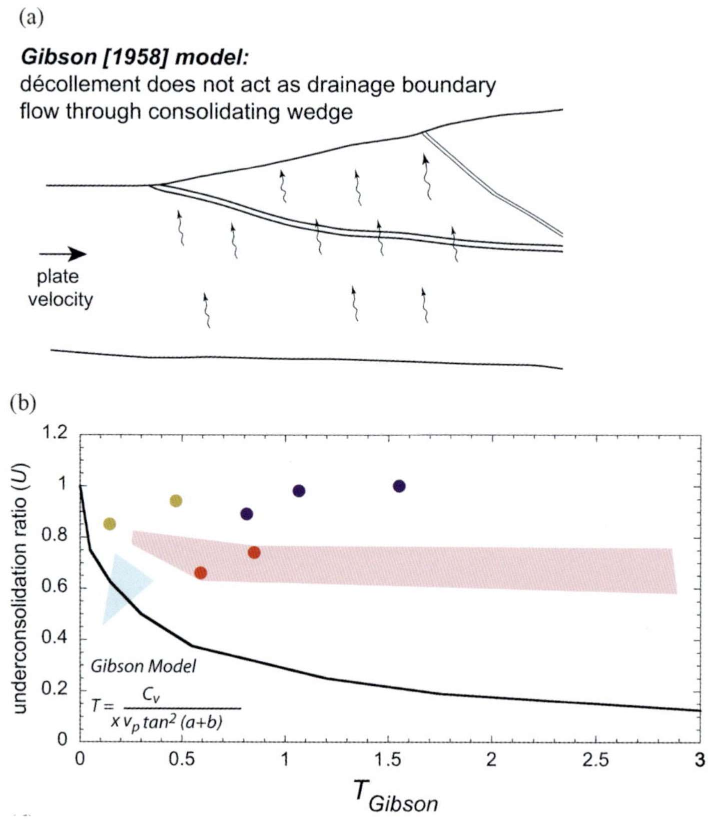

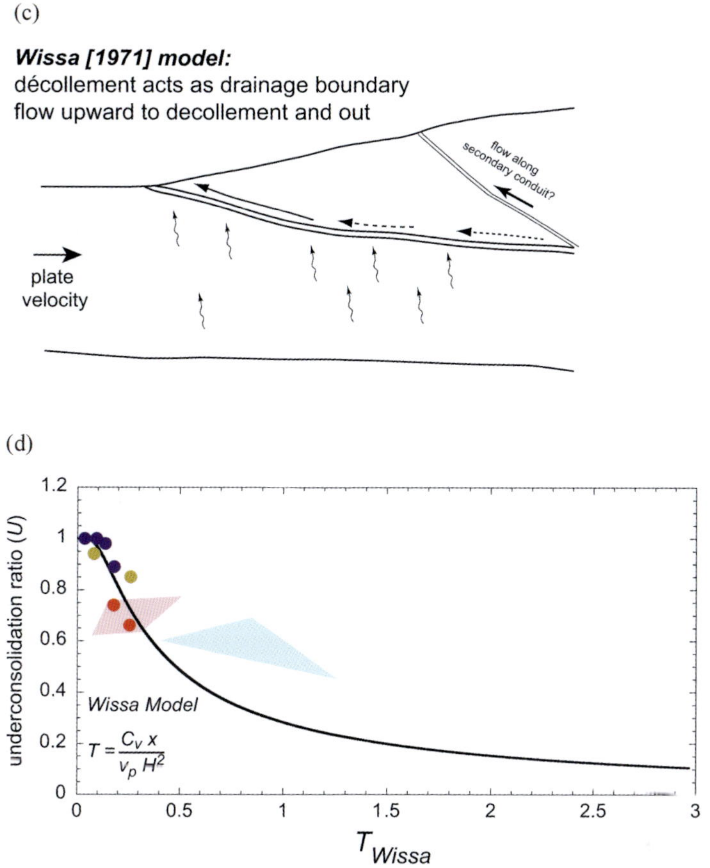

Figure 7.9 (a) Schematic showing drainage pattern for the *Gibson* [1958] consolidation model. (b) Modeled underconsolidation ratio U versus dimensionless time factor for the *Gibson* [1958] sedimentation model (solid black curve). (c) Schematic showing drainage pattern for the *Wissa et al.* [1971] consolidation model. (d) Modeled underconsolidation ratio U versus dimensionless time factor for the *Wissa et al.* [1971] model (solid black curve). In figures 7.9b and 7.9d, inferred mean values of U for boreholes at Barbados (blue), Nankai (red) and Costa Rica (tan) are shown for comparison. Values of U inferred from seismic velocity data are shown for Nankai (pink shaded region; 0–10 km from the trench) and north Barbados (light blue shaded area; 10–30 km from the trench [*Westbrook*, 1991]). The value of T_{gibson} for Site 1047 (Barbados) is 4.02; this is not shown in the range of values plotted in figure 7.9b but falls well above the Gibson [1958] solution with a value of $U = 1.0$. For all data shown in figures 7.9b and 7.9d, values of dimensionless time factor incorporate laboratory-derived values of C_v for samples from Nankai [*Gamage and Screaton*, 2003; Saffer, unpublished data], Costa Rica [*McKiernan and Saffer*, 2005], and Barbados [*Zwart et al.*, 1997]. As discussed in the text, the constant rate of strain model of *Wissa et al.* [1971] provides a better match to the data than that of *Gibson* [1958].

consolidation in a top-drained column resulting from a constant rate of strain [*Wissa et al.*, 1971] or from a constant rate of loading if the deformation behavior of the material is approximated as linear [*von Huene and Lee*, 1982]. This model applies to underthrust sediment if fluids escape vertically through the consolidating sediment to a free drainage (hydrostatically pressured) boundary at the décollement (figs. 7.9c and 7.9d).

As noted by *von Huene and Lee* [1982], neither model provides a perfect analog for underthrusting sediment. The *Gibson* [1958] model is similar to the subduction zone in that burial is continuous, and burial depth increases systematically with time (fig. 7.9a). However, the subduction zone setting differs from the *Gibson* [1958] model in that fluids escaping from the consolidating underthrust section to the seafloor may flow along high-permeability fault pathways, either the décollement at the top of the section or faults within the overriding wedge. Geochemical, thermal, and hydrologic investigations suggest that this is the case at many locations [e.g., *Kimura et al.*, 1997; *Moore*, 1989; *Moore and Vrolijk*, 1992; *Bangs et al.*, 1999; *Bekins et al.*, 1995]. Thus the *Gibson* [1958] model provides a probable upper bound on underconsolidation because it does not allow for fluid escape along fault conduits. In contrast, the *Wissa et al.* [1971] model is similar to the subduction zone setting in that the underthrust sediment is loaded at an approximately constant rate, and evidence suggests that fluid escapes upward to a highly permeable décollement and then seaward along it [e.g., *Moore*, 1989; *Moore and Vrolijk*, 1992]. However, the *Wissa et al.* [1971] model differs from the subduction system in that fluids probably do not escape to a free drainage (i.e., hydrostatically pressured) boundary at the top of the underthrusting sediment but must escape along either a décollement or through an overriding wedge characterized by finite permeability. Thus the *Wissa et al.* [1971] model provides a probable lower bound on underconsolidation because it allows free drainage at the top of the section, even at great distances from the trench.

The underconsolidation ratio U (defined as the ratio of excess pore pressure at a given time (t) to the total load emplaced by underthrusting) for both solutions is described as a function of a dimensionless time parameter, shown here as adapted for subduction loading by *von Huene and Lee* [1982]:

$$T_{\text{gibson}} = \frac{C_v}{x \cdot v_p \cdot \tan^2(\alpha + \beta)} \tag{7.6a}$$

$$T_{\text{wissa}} = \frac{C_v \cdot x}{v_p \cdot H^2} \tag{7.6b}$$

where x is distance from the trench (m), v_p is subduction rate (m s^{-1}), α and β are the surface slope of the wedge and subduction angle, respectively, and H is the thickness of the underthrust sediment (m). C_v is the coefficient of consolidation ($m^2\ s^{-1}$)

$$C_v = \frac{K}{\gamma_w \cdot m_v} \tag{7.7}$$

where K is sediment hydraulic conductivity ($m\ s^{-1}$), γ_w is specific weight of water ($kg\ m^{-2}\ s^{-2}$), and m_v is sediment volumetric compressibility (Pa^{-1}). Hydraulic conductivity (K) is related to intrinsic permeability (k) by

$$K = \frac{k\rho g}{\mu} \tag{7.8}$$

where ρ is water density ($kg\ m^{-3}$), g is gravitational acceleration ($m\ s^{-2}$), and μ is dynamic viscosity of water (Pa s).

Solutions for U as a function of dimensionless time factor for the *Gibson* [1958] and *Wissa et al.* [1971] models are shown in figures 7.9b and 7.9d, respectively (black solid lines). For the *Wissa et al.* [1971] model, the average value of U within the consolidating underthrust layer is shown (fig. 7.9d). For the *Gibson* [1958] model, U is shown for a depth at 90% of the total depth at any given time (fig. 7.9b). This solution provides an approximation for the consolidation state of underthrust sediment because the underthrust section initially comprises most of the total thickness and represents a decreasing fraction of the total thickness with time (or distance from the trench). However, the approximation introduces negligible error because the degree of underconsolidation predicted by the *Gibson* [1958] model varies only slightly within the deepest ~50% of the sediment section, regardless of time or distance from the trench.

Moreover, the behavior of the two solutions is fundamentally different, regardless of the choice of depth in the *Gibson* [1958] model. Both models predict decreasing underconsolidation with increased dimensionless time factor. However, for the *Gibson* [1958] solution the dimensionless time factor (T_{gibson}) decreases with distance from the trench, and the degree of underconsolidation increases. For the *Wissa et al.* [1971] solution, the time factor (T_{wissa}) increases with distance from the trench, and the degree of underconsolidation decreases.

The two models can be tested using existing data by combining inferred pore pressures with direct laboratory measurements of C_v (or of K and m_v) (fig. 7.9). When inferred underconsolidation ratios from Costa Rica, Nankai, and Barbados are plotted as a function of the two time factors, it is apparent that observations are not consistent with the *Gibson* [1958] model (fig. 7.9b) but are generally consistent with the *Wissa et al.* [1971] model (fig. 7.9d). The *Gibson* [1958] model predicts increasing values of U with increased distance from the trench, whereas inferred values exhibit the opposite trend (fig. 7.9b, table 7.2). The general agreement between inferred values of *U* and the *Wissa et al.* [1971] model implies that in underthrust sediment, at least over the front ~10 km of subduction zones, fluid escape is dominantly upward to a highly permeable décollement zone or to highly permeable pathways through the wedge. This further implies that the décollement (or faults within the wedge) must comprise a very small fraction of the total hydraulic impedance to the seafloor, allowing consolidation to progress as a free boundary at the top of the section would. Recent two-dimensional numerical modeling results for the Costa Rican subduction zone corroborate this inference in that they demonstrate that consolidation within the underthrust section generally proceeds as expected for drainage at a free upper boundary, provided that décollement permeability is at least an order of magnitude greater than sediment permeability [e.g., *Screaton and Saffer*, 2005]. It is noteworthy that data from multiple margins provide a reasonable match to this generalized model, despite large differences in values of sediment properties, v_p, and *H* (see additional discussion in the following section).

Factors That Control Pore Pressure in Subduction Zones

The maintenance of excess fluid pressures can be viewed as the result of a dynamic balance between fluid-pressure diffusion and geologic processes that act as distributed sources of fluid pressure (termed "geologic forcing") [e.g., *Neuzil*, 1995]. Geologic forcing includes processes that act as actual sources of fluid volume (e.g., mineral dehydration reactions), and those that act as virtual fluid sources (e.g., disequilibrium compaction). In subduction zones, geologic forcing is dominated by disequilibrium compaction caused by rapid burial and (in the overriding wedge) lateral tectonic stresses. As sediment is exposed to higher temperatures and pressures, the contribution of mineral dehydration reactions becomes increasingly significant [e.g., *Moore and Vrolijk*, 1992; *Bekins et al.*, 1995].

One- and two-dimensional numerical models have been used to investigate factors influencing pore pressure in trench sediment and accretionary complexes [e.g., *Screaton et al.*, 1990; *Screaton and Ge*, 2000; *Bekins and Dreiss*, 1992; *Saffer and Bekins*, 2002]. In numerical simulations of one-dimensional loading of a sediment column by trench sedimentation and underthrusting, geologic forcing is controlled mainly by the rate of loading (sedimentation rate or taper angle and plate convergence rate); sediment compressibility also plays a role [e.g., *Screaton and Ge*, 2000; *Gamage and Screaton*, 2003]. Sediment permeability is the dominant control on fluid escape, and thus modeled pore pressures.

In two-dimensions, geologic forcing in accretionary complexes is controlled primarily by the thickness of incoming sediment, plate convergence, and the rate of porosity loss with depth and distance from the trench [e.g., *Bekins and Dreiss*, 1992; *Saffer and Bekins*, 2002]. Factors controlling fluid escape include bulk sediment permeability, fault zone permeability, and drainage path length (which is controlled by sediment thickness and taper angle). Using numerical models, *Saffer and Bekins* [2002] demonstrated that sediment permeability, incoming sediment thickness, and plate convergence rate are the dominant controls on pore pressure. Accretionary complexes characterized by low sediment permeability or high rates of geologic forcing should sustain high pore pressures and thus small stable taper angles because critical taper angle depends on temporally and spatially averaged fluid pressure [e.g., *Davis et al.*, 1983]. Conversely, complexes characterized by high sediment permeability or low rates of geologic forcing should maintain larger taper angles. This general relationship between taper angle and sediment permeability is consistent with available lithologic data from active accretionary complexes [*Saffer and Bekins*, 2002]. Recent work has illustrated that for some cases, basement permeability may also play a role in affecting pore pressure along the base of accretionary complexes, particularly for wedges characterized by large taper angles and thin incoming sediment [*Matmon and Bekins*, 2006].

Inferred pore pressures from underthrust sediment at the toes of the Costa Rican, Nankai, and north Barbados subduction zones are generally consistent with the conceptual model that pore pressures reflect a dynamic balance between geologic forcing and fluid escape. For example, sediment underthrust at the Costa Rican margin is poorly drained, despite the fact that permeabilities are 1–3 orders of magnitude higher than at the north Barbados or Nankai margins [*Taylor and Leonard*, 1990; *Saffer et al.*, 2000]. This, in large part, is due to the extremely rapid loading rate at Costa Rica when compared with Nankai or north Barbados (8.5 cm yr^{-1} convergence rate and a margin wedge taper angle of ~14°). Conversely, pore pressures at north Barbados reflect nearly undrained conditions, despite slow loading (2.8 cm yr^{-1} convergence rate and ~4° taper angle). The high pore pressures at Barbados can be attributed to extremely low sediment permeability there.

Although it is clear that increased geologic forcing, increased path length, and decreased permeability should all lead to larger overpressures [e.g., *Neuzil*, 1995], the exact nature of the relationship between fluid pressure (or underconsolidation) and these quantities depends on the drainage geometry and boundary conditions. When the geologic forcing and drainage path length are normalized by the sediment permeability (e.g., the dimensionless time factors from *Wissa et al.* [1971] or *Gibson* [1958]), a clear relationship between fluid pressure within underthrust sediment and these parameters emerges (figs. 7.9a and 7.9b). As noted in the previous section, observed pore pressures for three different margins are most consistent with the model of *Wissa et al.* [1971] [e.g., *von Huene and Lee*, 1982], in which underthrust sediment undergoes a constant rate of loading and drains upward to a high-permeability décollement.

As an illustration of this concept, the solution of *Wissa et al.* [1971] can be used to quantify, to first order, the effect of sediment permeability on pore pressure (Figure 7.10a). In solving for layer-averaged excess fluid pressure in the underthrust sediment, I assume that the loading rate ($\partial\sigma_v/\partial t$) and time coefficient of consolidation (C_v) are constant with distance from the trench (and therefore with time since underthrusting). If k decreases more rapidly with underthrusting than m_v, C_v will decrease. This, in turn, would result in higher fluid pressures and larger underconsolidation ratios than shown in figure 7.10. Likewise, as the décollement itself comprises an increasing fraction of the total hydraulic impedance to the seafloor (as the drainage path length along it increases or if its permeability decreases with distance from the trench), underconsolidation ratios would be higher than shown. Finally, I do not account for fluid sources from mineral dehydration reactions occurring at distances of greater than ~20 km from the trench, which would also lead to higher underconsolidation ratios than shown. Nonetheless, the significant effect of C_v on pore pressures is evident from the results of this simple model. As C_v is decreased from 10^{-8} to 10^{-10} m^2 s^{-1}, underconsolidation ratios at all distances from the trench increase and effective stresses along the wedge base decrease. Higher values of C_v result in lower pore pressures, and significant underconsolidation does not persist as far from the trench as for lower values of C_v. Direct laboratory measurements of C_v over a wide range of effective stresses for sediment underthrust at Costa Rica, Nankai, and Barbados yield values from 6×10^{-9} to 4×10^{-8} m^2 s^{-1}. Interestingly, these values of C_v result in diminished underconsolidation of underthrust sediment by ~20–40 km from the trench (fig. 7.10, red and orange curves), consistent with results of recent seismic reflection work at Nankai [*Bangs et al.*, 2004] and consolidation trends assumed in numerical models of dewatering [e.g., *Saffer and Bekins*, 1998].

Mechanical Implications

One key implication of poorly drained behavior within underthrust sediment is that effective stress, and thus shear strength, do not increase appreciably with increased depth, at least near the toe of many subduction systems [e.g., *Brown et al.*, 2003]. This poorly drained condition provides a plausible and quantifiable mechanism explaining plate boundary fault weakness. A second implication is that changes in effective stress caused by pore pressure diffusion result in zones of minimum mechanical strength downsection (cf. fig. 7.4). These zones, in turn, may control the development and downstepping of faults [e.g., *Saffer*, 2003]. Zones of minimum strength likely migrate downsection with increased burial and drainage, potentially resulting in downward thickening of fault systems or mélange zones [e.g., *Moore and Byrne*, 1987]. A third implication of inferred pore pressures at the Costa Rican, Nankai, and Barbados margins is that the pore-pressure ratio, λ, is spatially variable. Inferred values of λ increase systematically with distance from the trench as burial progresses more rapidly than pore pressures diffuse (e.g., fig. 7.7) and also vary with depth (table 7.2, figs. 7.4 and 7.6) [*Saffer*, 2003]. This spatial variability is important in considering the critical taper angle of accreting wedges [e.g., *Breen and Orange*,

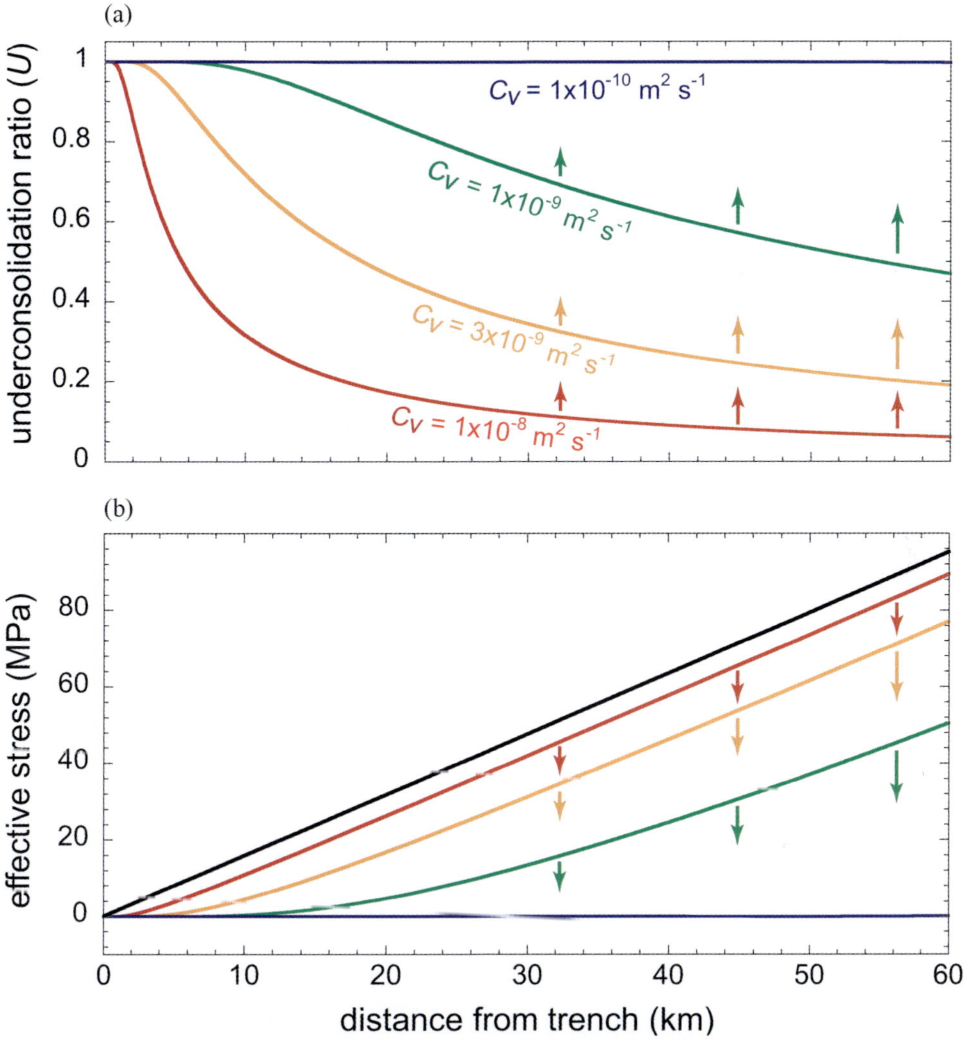

Figure 7.10 (a) Underconsolidation ratio *U* and (b) effective stress versus distance from the trench predicted by the constant rate of strain model of *Wissa et al.* [1971] for values of C_v as labeled in figure 7.10a. Lower values of C_v, which result from lower permeability or higher sediment compressibility, lead to larger underconsolidation ratios (i.e., higher pore pressures) and lower effective stresses. Arrows in both panels show (schematically) the expected deviation from the constant rate of strain solution caused by hydraulic impedance along the décollement, decreasing C_v, and/or additional fluid generation from dehydration reactions (see detailed discussion in text).

1992; *Zhao et al.*, 1986]. The weakening effect of elevated pore pressure may be at least partially counteracted by cementation or other diagenetic processes, which can increase sediment strength (likely associated with a threshold temperature) [e.g., *Morgan and Ask*, 2004], especially in subduction zones characterized by high thermal gradients, such as Nankai.

In a more general sense, through its control on effective stress, the balance between geologic forcing and fluid escape mediates mechanical failure of rocks and sediment. For accretionary complexes, *Saffer and Bekins* [2002] demonstrated that pore pressure (and sediment shear strength), manifested by accretionary wedge taper angle, result from a dynamic balance between factors driving geologic forcing (sediment thickness and plate convergence rate) and those limiting fluid escape (bulk sediment and fault permeability). In an end-member example for intraplate regions characterized by small geologic forcing and high permeability, *Townend and Zoback* [2000] demonstrate that hydrostatic pore pressures should be maintained throughout, leading to high strength. Taken together, these recent results indicate that failure of the brittle crust may be controlled, at least in part, by the relative magnitudes of geologic forcing and hydraulic conductivity. For example, one family of conceptual models explaining the inferred weakness of major transform faults such as the San Andreas suggest a combination of low permeability and fluids sourced from dehydration [e.g., *Irwin and Barnes*, 1975] or the mantle [e.g., *Rice*, 1992].

Spatial variation in pore pressure is also a potential control on the updip limit of the seismogenic zone along subduction megathrusts through its control on effective stress [*Scholz*, 1998; *Moore and Saffer*, 2001]. With increasing effective stress, the tendency for unstable sliding in materials characterized by velocity-weakening frictional behavior increases [e.g., *Scholz*, 1998]. Declining excess fluid pressure, and declining underconsolidation, with distance landward from the trench likely affects the onset of unstable slip at shallow depths. Differences in hydrologic behavior between margins clearly affects the evolution of effective stress with increased burial (fig. 7.10); therefore differences in fluid-pressure generation and sediment permeability are potentially important factors influencing the updip limit of the seismogenic zone.

Summary

Fluid pressure plays an important role in controlling fault and sediment strength; fluids are also potentially significant agents of chemical and heat transport in subduction zones. Estimation of in situ fluid pressures within underthrust sediment at shallow depths at the Costa Rican and Nankai margins, using several techniques, indicates significant overpressures. Estimated fluid pressures obtained from (1) laboratory consolidation tests, (2) projection of effective stress based on porosity loss, and (3) using seismic velocity as a proxy for consolidation state are all in good agreement (table 7.2). These results are also in good agreement with direct measurements where available. Taken together, these results suggest that, at least in some cases, laboratory consolidation test results can be successfully applied to in situ consolidation processes.

Elevated fluid pressures within underthrust sediment reported here as well as by others (see table 7.2 for summary) indicate limited drainage in response to rapid burial by subduction. Increasing excess pore pressure with depth also indicates a component of upward fluid escape, which increases with distance from the trench (cf. figs. 7.4, 7.6, and 7.8). Inferred fluid pressures at several locations are consistent with upward fluid escape to a permeable décollement zone, at least within a few kilometers of the trench (fig. 7.9). It is likely that with increased distance from the trench, the décollement comprises an increasing fraction of the hydraulic impedance to the seafloor (décollement permeability likely decreases with depth, and the path length to the seafloor increases); this should result in excess pore pressures larger than approximated for one-dimensional dewatering of top-drained underthrust sediment (fig. 7.10).

In general, the maintenance of elevated fluid pressures within subducted and accreted sediment can be attributed to a dynamic balance between geologic forcing (rapid burial), and fluid escape (controlled by sediment permeability and drainage path length) (cf. fig. 7.9) [*Neuzil*, 1995; *Saffer and Bekins*, 2002]. Better understanding the controls on pore-pressure development within underthrust and accreted sediment is critical toward constraining sediment and fault strength with increasing depth, driving forces for and rates of fluid expulsion, and may ultimately hold implications for the location of the updip limit of the seismogenic zone [e.g., *Moore and Saffer*, 2001]. However, a number of key parameters remain poorly constrained. These unknowns include porosity and porosity-effective stress relationships at depths >2–3 km, permeability of both fault zones and sediment and their variability with depth, sediment compressibility and mechanical anisotropy, and the magnitude and distribution of geologic forcing from mineral dehydration reactions and from lateral stressing in the fore arc. Future drilling, long-term monitoring, downhole experiments, and laboratory studies will yield improved constraints on these parameters and their spatial distribution, allowing more sophisticated analyses of pressure generation and maintenance.

Acknowledgments

Saffer was supported by NSF grant OCE-0241482 and a JOI-USSAC grant. This research used samples provided by the Ocean Drilling Program (ODP). ODP is sponsored by the U.S. NSF and participating countries under management of Joint Oceanographic Institutions (JOI), Inc. The clarity and presentation of this manuscript have been greatly improved by constructive comments from Glenn Spinelli, Greg Moore, Tim Dixon, and Peter Flemings.

References

Athy, L. F. (1930), Density, porosity, and compaction of sedimentary rocks, Am. Assoc. Petrol. Geol. Bull. 14, 1–24.

Bangs, N., G. Westbrook, J. Ladd, and P. Buhl (1990), Seismic velocities from the Barbados Ridge complex: Indicators of high pore fluid pressures in an accretionary complex, J. Geophys. Res., 95, 8767–8782.

Bangs, N., T. Shipley, J. C. Moore, and G. Moore (1999), Fluid accumulation and channeling along the northern Barbados Ridge decollement thrust, J. Geophys. Res., 104, 20, 399–20,414.

Bangs, N., et al. (2004), Evolution of the Nankai Trough décollement zone from the trench into the seismogenic zone: Inferences from 3-D seismic reflection imaging, Geology, 32, 273–276.

Becker, K., A. T. Fisher, and E. E. Davis (1997), The CORK experiment in hole 949C: Long-term observations of pressure and temperature in the Barbados accretionary prism, Proc. Ocean Drill. Program Sci. Results, 156, 247–252.

Bekins, B. A., and S. J. Dreiss (1992), A simplified analysis of parameters controlling dewatering in accretionary prisms, *Earth Planet. Sci. Lett., 109*, 275–287. Bekins, B. A., A. McCaffrey, and S. J. Dreiss (1995), Episodic and constant flow models for the origin of low-chloride waters in a modern accretionary complex, *Water Resour. Res., 31*, 3205–3215.

Bowers, G. L. (2001), Determining an appropriate pore-pressure estimation strategy,*Proc. Offshore Technol. Conf., 33*, 921–934.

Bray, C. J., and D. E. Karig (1985), Porosity of sediment in accretionary prisms and some implications for dewatering processes, *J. Geophys. Res., 90*, 768–778.

Bray, C., and D. E. Karig (1988), Dewatering and extensional deformation of the Shikoku Basin hemipelagic sediment in the Nankai Trough, *Pure Appl. Geophys., 128*, 725–747.

Breen, N., and D. L. Orange (1992), The effects of fluid escape on accretionary wedges. Variable porosity and wedge convexity, *J. Geophys. Res., 97*, 9265–9275.

Brown, K. M., et al. (2003), Compositional and fluid pressure controls on the state of stress on the Nankai subduction thrust: A weak plate boundary, *Earth Planet. Sci. Lett., 214*, 589–603.

Casagrande, A. (1936), Determination of preconsolidation load and its practical significance, *Proc. Int. Conf. Soil Mech. Found. Eng., 3*, 60–64.

Cochrane, G. R., J. C. Moore, M. E. MacKay, and G. F. Moore (1994), Velocity and inferred porosity model of the Oregon accretionary prism from multichannel seismic data: Implications on sediment dewatering and overpressure, *J. Geophys. Res., 99*, 7033–7043.

Cochrane, G., J. C. Moore, and H. J. Lee (1996), Sediment pore-fluid overpressuring and its effect on deformation at the toe of the Cascadia accretionary prism from seismic velocities, in *Subduction Top to Bottom, Geophys. Monogr. Ser.*, vol. 96, edited by G. Bebout et al., pp. 57–64, AGU, Washington, D. C.

Crawford, C. B. (1986), State of the art: Evaluation and interpretation of soil consolidation tests, in *Consolidation of Soils: Testing and Evaluation, ASTM STP 892*, edited by R. N. Yong and F. C. Townend, pp. 71–103, Am. Soc. for Test. and Mater., Philadelphia, Pa.

Davis, D., J. Suppe, and F. A. Dahlen (1983), Mechanics of fold-and-thrust belts and accretionary wedges, *J. Geophys. Res., 88*, 1153–1172.

Dugan, B., and P. B. Flemings (2000), Overpressure and fluid flow in the New Jersey continental slope; implications for slope failure and cold seeps, *Science, 289*, 288–291.

Erickson, S. N., and R. D. Jarrard (1998), Velocity-porosity relationships for water-saturated siliciclastic sediment, *J. Geophys. Res., 103*, 30,385–30,406.

Flemings, P. B. (2002), Overpressures above the decollement at Nankai: ODP Sites 808 and 1174, *Eos Trans. AGU, 83*(47), Fall Meet. Suppl., abstract H22E-08.

Foucher, J-P., P. Henry, and F. Harmegnies (1997), Long-term observations of pressure and temperature in hole 948D, Barbados accretionary prism, *Proc. Ocean Drill. Program Sci. Results, 156*, 239–245.

Fowler, S. R., R. S. White, and K. E. Louden (1985), Sediment dewatering in the Makran accretionary prism, *Earth Planet. Sci. Lett., 75*, 427–438.

Gamage, K., and E. J. Screaton (2003), Development of overpressures at Nankai accretionary prism, Ocean Drilling Program Sites 1173 and 1174, *Eos Trans. AGU, 84*(46), Fall Meet. Suppl., Abstract T52C-0276. Gibson, R. E. (1958), The progress of consolidation in a clay layer increasing in thickness with time, *Geotechnique, 8*, 171–182.

Hart, B. S., P. B. Flemings, and A. Deshpande (1995), Porosity and pressure: Role of compaction disequilibrium in the development of overpressures in a Gulf Coast Pleistocene basin, *Geology, 23*, 45–48.

Henry, P., et al. (2003), Anisotropy of electrical conductivity record of initial strain at the toe of the Nankai accretionary wedge, *J. Geophys. Res., 108*, 2407, doi:10.1029/2002JB002287.

Hills, D. J., et al. (2001), Preliminary results from integration of 2D PSDM and ODP Leg 196 LWD velocity data in the Nankai accretionary prism, *Eos Trans. AGU, 82*(47), Fall Meet. Suppl., Abstract T42D-0968.

Holtz, R. D., and W. D. Kovacs (2001), *An Introduction to Geotechnical Engineering*, Prentice Hall, Upper Saddle River, N. J.

Housen, B. A., et al. (1996), Strain decoupling across the decollement of the Barbados accretionary prism, *Geology, 24*, 127–130.

Hubbert, M. K., and W. W. Rubey (1959), Role of fluid pressure in mechanics of overthrust faulting, *Geol. Soc. Am. Bull., 70*, 115–166.

Irwin, W. P., and I. Barnes (1975), Effect of geologic structure and metamorphic fluids on seismic behavior of the San Andreas fault system in central and northern California, *Geology,* 3(12), 713–716.

Johnson, P. A., and T. V. McEvilly (1995), Fluids and faulting; suggestive evidence and crucial tests at Parkfield, in *Proceedings of Workshop LXIII;USGS Red-Book Conference on the Mechanical Involvement of Fluids in Faulting*, edited by S. Hickman, U.S. Geol. Surv. Open File Rep., 39–71.

Karig, D. E., and M. V. S. Ask (2003), Geological perspectives on consolidation of clay- rich marine sediment, *J. Geophys. Res.*, 108(B4), 2197, doi:10.1029/2001JB000652.

Karig, D. E., and G. Hou (1992), High-stress consolidation experiments and their geologic implications, *J. Geophys. Res., 97*, 289–300.

Karig, D. E. (1993), Reconsolidation tests and sonic velocity measurements of clay-rich sediment from the Nankai trough, *Proc. Ocean Drill. Program Sci. Results*, 131, 247–260.

Kimura, G., et al. (Eds.) (1997), *Proceedings of the Ocean Drilling Program, Initial Reports*, vol. 170, 458 pp., Ocean Drilling Program, Texas A&M Univ., College Station, Tex.

Lambe, T. W., and R. V. Whitman (1969), *Soil Mechanics*, 553 pp., John Wiley, Hoboken, N. J.

Lee, H. J. (1985), State of the art: Laboratory determination of the strength of marine soils, in *Strength Testing of Marine Sediment and in Situ Measurements, ASTM STP 883*, edited by R. C. Chaney and K. R. Demars, pp. 181-250, Am. Soc. for Test. and Mater., Philadelphia, Pa.

Matmon, D., and B. A. Bekins (2006), Hydromechanics of a high taper angle, low permeability prism: A case study from Peru, *J. Geophys. Res., 111*, B07101, doi:10.1029/2005JB003697,

McKiernan, A. W., and D. M. Saffer (2005), Data report: Permeability and consolidation characteristics of sediments collected during ODP Leg 205, Costa Rica, *Ocean Drill. Program Sci. Results*, in press.

Mitchell, J. K. (1993), *Fundamentals of Soil Behavior*, 437 pp., John Wiley, Hoboken, N. J.

Moore, G. F., et al. (2001), New insights into deformation and fluid flow processes in the Nankai Trough accretionary prism: Results of Ocean Drilling Program Leg 190, *Geochem., Geophys., Geosyst.,* 2(10), doi:10.129/2001GC000166.

Moore, J. C., and T. Byrne (1987), Thickening of fault zones: A mechanism of melange formation in accreting sediment, *Geology, 15*(11), 1040–1043.

Moore, J. C., and D. M. Saffer (2001), Updip limit of the seismogenic zone beneath the accretionary prism of southwest Japan: An effect of diagenetic to low-grade metamorphic processes and increasing effective stress, *Geology, 29*(2), 183–186.

Moore, J. C., and H. Tobin (1997), Estimated fluid pressures of the Barbados accretionary prism and adjacent sediment, *Proc. Ocean Drill. Program Sci. Results, 156*, 229–238.

Moore, J. C., and P. Vrolijk (1992), Fluids in accretionary prisms, *Rev. Geophys., 30*, 113–135.

Moore, J. C. (1989), Tectonics and hydrogeology of accretionary prisms: Role of the décollement zone, *J. Struct. Geol., 11*, 95–106.

Morgan, J. K., and D. E. Karig (1993), Ductile strains in clay-rich sediment from Hole 808C; preliminary results using X-ray pole figure goniometry, *Proc. Ocean Drill. Program Sci. Results, 131*, 141–155.

Morgan, J. K., and M. V. S. Ask (2004), Consolidation state and strength of underthrust sediment and evolution of the décollement at the Nankai accretionary margin: Results of uniaxial reconsolidation experiments, *J. Geophys. Res., 109*, B03102, doi:10.1029/2002JB002335.

Morgan, J. K., E. B. Sunderland, and M. V. S. Ask (2007), Deformation and diagenesis at the Nankai subduction zone: Implications for sediment mechanics, décollement initiation and propagation, this volume.

Neuzil, C. E. (1995), Abnormal pressures as hydrodynamic phenomena, *Am. J. Sci., 295*, 742–786.

Rice, J. R. (1992), Fault stress states, pore pressure distributions, and the weakness of the San Andreas fault, in *Fault Mechanics and Transport Properties of Rocks*, edited by B. Evans and T. F. Wong, pp. 475-503, Elsevier, New York.

Rubey, W. W., and M. K. Hubbert (1959), Role of fluid pressure in mechanics of verthrust faulting II: Overthrust belt in geosynclinal area of western Wyoming in light of fluid-pressure hypothesis, *Geol. Soc. Am. Bull.*, 70, 167–206.

Saffer, D. M., and B. A. Bekins (1998), Episodic fluid flow in the Nankai accretionary complex: Timescale, geochemistry, flow rates, and fluid budget, *J. Geophys. Res., 103*, 30,351–30,371.

Saffer, D. M. (2003), Pore pressure development and progressive dewatering in underthrust sediment at the Costa Rican subduction margin: Comparison with Northern Barbados and Nankai, *J. Geophys. Res., 108*(B5), 2261, doi:10.1029/2002JB001787.

Saffer, D. M., and B. A. Bekins (2002), Hydrologic controls on the mechanics and morphology of accretionary wedges and thrust belts, *Geology, 30* (30), 271–274.

Saffer, D. M., et al. (1997), Consolidation, pore fluid pressures, and implications for dewatering pathways of subducted sediment, offshore Costa Rica, *Eos Trans., AGU, 78*, 685.

Saffer, D. M., E. A. Silver, A. T. Fisher, H. Tobin, and K. Moran (2000), Inferred pore pressures at the Costa Rica subduction zone: Implications for dewatering processes, *Earth Planet. Sci. Lett., 177*, 193–207.

Scholz, C. H. (1998), Earthquakes and friction laws, *Nature, 391*, 37–42.

Screaton, E. J., D. R. Wuthritch, and S. J. Dreiss (1990), Permeabilities, fluid pressures, and flow rates in the Barbados ridge complex, *J. Geophys. Res., 95*, 8997–9007.

Screaton, E. J., and S. Ge (2000), Anomalously high porosities in the proto-decollement zone of the Barbados accretionary complex: Do they indicate overpressures?, *Geophys. Res. Lett., 27*, 1993–1996.

Screaton, E. J., D. M. Saffer, P. Henry, S. Hunze, and Leg 190 Shipboard Scientific Party (2002), Porosity loss within the underthrust sediments of the Nankai accretionary complex: Implications for overpressures, *Geology, 30*(1), 19–22.

Screaton , E. J., and D. M. Saffer (2005), Fluid expulsion and overpressure development during initial subduction at the Costa Rica convergent margin, *Earth Planet. Sci. Lett., 233*, 361–374.

Shi, Y., and C-Y. Wang (1988), Generation of high pore pressure in accretionary prisms: Inferences from the Barbados ridge complex, *J. Geophys. Res., 93*, 8893–8909.

Taylor, E., and J. Leonard (1990), Sediment consolidation and permeability at the Barbados forearc, *Proc. Ocean Drill. Program Sci. Results, 110*, 289–308.

Tobin, H. J., and D. Saffer (2003), Fluid pressure in the shallow plate interface at the Nankai Trough subduction zone, *Eos Trans. AGU, 84*(46), Fall Meet. Suppl., Abstract T41A-04.
Townend, J., and M. D. Zoback (2000), How faulting keeps the crust strong, *Geology, 28*(5), 399–402.
von Huene, R., and H. Lee (1982), The possible significance of pore fluid pressures in subduction zones, in *Studies in Continental Margin Geology*, edited by J. S. Watkins and C. L. Drake, *AAPG Mem., 34,* 781–791.
von Huene, R., et al. (1998), Mass and fluid flux during accretion at the Alaska margin, *Geol. Soc. Am., Bull., 110,* 468–482.
Vrolijk, P., T. Miller, and M. J. Gooch (1998), Hydrostatic consolidation tests of undeformed, clay-rich samples from the Barbados accretionary prism, Leg 156,
Proc. Ocean Drill. Program Initial Rep., 171A., 107–116.
Wang, K. (1994), Kinematics of dewatering accretionary prisms, *J. Geophys. Res., 99,* 4429–4438.
Westbrook, G. K. (1991), Geophysical evidence for the role of fluids in accretionary wedge tectonics, *Philos. Trans. R. Soc. London, Ser. A, 335,* 227–242.
Wissa, A. E. Z., et al. (1971), Consolidation at a constant rate of strain, *J. Soil Mech. Eng. Div., Am. Soc. Civ. Eng., 97,* SM10.
Zhao, W-L., et al. (1986), Origin of convex accretionary wedges: Evidence from Barbados, *J. Geophys. Res., 91,* 10,246–10,258.
Zwart, G., et al. (1997), Evaluation of hydrogeologic properties of the Barbados accretionary prism: A synthesis of Leg 156 results, *Proc. Ocean Drill. Program Sci. Results, 156,* 303–310.

8 Deformation and Mechanical Strength of Sediments at the Nankai Subduction Zone

Implications for Prism Evolution and Décollement Initiation and Propagation

Julia K. Morgan, E. Blanche Ramsey*, and Maria V. S. Ask

Abstract

As in many depositional settings, marine sediments at convergent margins undergo diagenetic changes before, during, and after mechanical consolidation and deformation. These changes influence mechanical behavior within and beneath the prism and along the décollement. To illustrate the interrelations between sediment diagenesis and deformation, we review physical properties and the types and distributions of deformation structures at several Ocean Drilling Program (ODP) and Deep Sea Drilling Project (DSDP) drill sites from the frontal regions of the Nankai accretionary prism. Both compactive and dilative deformation structures and fabrics are documented, denoting complicated stress paths during consolidation and tectonic deformation. Laboratory deformation experiments conducted on selected samples from these sites also demonstrate enhanced sediment strengths relative to their preconsolidation stress, both above and below the décollement horizon. This mechanical response indicates the presence of intergranular bonding or cementation that allows sediments to resist consolidation and deformation to relatively high stresses. Once their shear strengths are exceeded, however, cemented sediments can undergo rapid failure, leading to transient increases in pore pressures followed by consolidation. This deformation history may account for the localized compactive deformation bands within the prism. An analogous sequence may develop at depth within the underthrust sediments. Stress perturbations, e.g., near the updip limit of the seismogenic zone, may locally exceed the enhanced shear strengths of the underthrust sediments, leading to compactive failure and release of trapped pore fluids. Associated increases in pore-fluid pressures may enable décollement down-cutting and tectonic un-

*Formerly E. Blanche Sunderland

derplating. The resulting changes in structural and physical properties of the sediments may favor the onset of seismogenic slip along the décollement.

Introduction

Active convergent margins are the locus of some of the largest earthquakes on Earth, a consequence of the combined effects of low-angle fault geometries, cold subducting lithosphere, and the compressional stress environment, that favor large seismic rupture areas of the fault. The extent of the seismically active portions depends on temperature and pressure distributions within the crust and also, the compositional, elastic, and frictional properties of the fault and surrounding rocks [e.g., *Scholz*, 1990; *Tichelaar and Ruff*, 1993; *Hyndman et al.*, 1995, 1997; *Marone*, 1998; *Oleskevich et al.*, 1999]. Earthquake rupture, however, can extend updip along the fault into regions capable of supporting coseismic slip. Coseismic rupture may propagate onto splay faults and finally to the seafloor, causing rapid ground motions capable of generating devastating tsunami [e.g., *Ando*, 1991; *Park et al.*, 2000, 2002; *Tanioka and Satake*, 2001]. The frontal portions of the plate boundary fault may not sustain coseismic sliding but rather undergo aseismic slip as the overriding accretionary prism deforms during interseismic periods.

Significant changes must take place along the plate-boundary fault to account for the transition from seismic to aseismic slip. At the most frontal portions of accretionary prisms, porous sediments are partitioned by the décollement fault into accreted and underthrust packages that follow separate and distinct trajectories through the system [e.g., *Karig and Morgan*, 1994]. Such porous sediments cannot support high elastic shear strains and stress drops associated with earthquakes. At depth, progressive lithification, alteration, and probably pressurization, convert the marine sediments into rocks capable of unstable, seismogenic sliding. The detailed evolution in mechanical properties responsible for this change in behavior is not well understood; it is likely that many factors come into play, including evolving stress conditions, pore-fluid pressures, physical properties, mineral fabric, and pressure- and temperature-induced diagenetic changes, and rheologies [e.g., *Byerlee*, 1990; *Logan and Rauenzahn*, 1987; *Hyndman et al.*, 1995; *Marone*, 1998; *Moore and Saffer*, 2001].

The Nankai accretionary margin, located southeast of Japan (fig. 8.1), is a prime location to investigate transitions in physical and mechanical properties along the décollement and the surrounding wall rocks, and their influence of sliding behavior. The margin has a long history of damaging earthquakes with a mean recurrence interval of ~180 yrs [*Ando*, 1975, 1991]. Ongoing microearthquakes along a well-instrumented transect offshore of Shikoku Island, cluster close to the updip limit of the 1946 Nankaido earthquake rupture [*Obana*, 2001], thereby constraining the position of the seismic-aseismic transition or at least its lower bound [e.g., *Norabuena et al.*, 2004]. A recent three-dimensional

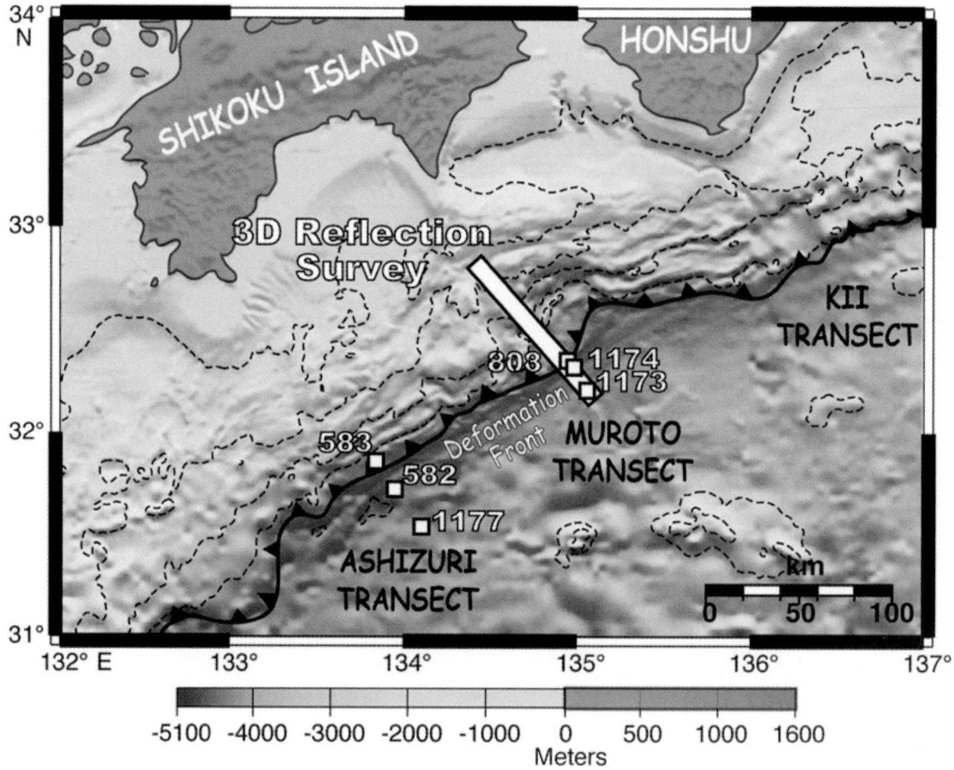

Figure 8.1 Location map for Deep Sea Drilling Project (DSDP) and Ocean Drilling Program (ODP) drill sites in the Nankai Trough and frontal accretionary prism. Sites 808, 1174, and 1173 lie along the Muroto transect; the Ashizuri transect is located ~100 km to the southwest. A three-dimensional (3-D) seismic reflection survey parallels the Muroto transect.

(3-D) seismic reflection survey along a transect parallel to Muroto Peninsula on Shikoku Island provides unprecedented imaging of the décollement in this area, as well as the frontal accretionary prism and incoming sedimentary section [*Moore et al.*, 2001a, 2001b]. As discussed below, several key changes in deformation structure, seismic properties, and fluid pressures converge near the inferred updip limit of the seismogenic zone, which in combination, may enable the onset of unstable sliding along the fault. Additionally, seismogenic sliding may trigger rapid changes in physical properties and mechanical state that may account for the features described above.

To date, no samples have been acquired from the seismogenic zone along the Nankai margin, although this is the target of future deep drilling efforts [*Kimura et al.*, 2003]. In the meantime, we must infer deep processes and properties based on materials and measurements collected from more accessible portions of the prism. Fortunately, the Nankai margin has been the site of many drilling ventures by the Ocean Drilling Program (ODP) and its predecessor, the Deep Sea Drilling Project (DSDP). Three ODP drill sites penetrate the frontal portion of the

Nankai accretionary prism in line with the 3-D seismic reflection survey, referred to here as the Muroto transect (fig. 8.1), and these sites can be compared with other sites that lie along the parallel Ashizuri transect, ~100 km to the southwest. In combination with the seismic data, the drill sites provide basic information about the physical properties, structure, and stratigraphy of the prism and the incoming sedimentary section [*Shipboard Scientific Party* (hereinafter referred to as SSP), 1991, 2001a, 2001b]. These observations are augmented by results of laboratory experiments and analyses carried out to better constrain the physical and mechanical properties of the prism materials.

Here, we review the characteristics of the incoming sedimentary sections and synthesize observations and results relating to structural and mechanical evolution of sediments within and below the Nankai accretionary prism along the Muroto transect. We develop a conceptual understanding of how sediments evolve near the toe of the prism and consider the consequences of ongoing sediment diagenesis on deformation behavior within the prism and along the plate-boundary décollement fault.

Large-Scale Characteristics of the Nankai Margin

The Nankai accretionary prism lies at the boundary between the Eurasian and Philippine Sea plates, off the southwest coast of Japan. The inactive Shikoku back-arc basin and overlying sediments are presently being subducted northwestward at a relative convergence rate of ~2–4 cm/yr, normal to the margin [*Seno et al.*, 1993]. A broad trench, the Nankai Trough, has formed above the downgoing plate and gives way to a landward sequence of ridges and benches defining the imbricate stack of thrust sheets forming at the toe of the Nankai accretionary prism (fig. 8.1). The margin is the site of several Deep Sea Drilling Project and Ocean Drilling Program surveys, including Legs 87, 131, 190, and 196 [e.g., *Kagami et al.*, 1986; *Hill et al.*, 1991; *Moore et al.*, 2001b; *Mikada et al.*, 2002]. A suite of six drill sites was cored during Legs 131 and 190 along a transect online with Muroto Point on Shikoku Island (fig. 8.1). We will focus on the drill sites penetrating the frontal portions of the prism and the seaward Shikoku Basin: Site 1173 samples a reference site through the incoming Shikoku Basin sequence; Site 1174, penetrates the protothrust zone and an incipient thrust fault; and the older Site 808, which passes through the frontal thrust fault, active décollement zone, and the underthrust section below. Both Sites 1173 and 808 were revisited for logging and long-term fluid monitoring operations by ODP Leg 196 in 2001 [*Mikada et al.*, 2002]. A comparative reference Site 1177 was drilled ~100 km to the southwest along the parallel Ashizuri transect (fig. 8.1), penetrating the deep incoming sedimentary section. Site 582 and 583, drilled during DSDP Leg 87, sampled shallower strata along this transect, within the Nankai Trough and the frontal accretionary prism, respectively.

The large-scale architecture of the Nankai accretionary margin along the Muroto Transect has been revealed by seismic reflection surveys [*Moore et al.*, 2001a, 2001b; *Gulick et al.*, 2004; *Bangs et al.*, 2004]. The basal décollement lies within landward dipping Shikoku Basin strata at the deformation front and is traced along a common acoustic reflection beneath the toe of the prism. At ~30–40 km from the deformation front, the taper of the wedge increases, reflecting a large offset out-of-sequence thrust fault that rises from the décollement horizon; the décollement appears to step down to the top of the igneous crust in the landward direction; a large-offset out-of-sequence thrust fault splays from the décollement at this point (fig. 8.2). The décollement step-down approximately coincides with the updip limit of the seismogenic zone, on the basis of the modeled rupture zone of the 1946 Nankaido earthquake [*Ando*, 1991; *Tanioka and Satake*, 2001], and the locked region of the fault [e.g., *Hyndman et al.*, 1995]. Increased microseismicity within 30–40 km from the deformation front falls close to the updip limit of seismogenic zone defined by thermal models [*Obana et al.*, 2001]. The strong reflectivity of the décollement horizon beneath the frontal prism is attributed to a high impedance contrast, due to the downward increase in porosity and possibly pore pressures [e.g., *Moore et al.*, 1990, 1991; *Moore and Shipley*, 1993; *SSP*, 1991, 2001a, 2001b]. A decrease in reflection amplitude near the step-down has been interpreted to denote dewatering of the underthrust section [*Bangs et al.*, 2004]. Hydrologic modeling of steady-state fluid flow indicates that the highest excess pore pressures should lie 20–30 km landward of the deformation front [*Saffer and Bekins*, 1998], which is close to the updip limit of the seismogenic zone [*Moore and Saffer*, 2001].

The Frontal Nankai Prism, Muroto Transect

The incoming sedimentary section on the Philippine Sea plate above the oceanic crust is dominated by bioturbated hemipelagic sediments of the Shikoku Basin sequence, deposited between the early-mid Miocene until the Quaternary (i.e., ~20–0.26 ma) [*SSP*, 1991, 2001a, 2001b; *Moore et al.*, 2001b; *Underwood*, this volume]. The Upper Shikoku Basin (USB) facies is distinguished by interbedded ash layers, which are absent within the lower Shikoku Basin (LSB) facies. The broad Nankai Trough is filled by a northwestward thickening wedge of interbedded turbidites and hemipelagic sediments accumulating today (fig. 8.3) [*Underwood et al.*, 1993; *Underwood*, this volume]. Onset of tectonic deformation is marked by a slight upward deflection of the flat-lying trench-fill strata and stratal thickening and incipient thrust faulting within the protothrust zone [*Moore et al.*, 1990, 1991]. An orderly stack of imbricate thrust sheets [e.g., *Gulick et al.*, 2004] marks the frontal portion of the prism. The basal décollement extends beneath the prism and protothrust zone and lies entirely within the LSB facies; its seaward projection along the age-equivalent horizon is referred to as the protodécollement. Three ODP drill sites intersect the

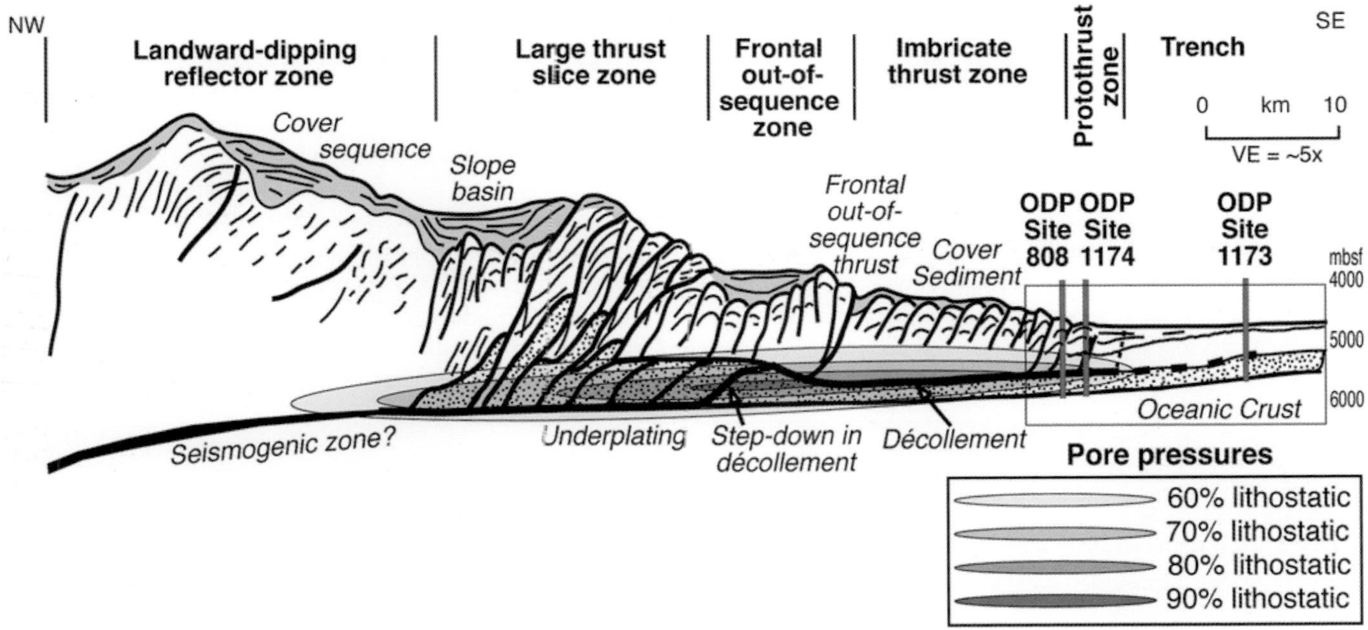

Figure 8.2 Interpretive cross section across the Muroto transect of the Nankai accretionary margin, showing structural domains (modified after *Moore et al.* [2001b]). A step down in the décollement near the frontal out-of-sequence thrust coincides with modeled peak pore-fluid pressures (shaded ellipses) from *Saffer and Bekins* [1998]. The aseismic to seismic transition lies just landward of this zone. Boxed region defines frontal prism, enlarged in figure 8.3. Stippled region denotes underthrust sediments that become underplated landward of the out-of-sequence thrust.

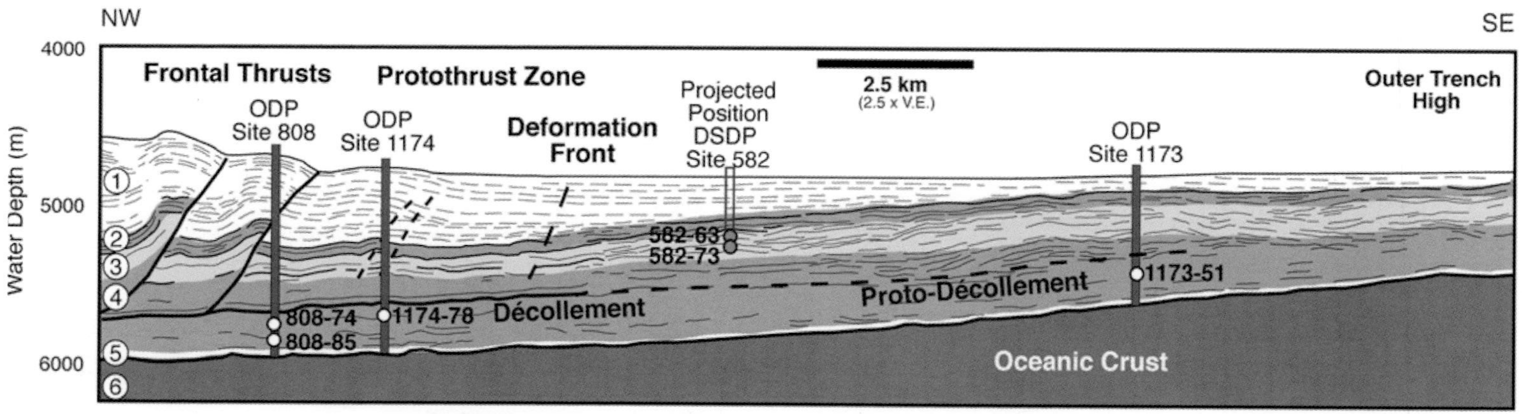

Figure 8.3 Interpreted cross section across frontal portion of the Nankai accretionary prism, along the Muroto Transect. Simplified stratigraphy is indicated by numbered units: landward thickening trench wedge (1), on-laps the trench-to-basin transition (2). The Upper Shikoku Basin (i.e., USB, 3) and Lower Shikoku Basin (i.e., LSB, 4) facies are underlain by thin volcaniclastic unit (5) and basaltic oceanic crust (6). The décollement zone lies within the Lower Shikoku Basin facies, and is correlated seaward of the deformation front along the age-equivalent horizon referred to as the protodécollement. The locations of the three ODP drill sites are shown; reference Site 1173 lies within the Shikoku Basin, Site 1174 passes through the protothrust zone, and Site 808 penetrates the frontal thrust fault. DSDP Site 582 to the southwest is projected into this transect. Locations of samples discussed in this study are indicated by white circles (gray circles for Site 582).

trench and frontal portion of the prism (fig. 8.3). The full stratigraphic section was penetrated at each site.

There are four main structural domains at the toe of the prism, each exhibiting distinct characteristics indicative of their compactive, diagenetic, and tectonic histories. These domains are distinguished as (1) reference section at Site 1173, presumed to have experienced no tectonic deformation, including units both above and below the protodécollement horizon; (2) tectonically deformed strata found above the décollement zone within the protothrust zone and frontal thrust sheets; (3) the underthrust section found beneath the décollement, landward of the deformation front; and (4) the décollement zone itself, which separates the accreted and underthrust packages.

Physical Properties

Shipboard porosity measurements from the three sites [*SSP*, 1991, 2001a, 2001b; *Screaton et al.*, 2002] provide a general overview of spatial changes in physical properties (fig. 8.4). The downward decreasing trend in porosity, η (see table 8.1 for all symbol definitions) at Site 1173 reverses at the top of the USB, at ~100 meters below sea floor (mbsf), and maintains a nearly constant porosity range of 65–68% to a depth of ~325 mbsf (fig. 8.4c). At this depth, porosities decrease to ~52 % over <100 m, and subsequently resume their steady downward decline. The sharp decrease in porosities coincides with the USB-LSB boundary and accompanies the downward loss of interbedded ash layers [*SSP*, 2001a]. Acoustic compressional velocities V_p increase steadily across the USB and its lower boundary, rising to ~1800 m/s at the protodécollement horizon [*SSP*, 2001a], indicating steadily increasing particle coupling downhole. The porosity step in the absence of the velocity change at the USB-LSB boundary suggests some type of diagenetic transformation that enables renewed consolidation without loss of intergranular contact [e.g., *SSP*, 2001a]. Acoustic velocities step down to ~1750 m/s below the protodécollement, while porosities show a very subtle step up, patterns again suggestive of an as yet undetermined diagenetic boundary.

Sites 1174 and 808 also show slight anomalies in porosities across the USB unit, but these are muted by comparison (figs. 8.4a and 8.4b). Slight downward increases in porosities occur across the frontal thrust fault at Site 808, ~410 mbsf, and across the protothrust fault at Site 1174, ~470 mbsf, probably reflecting incomplete consolidation of the footwalls following overthrusting [e.g., *SSP*, 1991]. More pronounced steps up in porosity occur across the basal décollement at both sites, ~964 mbsf at Site 808 and ~840 mbsf at Site 1174 (figs. 8.4a and 8.4b). Acoustic velocity profiles at these two sites are also complicated. Generally, velocities decrease in concert with porosity increases. Exceptions to this rule at Site 1174 include the lower trench wedge facies (i.e., outer trench wedge) and the top of the USB facies, where velocities are

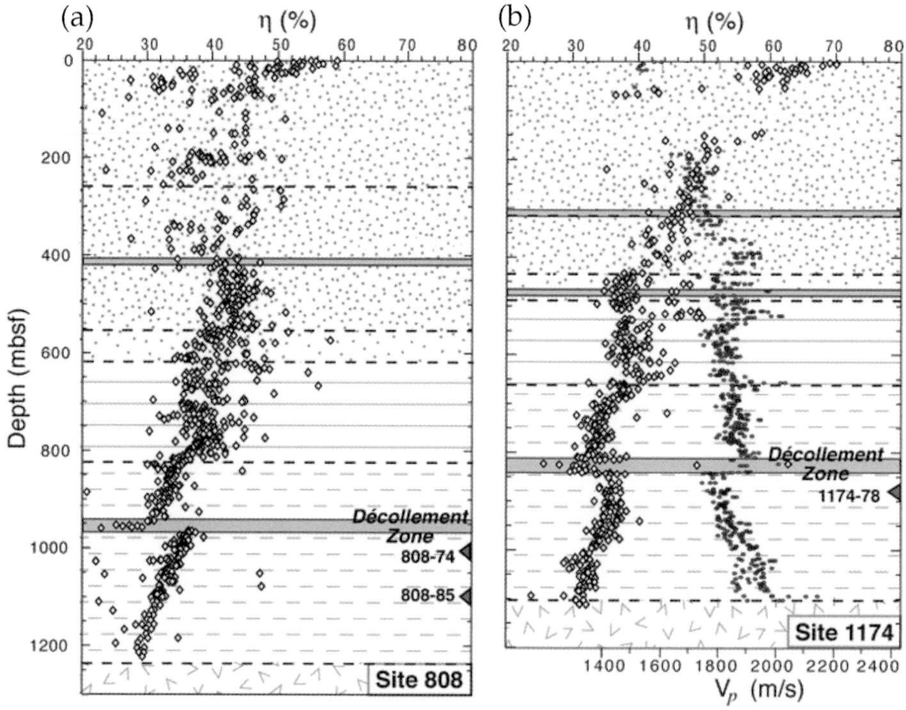

Figure 8.4 Shipboard porosities for the three ODP drill sites, and acoustic velocities for two sites: (a) Site 808, (b) Site 1174, (c) Site 1173 [*SSP*, 1991, 2001a]. Porosity values plotted here are not corrected for smectite interlayer water released during oven drying [*Brown and Ransom*, 1996]; the low smectite abundance in the bulk sediments (~10 wt %) at Nankai results in corrections of <1%. The locations and approximate thicknesses of the décollement zone and faults are indicated. Heavy dashed line shows calculated effective hydrostatic vertical stress with depth, σ_{vh}'. Sample positions and labels are denoted to right of each plot.

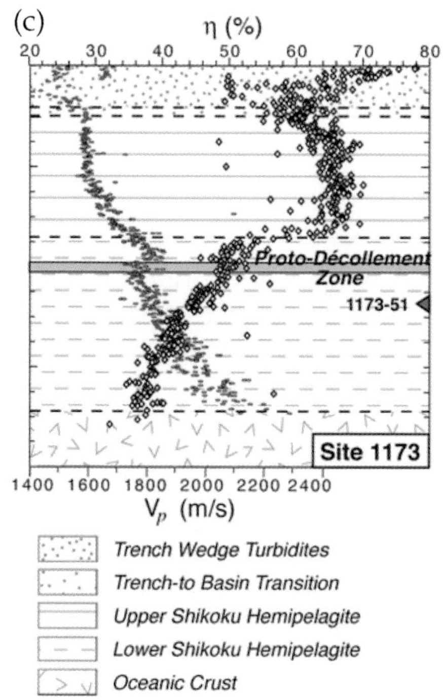

TABLE 8.1 **Common Symbols Used in Text**

Symbol	Definition
η	sediment porosity (%), volume of pore fluid/total volume
P_*	excess pore-fluid pressure (MPa)
σ_v	total vertical stress (MPa)
σ_v'	effective vertical stress (MPa)
σ_h'	effective horizontal stress (MPa)
σ_{vh}'	in situ effective vertical stress for hydrostatic pore pressure (MPa)
K_0	stress ratio during uniaxial strain, $K_0 = \sigma_h'/\sigma_v'$
$\Delta\sigma$	differential stress in tests (MPa), $\Delta\sigma = \sigma_v' - \sigma_h'$
σ_m'	effective mean stress in tests (MPa), $\sigma_m' = (\sigma_v' + 2\sigma_h')/3$
σ_y'	effective vertical yield stress in tests (MPa)
$\Delta\sigma_{peak}$	peak strength in tests (MPa)
σ'_{peak}	effective vertical peak stress at $\Delta\sigma_{peak}$ in tests, (MPa)
σ_c'	effective preconsolidation stress of sample (MPa), estimated from η and figure 8.3
σ_{v*}'	in situ effective vertical stress (MPa)
$C_{c\eta}$	modified compression index, $C_{c\eta} = -\eta/\Delta\log(\sigma_v')$
V_p	compression velocity

anomalously high. The former anomaly terminates at the base of the trench wedge facies, where pore-water profiles of both SiO_2 and Cl^+ begin gradual downhole declines [*SSP*, 2001b]. Again, these distinctive patterns may reflect diagenesis within the sedimentary column.

Some of the physical properties distributions documented in the Shikoku Basin units and overlying strata along the Muroto transect are mirrored at comparable drill sites along the Ashizuri transect ~100 km to the southwest, although with slight variations. A downhole increase in porosity at the top of the USB facies was noted at DSDP Site 582 [*Bray and Karig*, 1986], which projects along strike between Site 1173 and the deformation front (fig. 8.3). Sampling at ODP Site 1177 began in the lower USB facies farther seaward of Site 582 but also documented anomalous acoustic velocities in that unit and a downhole decrease in porosity at the USB-LSB boundary [*SSP*, 2001c]. The high porosities observed in the USB unit at Site 582 were previously interpreted to result from overpressures [e.g., *Bray and Karig*, 1986, 1988], but reconsolidation experiments on sediment cores showed unusually high yield strengths, more consistent with cementation [e.g., *Karig*, 1993]. In sharp contrast to the monotonous hemipelagic sediments that make up the LSB facies along the Muroto transect,

Site 1177 penetrated a ~330 m thick package of interbedded turbidites, which are characterized by porosities of 40% and higher [*SSP*, 2001c]. Although not the immediate focus of the present study, the along-strike variations in lithostratigraphy, and associated mineralogy and hydrogeology, must play an important role in the regional evolution of the Nankai prism and underlying décollement zone [e.g., *Moore et al.*, 1990; *Underwood*, this volume].

Deformation Structures and Fabrics

Ductile Deformation

Deformation within the Nankai prism sediments can take many forms, dependent on the stress history of the given domain [e.g., *Karig and Morgan*, 1994]. All units experience similar initial stages associated with deposition and subsequent burial at the seafloor. This normal compaction phase results in vertical shortening and consolidation, producing a downhole reduction in porosity, accompanied by the rotation of platy clay minerals into axisymmetric arrangements subparallel to bedding [e.g., *Behrmann and Kopf*, 1993; *Morgan and Karig*, 1993]. Both deformation modes are ductile because they impart bulk changes in physical properties and microfabric. Tectonic loading can also cause ductile deformation of the sediment matrix, and potentially, further consolidation. This results from increased horizontal stresses that cause horizontal flattening of pore spaces and rotations of clay minerals away from the bedding planes. Using the method of X-ray pole figure goniometry, *Morgan and Karig* [1993, 1995b] quantified the clay mineral orientation distributions at Site 808, demonstrating axisymmetric arrangements of clay minerals parallel to bedding below the décollement consistent with uniaxial consolidation history (fig. 8.5).

Figure 8.5 Results of X-ray pole figure goniometry analyses for Site 808 samples [*Morgan and Karig*, 1995b]. (a) Clay mineral orientation distribution reveals a moderate total anisotropy (A_3) between 20 and 30%, reflecting vertical flattening of sediments during consolidation. Lateral anisotropy (A_3) is ~10% above the décollement, denoting rotation of clay minerals away from bedding planes in response to horizontal tectonic shortening. A_3 decreases to 0–5% below the décollement, and A_2 increases slightly, consistent with uniaxial strain state. (b) Strain ellipsoid shape factor distinguishes the distortion of the sediment in response to tectonic and compaction strains. Nearly all samples lie within the field of flattening, reflecting the persistence of the bedding-planar clay fabric, but increasing constriction with depth implies increasing horizontal shortening. Again, the underthrust sediments exhibit almost pure flattening. Breccia fragments from within the décollement exhibit a gradation from flattening at the base, to modest constriction near the top, suggesting progressive incorporation of increasingly ductilely deformed sediments as the décollement grows.

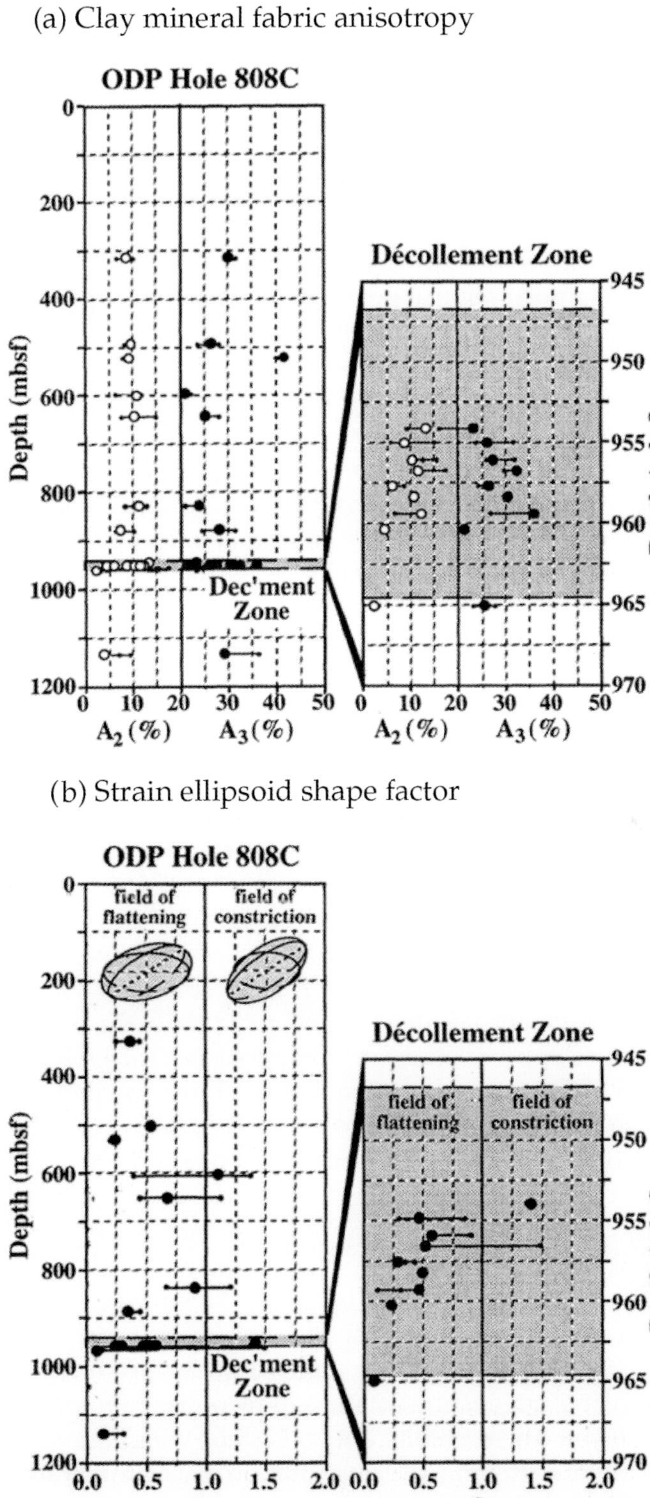
(a) Clay mineral fabric anisotropy
ODP Hole 808C
Décollement Zone
Dec'ment Zone
Depth (mbsf)
A2 (%)
A3 (%)
(b) Strain ellipsoid shape factor
ODP Hole 808C
field of flattening
field of constriction
Décollement Zone
Dec'ment Zone
Depth (mbsf)
k

Up to 10% lateral anisotropy was evident above the décollement at Site 808, based on orthorhombic clay mineral distributions within the prism above the décollement (fig. 8.5b). If clay minerals are assumed to rotate passively within the sediment matrix [*March*, 1932], then the shape of the bulk strain ellipsoid can be calculated [*Oertel*, 1983, 1985], documenting the ductile components of horizontal and vertical shortening. All samples measured within the accreted section at Site 808 yielded lateral anisotropies ~10%, which equates to nominal ductile horizontal shortening of ~10% [*Morgan and Karig*, 1993].

More detailed information about sediment microfabrics has been extracted from scanning electron microscopy (SEM) images [*Ujiie et al.*, 2003; *Sunderland and Morgan*, 2004]. Above the protodécollement zone at the reference site, sediments exhibit a wide range of grain sizes and open, randomly oriented microstructures (fig. 8.6a). A fine-grained clay phase is dispersed throughout the samples, adhering to the surfaces of larger grains. The larger grains define subhorizontal mineral preferred orientations, although local deviations up to ~30° were recognized [*Sunderland and Morgan*, 2004]. Below the protodécollement horizon, clay minerals are typically larger than above and more uniform in size and shape (fig. 8.6b). The distinctive fine-grained phase is absent. At these depths, clay mineral preferred orientations are dominantly subhorizontal; distinctive large pores, some of which might derive from ash dissolution [*Tribble and Wilkens*, 1994], persist throughout the core, i.e., to depths of at least 700 mbsf. *Ujiie et al.* [2003] noted the presence of fine-grained clay aggregates above and within the décollement horizon at Site 1173, which they interpreted to represent a bonding phase; similarly, these aggregates are absent in the deeper sediments.

SEM images of sediment microfabrics at Site 1174 show evidence for primary vertical consolidation followed by a distinct tectonic overprint [*Sunderland and Morgan*, 2004]. Similar to Site 1173 above the protodécollement, sediments in the shallowest units at Site 1174 exhibit open microstructures and the presence of the dispersed fine-grained clay phase (fig. 8.7a). This fine-clay phase, however, disappears about midway through the USB facies by at least 690 mbsf (fig. 8.7b); this is more than 100 m above the décollement horizon and therefore shallower than at the reference site. Below this depth, clay minerals exhibit more uniform shapes and sizes, reminiscent of the subdécollement grains at the reference site (figs. 8.7c and 8.7d). Sediment microfabrics within the protothrust zone, particularly in the shallower reaches, display a subtle domainal structure, defined by packets of subparallel clay platelets separated by less well-oriented grains (fig. 8.7a). This fabric is most apparent in samples close to faults and other discrete deformation structures [*Sunderland and Morgan*, 2004]. Furthermore, the aligned grains are inclined up to 70° to the horizontal plane, suggesting rotation in response to horizontal shortening (figs. 8.7a and 8.7b). These results are consistent with X-ray goniometry evidence for horizontal flattening, causing rotation of clay minerals away from the

(a)

(b)

Figure 8.6 Representative simultaneously extracted metals (SEM) images of LSB sediments from Site 1173. (a) Sample 190-1173A-35X-6, 114–118 cm (330.2 mbsf), taken from above the protodécollement horizon at the seaward reference site, shows wide size range of clay particles and random microfabric. (b) Sample 190-1173A-51X-2, 42–64 cm (476.3 mbsf), collected below the protodécollement zone at the reference site, shows more uniform grain size, moderate clay preferred orientation, and moderate sized secondary pores.

bedding plane (fig. 8.5); the SEM images demonstrate the heterogeneous, domainal nature of this grain rotation. Underthrust sediments at Sites 1174 and 808 (fig. 8.7d) lack the evidence for tectonic rotation observed above. Clay mineral preferred orientation is axisymmetric defining pretectonic bedding-parallel alignment. Both large and small pores exist throughout the Site 1174 cores, even below the décollement, despite the greater overburden.

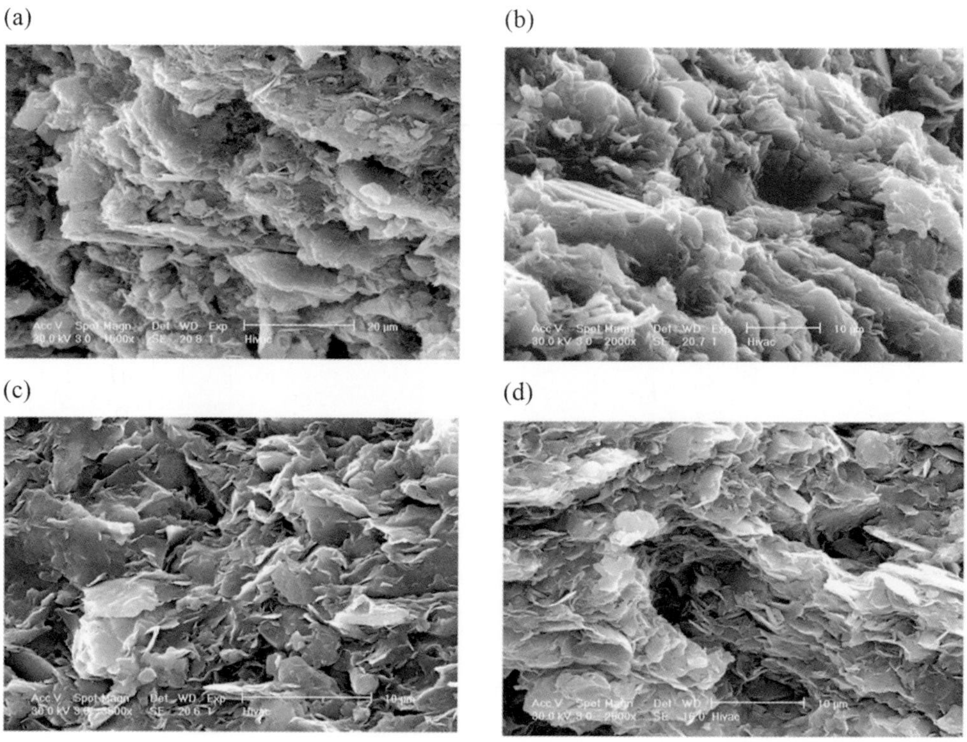

Figure 8.7 Representative SEM images of sediments from Site 1174. (a) Sample 1174B-21R-2W, 46–49 cm (335.1 mbsf) derives from the Outer Trench Wedge facies, and reveals domainal microstructure defined by stacks of clay minerals separated by more open microstructure. A wide range of particle sizes is evident, with a distinctive dispersed fine-grained clay phase. (b) Sample 1174B-58R-3W, 11–15 cm (691.1 mbsf) is from the LSB facies, with well-developed, slightly inclined stacks, and a range of particle sizes. (c) Sample 1174B-67R-2W, 101–120 cm (775.4 mbsf) from the LSB facies is characterized by more uniform, dispersed clay minerals of uniform size and shape. Bedding planar preferred orientation is subtle. (d) Sample 190-1174B-78R-1, 82–104 cm (880.8 mbsf) comes from below the décollement and exhibits uniform size and shape large grains with a well-developed clay preferred orientation and large secondary pores; up is toward top of page for all samples.

Brittle Deformation

Evidence for brittle deformation is found throughout the prism toe [*Taira et al.*, 1992; *Moore et al.*, 2001b], although it is most intense at Site 808 [*SSP*, 1991] and quite minimal at the reference Site 1173 [*SSP*, 2001a]. The microfaults recognized at Site 1173, with few exceptions, are high-angle normal faults with offsets of only a few centimeters [*SSP*, 2001a]. In one instance, the footwall of a small fault is also brecciated (fig. 8.8a), but all of the fractures are

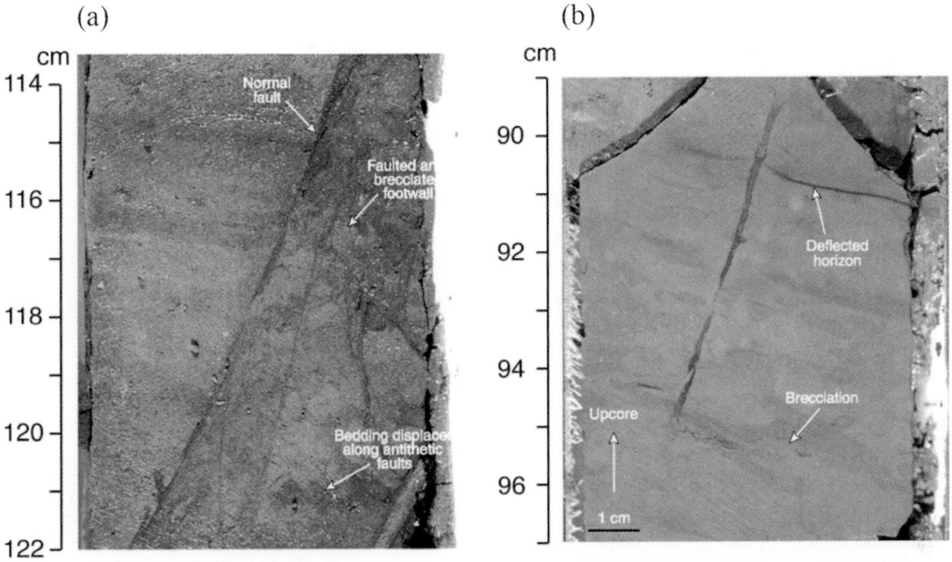

Figure 8.8 Representative deformation structures from Site 1173. (a) Healed high-angle normal fault showing brecciation in the footwall (interval 190-1173A-33X-4, 111–122 cm; 308.2 mbsf). (b) Mineralized fluid escape structure, showing zone of sediment brecciation near the base, deflected sediment horizon, and funnel-shaped top (interval 190-1173A-53X-5, 89–97 cm; 500.6 mbsf). Modified from *SSP* [2001a].

mineralized and healed and clearly inactive. A thin zone of foliated breccia occurs below the protodécollement horizon but correlates with no other obvious features. Site 1173 also reveals rare dewatering features, one of which deflected a bedding horizon before apparently venting to the old seafloor (fig. 8.8b). Occasional high-angle normal faults are also present within the Shikoku Basin units in the prism at Sites 1173 and 808. In all cases, these structures reveal no preferred orientations, consistent with pretectonic extension of the sediments due to compaction or plate bending prior to subduction [*SSP*, 2001b].

Deformation bands occur at shallow depths in the drill holes at Sites 1174 and 808, preferentially in the trench wedge hemipelagites and the USB sediments [e.g., *SSP*, 1991, 2001b; *Maltman et al.*, 1993; *Ujiie et al.*, 2004]. These tabular structures are characterized by porosity reductions and anastamosing zones of rotated clay minerals (fig. 8.9a). The deformed zone is generally continuous with the adjacent wall rocks, such that the deformation bands are better classified as brittle-ductile, rather than purely brittle, structures [*Karig and Lundberg*, 1990]. Anastamosing shear planes and rotated clay minerals within the deformation bands most commonly reveal a reverse sense of slip, consistent with subhorizontal maximum compressive tectonic stress. Deformation bands strike preferentially NE-SW, subparallel to the Nankai plate margin; they exhibit both northwest and southeast dips and occasionally occur in conjugate

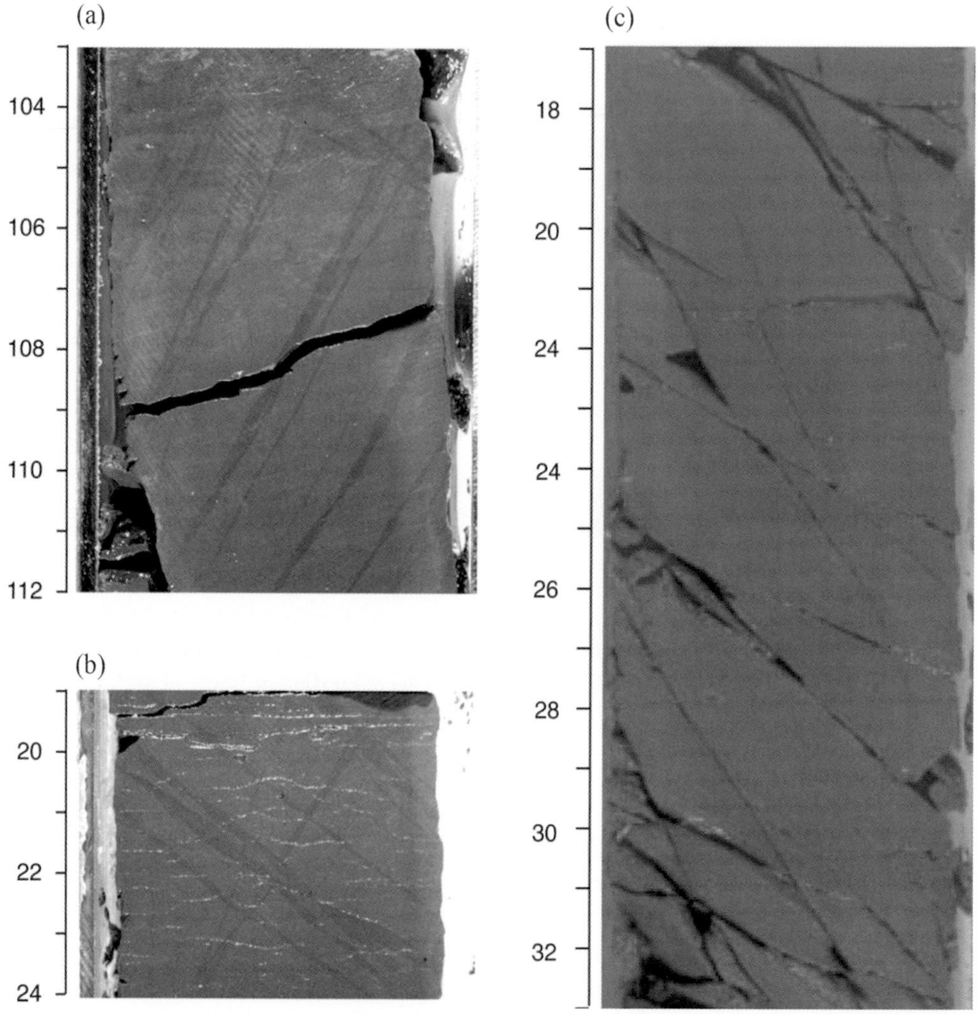

Figure 8.9 Representative deformation structures from Site 1174. (a) Deformation bands from interval 190-1174B-17R-2, 103–112 cm (297.1 mbsf), show characteristic tabular form, with anastamosing strands of varying widths. Note darker color within the bands, which correlates with decreased porosities, and also high dip angles. (b) Conjugate sets of deformation bands, again with very variable widths from interval 190-1174B-15R-2, 19–24 cm (277.2 mbsf); one set dips more steeply than the other. (c) Inclined fracture sets from interval 190-1174B-18R-2, 17–33 cm (306.0 mbsf), possibly contained within a prism backthrust. Modified from *SSP* [2001a].

pairs (fig. 8.9b) with a mean subvertical interband angle of ~62° [*SSP*, 1991, 2001b; *Karig and Lundberg*, 1990; *Ujiie et al.*, 2004]. The high dips associated with the deformation bands differ from those typically associated with conjugate brittle Coulomb shears, which form at ~30° to the subhorizontal maximum compressional stress field. To explain this angular discrepancy, it has been suggested that the deformation bands formed early during tectonic deformation and were subsequently rotated to higher angles by bulk horizontal shortening [*Karig and Lundberg,* 1990; *SSP,* 2001b]. Alternatively, the deformation bands initiated as high-angle reverse structures [e.g., *Ujiie et al.,* 2004]. *Karig and Lundberg* [1990] described a temporal sequence of deformation preserved near the frontal thrust fault, where high-angle deformation bands are overprinted by low-angle deformation bands, and finally, brittle microfaults that lie close to the 30° dips predicted for Coulomb shears.

Open fractures and microfaults, as observed in the recovered cores, become more common with depth at both Sites 1174 and 808; several show measurable offset,. Most distinctive are zones of intersecting sets of subparallel inclined fractures (fig. 8.9c), interpreted to define conjugate shear fractures [*SSP,* 2001b]. Fracture orientations vary with position in the prism but are generally consistent with subhorizontal maximum compressive stress orientations. At Site 1174, several of the inclined fracture sets coincide with interpreted positions of seismic scale thrust faults, suggesting that the fractures represent deformation localized along or adjacent to discrete forethrusts or backthrusts within the protothrust zone [*Moore et al.,* 2001a]. Commonly, the most fractured units are associated with foliated breccias and unusually poor core recovery, hinting at the loss of highly damaged fault rock during rotary drilling [*Sunderland,* 2003]. The most intense fracturing occurs just above the décollement zone at Site 1174 (fig. 8.10). This may reflect the presence of a deeper splay fault recognized on the seismic profiles through the drill site [*Moore et al.,* 2001b]. As with the deformation bands, microfractures demonstrate a preferred strike subparallel to the plate margin, consistent with a tectonic origin for these features [*SSP,* 2001b].

Relatively few deformation structures were observed within the underthrust section. Those that do occur are almost exclusively healed, high-angle normal faults showing nominal offset [*SSP,* 1991, 2001b], all consistent with pretectonic origin. The difference in deformation structures across the décollement shows that the fault is a zone of mechanical decoupling [e.g., *Taira et al.,* 1992].

Décollement Zone

The décollement zone stands out in cores from both prism sites as a 20–30 m thick zone of intense fracturing and brecciation, defined by a web of anastamosing shear surfaces [*Taira et al.,* 1992; *Moore et al.,* 2001b]. The décollement shows the most intense shattering where penetrated at Site 1174 (figs. 8.10d and 8.10e), although the intensely deformed fault zone yielded nearly 100% core recovery [*SSP,* 2001b]. The intensity of deformation and fragmentation varies across the fault zone (fig. 8.10) but generally increases down-section, as measured by the spacing of fractures and the size of breccia fragments (fig. 8.10a).

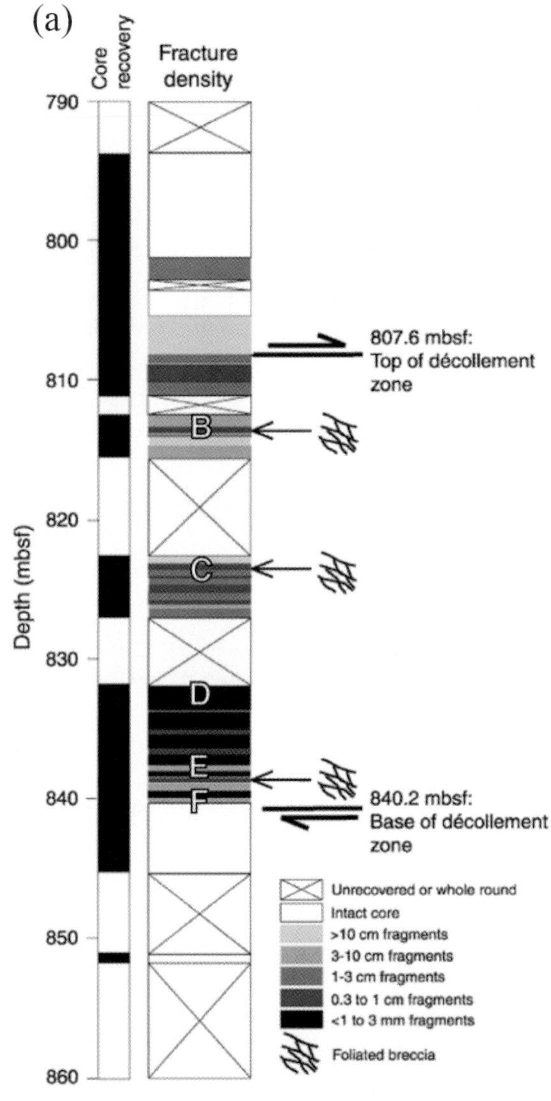

Figure 8.10 (a) Fracture intensity across the shipboard defined décollement fault zone based on breccia fragment size, from *SSP* [2001a]. Intensity of deformation increases downhole within the recovered intervals. Narrow zones of foliated breccias reflect the most intensely deformed zones; zones of severe shattering are recognized in several locations. *(continued on next page)*

The base of the décollement at Site 1174 is very sharply defined; a thin layer of foliated breccia marks the base of the fault (fig. 8.10f). Immediately below, sediments are unfractured and porosities are noticeably higher, representative of the underthrust section (figs. 8.4a and 8.4b). The top of the décollement is less easily discerned: at Site 1174, closely spaced inclined fractures extend down to at least 824 mbsf (fig. 8.10c); this package was originally interpreted to lie within the upper décollement zone [*SSP*, 2001b], although the inclined fracture

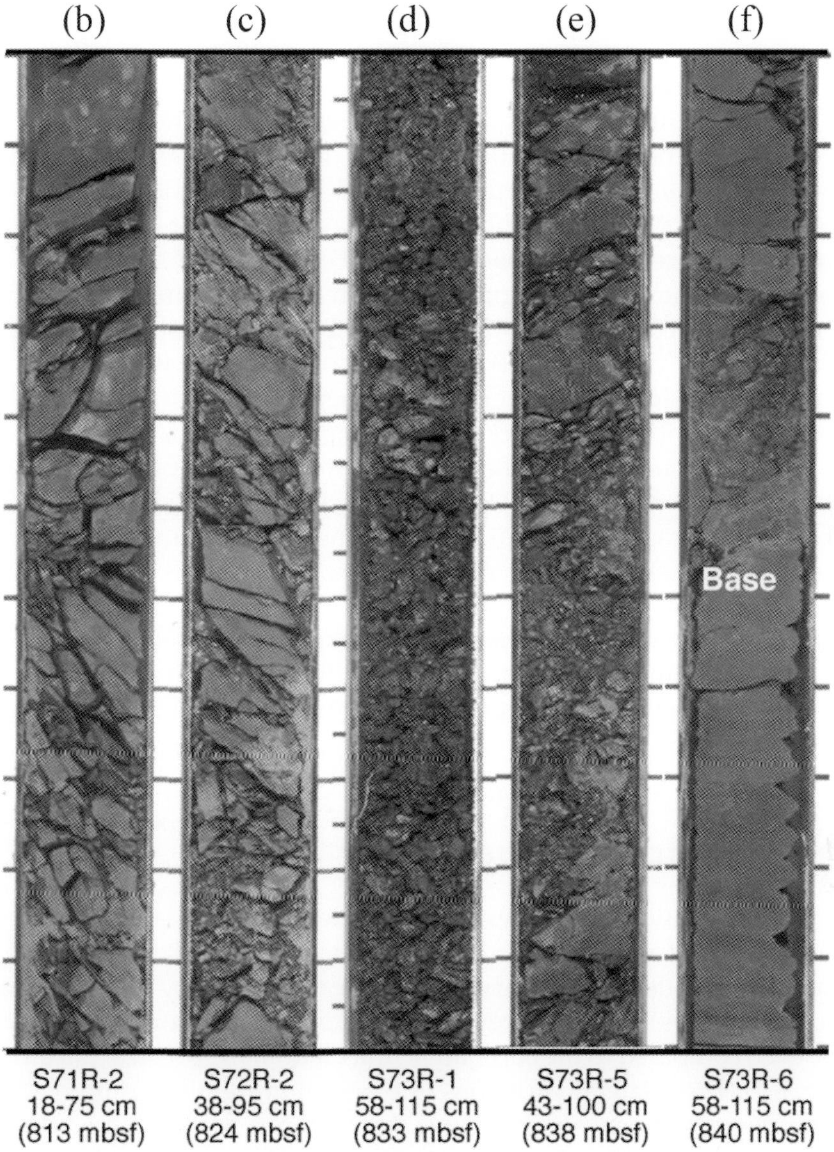

Figure 8.10 *(continued)* (b)–(f) Intervals reflecting types of deformation from within the décollement zone. Base of the décollement is sharply defined and immediately overlain by a layer of highly foliated breccia. The top of the décollement zone is characterized by high-angle inclined fracture sets that may correspond to the base of incipient thrust zones within the protothrust zone.

fabric suggests that it may denote the base of an intersecting protothrust which was imaged on the seismic reflection profiles at this depth [*Moore et al.*, 2001a].

Breccia fragments taken from within the décollement have porosities lower than sediments both above or below (figs. 8.4a and 8.4b), which could be interpreted as evidence for shear related dewatering [e.g., *Moore et al.*, 1986]. X-ray goniometry measurements of clay mineral fabrics [*Morgan and Karig*, 1995b], however, show a gradation in lateral anisotropies across the décollement zone (fig. 8.5), indicating a downward decrease in horizontal tectonic strain with depth, independent of the low porosities [*Morgan and Karig*, 1995b]. SEM images corroborate this interpretation; the clay microstructure within the breccia fragments shows no evidence for enhanced preferred orientation, although the sediments are very dense and lack the large pores that persist throughout the surrounding sediments [*Ujiie et al.*, 2003; *Sunderland and Morgan*, 2004]. It appears that the breccia fragments within the décollement zone have experienced enhanced consolidation under isotropic (hydrostatic) loading. *Morgan and Karig* [1995b] speculated that this consolidation was a result of the mechanical breakdown of intergranular cement within the décollement zone, which the surrounding wall rocks did not experience. *Ujiie et al.* [2003] recognized collapsed fine-grained clay aggregates that they suggested may have served as the cementing agent. Despite the lack of internal shear fabrics within the décollement fragments, most of the fragments exhibit thin layers of highly oriented clay minerals at their surfaces (figs. 8.11a and 8.11b), indicative of sliding along shear planes within the fault. The thinness of these layers and lack of internal shear fabric argue for shear displacement occurring under relatively high pore-fluid pressure conditions, reducing the effective normal stress and shear coupling across the slip surfaces and preventing penetrative shear strain within the fragments [*Morgan and Karig*, 1995b; *Ujiie et al.*, 2003].

Sediment Diagenesis

Drill core observations and physical properties measured at Site 1173, within the Shikoku Basin, are interpreted to define the reference state of the sediments that have entered the accretionary prism, at least at the very toe. Yet, measured physical properties at Site 1173 reveal distinct excursions from typical reference trends, even compared to the correlative units within the prism at Sites 1174 and 808 (fig. 8.4). These lateral variations in physical properties result in part from differing degrees of consolidation and changing stress states across the prism [e.g., *Bray and Karig*, 1985; *Morgan and Karig*, 1995a] but also probably reflect diagenetic changes in sediment state. For example, the sharp downhole decrease in porosities at the USB-LSB boundary at Site 1173 coincides with a downhole decline in dissolved silica, suggesting it denotes a silica diagenetic boundary, e.g., opal-A transforming to opal-CT [e.g., *SSP*, 2001a]. At first appearance the downhole densification and dissolved silica profile at Site 1173 resemble those correlated with progressive silica diagenesis, albeit in sediments initially much richer in biogenic silica [*O'Brien et al.*, 1989; *Nobes et al.*, 1992; *Tribble et al.*, 1992]. For this reason, we have tentatively identify the USB-LSB boundary at Site 1173 as the opal-A to opal-CT transition, a boundary that is likely to migrate with changing thermal state of the accreting sedimentary package.

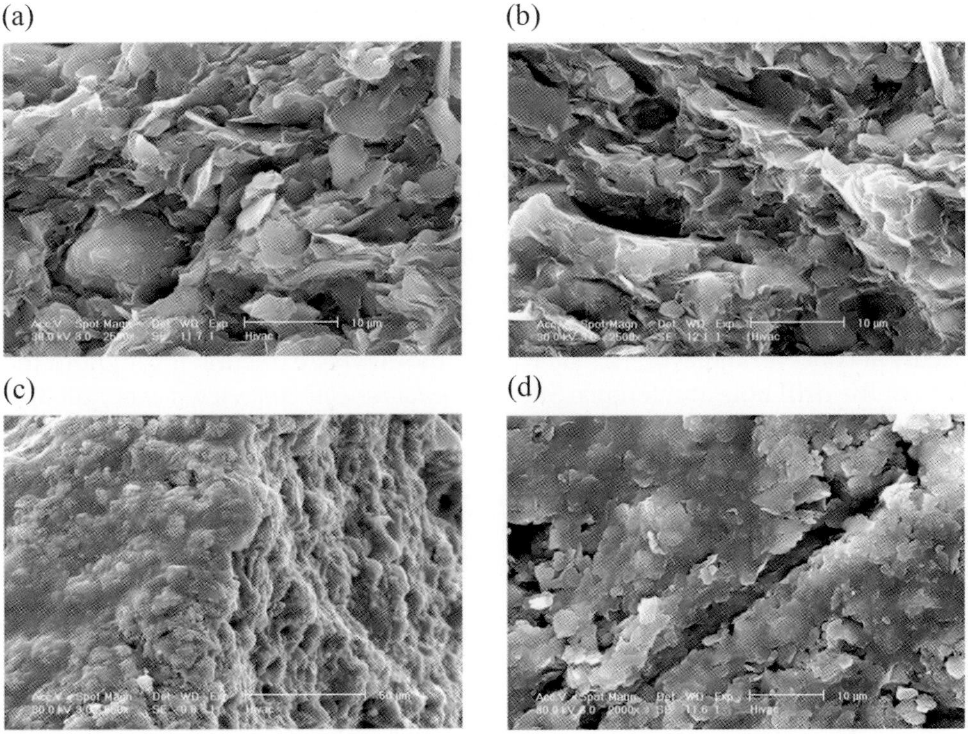

Figure 8.11 Representative SEM images of breccia fragments from the décollement zone at Site 1174. (a) Sample 1174B-71R-2W, 69–73 cm (813.57 mbsf) shows dense arrangement of large, dispersed, uniform size clay minerals; note lack of well-developed clay preferred orientation. Bedding orientation is unknown. (b) Sample 1174B-71R-3W, 23–28 cm (814.39 mbsf) shows similar dense packing of uniform clay minerals. (c) Sample 1174B-72R-3W, 103–105 cm (826.23 mbsf) shows thin zone of well-oriented clays polishing the surface of a breccia fragment; note the dense internal fabric of the breccia fragment. (d) Sample 1174B-71R-2W, 69–73 cm (813.57 mbsf) is close-up of thin oriented clay polish, with deep groove that exposes the less well-developed internal fabric.

Diagenetic changes in clay mineralogy have been better characterized along the Nankai margin. Semiquantitative clay mineral analyses [*Underwood et al.*, 1993; *Underwood and Steurer,* 2003; *Steurer and Underwood,* 2003] at all of the Nankai drill sites reveal downhole variations in smectite abundance, the combined result of variations in detrital contributions and ash alteration. At Site 1173, smectite abundance in the clay-size fraction generally increases with depth above the protodécollement, locally reaching values as high as 50–60 wt % of the clay-size fraction at ~390 mbsf, then drops markedly to ~40 wt % of the clay-size fraction below the protodécollement [*Steurer and Underwood,* 2003]; illite content increases slightly across the same boundary. Clay diagenesis is more advanced at the prism sites; at Site 1174 the onset of the smectite to illite transformation appears to have shifted ~100 m up-section relative to the age-equivalent décollement horizon, to a subbottom depth of ~700 m, where smectite abundance reaches ~40–50 wt % of the clay-size fraction. The smectite to illite transformation, however, has not gone to completion within the deeper sediments [*Steurer and Underwood,* 2003]. At both sites, the interpreted onset of the smectite diagenesis corresponds to the downhole loss of the dispersed fine-grained clay phase, which *Sunderland and Morgan* [2004] tentatively correlate with smectite (figs. 8.6 and 8.7). Along the Ashizuri transect, smectite diagenesis is not as far advanced owing to the lower geothermal gradient; smectite occurs at concentrations of ~40–60 wt % of the clay-size fraction throughout the base of the drill hole at reference Site 1177 [*Steurer and Underwood et al.*, 2003].

Although additional analyses remain to be conducted, the limited evidence to date suggests that significant diagenetic changes, involving both clay and silica phases, accompany sediment burial, accretion, and presumably subduction along the Nankai margin. No doubt, these changes introduce complicating influences on sediment strength [e.g., *Vrolijk*, 1990; *Saffer et al.*, 2001]. Of particular relevance to this study, opal and authigenic clays can serve as cementing agents that prevent sediment consolidation under ambient stress conditions [*O'Brien et al.*, 1989; *Dadey et al.*, 1991; *Tribble et al.*, 1992; *Tribble and Wilkens*, 1994]. The breakdown of these cements, either by mechanical processes or diagenic alteration, can have dramatic effects on physical properties and deformation behavior, influencing the structural evolution of the Nankai prism.

Sediment Strength and in Situ Stress State

We can gain first-order estimates of downhole effective stress conditions at the drill sites by comparing sediment porosities to the load imposed by the sediment overburden [e.g., *Lambe and Whitman*, 1979]. The dearth of deformation structures in cores recovered at Site 1173, and the downhole decrease in porosities in the LSB facies at Site 1173, suggest that the deeper sediments are normally consolidated [e.g., *Screaton et al.*, 2002; *Saffer*, 2003]. Therefore we can use the characteristics at this site to establish the reference state for sediments that will be incorporated into the accretionary prism. We estimate in situ effective vertical stresses throughout the hole by integrating the densities of the overburden at a given depth [e.g., *Morgan and Ask*, 2004], assuming there are no excess pore pressures acting on the sediment, i.e., $P_* = 0$. We define this as hydrostatic vertical stress, σ_{vh}'. If pore pressures in excess of hydrostatic occur, the effective vertical stress is reduced, i.e., $\sigma_v' = \sigma_{vh}' - P_*$. These definitions are explained in further detail elsewhere [*Morgan and Ask*, 2004]. In the upper portion of Site 1173, i.e., over the lower range of σ_{vh}', corrected porosities are very scattered, a result of the sandier lithologies in the outer trench wedge unit

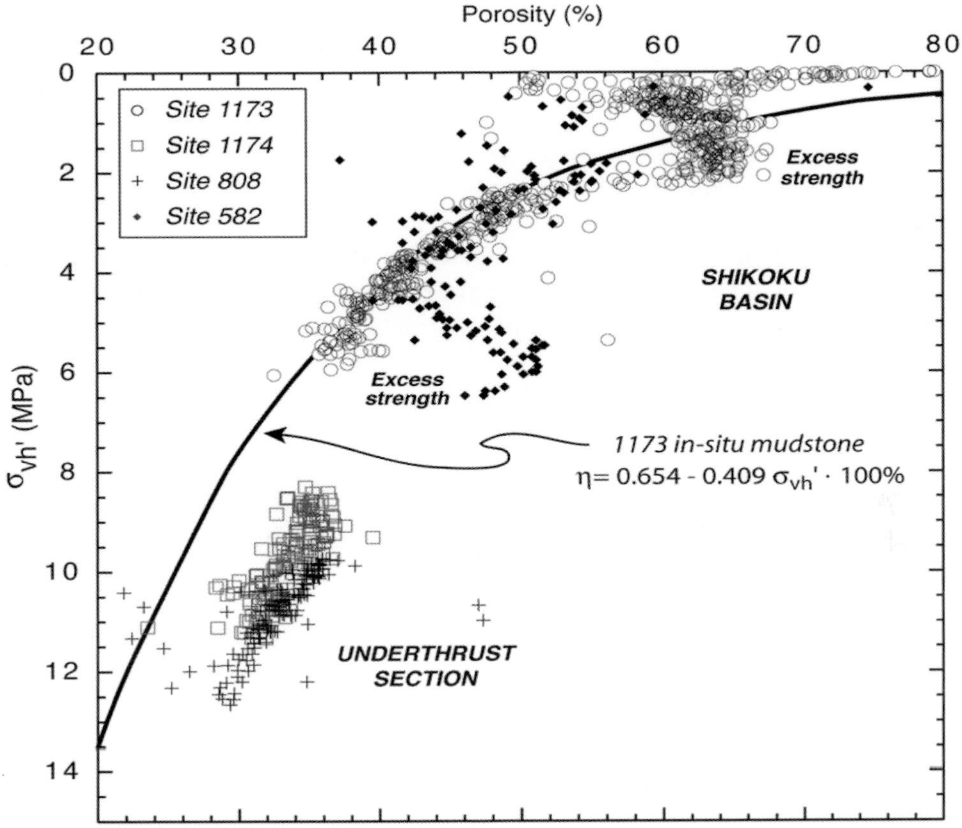

Figure 8.12 Shipboard porosities plotted against effective hydrostatic vertical stress σ'_{vh}. Only uniaxially consolidated sediments are shown: Site 1173, open circles; Site 582, solid diamonds; Site 1174 below décollement, open squares; Site 808 below décollement, crosses. A best-fit porosity-stress relationship (solid line and equation) is derived for porosities from the deeper LSB facies at Site 1173. This relationship predicts the effective vertical preconsolidation stress σ'_c, which serves as an estimate for in situ effective stress σ'_{v*}. See text for discussion.

(fig. 8.12). Between σ'_{vh} of 0.5 and 2.0 MPa, corresponding to the USB facies, porosities are deflected to anomalously high values, probably owing to the presence of some matrix cement that resists sediment consolidation [e.g., *SSP*, 2001a, *Karig*, 1993]. The rapid reduction in porosity across the USB-LSB facies boundary at Site 1173 (fig. 8.4c) reflects a downward decrease in mechanical strength of the sediment. In the deeper LSB facies at Site 1173, between σ'_{vh} of 2.0 and 6.0 MPa, sediment porosities lie along a distinct curvilinear trend, which can be fit to a simple logarithmic relationship,

$$\eta = 65.4 - 0.409 \log(\sigma'_{vh})\ 100\% \tag{8.1}$$

This $\eta - \sigma_{vh}'$ relationship serves to predict for a given porosity, the maximum in situ σ_v' that the sediment has experienced in the past, referred to as the preconsolidation stress, σ_c' (fig. 8.12). Porosities in the underthrust packages at the two prism Sites 1174 and 808 are clearly offset to higher values relative to the reference curve (fig. 8.12), suggesting a change in stress state or strength across the décollement. Several explanations for the offset in underthrust porosities have been proposed:

1. **Change from tectonic to uniaxial stress state across the décollement:** *Bray and Karig* [1985] documented a slight decrease in porosities between the reference and prism sites to the southwest, which they interpreted to reflect the increase in mean stress owing to tectonic loading. Similarly, the step up in porosities observed below the décollement zone at Sites 1174 and 808 could reflect the reduced horizontal stress state in the underthrust package [*Morgan and Karig*, 1995b]. The general coincidence of porosities at Sites 1174 and 808 with the reference consolidation curve, however, suggests that the accreted sediments have not experienced significantly greater tectonic consolidation (fig. 8.4).

2. **Excess pore pressures within the underthrust sediments, preventing sediment consolidation:** In this case, a lower effective stress state within the underthrust package would limit the amount of porosity reduction [e.g., *Taira et al.*, 1992]. If so, $\sigma_v' < \sigma_{vh}'$, and the plotted porosities should be shifted upward to lower values of σ_v' coincident with the reference line. The difference between σ_{vh}' and σ_v' serves as an estimate of P_* [e.g., *Screaton et al.*, 2002; *Saffer*, 2003]. Sediment yield strengths therefore could be predicted by the reference consolidation curve for a given value of η.

3. **Enhanced matrix strength, e.g., due to matrix cementation, allowing sediments to resist consolidation:** The enhanced strengths would allow the sediments to maintain anomalously high porosities despite increased burial loads, as has been suggested for the USB sediments within the Shikoku Basin [*Karig*, 1993; *SSP*, 2001b]. In this case, the in situ strength and associated yield stress during reconsolidation could be as high or higher than that predicted by either the in situ consolidation curve, or hydrostatic vertical stress σ_{vh}'.

The last two possibilities are very different and have distinct implications for the deformation of sediments with increased burial and subduction. Higher P_* and lower σ_v' beneath the décollement would imply weak underthrust sediments, which might be susceptible to décollement down-cutting and deformation. In contrast, if the underthrust sediments are stronger, they can resist such deformation to great depth, until subjected to increased tectonic stresses or pore-fluid pressures.

Laboratory experiments provide a means to discriminate between overpressured sediments that have undergone normal consolidation (option 2) and sediments that have been hardened by cementation or related diagenetic processes (option 3) [e.g., *Karig*, 1993; *Morgan and Ask*, 2004]. During normal consolidation, sediment porosities, η are assumed to be in equilibrium with the effective vertical stress, σ_v' [*Lambe and Whitman*, 1979; *Jones*, 1994]. This relationship is represented by an essentially straight line in $\log(\sigma_v') - \eta$ space (fig. 18.3a) over the higher stress ranges considered here [e.g., *Karig and Ask*, 2003]. The initial phase of uniaxial reconsolidation is characterized by elastic reloading with relatively little porosity change (fig. 8.13a), as indicated by the low value for the modified compression index, defined as $C_{c\eta} = -\Delta\eta/\Delta\log(\sigma_v')$ [*Karig and Ask*, 2003]. At the point of sediment yield, the sample experiences accelerated vertical strain and porosity loss, marked by a higher plastic $C_{c\eta}$ as the path returns to the normal consolidation trend. Uncemented sediments (path 1) generally experience sediment yield near the intersection of the reload path with the normal consolidation line (fig. 8.13a). The sample's preconsolidation stress σ_c' denotes the maximum value of σ_v' that the sample has experienced in the past [e.g., *Lambe and Whitman*, 1979]. In contrast, if sediments have

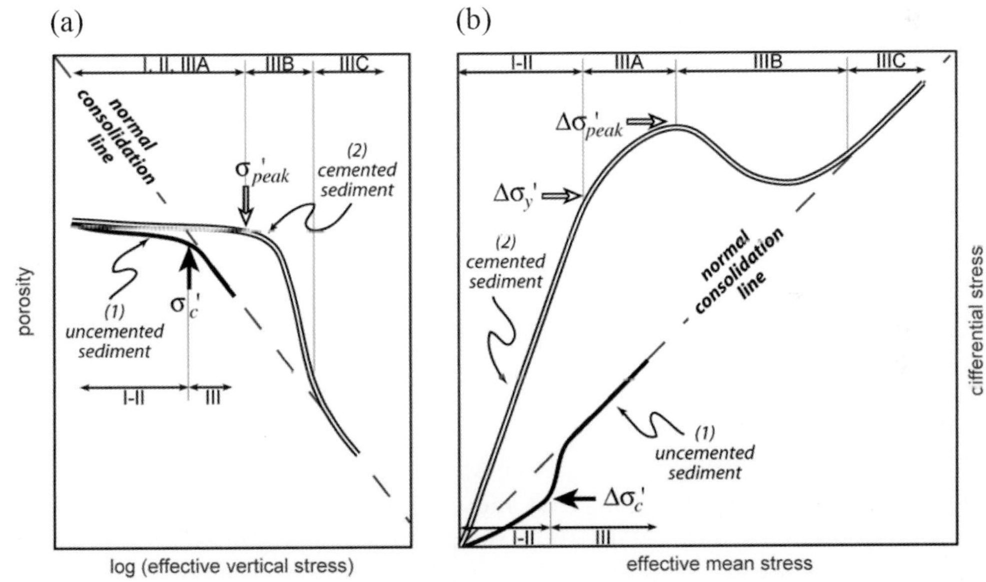

Figure 8.13 Schematic plots of reconsolidation experiments for normally consolidated (1) and cemented (2) sediment. (a) Plot of η versus $\log(\sigma_v')$ transforms the porosity stress relationship into a straight line. Experiment 1 shows initial elastic reconsolidation, characterized by a low initial slope, followed by sediment yield, reduction in η, and increase in slope, marking a return to normal consolidation; σ_c' is estimated at the intersection of tangent lines. Experiment 2 exhibits similar elastic reconsolidation but overshoots the predicted σ_c' due to enhanced matrix strength imparted by intergranular cementation. (b) Plot of $\Delta\sigma$ versus σ_m' is more sensitive to changes in sediment stress state. Path 1 is marked by initially low elastic slope followed by yield and return to normal consolidation trend; path 2 shows more rapid rise in $\Delta\sigma$, precursory yield at σ_y', leading to peak strength, $\Delta\sigma_{peak}$ at σ'_{peak}, and finally strain softening as the sample returns to normal consolidation trend – yield and peak stresses are shown in figure 8.13a.

been hardened by secondary consolidation or cementation, they may support vertical loads in excess of their past maximum σ_v' (path 2) [e.g., *Mitchell*, 1993; *Jones*, 1994; *Karig and Morgan*, 1994].

A plot of differential stress, $\Delta' = \sigma_v' - \sigma_h'$, against effective mean stress, σ_m' $(\sigma_h' + 2\sigma_h')/3$ (σ_h' is effective horizontal stress) may provide a more sensitive measure of sediment stress history [e.g., *Jones*, 1994; *Karig*, 1993]. As shown in Figure 8.13b, at low $\Delta\sigma$, uncemented sediments (path 1) undergo elastic reloading marked by relatively low slopes (phase II); increased slopes denote initial yield as the stress path regains the normal (i.e., primary or K_0) consolidation line (phase III). Owing to their greater bulk moduli, cemented materials initially exhibit steeper reload paths (path 2) and are characterized by three subphases following sediment yield (fig. 8.13b): Phase IIIA shows increasing $\Delta\sigma$ with σ_m' but with decreasing slopes as sediment cohesion is lost; phase IIIB initiates once the peak strength, $\Delta\sigma_{peak}$ is attained and is marked by a decline in $\Delta\sigma$ with σ_m' as intergranular cement rapidly breaks down. The sample eventually returns to K_0 consolidation path during phase IIIC. The effective vertical stress at the point of $\Delta\sigma_{peak}$ is termed σ_{peak}', and marks the onset of rapid porosity loss on the $\eta - \log(\sigma_v')$ plot (fig. 8.13a).

In order to determine the correct cause for the porosity step across the décollement horizon, we examine the results of several laboratory reconsolidation experiments carried out on samples collected from the frontal region of the Nankai Trough and prism.

Experimental Deformation Studies

Reconsolidation tests were carried out on whole-round cores from the USB facies at reference Site 582 to the southwest of the Muroto transect and on the underthrust sediments at the three Muroto sites [*Karig*, 1993; *Morgan and Ask*, 2004], that is, both regions that show anomalously high porosities. The sample locations and intervals, and results of the experiments, are shown in figure 8.3 and listed in table 8.2. (For simplicity, sample names are abbreviated by Site-Core throughout the text). Subcored samples had diameters of ~20 mm and heights of ~50–55 mm (i.e. ~2.5 times the diameter). Laboratory porosities were recalculated from the wet volume and weight of the sample and bulk densities, assuming shipboard values for grain and water densities [*SSP*, 2001a, 2001b]. All of the experiments were carried out in a triaxial rig located at Cornell University, under steadily increasing vertical stress with no lateral strain, designed to reproduce uniaxial consolidation and unloading paths assumed in both settings. Vertical and horizontal stresses, pore pressures, and dimensional and volumetric strains were closely monitored during the tests. Details of the experiments are presented in the references cited above; the laboratory facilities are described in more detail elsewhere [e.g., *Karig*, 1996; *Ask*, 2001].

TABLE **8.2** **Tests**[a]

	Sample, cm					
	190-1173A-51X-2, 57–64	190-1174A-78R-1, 98–104	131-808C-74R-2, 114–130[b]	131-808C-85R-1, 22-38	87-582-63-2, 130-150	87-582-73-1, 68-84
Depth, mbsf	476	881	1004	1098	644	744
Depth below déc'mt, m	87	41	41	134	—	—
σ_{vh}', MPa	3.3	8.6	9.9	10.8	5.6	6.4
σ_y', MPa	2.95 ± 0.03	4.97 ± 0.03	3.03 ± 0.05	6.40 ± 0.80	6.90 ± 0.20	9.0 ± 0.50
σ_c', MPa	3.20	5.20	6.40	7.70	1.8	2.4
σ_{peak}', MPa	4.0 ± 0.05	13.8 ± 0.2	>13.0	>13.5	8.01 ± 0.20	11.0 ± 0.20
P_f^{max}, MPa	0.35 ± 0.03	3.57 ± 0.03	6.87 ± 0.05	4.40 ± 0.80	—	—
S	1.25	2.65	>2.03	>1.75	4.45	4.58

[a] From Karig [1993] and Morgan and Ask [2004].
[b] Sample did not reach state of primary consolidation.

Summary results of the six reconsolidation experiments conducted on sediments from the Nankai margin are plotted in figures 8.14 and 8.15. A plot of $\Delta\sigma$ against σ_m' (fig. 8.14) demonstrates that all six samples exhibit similar elastic reloading paths (phase II) and undergo yield. Only four of the samples actually reached their peak strengths, $\Delta\sigma_{peak}$: samples 582-63 and 582-73 (shown in gray) from the USB facies, and samples 1173-51 and 1174-78 from beneath the décollement horizon. All of the deeper samples from the underthrust section exhibited distinctive precursory yield prior to reaching their peak strengths. Samples 1174-78, 808-74, and 808-85 all exhibit a transient decrease in $\Delta\sigma$ with σ_m', followed by renewed increase, often at slightly lower slope. Unfortunately, both tests on Site 808 samples were terminated before reaching $\Delta\sigma_{peak}$, and the test results for the former are very noisy and difficult to interpret (fig. 8.14). In all cases, the tests exhibit stress paths that resemble those of cemented materials that undergo mechanical breakdown in cementation at $\Delta\sigma_{peak}$. Samples 582-63, 1174-78, and 1173-51 also showed a return to a nearly linear path consistent with the normal consolidation path shown in figure 8.13b.

The response of sample porosity to increasing vertical stress is also very characteristic of cemented samples (fig. 8.13). All of the tests show typical elastic reloading phase, with relatively low modified compression index $C_{c\eta}$, denoting little change in η with increasing σ_v' (fig. 8.15). The four samples that reached their $\Delta\sigma_{peak}$ underwent more rapid consolidation, referred to as porosity yield

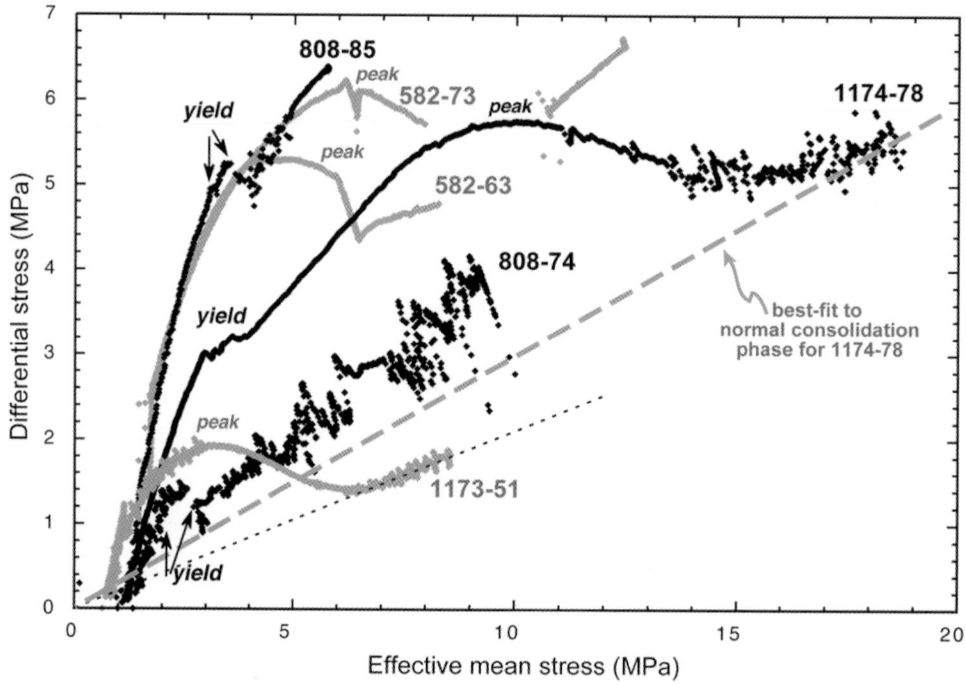

Figure 8.14 Differential stress, $\Delta\sigma$ plotted against effective mean stress σ_m' for all six tests; Site 582 test results are shown in gray [*Karig*, 1993; *Morgan and Ask*, 2004]. All tests show similar stress paths with initially steep elastic reloading, followed by yield and reduced rate of loading. Samples 1173-51 and 1174-78, as well as 582-63 and 582-73, both reached peak strength σ_{peak}' conditions, followed by strain softening; three of the tests return to normal consolidation paths. Samples 808-74 and 808-85 appear to have never reached peak strength conditions.

(small white arrows), defined by the intersecting tangents for the two linear portions of the $\eta - \log(\sigma_v')$ curves. The corresponding values of σ_{peak}' (large white arrows), which correlate with the $\Delta\sigma_{peak}$ values from figure 8.14, do not always correspond to the porosity yield stresses. The two samples from Site 808 appear to have never reached porosity yield. In all cases, however, the samples maintained porosities to σ_v' values in excess of their hydrostatic vertical stresses, σ_{vh}' (white bars), against which the shipboard porosity data are plotted (fig. 8.15). Clearly, the sediments are capable of resisting consolidation at all possible in situ σ_v', at least at laboratory strain rates.

The enhanced strengths indicated by the laboratory experiments do not necessarily preclude the presence of excess pore pressure in the reference or underthrust sections. The experimental results indicate, however, that porosities alone do not uniquely estimate the in situ σ_v'. Enhanced pore-fluid pressures could serve to maintain a stable stress state in which intergranular

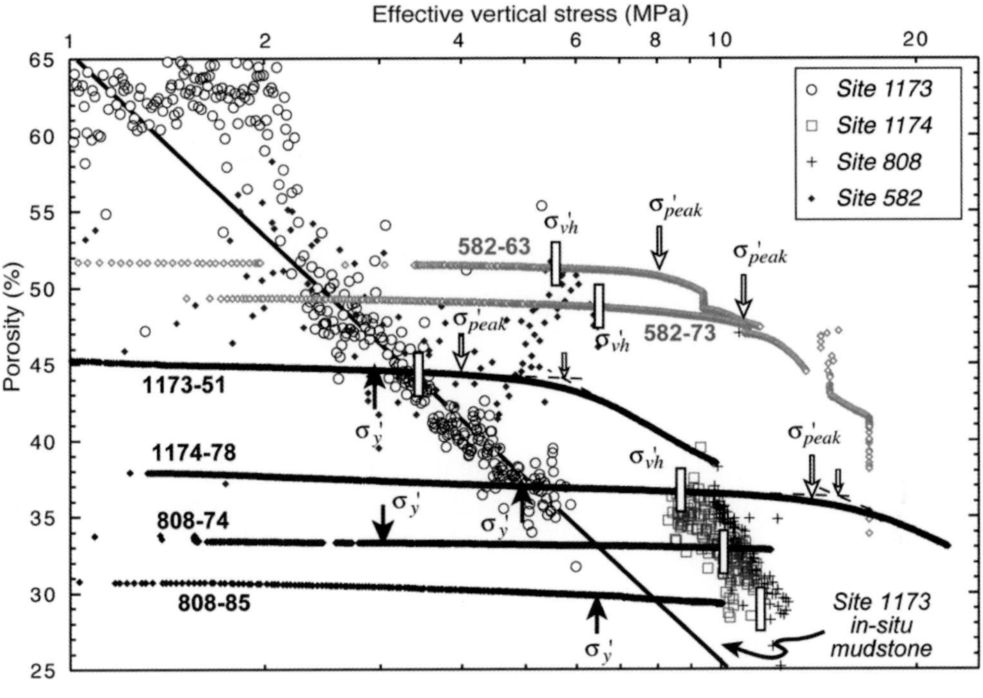

Figure 8.15 Plot of η versus $\log(\sigma_v')$ for all four tests; shipboard porosities are also plotted versus $\log(\sigma_{vh}')$ for uniaxially deformed sediments: Site 1173 (open circles), Site 582 (solid diamonds), Site 1174 (open squares), Site 808 (crosses). Solid line shows linear fit to Site 1173 LSB porosities, defining the in situ consolidation line and predictor for σ_c' from figure 8.12. Initial yield stress σ_y' from figure 8.14 is indicated for each experiment (black arrows); σ_y' is observed to approximate σ_c' (black line) for all but Sample 808-74. All samples maintained porosities well beyond their predicted σ_c' and also beyond σ_{vh}' (white bars), indicating substantial matrix strength. Vertical effective stress at peak strength, σ_{peak}' is picked for Samples 1173-51, 117478, as well as 582-63 and 582-73 (large white arrows). An alternative estimate for peak strength (small white arrows) is obtained from the intersection of tangents to the linear portions of η versus $\log(\sigma_v')$ data.

cement can form and develop, thereby reducing the likelihood of further consolidation. *Morgan and Ask* [2004] consider the possibility that the precursory yield recognized for all of the LSB samples at values of $\Delta\sigma < \Delta\sigma_{peak}$ denote the in situ σ_v'. If true, then all but one of the LSB samples may exist at or close to the predicted σ_{vh}', i.e., subject to minimal values of P_*. Several repeat experiments have been carried on subcores from the same samples presented here, and all show the same general response and precursory yield strengths; these data are the subject of continuing study (e.g., M. V. S. Ask and J. K. Morgan, manuscript in preparation, 2006).

The two suites of experiments, using samples from the USB at Site 582 and from the LSB below the protodécollement, yield very distinct results. Despite their much higher porosities, the two Site 582 samples showed initial slopes in $\Delta\sigma - \sigma'_m$ space as high as the most consolidated of the LSB samples, sample 808-85 (fig. 8.14). The high elastic moduli for these samples must arise from matrix cementation [*Karig*, 1993]. The USB samples also lack the precursory yield points detected in the deeper LSB samples. The two suites of samples are also distinguished by their sensitivities to applied load, defined as the ratio of sediment strength and preconsolidation stress predicted from the in situ consolidation curve [e.g., *Mitchell*, 1993]:

$$S = \sigma'_{\text{peak}} / \sigma'_c \tag{8.2}$$

The two USB samples from Site 582 exhibit moderate sensitivities of 4.45 and 4.58 respectively (fig. 8.15). The LSB samples also have measurable, but lower sensitivities of 1.36 for sample 1173-51, and 2.78 for sample 1174-78. The limited data for the Site 808 samples suggest that these sediments continue to harden downdip (fig. 8.15), implying increased sensitivity in the underthrust sediments with depth, which is likely to influence their deformation behavior within and beneath the prism.

Discussion

Sediments within several zones of anomalous porosity beneath and seaward of the Nankai prism exhibit mechanical evidence for some degree of cementation, which allows them to resist consolidation. This property is typical of many natural sediments, which are observed to harden by the combined processes of creep and chemical diagenesis [e.g., *Burland*, 1990; *Mitchell*, 1993; *Jones*, 1994]. The active cementing agents for the Nankai sediments are not known specifically but appear to be different within the shallower USB sediments (e.g., silica) and the deeper underthrust sediments (e.g., authigenic clay). Changes in stress conditions with the onset of tectonic loading within and beneath the prism, however, can overcome this resistance to deformation, leading to the mix of deformation structures observed within the prism drill cores, and interpreted from seismic reflection data. Below, we explore plausible mechanical histories of the accreting and underthrusting sediments, and their resulting tectonic expressions.

Deformation Within the Prism

Upon accretion, uniaxially deformed sediments are subject to a change in direction of the maximum principal stress, σ_1 from subvertical to subhorizontal. The stress state also changes from axisymmetric to triaxial, defined by three distinct principal stress magnitudes [e.g., *Karig and Morgan*, 1994]. Consequently, the deformation history of the sediments must now be tracked through three independently varying mechanical quantities: η (or alternatively, void ratio, e), $\Delta\sigma$, and σ_m'. As summarized by the theory of critical state soil (or sediment) mechanics [e.g., *Schofield and Wroth*, 1968; *Atkinson and Bransby*, 1978; *Wood*, 1990; *Jones*, 1994], the allowable states in which most sediments exist are represented by a 3-D yield surface defined in terms of these three quantities (fig. 8.16a). This surface is subdivided into a ductile (compactive) yield, or Roscoe, surface, and the brittle (dilative) failure, or Hvorslev, surface. The boundary between the two surfaces defines the critical state line (CSL), to which sediments will evolve with increasing $\Delta\sigma$ and σ_m'. For clarity, example stress paths are typically projected onto one of two planes of the 3-D volume: $\eta - \sigma_m'$ (fig. 8.16b) or $\Delta\sigma - \sigma_m'$ (fig. 8.16c).

Initial uniaxial (vertical) consolidation within the Shikoku Basin leads to increasing compaction with increasing burial. This stress path 1 tracks along the ductile compactive yield surface (figs. 8.16a and 8.16b). The anomalous porosities within the USB (figs. 8.4c and 8.13) reflect postburial cementation and strengthening, denoted by enhanced porosities and $\Delta\sigma$ relative to σ_m', i.e., stress path 2, which lies above the ductile yield surface. With increased $\Delta\sigma$ and changing diagenetic conditions, the enhanced strength of this unstable material can break down, leading to renewed consolidation as stress path 2 returns to the ductile yield surface (fig. 8.16). Both stress paths 1 and 2 are consistent with the general decrease in porosities with depth at Nankai (fig. 8.4), and with penetrative ductile deformation resulting in increasing bedding parallel alignment of clay minerals documented through X-ray goniometry and SEM techniques [*Morgan and Karig*, 1993; *Sunderland and Morgan*, 2004].

Within the prism, there is evidence of further distributed ductile deformation, as well as localized deformation bands and microfaults. At first glance, this deformation history implies stress path migration away from the normal consolidation curve, and toward critical state—stress path 3 or brittle failure—stress paths 4 (fig. 8.16). In fact, the picture is slightly more complicated, due to the change in orientation of σ_1 from subvertical within the basin, to subhorizontal within the prism. This leads to a reversal in the sign of the differential stress, defined as $\Delta\sigma = \sigma_v' - \sigma_h'$. Consequently, $\Delta\sigma$ is positive within the basin and negative within the prism. Laboratory experiments [e.g., *Wong et al.*, 1997] and theory [e.g., *Wood*, 1990] suggest that yield and failure conditions within the prism also will be sensitive to this reversal in sign. The yield surface is thought to be elliptical in shape and oriented symmetrically about the initial loading path. Thus the yield strength for uniaxially consolidated sediments will be lower within the prism than in the basin for the same σ_m' (fig. 8.17a). Given this knowledge, we can consider hypothetical stress paths that will lead to the observed deformation structures within the accreted sediments.

The active mode of deformation within the prism will be highly dependent on the initial cementation state and the rates of buildup and dissipation

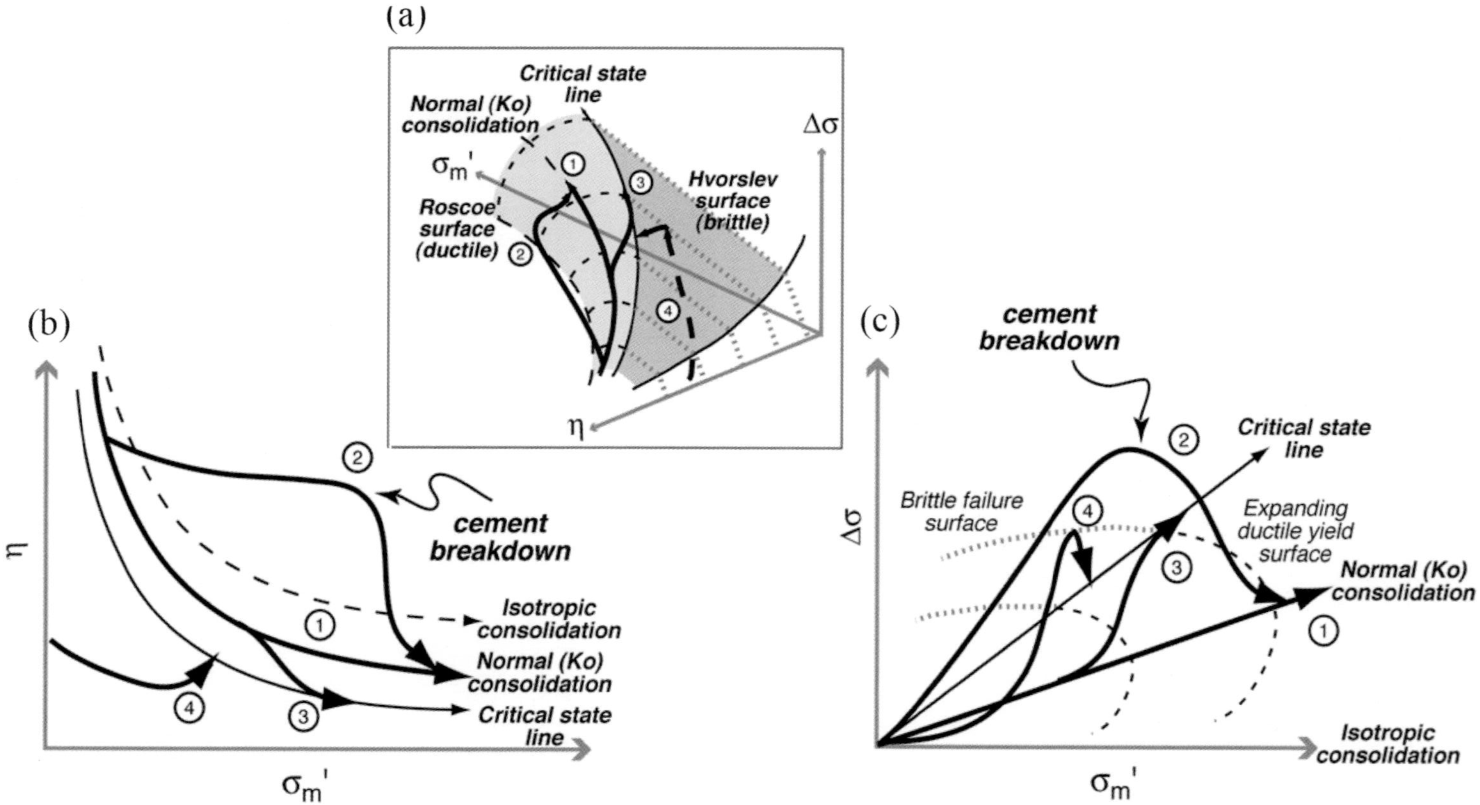

Figure 8.16 (a) A 3-D surface defining yield and failure conditions in critical state "soil" mechanics; sediment behavior varies as a function of initial consolidation state and its stress path in $\eta - \Delta\sigma - \sigma_m'$ space. The light gray surface represents ductile (compactive) yield, the darker gray surface represents brittle (dilative) failure. The junction of the two surfaces defines the critical state line (CSL), which defines pure plastic deformation. Four hypothetical stress paths through this 3-D volume are projected onto the $\eta - \sigma_m'$ plane (b), and the $\Delta\sigma - \sigma_m'$ plane (c): stress path (1) - normal uniaxial consolidation; stress path (2) - cemented sediments; stress path (3) - ductile (compactive) yield to failure along the CSL; stress path (4) - "overconsolidated" elastic sediments, reloaded to brittle failure. See text for discussion.

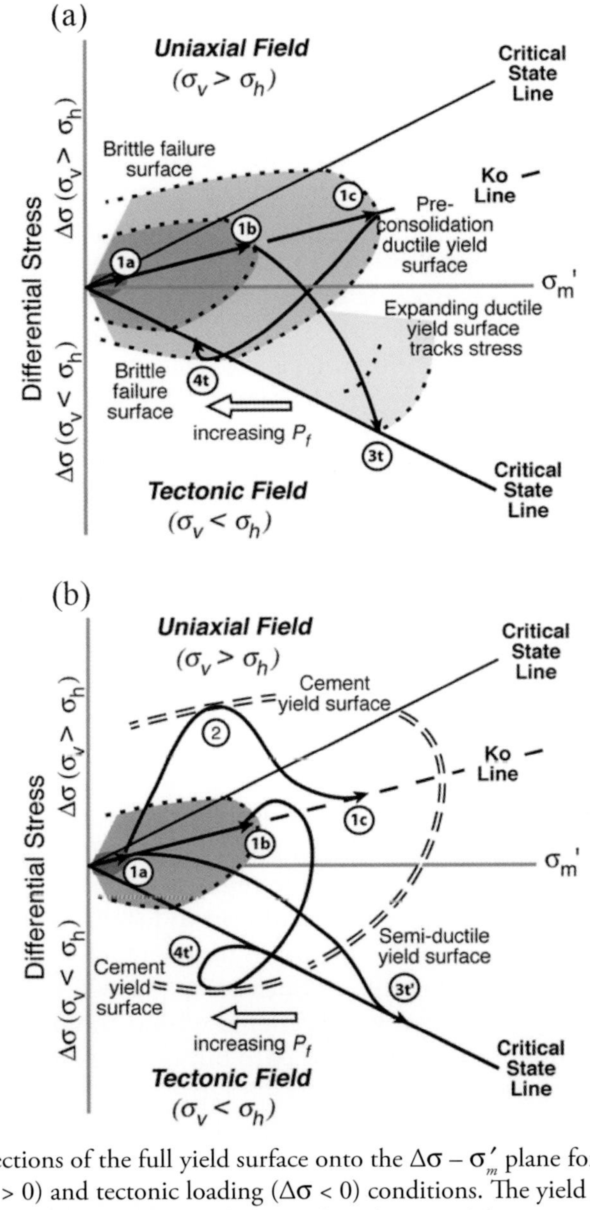

Figure 8.17 Projections of the full yield surface onto the $\Delta\sigma - \sigma'_m$ plane for both uniaxial ($\Delta\sigma > 0$) and tectonic loading ($\Delta\sigma < 0$) conditions. The yield surface is thought to be elliptical and centered on the consolidation stress path. Prism sediments exhibit lower shear strengths under tectonic, subhorizontal σ_1 than under subvertical σ_1. Arrowed paths denote possible stress paths within and beneath the prism. (a) Representative stress paths for uncemented, normally consolidated sediments: stress path 3t – compactive, ductile yield to critical state; stress path 4t – brittle, dilative failure. (b) Representative stress paths for cemented, normally consolidated sediments: stress path 2 – progressive breakdown of cementation bonds during renewed uniaxial consolidation; stress path 3t′ – compactive, ductile yield to critical state; stress path 4t′ – brittle failure. See text for discussion.

of excess pore-fluid pressures, P_*. The rotation of clay minerals away from the bedding planes implies tectonically induced ductile deformation [*Morgan and Karig*, 1993; *Sunderland and Morgan*, 2004], consistent with ductile yield under steadily increasing σ'_m, or "drained" conditions, i.e., low P_*. In this case, stress path 3t diverges from the K_0 consolidation line and approaches the critical state line in the tectonic quadrant ($\Delta\sigma < 0$). The ductile yield surface must grow outward with stress path 3t, leading to penetrative ductile strain associated with porosity reduction (fig. 8.17a). If tectonic stress is built up more rapidly than P_* can dissipate, however, deformation will occur under decreasing σ'_m, or undrained, conditions. For example, stress path 4t retreats from the uniaxially consolidated state, staying within the yield surface and elastic field, bypassing the critical state line to undergo brittle failure (fig. 8.17a). Associated deformation structures include the low-angle microfaults and fractures found at depth within the protothrust zone (e.g., fig. 8.9c).

Cemented sediments as documented by our mechanical experiments will be subject to even more complicated stress paths during tectonic loading (e.g., fig. 8.17b). Sediments consolidated to state 1a and cemented in situ will sustain high $\Delta\sigma$, under both subvertical and subhorizontal σ_1. Upon yield, however, these sensitive sediments will be subject to the sudden loss of cohesive strength, leading to rapid increases in P_* and transient low σ'_m conditions. If P_* dissipation keeps up with the loading rate, the steady increase in σ'_m will lead to penetrative ductile deformation and consolidation, i.e., stress path 3t′ (fig. 8.17b). If high strain rates and low permeabilities inhibit P_* dissipation, however, the decrease in σ'_m will favor localized failure, as shown by stress path 4t′. <Per Cammie this should be a prime.> In contrast to the uncemented analog (fig. 8.17a), the dissipation of pore pressure following undrained failure of the sensitive cemented sediments will lead to localized compaction, rather than dilation. This offers a plausible origin for compactive deformation bands found within the USB sediments at Sites 1174 and 808 (figs. 8.9a and 8.9b). More consolidated, less sensitive sediments within the underthrust package may follow a similar undrained stress path but undergo less total compaction due to their lower initial porosities (fig. 8.17b). This last stress path offers a plausible scenario for underplating of the underthrust sediments, due to the sudden imposition of subhorizontal σ_1 following seismogenic slip.

Structural Progression Within the Frontal Prism

The distribution of deformation structures at the three drill sites along the Muroto transect: Site 1173 within the Shikoku Basin, Site 1174 within the protothrust zone, and Site 808 within the frontal prism, demonstrates a progression in deformation mode and intensity with accretion. Bedding parallel clay mineral fabrics develop early during burial, and undergo progressive rotation perpendicular to the convergence direction within the prism [e.g., *Morgan and Karig*, 1993; *Sunderland and Morgan*, 2004]. At some distance inboard of the deformation front, compactive deformation bands develop, initially at high angles to the subhorizontal σ_1, but subsequently at lower angles, to be finally overprinted by brittle deformation structures, particularly at depth, and close to throughgoing thrust faults [e.g., *Maltman et al.*, 1993; *Morgan and Karig*, 1995b; *Ujiie et al.*, 2004]. This spatial distribution is indicative of intensification of tectonic loading with distance from the deformation front but may also denote progressive changes in shear strength and rheology as sediments experience mechanical and diagenetic changes.

One line of evidence in support of evolving material strength and rheology during sediment accretion lies in the occurrence of steeply dipping, reverse sense compactive deformation bands, commonly overprinted by more shallowly dipping deformation bands and microfaults [e.g., *Karig and Lundberg*, 1990; *Maltman et al.*, 1993; *Ujiie et al.*, 2004]. Previous explanations for the steep dips of deformation bands in the prism sediments include postformation ductile rotation [*Karig and Lundberg*, 1990; *Sunderland and Morgan*, 2004] or alternating dilative and compactive deformation during with sediment dewatering [*Vannucchi and Tobin*, 2000]. Alternatively, compactive deformation bands lie along a continuum of allowable deformation states and geometries, ranging from pure compaction bands oriented normal to σ_1, to dilative shear bands oriented ~30° to σ_1 [e.g., *Rudnicki and Rice*, 1975; *Issen and Rudnicki*, 2000]. Pure plastic deformation, with neither dilation nor compaction, i.e., critical state, lies between the two end-members at ~45° to σ_1. Laboratory experiments on sand packs and sandstones have reproduced the full range of predicted localized deformation structures and orientations, dependent upon experimental confining pressures and initial porosity [e.g., *Zhang et al.*, 1993; *Wong et al.*, 1997; *Olsson*, 1999; *Bésuelle*, 2001; *Fortin et al.*, 2004]. It is likely therefore that the abundance of high-angle compactive deformation bands within the shallower sediments at Sites 1174 and 808 is a result of the initial cemented state of the USB and shallower facies within the Shikoku Basin (fig. 8.4c). The first deformation bands initiate through ductile compactive yield at high angles (up to 60°) to the subhorizontal σ_1. The overprinting of increasingly shallower dipping deformation bands noted at Site 808 [*Karig and Lundberg*, 1990] implies a progressive decrease in sediment strength and porosity induced by mechanical or diagenetic breakdown of cement. With increasing consolidation, yield is accommodated by increasing components of shear relative to compaction, culminating in brittle and dilative faulting.

Spatial variations in deformation structures and fabrics correspond to an inferred temporal progression of deformation mode during accretion (fig. 8.18). Bedding parallel clay mineral fabrics define the reference state within the Shikoku Basin (fig. 8.18a). Near the deformation front, tectonic loading exceeds the shear strength of the cemented USB sediments, and intergranular cement breaks down heterogeneously along planar instabilities that form deformation bands at high angles to σ_1 (fig. 8.18b); localized compaction causes strengthening within the bands, arresting their propagation and transferring

Figure 8.18 Progression of coupled sediment deformation and diagenesis along the Muroto transect, modified from *Sunderland* [2003]. (a) Site 1173 is characterized by enhanced sediment porosities throughout the USB facies due to cementation (e.g., opal-A); diagenetic phase change (e.g., to opal-CT) at the base of USB facies allows for renewed vertical consolidation of the deeper sediments, resulting in increased bedding parallel clay mineral alignment. Smectite is found in increasing abundance above the protodécollement horizon, and drops significantly below, suggesting that the future fault plane coincides with the top of the gradual smectite-illite transition. (b) Deformation front, with facies and diagenetic boundaries interpolated from adjacent drill sites. Both the inferred opal-A to opal-CT and smectite-illite transitions have shifted up-section relative to the décollement horizon, due to increased burial beneath the trench wedge sediments. Tectonic deformation is manifested initially through high-angle compactive deformation bands at low angles to subhorizontal σ_1 within cemented USB and shallower facies, and through incipient microfaults within more consolidated LSB sediments. The upward migration of the diagenetic transitions and consolidation of the USB leads to enhanced domainal clay mineral preferred orientations. Displacement along the décollement horizon decouples the deforming sediments above from undisturbed strata below, and allows for continued diagenesis and cementation within the underthrust section. (c) Site 1174, within the protothrust zone, shows more advanced stages of deformation and diagenesis. The opal-A to opal-CT boundary has shifted upward to the base of the trench wedge facies; the top of the smectite-illite transition is now ~140 m above the décollement horizon. Diagenetic and mechanical breakdown of intergranular cement within the USB and trench wedge facies leads to distributed horizontal shortening, rotating clay minerals to steeper orientations. Lower-angle deformation bands overprint earlier generation; brittle protothrust faults splay from the décollement zone, overprinting ductile and brittle-ductile deformation structures. Underthrust sediments maintain comparatively high porosities and resist décollement downcutting owing to their enhanced strengths from ongoing diagenesis.

deformation onto adjacent bands [*Rudnicki and Rice,* 1975; *Aydin and Johnson,* 1983]. This produces zones of parallel bands, each with minimal displacement, as is observed in the Nankai prism toe [*Karig and Lundberg,* 1990; *Ujiie et al.,* 2004]. Deeper more consolidated sediments respond to tectonic loading by the formation of brittle microfaults at low angles to σ_1 (figs. 18.9b and 18.9c). Concurrently, sediments throughout the accreted sediment column are undergoing bulk subhorizontal consolidation, inducing penetrative clay mineral rotations documented by X-ray goniometry and SEM images [*Morgan and Karig,* 1993; *Sunderland and Morgan,* 2004]. In response, later stage deformation bands form at lower angles to σ_1. Eventually, reduced matrix permeability leads to

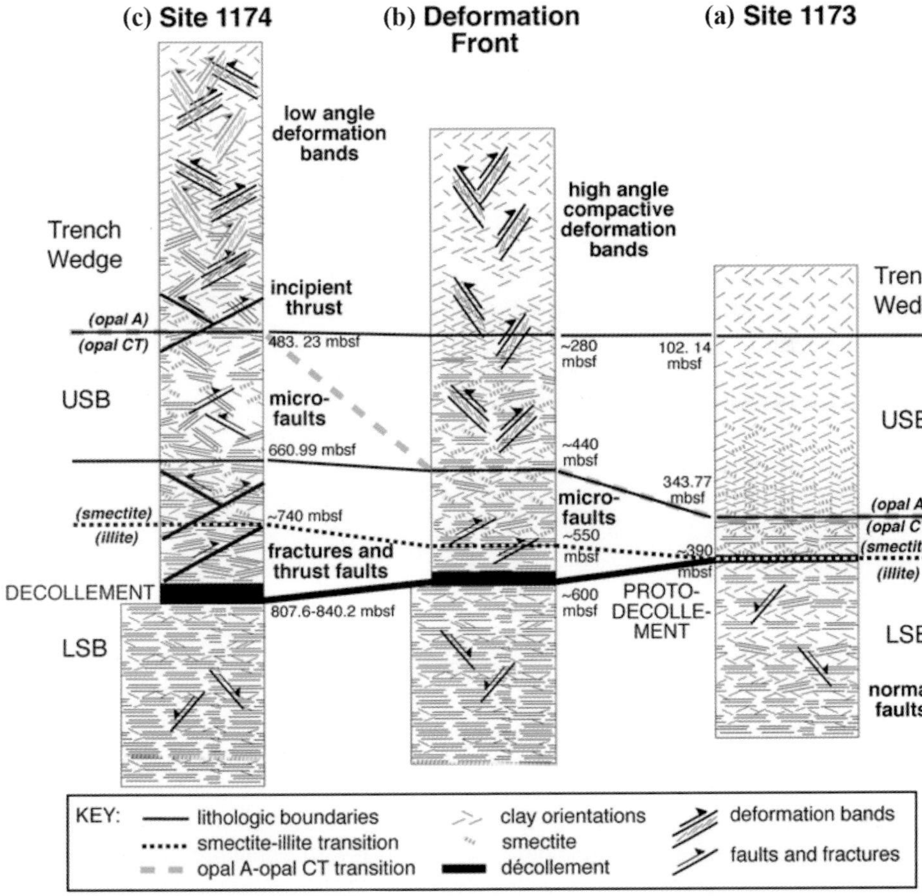

increases in P_*, driving deformation into the brittle field to form microfaults and throughgoing thrust faults observed to cross-cut the ductile and brittle-ductile structures and fabrics (fig. 8.18c) [*Karig and Morgan*, 1994; *Morgan and Karig*, 1995b].

Implications for Décollement Processes

The histories of the accreted and underthrust structural domains, separated by the décollement, are distinctly different. The entire accreted package has been deformed in some way, as reflected by the abundance of brittle deformation structures, and bulk ductile deformation fabrics (fig. 8.19). Therefore the prism sediments are undergoing constant tectonic "reworking." Their cohesive strengths are lost through mechanical and diagenetic breakdown of intergranular cementation, and through brittle faulting. To date, however, the sediments beneath the décollement at both Sites 1174 and 808 have not encountered significant tectonic shear stresses, and therefore remain within the

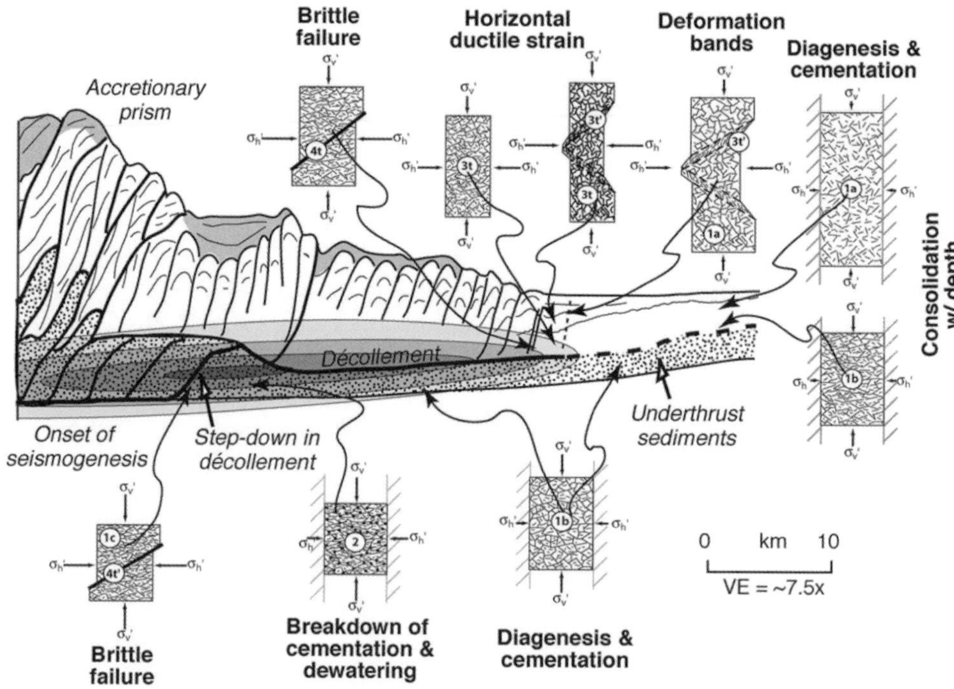

Figure 8.19 Spatial and temporal variations in mode of deformation and tectonic strain that develops during sediment accretion and underthrusting. The progression summarized in figure 8.18 produces ductile, brittle-ductile, and brittle deformation above the décollement zone. Sediment properties in the underthrust section are preserved beneath the toe of the prism, owing to ongoing sediment diagenesis and enhanced sediment strengths. Continuous tectonic reworking of the overthrust sediments maintains weak prism sediments by comparison, driving deformation upward into the accretionary prism. Stress perturbations near the updip limit of the décollement seismogenic zone ultimately exceed the strength of the underthrust sediments, leading to rapid breakdown of cementation, increase in pore-fluid pressures, and associated decrease in effective stress, allowing for décollement down-cutting and underplating. Subsequent consolidation and dewatering decreases the physical properties contrast across the décollement zone. Lateral changes in physical properties may result in a change in frictional behavior along the fault, leading to the onset of unstable seismogenic sliding along the dewatered décollement zone. Circled numbers refer to hypothetical stress paths discussed in the text.

upper quadrant shown in figures 8.17b and 8.17c; presumably, they lie along the normal consolidation line or at slightly higher $\Delta\sigma$ conditions. The underthrust sediments remain coherent, and even strengthen with increasing burial (fig. 8.19). Owing to their high strengths, these sediments can be carried 5 km or more landward without undergoing further consolidation [*Morgan and*

Ask, 2004]. These underthrust sediments can maintain high porosities to even greater depths if ongoing diagenesis allows the mechanical strength to stay ahead of the increased burial loads.

The implied contrast in strength across the décollement is the inverse of that predicted by sediment porosities alone (e.g., figs. 8.4a and 8.4b) but can explain several key observations at the frontal Nankai prism. First, the décollement is mechanically constrained to lie at the boundary between these two domains; it cannot cut downward into the stronger underthrust package, at least beneath the prism toe. Second, the fault must thicken by progressive incorporation of the weaker accreted sediments. This requirement is consistent with a downward intensification of fracture and fragmentation of décollement sediments (fig. 8.10), and the gradational decrease in lateral anisotropy of the internal clay preferred orientation thought to be inherited from ductile deformation in the prism. These observations suggest that the deepest sediments within the décollement were incorporated into the fault early, undergoing substantial brecciation before significant ductile strain accrued [*Morgan and Karig,* 1995b].

The intense fracturing and the persistence of very thin shear surfaces of décollement breccia fragments suggest that décollement slip occurs under high pore-pressure conditions [*Ujiie et al.,* 2003; *Sunderland and Morgan,* 2004]. The décollement fragments also appear to have undergone consolidation under essentially isotropic conditions, as might occur under fluctuating pore-fluid pressure conditions if low permeabilities prevent the fluids from permeating into the dense fragments [*Morgan and Karig,* 1995b]. The confinement of the fragments by high pore pressures may have caused the mechanical breakdown of authigenic clays that cemented the matrix, thereby allowing further consolidation [*Morgan and Karig,* 1995b; *Ujiie et al.,* 2003].

If the underthrust sediments are anomalously strong beneath the toe of the Nankai prism, then something must change downdip beneath the décollement to explain the structural and hydrologic changes that we recognize at the updip limit of the seismogenic zone (fig. 8.2). No doubt, continued mineral dehydration, along with low-grade diagenetic and metamorphic reactions, contribute to these changes [e.g., *Moore and Saffer,* 2001]. Furthermore, intergranular cements may dissolve away under more favorable pressure, temperature, and fluid chemistry conditions at depth (fig. 8.19). We propose that in addition, sudden stress perturbations, for example, generated in response to seismogenic slip along the downdip décollement, may help to overcome the cemented strength of the underthrust sediments, leading to rapid loss of shear strength, and rapid increase in pore fluid pressures (fig. 8.19). The lower effective stress state would favor brittle deformation within the underthrust package, enabling décollement down-cutting, ultimately, to the top of the oceanic crust, and underplating of the underthrust package (fig. 8.19). With time, the underthrust package would consolidate, reducing the porosity contrast across the décollement, and therefore the impedance contrast, as suggested by *Bangs et al.* [2004]. The lateral changes in physical properties and strength, and associated geochemical and mineralogic changes, could alter the mechanical behavior and frictional properties along the fault [e.g., *Zhang et al.*, 1993; *Moore and Saffer*, 2001a, b; *Brown et al.*, 2003; *Kopf and Brown*, 2003], enabling out-of-sequence faults to splay off of the new décollement surface, and possibly favoring unstable, and seismogenic sliding.

This hypothesis is based on characteristics and properties of remote sediments collected from the frontal portion of the Nankai prism along the Muroto transect. Along-strike differences in mineralogy, lithostratigraphy, and thermal and pore-fluid conditions, as documented along the Ashizuri transect [e.g., *Moore et al.*, 1990, 2001; *Underwood*, this volume], will very certainly influence sediment state and deformation behavior beneath and along the décollement zone. In particular, the presence of sandy turbidites within the deeper USB section elsewhere along the margin is presumed to decrease the potential for pore-pressure buildup within the underthrust package [e.g., *Moore et al.*, 1990]; diagenetic processes within these sandy units, and their effects on sediment strength and sliding behavior, are very poorly constrained at this time but may prove to be as complex or more so, than those we have elaborated here. The model presented here therefore still needs to be tested directly, e.g., through laboratory studies on materials from different locations along the Nankai margin, and ultimately through direct drilling and sampling of the seismogenic zone and adjacent fault rocks, an objective that will come to fruition as part of the NanTroSEIZE IODP drilling initiative [*Kimura et al.*, 2003].

Conclusions

Clay-rich sediments at the toe of the Nankai accretionary prism, southwest of Japan, undergo diagenetic changes before, during, and after consolidation and deformation within and beneath the prism, leading to spatial and temporal variations in mechanical behavior and deformation modes. Sediments deposited and locally cemented within the Shikoku Basin are partitioned by the basal décollement into accreted and underthrust packages, and subsequently follow distinctive deformation pathways across the margin (e.g., fig. 8.19). The modes of sediment deformation depend on their initial consolidation and diagenetic state, and locally, their enhanced strengths due to intergranular cementation. Sediments above the décollement encounter increased horizontal tectonic stresses, which induce discrete brittle deformation in the form of microfaults and fractures, penetrative horizontal ductile deformation, and localized brittle-ductile deformation bands. Concurrent with tectonic deformation, sediments undergo progressive diagenetic transformations, causing a breakdown in sediment cementation. Accordingly sediments undergo changes in mechanical behavior, as ductile horizontal shortening associated with clay mineral rotations and porosity reduction overprints previous deformation structures.

Below the active décollement zone along the Muroto transect, slightly enhanced porosities indicate delayed consolidation owing to sediment overpressures or cementation. Deformation experiments confirm the high yield strengths of the underthrust sediments, probably reflecting ongoing smectite-illite transformations within the deep sediments, and incipient matrix cementation through growth of authigenic clays. The degree of underconsolidation beneath the décollement increases in the downdip direction, leading to enhanced sediment sensitivity. The enhanced strengths of the underthrust sediments beneath the toe of the Nankai accretionary prism contrast with those of the deformed and continually remolded accreted sediments above, helping to guide the décollement horizon and evolution. At Site 1173, mineralogy and sediment fabric differences across the décollement suggest that the fault may ride near the top of a gradual diagenetic transition, possibly in the presence of locally enhanced pore fluid pressures. Beneath the frontal portion of the prism, the stronger underthrust sediments appear to resist décollement down-cutting, forcing the décollement zone to thicken by preferential incorporation of accreted strata during shear deformation.

Downdip, deep beneath the prism, higher pressures, temperatures, and chemical activities enable diagenetic alteration of the strong sediments. In addition, high shear stresses may develop near the updip limit of the seismogenic zone, breaking down the strength of the underthrust sediments, leading to loss of cohesion, shear failure, and release of trapped pore fluids. Associated rapid increases in pore pressure and reductions in shear strength along the fault zone will enable décollement down-cutting and the onset of underplating. Subsequent sediment consolidation will reduce contrasts in physical properties across the fault zone. This scenario provides a plausible explanation for the convergence of several structural and hydrological characteristics that have been interpreted close to the updip limit of the seismogenic zone beneath the Nankai prism, suggesting that the physical properties changes result in lateral contrasts in décollement strength and sliding behavior responsible for the onset of seismogenesis.

The features and interpretations presented here are specific to the Muroto transect, for which the greatest abundance of data are available. Significant along-strike variations in thermal structure, lithostratigraphy, and basement geometry will certainly influence the distribution and progression of deformation within the prism and along the décollement. Elsewhere along the margin, other factors, such as the presence of interbedded turbidites within the underthrust section along the Azshizuri transect, will play important, and potentially, controlling roles in décollement placement and properties. Future drilling investigations will shift to the northeast, along a new transect that lies offshore of the Kii Peninsula (fig. 8.1), where these mechanical models for the partitioning and evolution of deformation can be put to the test, as sediments enter the accretionary system, and ultimately the deep seismogenic zone.

Acknowledgments

This research used samples and data provided by the Ocean Drilling Program (ODP), collected during ODP Legs 131 and 190, and Deep Sea Drilling Project samples from Leg 87. ODP is sponsored by the U.S. National Science Foundation (NSF) and participating countries under management of Joint Oceanographic Institutions (JOI), Inc. Shore-based experiments were conducted at Cornell University. Dan Karig provided thoughtful advice and throughout the study. George Hade is gratefully thanked for engineering assistance that formed the basis for the well-functioning test equipment. X-ray goniometry analyses were carried out at the Cornell Materials Science Center; Maura Weathers provided technical advice and assistance. We also thank Angelo Benedetto for training and assistance with SEM imaging at Rice University, Kitty Milliken for use of the University of Texas Electron Imaging Facility, and Scott Stookey at CoreLab, Houston. We thank Mike Underwood and Paola Vannucchi for very complete and constructive reviews of an earlier version of this manuscript. Support for the research presented here came from JOI-USSSP grant 190-F001330 to Morgan at Rice University, and past NSF and JOI funding.

References

Ando, M. (1975), Source mechanisms and tectonic significance of historical earthquakes along the Nankai Trough, Japan, *Tectonophysics, 27*, 119–140.

Ando, M. (1991), A fault model of the 1946 Nankaido earthquake derived from tsunami data, *Phys. Earth Planet. Inter., 28*, 320–336.

Ask, M. V. S. (2001), Mechanical tests on claystone from Ocean Drilling Program Hole 1070A (Leg 173); Implications for elevated pore-fluid pressure in sediments within the ocean-continent transition zone, West Iberia, *Mar. Geol., 177*, 395–410.

Atkinson, J. H., and P. L. Bransby (1978), *The Mechanics of Soils – An Introduction to Critical State Soil Mechanics*, McGraw-Hill, New York.

Aydin, A. (1978), Small faults formed as deformation bands in sandstones, *Pure. Appl. Geophys., 116*, 913–930.

Aydin, A., and A. M. Johnson (1978), Development of faults as zones of deformation bands and as slip surfaces in sandstone, *Pure. Appl. Geophys., 116*, 931–942.

Antonellini, M., A. Aydin, and D. D. Pollard (1994), Microstructure of deformation Bands in porous sandstones at Arches National Park, Utah, *J. Struct. Geol., 16*, 941–959.

Bangs, N. L., T. Shipley, G. F. Moore, S. Gulick, S. Kuromoto, and Y. Nakamura (2004), Evolution of the Nankai Trough décollement from the trench into the seismogenic zone: Inferences from three-dimensional seismic reflection imaging, *Geology, 32*, 273–276, doi:10.1130/G20211.1.

Bésuelle, P. (2001), Compacting and dilating shear bands in porous rock, *J. Geophys. Res., 106*, 13,435–13,442.

Bray, C. J., and D. E. Karig (1985), Porosity of sediments in accretionary prisms and some implications for dewatering processes, *J. Geophys. Res., 90*, 768-778.

Bray, C. J., and D. E. Karig (1986), Physical properties of sediments from the Nankai Trough, DSDP Leg 87A, Sites 582 and 583, Deep Sea Drill. Program Initial *Rep., 87*, 817–842.

Bray, C. J., and D. E. Karig (1988), Dewatering and extensional deformation of the Shikoku Basin Hemipelagic Sediments in the Nankai Trough, *Pure. Appl. Geophys., 128*, 724–739.

Behrmann, J. H., and A. Kopf (1993), Textures and microfabrics in fine-grained muds and mudstones from Site 808, Nankai accretionary prism, *Proc. Ocean Drill. Program Sci. Results, 131*, 45–56.

Brown, K. M., A. Kopf, M. B. Underwood, and J. L. Weinberger (2003), Compositional and fluid pressure controls on the state of stress on the Nankai subduction thrust, a weak plate boundary, *Earth Planet. Sci. Lett., 214*, 589–603.

Brown, K. M., and B. Ransom (1996), Porosity corrections for smectite-rich sediments: Impact on studies of compaction, fluid generation, and tectonic history, *Geology, 24*(9), 843–846.

Burland, J. B. (1990), On the compressibility and shear strength of natural clays, *Geotechnique, 40*, 329–378.

Byerlee, J. (1990), Friction, overpressure and fault normal compression, *Geophys. Res. Lett., 17*, 2109–2112.

Cutillo, P. A., E. J. Screaton, and S. Ge (2003), Three-dimensional numerical simulation of fluid flow and heat transport within the Barbados Ridge accretionary complex, *J. Geophys. Res., 108*(B12), 2555, doi:10.1029/2002JB002240.

Dadey, K. A., M. Leinen, and A. J. Silva (1991), Anomalous stress history of sediments of the Northwest Pacific: The role of microstructure, in *Microstructure of Fine-Grained Sediments, From Mud to Shale*, , edited by R.H. Bennett, W. R. Bryant, and M. H. Hulbert, pp. 229–236, Springer, New York.

Fortin, J., A. Schubnel, and Y. Guéguen (2005), Elastic wave velocities and permeability evolution during compaction of Bleurwiller sandstone, *Int. J. Rock Mech. Mining Sci., 42*, 873–899.

Gulick, S. P., N. L. Bangs, T. H. Shipley, Y. Nakamura, G. F. Moore, and S. Kuramoto (2004), Three-dimensional architecture of the Nankai accretionary prism's imbricate thrust zone off Cape Muroto, Japan: Prism reconstruction via en echelon thrust propagation, *J. Geophys. Res., 109*, B02105, doi:10.1029/2003JB002654.

Hill, I. A., A. Taira, A. Klaus, and Shipboard Scientific Party (1991), *Proceedings of the Ocean Drilling Program, Initial Reports*, vol. 131, Ocean Drill. Program, College Station, Tex.

Hyndman, R., K. Wang, and M. Yamano (1995), Thermal constraints on the seismogenic portion of the southwestern Japan subduction thrust, *J. Geophys. Res., 100*, 15,373–15,392.

Issen, K.A., and J.W. Rudnicki (2000), Conditions for compaction bands in porous rock, *J. Geophys. Res., 105*, 21,529–21,536.

Jones, M. (1994), Mechanical principles of sediment deformation, in *Geological Deformation of Sediments*, edited by A. Maltman, pp. 37–72, CRC Press, Boca Raton, Fla.

Kagami, H., D. E. Karig, and Shipboard Scientific Party (1986), *Initial Reports of the Deep Sea Drilling Project*, vol. 87, 985 pp., U.S. Govt. Print. Off., Washington, D.C..

Karig, D. E. (1993), Reconsolidation tests and sonic velocity measurements of clay-rich sediments from the Nankai Trough, *Proc. Ocean Drill. Program Sci. Results, 131*, 247–260.

Karig, D. E. (1996), Uniaxial reconsolidation tests on porous sediments; mudstones from Site 987, *Proc. Ocean Drill. Program Sci. Results, 149*, 363–373.

Karig, D. E., and M. V. S. Ask (2003), Geological perspectives on consolidation of clay- rich marine sediments, *J. Geophys. Res., 108*, 2197, doi:10.1029/2001JB000652.

Karig, D. E., and N. Lundberg (1990), Deformation bands from the toe of the Nankai accretionary prism, *J. Geophys. Res., 95*, 9099–9109.

Karig, D. E., and J. K. Morgan (1994), Stress paths and strain histories during tectonic deformation of sediments, in *Geological Deformation of Sediments*, edited by A. Maltman, pp. 167–204, CRC Press, Boca Raton, Fla.

Kimura, G., H. Tobin, and Proponents (2003), NanTroSEIZE: The Nankai Trough Seismogenic Zone Experiment – A proposed IODP Complex drilling project, proposal, Integrated Ocean Drill. Program, College Station, TX.

Kopf, A., and K. M. Brown (2003), Friction experiments on saturated sediments and their implications for the stress state of the Nankai and Barbados subduction thrusts, *ar. Geol., 202*,193–210.

Lambe, W. T., and R. V. Whitman (1979), *Soil Mechanics*, 553 pp., John Wiley, New York.

Logan, J. M., and K. A. Rauenzahn (1987), Frictional dependence of gouge mixtures of quartz and montmorillonite on velocity, composition and fabric, *Tectonophysics, 144*, 87–108.

Maltman, A. J., T. Byrne, D. E. Karig, and S. Lallemant (1993), Deformation at the toe of an active accretionary prism: Synopsis of results from ODP Leg 131, Nankai, SW Japan, *J. Struct. Geol., 15*, 949–964.

Marone, C. (1998), Laboratory-derived friction laws and their application to seismic Faulting, *Annu. Rev. Earth Planet. Sci., 26*, 643–696.

March, A. (1932), Mathematische Theorie der Regelungen nach der Korngestalt bei affiner Deformation, *Z. Kristallogr., I*, 285–297.

Mikada, H., K. Becker, J. C. Moore, A. Klaus, and Shipboard Scientific Party (2002), *Proceedings of the Ocean Drilling Program, Initial Reports*, vol. 196, Ocean Drill. Program, College Station, Tex.

Mitchell, J. K. (1993), *Fundamentals of Soil Behavior*, 2nd ed., 437 pp., John Wiley, New York.

Moore, G. F., and T. H. Shipley, (1993), Character of the décollement in the Leg 131 area, Nankai Trough, *Proc. ODP, Sci. Results, 131*, Ocean Drilling Program, College Station, TX, 73–82.

Moore, G. F., T. H. Shipley, P. L. Stoffa, D. E. Karig, A. Taira, S. Kuramoto, H. Tokuyama, and K. Suyehiro (1990), Structure of the Nankai Trough accretionary zone from multichannel seismic reflection data, *J. Geophys. Res., 95*, 8753–8765.

Moore, G. F., Karig, D. E. Shipley, T. H. Taira, A., Stoffa, P. L., and Wood, W. T. (1991), Structural framework of the ODP Leg 131 area, Nankai Trough, *Proc. ODP, Init Repts., 131*, Ocean Drilling Program, College Station, TX, 15–20.

Moore, G. F., et al. (2001a), Data report: Structural setting of the Leg 190 Muroto Transect, *Proc. Ocean Drill. Program Initial Rep.* [CD-ROM], *190*, 1–14.

Moore, G. F., et al. (2001b), New insights into deformation and fluid flow processes in the Nankai Trough accretionary prism: Results of Ocean Drilling Program Leg 190, *Geochem. Geophys. Geosyst., 2*(10), doi:2001GC000166.

Moore, J. C., and D. M. Saffer (2001), Up-dip limit of the seismogenic zone beneath the prism of southwest Japan: An effect of diagenetic to low-grade metamorphic processes and increasing effective stress, *Geology, 29*, 183–186.

Moore, J. C., S. Roeske, N. Lundberg, J. Schoonmaker, D. S. Cowan, E. Gonzales, and S. E. Lucas (1986), Scaly fabrics from Deep Sea Drilling Project cores from forearcs, in *Structural Fabrics in Deep Sea Drilling Project Cores From Forearcs*, edited by J. C. Moore, *Mem. Geol. Soc. Am., 166*, 55–73.

Morgan, J. K., and M. V. S. Ask (2004), Consolidation state and strength of underthrust sediments and evolution of the décollement at the Nankai accretionary margin: Results of uniaxial reconsolidation experiments, *J. Geophys. Res., 109*, B03102, 10.1029/2002JB002335.

Morgan, J. K., and D. E. Karig (1993), Ductile strains in clay-rich sediments from Hole 808C: Preliminary results using X-ray pole figure goniometry, *Proc. Ocean Drill. Program Sci. Results, 131*, 141–155.

Morgan, J. K., and D. E. Karig (1995a), Diffuse deformation within the toe of the eastern Nankai accretionary prism: Balanced and restored section based on numerical strain calculations, *J. Struct. Geol., 17*, 31–45.

Morgan, J. K., and D. E. Karig (1995b), Décollement processes at the Nankai accretionary margin, southeast Japan: Propagation, deformation, and dewatering, *J. Geophys. Res., 100*, 15,221–15,231.

Nobes, D. C., R. W. Murray, S. Kuramoto, K. A. Pisciotto, and P. Holler (1992), Impact of silica diagenesis on physical property variations, *Proc. Ocean Drill. Program Sci. Results, 127–128*, 3–31.

Norabuena, E., et al. (2004), Geodetic and seismic constraints on some seismogenic zone processes in Costa Rica, *J. Geophys. Res., 109*, B11401, doi:10.1029/2003JB002931.

Obana, K., A. Kodaira, K. Mochizuki, and M. Shinohara (2001), Micro-seismicity around the seaward up-dip limit of the 1946 Nankai earthquake dislocation area, *Geophys. Res. Lett.*, *28*, 2333–2336.
O'Brien, D. K., M. H. Manghnani, and J. S. Tribble (1989), Irregular trends of physical properties in homogeneous clay-rich sediments of DSDP Leg 87 Hole 584, midslope terrace in the Japan Trench, *Mar. Geol.*, *87*, 183–194.
Oertel, G. (1983), The relationships of strain and preferred orientation of phyllosilicates grains in rock – A review, *Tectonophysics*, *100*, 413–447.
Oertel, G. (1985), Reorientation due to grain shape, in *Preferred Orientation in Deformed Metals and Rocks: An Introduction to Texture Analysis*, edited by H-R. Wenk, pp. 259–266, Elsevier, New York.
Oleskevich, D. A., R. D. Hyndman, and K. Wang (1999), The up-dip and down-dip limits to great subduction earthquakes: Thermal and structural models of Cascadia, south Alaska, SW Japan, and Chile: *J. Geophys. Res.*, *104*, 14,965–14,991.
Olsson, W. (1999), Theoretical and experimental investigation of compaction bands in porous rocks, *J. Geophys. Res.*, *104*, 7219–7228.
Park, J-O., T. Tsuru, S. Kodaira, A. Nakanisi, S. Miura, Y. Kaneda, and Y. Kono (2000), Out-of-sequence thrust faults developed in the coseismic slip zone of the 1946 Nankai earthquake (Mw = 8.2) off Shikoku, southwest Japan, *Geophys. Res. Lett.*, *27* 1033–1036.
Park, J-O., T. Tsuru, S. Kodaira, P.R. Cummins, and Y. Kaneda (2002), Splay fault branching along the Nankai subduction zone: *Science*, *297*, 1157–1160.
Rudnicki, J. W., and J. R. Rice (1975), Conditions for the localization of deformation in pressure sensitive dilatant materials, *J. Mech. Phys. Solids*, *23*, 371–394.
Saffer, D. M. (2003), Pore pressure development and progressive dewatering in underthrust sediments at the Costa Rican subduction margin: Comparison with northern Barbados and Nankai, *J. Geophys. Res.*, *108*(B5), 2261, doi:10.1029/2002JB001787.
Saffer, D. M., and B. A. Bekins (1998), Episodic fluid flow in the Nankai accretionary complex: Timescale, geochemistry, flow rates, and fluid budget, *J. Geophys. Res.*,*103*, 30,351–30,370.
Saffer, D. M., K. M. Frye, C. J. Marone, and K. Mair (2001), Laboratory results indicating complex and potentially unstable frictional behavior of smectite clay, *Geophys. Res. Lett.*, *28*, 2297–2300.
Schofield, A. N., and C. P. Wroth (1968), *Critical State Soil Mechanics*, 310 pp., McGraw-Hill, New York.
Scholz, C. H. (1990), *The Mechanics of Earthquakes and Faulting*, Cambridge University Press, Cambridge.
Screaton, E., D. Saffer, P. Henry, S. Hunze, and Leg 190 Shipboard Scientific Party (2002), Porosity loss within the underthrust sediments of the Nankai accretionary complex: Implications for overpressures, *Geology*, *30*(1), 19–22.
Seno, T., S. Stein, and A. A. Gripp, (1993), A model for the motion of the Philippine Sea plate consistent with NUVEL-1 and geological data, *J. Geophys. Res.*, *98*, 17,941–17,948.
Shipboard Scientific Party (1991), Site 808. *Proc. Ocean Drill. Program, Init. Rep.*, *131*, 71–269.
Shipboard Scientific Party (2001), Site 1173 [online]. In Moore, G. F., Taira, A., Klaus, A., et al., *Proc. Ocean Drill. Program Init. Rep.*, *190*. (Available at http://www-odp.tamu.edu/publications/190_IR/VOLUME/CHAPTERS/IR190_04.PDF)
Shipboard Scientific Party (2001), Site 1174 [online]. In Moore, G. F., Taira, A., Klaus, A., et al., *Proc. Ocean Drill. Program Init. Rep.*, *190*. (Available at http://www-odp.tamu.edu/publications/190_IR/VOLUME/CHAPTERS/IR190_05.PDF)
Shipboard Scientific Party (2001), Site 1177 [online]. In Moore, G. F., Taira, A., Klaus, A., et al., *Proc. Ocean Drill. Program Init. Rep.*, *190*. (Available at http://www-odp.tamu.edu/publications/190_IR/VOLUME/ CHAPTERS/IR190_08.PDF)
Steurer, J. F., and M. B. Underwood (2003), Clay mineralogy of mudstones from the Nankai Trough reference sites and frontal accretionary prism [online], *Proc. Ocean Drill.*

Program Sci. Results, 190/196. (Available at http://www-odp.tamu.edu/publications/190196SR211/211.htm)

Sunderland, E. B. (2003), Sediment microstructures and deformation at the toe of the Nankai accretionary prism: Effects of accretion and diagenesis and evolution of the décollement, M.S. thesis, 240 pp., Rice Univ., Houston, Tex.

Sunderland, E. B., and J. K. Morgan (2004), Microstructural variations in sediments from the toe of the Nankai accretionary prism: Results of scanning electron microscope analysis [online], *Proc. Ocean Drill. Program Sci. Results, 190/196.* (Available at http://www-odp.tamu.edu/publications/190196SR212/212.htm)

Taira, A., I. Hill, J. Firth, and Shipboard Scientific Party (1992), Sediment deformation and hydrogeology of the Nankai Trough accretionary prism: Synthesis of shipboard results of ODP Leg 131, *Earth. Planet. Sci. Lett., 109,* 431–450.

Tanioka, Y., and K. Satake (2001), Detailed coseismic slip distribution of the 1944 Tonankai earthquake estimated from tsunami waveforms, *Geophys. Res. Lett., 28,* 1075–1078.

Tichelaar, B. W., and L. J. Ruff (1993), Depth of seismic coupling along subduction zones, *J. Geophys. Res., 98,* 2017–2037.

Torrance, J. K. (1983), Towards a general model of quick-clay development, *Sedimentology, 30,* 547–555.

Torrance, J. K. (1984), A comparison of marine clays from Ariake Bay, Japan, and the South Nation River landslide site, Canada, Soils and Foundations, *Jpn. Soc. Soil Mech. Found. Eng., 24,* 75–81.

Tribble, J., and R. H. Wilkens (1994), Microfabric of altered ash layers, ODP Leg 131, Nankai Trough, *Clays Clay Miner., 42,* 428–436.

Tribble, J. S., F. T. MacKenzie, J. Urmos, D. K. O'Brien, and M. H. Manghnani (1992),
Effects of biogenic silica on acoustic and physical properties of clay-rich marine sediments, *Bull. Am. Assoc. Petrol. Geol., 76,* 792–804.

Ujiie, K., T. Hisamitsu, and A. Taira (2003), Deformation and fluid pressure variation during initiation and evolution of the plate boundary décollement zone in the Nankai accretionary prism, *J. Geophys. Res., 108*(B8), 2398, doi:10.1029/2002JB002314.

Ujiie, K., A. J. Maltman, and M. Sánchez-Gómez (2004), Origin of deformation bands in argillaceous sediments at the toe of the Nankai accretionary prism, southwest Japan, *J. Struct. Geol., 26,* 221–231.

Underwood, M. B. (2007), Sediment inputs to subduction zones: Why lithostratigraphy and clay mineralogy matter, this volume.

Underwood, M. B., and J. F. Steurer (2003), Composition and sources of clay from the trench slope and shallow accretionary prism of Nankai Trough [online], *Proc. Ocean Drill. Program Sci. Results, 190/196.* (Available at http://www-odp.tamu.edu/publications/190196SR206/206.htm)

Underwood, M. B., K. Pickering, J. M. Gieskes, M. Kastner, and R. Orr (1993), Sediment geochemistry, clay mineralogy, and diagenesis: A synthesis of data from Leg 131, Nankai Trough, *Proc. Ocean Drill. Program Sci Results, 131,* 343–364.

Vannucchi, P., and H. Tobin (2000), Deformation structures and implications for fluid flow at the Costa Rica convergent margin, ODP Sites 1040 and 1043, Leg 170, *J. Struct. Geol., 22,* 1087–1103.

Vrolijk, P. (1990), On the mechanical role of smectite in subduction zones, *Geology, 18*(8), 703–707.

Wong, T.-F., C. David, and W. Shu (1997), The transition from brittle faulting to cataclastic flow in porous sandstones: Mechanical deformation, *J. Geophys. Res.,102,* 3009–3025.

Wood, D. M. (1990), *Soil Behaviour and Critical State Soil Mechanics,* 462 pp., Cambridge Univ. Press, New York,

Zhang, J., D. M. Davis, and T. F. Wong (1993), The brittle-ductile transition in porous sedimentary rocks: Geological implications for accretionary wedge aseismicity, *J. Struct. Geol., 15,* 819–830.

9

The Nicaragua Convergent Margin

Seismic Reflection Imaging of the Source of a Tsunami Earthquake

Kirk D. McIntosh, Eli A. Silver, Imtiaz Ahmed, Arnim Berhorst, Cesar R. Ranero, Robyn K. Kelly, and Ernst R. Flueh

Abstract

Deep-penetration seismic reflection data acquired across the Pacific margin of Nicaragua reveal the regional structure of the margin and provide direct evidence for the involvement of subducted seamounts in the 1992, $M = 7.2$ tsunami earthquake. Prior to this survey, seismic reflection data sets were limited to shallow-penetration exploration data across the shelf and a single relatively low penetration profile across the trench slope and subducting Cocos plate. The new data characterize the structure of the Cocos plate and provide images of its morphology during the subduction process. Using the seismic reflection data and nearly full swath bathymetric coverage, we find that the Cocos plate preferentially forms half-graben structures as it bends into the trench, with fault throw down toward the trench on nearly all the faults. Fault throw is at least 500 m in some cases with basement of the footwall block occasionally exposed owing to the relatively thin 300–400 m thick oceanic sedimentary section. As these structures are subducted, the lower slope portion of the upper plate conforms to their shape undergoing nearly continuous deformation. Images of the plate boundary show that the half-graben structures are maintained during subduction beneath the trench slope and outer shelf to depths of at least 20 km. We have also imaged at least three moderate size (1–2 km high) seamounts subducted beneath the margin wedge. These have all left evidence of their passage in the seafloor morphology, and they are currently deforming mid- to upper-slope areas. These seamounts also appear to have been involved in the rupture pattern of the 1992 earthquake because they are within the two widely separated zones of enhanced moment release and occur where these zones extend anomalously close to the trench. We suggest that the shallow rupture was likely influenced by these seamounts, and as a result, an anomalously large tsunami was excited. Our seismic data show for the first time that off northern Nicaragua the Sandino fore-arc Basin sedimentary section has a minimum thickness of 14 km and could easily be as thick as 22 km. Much of this material is Cretaceous or Paleogene in age, but sedimentation has continued through the Neogene. Subsidence to accommodate this material has centered on the landward part of the basin, with an outer high present through nearly all the basin history. During Neogene-Recent there has been subsidence of the basin and the outer high off northern Nicaragua. We have identified normal faults cutting the upper plate, many of which are seaward dipping. This suggests margin wedge extension, probably as a response to tectonic erosion. However, with net progradation of the shelf edge through this time, rather than retreat, tectonic erosion must be limited, so steepening slab dip and sediment loading are probably important factors in the observed subsidence.

Introduction

The *M* 7.2 tsunami earthquake of 1992 (fig. 9.1) dramatically increased the interest in Nicaragua margin structure and pointed to its inherent contrasts with neighboring Costa Rica. This earthquake was characterized by long, slow rupture along most of the Nicaragua margin, significant moment release anomalously close to the trench, and generation of a deadly tsunami [*Satake*, 1994; *Velasco et al.*, 1994; *Ihmlé*, 1996; *Piatanesi et al.*, 1996]. No similar earthquake/tsunami events are historically known for Costa Rica, and the segmentation of its margin make similar wide-reaching events unlikely [*Ambraseys and Adams*, 1996; *Protti et al.*, 1995]. With moment release close to the trench, this event made a favorable target for marine geophysical investigations, but the available data were very limited. Aside from industry exploration of the Sandino Basin area, much of what was previously known about the structure off Nicaragua was interpreted from MCS profile NIC-1, a seismic line acquired in the late 1970s. *Ranero et al.* [2000] combined this profile (which is nearly coincident with NIC-80 acquired in the current project; fig. 9.1) with higher-quality industry data across the Sandino Basin and used information from exploration drill holes to establish a general tectonic history of the Nicaragua margin. In addition to the reflection profile the deeper structure of the margin was investigated by *Walther et al.* [2000] using refraction data recorded by ocean bottom hydrophones (OBH) and land-based seismographs. They found upper-plate margin wedge velocities of 2.2 km/s near the trench rapidly increasing to 4.1 km/s beneath the upper slope. *Walther et al.* [2000] interpreted that the upper-plate structure was largely composed of oceanic igneous material, which was configured in Late Cretaceous as a result of subduction initiation on the western margin of the Caribbean Large Igneous province.

In this paper we present results of multichannel seismic (MCS) reflection profiling across the Middle America convergent margin offshore Nicaragua, which characterize the subducting Cocos plate, illustrate its interactions with

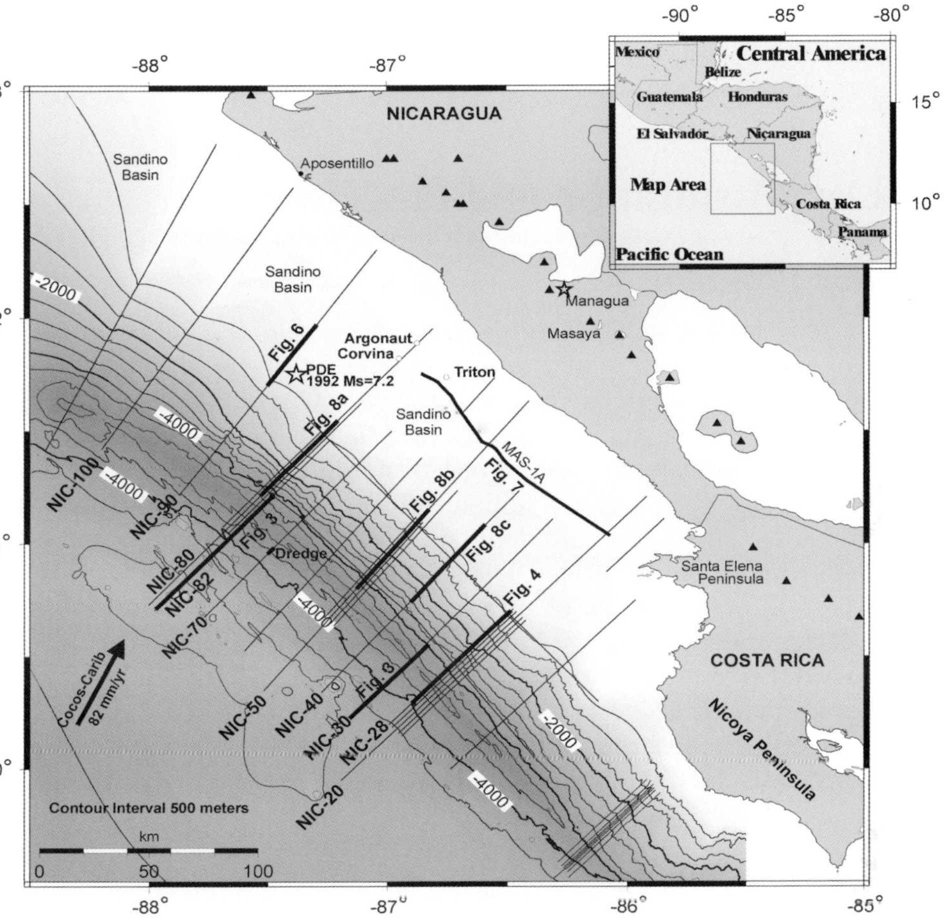

Figure 9.1 Map of study area offshore Nicaragua. Black lines, except off Nicoya Peninsula, show locations of seismic profiles acquired in 2000 cruise aboard the R/V *Maurice Ewing* (EW00-05). Small solid triangles mark volcanoes of the Central American arc. Three offshore exploration wells are labeled and marked, as is the epicenter of the 1992 earthquake. Plate motion vector estimated using poles of *DeMets et al.* [1994]. Heavy lines with labels overlying seismic line locations indicate portions of the lines shown in the labeled figures.

the overriding Caribbean plate, and provide images of the plate-boundary zone and its geometry well into the seismogenic zone. We first present two long MCS profiles to establish some of the regional characteristics of the margin. We then focus on the Cocos plate and trench slope structure. The seismic data and an expanded swath bathymetry data set provide important clues about the 1992 tsunami earthquake, establishing a structural framework to help understand the updip limit of seismogenic rupture here and the role of subducted seamounts. We also present data showing structure and stratigraphy of the Sandino fore-arc Basin, which document the tectonic history of the margin. Here we find evidence for vertical motions of the margin, including long-term subsidence in some areas and uplift in other areas.

Data Acquisition and Processing

Seismic Program

From 26 May to 27 June 2000 we used the R/V *Maurice Ewing* to collect an extensive seismic reflection and refraction data set off the Pacific margin of Nicaragua and northwestern Costa Rica (fig. 9.1). Our data set spans the Sandino fore-arc Basin, the trench slope, the Middle America Trench, and the Cocos plate to the outer bulge. We recorded more than 2800 km of MCS data and ~580 km of OBH and land station refraction data. The wide-angle data set includes 16+ successful land deployments on two lines and 39 successful OBH deployments on three lines. For reflection data acquisition we used the 6-km-long hydrophone streamer on the Ewing configured to record 240 channels of data with samples every 2 ms. Our nominal record length was 16 s, but we adjusted this to record 20 s in the landward parts of long profiles to make sure we recorded the deepest possible plate boundary reflections. In both cases we aimed to shoot the airgun array every 50 m. The airgun array consisted of 20 airguns with a total volume of 6817 cubic inches towed at ~7 m depth.

Data Processing

For the images presented in this paper we performed the following processing sequence: After initial trace editing to eliminate noisy traces in the shot gathers, we applied a broad 6–80 Hz bandpass filter, spherical divergence gain correction, and deconvolution. We used a multichannel predictive deconvolution algorithm, which designed the filter operator with 81 traces in two time windows. The prediction distance was 16 ms for the upper window, generally applied to the sedimentary section, 60 ms for the lower window, and operator lengths were 240 and 360 ms for these respective windows. Following deconvolution we removed the spherical divergence correction and sorted to 60-fold common midpoint gathers (CMPs). We typically picked stacking velocities every 100 CMPs (1.25 km), using a combination of semblance analysis and sets of gathers and stacks corrected with trial velocity functions. Before stack we applied a far-trace mute to remove direct arrivals and refractions, and we applied a near trace mute, starting at just above twice the water-bottom time, to suppress the seafloor multiple reflection. After stack we applied a two or three window time variant filter and gain, typically 1 s AGC, before display. We used poststack time migration in the frequency-space (FX) domain to obtain our final images, with migration velocity functions derived from simplified and smoothed stacking velocity functions. One of the difficulties we had was obtaining consistently good images of the slope structure, particularly in imaging the base of the sedimentary slope section and some of the structures above and below it. Part of the problem here may be due to streamer feathering during acquisition due to wind and currents. A postcruise examination of feathering suggests that it was typically worst in the slope area.

Regional Structure

Our data acquisition plan for this project focused on the Cocos plate and trench slope areas, but we also recorded six MCS dip profiles across the entire margin to image the deeper portions of the subduction plate boundary and to investigate the structural development of the Sandino fore-arc Basin. Two of these profiles, NIC-50 and NIC-100 (location in fig. 9.1), are shown in plates 9.1 and 9.2 to provide a regional perspective and to show some of the dramatic structural variations along strike.

NIC-50

Profile NIC-50 is located off central Nicaragua and extends from the outer bulge of the subducting Cocos plate to near the coast approximately in line with the Masaya volcano south of Managua (fig. 9.1). Plate 9.1 shows all of NIC-50 except the most seaward portion. Plate 9.2 shows this same section converted approximately to depth using velocities from our unpublished refraction model. This view gives a better representation of the true geometry of the Nicaragua margin, although some distortion remains owing to the simplified velocity model used for the depth conversion. On the Cocos plate portion of NIC-50, no seamounts are present and a relatively consistent stratigraphic section is visible despite the recent normal faulting. Landward of the trench axis and beneath the lower continental slope, the plate boundary and Cocos plate are represented by a complex reflection configuration at ~7 s (two-way travel time). This set of reflections is present to CMP 7800, where it is largely obscured by energy from the seafloor multiple reflection and associated later arrivals. Fortunately, the plate-boundary reflection emerges from the zone of multiples below CMP 8800 still at ~7 s travel time. We note that this apparent horizontal position of the plate boundary in plate 9.1 is a distortion owing to the display in time rather than depth. The increasing landward thickness of high seismic velocity material in the upper-plate margin wedge reduces the travel time to the plate boundary relative to paths through the water column (plate 9.1, see insets showing velocity, depth, and two-way travel time relationships at two sites). In fact, as shown in plate 9.2, the plate-boundary dip actually gets progressively steeper throughout this image. The distortion is reduced in the time section (plate 9.1) beneath the upper slope and shelf as the

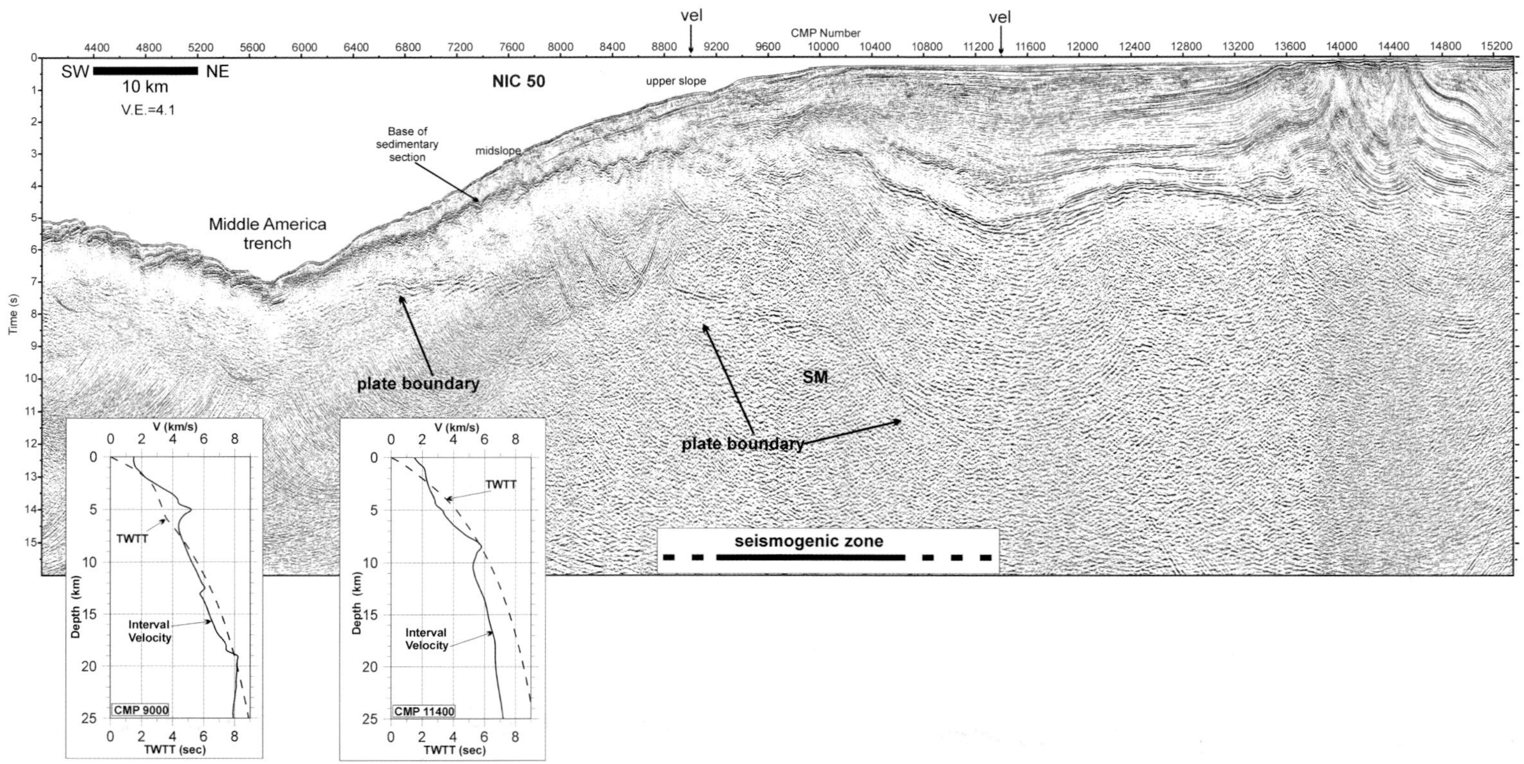

Plate 9.1 Regional seismic reflection profile NIC-50 across the Middle America Trench, trench slope, and Sandino fore-arc Basin. Except for interference with multiple reflections, the plate-boundary zone reflection is visible from the trench to CMP 11200, a distance of 67 km, with the depth at this point ~33 km. The Sandino Basin sedimentary section includes Cretaceous-Holocene units. The seaward two thirds of the basin is characterized by normal faults, while Late Miocene-Pliocene compressional deformation dominates at the landward end of the section. The expected seismogenic portion of the plate boundary is indicated at the base of the figure. Insets show interval velocity versus depth and two-way travel time (TWTT) versus depth for CMP 9000 and CMP 11400 extracted from our unpublished tomographic velocity model along this profile. Letters SM mark a probable subducted seamount along the plate boundary and V.E. is vertical exaggeration at the seafloor.

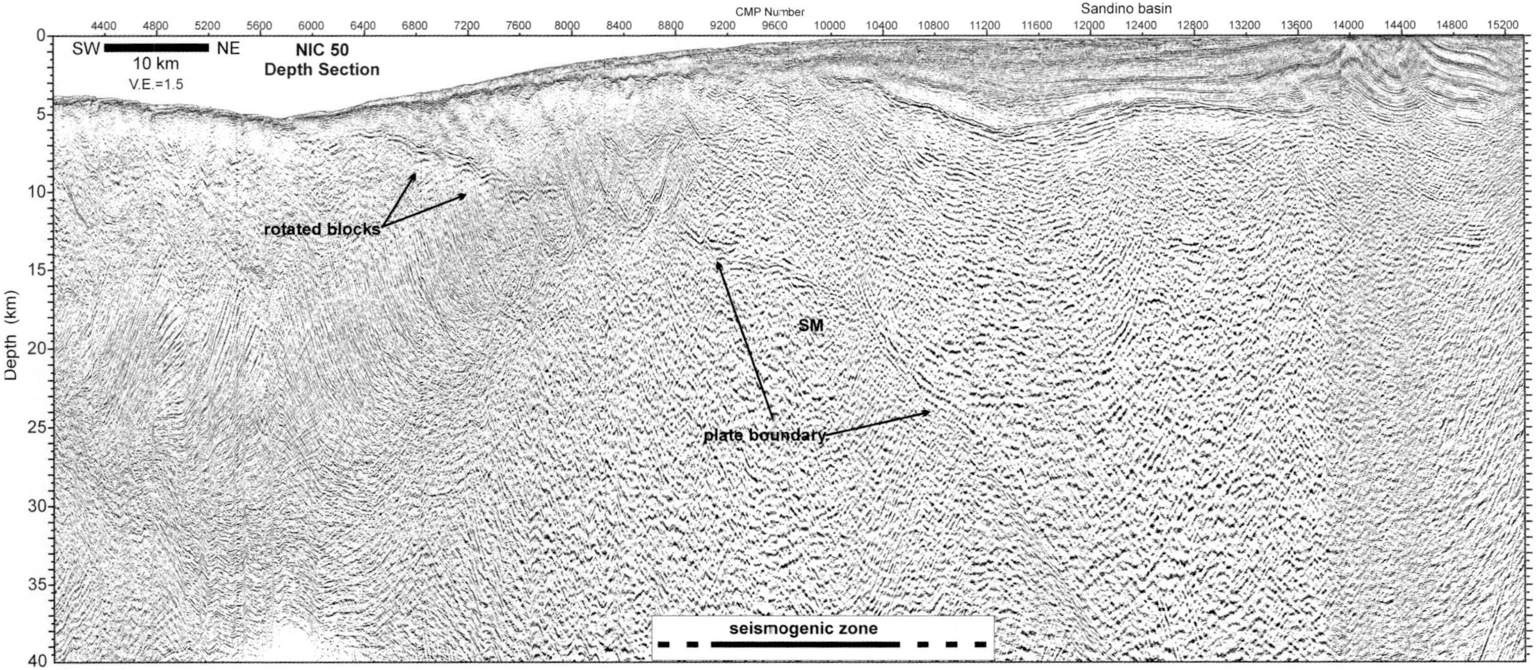

Plate 9.2 Depth-converted version of NIC-50 plotted at the same horizontal scale as in plate 9.1 and a vertical exaggeration of 1.5. The velocities used for depth conversion come from a simple layered model (sedimentary layer, upper crust, lower crust, mantle) with velocity at the top of each layer and intralayer velocity gradients estimated from our unpublished refraction model.

plate-boundary reflection dips steeply below the Sandino Basin. Some complexity in this reflection occurs below CMP 10000, where a seaward-dipping reflection segment appears. Although not well resolved on this section, we believe this is a small, subducted seamount because it is imaged on strike line NIC-125 at the intersection of these two lines. Just landward of the interpreted seamount, there is a steeper arrival; it may be partially generated by the landward flank of the seamount, but it is most likely a migrated, peg leg multiple generated by multiple paths between the high impedance base of the sedimentary basin and the seafloor or free (sea) surface. Another reflection segment extends at shallower dip from 10.5 s (CMP 10600) to 12 s (CMP 11000). This segment is also likely from the plate-boundary zone and corresponds to depths of 27–33 km using the velocities of *Walther et al.* [2000] and our unpublished refraction data along this profile (plate 9.1, inset). The depth-converted section (plate 9.2) shows this segment of the plate boundary as well, but the simplified gradient velocity function applied across the section apparently underestimated the velocity and resulting depth in this zone.

At shallower levels we observe landward-thickening sedimentary cover overlying the margin wedge. On NIC-50 this sedimentary section thickens gradually to the upper slope in the vicinity of CMP 8800. Here, the high-amplitude reflection at the base of the sedimentary section becomes unclear but probably deepens, as a thicker section of more coherent stratigraphic layering is visible beneath the uppermost slope. This trend is reversed at ~CMP 9900 as the high-amplitude, top-wedge reflection reappears at a higher level (~3 s), possibly owing to early compressional deformation (thrust faulting) within the basement. This uplifted basement marks the outer fore-arc high, which has separated the Sandino Basin from the trench-slope since Eocene or earlier, possibly even Late Cretaceous [*Ranero et al.*, 2000]. Landward of the outer high the basin reaches a time thickness of 5 s (~6–7 km) before shallowing to 4 s (~4.5 km) at CMP 12600. This modest basement high also marks a divide in extensional deformation in the seaward direction and compressional deformation landward. The two large anticlinal structures involve most of the stratigraphic section present in the rest of the Sandino Basin. Thus, although compression within the basin may have occurred as early as Eocene time [*Ranero et al.*, 2000], the primary development of these structures is probably Late Miocene to Early Pliocene in age [*Weinberg*, 1992], with some continued deformation perhaps to the present [*Gerth et al.*, 1999]. Interestingly, the thick sedimentary pile involved in these structures indicates that the deepest part of the basin was likely landward from the end of this profile, which comes to within 8 km of the present coast. Note that the apparently flat-lying events at shallow levels (<1 s) overlying the anticlines are artifacts, reverberating seafloor multiple reflections that were not successfully removed during processing. We also note that the compressional structures on NIC-50 are part of a nearly coast-parallel trend extending the length of the Nicaragua margin.

NIC-100

Profile NIC-100 is located near the western limit of our study area (fig. 9.1), extending from just seaward of the trench to near the coast at Aposentillo. NIC-100 (plate 9.3) is similar to NIC-50 in many ways, but there are also major differences. For example, the trench here is significantly broader, with lower relief on the Cocos plate side of the trench, the distance from the trench to shelf is nearly 20 km greater on NIC-100, and the Sandino Basin is phenomenally deep here. We can follow the plate-boundary reflection beneath the lower slope to CMP 4100 at 8 s, where it crosses the remnants of the seafloor multiple reflection. Landward the plate-boundary reflection emerges and is clear to CMP 5800 at 9 s (~20–22 km depth). Beyond this point the boundary is unclear, but a consistent band of reflections is present from CMP 7500 at 11.5 s to CMP 9500 at 16 s. We interpret these deep arrivals as probable plate-boundary zone reflections and mark its position between ~ 29 and 40–45 km depth (at CMPs 7500 and 9500, respectively).

As we observed on NIC-50 (plate 9.1), the sedimentary section overlying the margin wedge gradually thickens landward, but the top-basement reflection becomes uncertain beneath the upper slope (beyond CMP 5100). Here, a set of normal faults, dipping in both the landward and seaward directions, offsets the sedimentary section and may extend into the underlying wedge. This complex structural domain extends to just beneath the shelf edge, where the Sandino Basin begins. The seaward boundary of the Sandino Basin is not imaged, perhaps indicating a high-angle basin-bounding fault (near CMP 7300). It is apparent, however, that sedimentary units are probably present to 5.5 s at CMP 7900 and the high-amplitude reflection at this point extends basinward to at least 9.5 s and a possible maximum of ~11 s. Stacking velocities near the center of the basin at 8 s were 3.7 km/s, so we can use that as a rough approximation (velocity discrimination is generally poor at depths greater than the streamer length, here 6 km). In comparison, velocities extracted from the NIC-50 tomographic model (unpublished) lead to an estimated average velocity of 2.8 km/s at 5.0-s two-way travel time, just above basement at the deepest part of the sedimentary section on that profile (plate 9.1, inset). If the sedimentary velocity-depth trend from the NIC-50 model is extrapolated to 8.0 s, then the average velocity is 4.2 km/s. Assuming a minimum average velocity of 3 km/s for the entire sedimentary section, travel times of 9.5 s to 11 s indicate a sedimentary thickness of 14–16.5 km in the Sandino Basin. However if the average velocity were 4 km/s, as seems likely, then the basin thickness would be 22 km. The compressional deformation visible on NIC-50 and along most of the margin is not manifested on this profile. However, the seaward dips near the end of the profile are likely the flank of a related anticline farther to the northeast, on trend with the San Cayetano anticline mapped onshore to the southeast [*Weinberg*, 1992]. We note that even very young strata appear to be uplifted and erosionally truncated (plate 9.3, inset), suggesting recent or ongoing deformation here.

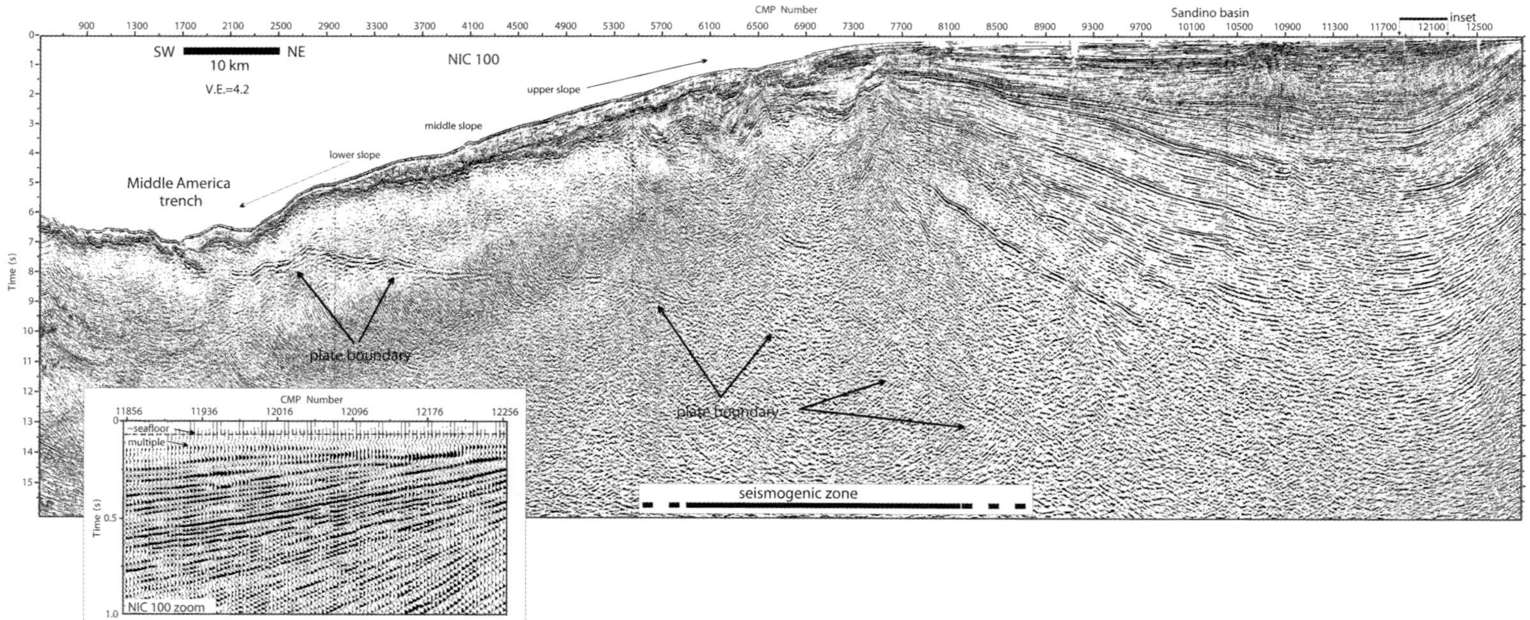

Plate 9.3 Regional seismic reflection profile NIC-100 across the Middle America Trench, trench slope, and Sandino fore-arc Basin. This profile demonstrates the northwestward thickening of the Sandino Basin. We estimate the sediment thickness here to be at least 16 km. The margin wedge is thicker here than along NIC-50 at similar distances from the trench. The plate-boundary reflection is clear to beneath the upper slope, and we interpret a deeper set of arrivals mark the plate boundary beneath the Sandino Basin. The inset at lower left is an expanded view of young, shallow strata, which have been uplifted and truncated; location of this data segment is marked in the upper right of the profile.

Cocos Plate

A primary goal of our project is to characterize the structure and morphology of the subducting Cocos plate and investigate how variations in these parameters may also relate to variations in seismogenic character and sediment dynamics. Accordingly, we added to the existing swath bathymetry database [*von Huene et al.*, 2000] to cover nearly the entire trench and slope area offshore Nicaragua. In addition, to trace the development of Cocos plate normal faults we extended four seismic lines to the crest of the outer bulge (fig. 9.1). Figure 9.2 shows a perspective view of the swath bathymetry data along the Middle America Trench, extending from just northwest of NIC-90 to southeast of NIC-20. Seamounts are abundant in this area, with sizes up to 1.25 km high and ~10 km basal diameter and including a full range of smaller sizes. Figure 9.2 also shows the prominent structural fabric, which develops as the Cocos plate bends in to the Middle America Trench and is broken by normal faults. Although the bathymetric coverage is limited to only a few swaths farther seaward from the trench, it is clear that normal faults are poorly developed or absent across the outer bulge and form as the plate approaches the trench. The Cocos plate normal faults form along the entire area offshore Nicaragua, but the greatest structural relief occurs where the outer bulge is best developed, as shown by the contours in figure 9.1, and the relief tapers off in particular to

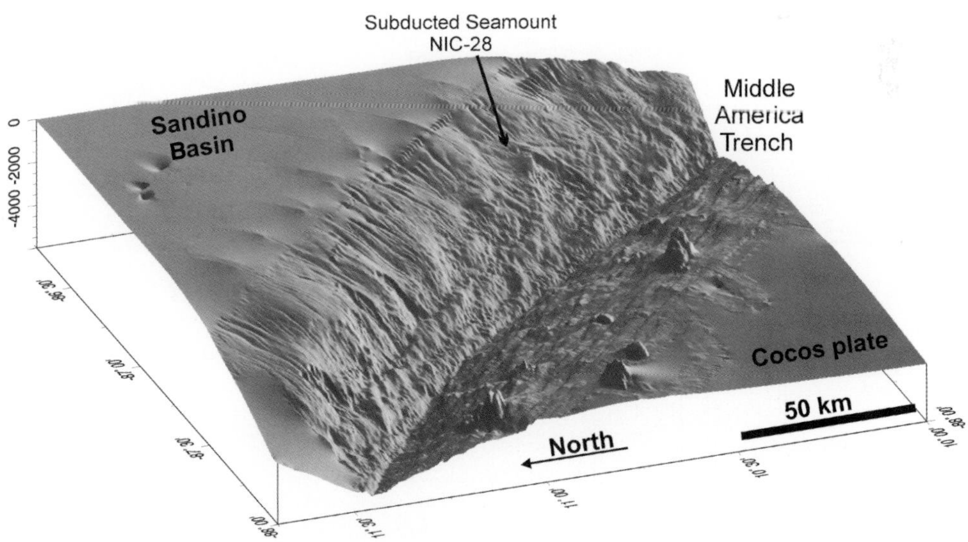

Figure 9.2 Perspective view of swath bathymetry looking southeast along the Middle America Trench offshore Nicaragua. Smooth areas in the extreme northeast and southwest are constrained only by satellite-derived data. Note the linear structural fabric produced by Cocos plate normal faults and numerous small- to medium-size seamounts. The relatively steep middle slope is marked by numerous slumps, some of which are directly associated with subducting seamounts.

the southeast toward Nicoya Peninsula, Costa Rica, where the bulge is poorly developed and the depth to the trench is substantially reduced.

Although the swath bathymetry clearly shows the presence and intensity of Cocos plate normal faulting, the sedimentary section overlying the igneous crust smoothes the structure somewhat. Seismic reflection profiles, however, clearly show fault offsets at the seafloor and at the top of the basement. Two representative profiles are shown in figure 9.3. Profile NIC-82 crosses one of the larger seamounts in the area and shows the dramatic development of the normal faults as the oceanic plate approaches the trench. Some fault offsets shown here and detected on other profiles are nearly 500 m at the seafloor. The other outstanding feature of this fault pattern is that nearly all the faults have throws down to the northeast, or toward the trench. This pattern also features fault-block rotation, which may lead to enhanced sediment subduction efficiency [*Kelly et al.*, 2001]. Profile NIC-30, located ~100 km southeast of NIC-82, shows a similar but subdued pattern of normal faulting, including fault-block rotation and consistent down to the trench offsets (fig. 9.3). Here, the incoming sedimentary section is much more uniform in thickness and the fault offsets are substantially smaller. The uniform sedimentary cover is apparently

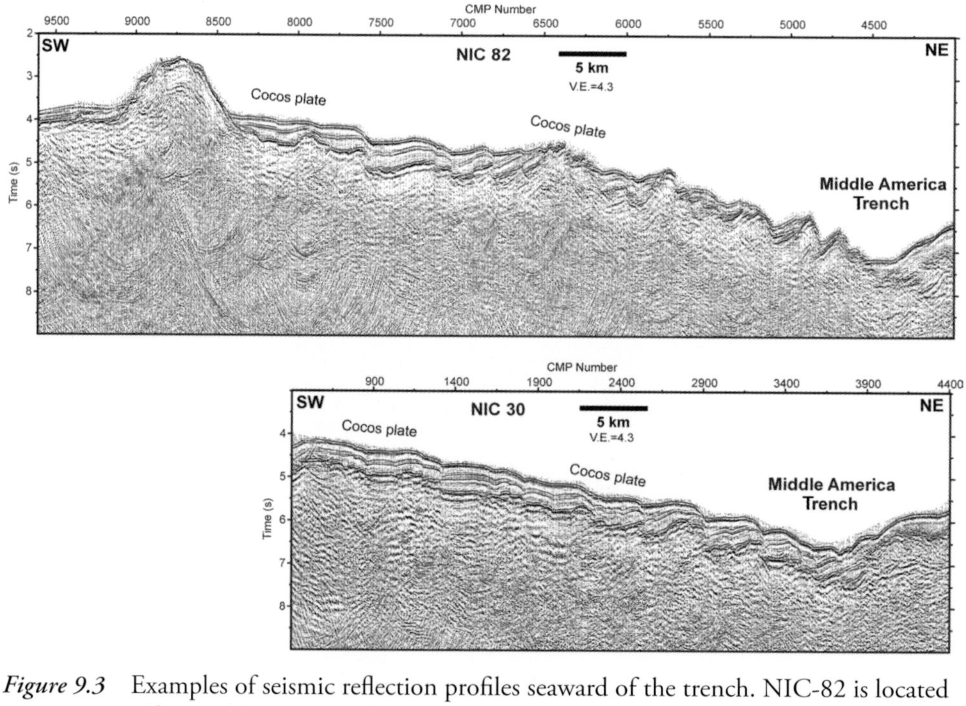

Figure 9.3 Examples of seismic reflection profiles seaward of the trench. NIC-82 is located off central Nicaragua where the outer bulge is well developed and relief into the trench is near the maximum. The seamount imaged here is typical of the larger seamounts detected offshore Nicaragua. NIC-30, farther to the southeast, shows a more continuous sedimentary section and reduced offset on the normal faults.

due to the lack of seamounts and other positive bathymetric features, both of which characterize NIC-82. The two profiles are essentially end-members representing the variation in Cocos plate structures offshore Nicaragua. The trend of lower relief and reduced normal fault offset to the southeast indicated by the swath bathymetry and our entire seismic data set continues to a structural minimum offshore of the Nicoya Peninsula where the Cocos plate is the smoothest [*von Huene et al.*, 2000; *Kelly et al.*, 2001].

Continental Slope and Margin Wedge

The morphology of the trench slope includes a lower slope domain rising from the trench typically with several moderate slopes and intervening benches, a rapidly steepening midslope zone and an upper slope, which shallows to the shelf edge (fig. 9.2). The lower slope shape, in some cases, is directly related to fault blocks subducting beneath. The larger slope/bench combinations are composite features created by deposition, mostly slumping, internal wedge deformation, and subducting plate structures. The midslope area, much steeper than upper or lower slopes, has pervasive scars (fig. 9.2) caused by slumping, in some cases directly related to seamount subduction. The transition to the upper slope is marked by a reduced gradient, which ultimately flattens at the shelf edge. The upper slope and shelf edge appear to be much more stable than the middle and lower slope areas as small submarine canyons or rills are well developed along much of the margin.

Our seismic reflection data reveal some of the internal structure beneath the trench slope and also confirm the effects of the subducting Cocos plate structures. The trench-slope offshore Nicaragua has a sedimentary section of variable thickness, which is separated from the underlying margin wedge by an irregular but highly reflective seismic boundary, the rough surface described for offshore Costa Rica by *Shipley et al.* [1992]. Typically, this sedimentary section thins rapidly from the upper slope to the midslope and remains thin across the lower slope. Seismic profile NIC-28 (fig. 9.4) shows an example of how the lower slope can conform to the shape of tilted fault blocks subducting beneath it. In this case the two slope/bench combinations nearest the trench can be tied directly to two subducting fault blocks. The lowermost bench is essentially level with the top of the basement reflection underlying the slope and may also indicate basement outcrop. The second bench, just upslope, results from an upturn of the lowermost part of the slope apron, coinciding with the location of a tilted fault block in the subducting Cocos plate.

As seen in figure 9.4, the slope section is apparently thinned owing to slumping, which in this case is due to a subducting seamount. Here, the slump scarp at the seafloor is directly above the trailing edge of the seamount, a relationship that is typical of seamount subduction [*Lallemand and Le Pichon*, 1987; *Collot and Fisher*, 1989; *Lallemand et al.*, 1994]. The scar resulting from this seamount

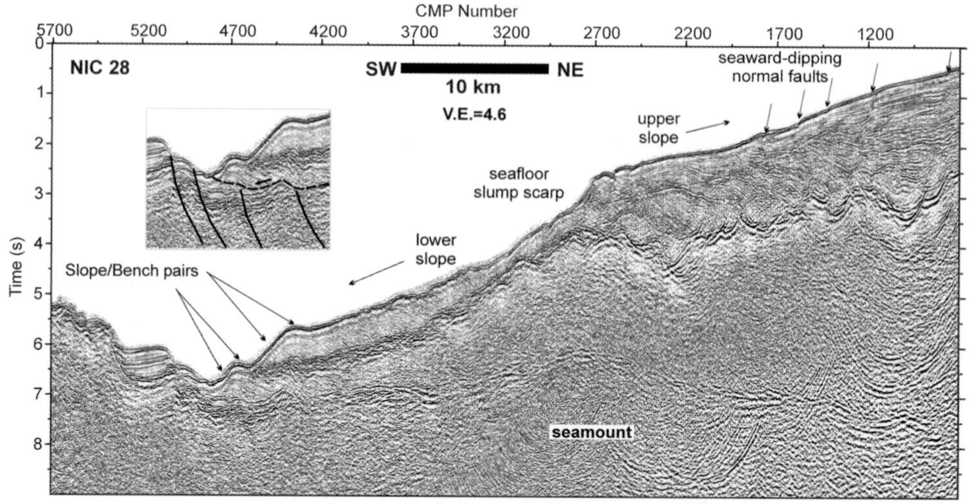

Figure 9.4 Seismic reflection profile NIC-28 across the landward trench slope. This profile shows how the upper plate conforms to the structure of the subducting plate. Near the trench the two seafloor slope/bench combinations correspond directly to subducted fault blocks. Beneath the midslope, we image a subducted seamount, which uplifts the upper plate and results in a seafloor slump scarp above its trailing edge.

is also well displayed in the seafloor image (fig. 9.2), which shows the fresh scarp and the uplifted seafloor just landward where the seamount is elevating the margin wedge. This seamount is 1.5–2.0 km high (assuming velocities of 3–4 km/s as indicated by refraction models and stacking) and ~10 km across in the dip direction; thus it is similar in size to several other seamounts seaward of the trench visible in figure 9.2, which will subduct in the next 100–500 kyr. On the basis of this observation and other subducted seamounts visible on profiles NIC-80 and NIC-90, we suggest that subducting seamounts probably play a key role in the slope morphology here, especially in development of the steep midslope zone.

As noted above, the sedimentary section overlying the margin wedge thins dramatically at a midslope transition (CMP 2800). This also marks the seaward limit of coherent stratigraphic layering in the sedimentary section. In this case the transition is directly influenced by the subducting seamount, but a similar morphological change occurs all along the margin, with or without currently subducting seamounts. We interpret that the coherent strata generally present across the upper slope may indicate a threshold of wedge thickness and competence at which subducting seamounts have a muted, perhaps more regional effect. As we have seen, the thin wedge below the lower slope tends to conform with the subducting plate structures, so it must be relatively weak. We further suggest that the midslope morphological change reflects a transition in structural integrity of the margin wedge to a bulk material capable of storing elastic strain. The position of the transition depends on wedge thickness, subsequent erosional processes, and wedge composition. Several lines of evidence indicate that the wedge present beneath the rough surface here is composed of oceanic igneous rock. Reflection (MCS) processing and refraction data show that the margin wedge has *P*-wave velocities > 4 km/s at the top and generally increases with depth and distance from the trench to reach velocities of 5 to >6 km/s at deeper levels beneath the upper slope and shelf [*Ranero et al.*, 2000; *Walther et al.*, 2000; K. McIntosh, unpublished data, 2006]. We expect that velocities are substantially reduced owing to deformation and fluid-related alteration beneath the trench slope, but these processes likely have decreasing effect landward as indicated by the increasing velocities. Direct evidence supporting the igneous nature of the margin wedge is supplied by samples dredged from a midslope seafloor exposure of the rough surface. We recovered basalts, Late Cretaceous limestones, red chert, and basaltic sandstone [*Silver et al.*, 2000], providing strong evidence for the composition of the margin wedge as well as a date for its probable cover assemblage. The sedimentary rocks yielded ages of Santonian-Maestrichtian, similar to sedimentary rocks associated with the Nicoya Complex of Costa Rica, based on both nanofossil (S. Kling, written commununication, 2000) and foraminiferal evidence (E. Thomas, written communication, 2000)

While the upper slope has smoother morphology than the mid- and lower-slope domains, we recognize significant deformation in the subsurface there, primarily in the form of normal faults. For example, in figure 9.4, there are several seaward-dipping normal faults, which cut through the entire sedimentary section and offset the basement and the seafloor. Thus they appear to be active and apparently involve the underlying margin wedge, although fault-plane reflections associated with these faults have not been clearly identified within the basement wedge. If this analysis is correct, then the upper plate is extending seaward, most likely as a result of tectonic erosion (we have not determined whether erosion is frontal, basal, or both). Another view of the upper slope is shown on profile NIC-80 (fig. 9.5). This transect is also characterized by normal faults, but there is a clear change in degree of deformation through time. The lower section, just overlying basement, is offset by seaward-dipping faults, some of which appear to penetrate the basement. Overlying this poorly reflective unit some strata thicken toward the normal faults representing syndeformational growth strata. However, this style of deformation stops abruptly up section, such that in the upper 1 s of the record (~1 km) fault offsets are quite small. In the landward quarter of figure 9.5 the normal faults are more visible throughout the sedimentary section and become dominantly landward dipping. This zone is landward of outer basement high and marks a transition to faulting toward the interior of the Sandino Basin.

Profile NIC-80 is nearly colocated with the composite profile interpreted by *Ranero et al.* [2000], so it is straightforward to identify the corresponding

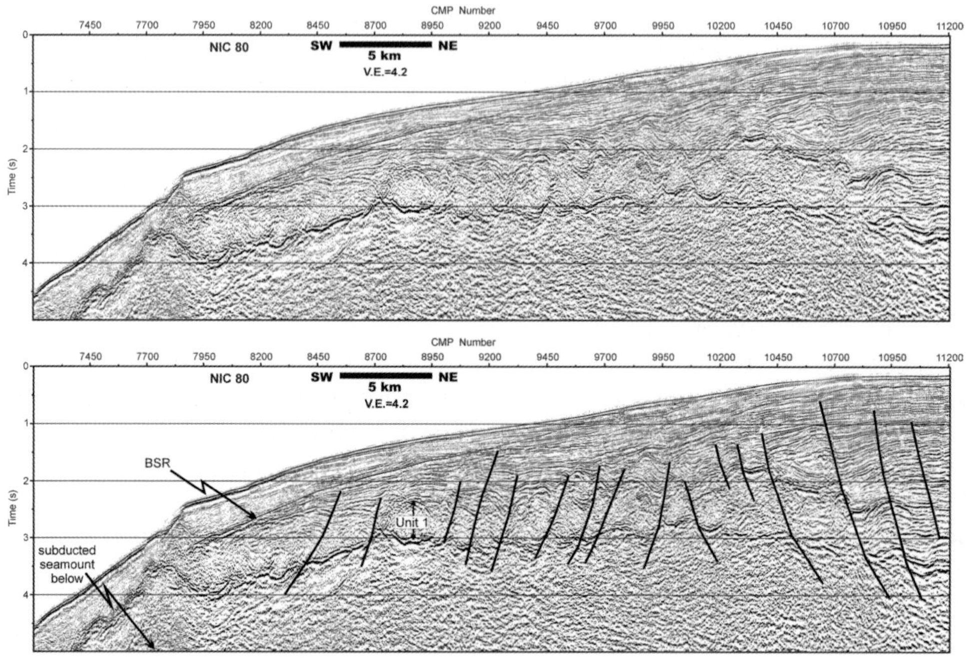

Figure 9.5 A view of the upper slope from profile NIC-80. The seafloor scarp marks the underlying subducted seamount. Normal faults change dominant dip direction at ~CMP 10000, which marks the fore-arc basement high. The lowermost sedimentary unit (unit 1 of *Ranero et al.* [2000]) is likely Cretaceous to Eocene in age. Fault offset generally decreases up section. BSR, bottom simulating reflector, which marks the base of the gas hydrate stability zone.

stratigraphic units. We interpret that the lower unit is their unit 1, Rivas and Brito Formations, age Late Cretaceous - Middle Eocene. This unit extends seaward at least to the basement high associated with the subducting seamount (CMP 7800, fig. 9.5) and on adjacent profiles appears to extend at least to the midslope area and possibly across the lower slope. This interpretation is consistent with the Late Cretaceous age of our sedimentary dredge samples, which are likely from the base of this section [*Silver et al.*, 2000].

Sandino Fore-Arc Basin

The Sandino Basin developed as a fore-arc basin at least by late Cretaceous time and has been a site of sedimentary deposition throughout the subsequent period [*Winsemann*, 1994; *Walther et al.*, 2000; *Ranero et al.*, 2000]. Although Cenomanian siliciclastic turbidite deposits are known in southwestern Nicaragua, establishing a possible early Cretaceous origin [*Winsemann*, 1994], compressional deformation during Campanian time [*Kuijpers*, 1980; *Bourgois et al.*, 1984] probably caused major basement structural modification in Nicaragua and northern Costa Rica. The sedimentary units in the Sandino Basin are known from onshore outcrops and have been described by numerous workers [e.g., *Weyl*, 1980]. As noted above, *Ranero et al.* [2000] correlated stratigraphy encountered in onshore and offshore petroleum exploration wells to a seismic transect across the basin to establish approximate dates and characteristics for the various seismic units. Both the wells and the onshore sections show that there has been a general trend of shallowing depositional environment through time progressing from basinal, to fan/channel systems, to prograding wedges, and outcrops younger than late Oligocene show shallow water sedimentary facies [*Winsemann*, 1994].

As depicted by the regional profiles NIC-50 and NIC-100 (plates 9.1 and 9.2), the Sandino Basin deepens dramatically along strike to the northwest and becomes generally shallower to the southeast. Both of these profiles show an outer basement high, separating the fore-arc basin from the trench slope. This has been an enduring feature of the margin as sedimentary units of nearly all ages, Cretaceous-Miocene, thin or pinch-out toward it. While the long history of an outer high is clear, the huge accumulation of strata requires that both the basement high and, especially, the adjacent basin have subsided through time. Our data set also provides evidence of how depositional systems have varied through time reflecting the trade-offs between subsidence, sediment supply, and deformation rates. One of the best examples of this is shown on NIC-90 (fig. 9.6), where progradational sedimentary structures mark the paleoshelf edge from early Miocene to the present. Two important observations that can be made from figure 9.6 are that subsidence is nearly uniform across the margin over this time, and there has been a general seaward advance in shelf edge through time. Uniform subsidence is indicated by the almost horizontal sequence boundaries (bold black lines) across this outer half of the Sandino Basin. The seaward advance of the shelf edge is demonstrated by seaward-dipping clinoforms in the Lower Miocene section, CMPs 4400–5400, in the lower right part of figure 9.6, to the current shelf edge, and several other progradational sedimentary packages are apparent in between. The deformation at the seaward end of figure 9.6 is quite recent, given the offsets of strata shallow in the section The uplift at deeper levels at CMP 6800 may be due to diapiric mass movement.

Although fore-arc subsidence is clear in the central and northern Sandino Basin, substantial uplift has occurred in its southeastern portions. This is reflected in the onshore surface geology with progressively older units exposed to the southeast [*Weyl*, 1980; *Ranero et al.*, 2000], and an older vintage (1977) seismic strike profile, MAS-1A, shows how the uplift is manifested in the landward portion of the Sandino Basin, approaching the Santa Elena and Nicoya peninsulas of Costa Rica (fig. 9.7; location in fig. 9.1). We have attempted to jump correlate the *Ranero et al.* [2000] stratigraphic units from the northwest continuation of this profile, which comes close to the Corvina and Triton exploration wells. Owing to substantial along-strike deformation the correlation is tenuous, but because the section is apparently anchored by a basement

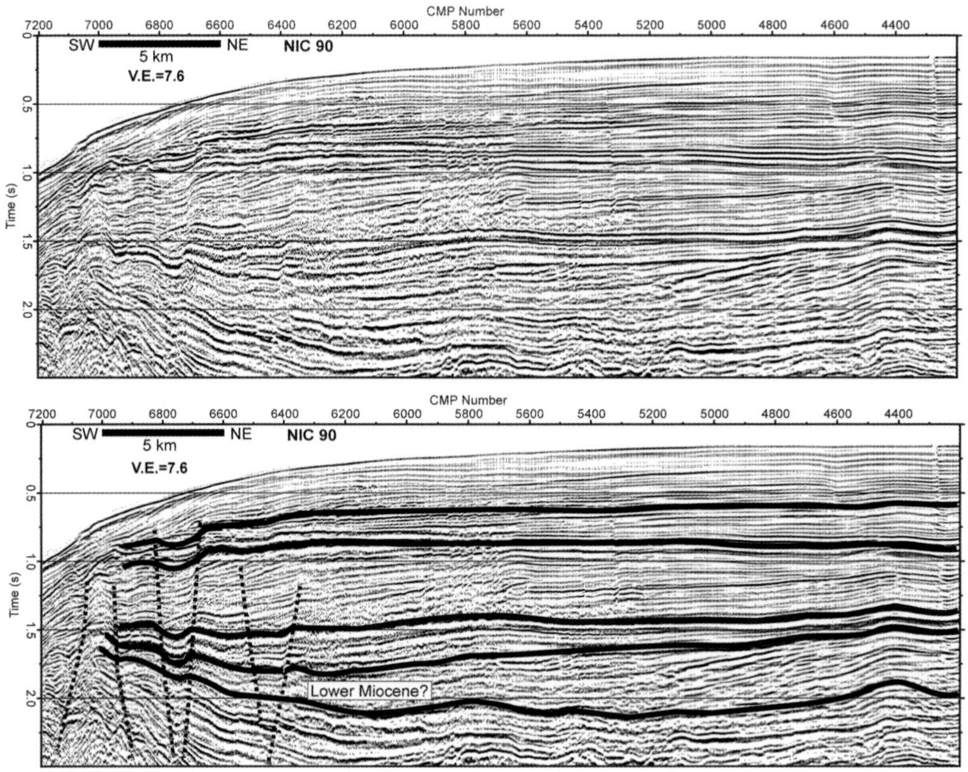

Figure 9.6 Profile NIC-90 across upper slope and outer shelf. The portion shown is interpreted as the Miocene to Recent sedimentary section. Major sequence boundaries, marked with the heavy black curves, bound progradational sedimentary packages showing seaward advancement over the last 20+ myr. These shallow water units also require continuing subsidence, but lack of tilting suggests a regional process of accommodation.

horizon, our most likely error is an age underestimate of the shallower units. The first-order observation on MAS-1A (fig. 9.7) is that all the strata are tilted, including the shallowest units. The sediment package labeled unit 5 thins toward the uplifted area but unit 6 strata above it are nearly uniform in thickness and also tilted. Thus it appears that at least some of the uplift is as recent as Pliocene or later (unit 6) with possible earlier uplift in Middle to Late Miocene (unit 5). The lower sedimentary unit is undifferentiated and may include units 1–3, or Cretaceous-Oligocene strata. The deeper and older portion of this unit apparently thickens significantly to the southeast, suggesting a Cretaceous depocenter near the current Nicaragua-Costa Rica border. Within the lower sedimentary unit we observe a continuous reflection with a minimum of 3 s of time change across the section. Using an average seismic velocity of 3 km/s, this equates to an uplift of 4.5 km. This uplift continues to the southeast as basement, shown here at 3-s two-way time, crops out at the surface in the Nicoya and Santa Elena peninsulas.

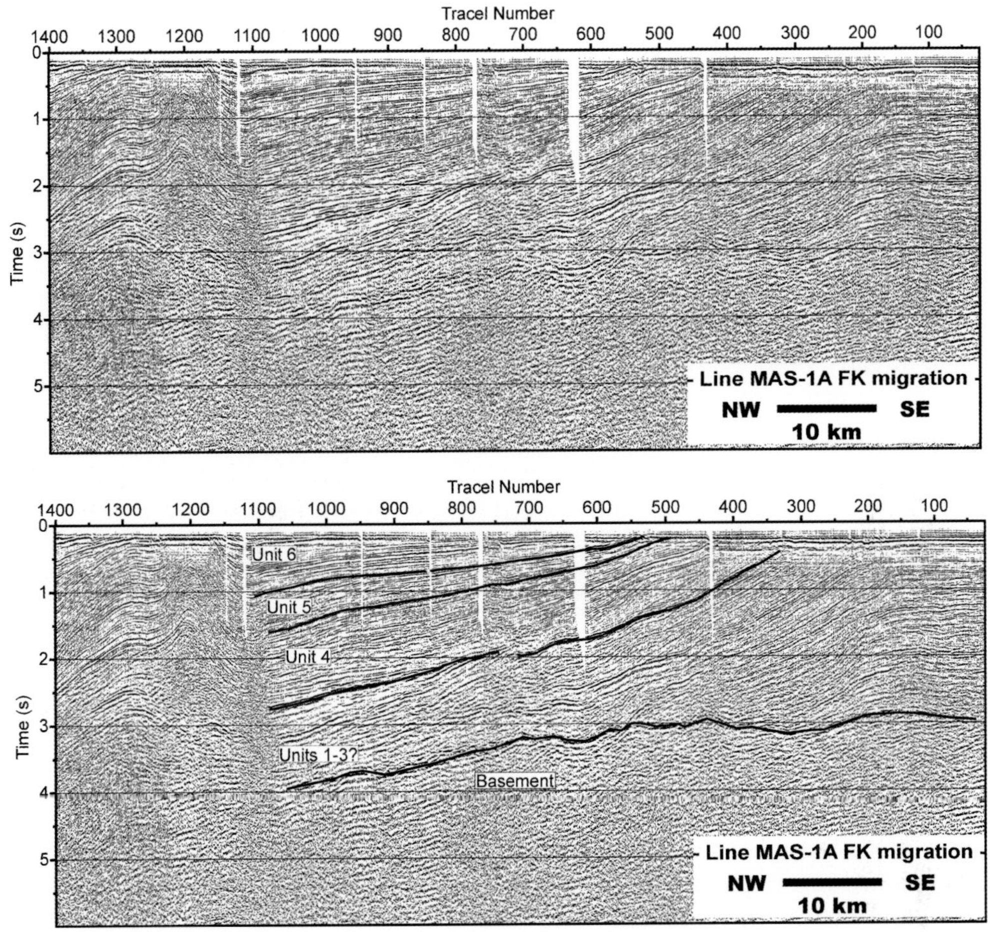

Figure 9.7 Profile MAS-1A, a strike profile along the southeastern part of the Sandino Basin. The southeastern part of this profile documents strong uplift of the Sandino Basin sedimentary section, apparently involving relatively young units. Our correlation is approximate owing to a jump across the deformed zone to the northwest. Units as defined by *Ranero et al.* [2000]: unit 1, Rivas/Brito formations, Late Cretaceous-Middle Eocene; unit 2, Brito Formation, Late Eocene-Early Oligocene; unit 3, Brito/Masachapa formations., Early-Late Oligocene; Unit 4, Masachapa Formation, Late Oligocene-Middle Miocene; unit 5, El Fraile Formation, Middle-Late Miocene; unit 6, El Salto Formation, Pliocene-Holocene.

Plate-Boundary Structures

Understanding the dimensions, physical properties, and structure of a subduction plate- boundary zone is critically important to all the disciplines investigating convergent margins. Unfortunately, the plate-boundary zone is a very difficult target using almost any geological or geophysical tool. We set out to image the plate-boundary zone to identify evidence for sediment accretion or tectonic erosion and to document the interaction between structures of the Cocos plate and the overriding margin wedge, which may affect seismogenic character of the plate boundary. The easiest part of this effort was our examination of the Cocos plate seaward of the trench; the seafloor is essentially the future plate boundary as the limited sedimentary section mimics the basement structure and is probably mostly subducted (figs. 9.2 and 9.3). However, we must look beneath the margin wedge to see whether the rotated normal faults and the scattered seamounts persist to seismogenic depths. We have already demonstrated that the plate boundary tends to conform to the basement structure near the trench but then becomes smoother landward (fig. 9.4). The exception to this trend is a subducting seamount. Profile NIC-28 (fig. 9.4) shows that even where a seamount is subducted beneath the midslope area, the upper plate still undergoes significant deformation. Here, the rough surface boundary is substantially elevated, and, at a height of 1.5–2.0 km, the seamount has apparently been subducted intact to this point 25 km from the trench and ~7 km below the seafloor.

We also have evidence that the half-graben structures of the Cocos plate are subducted intact. Figure 9.8 shows parts of three profiles, focusing on the plate boundary. The subset of NIC-80 (fig. 9.8a) shows the flanks of a subducted seamount beneath CMP 7800, and between CMPs 9200 and 10000, the plate-boundary zone is marked by sets of dipping and nearly flat reflections. The plate boundary is marked by landward (northeast) dipping reflections, steepening to the northeast, while the flatter reflection segments below delineate the top of basement in rotated Cocos plate fault blocks. We are confident in this interpretation because it is confirmed on numerous other profiles in the data set including NIC-50 and NIC-40, shown in figures 9.8b and 9.8c. On NIC-50 the subducted half-grabens are best imaged in the range CMP 7000 to 7600. Over at least part of this range the plate-boundary reflection is apparent as a nearly flat arrival. NIC-40 (fig. 9.8c) shows at least two more examples of reflection segments oblique to the general trend of the plate-boundary fault zone. We show a simplified diagram in figure 9.8d to illustrate our interpretation of the observations made above. We do not actually expect such regular fault-block size and offset, and as shown in figure 9.8a, seamounts or smaller igneous bodies are commonly present. The key point to make is that our images of the plate-boundary zone indicate that the general morphology of the Cocos plate, with seamounts and rotated fault blocks, is maintained during subduction, including the passage through the seismogenic zone. The image of NIC-80, for example, shows the fault blocks at more than 20-km depth (5+km/s average velocity), which is well within the seismogenic zone.

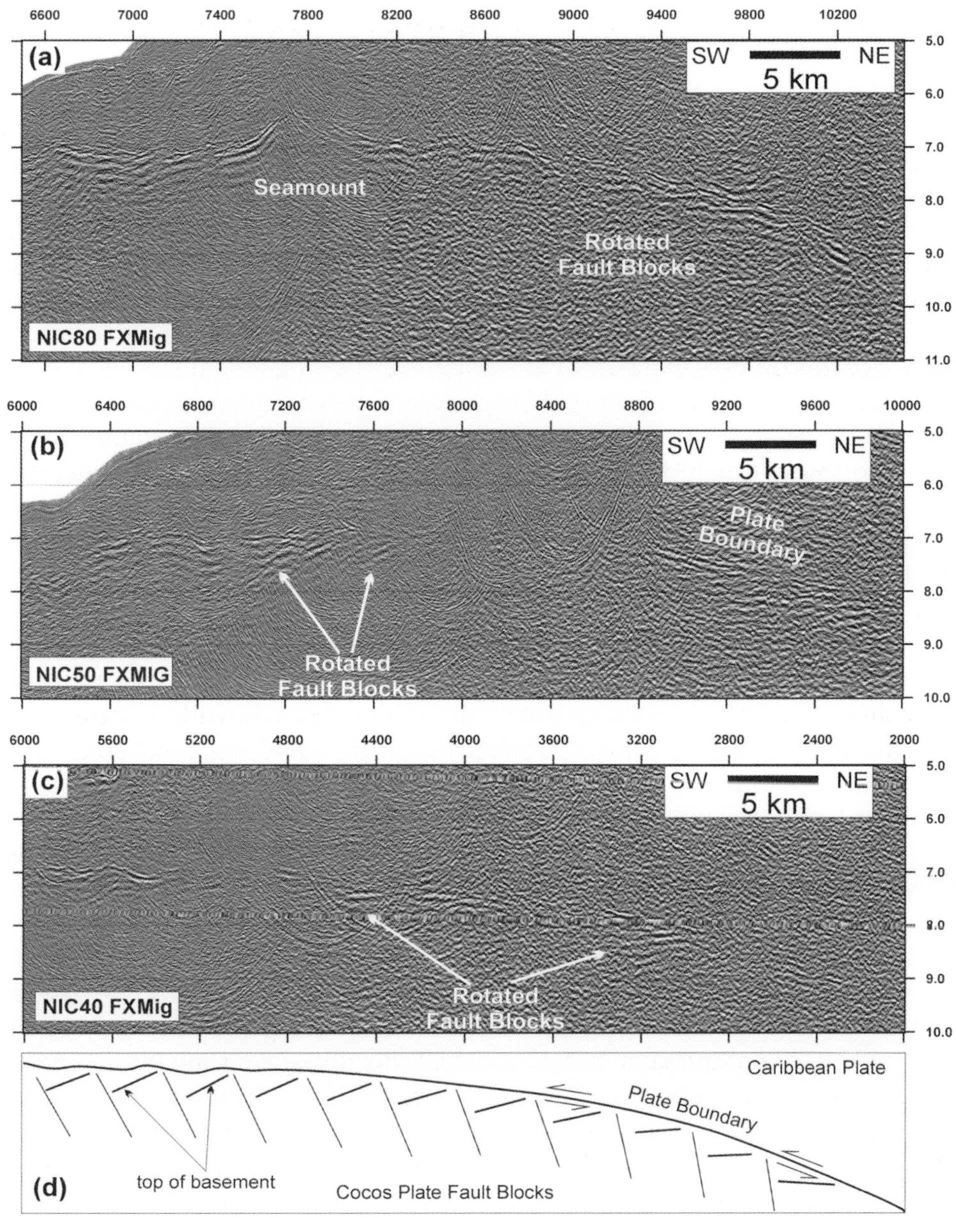

Figure 9.8 Three images of the plate boundary from the midslope to the outer shelf: (a) NIC-80, (b) NIC-50, and (c) NIC-40. Reflection segments oblique to the plate-boundary trend are a persistent feature we interpret as top-basement reflections in rotated fault blocks. The landward half of these boundary segments is within the seismogenic rupture domain. (d) A simplified diagram illustrating our interpretation of the plate-boundary structures. See text for further discussion. Locations shown in figure 9.1.

Discussion

Deformation and Tectonic Erosion

Our observations suggest that most or this entire margin is undergoing tectonic erosion, but the erosion may be limited. From figures 9.4 and 9.5, it is certain that the sedimentary cover is being removed from the slope areas. This material is redeposited in the trench, lower slope, and onto fault blocks of the Cocos plate, so in the absence of a large trench wedge, it is apparently subducted. Such material could be accreted at the trench or underplated, but we see no evidence of this process in the margin wedge We also note that the rough surface reflector (consisting of basalt and Late Cretaceous sediments where dredged) extends nearly to the trench (fig. 9.4; plates 9.1 and 9.2) throughout our survey area, indicating that the igneous basement probably occupies nearly the entire margin wedge. We also have evidence on many profiles (e.g., fig. 9.4) that seaward-dipping normal faults cut the sedimentary apron and offset the margin wedge. This indicates an active process of wedge extension and thinning, most likely owing at least in part to tectonic erosion near the seaward edge of the upper plate. However, the allowable tectonic erosion here is limited because there has been a net seaward advancement of the shelf edge since early Miocene times despite continuing margin subsidence (fig. 9.6). Other factors in the margin subsidence are sediment loading on the shelf and slope and steepening dip of the Cocos plate slab. An expected companion to subduction erosion is landward progression of the volcanic arc [*von Huene and Scholl*, 1991], but it is clear from geologic fieldwork on land that the Nicaragua volcanic arc has moved seaward through time. *Weinberg* [1992] noted the general seaward arc migration through the Miocene and suggested steepening slab dip and upper plate extension were important and related factors. Work by *Ehrenborg* [1996] showed that at the beginning of the Miocene, the arc was as much as 100 km northeast of its present position, and *Plank et al.* [2002] analyzed and dated numerous additional Miocene samples, which documented seaward arc migration throughout the Miocene. The migration of the arc, in addition to marking slab steepening or roll-back, likely also acted as a source for seaward migration of clastic sedimentation, so margin subsidence is an expected result from both a deeper slab and sediment loading, independent of subduction erosion.

Comparisons to Costa Rica

As we have demonstrated, the ubiquitous fault blocks and common seamounts on the Cocos plate have a strong effect on the upper-plate morphology as they subduct along the Nicaraguan margin. We also noted that the dip of the subducting Cocos plate and the depth to the trench axis decreases to the southeast. This trend continues offshore Costa Rica along most of the Nicoya peninsula, where the Cocos plate is the smoothest [*von Huene et al.*, 1995, 2000]. Past the central

Nicoya Peninsula to the southeast, the Cocos plate lithosphere was formed at the Cocos-Nazca spreading center rather than the east Pacific Rise [*Barckhausen et al.*, 2001] and a number of seamount chains occur because of interaction with the Galapagos hot spot [*Werner et al.*, 1999], culminating with the Cocos Ridge hot spot track. The seamounts here are significantly larger than offshore Nicaragua, with sizes commonly 2–4 km high and 20 km in diameter [*von Huene et al.*, 2000, 2004] as opposed to seamounts such as the ones in figures 9.2 and 9.3, which are 1–2 km high and ~10 km in diameter. As shown by *Ranero and von Huene* [2000] and *Walther* [2002], this variation in crustal morphology is associated with thicker crust, it was modified somewhat after original oceanic crust was formed, and ages of subducting lithosphere decrease steadily along the trench off southern Costa Rica [*Barckhausen et al.*, 2001]. The end result is that offshore central and southern Costa Rica the Cocos plate is young, shallowly dipping, and highly erosive to the continental margin.

The upper plate along the Costa Rica margin has a similar basic appearance to offshore Nicaragua, with the high-velocity margin wedge and an overlying seaward-thinning sedimentary section separated by a high-impedance seismic reflector. However, one common feature of the Costa Rican margin is a small frontal prism, typically 5–8 km wide [*von Huene et al.* 2000], frequently with few coherent internal reflections. In contrast, a frontal prism is often absent off Nicaragua, as the rough surface boundary typically extends to <1 km of the trench axis. Another feature of many Costa Rica transects is a set of landward-dipping reflections in the lower slope area beneath the rough surface reflection [*Shipley et al.*, 1992; *von Huene et al.*, 2000]. The interpretation of these reflector packages is not clear, but possibilities include the following: (1) the thin seaward edge of the igneous margin wedge compressionally telescoped; (2) a paleo accretionary prism of Eocene to Miocene age [*Baumgartner et al.*, 1989; P. Baumgartner, personal communication, 2002]; (3) active accretion by underplating [*Shipley et al.*, 1992; *McIntosh and Sen*, 2000]. Offshore Nicaragua we observe no similar reflection patterns. This suggests that there has been no current or recent sediment accretion here, and, if there was an older accretionary prism, then it appears to have been removed. Farther upslope, along the middle- and upper-slope areas of both Costa Rica and Nicaragua, normal faults are commonly observed. However, landward-dipping normal faults are more prevalent off Costa Rica [*McIntosh et al.*, 1993; *Ranero and von Huene*, 2000], and Nicaragua shows a mix of normal fault dip, but seaward-dipping normal faults appear to dominate.

As discussed above, some degree of tectonic erosion is likely along the Nicaraguan margin on the basis of shelf subsidence, common seaward-dipping normal faults, and the presence of many subducting seamounts. Off central to southern Costa Rica, seamounts cover as much as 40% of the Cocos plate [*von Huene et al.*, 2004] and promote widespread tectonic erosion of the upper plate [*von Huene et al.*, 1995; *Ranero and von Huene*, 2000]. In this area, there are local submarine embayments and apparent retreat of the shelf edge or truncation of the trench slope. Where the smooth segment of the Cocos plate subducts beneath the Nicoya Peninsula, it is uncertain to what degree subduction erosion has been active. *Ranero and von Huene* [2000] noted that the seaward 50 km of the margin wedge is thinner off Nicoya than off Nicaragua, which may suggest erosional thinning. The most extreme interpretation of tectonic erosion proposed for the Costa Rica margin involves subsidence of the outermost part of the high-velocity margin wedge from near sea level to its present depth of ~3900 m, at a location that is now only ~7 km landward of the trench, over the last ~16 Ma [*Vannucchi et al.*, 2001]. This is based on recovery of a lower-middle Miocene carbonate breccia, interpreted to be *in situ*, but of shallow-water origin, at ODP Site 1042 [*Kimura et al.*, 1997; *Vannucchi et al.*, 2001]. Additional supporting evidence for massive erosion comes from apparent subsidence trends derived through paleobathymetric analysis of fauna recovered in the sedimentary apron during ODP Leg 170 and DSDP Leg 84 [*Vannucchi et al.*, 2003]. It is possible here, in the absence of significant subducting structure, that basal erosion through a hydrofracturing process may occur [*von Huene et al.*, 2003]. However, this is also where landward-dipping reflections have been observed and possible underplating has been interpreted [*Shipley et al.*, 1992; *McIntosh and Sen*, 2000], so deep drilling to sample the margin wedge and this reflector package is probably necessary to determine the true mode and history of the subduction processes.

Seismogenic Processes

According to the catalog compiled by *Ambraseys and Adams* [1996], there have been five $Ms \geq 7.0$ earthquakes off the Pacific coast of Nicaragua since 1898. Of these events, two have been located near the coast, >110 km from the trench, and one is located approximately at the trench and may be a lower-plate event. The remaining two events, from 1901 and 1992, had epicenters near the shelf edge and both generated tsunamis, although the 1901 tsunami was apparently significantly smaller [*Ambraseys and Adams*, 1996].

The 1992 event is the best-studied large earthquake in the Nicaragua area and likely provides important clues about the seismogenic character of the subduction boundary in this area. The earthquake itself ($Ms = 7.2$) caused little damage, but the associated tsunami killed 170 people and left 13,000 people homeless [*Imamura et al*, 1993; *Satake et al.*, 1993]. The epicenter of this event [National Earthquake Information Center (NEIC), 1992] was located beneath the uppermost slope off north central Nicaragua (PDE in fig. 9.1), but the Harvard CMT location was located nearly at the trench [*Dziewonski et al.*, 1993]. Furthermore, subsequent research has shown that rupture occurred along nearly the entire Nicaraguan margin, and critical for tsunami generation, a large amount of seismogenic slip occurred close to the trench at shallow levels along the subduction thrust boundary [*Velasco et al.*, 1994; *Satake*, 1995; *Ihmlé*, 1996]. Rupture associated with this event is also described as slow (rupture velocities ~1–2 km/s), smooth, and long lasting (duration 100–120 s) [*Kanamori and Kikuchi*, 1993; *Velasco et al.*, 1994; *Ihmlé*, 1996]. Thus this event is a prime example of a tsunami earthquake as discussed by *Kanamori* [1972] and *Pelayo and Wiens* [1992].

An important point about the 1992 earthquake is that rupture extended to near the trench. This is typical of tsunami earthquakes, which also commonly have shallowly dipping fault planes and shallow source depth [*Pelayo and Wiens*, 1992]. All these observations indicate seismogenic rupture seaward of the expected seismogenic front, the point at which materials in and around the plate boundary have sufficient strength to store elastic strain. This paradox is confirmed by the slow rupture velocities typically observed. Rupture velocity is limited by the shear velocity of the rupture zone material [*Fossum and Freund*, 1975; *Geller*, 1976], so rupture velocities estimated at ~1 km/s during the 1992 event have led to the conclusion that there must be a weak, low-velocity sediment layer along the plate boundary [*Kanamori and Kikuchi*, 1993; *Velasco et al.*, 1994; *Ihmlé*, 1996]. To account for the seismogenic rupture capability then the sediment layer may be conditionally stable, as described by *Scholz* [1990], such that slip may occur aseismically, but can become unstable if triggered by a velocity jump owing to rupture in an adjacent region [*Ihmlé*, 1996].

Our results depict a plate-boundary megathrust zone between an upper plate composed of oceanic igneous rock and a lower plate featuring sediment-filled half grabens separated by oceanic basement exposures. In addition to the Cocos plate fault blocks, which are continuous for tens of kilometers along strike (fig. 9.2), small- to medium-sized seamounts are scattered throughout. We suggest that typical depth-dependent processes of sediment compaction, dewatering, and mineral dehydration largely control the seismogenic character of the margin, but the exposed basement fault blocks and, especially, subducted seamounts also appear to significantly affect rupture patterns. Figure 9.9 is a map showing earthquake locations over the period of 1975 to 1982, recorded by a U.S. Geological Survey/Nicaraguan local seismic network. The events displayed are limited to depths <100 km and are not scaled for magnitude. The general pattern is a band of seismicity along the volcanic arc, another band near the coast associated with Wadati-Benioff seismicity, and a less continuous band, which is densest beneath the shelf edge off central Nicaragua associated with slip on the plate boundary. In this data set, there is a clear lack of interplate events in the region spanned by NIC-90 and a reduced density of events in the southeastern margin segment. Seismicity recorded after the 1992 earthquake (INETER Web site) also shows reduced activity in the northwest zone relative to the adjacent segments. However, the apparent lack of seismicity to the southeast may be mostly a function of distance from the Nicaraguan network; data recorded by a northern Costa Rican network operated by Instituto Costarricense de Electricidad (ICE) and University of Costa Rica show relatively widespread seismicity in this area (M. Mora, written communication., 2004). On the basis of the seismicity data, *Harlow et al.* [1981] identified the northwestern segment as a seismic gap, likely capable

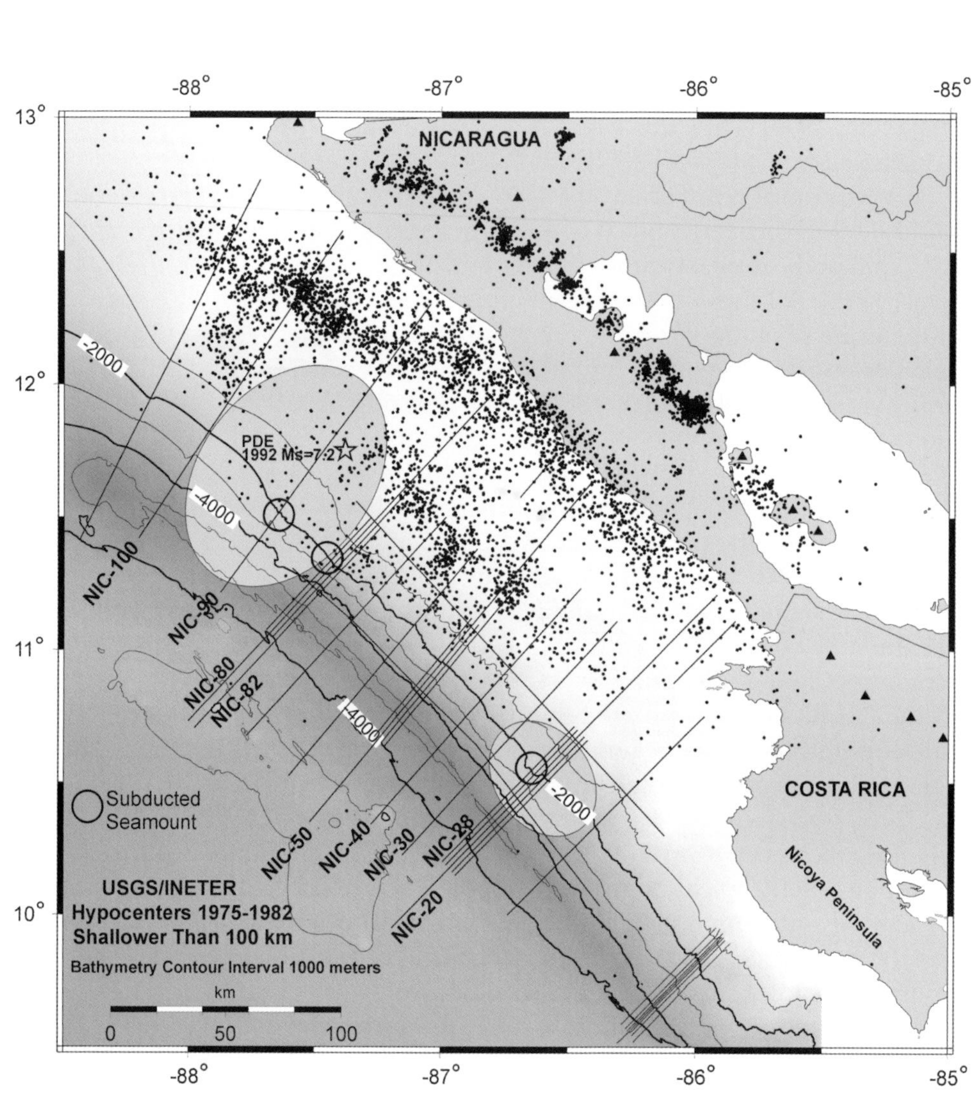

Figure 9.9 Map showing events (shallower than 100 km) recorded by the local seismic network from 1975 to 1982. The superimposed oval zones mark areas where significant moment was released during the 1992 tsunami earthquake and where we have imaged subducting seamounts. Both areas were conspicuously inactive during this period 10 years prior to the rupture. Approximate seamount locations within the rupture areas are marked by the three smaller circles.

of an $M > 7$ earthquake. The shaded oval area here approximates the region of large moment release in the subsequent 1992 earthquake as determined by *Ihmlé* [1996], which includes the epicenter beneath the shelf edge and extends to near the trench. The smaller shaded oval area in the southeast is a secondary area of significant moment release determined in the same analysis. Our data set now shows the presence of seamounts in both of these primary rupture areas of the 1992 earthquake (e.g., NIC-28 on fig. 9.4, NIC-80 on fig. 9.8, and

NIC-90). In each case the seamounts are found beneath the middle slope, seaward of generally expected seismogenic rupture. We suggest that these seamounts have been subducted to the point where the upper plate has become thicker, somewhat less deformed, and is capable of storing elastic strain. If so, the key ingredient for seismogenic rupture in this seaward position is apparently the ability of seamounts to develop enhanced levels of interplate normal stress, as proposed by *Scholz and Small* [1997].

Models of tsunami excitation typically identify shallow rupture as a key factor, so it seems likely that the seamounts we have identified were instrumental in producing the destructive tsunami associated with the 1992 event. Our interpretation is similar to that proposed by *Bilek and Lay* [2002] in which there is an unstable zone of the plate boundary at typical seismogenic depths, e.g., >10–15 km but unstable patches surrounded by conditionally stable material updip. In the case of Nicaragua the nominal seismogenic zone in Nicaragua appears to occur beneath the upper slope to outer shelf. If seamounts beneath the middle slope represent the unstable, updip patches, then rupture may be triggered there and spread into the conditionally stable zones. Owing to the shallow rupture position, efficient tsunami excitation is likely. We note that rough, horst and sediment-filled graben structure on the subducting Pacific plate has been linked to tsunami earthquakes at the Japan Trench [*Tanioka et al.*, 1997]. We originally suspected this mechanism in the 1992 Nicaragua event, and it may be a factor; however; the coincidence of large moment release and subducted seamounts appears to demonstrate cause and effect.

As noted above and shown in figure 9.9, the interplate seismogenic zone occurs largely beneath the upper slope and outer shelf. Downdip is an apparent gap in seismicity and then an increase in intraslab earthquakes. The gap may be an artifact, at least in part, due to an oversimplified velocity model used for earthquake relocation (J. M. Protti, personal communication, 2001), but events located in the landward band are primarily at depths of >45 km so are almost certainly intraslab. We also note that more recent data recorded by the local network, now operated by INETER, show patterns very similar to the older data set. Plates 9.1, 9.2, and 9.3 show that the outer fore-arc basement high overlies the seismogenic portion of the plate boundary, while the deeper Sandino Basin overlies the deeper, aseismic portion of the boundary. This structural/seismic relationship is consistent with cumulative, permanent deformation expected from repeated earthquake cycles with similar seismogenic properties. Geodetic work in Japan by *Thatcher* [1984] showed that coseismic uplift near the seismogenic rupture zone of great earthquakes and subsidence at positions farther from the trench were representative of cumulative deformation patterns over a complete earthquake cycle and also similar to the Quaternary pattern displayed by uplifted marine terraces. If this approach is valid in the Central American environment, then the long-lived structural configuration seen here also supports relatively consistent subduction and seismogenic processes through time.

Conclusions

Prior to the acquisition and analysis of the seismic reflection data set described here the convergent margin structure offshore Nicaragua was poorly known. We have now been able to characterize the subducting Cocos plate morphology, the upper plate structure, and aspects of the plate boundary megathrust with high-quality seismic reflection data spanning ~250 km of the margin. Among the exciting discoveries in our data analysis are the ultra-thick sedimentary section in the NW Sandino forearc basin and the presence of subducting seamounts in high seismic moment release zones of the 1992 tsunami earthquake. The specific conclusions from our analysis are listed below:

1. Our data set shows the plate boundary well into the seismogenic zone and in some cases probably through the entire zone. Normal faults produced by Cocos plate bending persist to seismogenic depths and result in a lower plate with alternating sediment-filled half grabens and exposed basement blocks.

2. As indicated by frequent and active seaward-dipping normal faults, the upper plate here is undergoing extension, most likely as a result of moderate subduction erosion.

3. Seamounts play an important role in the structural development of the upper plate. They subduct intact beneath the lower and middle slope areas, promote upper-plate deformation, and are probably a key factor in subduction erosion. Several of the slump scarps present in the middle slope area have been caused by recently subducted seamounts.

4. Seamounts appear to also act as zones of high moment release during subduction zone earthquake events and may allow rupture at abnormally shallow depths. During the 1992 tsunami earthquake the areas of large moment release occurred in two areas, which extended to near the trench and also contained subducted seamounts. By allowing large rupture near the trench, the seamounts likely promoted much greater tsunami excitation than expected for a $M = 7.2$ earthquake.

5. The Sandino Basin and the outer fore-arc basement high have maintained a nearly stable configuration since at least Late Cretaceous. While there has been gradual subsidence along much of the margin, it has been focused landward of the outer high. Seaward progradation of the shelf edge at least since Early Miocene apparently precludes any large-scale tectonic erosion of the margin, which would have resulted in margin retreat. While some tectonic erosion is almost certainly taking place here, we conclude that steepening slab dip and sediment loading are also very important factors in the observed margin subsidence.

6. The outer fore-arc high appears to occur above the interplate seismogenic zone, while the deeper parts of the Sandino Basin overlie a more stable sliding regime of the plate boundary. The longevity of the fore-arc/fore-arc basin structure here suggests that seismogenic characteristics may also have been relatively consistent through time.

Acknowledgments

We thank the captain, crew, and science staff of the R/V *Maurice Ewing* for their excellent performance during the EW00-05 cruise. The Ewing airgun source and 6 km streamer were the keys to our success. Ming Zeng was a big help in processing the seismic data. We thank Wilfried Strauch, Carlos Guzman, and fellow workers from INETER. We also thank Mauricio Darce and Manuel Duarte for helpful discussions and access to the seismic data set at INE. We thank Sue Bilek and Andy Goodliffe for very constructive reviews that improved this paper significantly. McIntosh was supported by NSF OCE-9905355 and Silver was supported by NSF OCE-9909859 through the MARGINS Program. Participation by GEOMAR scientists was funded by Deutsche Forschungsgemeinschaft grant F1147/13. This is UTIG contribution 1695.

References

Ambraseys, N. N., and R. D. Adams (1996), Large-magnitude Central American earthquakes, 1898–1994, *Geophys. J. Int.*, *127*, 665.

Barckhausen, U., et al. (2001), Revised tectonic boundaries in the Cocos Plate off Costa Rica; implications for the segmentation of the convergent margin and for plate tectonic models, *J. Geophys. Res.*, *106*, 19,207.Baumgartner, P., et al. (1989), Paleogene accretion and suspect terranes in southern Costa Rica (Osa, Burica, Central America), paper presented at Transactions of the 12th Caribbean Geological Conference, Miami Geol. Soc., St. Croix, U.S. Virgin Islands.

Bilek, S. L., and T. Lay (2002), Tsunami earthquakes possibly widespread manifestations of frictional conditional stability, *Geophys. Res. Lett.*, *29*(14), 1673, doi:10.1029/2002GL015215.

Bourgois, J., et al. (1984), The geologic history of the Caribbean-Cocos Plate boundary with special reference to the Nicoya ophiolite complex (Costa Rica) and D.S.D.P. results (Legs 67 and 84 off Guatemala); a synthesis, *Tectonophysics*, *108*, 1.

Collot, J.-Y., and M. A. Fisher (1989), Formation of forearc basins by collision between seamounts and accretionary wedges; an example from the New Hebrides subduction zone, *Geol.*, *17*(10), 930.

DeMets, C., et al. (1994), Effect of recent revisions to the geomagnetic reversal time scale on estimates of current plate motions, *Geophys. Res. Lett.*, *21*, 2191.

Dziewonski, A. M., et al. (1993), Centroid-moment tensor solutions for July-September 1992, *Phys. Earth Planet. Inter.*, *79*, 287.

Ehrenborg, J. (1996), A new stratigraphy for the Tertiary volcanic rocks of the Nicaraguan Highland, *Geol. Soc. Am. Bull.*, *108*, 830.

Fossum, A. F., and L. B. Freund (1975), Nonuniformly moving shear crack model of a shallow focus earthquake mechanism, *J. Geophys. Res.*, *80*, 3343.

Geller, R. J. (1976), Scaling relations for earthquake source parameters and magnitudes, *Bull. Seismol. Soc. Am.*, *66*, 1501.

Gerth, R., et al. (1999), La Boquita Terrace, Nicaragua; evidence for uplift of the Central Pacific coast, *Eos Trans. AGU*, *80*, 1033.

Harlow, D. H., et al. (1981), Quiet zone within a seismic gap near western Nicaragua; possible location of a future large earthquake, *Science*, *213*, 648.

Ihmlé, P. F. (1996), Frequency-dependent relocation of the 1992 Nicaragua slow earthquake; an empirical Green's function approach, *Geophys. J. Int.*, *127*, 75.

Imamura, F., et al. (1993), Estimate of the tsunami source of the 1992 Nicaraguan earthquake from tsunami data, *Geophys. Res. Lett.*, *20*, 1515.

Kanamori, H. (1972), Mechanism of tsunami earthquakes, *Phys. Earth Planet. Inter., 6*, 346.

Kanamori, H., and M. Kikuchi (1993), The 1992 Nicaragua earthquake; a slow tsunami earthquake associated with subducted sediments, *Nature, 361*, 714.

Kelly, R. K., et al. (2001), A quantitative analysis of flexural faulting in the Cocos Plate at the Middle America Trench from Nicaragua to Costa Rica, *Eos Trans. AGU, 82*(47), Fall Meet. Suppl., Abstract T22C-0929.

Kimura, G., et al. (1997), Costa Rica accretionary wedge; covering Leg 170 of the cruises of the drilling vessel JOIDES Resolution, San Diego, California, to Balboa, Panama, sites 1039–1043, 16 October–17 December, *Proc. Ocean Drill. Program Initial Rep., 170*, 458 pp.

Kuijpers, E. P. (1980), The geological history of the Nicoya ophiolite complex, Costa Rica, and its geotectonic significance, *Tectonophysics, 68*, 233.

Lallemand, S., and X. Le Pichon (1987), Coulomb wedge model applied to the subduction of seamounts in the Japan Trench, *Geology, 15*(11), 1065.

Lallemand, S. E., et al. (1994), Coulomb theory applied to accretionary and nonaccretionary wedges; possible causes for tectonic erosion and/or frontal accretion, *J. Geophys. Res., 99*, 12,033.

McIntosh, K., et al. (1993), Evidence and mechanisms for forearc extension at the accretionary Costa Rica convergent margin, *Tectonics, 12*, 1380.

McIntosh, K. D., and M. K. Sen (2000), Geophysical evidence for dewatering and deformation processes in the ODP Leg 170 area offshore Costa Rica, *Earth Planet. Sci. Lett., 178*, 125.

National Earthquake Information Center (1992), Preliminary determination of epicenters monthly listing, report, U. S. Dep. of the Inter., Washington, D. C.

Pelayo, A. M., and D. A. Wiens (1992), Tsunami earthquakes; slow thrust-faulting events in the accretionary wedge, *J. Geophys. Res., 97*, 15,321.

Piatanesi, A., et al. (1996), The slip distribution of the 1992 Nicaragua earthquake from tsunami run-up data, *Geophys. Res. Lett., 23*, 37.

Plank, T., et al. (2002), Nicaraguan volcanoes record paleoceanographic changes accompanying closure of the Panama gateway, *Geology, 30*(12), 1087.

Protti, M., et al. (1994), The geometry of the Wadati-Benioff zone under southern Central America and its tectonic significance; results from a high-resolution local seismographic network, *Phys. Earth Planet. Inter., 84*, 271.

Protti, M., et al. (1995), Correlation between the age of the subducting Cocos plate and the geometry of the Wadati-Benioff zone under Nicaragua and Costa Rica, *Spec. Pap. Geol. Soc. Am., 295*, 309–326.

Ranero, C. R., and R. von Huene (2000), Subduction erosion along the Middle America convergent margin, *Nature, 404*, 748.

Ranero, C. R., et al. (2000), A cross section of the convergent Pacific margin of Nicaragua, *Tectonics, 19*, 335.

Satake, K. (1994), Mechanism of the 1992 Nicaragua tsunami earthquake, *Geophys. Res. Lett., 21*, 2519.

Satake, K. (1995), Linear and nonlinear computations of the 1992 Nicaragua earthquake tsunami, *Pure Appl. Geophys., 144*, 455.

Satake, K., J. Bourgeois, K. Abe, Y. Tsuji, F. Imamura, Y. Iio, H. Katao, E. Noguera, and F. Estrada (1993), Tsunami field survey of the 1992 Nicaragua earthquake, *Eos Trans. AGU, 74*, 145, 156.

Scholz, C. H. (1990), *The Mechanics of Earthquakes and Faulting*, 439 pp., Cambridge Univ. Press, New York.

Scholz, C. H., and C. Small (1997), The effect of seamount subduction on seismic coupling, *Geology, 25*(6), 487.

Shipley, T. H., et al. (1992), Three-dimensional seismic imaging of the Costa Rica accretionary prism; structural diversity in a small volume of the lower slope, *J.Geophys. Res., 97*, 4439.

Silver, E. A., et al. (2000), Influence of lower plate structure on the overriding slope offshore Nicaragua; new geophysical observations and the first dredge samples of basement rocks, *Eos Trans. AGU, 81*, 1204.

Tanioka, Y., et al. (1997), What controls the lateral variation of large earthquake occurrence along the Japan Trench? *Island Arc, 6*, 261.

Thatcher, W. (1984), The earthquake deformation cycle at the Nankai Trough, southwest Japan, *J. Geophys. Res., 89*, 3087.

Vannucchi, P., et al. (2001), Tectonic erosion and consequent collapse of the Pacific margin of Costa Rica; combined implications from ODP Leg 170, seismic offshore data, and regional geology of the Nicoya Peninsula, *Tectonics, 20*, 649.

Vannucchi, P., et al. (2003), Fast rates of subduction erosion along the Costa Rica Pacific margin: Implications for nonsteady rates of crustal recycling at subduction zones, *J. Geophys. Res., 108*(B11), 2511, doi:10.1029/2002JB002207.

Velasco, A. A., C. J. Ammon, T. Lay, and J. Zhang (1994), Imaging a slow bilateral rupture with broadband seismic waves; the September 2, 1992 Nicaraguan tsunami earthquake, *Geophys. Res. Lett., 21*, 2629.

von Huene, R., and D. W. Scholl (1991), Observations at convergent margins concerning sediment subduction, subduction erosion, and the growth of continental crust, *Rev. Geophys., 29*, 279.

von Huene, R., et al. (1995), Morphotectonics of the Pacific convergent margin of Costa Rica, *Spec. Pap. Geol. Soc. Am., 295*, 291.

von Huene, R., et al. (2000), Quaternary convergent margin tectonics of Costa Rica, segmentation of the Cocos Plate, and Central American volcanism, *Tectonics, 19*, 314.

von Huene, R., et al. (2003), Low friction subduction erosion, a model to test with ocean drilling, *Eos Trans. AGU, 84*(46), Fall Meet. Suppl., Abstract T41A-08.

von Huene, R., et al. (2004), Tsunamigenic slope failure along the Middle America Trench in two tectonic settings, *Mar. Geol., 203*, 303.

Walther, C. (2002), Crustal structure of the Cocos Ridge northeast of Cocos Island, Panama Basin, *Geophys. Res. Lett., 29*(20), 1986, doi:10.1029/2001GL014267.

Walther, C. H. E., et al. (2000), Crustal structure across the Pacific margin of Nicaragua; evidence for ophiolitic basement and a shallow mantle sliver, *Geophys. J. Int., 141*, 759.

Weinberg, R. F. (1992), Neotectonic development of western Nicaragua, *Tectonics, 11*, 1010.

Werner, R., et al. (1999), Drowned 14-m.y.-old Galapagos Archipelago off the coast of Costa Rica; implications for tectonic and evolutionary models, *Geology, 27*(6), 499.

Weyl, R. (1980), *Geology of Central America*, translated from German, 371 pp., Gebrueder Borntraeger, Berlin, Germany.

Winsemann, J. (1994), Origin and tectonic history of the western Caribbean region, in *Profil*, edited, p. 1, Inst. für Geol. und Paläontol., Stuttgart, Germany.

10

How Accretionary Prisms Elucidate Seismogenesis in Subduction Zones

J. Casey Moore, Christie Rowe, and Francesca Meneghini

Abstract

Earthquakes occur along the plate-boundary thrusts underlying accretionary prisms and along out-of-sequence thrusts that cut through prisms. Thermal models suggest that the earthquakes on the plate-boundary thrusts initiate in a temperature range of 125°C to ~350°C. Because syndeformational diagenetic and metamorphic alterations recorded in accretionary prisms have specific temperature ranges, the alterations and the associated deformation can be correlated to the temperature range that accretionary prisms are seismogenic. Comparison of accreted rocks deformed above, within, and below the seismogenic zone suggests characteristics of rocks at seismogenic depths that may make them earthquake prone. During passage through temperatures from 50° to 150°C, accretionary prism sediments become rocks, undergoing diagenetic reactions including the transformation of smectite to illite, albitization of detrital feldspar, dehydration of opal, and the generation of hydrocarbons. Although the smectite to illite transition does not change the frictional properties of the prism so that it becomes seismogenic, water and cations (calcium, magnesium, iron) released during this transition and the albitization process foster cementation. Cementation and veining by carbonates becomes common by 125°C, perhaps due to the above-mentioned release of cations. Pressure-solution fabrics begin to be apparent at ~150°C, with well-developed cleavages and quartz veining common by 200°C. Pressure solution may be facilitated by the diagenetic formation of illite. Quartz veining and cementation in the 150°–300°C range facilitates the change from a velocity-strengthening, clay-dominated to a velocity-weakening, quartz-influenced, earthquake-prone rheology. The diagenetic-metamorphic reactions occurring at temperatures from 125° to ~300°C cement and add rigidity to the thickening upper plate of the accretionary prism. This developing elastic strength of the upper plate is required to store the elastic strain energy required for an earthquake. In accretionary prisms, brittle fabrics are progressively replaced by ductile fabrics through a temperature range of ~150°–325°C. Although rocks in the seismogenic zone have lost most of their intergranular fluid through consolidation, vein geometries and fluid inclusions suggest high fluid pressures, approaching lithostatic. Strain localization in the form of discrete shear surfaces occurs across the lower aseismic to seismic transition. Strain localization is observed both at outcrop and map scale. At map scale, the seawardmost occurrence of out-of-sequence thrusts defines the leading edge of the rigidified accretionary prism that is capable of storing elastic energy.

Introduction

Accretionary prisms comprise materials that have been transferred from the oceanic plate to the overriding plate in subduction zones. At all but the shallowest levels, near the front of the prism, these sediments and rocks have been underthrust with the lower plate and accreted to the upper plate (fig. 10.1). This accretion process involves a downward migration of the subduction thrust. Thus accretionary prisms constructed by this "underplating" process include faults that were for a period of time subduction thrusts. In addition to the faults active during the initial emplacement of their rocks, accretionary prisms are cut by later faults, or out-of-sequence thrusts. In combination, the basal subduction thrusts and the later faults encompass faults active in the great subduction earthquakes [*Plafker*, 1972], the largest known seismic

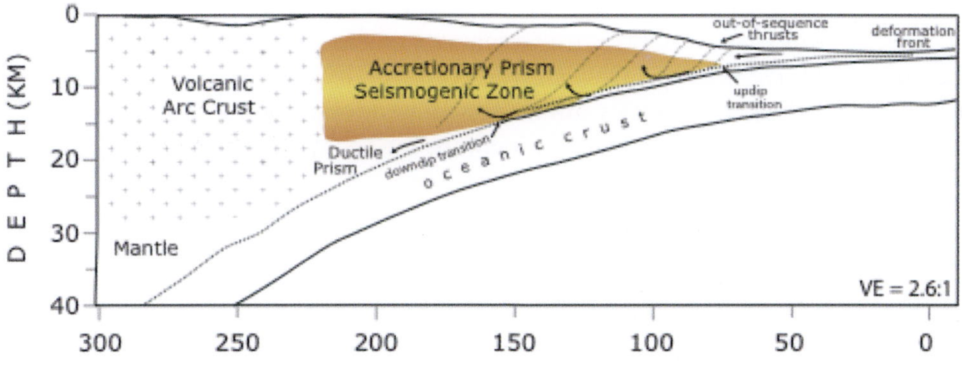

Figure 10.1 Cross section showing transport paths of rocks through the accretionary prism. Note that out-of-sequence thrusts begin at about seaward limit of prism seismogenic zone. Seismogenic zone defined by temperature limits of 125°–350°C; limits would shift with variations in geothermal gradient. Arrows show material transport paths in prism.

events [*Lay and Bilek*, this volume). The history of deformation through the earthquake cycle is recorded in the rocks of accretionary prisms. To the degree that we can decipher this history, we can learn much about the inception and evolution of the deformational processes associated with earthquakes.

Investigation of old rocks from the seismogenic realm provides windows on processes that cannot be observed short of deep drilling. Terranes of old rocks allow direct investigation of three-dimensional variations, which will be useful in interpreting any results from more physically limited boreholes. However, investigations of old rocks are constrained in that they offer an interpretable record of past activity but do not allow direct observation of active processes.

Defining the Seismogenic Zone at Convergent Plate Boundaries

The seismogenic zone is most properly defined by a depth interval of unstable (stick-slip) frictional behavior where earthquakes can initiate [*Scholz*, 2002, p. 152]. In practice, this depth interval can best be identified by aftershock zones of large earthquakes. In the absence of well-resolved aftershock distributions, coseismic rupture areas estimated from earthquake and/or tsunami wave inversions have been used to infer the limits of the seismogenic zone. Because coseismic rupture may propagate into a zone of conditional stability or even marginally into the unstable regime [*Scholz*, 2002, p. 88], coseismic rupture may overestimate the seismogenic zone as defined above. Geodetic measurements indicating the distribution of locking or partial locking can define the seismogenic zone [*Norabuena et al.*, 2004]. Interseismic microearthquakes have been used to estimate the limits of the seismogenic zone [e.g., *Obana et al.*, 2003] but may only be active over part of the seismogenic zone of large earthquakes [*Schwartz and Deshon*, this volume]. In summary, aftershock distribution and locked intervals inferred from geodetic measurements probably provide the best practical means of locating the seismogenic zone.

Recognizing Faults Rocks from the Seismogenic Zone

Progressive addition or underplating of material at the base of accretionary prisms results in their uplift. Therefore the oldest rocks, exhumed from the deepest level, are the structurally highest, and commonly innermost parts of the continually accreted prism, with less deeply buried rocks progressively exposed in a seaward direction [*Brandon and Vance*, 1992; *Plafker et al.*, 1994; *Taira et al.*, 1988]. Typically, packages of rocks with internally similar sedimentary, deformational, and metamorphic histories are separated from each other by prominent out-of-sequence thrust faults. The best-exposed packages of ancient accreted rocks occur where subduction was characterized by a generally high input of sediment and consequent creation and uplift of a volumetrically significant accretionary prism, for example, prisms in California, Washington, Alaska, and southwestern Japan. Even though these accretionary prisms are dominated by clastic rocks of the high sediment input intervals, selected sections include units formed during periods of low sediment supply, in which the subduction zone was more sediment starved. Therefore studies of these ancient rocks can provide information on seismogenesis in both sediment-dominated and sediment-poor systems.

Given exhumed prism rocks of a range of ages, metamorphic grades, and structural styles, how can we recognize rocks that have deformed at the seismogenic zone? Pseudotachylytes, or rocks formed by frictional melting on fault surfaces, are most explicit evidence of past seismogenic behavior [*Cowan*, 1999; *Sibson*, 1975]. However, pseudotachylytes are rare in accretionary prisms [*Ikesawa et al.*, 2003], and there is much disagreement on what structural features may actually form during a seismic event [e.g. *Cowan*, 1999]. A more inclusive approach to understanding rocks that have behaved seismogenically is to simply investigate faults in any volume of rock that has deformed through the P-T conditions where modern earthquakes occur. We can reasonably infer that these rocks have in part deformed seismogenically and that this deformation occurred along brittle structural features. Thus we are asking not whether a particular structure represents a seismogenic, or high velocity, displacement [see *Cowan*, 1999] but what features of rocks in the seismogenic zone may lead to stick-slip behavior. Accordingly, comparison of these rocks to those that behaved aseismically at shallower and greater depths provides insights on the rock fabrics and material properties that control seismogenic behavior.

During the last decade, thermal models [*Currie et al.*, 2002; *Hyndman et al.*, 1997; *Oleskevich et al.*, 1999; *Wang*, 1995] have defined the temperature range at which modern interplate earthquakes occur at seismogenic zones. The temperature range of large interplate thrust earthquakes is from 100°–150°C to 350°–450°C, with the depth limits varying depending on the thermal gradient of the associated subduction zone. The extreme deeper limit of the seismogenic zone may be due to earthquakes initiating at temperatures <350°C but propagating in to a conditionally stable region at temperatures perhaps as high as 450°C. Although there are undoubtedly significant errors in the thermal models, the consistency of temperature ranges for interplate thrust earthquakes at modern subduction zones provides confidence that the above-mentioned temperature limits for the seismogenic zone can be used to define the temperature limits of seismogenic activity in ancient rocks (figs. 10.2 and 10.3). For example, it may be difficult to identify exact temperatures of the boundaries, we can be confident at 50°C we have not reached the seismogenic zone, at 200°–300°C we are within the seismogenic zone, and at 450°C we are beyond the seismogenic zone.

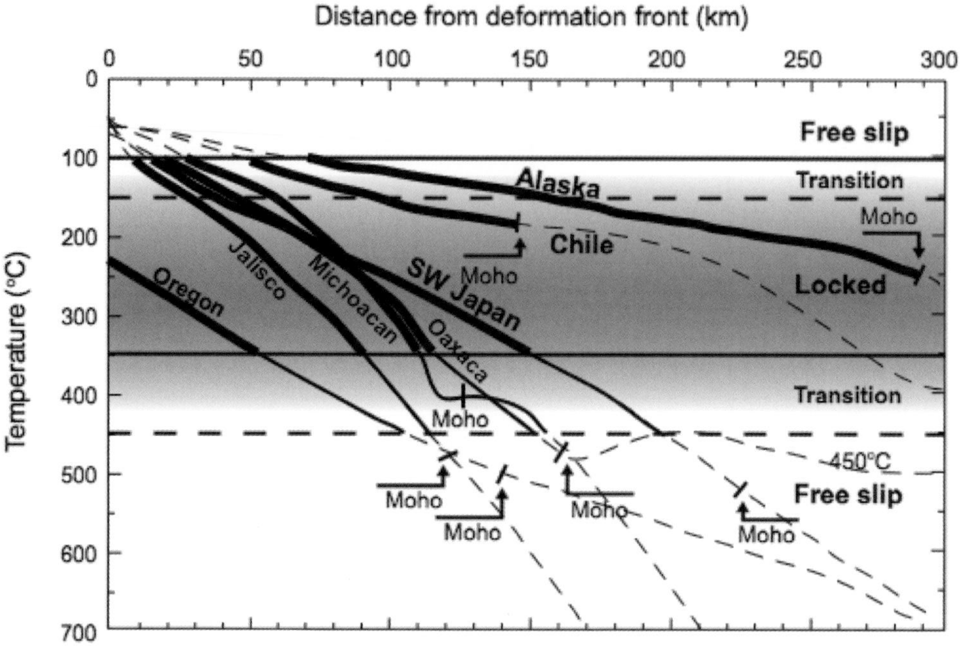

Figure 10.2 Compilation of modeled temperatures of subduction thrust interfaces verses distance landward of deformation front. Temperatures estimated from two-dimensional thermal models with upper and lower bounds of seismogenic zone estimated from aftershocks or modeled fault slip from seismic and/or tsunami waves [*Currie et al.*, 2002; *Hyndman*, 1997]. Note the transitional nature of both the upper and lower boundaries to the locked seismogenic zone.

Pinpointing fault rock units deforming in the seismogenic temperature range requires identification of syndeformational metamorphic minerals [*Ernst*, 1990], or the use of fluid inclusion P-T estimates derived from syndeformational minerals [*Vrolijk et al.*, 1988] that formed under seismogenic P-T conditions (fig. 10.3). Plotting the P-T limits of the seismogenic zone on a petrogenetic grid then ties the seismogenic zone to observable minerals formed syntectonically (fig. 10.3). We have used 350°C as the higher-temperature cutoff of the seismogenic zone overlain on the petrogenetic grid. By that temperature, quartz is ductile and most accreted sediments are highly siliceous, so we expect aseismic behavior above 350°C.

Within rock units deformed at seismogenic depths, we can recognize ductile deformation that would be occurring during interseismic intervals. We can also identify brittle failure surfaces of high strain that may or may not represent deformation at seismogenic strain rates [*Cowan*, 1999]. We agree that unambiguous identification of seismic slip surfaces, except pseudotachylyte, is ambiguous. However, we are studying rocks deformed in the accretionary prism realm that have suffered the effects of the largest earthquakes in the

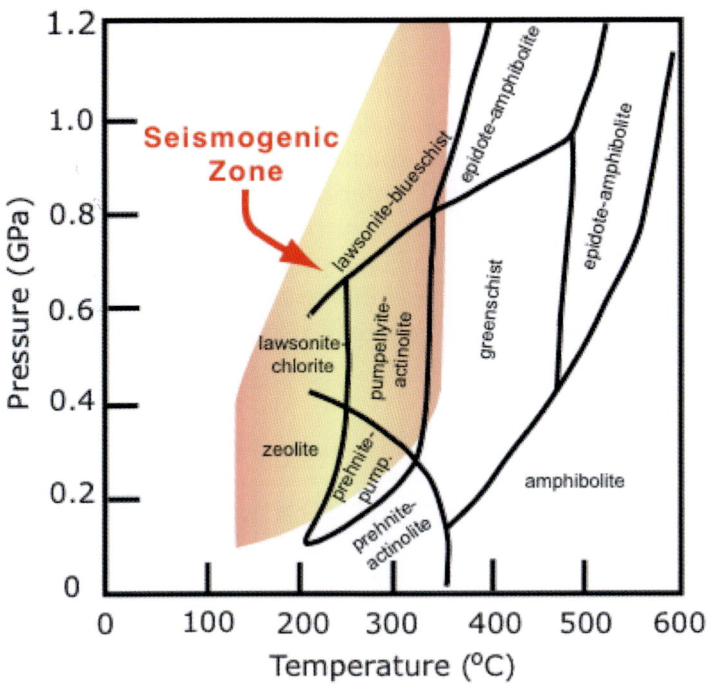

Figure 10.3 P-T limits of seismogenic zone estimated from modeled temperature limits of modern systems (125°–350°C) limited by thermal gradients from ancient prisms. Upper pressure limit constrained by extremely cold subduction systems such as ancient Franciscan Complex [*Ernst*, 1990] and lower limits by very warm systems such as the Shimanto Complex of southwestern Japan [*Lewis and Byrne*, 2003] and Cascadia subduction zone [*Hyndman and Wang*, 1993]. Limits of metamorphic mineral facies from *Peacock* [1993].

world. Thus it is statistically likely that some of the brittle failure surfaces in these rocks record coseimic slip. This must be particularly true at subduction zones similar to the Cascadia margin where nearly 100% of interplate motion must be accommodated coseismically at the depths the seismogenic zone is completely locked [*Wang et al.*, 2003].

Distribution of Deformation in the Seismogenic Zone

As depicted in figure 10.2, the seismogenic zone comprises a rock volume, not just a two-dimensional interface along the basal subduction thrust. Although major slip on large earthquakes is concentrated on the basal thrust and out-of-sequence thrusts [*Plafker*, 1972], small earthquakes occur through the seismogenic portion of the accretionary prism [*Lewis et al.*, 2003]. Structural studies suggest that the major deformation of the prism rocks occurs along the subduction thrust. Structural analyses indicate that accretionary prism deformational fabrics are largely acquired during the initial emplacement of the rocks [*Fisher and Byrne*, 1987; *Kusky and Bradley*, 1999], However, faulting [e.g., *Lewis and Byrne*, 2003] and interseismic ductile deformation [e.g., *Fisher and Byrne*, 1992] continue in the seismogenic interior of the accretionary prism. The sequence of structural superpositions is normally clear. Our summary attempts to encompass all the mineralogical changes and structural fabrics developed in the seismogenic zone of the accretionary prism, whether along the high-strain basal region, on out-of-sequence thrusts, or in lower-strain rate interior regions.

What Progressive Changes Occur in Mineralogy and Fault Rock Fabrics Into and Through the Seismogenic Zone?

Our discussion is based on cataloging and organizing the mineralogical and microscopic to outcrop-scale deformational features observed in accretionary prism rocks that have been deformed at P-T conditions above, within, and below the seismogenic zone (table 10.1). Although related, evidence for fluids, fluid pressure, and strain localization are discussed separately. Metamorphic and deformational phenomena tabulated in figure 10.4 illustrate the broad trends from the upper aseismic zone through the seismogenic zone into the lower aseismic zone. Although the boundaries between these realms are transitional, major changes are apparent from one to another.

As used in this paper and in figure 10.4, "brittle deformation" is associated with loss of cohesion along fractures and faults, and "ductile deformation" refers to coherent nonrecoverable deformation without fracturing at scale of crystal grains or larger [*Twiss and Moores*, 1992; *van der Pluijm and Marshak*, 1997, 2004]. Following these authors, we include cataclasis and granular flow in brittle deformation because cataclasis involves brittle fracture and they are both controlled by frictional rheology. Optical microscopic investigations clearly distinguish brittle and ductile phenomena, as defined above, and these observations can be extended to hand specimen and outcrop scales to distinguish brittle and ductile fabrics.

Observations

Diagenesis to Low-Grade Metamorphism Our discussion of metamorphic mineral transitions occurring in subducted rocks [e.g., *Ernst*, 1990] only focuses on a few alterations that apparently strongly influence the structural

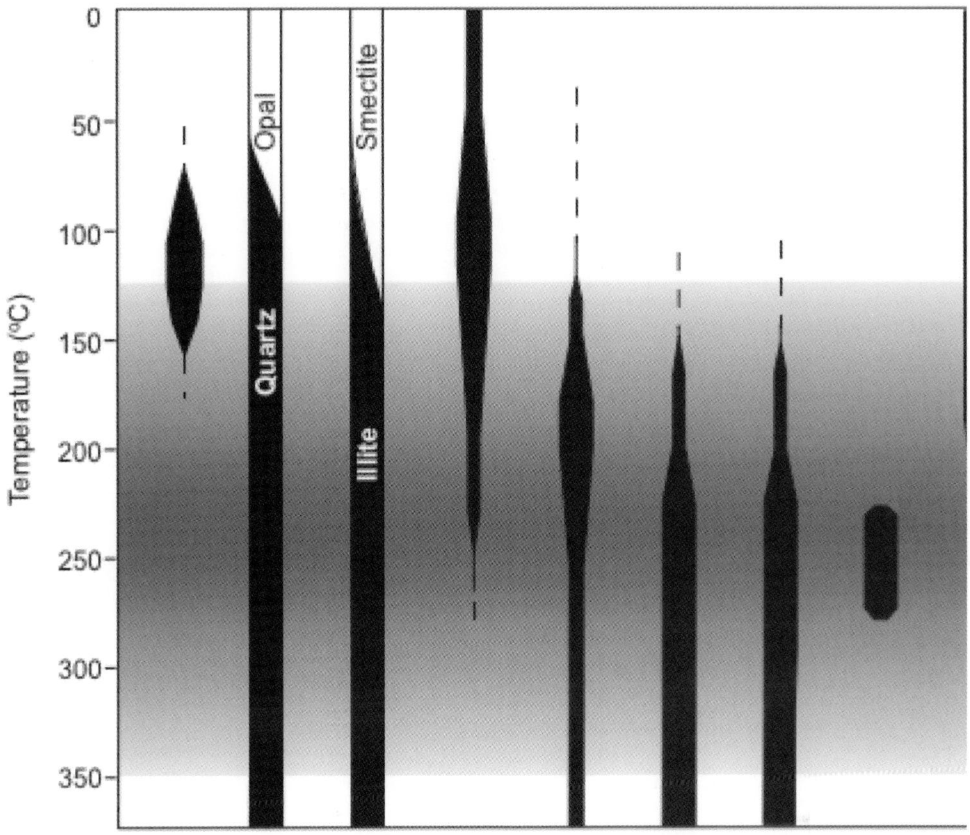

Figure 10.4 Mineralogical transitions and fabric changes in the 0°–375°C temperature range in accretionary prism. Abundance of minerals as well as intensity of metamorphic and structural processes is shown schematically by width of bars. Temperature range of pseudotachylyte (frictional melt) is that of country rocks in which melting occurred. Information on this diagram is generalized from sources in table 10.1 and references cited in discussion section.

evolution of the rocks. Of these, the formation of metamorphic albite releases calcium from more calcic-rich detrital plagioclase [*Boles,* 1982]. The transition from smectite to illite releases water, calcium, iron, and magnesium [*Boles and Franks,* 1979]. Sediments incoming to the subduction zone may include biogenic opal (diatoms or radiolarians) or opal in volcanic glass. Both transform to quartz below 100°C, causing a decrease in solid volume, and water production [*Behl and Garrison,* 1994; *Isaacs et al.,* 1983]. Hydrocarbon production begins near the surface peaks around 100°C and declines with the production of methane to above 200°C [*Hunt,* 1996]. Hydrocarbon production not only provides fluid by various carbon species that can influence mineral precipitation but also low density fluids that can lead to overpressuring.

TABLE 10.1 **Observed Cements, Veins, and Relevant Structural Fabrics in Various Temperature and Metamorphic Facies Ranges**

Temperature and Pressure Range	Observations and Source Localities	References
<100°C Unmetamorphosed	No veins and cements through nearly all cored rocks in ODP. Minor carbonates in Barbados drill cores. Extremely rare quartz, carbonate, and sulfides in Nankai Trough. Accreted terrigenous deposits on Barbados Island lack any carbonate/quartz veins or cements. Faulting and stratal disruption includes cataclasis of sand sized particles widespread development of scaly fabric in mudstones.	*Byrne et al.*, 1993; *Labaume et al.*, 1997; *Maltman et al.*, 1993, 1997; *Moore and Allwardt*, 1980; *Speed*, 1990; *Vrolijk and Sheppard*, 1991.
100°–150°C; 5 km Typically Zeolite Facies	Evidence for cementation and veining variable. Locally significant carbonate cementation along fault zones. Quartz cementation/veins rare and minor. Cataclasis of sand-sized particles widespread. Development of scaly fabric in mudstones. Incipient pressure solution cleavage.	*Geddes*, 1993; *Moore*, 1980; *Moore and Allwardt*, 1980; *Orange et al.*, 1993; *Orange and Underwood*, 1995; *Rowe et al.*, 2002.
150°–300°C; 5–15 km Typically Prehnite-Pumpellyite Facies	Quartz- and carbonate-veining common. Cataclasis in sands and veined materials temporally alternates with pressure solution. Cleavages common, and especially visible in more coherent units. Quartz veins/cements common in the 200°–300°C range. Rare pseudotachylyte.	*Byrne*, 1984; *DiTullio and Byrne*, 1990; *DiTullio et al.*, 1993; *Fisher*, 1996; *Hibbard et al.*, 1993; *Ikesawa et al.*, 2003; *Kimura and Mukai*, 1990; *Onishi and Kimura*, 1995; *Sample and Moore*, 1987; *Sample*, 1990; *Vrolijk et al.*, 1988.
150°–350°C; 15–27 km Classic HP-LT (e.g., Franciscan Blueschists of eastern belt)	Often coherent layering with multiple fold generations, local stratal disruption predating metamorphism. Pressure solution common.	*Jayko et al.*, 1986; *Ring and Brandon*, 1999.
>300°C; 8-27 km Prehnite-Actinolite, and Blueschist-Greenschist Facies	Widespread foliation development, pressure solution, microscopically apparent recrystallization during development of metamorphic assemblages. Many terranes surprising coherent with multiple phases of folding. Brittle deformation during peak metamorphism not obvious, but a previous history is suggested by the stratal disruption in some areas.	*Helper*, 1986; *Norris and Bishop*, 1990; *Ring and Brandon*, 1999; *Roeske*, 1986; *Schwartz and Stockhert*, 1996.

[a] Temperature ranges are associated with typical metamorphic facies with actual temperature conditions determined from metamorphic minerals, fluid inclusions, and vitrinite reflectance. The pressure and temperature range encompass those of the cited examples with the associated metamorphic facies being identified. The P-T ranges do not include the maximum range of the particular metamorphic facies. Surface and near-surface carbonate precipitates associated with cold seeps not included as their formation is too shallow to be involved in seismicity.

Cements and Veins We focus on veins and cements that can be observed petrographically, excluding subtle clay cements that may be easily obscured by mechanical deformation. Observations from Ocean Drilling Program/Deep Sea Drilling Project (ODP/DSDP) cores and the few exposures of shallowly buried accretionary prisms indicate that they are generally uncemented and sparsely veined below 125°C, excluding carbonates formed by surface seepage [*Kulm and Suess,* 1990] and are not involved in seismogenesis. Above these temperatures, calcite is locally abundant to the 150°–250°C range. Quartz veining is not common until ~200°C. In prism rocks exposed on land, veins and cements occur both distributed through the rocks (fig. 10.5) and in locally high concentrations in fault zones (fig. 10.6). The distributed veins occur in extensional fractures in boudins or in fractures in coherent rock units and also coat shear surfaces of boudins and distributed faults. The extensional veins inside the boudins are protected from the destructive slip processes occurring along shear surfaces coating boudins, perhaps accounting for their abundance over shear veins at some localities.

Figure 10.5 Photograph of quartz-veined block surface of stratally disrupted zone, Paleocene accretionary prism, Kodiak Islands, Alaska [*Vrolijk et al.*, 1988]. Shear veins coating surface of this block probably developed during interseismic phase of deformation but would provide a quartz-influenced rheology during seismic deformation.

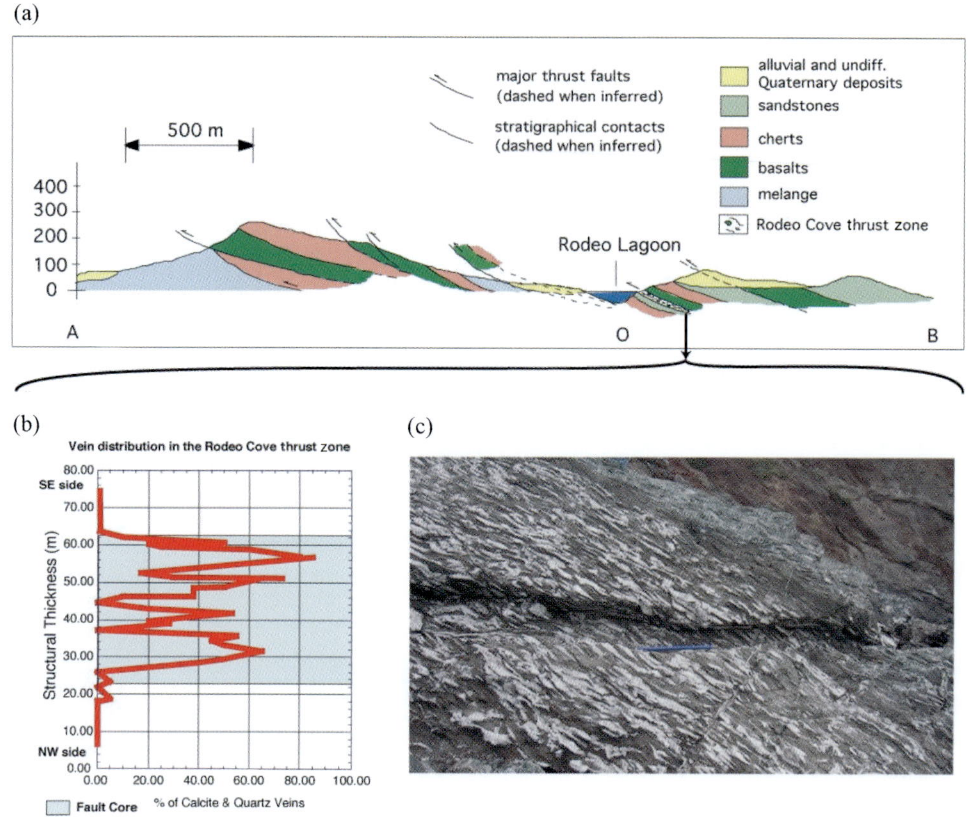

Figure 10.6 Evidence for localized fluid flow on a subduction thrust system. (a) Cross section viewed to the ENE through Rodeo Cove area, Marin Headlands [*Blake et al.*, 2000]. (b) Concentration of carbonate and quartz veins in Rodeo Cove thrust zone [*Meneghini*, 2003]. Percentage of veins reflects average over decimeter intervals. (c) Outcrop of carbonate and quartz veins in Rodeo Cove thrust zone. Note parallelism of veins to dip of fault zone in fig. 10.6a. Veins are also parallel to pressure-solution foliation in fault zone [*Meneghini*, 2003]

Brittle Deformation Brittle deformation includes fabrics associated with faulting in subduction zones: scaly fabric (mudstone), stratal disruption, and the associated development of complex cataclastic shear bands as well as macroscopically discrete shear surfaces [*Moore,* 1986]. In subduction zones, cataclastic phenomena are commonly localized into millimeter-wide shear bands [*Aydin,* 1978] of complex three-dimensional geometry, or "web structure" [*Byrne,* 1984]. Apparently, the shear bands achieve their complex three-dimensional structure because of the ongoing rotational deformation and high strain that ultimately result in stratal disruption. Granular flow and distributed cataclastic flow occur at the earliest phases of development of subduction shear zones [*Lucas and Moore,* 1986]. Pseudotachylyte is rare and developed in rocks with background temperatures of 230°–270°C prior to melting at much higher temperatures [*Ikesawa et al.*, 2003; *Kitamura et al.*, 2005; *Rowe et al.*, 2005].

Ductile Deformation Pressure solution is the initial ductile deformation mechanism in accretionary prisms. At temperatures <140°–150°C pressure solution is subtle, characterized by local indentation and suturing of grains. At higher temperatures, solution processes begin to produce cleavages of increasing intensity. Pressure solution is the most dominant ductile deformation mechanism in accretionary prisms [*Feehan and Brandon*, 1999; *Fisher and Byrne*, 1992; *Fisher and Brantley*, 1993; *Ring and Brandon*, 1999; *Schwartz and Stockhert*, 1996]. Because pressure solution is commonly superimposed on previously stratally disrupted units, strong planar cleavages are not always obvious. Pressure solution apparently remains the dominant ductile deformation mechanism to temperatures exceeding 350°C. Intracrystalline deformation along dislocations is relatively rare [*Schwartz and Stockhert*, 1996], but the rocks undergo syntectonic recrystallization during growth of new phases, for example [*Jayko et al.*, 1986].

Discussion

Above we subdivide the critical diagenetic-metamorphic and deformational processes operating in both aseismic and seismic realms of accretionary prisms. Linkages exist between many of these processes and together control deformation and earthquake generation in this environment. Below we provide some interpretations on how these processes are interrelated and foster seismogenesis.

Smectite-Illite Transition and Albitization The smectite to illite transition, which is virtually complete at 150°C, was proposed as a possible cause of the upper aseismic to seismic transition in subduction zones (fig. 10.1) [*Hyndman*, 1997; *Vrolijk*, 1990] because this phase transition causes an increase in coefficient of friction (or strengthening of fault surfaces). However, recent experimental work on the rate-dependent frictional properties of illite shows that it is velocity strengthening [*Morrow et al.*, 1992; *Saffer and Marone*, 2003], which is not consistent with accelerating runaway slip behavior associated with an earthquake [*Scholz*, 2002].

Although the smectite to illite transition may not have a direct physical effect on seismogenesis, the chemistry of the phase change may be significant. The smectite to illite transition produces calcium, iron, and magnesium, which are expelled into pore waters. The albitization of plagioclase also produces calcium over about the same temperature range as the smectite to illite transition. The influx of these cations is thought to cause carbonate cementation during burial of sedimentary rocks [*Boles and Franks*, 1979] and also in accretionary prisms [*Sample*, 1990].

Carbonate Cementation Carbonate cementation/veining occurs in significant amounts starting at ~100°–150°C, which may reflect an influx of calcium, iron, and magnesium causing the concentrations of these cations in pore waters to exceed the solubility of the various carbonate minerals. CO_3^{-2} in the carbonates is probably ultimately derived from organic matter [*Mullis et al.*, 1994], especially because limestones are rare in accretionary prisms. The presence of interstitial carbonate can add to the cohesion of the rock volume and perhaps change its response to deformation [*Muhuri et al.*, 2003]. It is not clear whether carbonate is velocity weakening or velocity strengthening [*Lockner and Byerlee*, 1986], and therefore its significance in modifying the frictional response of discrete fault zones is underdetermined. Carbonate cementation is one process that would rigidify the accretionary prism and therefore contribute to its elastic strength and earthquake potential (see below). Carbonate cementation would also reduce porosity and permeability and foster high fluid pressure.

Pressure Solution and Quartz Cementation Pressure solution is normally considered as a component of diffusive mass transfer (dissolution, diffusion, and precipitation). Because fluid advection and diffusion are both important in the transport of solutes in accretionary prisms, we focus our discussion on the dissolution and reprecipitation of materials. The onset of pressure solution appears approximately when the smectite to illite transition is complete. Research by sedimentary petrologists indicates that dissolution is strongly is facilitated by the presence of an illite-muscovite phase under very low stresses [*Bjorkum*, 1996]. Therefore the combination of increasing temperature, increasing effective stress, and the increase in abundance of illite from the smectite-illite transition may cause the onset of pressure solution observed at ~150°C in accretionary prisms. This interpretation suggests additional importance of the smectite-illite transition as a chemical, if not a physical, transition increasing rock coherence and possibly contributing to seismogenic potential.

Byrne [1998] has suggested diffusive mass-transfer processes homogenize rock fabrics and therefore allow generation of recordable earthquakes. *Rowe and Moore* [2003] further pointed out that the common quartz veining and cements occurring in areas of pressure solution change the frictional properties of the rock, leading to a velocity-weakening slip behavior. Apparently, silica is being mobilized by dissolution and redeposited at temperatures >150°C, well within the 100°–300°C temperature range where quartz demonstrates velocity-weakening behavior [*Blanpied et al.*, 1995] (fig. 10.7). Other experiments suggest quartz is velocity weakening even at lower temperatures [*Di Toro et al.*, 2004]. However, faults in accretionary prisms are clay dominated at lower temperatures, so the onset of quartz deposition in faults above 150°C is an important control on their stick-slip behavior.

Schwartz and Stockhert [1996] argued that the abundance of pressure solution and paucity of intracrystalline deformation is due to the low stresses in accretionary prisms, relative to continental deformation. Because accretionary prisms are thrust systems, higher mean stresses might be expected for a given

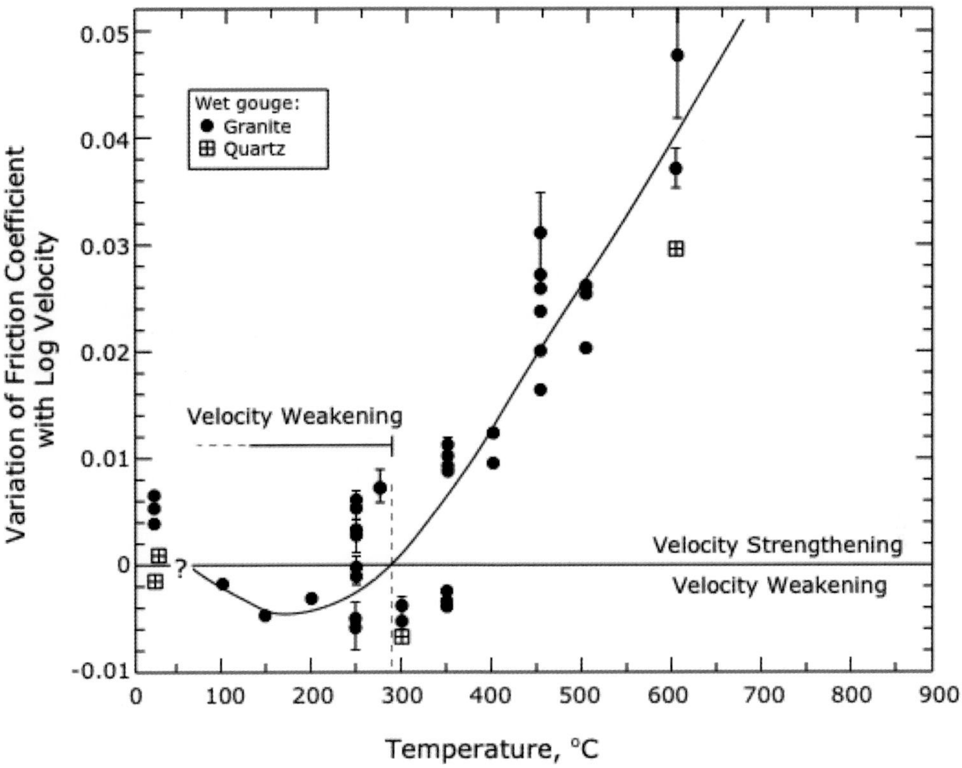

Figure 10.7 Velocity dependence of granite and quartz gouge samples [*Blanpied et al.*, 1995]. Note that velocity weakening occurs below ~300°C. The lower limit of velocity weakening is poorly determined. However, fault zones in accretionary prisms are clay dominated below ~150°C, which would favor stable sliding. The infusion of quartz into faults above 150°C and the tendency of quartz to be velocity weakening to ~300°C argues for unstable slip along fault zones in the 150°–300°C range, approximating the seismogenic zone. Quartz and granite frictional behavior are plotted together because quartz is a framework silicate and granite is dominated by framework silicates or quartz and feldspar.

depth. However, the ubiquitous presence of fluids in accretionary prisms may mediate the stress levels (see fluid pressures below).

In subduction zones, brittle deformation dominates to 150°–200°C. Brittle structures in accretionary prisms begin to be overprinted by ductile structures at ~150°–200°C. Rocks deformed in the 200°–300°C range commonly show ductile deformation, acting as pressure solution, repeatedly overprinted by brittle failure [*DiTullio and Byrne*, 1990; *Lewis and Byrne*, 2001; *Meneghini*, 2003; *Onishi and Kimura*, 1995; *Rowe and Moore*, 2003]. Microscopically, this alternation is seen as a pressure-solution fabric and as syntectonic growth of quartz fibers and other minerals that are cross cut by brittle fractures. Subsequently, the brittle-fracture phenomena are redeformed by pressure-solution fabrics (fig. 10.8). In rocks deformed at 300°C and above, there is little evidence for brittle deformation.

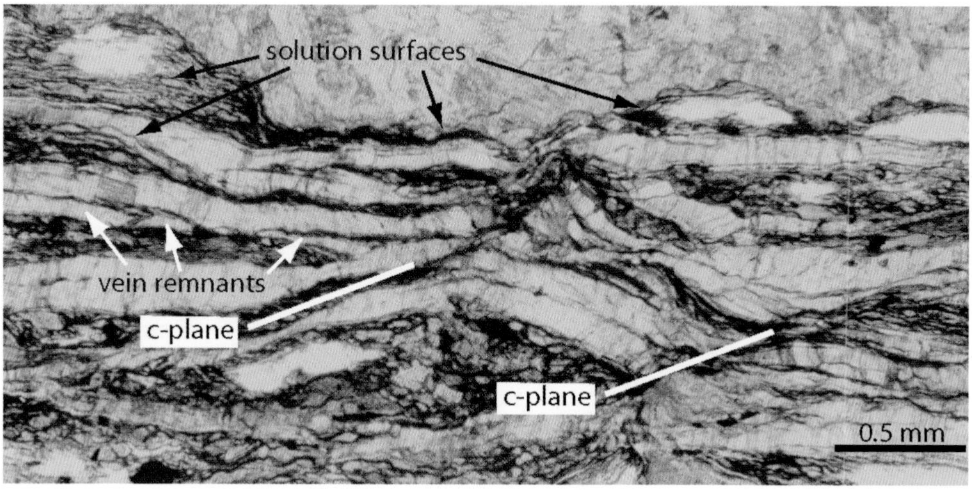

Figure 10.8 Veined fault surface showing interplay of solution and brittle fracture as vein emplacement. Light grey represents carbonate veins, whereas greenish layers are dominantly chlorite. Vein boundaries, especially large vein at top of figure, commonly modified by irregular pressure solution surfaces. Chloritic layers also show internal pressure solution surfaces. Note remnants of vein in upper right showing extreme solution effects resulting in boudinage of vein. Veins show varying degrees of disruption and modification by solution, suggesting veins of differing ages. Veins and pressure-solution fabric are approximately horizontal in photomicrograph. Pressure-solution fabric, in combination with shear surfaces, or "c-planes" define S-C fabric characteristic of shear zones. Photomicrograph from veined core of fault shown in figure 10.6 [*Meneghini*, 2003].

The apparently widespread alternation of pressure solution and brittle deformation in the 200°–300°C range suggests an alternation of fluid pressure and/or differential stress. A low fluid pressure, low differential stress regime would favor pressure solution, whereas high fluid pressures and/or high differential stress could cause brittle failure. The pressure-solution regime represents relatively low strain rate interseismic deformation, as has been suggested elsewhere [*Duebendorfer et al.*, 1998; *Gratier and Gamond*, 1990]. The brittle deformation may be coseismic, although it could also represent fracture at less than seismic strain rates [*Cowan*, 1999].

How Does Fluid Pressure Vary Above and Into the Seismogenic Zone?

Background

Because pore pressure is one of the most critical controls on rock deformation, a principal goal of the proposed deep drilling into seismogenic zones is to measure pore pressure [*Kimura et al.*, 2003]. Shallow direct measurements of fluid pressures at the frontal part of accretionary prisms have yielded modest overpressuring [*Becker et al.*, 1997; *Screaton et al.*, 1995], and modeling suggests development of higher pressures at somewhat greater depths [*Bekins and Dreiss*, 1992; *Saffer and Bekins*, 1998]. Moreover, both frontal thrusts [*Le Pichon et al.*, 1987; *Moore et al.*, 1990] and out-of-sequence thrusts [*Shipboard Scientific Party*, 1994] are known to seep fluid at the surface, thus indicating they are conduits of active fluid flow that are be overpressured. The onset of high fluid pressures at depths <3–4 km is due to the rapid consolidation of sediments at shallow depths [e.g., *Screaton et al.*, 2002]. Also, mineral phase transitions including smectite to illite and opal to quartz transformations may contribute to overpressuring [*Moore and Vrolijk*, 1992]. The wide distribution of high fluid pressures suggested by the models coupled with the low state of consolidation and lithification of the sediments may explain the lack of seismicity at the depths of <4 km in sediment-dominated accretionary prisms. The disappearance of the décollement as a clear reflector at seismogenic depths in the Nankai accretionary prism has been attributed to dewatering and presumably depressurization of the sediment underthrust below the décollement [*Bangs et al.*, 2004].

Observations from Subaerially Exposed Accretionary Prisms

Uplifted accretionary prism rocks provide a record of elevated fluid pressures preserved in both the fluid inclusions and geometry of veins and cements. Studies of fluid inclusions in shear and extensional veins of accretionary prisms show high and variable fluid pressures that are interpreted to be near lithostatic [*Lewis et al.*, 2000; *Vrolijk*, 1987]. *Vrolijk* [1987] suggested that the quartz vein growth was associated with a phase of major fluid movement along the décollement of the accretionary prism and that it initiated at the high range of observed fluid pressures and cyclically decreased through time. On the basis of low wall rock porosities and the discontinuous and limited occurrence of veins, *Lewis and Byrne* [2003; *Lewis et al.*, 2000] suggest that the observed veins reflect a small amount of fluid trickling through a previously emplaced accretionary prism, albeit at fluid pressures 75% of lithostatic or higher.

Extensive veining paralleling a shallowly dipping latest Cretaceous accretionary thrust in the Marin Headlands of northern California (fig. 10.6) indicates that fluid pressures were sufficient to open cracks perpendicular to the minimum principal stress plus any tensional strength of the rock [*Meneghini*, 2003]. Because this was a shallowly dipping accretionary thrust and the veins parallel it, the veins would have required a fluid pressure at least equal to the overburden. The zone of extensional veining is ~40 m thick along the thrust surface (fig. 10.6). The extraordinary quantity of veins in the fault zone suggests copious fluid flow. Cross-cutting relationships suggest generations of veining, brittle fracture, and pressure solution. Together these structures indicate alternation of compression and dilation across the fault surface that may represent fault-valve behavior and may be tied to the seismic cycle [*Meneghini*, 2003]. Apparently, the 40-m-thick zone of veins developed cumulatively with only a portion being active at any time. A 40-m thickness of dilated fault might be seismically resolvable at this depth. A thin dilated zone would not be imaged if it were less than ~10 m thick. Thus this fault zone may be a good analogue for the area downdip along the Nankai décollement where the reflection disappears [*Bangs et al.*, 2004].

Overall, fluid pressures appear to be high and broadly distributed in time and space at shallow depths due to the intergranular fluid sources, mineral phase changes, and hydrocarbon generation that produce fluids. At greater depths the fluid pressures also apparently transiently high, and the conduits of flow are more localized along faults owing to the reduction of intergranular permeability resulting from the consolidation and low-grade metamorphism in the accretionary prism [*Lewis and Byrne*, 2003]. As prism rocks are buried deeper than ~5 km, porosity changes are small [*Bray and Karig*, 1985]; hence fluids are probably derived primarily from continuing mineral dehydration reactions [*Hacker et al.*, 2003].

Strain Localization

In accretionary prisms the early faults are characterized by broader zones of scaly mudstone and stratal disruption, shear bands, and minor faults [*Shipboard Scientific Party*, 1991]. Faults developed later in consolidated prism interiors are more discrete than earlier primary accretionary structures (fig. 10.9). Experimental studies suggest that localized deformation on discrete, throughgoing shear surfaces tends to be of lower strength and more likely to be velocity weakening than distributed deformation in broader shear zones [*Beeler et al.*, 1996; *Marone*, 1998].

On a large scale in accretionary prisms it appears that shear strain localization in the form of out-of-sequence thrusts approximately correlates with the seaward limit of modeled earthquake displacements and onset of recorded microearthquakes in the Nankai Trough (fig. 10.10) [*Obana et al.*, 2003; *Park et al.*, 2000]. This implies that the consolidation/lithification processes in the prism are allowing it to deform as discrete large-scale blocks with slip localized along the out-of-sequence thrusts [*Lewis and Byrne*, 2003]. Out-of-sequence thrusts are clearly defined in ancient prisms and tend to partition them into packages with differing sedimentary, structural, or metamorphic history (fig. 10.11). The demonstrated activity of these out-of-sequence thrusts in past great earthquakes indicates a significant role in releasing seismic energy [*Pflaker*, 1972]. Moreover, modern activity of out-of-sequence thrusts suggests major activity near the frontal part of the accretionary prism and quiescence in a landward portion of the prism (fig. 10.11).

(a)

(b)

Figure 10.9 Outcrop examples of distributed faulting verses shear localization in accretionary prisms. (a) Outcrop of a stratally disrupted fault zone, Eocene accretionary prism, Dominican Republic [*Witschard and Dolan*, 1990]. Broad zone of deformation continues at least 10 m right beyond the edge of the photo. Inclined fabric of disruption zone indicates stratified rocks, where Jim Dolan is standing, are up relative to the disrupted rocks. (b) Discrete out-of-sequence thrust fault cutting early steeply dipping stratally disrupted fabric (beneath Peter Vrolijk's feet, right); note white quartz vein in fault being pointed to by Tim Byrne (left). Paleocene accretionary prism, Kodiak Islands Alaska.

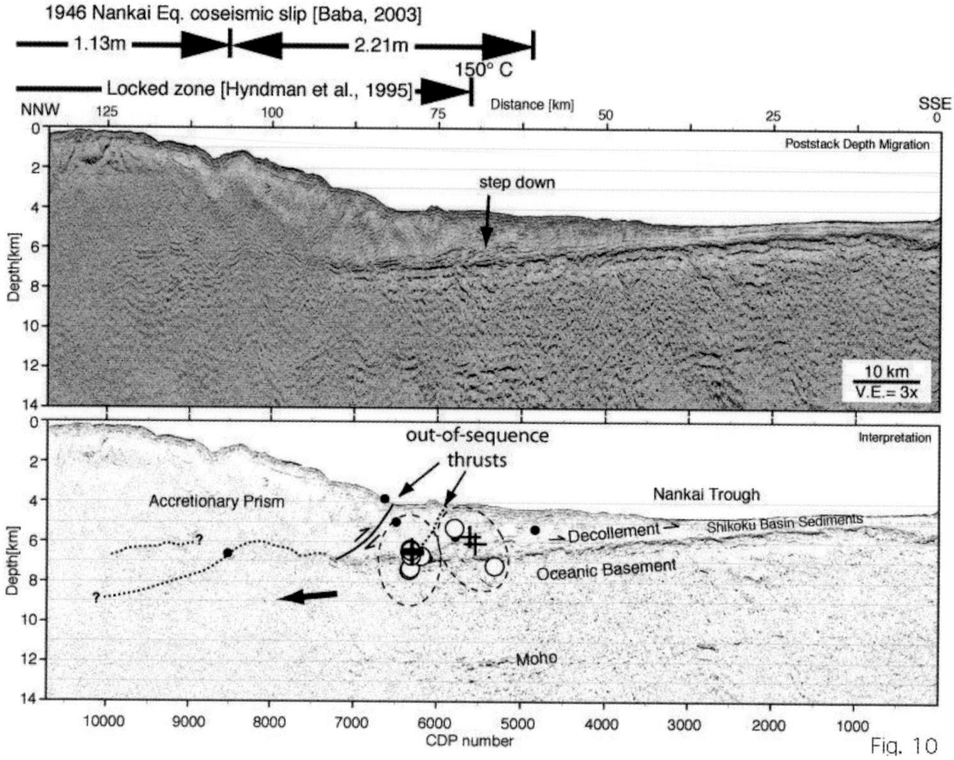

Figure 10.10 Seismic reflection depth section with nearby microearthquakes projected on to it [*Obana et al.*, 2003]. Note that microearthquakes die out just seaward of the initial development of out-of-sequence thrusts. Note also that the accretionary prism has a shallow even taper in its seaward extent, but the surface slope begins to steepen and become more irregular landward of the development of the out-of-sequence thrusts and the onset of seismicity.

Development of Upper-Plate Rigidity

The sudden slip during an earthquake results from the release of strain accumulated in rock masses on either side of the fault [e.g., *Scholz*, 2002, pp. 244–247]. Subduction thrusts beneath accretionary prisms are unique in that the upper plate begins as poorly consolidated, overpressured sediment with insufficient elastic strength to accumulate strain [*Byrne et al.*, 1988]. Even if the mechanical properties of the fault surface evolve to support unstable slip, the upper plate must develop sufficient elastic strength to produce the slip event. The accretionary prism thickens landward as required for its continued growth [*Davis et al.*, 1983]; owing to their progressive burial, the included rocks undergo consolidation, metamorphism, and cementation acquiring elastic strength. The onset of seismic activity typically occurs where the overlying prism has minimum thickness of 3–5 km and typically at localities with a high thermal

(a)

(b)

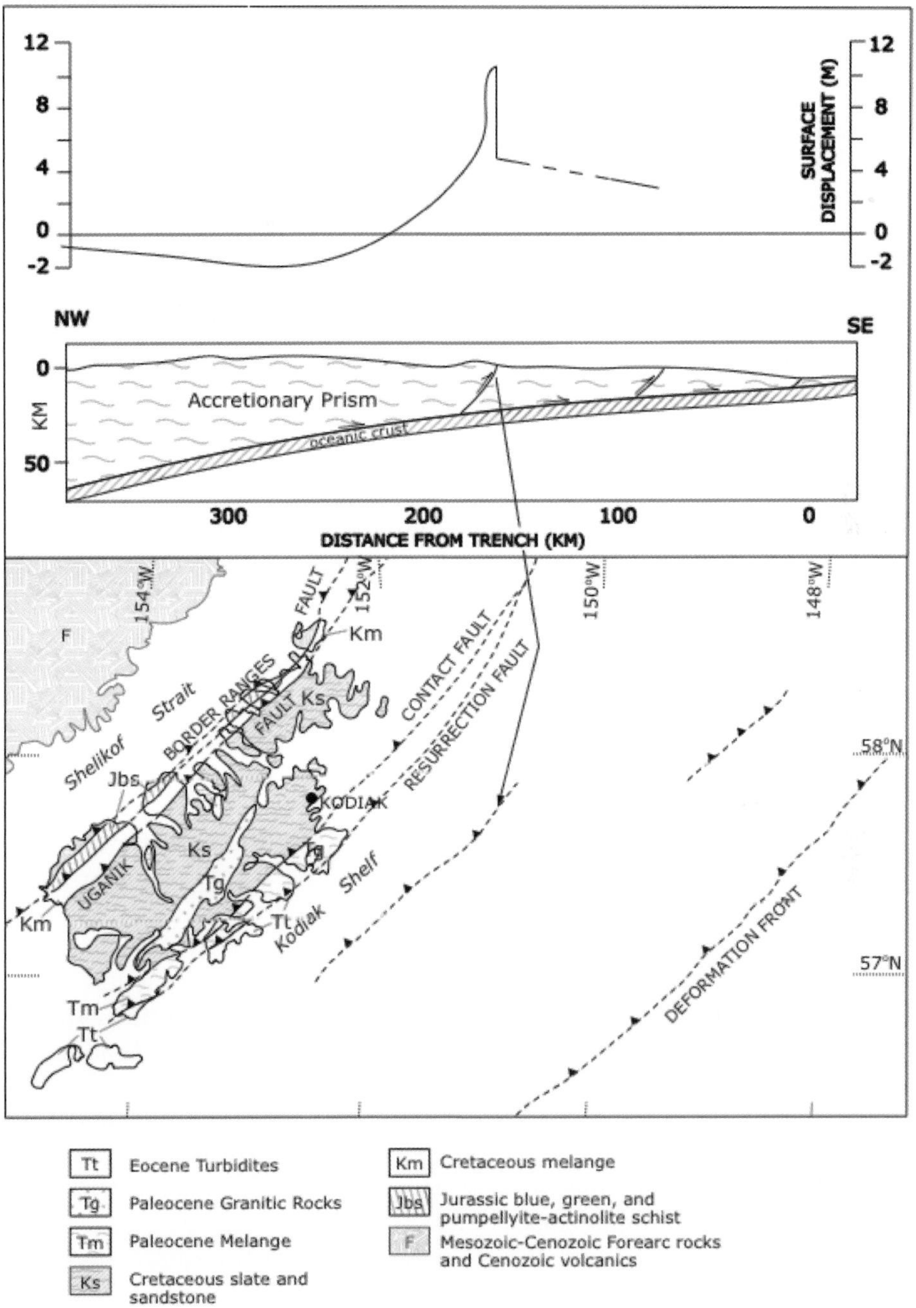

Figure 10.11 (a) Map showing thrust faults cutting accretionary prism [*Moore et al.*, 1991; *Plafker*, 1972; *Plafker et al.*, 1994]. All but frontal thrust is out-of-sequence thrusts. (b) Cross section from Prince William Sound, just north of the area shown in figure 10.11a, showing correlation to fault that was active during 1964 earthquake [*Plafker*, 1972]. Faults seaward of the correlated fault were inferred to be active during the 1964 earthquake. Seismic reflection data show that these faults deform young sediments at the seafloor.

gradient [*Obana et al.*, 2003; *Bilek and Lay*, 2000]. Apparently, the high thermal gradients accelerate metamorphism, cementation, and rigidification of the upper plate.

Summary

Accretionary prisms develop by the underthrusting and emplacement of rocks carried by the downgoing plate. Aside from those materials scraped off at the deformation front, all rocks that are underthrust more deeply must descend beneath the décollement, and experience the décollement step-down through or around them, thus achieving accretion to the upper plate. Thus, depending on their depth of accretion, these rocks preserve a record of deformation in the upper aseismic zone, the seismogenic zone, and the lower aseismic zone. Drawing from the consensus of thermal models of modern seismicity in subduction zones, we define prism rocks of the upper aseismic zone as those emplaced at temperatures of <125°C, of the seismogenic zone as those that have been subjected to temperatures of 125°–350°C, and those of the lower aseismic zone as those emplaced at temperatures >350°C. These temperature limits are approximate and transitional with errors of about ±15–20%. They apply to continental margin, not oceanic subduction zone settings. As sediments and rocks are emplaced at differing temperatures and depths, they acquire specific rheologically significant metamorphic and deformational imprints:

1. During initial burial to ~125°C, accretionary prism sediments become rocks, undergoing diagenetic reactions including of the transformation of smectite to illite, the albitization of detrital feldspar, dehydration of opal, and hydrocarbon generation. Although the smectite to illite transition does increase the coefficient of friction or strength, illite's rate-dependent friction is velocity strengthening and not subject to accelerating slip. However, the cations (calcium, magnesium, iron) released during the smectite to illite transition and the albitization process foster cementation.

2. Cementation and veining by carbonates becomes common above 125°C, perhaps owing to the above-mentioned release of cations. Pressure-solution fabrics begin to be apparent at ~150°C, with well-developed cleavages and quartz veining common by 200°C. Pressure solution may be facilitated by the diagenetic formation of illite and its catalyzing effect on quartz and feldspar dissolution.

3. Quartz veining and cementation in the 150°–300°C range facilitate the change from a velocity-strengthening to a velocity-weakening, earthquake-prone rheology.

4. Cementation by quartz and carbonates increases cohesion and may also make the rocks subject to unstable failure.

5. In accretionary prisms, brittle fabrics are progressively replaced by ductile fabrics through a temperature range of ~150°–325°C.

6. Although rocks in the seismogenic zone have lost most of their intergranular fluid through consolidation, vein geometries and fluid inclusions suggest high fluid pressures, approximately at lithostatic.

7. Strain localization in the formation of discrete faults occurs across the lower aseismic to seismic transition. Strain localization is observed both at microscopic, outcrop, and map scale. Active out-of-sequence thrusts occur in the frontal portions of accretionary prisms and represent a map-scale example of strain localization and seismic behavior.

The evolution of the accretionary prism from aseismic to seismic behavior involves the changes in material properties from fault zones dominated by clays to those in which quartz and carbonate are introduced by diffusion and fluid advection. Although both the quartz and carbonate increase cohesion, only quartz displays a velocity-weakening frictional response. The rate-dependent frictional behavior of carbonate is not well defined.

Consolidation, cementation, and other diagenetic-metamorphic transitions of the accretionary prism lead to the development of an upper plate of the subduction zone with increasing elastic strength. Development of this rigid upper plate is a prerequisite for plate-boundary seismicity because stick-slip behavior requires that both the hanging wall and foot wall of a fault store elastic strain energy. The seawardmost out-of-sequence thrusts define the seaward edge of this elastically rigid block.

Acknowledgments

We thank Peter Eichhubl and Jim Boles for helpful discussions on diagenesis and low-grade metamorphism. Sampling access to the outcrops in Marin Headlands provided by the National Park Service. We thank Jon Lewis and Tim Byrne for helpful reviews that clarified the text. Koichiro Obana provided a high-quality copy of figure 10.11 and American Geophysical Union granted permission for its reproduction here. Research supported by NSF grants OCE-9802264 and OCE-0203664 and Italian government grant M.I.U.R. COFIN 2003.

References

Aydin, A. (1978), Small faults as deformation bands in sandstone, *Pure Appl. Geophys., 116*, 913–930.

Bangs, N., S. Shipley, G. Moore, S. Gulick, S. Kuromoto, and Y. Nakamura (2004), Evolution of the Nankai Trough décollement from the trench into the seismogenic zone: Inferences from three-dimensional seismic reflection imaging, *Geology, 32*(4), 273–276.

Becker, K., A. T. Fisher, and E. E. Davis (1997), The CORK experiment in Hole 949C: long-term observations of pressure and temperature in the Barbados accretionary prism, *Proc. Ocean Drill. Program Sci. Results, 156,* 239–252.

Beeler, N. M., et al. (1996), Frictional behavior of large displacement experimental faults, *J. Geophys. Res., 101,* 8697–8715.

Behl, R. J., and R. E. Garrison (1994), The origin of chert in the Monterey Formation of California (USA), in *Siliceous, Phosphatic and Glauconitic Sediments of the Tertiary and Mesozoic,* edited by A. Iijima and Garrison, pp. 101–132, VSP, Int. Sci.,Utrecht, Netherlands.

Bekins, B., and S. Dreiss (1992), A simplified analysis of parameters controlling dewatering in accretionary prisms, *Earth Planet. Sci. Lett., 109,* 275–287.

Bilek, S. L., and T. Lay (2000), Depth dependent rupture properties in circum-Pacific subduction zones, in *Geocomplexity and the Physics of Earthquakes, Geophys. Monogr. Ser.*, vol. 120, edited by J. B. Rundle, D. L. Turcott, and W. Klein, pp. 165–186, AGU, Washington, D.C.

Bjorkum, P. A. (1996), How important is pressure in causing dissolution of quartz in sandstone? *J. Sediment. Res., 66,* 147–154.

Blake, M. C. J., R. W. Graymer, and D. L. Jones (2000), Geologic map and map database of parts of Marin, San Francisco, Alameda, Contra Costa, and Sonoma counties, California, *U.S. Geol. Surv. Miscell. Field Stud., MF-2337.*

Blanpied, M. L., D. A. Lockner, and J. D. Byerlee (1995), Frictional slip of granite at hydrothermal conditions, *J. Geophys. Res., 100,* 13,045–13,064.

Boles, J. R. (1982), Active albitization of plagioclase, Gulf Coast Tertiary, *Am. J. Sci., 282,* 165–180.

Boles, J. R., and S. G. Franks (1979), Clay diagenesis in Wilcox sandstones of southwest Texas—Implications of smectite diagenesis on sandstone cementation, *J. Sediment. Petrol., 49,* 55–70.

Brandon, M. T., and J. A. Vance (1992), Tectonic evolution of the Cenozoic Olympic subduction complex, Washington State, as deduced from fission track ages for detrital zircons, *Am. J. Sci., 292,* 564–636.

Bray, C. J., and D. E. Karig (1985), Porosity of sediments in accretionary prisms and some implications for dewatering processes, *J. Geophys. Res., 90,* 768–778.

Byrne, D. E., D. M. Davis, and L. R. Sykes (1988), Loci and maximum size of thrust earthquakes and the mechanics of the shallow region of subduction zones, *Tectonics, 7,* 833–857.

Byrne, T. (1984), Early deformation in melange terranes of the Ghost Rocks Formation, Kodiak Islands, Alaska, *Spec. Pap. Geol. Soc. Am., 198,* 21–52.

Byrne, T. (1998), Seismicity, slate belts and coupling along convergent plate boundaries, *Eos Trans. AGU, 79,* W-114.

Byrne, T., A. J. Maltman, E. Stephenson, and W. Soh (1993), Deformation structures and fluid flow in the toe region of the Nankai accretionary prism, *Proc. Ocean Drill. Program Sci. Results, 131,* 83–101.

Cowan, D. S. (1999), Do faults preserve a record of seismic slip? A field geologist's opinion, *J. Struct. Geol., 21,* 995–1001.

Currie, C. A., R. D. Hyndman, K. Wang, and V. Kostoglodov (2002), Thermal models of the Mexico subduction zone: Implications for the megathrust seismogenic zone, *J. Geophys. Res., 107*(B12), 2370, doi:10.1029/2001JB000886.

Davis, D. J., J. Suppe, and F. A. Dahlen (1983), Mechanics of fold-and-thrust belts and accretionary wedges, *J. Geophys. Res., 88,* 1153–1172.

Di Toro, G., D. L. Goldsby, and T. E. Tullis (2004), Friction falls towards zero in quartz rock as slip velocity approaches seismic rates, *Nature, 427,* 436–439.

DiTullio, L., and T. Byrne (1990), Deformation paths in the shallow levels of an acccretionary prism: The Eocene Shimanto belt of southwest Japan, *Geol. Soc. Am. Bull.*, *102*, 1420–1438.

DiTullio, L., M. Laughland, and T. Byrne (1993), Thermal and constraints on deformation from illite crystallinity and vitrinite relfectance in shallow levels of an acretionary prism Eocene-Oligocene Shimanto Belt, southwest Japan, in *Thermal Evolution of the Tertiary Shimanto Belt, Southwest Japan: An Example of Ridge-Trench Interaction*, edited by M. B. Underwood, *Spec. Pap. Geol. Soc Am.*, *273*, 63–82,

Duebendorfer, E. M., J. Vermilye, P. A. Beiser, and T. L. Davis (1998), Evidence for aseismic deformation in the western Transverse Ranges, southern California; implications for seismic risk assessment, *Geology*, *26*(3), 271–274.

Ernst, W. G. (1990), Thermobarometric and fluid expulsion history of subduction zones, *J. Geophys. Res.*, *95*, 9047–9053.

Feehan, J., and M. T. Brandon (1999), Contribution of ductile flow to exhumation of low-T-high-*P* metamorphic rocks: San Juan-Cascade nappes, NW Washington State, *J. Geophys. Res.*, *104*, 10,883–810,902.

Fisher, D., and T. Byrne (1987), Structural evolution of underthrusted sediments, Kodiak Islands, Alaska, *Tectonics*, *6*, 775–794.

Fisher, D., and T. Byrne (1992), Strain variations in an ancient accretionary complex: implications for forearc evolution, *Tectonics*, *11*, 330–347.

Fisher, D. M. (1996), Fabrics and veins in the forearc: A record of cyclic fluid flow at depths of <15 km, in *Subduction Top to Bottom*, *Geophys. Monogr. Ser.*, vol. 96, edited by J. P. Platt, pp. 75–89, AGU, Washington, D. C.

Fisher, D. M., and S. L. Brantley (1993), Models of quartz overgrowth and vein formation: Deformation and fluid flow in an ancient subduction zone, *J. Geophys. Res.*, *97*, 20,043–20,061.

Geddes, D. (1993), Carbonate cement in sandstone as an indicator of accretionary complex hydrogeology, M.S. thesis, 107 pp., Univ. of Calif., Santa Cruz.

Gratier, J. P., and J. F. Gamond (1990), Transition between seismic and aseismic deformation in the upper crust, in *Deformation Mechanisms, Rheology and Tectonics,* edited by R. J. Knipe and E. H. Rutter, *Spec. Publ. Geol. Soc. London*, *54*, 461–473.

Hacker, B. R., G. A. Abers, and S. M. Peacock (2003), Subduction factory: 1. Theoretical mineralogy, densities, seismic wave speeds, and H_2O contents, *J. Geophys. Res.*, *108*(B1), 2029, doi:10.1029/2001JB001127.

Helper, M. A. (1986), Deformation and high P/T metamorphism in the central part of the Condrey Mountain window, north-central Klamath Mountains, California and Oregon, in *Blueschists and Eclogites*, edited by B. W. Evans and E. H. Brown, *Mem. Geol. Soc. Am.*, *164*, 107–123.

Hibbard, J. P., M. M. Laughland, S. M. Kang, and D. Karig (1993), The thermal imprint of spreading ridge subduction on the upper levels of an accretionary prism, Southwest Japan, in *Thermal Evolution of the Tertiary Shimanto Belt, Southwest Japan: An Example of Ridge-Trench Interaction*, edited by M. B. Underwood, pp. 83–101, Geol. Soc. of Am., Boulder Colo.

Hunt, J. M. (1996), *Petroleum Geochemistry and Geology,* 2nd ed., 743 pp., W. H. Freeman, New York.

Hyndman, R. D. (1997), Seismogenic zone of subduction thrust faults, *Island Arc*, *6*, 244–260.

Hyndman, R. D., and K. Wang (1993), Thermal constraints on the zone of major thrust earthquake failure: The Cascadia subduction zone, *J. Geophys. Res.*, *98*, 2039–2060.

Hyndman, R. D., M. Yamano, and D. A. Oleskevich (1997), The seismogenic zone of subduction thrust faults, *Island Arc*, *6*, 224–260.

Ikesawa, E., A. Sakaguchi, and G. Kimura (2003), Pseudotachylyte from an ancient accretionary complex: Evidence for melt generation during seismic slip along a master décollement, *Geology*, *31*(7), 637–640.

Isaacs, C. M., K. A. Pisciotto, and R. E. Garrison (1983), Facies and diagenesis of the Miocene Monterey Formation, California: A summary, in *Siliceous Deposits in the Pacific Region, Developments in Sedimentology*, vol. 36, edited by A. Iijima, J. R. Hein, and R. Siever, pp. 247–282, Elsevier, New York.

Jayko, A. S., M. C. J. Blake, and R. N. Brothers (1986), Blueschist metamorphism of the Eastern Franciscan belt, northern California, *Mem. Geol. Soc. Am.*, *164*, 107–123.

Kimura, G., and A. Mukai (1990), Underplated units in an accretionary complex: Melange of the Shimanto Belt of eastern Shikoku, southwestern Japan, *Tectonics*, *10*, 31–50.

Kimura, G., H. Tobin, and N. W. Group (2003), NanTroSEIZE: The Nankai Trough Seismogenic Zone Experiment Complex Drilling Project, report, p. 33, N. M. Tech, Socorro.

Kitamura, Y., K. Sato, E. Ikesawa, K. Ikehara-Ohmori, G. Kimura, H. Kondo, K. Ujiie, C. T. Onishi, K. Kawabata, Y. M. H. Hashimoto, and H. Masago (2005), Melange and its seismogenic roof decollement: A plate boundary fault rock in the subduction zone- An example from the Shimanto Belkt, Japan, *Tectonics*, *24*, p. TC5012, doi:10.1029/2004TX001635, 2005.

Kulm, L. D., and E. Suess (1990), The relation of carbonate deposits to fluid venting processes: Oregon accretionary prism, *J. Geophys. Res.*, *95*, 8899–8915.

Kusky, T. M., and D. C. Bradley (1999), Kinematic analysis of Mélange fabrics: Examples and applications from the McHugh Complex, Kenai Peninsula, Alaska, *J. Struct. Geol.*, *21*, 1773–1796.

Labaume, P., M. Kastner, A. Trave, and P. Henry (1997), Carbonate veins from the decollement zone at the toe of the Northern Barbados accretionary prism: microstructure, mineralogy, geochemistry and relations with prism structures and fluid regime, *Proc. Ocean Drill. Program Sci. Results*, *156*, 79–96.

Lay, T., and S. Bilek (2007), Anomalous earthquake ruptures at shallow depths on subduction zone megathrusts, this volume.

Le Pichon, X., et al. (1987), Nankai Trough and Zenisu Ridge: A deep-sea submersible survey, *Earth Planet. Sci. Lett.*, *83*, 285–299.

Lewis, J. C., and T. Byrne (2001), Fault kinematics and past plate motions at a convergent plate boundary: Tertiary Shimanto Belt, southwest Japan, *Tectonics*, *20*, 548–565.

Lewis, J. C., and T. Byrne (2003), History of metamorphic fluids along outcrop-scale faults in a Paleogene accretionary prism, SW Japan: Implications for prism-scale hydrology, *Geochem. Geophys. Geosyst.*, *4*, 9007, doi:10.1029/2002GC000359.

Lewis, J. C., T. Byrne, J. D. Pasteris, D. London, and G. B. I. Morgan (2000), Early Tertiary fluid flow and pressure-temperature conditions in the Shimanto accretionary complex of south-west Japan: Constraints from fluid inclusions, *J. Metamorphic Geol.*, *18*, 319–333.

Lewis, J. C., J. R. Unruh, and R. J. Twiss (2003), Seismogenic strain and motion of the Oregon coast block, *Geology*, *31*(22), 183–186.

Lockner, D. A., and J. Byerlee (1986), Laboratory measurements of velocity-dependent friction strength, *U.S. Geol. Surv. Open File Rep., 86–417*.

Lucas, S. E., and J. C. Moore (1986), Cataclastic deformation in accretionary prisms: DSDP Leg 66, and onland examples from Barbados and Kodiak Islands, in *Structural Fabrics From Deep Sea Drilling Project Cores From Forearcs*, edited by J. C. Moore, *Mem. Geol. Soc. Am.* , *166*, 89–103.

Maltman, A. J., T. Byrne, D. E. Karig, S. J. Lallemant, R. Knipe, and D. Prior (1993), Deformation structures at Site 808, Nankai accretionary prism, Japan, *Proc. Ocean Drill. Program Sci. Results*, *131*, 123–133.

Maltman, A. J., P. Labaume, and B. Housen (1997), Structural geology of the decollement at the toe of the Barbados accretionary prism, *Proc. Ocean Drill. Program Sci. Results, 156*, 279–292.

Marone, C. (1998), Laboratory-derived friction laws and their application to seismic faulting, *Annu. Rev. Earth Planet. Sci., 26*, 643–696.

Meneghini, F. (2003), Hydrofracture, fluid flow and deformation in fossil accretionary prisms: The examples on the Internal Ligurian Units (northern Apennines, Italy) and the Franciscan Complex (California Coast Ranges, N. California, USA), Ph.D. thesis, 174 pp, Univ. of Pisa, Pisa, Italy.

Moore, G. W. (1980), Slickensides in deep sea cores near the Japan Trench, Leg 57, Deep Sea Drilling Project, *Initial Rep. Deep Sea Drill. Proj., 56–57*, part 2, 1107–1116.

Moore, J. C. (Ed.) (1986), *Structural Fabrics in Deep Sea Drilling Project Cores From Forearcs, Mem. Geol. Soc. Am., 166*, 160 pp.

Moore, J. C., and A. Allwardt (1980), Progressive deformation of a Tertiary trench slope, Kodiak Islands, Alaska, *J. Geophys. Res., 85*, 4741–4756.

Moore, J. C., and P. Vrolijk (1992), Fluids in accretionary prisms, *Rev. Geophys., 30*, 113–135.

Moore, J. C., D. L. Orange, and L. D. Kulm (1990), Interrelationship of fluid venting and structural evolution: Alvin observations from the frontal accretionary prism, Oregon, *J. Geophys. Res., 95*, 8795–8808.

Moore, J. C., F. Diebold, M., V. H. Sample, R., T. Brocher, B. Page, D. Stone, M. Talwani, J. Ewing, C. Rowe, and J. Davies (1991), EDGE deep seismic reflection transect of the eastern Aleutian arc-trench layered lower crust reveals underplating and continental growth, *Geology, 19*(5), 420–424.

Morrow, C. A., B. Radney, and J. Byerlee (1992), Frictional strength and the effective pressure law of montmorillonite and illite clays, in *Fault Mechanics and Transport Properties of Rocks*, edited by Evans, B., and T. F. Wong, pp. 69–88, Academic Press, San Diego.

Muhuri, S. K., T. A. Dewers, E. S. J. Thurmann, and Z. Reches (2003), Interseismic fault strengthening and earthquake-slip instability: Friction or cohesion, *Geology, 31*(10), 881–884.

Mullis, J., J. Dubessy, B. Poty, and J. O'Neil (1994), Fluid regimes during late stages of a continental collision: Physical, chemical, and stable isotope measurements of fluid inclusions in fissure quartz from a geotraverse through the Central Alps, Switzerland, *Geochim. Cosmochim. Acta, 58*, 2239–2267.

Norabuena, E., et al. (2004), Geodetic and seismic constraints on some seismogenic zone processes in Costa Rica, *J. Geophys. Res., 109*, B11403, doi:10.1029/2003JB002931.

Norris, R. J., and D. G. Bishop (1990), Deformed conglomerates and textural zones in the Otago Schists, South island, New Zealand, *Tectonophysics, 174*, 331–349.

Obana, K., et al. (2003), Microseismicity at the seaward updip limit of the western Nankai Trough seismogenic zone, *J. Geophys. Res., 108*(B10), 2459, doi:10.1029/2002JB002370.

Oleskevich, D. A., R. D. Hyndman, and K. Wang (1999), The updip and downdip limits to great subduction earthquakes: Thermal and structural models of Cascadia, south Alaska, SW Japan, and Chile, *J. Geophys. Res., 104*, 14,965–14,991.

Onishi, C. T., and G. Kimura (1995), Change in fabric of melange in Shimanto Belt, Japan: Change in relative convergence? *Tectonics, 14*, 1273–1289.

Orange, D. L., and M. B. Underwood (1995), Patterns of thermal maturity as diagnostic criteria for interpretation of melanges, *Geology, 23*(12), 1144–1148.

Orange, D. L., D. S. Geddes, and J. C. Moore (1993), Structural and fluid evolution of a young accretionary complex; the Hoh rock assemblage of the western Olympic Peninsula, Washington, *Geol. Soc. Am. Bull., 105*, 1053–1075.

Park, J. O., et al. (2000), Out-of-sequence thrust faults developed in the coseismic slip zone of the 1946 Nankai earthquake (M_w = 8.2) off Shikoku, southwest Japan, *Geophys. Res. Lett., 27*, 1033–1036.

Peacock, S. M. (1993), The importance of blueschist eclogite dehyration reactions in subducting oceanic crust, *Geol. Soc. Am. Bull.*, *105*, 684–694.

Plafker, G. (1972), Alaskan earthquake of 1964 and Chilean earthquake of 1960: Implications for arc tectonics, *J. Geophys. Res.*, *77*, 901–925.

Plafker, G., J. C. Moore, and G. R. Winkler (1994), Geology of the southern Alaska margin, in *The Geology of Alaska*, edited by P. G. and H. C. Berg, pp. 389–449, Geol. Soc. of Am., Boulder, Colo.

Ring, U., and M. T. Brandon (1999), Ductile deformation as mass loss in the Franciscan Subduction Complex: Implications for exhumation processes in accretionary wedges, in *Ehumation Processes: Normal Faulting, Ductile Flow, and Erosion*, edited by U. Ring et al., pp. 55–86, Geol. Soc., London.

Roeske, S. (1986), Field relations and metamorphism of the Raspberry Schist, Kodiak Islands, Alaska, in *Blueschist and Eclogites*, edited by B. W. Evans and E. H. Brown, *Mem. Geol. Soc. Am.*, *164*, 169–184.

Rowe, C., E. Thompson, and J. C. Moore (2002), Contrasts in faulting and veining across the aseismic to seismic transition, Kodiak Accretionary Complex, Alaska, *Eos Trans. AGU*, *83*(47), Fall Meet. Suppl., Abstract T21A-1060.

Rowe, C., J. C. Moore, F. Meneghini, and A. W. McKiernan (2005), Large-scale pseudotachylytes and fluidized cataclasites from an ancient subduction thrust fault, *Geology*, *33*(12), 937–940.

Rowe, C. D., and J. C. Moore (2003), The upper aseismic to seismic transition: A silica mobility threshold, *Eos Trans. AGU*, *84*(46), Fall Meet. Suppl., Abstract T41E-02.

Saffer, D., and C. Marone (2003), Comparisons of smectite and illite-rich gouge, frictional properties: applications to the updip limit of the seismogenic zone of subduction thrusts, *Earth and Planetary Science Letters*, *215*, 219–235.

Saffer, D. M., and B. A. Bekins (1998), Episodic fluid flow in the Nankai accretionary complex: Timescale, geochemistry, flow rates, and fluid budget, *J. Geophys. Res.*, *103*, 30,351–330,370.

Sample, J. C., and J. C. Moore (1987), Structural style and kinematics of an underplated slate belt, Kodiak Islands, Alaska, *Geol. Soc. Am. Bull.*, *99*, 7–20.

Sample, J. S. (1990), The effect of carbonate cementation of underthrust sediments on deformation styles during underplating, *J. Geophys. Res.*, *95*, 9111–9121.

Scholz, C. H. (2002), *The Mechanics of Earthquakes and Faulting*, 2nd ed., 471 pp., Cambridge Univ. Press, New York.

Schwartz, S., and B. Stockhert (1996), Pressure solution in siliciclastic HP-LT metamorphic rocks—Constraints on the state of stress in deep levels of accretionary complexes, *Tectonophysics*, *255*, 203–209.

Schwartz, S. Y., and H. R. DeShon (2007), Distinct updip limits to geodetic locking and microseismicity at the northern Costa Rica seismogenic zone: Evidence for two mechanical transitions, this volume.

Screaton, E. J., B. Carson, and G. P. Lennon (1995), Hydrologic properties of a thrust fault within the Oregon accretionary prism, *J. Geophys. Res.*, *100*, 20025–20035.

Screaton, E. J., et al. (2002), Porosity loss within the underthrust sediments of the Nankai accretionary complex; implications for overpressures, *Geology*, *30*(1), 19–22.

Shipboard Scientific Party (1991), Site 808, in *Proc. Ocean Drill. Program Initial Reports*, edited by Tiara, A., et al., v. 131, pp. 71–269, (Ocean Drilling Program), College Station, TX.

Shipboard Scientific Party (1994), Site 892, in *Proc. Ocean Drill. Program Initial Reports*, *146*, Part 1, 301–378.

Sibson, R. H. (1975), Generation of pseudotachylyte by ancient seismic faulting, *Geophys. J. R. Astron. Soc.*, *43*, 775–794.

Speed, R. (1990), Volume loss and defluidization history of Barbados, *J. Geophys. Res.*, *95*, 8983–8996.

Taira, A., J. Katto, M. Tashiro, M. Okamura, and K. Kodama (1988), The Shimanto Belt in Shikoku Japan: Evolution of a Cretaceous to Miocene accretionary prism, *Mod. Geol.*, *12*, 5–46.

Twiss, R. J., and E. M. Moores (1992), *Structural Geology*, 532 pp., W. H. Freeman, New York.

van der Pluijm, B. A., and S. Marshak (1997), *Earth Structure: An Introduction to Structural Geology and Tectonics*, 495 pp., McGraw-Hill, New York.

van der Pluijm, B. A., and S. Marshak (2004), *Earth Structure: An Introduction to Strutural Geology and Tectonics*, 656 pp., W.W. Norton, New York.

Vrolijk, P. J. (1987), Tectonically-driven fluid flow in the Kodiak accretionary complex, Alaska, *Geology*, *15*(5), 466–469.

Vrolijk, P. (1990), On the mechanical role of smectite in subduction zones, *Geology*, *18*(8), 703–707.

Vrolijk, P., and S. M. F. Sheppard (1991), Syntectonic carbonate veins from the Barbados accretionary prism (ODP Leg 110): Record of paleohydrology, *Sedimentology*, *38*, 671–690.

Vrolijk, P., G. Myers, and J. C. Moore (1988), Warm fluid migration along tectonic melanges in the Kodiak accretionary complex, Alaska, *J. Geophys. Res.*, *93*, 10,313–10,324.

Wang, K., R. Wells, S. Mazzotti, R. D. Hyndman, and T. Sagiya (2003), A revised dislocation model of interseismic deformation of the Cascadia subduction zone, *J. Geophys. Res.*, *108*, 2026, doi:10.1029/2001JB001227.

Wang, K. H., R. D. Yamano, and M. Makoto (1995), Thermal regime of the southwest Japan subduction zone, *Tectonophysics*, *248*, 53–69.

Witschard, M., and J. F. Dolan (1990), Contrasting structural styles in siliciclastic and carbonate rocks of an offscraped sequence: The Peralta accretionary prism, Hispaniola, *Geol. Soc. Am. Bull.*, *102*, 792–806.

PART IV

Laboratory Studies

11

Friction of the Smectite Clay Montmorillonite

A Review and Interpretation of Data

Diane E. Moore and David A. Lockner

Abstract

Previous frictional strength experiments on the smectite clay montmorillonite conducted under supposedly similar conditions have yielded apparent values of the coefficient of friction μ (the shear stress/effective normal stress) that can range from 0.06 to 0.78. We report some new laboratory data that illustrate that this wide variation is in part a function of the large difference in strength between thoroughly dried (μ ≥ 0.7) and water-saturated (μ ≤ 0.3) montmorillonite. Dry montmorillonite gouge is subject to standard frictional processes such as abrasion, wear, and fracture during shear. In contrast, shear of water-saturated montmorillonite gouge is concentrated in thin films of water that are adsorbed onto the (001) surfaces of the platy grains. The water-saturated shear strength of layer-structure minerals increases with the strength of the bonding of the polar water molecules to the (001) surfaces, and the relative weakness of water-saturated montmorillonite may be largely owing to its small layer charge. Published values of μ for montmorillonite that are considered to represent (deionized) water-saturated, equilibrated conditions increase regularly from 0.06 at effective normal stresses, $\sigma_n' < 1$ MPa to 0.30 at $\sigma_n' \approx 300$ MPa. The direct correlation of μ with effective stress is characteristic of sheet-silicate gouges, and it is attributed to decreasing thickness of the surface water films with increasing σ_n'. Increasing salinity of the pore fluid has the same effect on shear strength as increasing σ_n'. Most of the published strength data for montmorillonite fall outside the range of μ for water-saturated, equilibrated samples. Of these, the samples that are overly strong for a given set of experimental conditions may have been only partially saturated. The overly weak samples appear to be caused by inadequate drainage and consequent buildup of internal pore pressure. The velocity dependence of montmorillonite strength has not been extensively investigated. Water-saturated montmorillonite gouge is velocity strengthening over the range of conditions tested to date, whereas partially saturated montmorillonite gouge may be velocity weakening at some velocities. These results highlight the hazards of interpreting fault-zone behavior based on experiments that do not approximate natural conditions. For example, the use of unsaturated clay rheology in modeling subducted sediment wedges in megathrust environments could lead to erroneous predictions.

Introduction

The smectite clay montmorillonite is an important low-temperature alteration product of mafic igneous volcanic rocks, particularly tuffs and volcanic ash [*Grim and Güven*, 1978]. Because of the close association of volcanic arcs with subduction zones, montmorillonite can be an important constituent of the accretionary wedges and downgoing sediments in subduction zones (see Table 1 of *Vrolijk* [1990, and references therein]). According to *Vrolijk* [1990], airborne volcanic ash incorporated into the hemipelagic deposits of the ocean basins alters to produce weak, smectite-rich horizons, along which the décollement may preferentially form as the oceanic crust is subducted. Localization of the décollement within smectite-rich units has been observed in the North Barbados [e.g., *Pudsey*, 1984; *Capet et al.*, 1990; *Tribble*, 1990] and Nankai [e.g., *Chamley et al.*, 1986] troughs. Along-strike variations in smectite-clay abundance in the décollement of the Cascadia subduction zone may control the vergence direction of thrust faults in the overlying accretionary wedge [*Underwood*, 2002].

Montmorillonite is also commonly found in other fault-zone settings [e.g., *Vrolijk and van der Pluijm*, 1999]. In particular, montmorillonite-bearing fault gouge has been identified in core from shallow drill holes into strike-slip faults such as the San Andreas fault system [*Wu et al.*, 1975; *Liechti and Zoback*, 1979] and the Nojima fault of Japan [*Ohtani et al.*, 2000]. Clays have been postulated to be present at considerable depths in some faults [*Wu et al.*, 1975; *Wu*, 1978], although montmorillonite may not be stable at temperatures above ≈150°C [e.g., *Abercrombie et al.*, 1994].

Since its identification as a low-strength mineral [*Kenney*, 1967; *Summers and Byerlee*, 1977a], montmorillonite has received considerable attention from the rock mechanics community as a possible cause of weak faults. Unfortunately, many of the published results are contradictory with respect to both frictional strength and the velocity dependence of friction. The purpose of this paper is to offer explanations for the apparent contradictions among the existing data, based on our recent investigations of the dry and water-saturated frictional strengths of sheet-structure minerals [*Moore and Lockner*, 2004] combined with the earlier work of *Wang et al.* [1980]. Assuming our interpretation is correct, this provides a framework for understanding the apparently inconsistent measurements of montmorillonite strength.

Summary of Existing Friction Data on Montmorillonite

Figure 11.1 summarizes the published friction data for montmorillonite (data sources are identified in figures 11.8 and 11.10). Where necessary, μ was calculated using the reported experimental conditions and stress measurements. The gap in the data between ~0.5 and 5 MPa separates the soil mechanics investigations conducted at very low-stress conditions from the higher-stress rock mechanics studies. The reported values of coefficient of friction μ at a given value of apparent effective normal stress σ_n' vary widely. For example, between 100 and 150 MPa σ_n', μ ranges from 0.06 to 0.78, more than an order of magnitude variation. This wide range is too large to be explained by differences in sample composition or systematic errors in test procedures.

Fewer measurements have been made of the velocity dependence of montmorillonite strength, but these too are contradictory. *Logan and Rauenzahn* [1987], *Morrow et al.* [1992], *Brown et al.* [2003], and D. A. Lockner and H. Tanaka (manuscript in preparation, 2006) measured solely velocity-strengthening behavior (shear strength increases with increasing velocity), whereas *Saffer et al.* [2001]

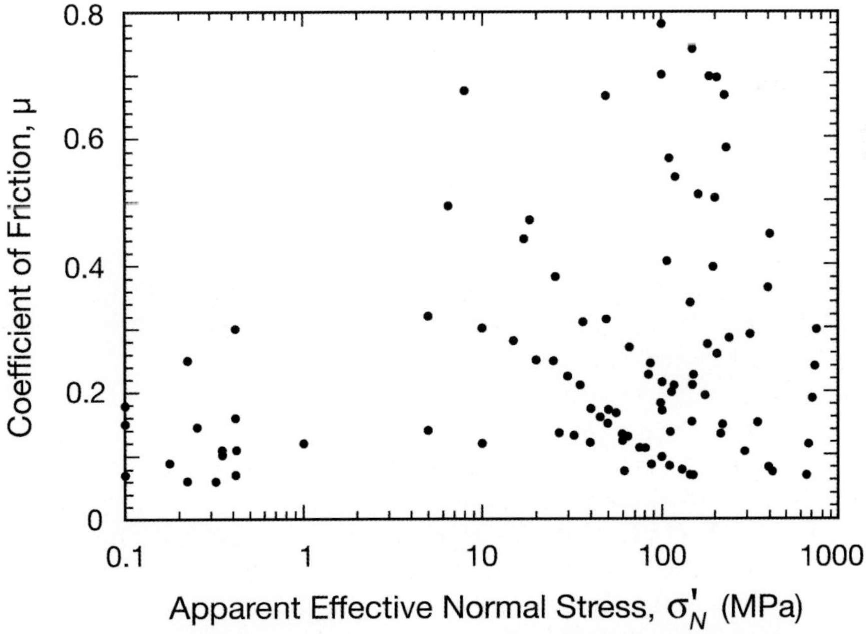

Figure 11.1 Compilation of room-temperature friction data for montmorillonite from soil- and rock-mechanics investigations (see fig. 11.8 and 11.10 for data sources). At a given apparent effective normal stress, reported coefficients of friction can vary by as much as 0.7.

reported that under certain conditions montmorillonite is velocity weakening (shear strength decreases with increasing velocity) and consequently capable of unstable slip.

If all these data are taken at face value, montmorillonite can be strong or weak, stable or unstable under reportedly similar conditions.

Experiments Conducted

We ran five triaxial friction experiments to illustrate some aspects of the frictional strength of montmorillonite. The starting material is a commercial montmorillonite clay, Volclay MPS-1, that has the composition: $Na_{0.3}(Fe^{+2},Mg)_{0.3}(Al,Fe^{+3})_{1.7}Si_4O_{10}(OH)_2 \cdot nH_2O$. The grains are thin plates with vaguely hexagonal outlines and diameters generally ≤0.5 μm (fig. 11.2). The very fine grain size is typical of montmorillonite as is a large ratio of (001) surface area to grain thickness [e.g., *Güven*, 1988; *Tournassat et al.*, 2003]. The montmorillonite sample required no preparation for use as a gouge material.

For the experiments a 1.1-mm-thick layer of the montmorillonite gouge was placed along a 30° sawcut between 25.4-mm-diameter cylindrical driving blocks (fig. 11.3). Westerly granite was used for the lower forcing block in all five experiments and either Berea sandstone or Westerly granite for the upper block, which is located at the pore-fluid inlet. As figure 11.3 illustrates, the permeability of the upper forcing block is critical to good fluid communication

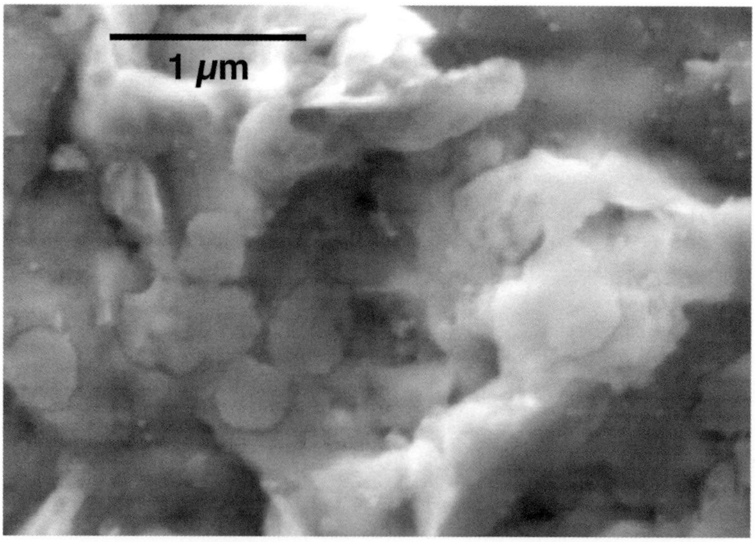

Figure 11.2 Secondary electron scanning electron microscope (SEM) image of the Na-montmorillonite MPS-1 starting material used in the experiments conducted for this study.

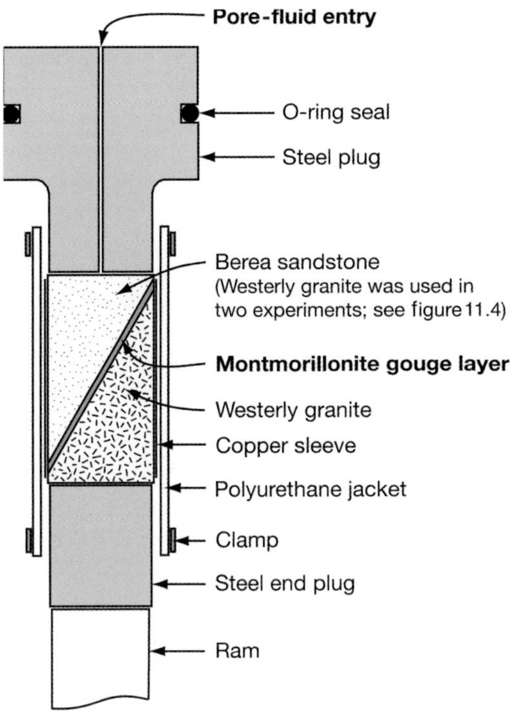

Figure 11.3 Sample assembly for triaxial friction experiments.

between the gouge layer and the external pore-pressure system, and the two rock types were chosen to demonstrate this. On the one hand, Berea sandstone has a porosity of 18% and a relatively high permeability on the order of 10^{-13} m^2 [*Zoback and Byerlee*, 1975]. On the other hand, the permeability of Westerly granite is in the approximate range 10^{-20}–10^{-19} m^2 [*Morrow et al.*, 1986], 6–7 orders of magnitude lower than that of Berea sandstone.

The gouge layer was applied to one sawcut surface as a thick slurry, and the other block was placed on top of the gouge layer. The sample was inserted into a 0.05-mm-thick copper sleeve to keep the assembly intact; the copper sleeve ruptures during the initial 1–2 mm of shear and therefore does not contribute to the reported strengths. With one exception, the samples were vacuum-dried overnight (~22 hours) at 125°C, and upon removal from the oven, the dried samples were immediately jacketed and placed in the pressure vessel. The sample assembly and the pore-pressure lines were evacuated for ≥20 min. Confining pressure was then raised to the desired level; if the experiment was to be conducted under water-saturated conditions, the fluid pressure was applied after the confining pressure. Room-temperature velocity-stepping experiments were conducted at axial velocities of 0.05, 0.5, and 5.0 μm/s. Constant σ_n' was maintained in each experiment by means of computer-controlled adjustments to the confining pressure. The exception was a sample that was

Figure 11.4 Friction experiments on montmorillonite conducted for this study. (a) Contrasting strengths of montmorillonite samples dried overnight at 125°C, one run dry and the other run at 10-MPa pore pressure of deionized water. Both samples used Berea sandstone as the forcing block adjoining the pore-pressure inlet (fig. 11.3). Experiments were run at a sequence of axial velocities, as indicated. (b) Velocity-stepping friction tests at 150 MPa σ_n', using sandstone/granite forcing blocks in one experiment and granite/granite forcing blocks in the other two. (c) Enlargement of the strength tests between 3.5- and 8.5-mm axial displacement, to better illustrate the changes in μ accompanying the velocity steps. Labels i–v identify the strength experiments from figures 11.4a and 11.4b.

dried at 50°C for 3.5 hours, evacuated for 3 hours in the pressure vessel, and allowed to sit overnight at the confining and pore-fluid pressure conditions of the experiment prior to testing.

A prominent feature of montmorillonite friction is the very large difference between dry and water-saturated strengths at the same effective normal stress (fig. 11.4a). At 100 MPa σ_n', μ of thoroughly dry montmorillonite is roughly 4 times as large as that of water-saturated montmorillonite. Raising σ_n' from 100 to 150 MPa increases the water-saturated value of μ by ≈0.03 (fig. 11.4b). The two experiments in figure 11.4b that used Westerly granite for the upper forcing blocks were designed to illustrate how the low-permeability granite increases the time needed to equilibrate pore-fluid pressure in the gouge layer. The equilibration times were even longer than expected; when the samples were disassembled immediately after the experiments, the sheared gouge layers showed no signs of dampness, even though the top of the upper granite piece, adjacent to the pore-pressure inlet (fig. 11.3), was wet.

The changes in strength accompanying the velocity steps are briefly summarized here. The two truly water-saturated samples (ii and v, the ones using sandstone/granite forcing-block pairs) are velocity strengthening over the range tested, as better illustrated in the enlarged plots in figure 11.4c. The more irregular strength behavior of the two samples with granite/granite blocks (figs. 11.4b and 11.4c) made it difficult to determine accurately the change in μ accompanying a velocity step. In all cases, however, the estimated range of values indicated velocity-strengthening behavior, despite the series of small stress drops that occurred during the 5-μm/s step of the sample dried at 125°C. The strength of the dry sample, i (fig.11. 4c) is essentially independent of velocity within the uncertainty of the measurements.

Different shear-induced textures developed during the dry and water-saturated experiments. Water-saturated gouge contains well-developed, shiny and slickensided shear surfaces (fig. 11.5a), and (001) faces of the montmorillonite grains in the shears are oriented parallel to the shear planes (fig. 11.5b). Shear surfaces of the dry gouge sample tested at 100 MPa σ_n' (fig. 11.4a) are

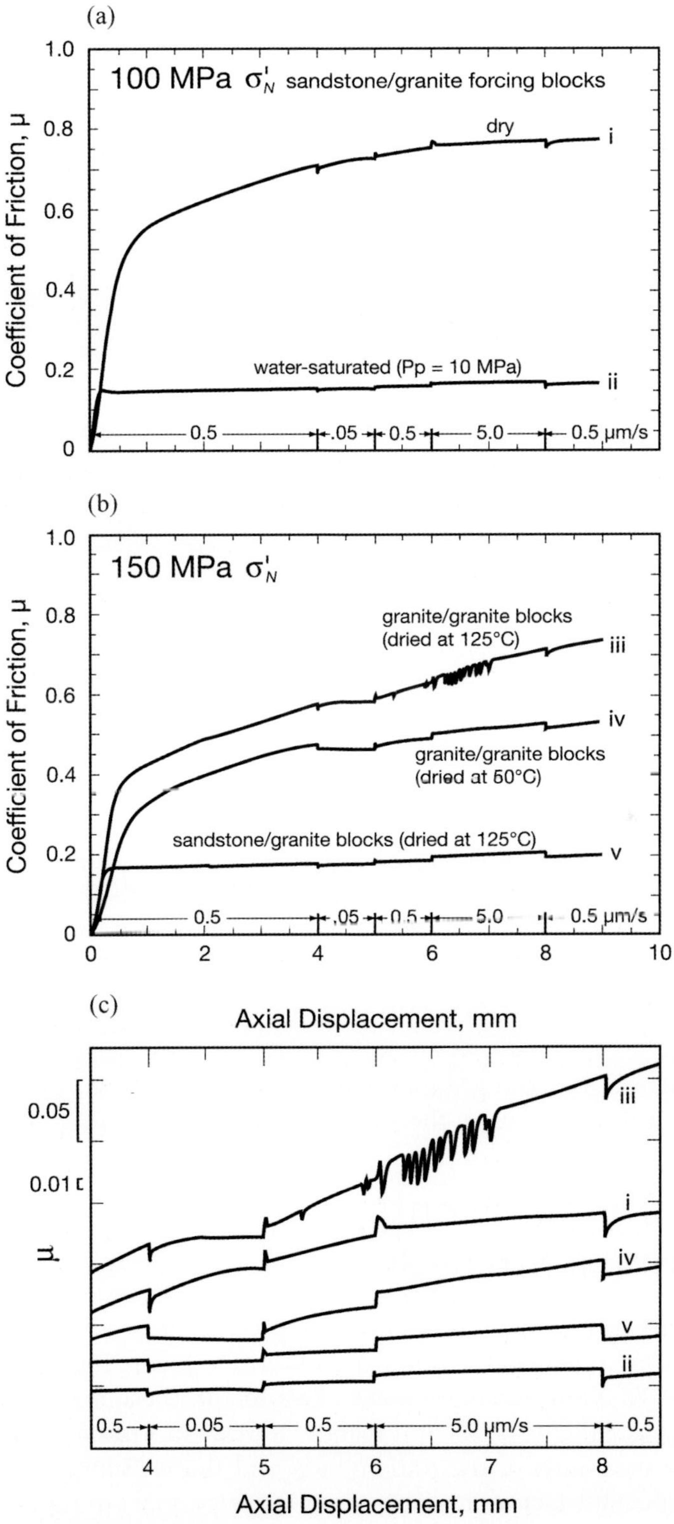

(a)
100 MPa σ'N sandstone/granite forcing blocks
dry
i
water-saturated (Pp = 10 MPa)
ii
0.5 .05 0.5 5.0 0.5 μm/s
Coefficient of Friction, μ
(b)
150 MPa σ'N
granite/granite blocks (dried at 125°C)
iii
iv
granite/granite blocks (dried at 50°C)
sandstone/granite blocks (dried at 125°C)
v
0.5 .05 0.5 5.0 0.5 μm/s
Coefficient of Friction, μ
Axial Displacement, mm
(c)
0.05
0.01
μ
iii
i
iv
v
ii
0.5 0.05 0.5 5.0 μm/s 0.5
Axial Displacement, mm

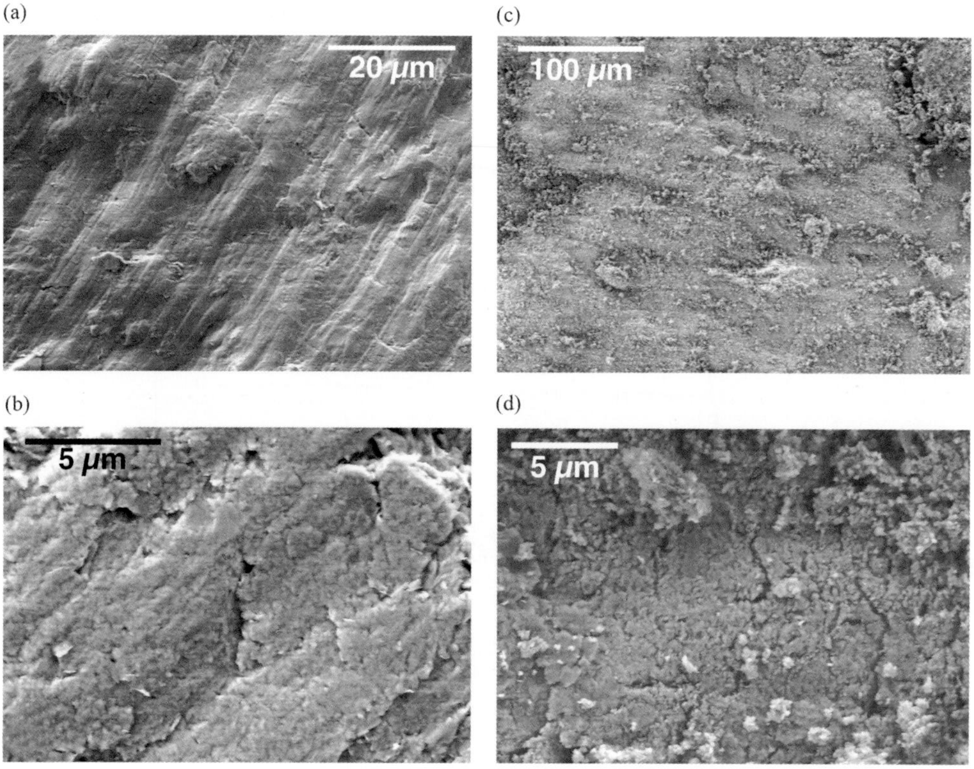

Figure 11.5 Secondary-electron SEM images of tested gouge samples. The water-saturated sample run at 150 MPa σ_n' shows good development of (a) shiny, slickensided shear surfaces characterized by (b) the strong preferred orientation of the platy montmorillonite grains parallel to the shear plane. The dry sample tested at 100 MPa σ_n' is characterized by (c) dull, grainy boundary-shear surfaces, although (d) grains in the shears have rotated into alignment with the shear surfaces.

considerably rougher and grainier in appearance (fig. 11.5c), and the preferred orientation of the platelets in the shears is only locally evident (fig. 11.5d).

Controls on Montmorillonite Frictional Strength

For layer-structured minerals such as the phyllosilicates, the bonding within a layer is strong whereas the bonds that hold the layers together (the (001) or interlayer bonds) are relatively weak. As a result, these minerals have a perfect cleavage parallel to (001) and usually a platy morphology (e.g., fig. 11.2). The relative weakness of the (001) bonds and the ensuing shape anisotropy are both important factors in controlling the frictional strengths of the layer-structure minerals. In addition, dry and water-saturated shear of montmorillonite are characterized by the operation of different frictional processes that can cause their strengths to differ significantly.

Water-Saturated Friction

Water lubricates the shear of montmorillonite (fig. 11.4) and that of all other sheet-silicate minerals [*Horn and Deere*, 1962; *Moore and Lockner*, 2004]. Water attracted to the surfaces of the platelets forms thin, structured films with an integral number of layers when constrained between the basal planes of two adjoining sheet-silicate crystals [e.g., *Israelachvili, et al.*, 1988]. The films have a measurable shear strength that arises because the water molecules are bonded to the (001) surfaces of the crystals and also to the water molecules in other layers [*Israelachvili et al.*, 1988]. As illustrated in figures 11.5a and 11.5b, shear of water-saturated montmorillonite is localized to well-defined shear planes, and the platy grains in the shears have rotated so that their basal planes are parallel to the shear surfaces. Such deformation textures are characteristic of gouges composed of layer-structure minerals, and *Moore and Lockner* [2004] concluded that the shear of the phyllosilicate-rich gouges during their triaxial friction experiments was concentrated in the films of water adsorbed onto the (001) surfaces. Similar conclusions have been reached for the shear of clay sediments [e.g., *Rosenqvist*, 1984].

The shear strength of the water films will be a function of the degree of attraction of the polar water molecules to the (001) surfaces; specifically, the strength is correlated with factors that contribute to the electrostatic component of the (001) interlayer bond [*Moore and Lockner*, 2004]. The smectite clays are 2:1 sheet silicates, in which a layer consists of an octahedral sheet sandwiched between two tetrahedral sheets. A net negative layer charge can arise in these minerals from substitutions in either the tetrahedral or octahedral sheet, and this charge is balanced by the addition of cations to the interlayer. On the basis of his calculations, *Giese* [1978] concluded that the layer charge is the major contributor to the (001) bond of the 2:1 sheet silicates. Because the tetrahedral sheet is always at the surface of a crystal in this mineral group, a water molecule should adhere more strongly to a 2:1 sheet silicate whose layer charge originates in the tetrahedral sheet [e.g., *Bleam*, 1990; *Skipper et al.*, 1993]. Consistent with this, the coefficient of friction correlates better with the tetrahedral-sheet charge (fig. 11.6) than with the total layer charge [*Moore and Lockner*, 2004]. Sheet-silicate minerals can be separated into two groups, trioctahedral and dioctahedral, based on the cations occupying the octahedral sheet. All of the octahedral cation sites of an ideal trioctahedral mineral are filled with divalent cations, whereas two thirds of the octahedral cation sites in a dioctahedral mineral are filled with trivalent cations and the other third are empty. A dioctahedral mineral always has somewhat stronger interlayer bonds than its trioctahedral equivalent [*Giese*, 1978], and the dioctahedral minerals in

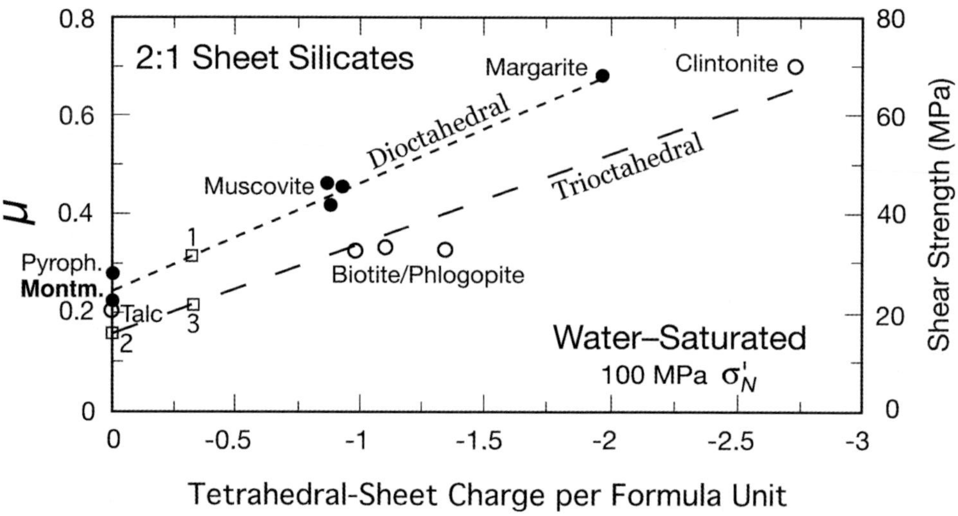

Figure 11.6 Correlation of water-saturated coefficient of friction of 2:1 sheet silicates at 100 MPa σ_n' with the charge on the tetrahedral sheets that form the surfaces of the (001) cleavage faces. Charge determinations were made using the measured composition of each mineral [*Moore and Lockner*, 2004]. For a given layer charge, μ is 0.10–0.15 larger for a dioctahedral mineral (solid circles) than for its trioctahedral (open circles) equivalent. The open squares labeled 1, 2, and 3 are the predicted values of μ for other smectite clays based on their ideal compositions: 1, beidellite and nontronite; 2, hectorite; 3, saponite.

figure 11.6 have consistently higher frictional strengths than their trioctahedral counterparts.

The coefficient of friction of water-saturated montmorillonite (fig. 11.6) is in accord with its 2:1 dioctahedral crystal structure, in which a small layer charge arises from substitutions in the octahedral rather than the tetrahedral sheet. Thus the strength of water-saturated montmorillonite gouge can be explained completely in terms of shear through the surface water films, as hypothesized for the other layer-structure minerals. The previous emphasis on the interlayer-water content of montmorillonite as the cause of its low strength [e.g., *Summers and Byerlee*, 1977a; *Bird*, 1984] may have arisen because the strengths of other very weak minerals such as the pyrophyllite and talc in figure 11.6 (both of them nonswelling minerals) had not yet been determined. Shear through surface water films of an expanded montmorillonite grain should always be easier than shear through its interlayer bonds, because the interlayer cations bind the layers together and generally limit the amount of expansion to <10 Å (1 nm). In contrast, the water films between separate platy grains can be considerably thicker [e.g., *Güven*, 1992; *Renard and Ortoleva*, 1997], particularly at $\sigma_n' < 100$ MPa (fig. 11.7a).

Dry Friction

The electrostatic components of the (001) bonds of many crystallographically well-characterized sheet-structure minerals were calculated by *Giese* [1978, 1980] and *Bish and Giese* [1981]. The coefficient of friction of thoroughly dried gouges of these minerals increases with increasing calculated (001) bond strength to values of ~70 kcal/mole [*Morrow et al.*, 2000; *Moore and Lockner*, 2004], at which point $\mu \approx 0.8$, satisfying "Byerlee's law" [*Byerlee*, 1978]. The coefficient of friction remains at ≈0.8 for minerals with bond strengths >70 kcal/mole. *Moore and Lockner* [2004] concluded that for those minerals with $\mu < 0.8$, dry shearing occurs by cleaving through the (001) planes of the grains that have rotated into alignment with the shear planes. For the minerals with $\mu \approx 0.8$, shearing occurs by means of the frictional processes characteristic of "strong" minerals such as abrasion and wear, brittle fracture, and rolling [*Paterson*, 1978].

Because of the lack of a crystallographic structural refinement, the interlayer bond strength of montmorillonite has not been determined. However, on the basis of its overall crystal structure, dry montmorillonite was expected to be weaker than muscovite and biotite, for which μ (dry) at 100 MPa σ_n' averages 0.59 and 0.54, respectively [*Moore and Lockner*, 2004]. Surprisingly, the dry montmorillonite sample in figure 11.4a has $\mu \approx 0.8$. The extremely fine grain size typical of montmorillonite clays and the very large area-to-thickness ratio may contribute to the unexpectedly high strength of dry montmorillonite. A montmorillonite grain that is only 1 unit cell thick [e.g., *Güven*, 1992; *Tournassat et al.*, 2003] cannot deform by cleaving through interlayer bonds. A minimum number of layers per crystal may be needed for the process to occur; thinner grains might preferentially flex or buckle rather than shear through (001). According to *Güven* [1992], montmorillonite lamellae only a few layers in thickness will often curl and fold. The grainy shear surfaces of the dry montmorillonite sample (figs. 11.5c and 11.5d) are similar to those of other phyllosilicates for which μ (dry) ≈ 0.8 [*Moore and Lockner*, 2004], suggesting that similar frictional processes were operative for all of them.

Interpretation of Published Strength Data on Montmorillonite

Water-Saturated, Equilibrated Strengths

To sort out the data in figure 11.1, we first identified those experiments that were conducted on water-saturated samples, with deionized water as the pore fluid, and that are considered to have maintained equilibrated fluid pressures during the tests. The data assigned to this group are shown in figure 11.7b. *Kenney* [1967], *Lupini et al.* [1981], and *Maio and Fenelli* [1994] conducted

Figure 11.7 (a) Thickness of water films between two platy crystal surfaces, calculated from Debye-Hückel/osmotic theory, as a function of σ_n', from figure 11.4 of *Renard and Ortoleva* [1997, p. 1968]. This curve is constructed for a mineral with a surface charge energy of −0.01 C/m^2, at room temperature (25°C). (b) Compilation of friction experiments that are considered to represent equilibrated, water-saturated conditions using deionized water as pore fluid. The vertical bar at the low-stress end is the range of μ from those experiments plotted in figure 11.8 that used deionized water. (c) Correlation of shear strengths calculated from the data in figure 11.7b with the inverse of the calculated water-film thicknesses from figure 11.7a at the effective normal stresses of the experiments.

direct-shear tests on montmorillonite at effective normal stresses in the range 0.1–0.5 MPa (fig. 11.8). They reported residual strengths, that is, the strengths of the clay samples following the development of localized shears containing aligned clay platelets [*Skempton*, 1964, 1985] similar to the one shown in figure 11.5a. Care was taken in all three studies to equilibrate the fluid pressure. The reported values of μ from their experiments using pure water range between 0.06 and 0.11, as indicated by the vertical bar in figure 11.7b.

All of the data at higher σ_n' are from triaxial friction experiments in which Berea sandstone was the forcing block at the pore-pressure inlet. The high permeability of Berea sandstone assures that the pore-pressure difference between the external control system and the sawcut surface of the gouge layer is vanishingly small. The two sandstone/granite forcing-block experiments from figure 11.4 are included in figure 11.7b, along with experiments from *Morrow et al.* [1992, 2000] and D. A. Lockner and H. Tanaka (manuscript in preparation, 2006). Most of the data points from *Morrow et al.* [1992] included in figure 11.7b are the averages of two to four tests conducted at different combinations of fluid and confining pressures. Although some of their experiments were run at high σ_n', thin sections of those samples show no evidence of damage to the Berea sandstone forcing blocks. The data point from the *Morrow et al.* [2000] paper is the final value of μ from an experiment that was run dry to 4-mm axial displacement, then at 10-MPa pore-fluid pressure (100 MPa σ_n') to 9-mm axial displacement. The reported values of μ from D. A. Lockner and H. Tanaka (manuscript in preparation, 2006) are for a velocity of 1 μm/s, which is closest to the velocities of the other high-σ_n' experiments in figure 11.7b.

In figure 11.7b, μ increases relatively rapidly with increasing σ_n' at low stress levels and then more slowly at higher stresses. The spread of values at a given σ_n' reflects differences in experimental conditions such as the sliding velocity and the amount of displacement. *Israelachvili et al.* [1988] determined experimentally that the shear strength of a water film between two mica plates increases as the number of water layers in the film decreases. This inverse relation between water-film thickness and shear strength can be tested by com-

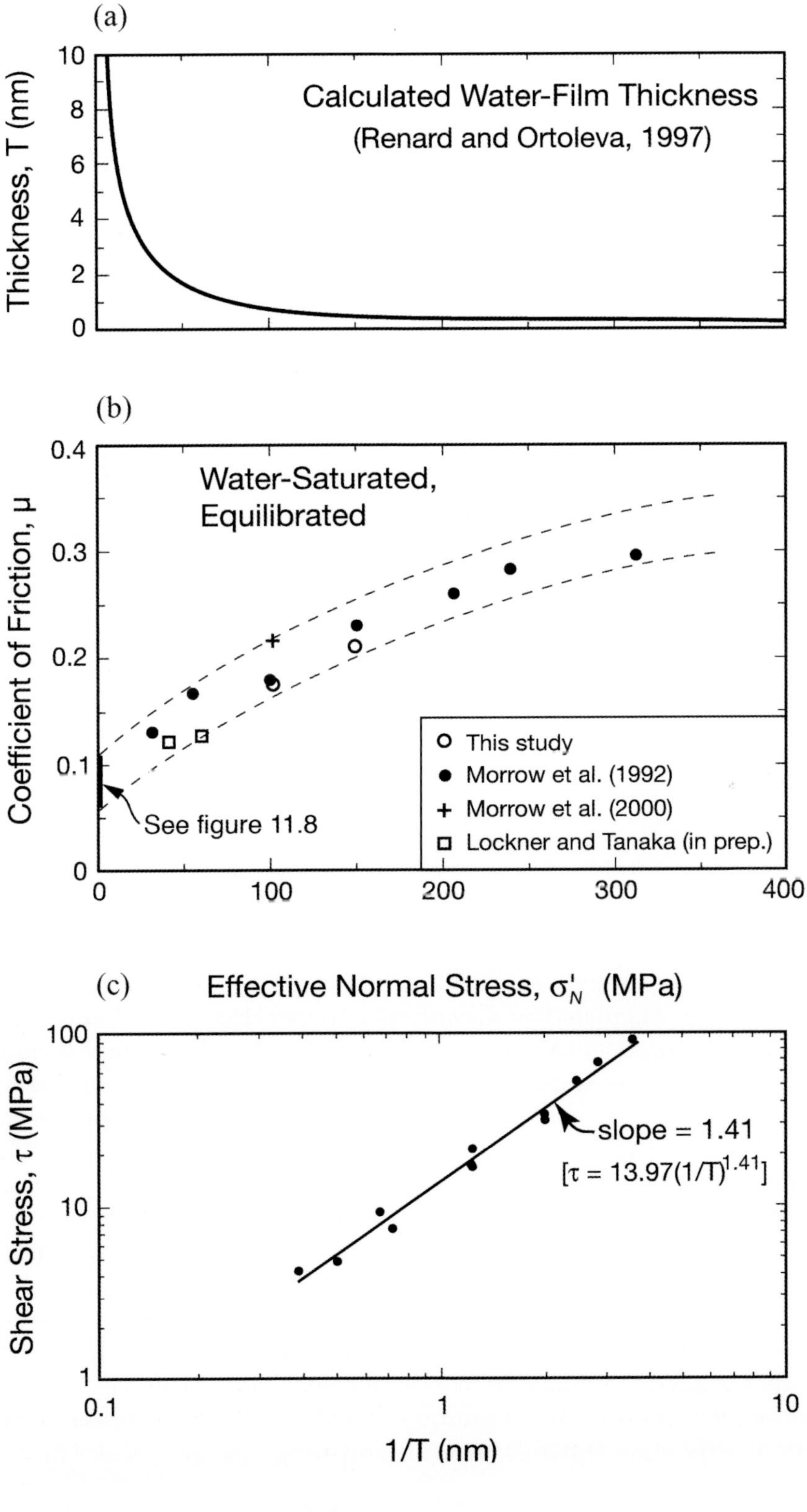
(a)
Calculated Water-Film Thickness
(Renard and Ortoleva, 1997)
Thickness, T (nm)
0
2
4
6
8
10
(b)
Water-Saturated, Equilibrated
Coefficient of Friction, μ
0
0.1
0.2
0.3
0.4
See figure 11.8
This study
Morrow et al. (1992)
Morrow et al. (2000)
Lockner and Tanaka (in prep.)
0
100
200
300
400
(c)
Effective Normal Stress, σ'N (MPa)
Shear Stress, τ (MPa)
1
10
100
slope = 1.41
[τ = 13.97(1/T)^1.41]
0.1
1
10
1/T (nm)

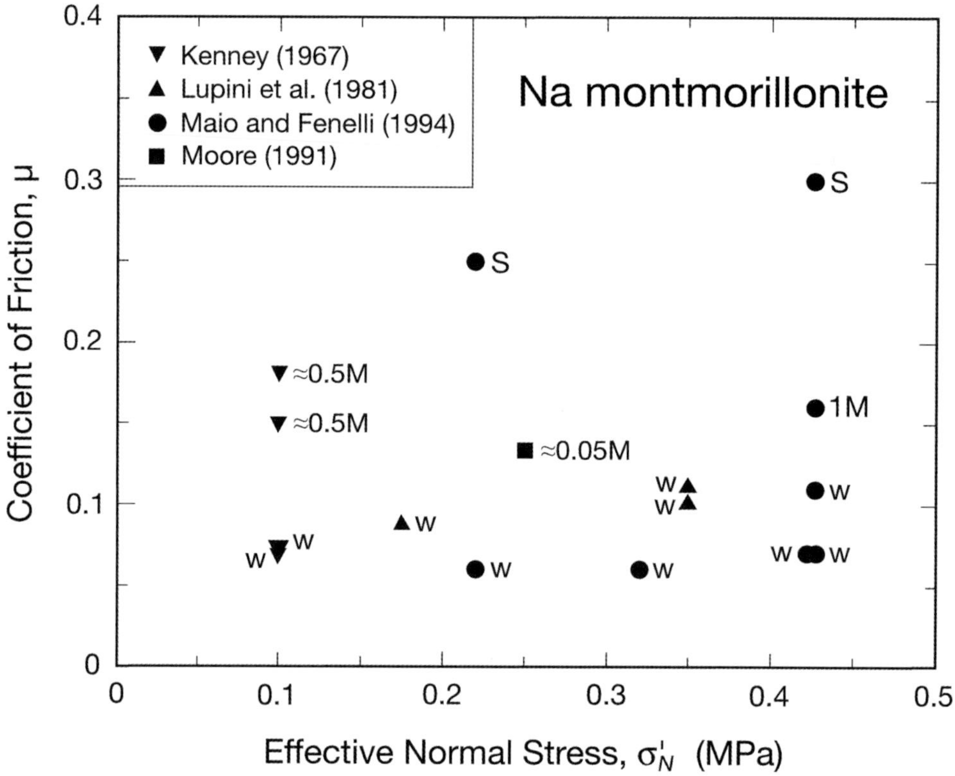

Figure 11.8 Coefficient of friction of Na-montmorillonite at $\sigma_n' < 0.5$ MPa, obtained from direct-shear and shear-box experiments. Labels: w, deionized or distilled water; ≈0.05 M, ≈0.5 M, 1 M = NaCl solutions of various concentrations; S, saturated NaCl solution.

paring clay shear strength ($\tau = \mu\sigma_n'$) in figure 11.7b with water-film thickness (figure 11.7a), as calculated by *Renard and Ortoleva* [1997]. The result, shown in figure 11.7c, is consistent with a model in which shear resistance increases at higher pressures where more weakly bound water layers are driven off. The effect of decreasing water-film thickness is thus to reduce the lubricating ability of the water film and, in consequence, to raise the frictional strength. This behavior contrasts to that of common rock-forming minerals such as quartz and feldspar (see fig. 11.7 of *Byerlee* [1978, p. 624]), for which μ decreases at effective pressures above ≈200 MPa, but it appears to be typical of water-saturated clay- and other phyllosilicate-rich gouges. For example, at room temperature the coefficient of friction of the serpentine mineral chrysotile increases from 0.17 to 0.32 with increasing σ_n' between 40 and 300 MPa [*Moore et al.*, 2004], a range that is nearly identical to that of montmorillonite (fig. 11.9). The differences between dry and water-saturated frictional strengths of the other two important serpentine minerals, lizardite and antigorite, are smaller than found

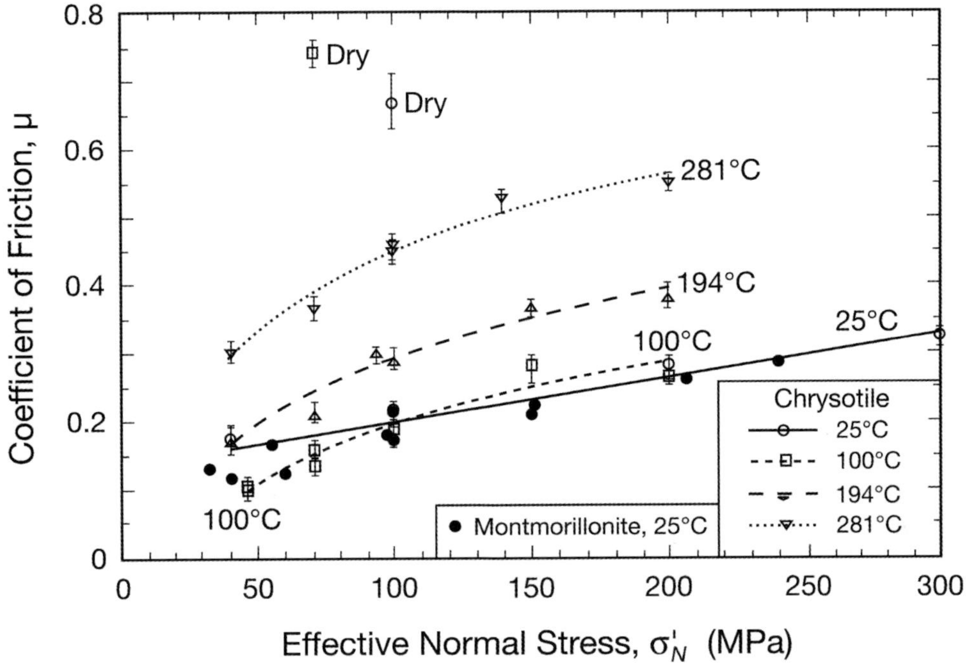

Figure 11.9 Variation of μ of the serpentine mineral chrysotile with temperature and σ_n', from *Moore et al.* [2004]. All but the two experiments labelled "Dry" were run under fluid-saturated conditions. The water-saturated, equilibrated montmorillonite friction data at $\sigma_n' > 30$ MPa from figure 11.7b are plotted for comparison. Room-temperature chrysotile and montmorillonite strengths are nearly identical over much of the range of effective normal stresses tested.

for chrysotile, and these two minerals are characterized by somewhat smaller rates of increase of μ with increasing σ_n' [*Moore et al.*, 1997]. The combined data of *Horn and Deere* [1962], *Kenney* [1967], and *Moore and Lockner* [2004] suggest that μ also increases with σ_n' for water-saturated muscovite, biotite, chlorite, kaolinite, and talc.

Other Experiments

The data set from figure 11.1 is replotted in figure 11.10a with the field of water-saturated, equilibrated montmorillonite strengths from figure 11.7b, labeled "Saturated," superimposed on it. Another field labeled "Dry Strength" at $\mu \geq 0.7$ is defined by the dry experiment from figure 4a and the dry value of μ for montmorillonite reported by *Morrow et al.* [2000]. The granite/granite sample from figure 11.4b that was dried at 125°C plots with the dry samples despite the introduction of water to the top of the upper granite cylinder (fig. 11.3). In compiling the available friction data on rocks and rock-forming

Figure 11.10 Interpretations of the data presented in figure 11.1, plotting (a) apparent coefficient of friction against apparent effective normal stress and (b) shear stress versus apparent effective normal stress. When thoroughly dried, montmorillonite is strong ($\mu \geq 0.7$). The water-saturated, equilibrated field from figure 11.7b is considered to define the true strength of wet montmorillonite. Those reported values of μ intermediate between the dry and wet strengths are labeled "Partially Saturated," and those with apparent friction coefficients lower than the equilibrium value at a given effective normal stress are labeled "Overpressured" (see text). In the overpressured field the reported effective normal stress may be higher than the true value, because of excess fluid pressures that are not accounted for. Key: solid green squares, this study; red and yellow vertical bars, see figure 11.8; D, experiments run on reportedly dried montmorillonite samples (drying procedures vary); solid circles, Na-montmorillonite; open circles, Ca-montmorillonite, with colors: green, *Morrow et al.* [2000]; dark blue, *Morrow et al.* [1992]; light green, *Morrow et al.* [1982]; aqua, *Bird* [1984], where I, II, and III experiments were run on samples equilibrated at relative humidities specified by the numbers in parentheses; brownish orange, *Summers and Byerlee* [1977a, 1977b], for which samples labeled G were dried for a longer time than those labeled S and samples labeled w were run wet; yellow, *Wang et al.* [1980]; black, *Saffer et al.* [2001]; dark gray, *Saffer and Marone* [2003]; red, *Brown et al.* [2003]; magenta, *Logan and Rauenzahn* [1987]; light gray, *Shimamoto and Logan* [1981]; lavender, D. A. Lockner and H. Tanaka (manuscript in preparation, 2006).

minerals, *Byerlee* [1978] found that, with the major exception of the sheet-silicate minerals, $\mu = 0.80–0.85$ at normal stresses ≤200 MPa, and μ tended to decrease somewhat at higher stresses. These trends, known as Byerlee's law, are included in figure 11.10a and they should provide the upper limit for dry montmorillonite strength. For reference, a horizontal line was added at $\mu = 0.06$, the lower limit of reported values. These data (excluding the experiments using saline pore fluids) are recast in figure 11.10b in a plot of shear stress versus apparent effective normal stress, with the dry and saturated fields represented by dashed green lines.

A few data points from studies not included in figure 11.7b plot in the saturated field of figure 11.10a, but most fall well outside it. As discussed below, the data that plot between the dry and water-saturated regions are interpreted to represent a "partially saturated" state in which the operative friction mechanisms may be a mixture of those characteristic of dry and water-saturated gouge. Those data plotting between the saturated field and the $\mu = 0.06$ line may have developed fluid overpressures ("overpressured") that were not alleviated over the duration of the experiments.

Partially Saturated Montmorillonite exposed to room air incorporates atmospheric water onto its crystal surfaces and interlamellar spaces. As the vapor

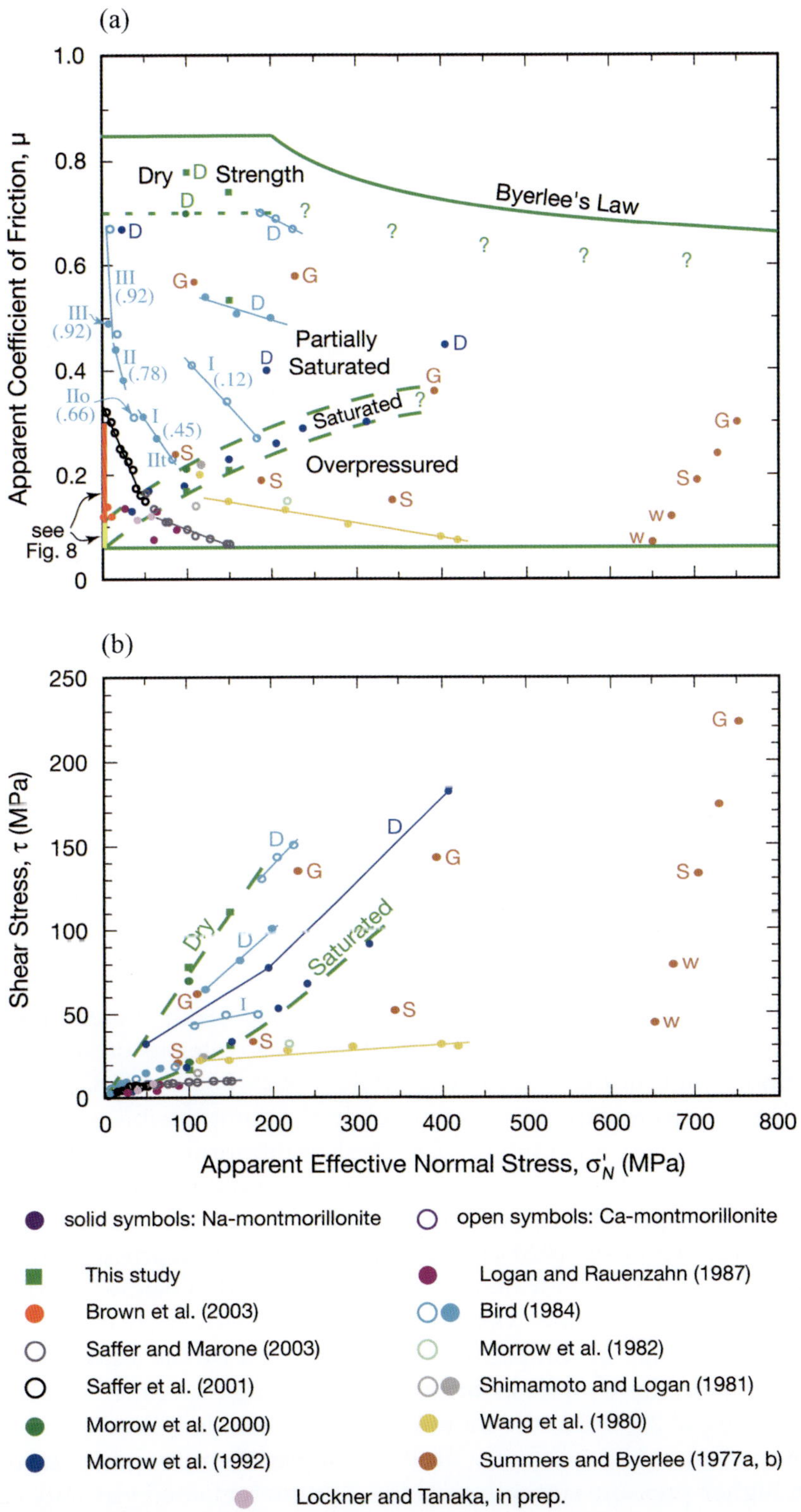
(a)
Apparent Coefficient of Friction, μ
Dry D Strength
Byerlee's Law
Partially Saturated
Saturated
Overpressured
see Fig. 8
(b)
Shear Stress, τ (MPa)
Dry
Saturated
Apparent Effective Normal Stress, σ'N (MPa)
solid symbols: Na-montmorillonite
open symbols: Ca-montmorillonite
This study
Brown et al. (2003)
Saffer and Marone (2003)
Saffer et al. (2001)
Morrow et al. (2000)
Morrow et al. (1992)
Logan and Rauenzahn (1987)
Bird (1984)
Morrow et al. (1982)
Shimamoto and Logan (1981)
Wang et al. (1980)
Summers and Byerlee (1977a, b)
Lockner and Tanaka, in prep.

pressure of air in contact with a montmorillonite grain rises, the interlayer-water content increases in discontinuous steps, whereas the amount of water adsorbed onto the crystal surfaces increases continuously up to its saturation level [*Newman*, 1987]. With respect to the interlayer water, the transition from a one- to a two-layer structure in Na-montmorillonite occurred at a relative humidity, p/p_o between 0.5 and 0.6, and a uniform three-layer structure at $p/p_o > 0.95$ [*Newman*, 1987]. *Ormerod and Newman* [1983] determined that the water sorption on the external surfaces of montmorillonite is similar to that on reference oxides [*Hagymassy et al.*, 1969]. The equivalent of 1 monolayer is adsorbed onto the external surfaces of those oxides at p/p_o between 0.1 and 0.25, 2 at $p/p_o \approx 0.6$, 3 at $p/p_o \approx 0.85$, and about six monolayers at saturation [*Hagymassy et al.*, 1969].

At a specific vapor pressure, the total amount of water taken on by a smectite clay is a function of the mineral composition and particle size, the size controlling the proportion of surface to interlamellar area [*Newman*, 1987]. Because of its typically very fine grain size [e.g., *Güven*, 1988; *Tournassat et al.*, 2003] and correspondingly large proportion of external surface area, much of the water adsorbed by montmorillonite is external rather than interlayer water [*Newman*, 1987]. The very high specific surface area of montmorillonite compared to other minerals (e.g., 700–800 m^2/g for montmorillonite compared to 100–200 m^2/g for illite and only 5–20 m^2/g for kaolinite [*Fripiat*, 1965]) also means that a room-dry montmorillonite sample will have a significantly larger adsorbed-water content than an illite or kaolinite sample stored at the same room humidity. As a result, montmorillonite and other smectite clays may be much more sensitive than other phyllosilicates to all features of the experimental assembly and procedures that can affect the pore-fluid system.

As a room-dry montmorillonite gouge is compacted, the amount of void space decreases and the water adsorbed onto the surfaces fills a greater proportion of the remaining space. At a sufficiently high normal stress the gouge may become fully saturated [*Wang et al.*, 1980]. Potentially, montmorillonite could also lose some of its interlayer water to unsaturated pore space under high normal loads [*Bird*, 1984, *Colten-Bradley*, 1987]. *Wang et al.* [1980] introduced this hypothesis to explain the compressibility and strength data they obtained on cylinders of packed montmorillonite. They found an abrupt decrease in the rate of compaction of the montmorillonite cylinders at ~200 MPa, corresponding to a volumetric strain of ~12%. On the basis of their estimates of initial porosity (≈20%) and measurements of adsorbed-water content (≈5–7 wt %) of the montmorillonite, they concluded that below 200 MPa much of the decrease in volume was because of the elimination of empty pores and that the change in compressibility was caused by the attainment of water saturation in the sample. In subsequent shearing experiments at confining pressures below 200 MPa, the montmorillonite cylinders showed a small net decrease in volume, consistent with the collapse of empty pores. In contrast, during higher-pressure strength tests the volume remained virtually constant to ~90% of peak strength [*Wang et al.*, 1980]. The exact frictional mechanisms operative in a partially saturated gouge are not known, but because of the presence of some unfilled void space, they may possibly be a mixture of those characteristic of dry and water-saturated conditions.

On the basis of the hypothesis of *Wang et al.* [1980], a montmorillonite sample equilibrated at high relative humidity should reach water saturation at lower applied normal stresses than one equilibrated at a lower humidity. Conversely, at a given normal stress in an experiment conducted under partially saturated conditions, the montmorillonite gouge equilibrated at a higher relative humidity should have somewhat lower frictional strength. This latter relation has been demonstrated experimentally for the serpentine mineral antigorite by *Reinen et al.* [1994], and the values of μ that they obtained for partially saturated antigorite are all intermediate between the dry and water-saturated antigorite-gouge strengths measured by *Morrow et al.* [2000]. *Horn and Deere* [1962] also found substantial decreases in μ with increasing water content between oven-dried (105°C) and water-saturated serpentine as well as muscovite, phlogopite, biotite, chlorite, and talc. In the same way, the difference in strength between the two granite/granite samples in figure 11.4b can be explained by the lower drying temperature of the weaker sample, although the possibility that trace amounts of water reached the weaker gouge layer during the experiment cannot be ruled out.

The various sets of experimental data that plot in the partially saturated region of figure 11.10a are all consistent with this hypothesis. The majority of the unsaturated data points come from the paper by *Bird* [1984], who tested a sodium-rich (solid circles in figure 11.10) and a calcium-rich (open circles) montmorillonite in a triaxial apparatus without applied fluid pressure. The gouge samples were equilibrated at humidity levels that he had determined from X-ray diffraction analysis should yield montmorillonite containing 0, 1, 2, or 3 water interlayers (labeled D, I, II for Na-montmorillonite and IIt and IIo for Ca-montmorillonite, and III, respectively, in fig. 11.10a). From comparison of figures 2, 3, and 9 in *Bird*'s [1984] paper, the relative humidities used to prepare specific samples were as follows: 0.12 = Ca-montmorillonite I; 0.45 = Ca-montmorillonite IIt and Na-montmorillonite I; 0.66 = Ca-montmorillonite IIo; 0.78 = Na-montmorillonite II; and 0.92 = Ca- and Na-montmorillonite III. His thesis was that shear of montmorillonite was concentrated in the interlayers, and strength decreased with increasing hydration of the interlayer. He did not consider the possibility of shear through surface-adsorbed water films. However, the data are completely consistent with the hypothesis proposed here. In all cases the coefficient of friction of montmorillonite gouge equilibrated at a specific humidity level decreases linearly with increasing σ_n'. The trend for the Group I Ca-montmorillonite samples ($p/p_o = 0.12$) intersects the saturated field at a higher normal stress than the Group I Na-montmorillonites and IIt Ca-montmorillonite equilibrated at $p/p_o = 0.45$. The three Group III samples ($p/p_o = 0.92$), both Ca- and Na-montmorillonite, also have comparable strength behavior, and the extrapolated linear trend of the Group III data would intersect the saturated field at the lowest normal stresses of the set. The dry Na-montmorillonite samples are unusual in that they are weaker than both the dry Ca-montmorillonites and the Na-montmorillonites that define the dry strength field in figure 11.10a. Even so, they are considerably stronger than the Group I Ca-montmorillonites ($p/p_o = 0.12$), which were tested over the same range of normal stresses. Na-montmorillonite takes on water more readily than does Ca-montmorillonite [e.g., *Keller et al.*, 1986], and *Bird*'s [1984] dry Na-montmorillonite may have contained traces of adsorbed water.

Saffer et al. [2001] ran direct-shear friction experiments on a room-dry Ca-smectite with 11 wt % water (presumably the nonstructural water content). Their strength data follow a linear trend of decreasing μ with increasing σ_n' similar to the trends of *Bird*'s [1984] data, and they noted a transition in velocity dependence from velocity weakening at velocities <20 μm/s for normal stresses below 30–40 MPa to exclusively velocity-strengthening behavior at higher normal stresses. Their 40–50 MPa results plot within or adjacent to the saturated field, which suggests that the transition in velocity dependence may be a function of the degree of water-saturation of the samples. A shift from velocity-weakening to velocity-strengthening behavior with increasing humidity was reported for antigorite by *Reinen et al.* [1994]. D. A. Lockner and H. Tanaka (manuscript in preparation, 2006) measured exclusively velocity-strengthening behavior for saturated montmorillonite gouge at 40 and 60 MPa σ_n' and velocities in the range 0.0001 to 10 μm/s.

Drying temperatures above 100°C are needed to completely remove the adsorbed water from montmorillonite [e.g., *Mackenzie*, 1957; *Deer et al.*, 1962], and the coefficients of friction of three samples of montmorillonite gouge that *Morrow et al.* [1992] dried at 100°C before testing without applied fluid pressure all plot in the unsaturated rather than the dry field in figure 11.10a. These three data points do not define a linear trend, perhaps reflecting differences in sample treatment after drying.

Summers and Byerlee [1977b] ran two sets of strength experiments on montmorillonite gouge at confining pressures of ~73, 155, 312, and 625 MPa. The samples were vacuum dried at an unspecified temperature, with one group being dried for an unspecified shorter period of time (labeled S) than the other (labeled G). In all cases, the sample with the greater drying time was stronger than the one with the shorter drying time tested at the same confining pressure. The two lowest-pressure G samples and the lowest-pressure S sample plot in the partially saturated field.

Overpressured The development of fluid overpressures was proposed by *Wang et al.* [1980] to explain the lack of change in the shear strength of cylinders of compacted montmorillonite with increasing confining pressure (fig. 11.10b). Their sample configuration did not allow any loss of fluid; brass disks were placed on the ends of the montmorillonite cylinders, and the whole assembly was then covered with liquid rubber. (Note that the data plotted in fig. 11.10 from the *Wang et al.* [1980] paper are the peak strengths of the experiments.) As described previously, the results of compaction tests on the montmorillonite cylinders suggested to *Wang et al.* [1980] that the samples reached fluid-saturated conditions at ≤200 MPa confining pressure. They concluded that further increases in confining pressure were matched by increases in pore pressure to maintain a constant effective pressure. On the basis of their interpretation, the reported effective normal stresses of the data in the overpressured field in figure 11.10a are higher than the true values. In consequence, the coefficients of friction calculated using those numbers (figure 11.10a) would be lower than the true values.

The sample assemblies and test conditions of all the experiments plotting in the overpressured field will favor the development of fluid overpressures. *Saffer and Marone* [2003] extended the velocity-stepping experiments of *Saffer et al.* [2001] to higher normal stresses, using the same Ca-montmorillonite starting material and sliding velocities of 1–200 μm/s. In their direct-shear experimental apparatus the room-dry clay was placed between steel forcing blocks. Although water could escape from the edges of the sample, the time constant for fluid-pressure equilibration in this sample geometry would probably exceed many of the test sequence times.

As a comparison, the decay times for reducing a 1-MPa pore-pressure transient in a clay layer were estimated for the triaxial apparatus in figure 11.3 and the direct-shear setup of *Saffer and Marone* [2003]. For both cases a storage capacity of 10^{-3} MPa^{-1} was assumed, so that an imposed volumetric strain of 10^{-3} would result in a 1-MPa pore-pressure change. For the sawcut triaxial geometry this pressure pulse is dissipated by flow of fluid across the gouge layer to the sandstone driving block (<0.1 cm). Using a permeability of 10^{-6} darcy ($\approx 10^{-18}$ m^2), the time needed to reduce the pore pressure transient across a 0.1-cm gouge layer by 80% is on the order of 1 s [*Carslaw and Jaeger*, 1984, p. 101]. *Saffer and Marone* [2003] did not provide the lateral dimensions of their gouge layers, but *Frye and Marone* [2002] reported a sample size of 10 cm × 10 cm for experiments conducted in the same direct-shear apparatus. For diffusion problems of this type, decay timescales approximately as $(\text{distance})^2$. In the direct shear apparatus, the gouge layer is open to air at the edges, and dissipation of overpressured fluid in the center of the sample would require a radial flow path of ~5 cm. In this case, the time required for a 1-MPa pressure pulse to decay by 80% [*Carslaw and Jaeger*, 1984, p. 200] would be >15 min. Consequently, rapid deformation tests could be complicated by the presence of unknown internal pore pressures. From figure 11.10a, the *Saffer and Marone* [2003] montmorillonite gouge is inferred to have reached saturation at 40–50 MPa σ_n' and to have entered the overpressured field above 60 MPa. The trend of their 60–150 MPa σ_n' data in figure 11.10b parallels that of the *Wang et al.* [1980] experiments at lower shear stresses consistent with the larger adsorbed-water content of their montmorillonite and the fact that *Wang et al.* [1980] reported peak rather than residual strengths.

Summers and Byerlee [1977a, 1977b] used granite forcing blocks for their montmorillonite-gouge strength experiments. The low granite permeability will significantly slow the rate of fluid-pressure equilibration, as illustrated in reverse by the two granite/granite experiments in figure 11.4b. Three of the S (fig. 11.10) samples that *Summers and Byerlee* [1977a, 1977b] tested and one of the G samples plot in the overpressured field. *Summers and Byerlee* [1977b] ran two additional experiments at ≈625 MPa confining pressure on wet (w) montmorillonite samples; that is, the gouge layer was placed on the sawcut surface as a slurry and not dried before use. In addition, *Summers and Byerlee* [1977a] tested an air-dried montmorillonite sample at 625 MPa confining pressure. Of the five 625-MPa experiments (fig. 11.10), the two wet samples are the weakest and the oven-dried sample with the greater drying time (G) the strongest. The overall increase in shear strength with increasing "dryness" of the samples (fig. 11.10b) is consistent with the generation of smaller fluid overpressures in the drier samples.

Logan and Rauenzahn [1987] used Tennessee sandstone forcing blocks for their triaxial friction experiments. Tennessee (Crab Orchard) sandstone has a porosity of 2–3% and a permeability of 4×10^{-19} m^2 at 20-MPa effective confining pressure (D. A. Lockner and C. A. Morrow, manuscript in preparation, 2006), which is close to the permeability of Westerly granite. *Logan and Rauenzahn* [1987] saturated their samples with water for 20 hours prior to conducting velocity-stepping experiments without an imposed fluid pressure. Two data points from experiments at 25- and 50-MPa confining pressure plot within the saturated field (fig. 11.10). Another 50-MPa test was run at such high velocities (maximum of 200 μm/s, which was also the maximum velocity of the *Saffer and Marone* [2003] experiments) that it was completed in ~15 min. This sample was considerably weaker than the other one run at 50 MPa, and *Logan and Rauenzahn* [1987] attributed the low strength to the lack of attainment of pore-pressure equilibrium. Their one experiment at 70 MPa confining pressure also plots in the overpressured field.

Two additional data points in the overpressured area are also from experiments conducted using forcing blocks of Westerly granite [*Morrow et al.*, 1982] and Tennessee sandstone [*Shimamoto and Logan*, 1981].

Effect of Pore-Fluid Composition

Most of the experiments investigating the effect of fluid chemistry on montmorillonite strength have been conducted at low-stress conditions (fig. 11.8). These experiments show an overall increase in μ with increasing salinity, although there is some variability in the data for concentrations between ≈0.05 M and 1 M NaCl that may reflect the very different procedures used to control pore-fluid chemistry in those studies. Saline pore fluids will tend to collapse the double layers of water surrounding the montmorillonite grains, which, in turn, will increase shear strength. A saturated NaCl solution raises μ fourfold to fivefold at these low stresses to a value as high as 0.3, the equivalent of increasing σ_n' to ≈250–300 MPa for a montmorillonite gouge equilibrated with deionized water (fig. 11.7b). *Brown et al.* [2003] ran direct-shear experiments at somewhat higher stresses of 2–30 MPa σ_n' (fig. 11.10a) using seawater as pore fluid, which is roughly equivalent to a 0.5 M NaCl solution. Their results for tests at 0.1 μm/s shear rate plot just above the saturated field in figure 11.10a. The effect of changing salinity is expected to be most pronounced at low effective normal stresses, where very low salinity fluids can result in thick films of structured water (fig. 11.7a).

Suggestions for Future Research

Temperature and Velocity Dependence of Montmorillonite Strength

Enough water-saturated, equilibrated friction experiments have been conducted on montmorillonite to determine the stress dependence of μ between 0.1 and 300 MPa σ_n' at room temperature. Although less work has been done on the velocity dependence of montmorillonite strength, the available evidence from this and previous studies [*Morrow et al.*, 1992; *Brown et al.*, 2003; D. A. Lockner and H. Tanaka, manuscript in preparation, 2006] also suggest that under water-saturated conditions at room temperature, montmorillonite strength consistently increases with increasing velocity. Those data indicating $\Delta\mu \leq 0$ with increasing velocity appear to be associated with dry (fig. 11.4) or unsaturated [*Saffer et al.*, 2001] experimental conditions. D. A. Lockner and H. Tanaka (manuscript in preparation, 2006) discuss the room-temperature velocity dependence of montmorillonite strength in more detail.

In contrast, no friction experiments have been conducted at elevated temperatures within the stability range of montmorillonite. As noted previously, montmorillonite and chrysotile have similar room-temperature strengths, particularly at high σ_n' (fig. 11.9), and *Moore et al.* [1997, 2004] have shown that heated chrysotile undergoes a minimum of strength at T ≈ 100°C and $\sigma_n' \leq 40$ MPa (fig. 11.9). If heated montmorillonite and chrysotile also behave similarly, then montmorillonite could potentially become even weaker at some depth interval in a fault zone. The velocity dependence of μ may also vary with temperature, as was observed with chrysotile [*Moore et al.*, 1997, 2004]. Because of the importance of montmorillonite in subduction- and other fault-zone settings, the temperature dependence of montmorillonite frictional strength over a wide range of velocities needs to be thoroughly investigated. If possible, such hydrothermal strength experiments should be conducted using realistic fluid compositions, as *Brown et al.* [2003] have done in their room-temperature experiments.

Effects of Fluid and Mineral Composition

All of the experiments in figure 11.7b were run on Na-montmorillonite. The nature of the interlayer cation might be expected to have little effect on montmorillonite strength, but the strength behavior of Ca- and Na-montmorillonite has been directly compared under water-saturated, equilibrated conditions only in the investigation conducted by *Kenney* [1967] at 0.1 MPa effective normal stress. In identical experiments run using pure water as pore fluid, he found that the coefficient of friction of Ca-montmorillonite was about twice the value for Na-montmorillonite [*Kenney*, 1967]. However, the strength of the Ca-montmorillonite was essentially unaffected by the use of a saturated solution of $CaSO_2$ as pore fluid [*Kenney*, 1967], whereas that of Na-montmorillonite doubled when deioned water was replaced by ≈0.5 M NaCl solution (fig. 11.8). The very low-stress conditions of the experiments may have influenced the effect of fluid chemistry on Na-montmorillonite strength, and experiments at higher effective stresses more appropriate for fault zones need to be conducted. For example, *Alcantar et al.* [2003] found that at normal loads >10 MPa, the thickness of a water film between two muscovite sheets increases with increasing salinity. In their experiments, NaCl solutions had a markedly different effect on water-film thickness than $CaCl_2$ solutions.

Kenney [1967] tested Ca-montmorillonite in a Ca-rich fluid and Na-montmorillonite in Na-rich fluid. Similarly, *Brown et al.* [2003] used Na-rich seawater for their strength experiments on Na-montmorillonite. Fluid chemistry, in particular the cation species in solution [e.g., *Alcantar et al.*, 2003], may have an effect on the strength of the smectite clays and should be investigated. Such studies would have application to montmorillonite-rich rock units cut by and incorporated into fault zones in which the migrating fluids have very different chemistry from those present when the clays formed. Changing fluid chemistry with time during the earthquake cycle may also influence frictional behavior.

Because of differences in their crystal structures and compositions [*Deer et al.*, 1962], the frictional strengths of other smectite clays may differ somewhat from that of montmorillonite. On the basis of their ideal compositions, the predicted values of μ at 100 MPa σ_n' for beidellite and (1) nontronite, (2) hectorite, and (3) saponite are indicated by the open squares in figure 11.6. Of these, the combination of negative tetrahedral-sheet charge and dioctahedral character should make beidellite and nontronite the strongest of the smectite minerals, whereas the trioctahedral character and lack of tetrahedral-sheet charge should make hectorite the weakest. Montmorillonite and saponite are expected to have similar strengths that at 100 MPa σ_n' may be ≈50% lower than for beidellite and nontronite (fig. 11.6). These predictions can be readily tested in laboratory friction experiments.

The Smectite-to-Illite Transition

The smectite-to-illite replacement reaction has been documented in numerous fault-zone settings [e.g., *Tribble*, 1990; *Vrolijk and van der Pluijm*, 1999; *Brown et al.*, 2003], and it has been hypothesized to influence fault-zone rheology in various ways. Laboratory experiments designed to investigate the implications of this reaction for fault strength typically are conducted on mixtures of montmorillonite and disaggregated illite-rich shales [e.g., *Morrow et al.*, 1992; *Brown et al.*, 2003; *Saffer and Marone*, 2003]. Although such mixtures may provide a reasonable estimate of the strength of a mixed-layer illite-smectite clay, at least one comparative study using natural mixed-layer clays should be conducted to verify this assumption. The reason for doing this is that some mechanisms proposed for the illitization of smectite clays involve the formation of intermediate clay structures with different chemical and crystallographic properties and, in consequence, potentially different frictional behavior. *Boles and Franks* [1979] considered the chemical environment for the replacement reaction to be a closed system with enough K available in detrital K-feldspar and mica grains for easy acquisition by the smectite, and K-Na exchange may have preceded Al-Si exchange. They proposed that the newly formed illite occurs as separate packets rather than in a sequence of alternating layers with smectite. Such reaction products should be adequately represented in strength experiments by mixtures of illite and smectite. In contrast, *Bell* [1986] proposed that Al from the octahedral sheet of smectite first replaces Si in the tetrahedral sheet to form an illite-like tetrahedral sheet. He also considered the reaction environment to be an open system in which many partly altered smectites contain high-charge layers but lack the K to convert them completely to illite. Such clays would have a high swelling potential indicating a large percentage of smectite, but their frictional strengths might be more similar to illite than smectite clays, because of the large tetrahedral-sheet charge (fig. 11.6).

The smectite-to-illite reaction has been proposed as the cause of the onset of seismicity at shallow depths in subduction zones [*Vrolijk*, 1990]. However, the smectite-clay content of sediments entering different subduction zones is highly variable [e.g., *Hyndman et al.*, 1997], and smectite abundance can also vary substantially along strike within a subduction zone [*Underwood*, 2002]. Because of this, a more general process than the replacement of smectite clay by illite is required to explain earthquake generation at relatively shallow depths in subduction zones. *Moore and Saffer* [2001] emphasized that the temperature range of the smectite-to-illite transition is characterized by a variety of diagenetic to low-grade metamorphic reactions and processes that will tend to lithify and thereby strengthen the sediments [see also *Brown et al.*, 2003]. As one of the possible reactions involved in this lithification process, understanding the overall smectite-to-illite reaction is important in terms of the changes in mineral assemblage, porosity structure, and fabric development and their combined effect on density, permeability, strength, and sliding stability of the sediments. For example, the formation of illite from smectite and K-feldspar yields SiO_2 as a reaction product [e.g., *Abercrombie et al.*, 1994] that likely will be deposited in the pore spaces of the sediment, affecting strength and perhaps sliding stability. Experiments designed to study the evolution of the rheological properties of marine sediments under hydrothermal conditions should provide important information about the onset of seismicity in subduction zones.

Conclusions

Our work on the frictional behavior of thoroughly dry and water-saturated montmorillonite, combined with the hypotheses of *Wang et al.* [1980], provides a possible framework for reconciling the disparate friction data published to date on that important mineral. These interpretations should be tested further under well-controlled experimental conditions to determine their validity. Nevertheless, this work demonstrates that in order to advance our understanding of the frictional properties of montmorillonite and other smectite clays under conditions applicable to fault zones, future research should be conducted only on fluid-saturated samples using experimental assemblies and procedures that will ensure that the fluid pressures remain equilibrated. Study of the interplay of temperature, mineral and fluid chemistry, and effective stress in influencing smectite-clay frictional behavior may be critical to proper understanding of the role of these minerals in fault zones.

References

Abercrombie, H. J., I. E. Hutcheon, J. D. Bloch, and P. de Caritat (1994), Silica activity and the smectite-illite reaction, *Geology, 22*(6), 539–542.

Alcantar, N., J. Israelachvili, and J. Boles (2003), Forces and ionic transport between mica surfaces: Implications for pressure solution, *Geochim. Cosmochim. Acta, 67*, 1289–1304.

Bell, T. E. (1986), Microstructure in mixed-layer illite/smectite and its relationship to the reaction of smectite to illite, *Clays Clay Miner., 34*, 146–154.

Bird, P. (1984), Hydration-phase diagrams and friction of montmorillonite under laboratory and geologic conditions, with implications for shale compaction, slope stability, and fault gouge, *Tectonophysics, 107*, 235–260.

Bish, D. L., and R. F. Giese Jr. (1981), Interlayer bonding in IIb chlorite, *Am. Mineral., 66*, 1216–1220.

Bleam, W. F. (1990), The nature of cation-substitution sites in phyllosilicates, *Clays Clay Miner., 38*, 527–536.

Boles, J. R., and S. G. Franks (1979), Clay diagenesis in Wilcox sandstones of southeast Texas: Implications of smectite diagenesis on sandstone cementation, *J. Sediment. Petrol., 49*, 55–70.

Brown, K. M., A. Kopf, M. B. Underwood, and J. L. Weinberger (2003), Compositional and fluid pressure controls on the state of stress on the Nankai subduction thrust: A weak plate boundary, *Earth Planet. Sci. Lett., 214*, 589–603.

Byerlee, J. (1978), Friction of rocks, *Pure Appl. Geophys., 116*, 615–626.

Capet, X., H. Chamley, C. Beck, and T. Holtzapffel (1990), Clay mineralogy of sites 671 and 672, Barbados Ridge accretionary complex and Atlantic abyssal basin: Paleoenvironmental and diagenetic implications, *Proc. Ocean Drill. Program Sci. Results, 110*, 345–363.

Carslaw, H. S., and J. C. Jaeger (1984), *Conduction of Heat in Solids*, 2nd ed., 510 pp., Clarendon, Oxford, U.K.

Chamley, H., J. P. Cadet, and J. Charvet (1986), Nankai trough and Japan trench late Cenozoic paleoenvironments deduced from clay mineralogical data, *Initial Rep. Deep Sea Drill. Proj., 87*, 633–641.
Colten-Bradley, V. A. (1987), Role of pressure in smectite dehydration — Effects on geopressure and smectite-to-illite transformation, *Am. Assoc. Petrol. Geol. Bull., 71*, 1414–1427.
Deer, W. A., R. A. Howie, and J. Zussman (1962), *Rock-Forming Minerals*, vol. 3, *Sheet Silicates*, 270 pp., John Wiley, Hoboken, N. J.
Fripiat, J. J. (1965), Surface chemistry and soil science, in Experimental Pedology, *Proceedings of the Eleventh Easter School in Agricultural Science*, edited by E. G. Hallsworth and D. V. Crawford, pp. 3–13, Butterworths, London.
Frye, K. M., and C. Marone (2002), Effect of humidity on granular friction at room temperature, *J. Geophys. Res., 107*(B11), 2309, doi:10.1029/2001JB000654.
Giese, R. F., Jr. (1978), The electrostatic interlayer forces of layer structure minerals, *Clays Clay Miner., 26*, 51–57.
Giese, R. F., Jr. (1980), Hydroxyl orientations and interlayer bonding in amesite, *Clays Clay Miner., 28*, 81–86.
Grim, R. E., and N. Güven (1978), Bentonites. Geology, mineralogy, properties and uses, *Dev. Sediment, 24*, 256 pp.
Güven, N. (1988), Smectites, in hydrous phyllosilicates (exclusive of micas), *Rev. Mineral, 19*, 497–559.
Güven, N. (1992), Molecular aspects of clay-water interactions, in *Clay-Water Interface and Its Rheological Implications*, edited by N. Güven, *Clay Min. Soc. Workshop Lect.*, *4*, 1–79.
Hagymassy, J., Jr., S. Brunauer, and R. S. Mikhail (1969), Pore structure analysis by water vapor adsorption I. t-curves for water vapor, *J. Colloid Interface Sci., 29*, 485–491.
Horn, H. M., and D. U. Deere (1962), Frictional characteristics of minerals, *Geotechnique, 12*, 319–335.
Hyndman, R. D., M. Yamano, and D. A. Oleskevich, (1997), The seismogenic zone of subduction thrust earthquakes, *Island Arc, 6*, 244–260.
Israelachvili, J. N., P. M. McGuiggan, and A. M. Homola (1988), Dynamic properties of molecularly thin liquid films, *Science, 240*, 189–191.
Keller, W. D., R. C. Reynolds, and A. Inoue (1986), Morphology of clay minerals in the smectite-to-illite conversion series by scanning electron microscopy, *Clays Clay Miner, 34*, 187–197.
Kenney, T. C. (1967), The influence of mineral composition on the residual strength of natural soils, in *Proceedings of the Geotechnical Conference on Shear Strength Properties of Natural Soils and Rocks, Oslo, Norway 1967*, pp. 123–129, Norwegian Geotechnical Institute, Oslo.
Liechti, R., and M. D. Zoback (1979), Preliminary analysis of clay gouge from a well in the San Andreas fault zone in central California, *Proceedings of Conference VIII, Analysis of Actual Fault Zones in Bedrock, U. S. Geol. Surv. Open File Rep., 79-1239*, 269.
Logan, J. M., and K. A. Rauenzahn (1987), Frictional dependence of gouge mixtures of quartz and montmorillonite on velocity, composition, and fabric, *Tectonophysics, 144*, 87–108.
Lupini, J. F., A. E. Skinner, and P. R. Vaughan (1981), The drained residual strength of cohesive soils, *Geotechnique, 31*, 181–213.
Mackenzie, R. C., ed. (1957), *The Differential Thermal Analysis of Clays*, 456 pp., Mineral Soc., London.
Maio, C. Di, and G. B. Fenelli (1994), Residual strength of kaolin and bentonite: The influence of their constituent pore fluid, *Geotechnique, 44*, 217–226.
Moore, D. E., and D. A. Lockner (2004), Crystallographic controls on the frictional behavior of dry and water-saturated sheet-structure minerals, *J. Geophys. Res., 109*, B03401, doi:10.1029/2003JB002582.
Moore, D. E., D. A. Lockner, M. Shengli, R. Summers, and J. D. Byerlee (1997), Strengths of serpentinite gouges at elevated temperatures, *J. Geophys. Res., 102*, 14,787–14,801.

Moore, D. E., D. A. Lockner, H. Tanaka, and K. Iwata (2004), The coefficient of friction of chrysotile gouge at seismogenic depths, *Int. Geol. Rev., 46*, 385–398.

Moore, J. C., and D. Saffer (2001), Updip limit of the seismogenic zone beneath the accretionary prism of southwest Japan: An effect of diagenetic to low-grade metamorphic processes and increasing effective stress, *Geology, 29*(2), 183–186.

Moore, R. (1991), The chemical and mineralogical controls upon the residual strength of pure and natural clays, *Geotechnique, 41*, 35–47.

Morrow, C. A., Z. Bo-Chong, and J. D. Byerlee (1986), Effective pressure law for permeability of Westerly granite under cyclic loading, *J. Geophys. Res., 91*, 3870–3876.

Morrow, C. A., L. Q. Shi, and J. D. Byerlee (1982), Strain hardening and strength of clay-rich fault gouges, *J. Geophys. Res., 87*, 6771–6780.

Morrow, C., B. Radney, and J. Byerlee (1992), Frictional strength and the effective pressure law of montmorillonite and illite clays, in *Fault Mechanisms and Transport Properties of Rocks*, edited by B. Evans and T.-F. Wong, pp. 69–88, Elsevier, New York.

Morrow, C. A., D. E. Moore, and D. A. Lockner (2000), The effect of mineral bond strength and adsorbed water on fault gouge frictional strength, *Geophys. Res. Lett., 27*, 815–818.

Newman, A. C. D. (1987), The interaction of water with clay mineral surfaces, in *Chemistry of Clays and Clay Minerals, Mineral. Soc. Monogr.*, vol. 6, edited by A. C. D. Newman, pp. 237–274, John Wiley, Hoboken, N. J.

Ohtani, T., K. Fujimoto, H. Ito, H. Tanaka, N. Tomida, and T. Higuchi (2000), Fault rocks and past to recent fluid characteristics from the borehole survey of the Nojima fault ruptured in the 1995 Kobe earthquake, southwest Japan, *J. Geophys. Res., 105*, 16,161–16,171.

Ormerod, E. C., and A. C. D. Newman (1983), Water sorption on Ca-saturated clays: II. Internal and external surfaces of montmorillonite, *Clay Miner., 18*, 289–299.

Paterson, M. S. (1978), *Experimental Rock Deformation—The Brittle Field*, 254 pp., Springer, New York.

Pudsey, C. J. (1984), X-ray mineralogy of Miocene and older sediments from Deep Sea Drilling Project Leg 78A, *Init. Rep. Deep Sea Drill. Proj., 78A*, 325–342.

Reinen, L. A., J. D. Weeks, and T. E. Tullis (1994), The frictional behavior of lizardite and antigorite serpentinites: Experiments, constitutive models, and implications for natural faults, *Pure Appl. Geophys., 143*, 317–358.

Renard, F., and P. Ortoleva (1997), Water films at grain-grain contacts: Debye-Hückel, osmotic model of stress, salinity, and mineralogy dependence, *Geochim. Cosmochim. Acta, 61*, 1963–1970.

Rosenqvist, I. T. (1984), The importance of pore-water chemistry on mechanical and engineering properties of clay soils, *Phil. Trans. R. Soc. London, Ser. A, 311*, 369–373.

Saffer, D. M., and C. Marone (2003), Comparison of smectite- and illite-rich gouge frictional properties: Application to the updip limit of the seismogenic zone along subduction megathrusts, *Earth Planet. Sci. Lett., 215*, 219–235.

Saffer, D. M., K. M. Frye, C. Marone, and K. Mair (2001), Laboratory results indicating complex and potentially unstable frictional behavior of smectite clay, *Geophys. Res. Lett., 28*, 2297–2300.

Shimamoto, T., and J. M. Logan (1981), Effects of simulated clay gouges on the sliding behavior of Tennessee sandstone, *Tectonophysics, 75*, 243–255.

Skempton, A. W. (1964), Long-term stability of clay slopes, *Geotechnique, 14*, 77–101.

Skempton, A. W. (1985), Residual strength of clays in landslides, folded strata and the laboratory, *Geotechnique, 35*, 3–18.

Skipper, N. T., K. Refson, and J. D. C. McConnell (1993), Monte Carlo simulations of Mg- and Na-smectites, in *Geochemistry of Clay-Pore Fluid Interactions*, edited by D. A. C. Manning, P. L. Hall, and C. R. Hughes, pp. 40–61, CRC Press, Boca Raton, Fla.

Summers, R., and J. Byerlee (1977a), A note on the effect of fault gouge composition on the stability of frictional sliding, *Int. J. Rock Mech. Min. Sci., 14*, 155–160.

Summers, R., and J. Byerlee (1977b), Summary of results of frictional sliding studies, at confining pressures up to 6.98 kb, in selected rock materials, *U.S. Geol. Surv. Open File Rep., 77–142*, 129 pp.
Tournassat, C., A. Neaman, F. Villéras, D. Bosbach, and L. Charlet (2003), Nanomorphology of montmorillonite particles: Estimation of the clay edge sorption site density by low-pressure gas adsorption and AFM observations, *Am. Mineral., 88*, 1989–1995.
Tribble, J. S. (1990), Clay diagenesis in the Barbados accretionary complex: Potential impact on hydrology and subduction dynamics, *Proc. Ocean Drill. Program Sci. Results, 110*, 97–110.
Underwood, M. B. (2002), Strike-parallel variations in clay minerals and fault vergence in the Cascadia subduction zone, *Geology, 30*(5), 155–158.
Vrolijk, P. (1990), On the mechanical role of smectite in subduction zones, *Geology, 18*(8), 703–707.
Vrolijk, P., and B. A. van der Pluijm (1999), Clay gouge, *J. Struct. Geol., 21*, 1039–1048.
Wang, C.-Y., N.-H. Mao, and F. T. Wu, (1980), Mechanical properties of clays at high pressure, *J. Geophys. Res., 85*, 1462–1468.
Wu, F. T. (1978), Mineralogy and physical nature of clay gouge, *Pure Appl. Geophys., 116*, 655–689.
Wu, F. T., L. Blatter, and H. Roberson (1975), Clay gouges in the San Andreas fault system and their possible implications, *Pure Appl. Geophys., 113*, 87–95.
Zoback, M. D., and J. D. Byerlee (1975), Permeability and effective stress, *Am. Assoc. Petrol. Geol. Bull., 59*, 154–158.

12

Fault Friction and the Upper Transition from Seismic to Aseismic Faulting

Chris Marone and Demian M. Saffer

Abstract

This paper discusses the processes and mechanical factors that define the seismogenic zone and the updip transition from seismic to aseismic faulting. We review friction laws for granular and clay-rich fault zones and discuss them in the context of the mechanics of stick-slip and the requirements for instability. Seismogenic faulting is driven by the unstable release of stored elastic energy, and thus explanations of the upper stability transition must be couched in terms of fault-zone rheology. We summarize several observations that define the seismogenic zone and its upper boundary, including microseismicity, earthquake afterslip, and the spatial distribution of seismic moment release during large earthquakes. We focus on two hypotheses for the updip limit of seismic faulting. The clay mineral hypothesis posits that a thermally driven transition from a smectite to an illite structure produces a transition from aseismic to seismic behavior. The fault gouge lithification hypothesis posits that the stability transition reflects a threshold consolidation, or lithification, state and a switch from pervasive to localized shear. Existing laboratory data, which cover a limited range of conditions, indicate that the smectite-illite transformation results in an increase in friction but not a transition from stable to unstable behavior. It appears that other depth- and temperature-dependent processes may play an important role in altering the frictional properties of subduction faults and establishing the updip seismic limit.

Introduction

Plate boundaries can be divided into three main zones as a function of increasing depth: an aseismic updip zone, the seismogenic zone, and a deep aseismic zone. Identifying the transitions between these zones and the processes controlling their locations are key goals in understanding the mechanics of slip along subduction zone megathrusts. In this paper, we focus on the updip limit of the seismogenic zone and discuss that stability transition in the context of fault-zone frictional rheology.

Subduction zone megathrusts host a wide range of fault behaviors including interseismic creep, slow earthquakes, earthquakes with normal (fast) rupture velocity, tsunamogenic earthquakes, postseismic slip, and fault healing [*Satake,* 1994; *Draggert et al.,* 2001, *Miller et al.,* 2002; *McGuire and Segall,* 2003]. These observations indicate that the limits of the seismogenic zone must be carefully defined in both space and time. For example, the upper limit of unstable faulting as defined by microearthquakes may not necessarily coincide with the updip limit of tsunamogenic faulting or the updip limit of seismic moment release in large earthquakes [e.g., *Marone and Scholz,* 1988; *Davis et al.,* 1994]. Comparison between the data sets that define the seismogenic zone represents an important opportunity to test hypotheses for the updip stability transition.

Two main hypotheses for the updip stability transition have been proposed. The clay mineral hypothesis posits that a thermally driven transition from a hydrous (smectite) to an anhydrous clay structure (illite) produces a transition from aseismic to seismic behavior. This hypothesis explicitly assumes that smectite exhibits stable, velocity-strengthening frictional behavior and that illite exhibits the necessary conditions for unstable behavior, which (in the context of rate/state friction) means velocity-weakening behavior. The fault gouge lithification hypothesis posits that the stability transition is the result of a change from pervasive shear accommodation within a fault zone, in which grain crushing and consolidation are important, to localized shear within highly consolidated (lithified) fault-zone material. We review and discuss each hypothesis in the context of laboratory friction data and constitutive laws for unstable faulting.

We summarize data from laboratory experiments on the frictional behavior of clays and granular fault gouge. Existing data show that the transformation of a smectite clay structure to an illite structure results in an increase in frictional strength but does not result in a transition from stable frictional behavior to potentially unstable behavior. Existing data support the fault gouge lithification hypothesis as an explanation for the stability transition at the updip limit of the seismogenic zone. However, the available data are limited, and at this time the clay mineral hypothesis cannot be ruled out. In addition, some of the potential consequences of smectite transformation, such as silica cementation and decreased fluid content, may contribute to the transition to consolidated behavior. We suggest that depth- and temperature-dependent processes, such as cementation, consolidation, and slip localization, may play an important role in changing the frictional properties of subduction zone faults and that these processes, in addition to clay mineralogy, should be the focus of future investigation.

Hypotheses for Stability Transitions That Define the Seismogenic Zone

While it is generally agreed that the downdip limit of interplate seismicity is controlled by either (1) the onset of crystal plasticity and associated creep behavior at ~350°–450°C, with the exact temperature depending upon mineralogy [e.g., *Brace and Kohlstedt,* 1980; *Sibson,* 1982; *Scholz,* 2002; *Hyndman et al.,* 1997; *Peacock and Hyndman,* 1999] or (2) the intersection of the subducting slab with the fore-arc mantle wedge in colder subduction zones [e.g., *Peacock and Hyndman,* 1999]; there is little consensus regarding processes that control the updip limit of seismogenic faulting [e.g., *Moore and Saffer,* 2001].

It is widely held that the updip onset of seismogenic behavior in subduction zones is tied to the transformation of stably sliding smectite clay to illite. The popularity of this hypothesis is due to (1) studies indicating that the thermally driven dehydration of smectite to illite is completed by temperatures of 120°–150°C [e.g., *Pytte and Reynolds,* 1988], (2) laboratory studies that demonstrate a difference in frictional strength between smectite and illite, and (3) correlation between the location of the 150°C isotherm along subduction faults predicted by thermal models and the observed seismic front along the Nankai, Cascadia, South Alaska, and Chile subduction zones [*Hyndman et al.,* 1997; *Oleskevich et al.,* 1999].

An alternative hypothesis, offered by *Marone and Scholz* [1988] to explain the onset of seismicity at depths of 3–5 km on strike-slip faults (and adapted by *Davis et al.* [1994] for accretionary prisms), is that the updip limit reflects a threshold lithification state within fault gouge. This model recognizes that fault gouge is unconsolidated at shallow depths and that shear is accommodated primarily by distributed deformation within the fault zone. At a critical depth, which depends on fault-slip rate, mineralogy, and geothermal gradient, fault gouge reaches a threshold lithification state and deformation becomes localized. Whereas a poorly consolidated fault gouge exhibits properties of a granular material, highly consolidated and lithified gouge behaves, essentially, as a frictional solid, which may be cut by subfaults. This hypothesis is supported by laboratory observations linking brittle deformation and localized slip with stick-slip behavior and distributed deformation with stable frictional sliding [e.g., *Marone et al.,* 1992; *Beeler et al.,* 1996; *Marone,* 1998].

As a third hypothesis, *Moore and Saffer* [2001] argue that the depth of the inferred 150°C isotherm corresponds to a wide range of diagenetic processes and increasing effective stress associated with declining fluid overpressure, which, collectively, may explain the transition to unstable slip. This hypothesis shares key features of the critical lithification concept. A fourth hypothesis can be invoked on the basis of the transition from stable to unstable frictional behavior as a function of increased normal stress. Many frictional systems tend to exhibit stable sliding at low normal load and a transition to stick-slip behavior above a critical load [*Bowden and Tabor*, 1950; *Rabinowicz*, 1956; *Dieterich*, 1986; *Scholz*, 1988]. *Dieterich* [1986] showed how this could be applied to faulting, in terms of loading stiffness and elastic interaction. However, several lines of evidence show that this is not a valid explanation of the upper seismic stability transition, including the depth-frequency distribution of seismicity and its variation with fault-slip rate and the mode of faulting [*Sibson*, 1982; *Marone and Scholz*, 1988; *Scholz*, 2002].

Friction Mechanics and Stability of Faulting

Fault Strength Versus Stability

The stability of tectonic faulting is determined by fault-zone rheology and continuum interactions between the fault zone and its surroundings. In the simplest form, seismogenic faulting is driven by the unstable release of stored elastic energy as a fault zone weakens. Thus fault weakening is a necessary condition for instability, and any hypothesis for a stability transition from stable to unstable faulting must include a mechanism for fault-zone weakening. A second requirement for seismic faulting is the existence of stored elastic energy. A fault cannot grow dynamically without an energy source, and this fact leads to a common misunderstanding about the role of frictional strength in earthquake faulting.

It is often assumed that strong faults, with greater elastic energy stored locally, will exhibit unstable slip, whereas weak faults will creep and exhibit aseismic slip. In fact, frictional instability is dictated by (often small) changes in strength rather than the absolute strength. To first order, instability is independent of fault strength (see discussion by *Beeler* [this volume]). While it is true that earthquake nucleation is unlikely on a fault with nearly zero strength, the converse is not true: strong faults are no more likely to be unstable than weak faults.

Friction Laws for Stick-Slip Versus Stable Sliding

It has long been recognized that stick-slip frictional instability represents a useful analog for shallow (<50 km) earthquakes [e.g., *Bridgeman*, 1936; *Bowden and Tabor*, 1960], and from these earliest times friction laws were sought to explain unstable sliding. The earliest friction laws lacked predictive power regarding the distinction between stable and unstable sliding, but the introduction of a slip-weakening distance by *Rabionowicz* [1951, 1956] with later important work by others [e.g., *Ida*, 1972; *Palmer and Rice*, 1973] changed that. Rabinowicz's experiments on the nature of static friction and the transition from static to dynamic friction led to a slip-weakening friction law (equation (12.1)) and the ability to link (1) weakening on a frictional contact and (2) elastic interaction between the contact and its continuum surroundings.

$$\begin{aligned} \mu &= \mu_s - (\mu_s - \mu_d)u/L \quad & u \le L \\ \mu &= \mu_d & u > L \end{aligned} \tag{12.1}$$

Here, μ_s is the coefficient of static friction, μ_d is the coefficient of dynamic friction, u is slip, and L is the slip-weakening distance.

Modern friction constitutive laws have moved beyond slip weakening to consider time- and slip-dependent restrengthening and variations in friction with slip rate. These laws can reproduce, in a simplified manner, the entire seismic cycle including earthquake nucleation, dynamic rupture, postseismic slip, and interseismic healing. As a result, testable hypotheses for the seismic cycle and earthquake data can be couched in terms of constitutive laws.

Spring-Slider Analog and a Critical Stiffness for Instability The essential elements of an earthquake-like frictional instability can be understood with reference to an elastic element coupled to a massless block that obeys a slip-weakening friction law at its sliding contact (fig. 12.1). We may assume unit

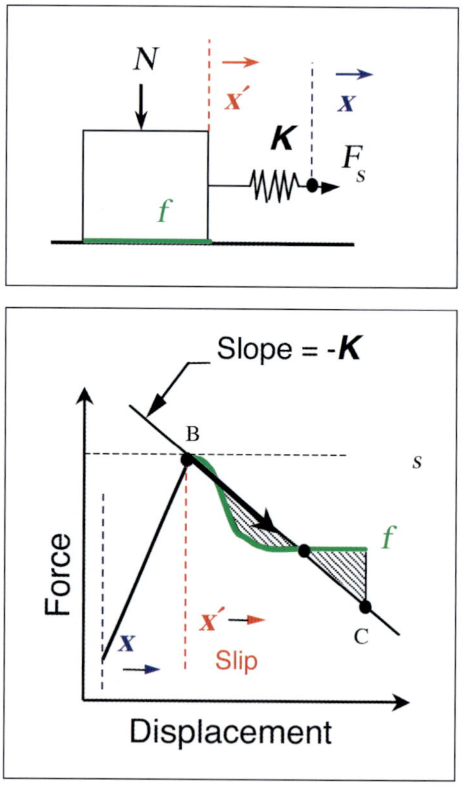

Figure 12.1 Spring-block model and continuum interactions for stick-slip friction. K is the spring constant and μ_s is the coefficient of static friction. Note that frictional resistance f decreases suddenly when the block begins to move at $x' = 0$. Energy is conserved, such that the block continues to slide for some distance beyond the point where f is greater than K.

contact area at the base of the slider, so that forces and stresses can be used interchangeably in the discussion that follows. By assumption of a static coefficient of friction, the fault is locked until the remotely applied force F_s exceeds the friction force (point B in fig. 12.1). At this point, slip may occur by stable sliding, for which the spring force is balanced by the friction force, or by stick-slip. The stability of each slip increment is dictated by the relative magnitude of the change in spring force relative to the change in frictional force. For a slip-weakening friction law (fig. 12.2), the friction force decreases with slip at a rate given by $N(\mu_s - \mu_d)/L$, where N is normal force. Regardless of the friction

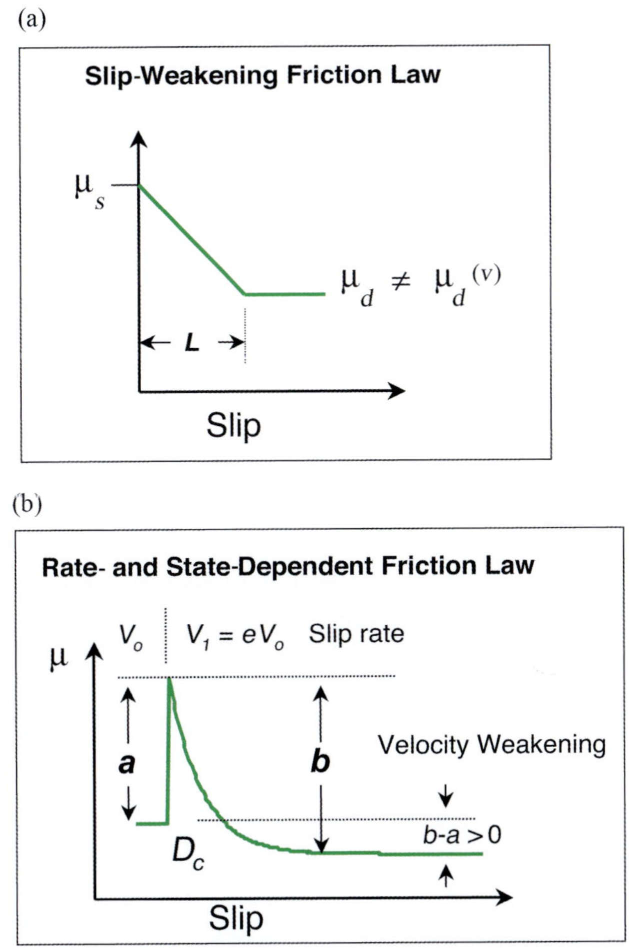

Figure 12.2 Displacement dependence of friction for slip weakening and rate-state friction laws. (a) L is the slip-weakening distance. Note that dynamic friction is independent of slip velocity. (b) Friction response to a sudden perturbation in slip velocity. The friction rate parameter (a-b) describes the dependence of steady-state sliding friction on slip velocity.

behavior, the spring force reduces with slip at a rate given by the spring constant K (fig. 12.1). If the magnitude of the reduction in frictional force exceeds the change in spring force, a force imbalance occurs and stick-slip behavior results.

It is convenient to define the rate of frictional weakening as a critical stiffness K_c:

$$K_c = N(\mu_s - \mu_d)/L \tag{12.2}$$

With the limitations imposed by the simplified friction law and by neglecting mass of the slider block, the spring-block system undergoes stick-slip if $K < K_c$ and stable sliding otherwise.

The spring-slider system of figure 12.1 is clearly woefully simplistic compared to seismogenic plate boundaries. Fault-zone frictional properties are more complex than can be described by a slip-weakening friction law, and elastic interactions between a fault patch that enlarges with slip and its surroundings are not well characterized by a one-dimensional spring constant. Nevertheless, the essential elements that determine the stability of the spring-slider system (e.g., equation (12.2)) form the basis for all other, more complex analyses. For descriptions of recent works on slip stability, including treatments and discussion of the role of inertia and pore fluids, readers are directed to the works and summaries of *Rice* [1983], *Rice and Ruina* [1983], *Rice* [1993], *Roy and Marone* [1996], *Segall and Rice* [1995], *Perfettini et al.* [2001], and *Scholz* [2002].

Rate-State Friction A major limitation of the slip-weakening friction model is that only a single earthquake can be studied. Slip weakening provides no basis for repetitive stick-slip failure, interseismic fault restrengthening or for modeling tsunamogenic earthquakes or creep events. While the updip limit of seismic faulting could be described by slip weakening (the transition from seismic to aseismic faulting would be ascribed to a change from slip weakening to slip strengthening, in which the dynamic friction exceeded the static value), testable, processes-based hypotheses for the limits of the seismogenic zone are beyond the scope of this class of friction laws.

Rate- and state-dependent friction laws (fig. 12.2) were developed to describe the dependence of steady-state sliding friction on slip rate and the variations of friction with contact time and slip [*Dieterich,* 1979; *Ruina,* 1983]. In the context of rate and state constitutive laws, there is no distinction between static and dynamic friction; both are simply special cases of a more generalized view of the variation of friction with slip rate and contact state. The state variable describes physicochemical changes of the frictional contacts (or the fault zone) as a function of contact time, slip history, normal stress, and chemical environment, in addition to other factors [e.g., *Marone,* 1998; *Scholz,* 2002].

Figure 12.2 shows the essential elements of rate and state friction behavior. The coefficient of friction is a function of slip rate V and a state variable.

Variations in friction are described by dimensionless constants a and b and critical slip distance Dc. The parameter a describes the instantaneous change in frictional strength for a sudden perturbation in slip rate, and b describes subsequent time- and slip-dependent evolution (fig. 12.2). The steady-state coefficient of sliding friction varies with the log of slip velocity V as

$$\Delta\mu_{ss} = (a - b)\,\Delta\ln V \tag{12.3}$$

If parameter a exceeds parameter b, steady-state friction increases with sliding velocity and friction is said to exhibit velocity strengthening. This is a type of viscous behavior, as can be seen by replacing velocity with strain rate and noting that shear strength increases with increasing strain rate. Velocity-weakening friction occurs when the evolution parameter b is larger than parameter a. As shown in figure 12.2, velocity weakening implies that the steady-state frictional strength decreases as sliding velocity increases.

For a massless spring-block system (e.g., fig. 12.1) that obeys rate and state friction, the stability of sliding is governed by a critical stiffness K_c given by

$$K_c = \sigma_n\,(b - a)/D_c \tag{12.4}$$

where σ_n is normal stress and we assume unit area for the slider block for comparison to equation (12.2). Unstable, earthquake-like, sliding of a spring-block system occurs if $K < K_c$ and stable sliding occurs otherwise. A discussion of stability analyses and additional background can be found in the work of *Rice and Ruina* [1983].

Equation (12.4) conveys two essential points. First, instability requires velocity weakening; sliding is always stable if parameter a is larger than b, because K is $> K_c$. Second, the sufficient condition for unstable motion involves only the velocity dependence of friction, the critical slip distance, normal stress, and the stiffness of the surroundings K. The degree of potential instability increases for larger values of K_c. As is evident from equations (12.2) and (12.4), in the context of slip-weakening friction laws the distinction between stable and unstable sliding does not depend on the absolute frictional strength.

Frictional Properties of Rock and Granular Fault Gouge Laboratory experiments indicate that the frictional properties of fault rocks range from strongly velocity strengthening to strongly velocity weakening. Their behavior depends on numerous conditions, including temperature, the presence and thickness of a gouge layer, rock or gouge composition, grain size (for gouge), normal stress, humidity, and shear strain. For quartzo-feldspathic rocks, temperatures higher than ~350° C generally result in velocity-strengthening behavior [e.g., *Blanpied et al.*, 1998]. At lower temperatures, many factors have subtle but important effects on the velocity dependence of steady-state friction, and it is difficult to predict the frictional behavior of materials a priori [e.g., *Tullis*,

1988]. Some clays and serpentinite minerals tend to exhibit velocity strengthening [e.g., *Reinen et al.,* 1994]. In general, smoother frictional surfaces tend to exhibit velocity weakening, whereas rougher surfaces and conditions that favor high rates of wear tend to exhibit velocity strengthening [*Dieterich,* 1981; *Marone et al.,* 1990]. Also, rock-on-rock sliding under stresses and temperatures appropriate for the upper crust tends to result in velocity-weakening frictional behavior [*Marone,* 1998].

In experiments on crushed rock and granular fault gouge, velocity strengthening is associated with pervasive strain and conditions that promote comminution, whereas shear localization tends to reduce the parameter (a-b) and result in velocity weakening [e.g., *Beeler et al.,* 1996; *Mair and Marone,* 1999]. Laboratory experiments show that compaction and dilation play a key role in determining the frictional behavior of fault gouge. Under nominally dry conditions the rate of dilatancy with shear strain varies systematically with sliding velocity (or strain rate) and influences the frictional strength and its velocity dependence [*Marone et al.,* 1990].

In the presence of fluids, pore-volume changes may have significant effects on fault strength and the seismic cycle [e.g., *Sleep and Blanpied,* 1992, 1994; *Segall and Rice,* 1995]. The degree of shear localization and the shear band width have been shown to influence the critical slip distance; such that D_c decreases as the degree of shear localization increases [*Marone and Kilgore,* 1993]. Because a reduction in D_c tends to promote instability (equation (12.4)), shear localization plays a key role in fault stability [*Marone,* 1998].

Nucleation of Instability and Minimum Earthquake Size A velocity-weakening fault zone may exhibit seismic slip if the local elastic stiffness K is less than K_c. The one-dimensional stiffness K of a region around a fault patch is proportional to the shear modulus G divided by the patch radius. This means that for a given set of friction parameters, slipping patches larger than a critical size (R_{crit}) will be unstable, where $R_{crit} = CGD_c/\sigma_n(b - a)$ and C is a proportionality constant. Thus the critical patch radius for instability can be related to K_c [*Dieterich,* 1986], and this dimension corresponds to a minimum earthquake size [*Richardson and Jordan,* 2002]. Note that a theoretical earthquake source of dimension R_{crit} will radiate zero seismic energy, and thus, to be detected seismically, an unstable fault patch would need to exceed R_{crit}. Theoretical and laboratory-based estimates of the minimum earthquake size range from magnitude –2 to 0 or slightly larger. While this value is highly uncertain for any given fault zone, it is likely to be much smaller than the earthquake detection threshold for most seismic arrays. Thus, when considering historical catalogs or large, high-resolution seismic data sets, it is reasonable to assume that recordable earthquakes will nucleate in any region that exhibits velocity-weakening frictional behavior, no matter how small the value of $(a - b)$.

Stability Transitions and the Seismogenic Zone

In the context of rate and state friction, a fault zone that exhibits velocity-strengthening frictional behavior will undergo aseismic deformation. This implies that the seismogenic zone as defined by earthquake nucleation maps the transition from stable, velocity-strengthening frictional behavior to unstable, velocity-weakening behavior (fig. 12.3).

Figure 12.3 shows a depth-frequency distribution of seismicity for an active, mature fault zone, in which the upper region is dominantly aseismic. The updip and downdip limits of the seismogenic zone are taken as the depths below which (and above which, respectively) 90% of the seismicity occurs. *Marone and Scholz* [1988] studied continental faults and showed that the upper-stability transition occurs only for mature fault zones. Young faults and those with long recurrence intervals or negligible gouge zones do not exhibit an upper stability transition [*Marone and Scholz,* 1988]. They argued that the updip limit of the seismogenic zone marks a transition from dominantly pervasive shear of poorly consolidated fault gouge to localized slip within indurated gouge. In this model, the transition depth reflects a balance between granulation, driven by fault slip, and lithification driven by pressure, fluids, and elevated temperature. This is the laboratory basis for the critical lithification hypothesis discussed above.

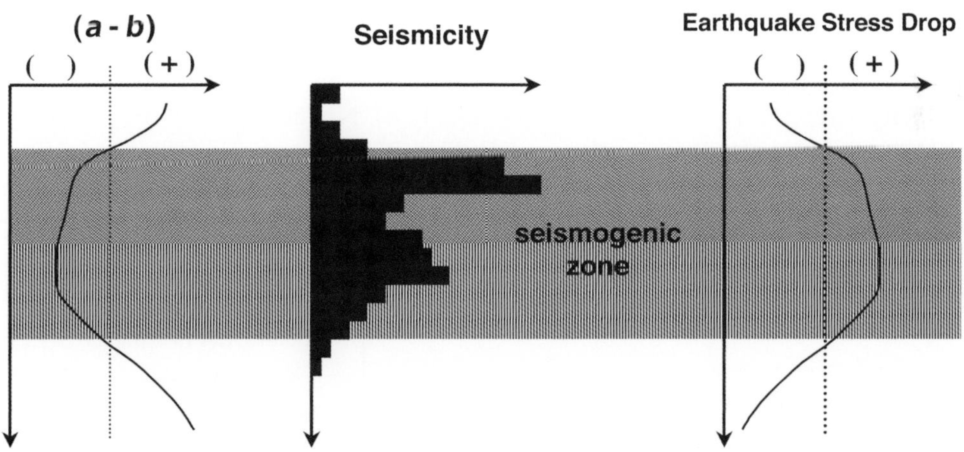

Figure 12.3 Definition of the seismogenic zone. Center panel shows a depth-frequency distribution of seismicity. The updip and downdip limits of the seismogenic zone are taken as the depths below which (and above which, respectively) 90% of the seismicity occurs. Left and right panels show schematic relationship between seismogenic zone and the parameters that dictate earthquake nucleation $(a - b)$ and seismic moment release for a propagating rupture (stress drop). Note that dynamic stress drop is proportional to slip speed, $(a - b)$, and normal stress; hence details of the shape are schematic only.

At a given depth, earthquake nucleation is dictated by the local friction-rate parameter $(a - b)$, normal stress, and the critical stiffness K_c. However, large earthquakes may propagate beyond the limits of the seismogenic zone defined by nucleation (fig. 12.4). Ruptures that are sufficiently energetic may extend outside the region of velocity weakening, and in some cases, such events may rupture completely through the upper stable region. Dynamic rupture of a velocity-strengthening material implies negative stress drop (fig. 12.3) because the local, dynamic fault strength exceeds the initial stress level. Events with significant regions of negative stress drop generally have distinctive characteristics, such as low-average rupture velocity.

The updip limit of seismicity on mature continental faults is generally at a depth of 3–5 km, and this depth is in rough agreement with the upper-stability transition along the southwest segment of the Nankai Trough, which is generally taken to be 4–5 km below the seafloor (fig. 12.4). Modeling of the 1946 Nankai earthquake and the associated tsunamogenic faulting indicates a strong reduction in coseismic slip above 5 km (fig. 12.5). Coseismic rupture along the more vertical, splay fault (within the potentially overconsolidated accretionary wedge sediments) is consistent with strong velocity strengthening along the main décollement (fig. 12.5).

Earthquake Afterslip The 1979 Imperial Valley, California, earthquake ruptured a section of the seismogenic fault overlain by a thick layer of unconsolidated sediment [*Fuis et al.*, 1982]. Both kinematic modeling of the seismic strong motion records [*Archuleta*, 1984] and dynamic modeling of the coseismic rupture [*Quin*, 1990] indicate that the region above 5 km exhibits strong velocity-strengthening characteristics (fig. 12.6). Negative stress drop implies that the dynamic fault strength exceeds the initial stress level; hence coseismic rupture into the upper region results in a sudden increase in stress there. As a consequence, the fault slips there during the period immediately following the

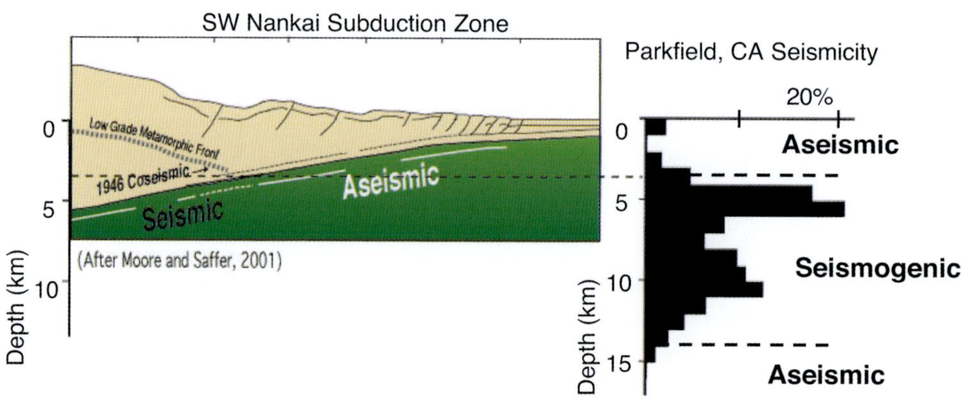

Figure 12.4 Data sets that define the seismogenic zone and the upper transition from unstable to stable frictional behavior for a subduction zone megathrusts (Nankai) and a continental strike-slip fault (Parkfield).

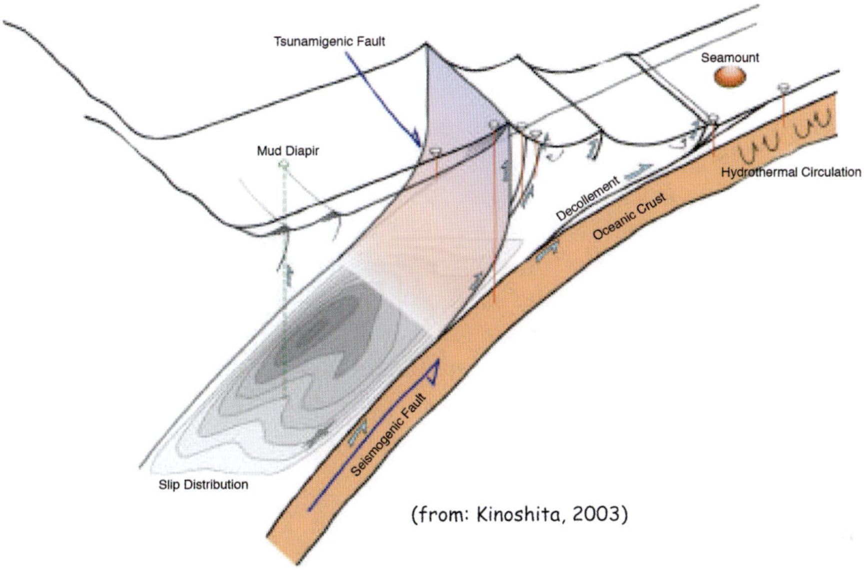

Figure 12.5 Geometry and major faults of the Nankai accretionary prism and subduction zone. Contours show slip distribution for the 1944 and 1946 Nakaido earthquakes as inferred from tsunami data and strong-motion seismic records. Courtesy of M. Kinoshita (personal communication, 2003).

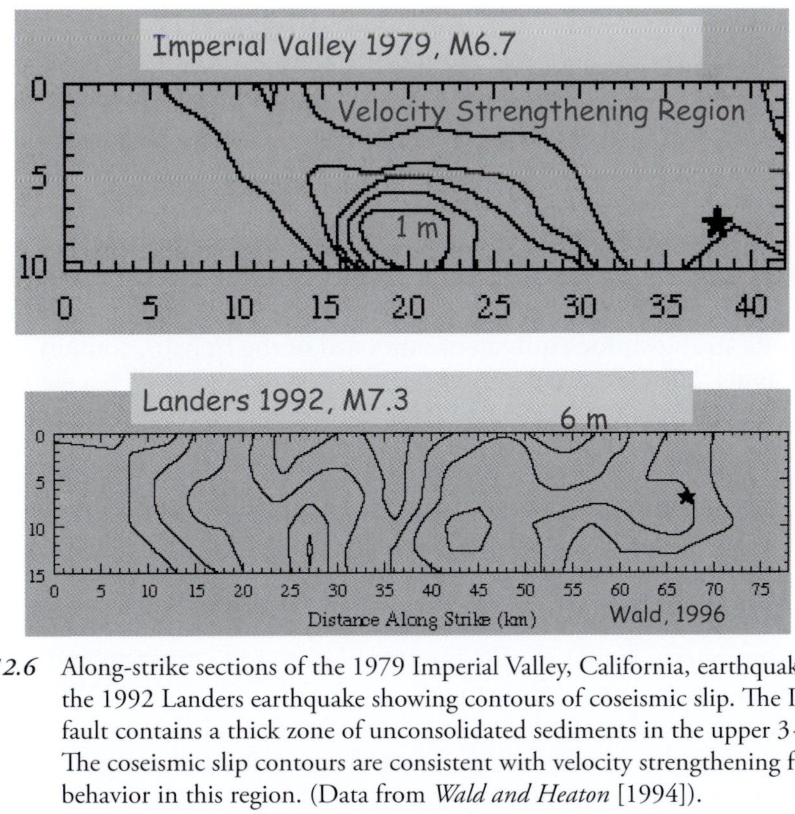

Figure 12.6 Along-strike sections of the 1979 Imperial Valley, California, earthquake and the 1992 Landers earthquake showing contours of coseismic slip. The Imperial fault contains a thick zone of unconsolidated sediments in the upper 3–4 km. The coseismic slip contours are consistent with velocity strengthening frictional behavior in this region. (Data from *Wald and Heaton* [1994]).

earthquake, which is known as earthquake afterslip. *Marone et al.* [1991] discussed a model for earthquake afterslip and found good agreement between the depth of the velocity-strengthening region inferred from rate-state friction laws, using laboratory-derived values of (*a-b*), and independent estimates of the upper-stability transition based on microseismicity [*Doser and Kanamori,* 1986] and crustal structure [*Fuis et al.,* 1982].

A prediction of the afterslip model is that ruptures on immature faults will not exhibit buried coseismic slip or shallow afterslip because the velocity-weakening zone extends to the surface, and no region of negative coseismic stress drop occurs. The 1992 *M* 7.3 Landers earthquake and the 1999 *M* 7.1 Hector Mine earthquake show exactly this behavior (fig. 12.6). Strong rupture extended to the surface and the coseismic slip contours indicate no updip reduction of slip.

Review of Relevant Laboratory Friction Data

Laboratory studies have investigated many factors that influence the frictional properties of fault-zone rocks. These factors include initial sediment composition and grain size distribution [e.g., *Lupini et al.,* 1981; *Logan and Rauenzahn,* 1987; *Bos and Spiers,* 2000, 2002], normal stress and slip velocity [*Marone et al.,* 1990; *Beeler et al.,* 1996; *Mair and Marone,* 1999], variations in mineralogy as might be expected with increasing fault plane depth [e.g., *Saffer and Marone,* 2003], increasing consolidation [*Zhang et al.,* 1993], and cementation.

At subduction zones, megathrusts initiate within sediment of the incoming plate, or are at least bounded by the sediment on one side; the fault gouge is typically derived from these sediments [e.g., *Tobin et al.,* 2001]. The sediment and fault zones dominantly contain a mixture of quartz, clay minerals (typically illite, smectite, chlorite, and kaolinite), and plagioclase. For example, sediments incoming to the southwest Nankai subduction zone, recovered by drilling on Ocean Drilling Program (ODP) Leg 190, sampled at the décollement (or its stratigraphic equivalent outboard of the trench), contain ~37–58% clay minerals, 30–42% quartz, 7–13% plagioclase, and 0–11% calcite [*Moore et al.,* 2001]. Of the clay minerals, 74–83% is illite-smectite; the remainder is dominantly chlorite. With progressive burial and heating, the smectite in these sediments transforms to illite. The clay fraction of some fault gouges is enriched in illite when compared to the mineralogy of wall rocks [*Vrolik,* 1990; *Vrolik and Van der Pluijm,* 1999]. However, the clay mineralogy of subduction zone faults, at least as observed at shallow depths, does not follow this pattern [e.g., *Underwood et al.,* 1993; *Moore et al.,* 2001; *Deng and Underwood,* 2001].

We summarize recent results on granular quartz, smectite and illite clay, and clay-quartz mixtures, focusing on works of *Saffer et al.* [2001], *Saffer and Marone* [2003], and *Mair and Marone* [1999]. The illite shale material discussed here is comparable in composition to some clay-rich fault gouges used in previous experimental work on illite gouges and also to the marine sediments that serve as source material for fault gouge in subduction zones [e.g., *Saffer and Marone*, 2003]. These studies employed the double-direct shear configuration. Layers were sheared under constant normal stress, ranging from 5 to 150 MPa, sliding velocities of 0.1 μm/s to a few millimeters per second, and a nominally dry condition or controlled humidity at room temperature. In all tests, normal stress across the layers was maintained by a fast, hydraulically driven servocontroller, and changes in porosity were monitored continuously as a function of slip and time. Materials consisted of Ca-Smectite clay (obtained as a powder from GSA Resources), illite-shale (obtained from Wards Geological Supply), and quartz powder (F-110, obtained from the US silica company). F-110 is a pure quartz sand (99.8% SiO_2) with minor amounts of Fe_20_3, Al_20_3 (< 0.1% each), and other oxides. Particles are rounded to subangular and the size distribution is as follows: 8% 53–75 μm, 25% 75–106 μm, 44% 106–150 μm, 18% 150–212 μm, 4% 150–212 μm, with <1% beyond the upper and lower limits. Clay particles ranged from 2 to 500 μm. Details of the XRD analyses can be found in the work of *Saffer and Marone* [2003].

Frictional Strength of Illite Shale, Smectite, and Quartz-Smectite Mixtures

Figure 12.7 shows basic stress-strain characteristics for four materials. Quartz and illite shale are frictionally strong and exhibit coefficients of sliding friction of 0.62 and 0.55, respectively. Smectite friction is low, ~0.25 at a normal stress of 20 MPa, and the 50/50 clay-quartz mixture has an intermediate strength, with a friction coefficient of ~0.4. Pure smectite shows a distinct peak stress with subsequent weakening over the shear strain range from 50 to 200% (fig. 12.7a). The other materials obtain a steady-state, residual friction value after shear strains of 3 to 4. The experiments shown in figure 12.7 include sections of constant loading velocity as well as numerous perturbations in the loading rate, discussed below, and these appear as transients and second-order variations in the normalized shear strength.

Frictional strength varies systematically with quartz content in smectite-quartz mixtures (fig. 12.7b). The mixture strengths are somewhat lower than predicted by simply averaging the strength of the end members; e.g., compare the 50/50 mixture with the 90/10 and 10/90 mixtures in figure 12.7b (see discussion of polyphase mixtures in the work of *Beeler* [this volume]). During initial loading at 25 MPa normal stress, layers with greater clay content exhibit a more distinct peak stress (fig. 12.7b). In these experiments the shear load was removed and reapplied at a shear strain of ~2. The reloading curves do not show a peak stress. This is consistent with the interpretation that the peak and subsequent weakening result from rotation and alignment of clay grains [e.g., *Lupini et al.*, 1981; *Saffer and Marone*, 2003]. Clay grains have random orientation when the layers are initially prepared. The application of normal load compacts the gouge layer, but clay grain alignment occurs primarily during shear and is not a reversible process.

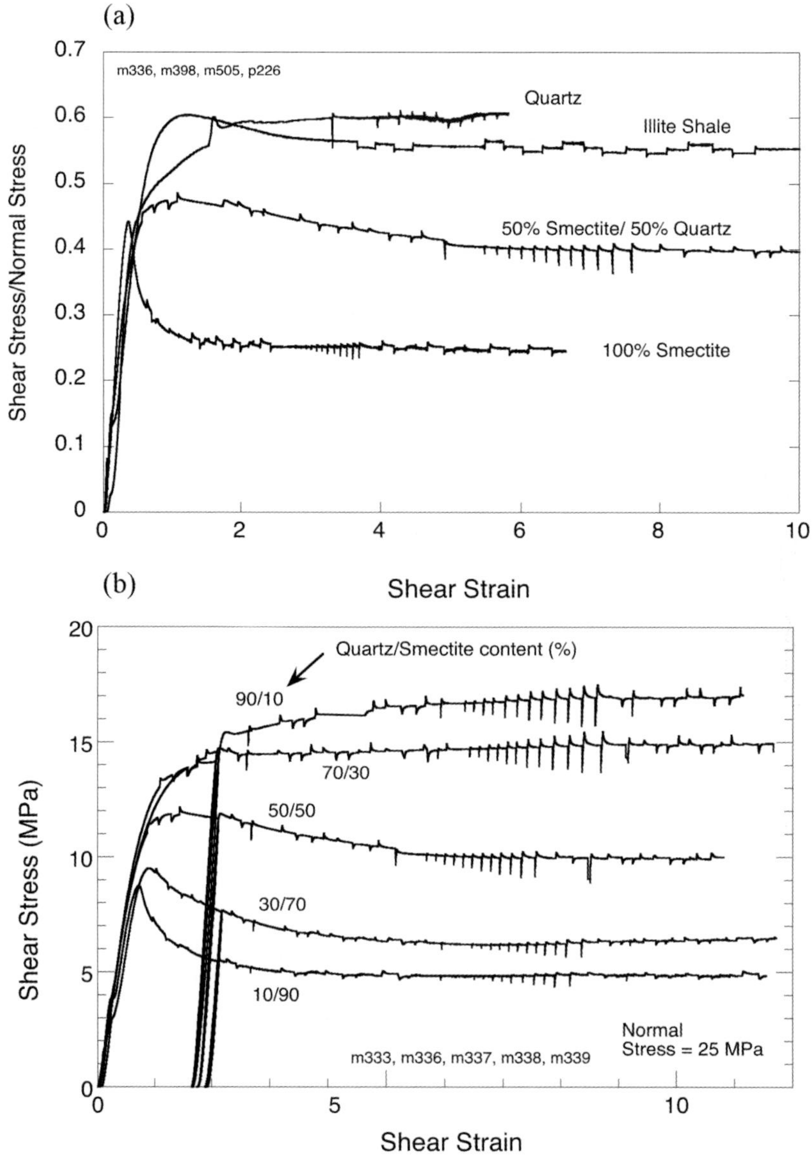

Figure 12.7 Laboratory friction data. (a) Friction (shear stress/normal stress) versus shear strain for four materials. (b) Stress-strain behavior as a function of quartz content in layers of smectite clay. [After *Saffer and Marone*, 2003]

The four materials exhibit a range of failure behaviors as a function of increasing normal stress from 5 to 150 MPa (fig. 12.8). As is typical for gouge, all have cohesion values near zero. For pure smectite gouge the coefficient of sliding friction ranges from 0.07 to 0.30, and decreases with increasing normal stress (fig. 12.8). Smectite gouge undergoes a transition from brittle to pressure insensitive deformation at a normal stress of 35 MPa; this value also marks a change in the angle of internal friction for the 50/50 smectite-quartz mixture.

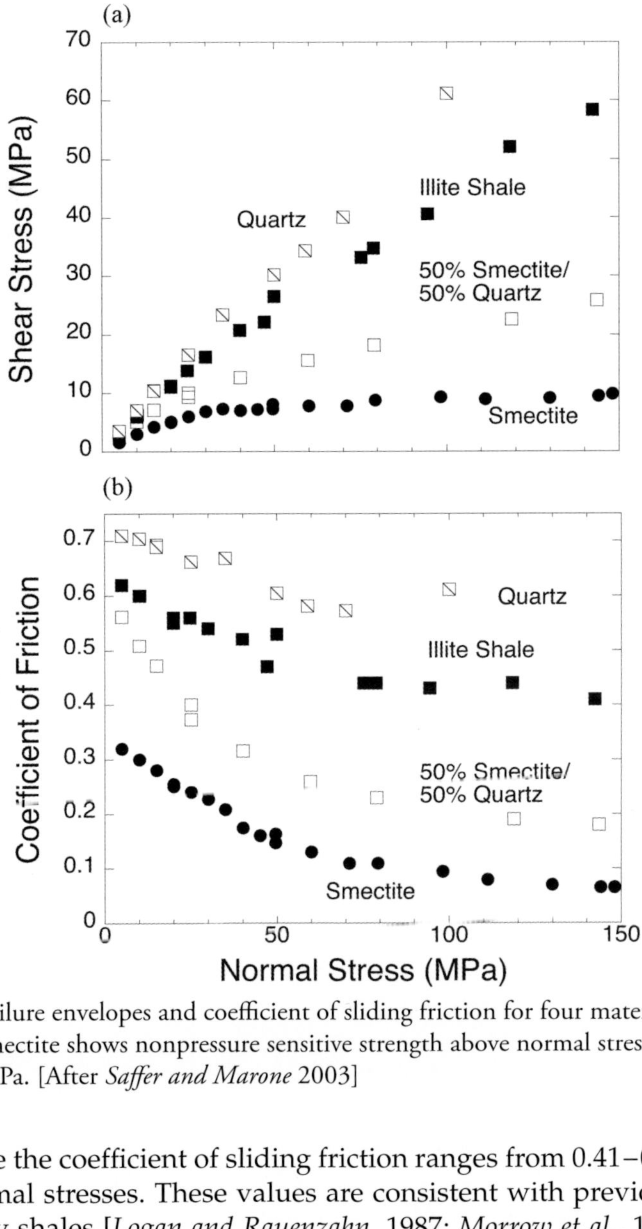

Figure 12.8 Failure envelopes and coefficient of sliding friction for four materials. Note that smectite shows nonpressure sensitive strength above normal stresses of 30–40 MPa. [After *Saffer and Marone* 2003]

For illite shale the coefficient of sliding friction ranges from 0.41–0.63 over the range of normal stresses. These values are consistent with previous work on clays and clay shales [*Logan and Rauenzahn*, 1987; *Morrow et al.*, 1982; *Morrow et al.*, 1992; *Morrow et al.*, 2000]. Illite shale has nearly the same failure envelope as quartz and each exhibits a constant angle of internal friction over this range of normal stresses (fig. 12.8).

Frictional Behavior of Illite Shale, Smectite, and Quartz-Smectite Mixtures

Velocity stepping experiments are done to assess the transient and steady-state frictional response (fig. 12.9). Figure 12.9b gives the full rate and state friction law and the relation to describe elastic coupling between the testing apparatus and shearing layers. Figure 12.9a shows results for velocity stepping tests on layers of illite-shale and indicates velocity-strengthening frictional behavior (i.e., the steady-state coefficient of sliding friction increases with increasing sliding velocity). Equations (12.1), (12.2), and (12.5) (fig. 12.9b) are solved to model the full transient and steady-state frictional response; these models provide good fits to the range of materials discussed here [*Mair and Marone*, 1999; *Saffer and Marone*, 2003].

As shown by *Saffer et al.* [2001], Ca-smectite exhibits velocity-weakening behavior under a limited set of conditions of low normal stress and sliding velocities. Here we focus only on the sliding velocities <20 μm/s (fig. 12.10). Smectite exhibits a clear transition from velocity-weakening at normal stresses

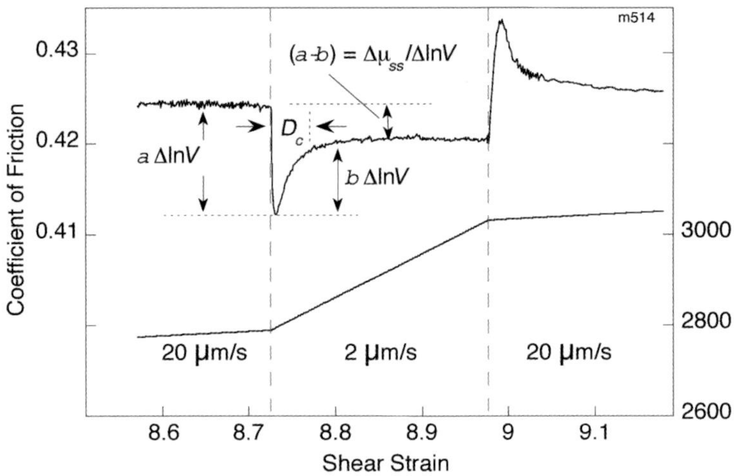

$$(1)\quad \mu(\theta,v) = \mu_0 + a\ln\left(\frac{v}{v_o}\right) + b\ln\left(\frac{v_o\theta}{D_c}\right)$$

$$(2)\quad \frac{d\theta}{dt} = 1-\frac{v\theta}{D_c}$$

$$(3)\quad \theta_{ss} = \frac{D_c}{v}$$

$$(4)\quad -\mu_{ss} = (a-b)\ln\left(\frac{v}{v_o}\right)$$

$$(5)\quad \frac{d\mu}{dt} = k'(v_{lp} - v)$$

Figure 12.9 (a) Rate state friction behavior, showing velocity-strengthening behavior. (b) Friction constitutive equations and relation for elastic interaction with the testing machine.

below 35–40 MPa to velocity strengthening for higher normal stresses. The minimum value of (*a-b*) is = –0.0025 at a normal stress of 5 MPa. Maximum values of (*a-b*) of 0.0015 to 0.002 are observed at normal stresses >75 MPa and are consistent with the value of 0.005 reported by *Morrow et al.* [1992] for similar normal stresses and sliding velocities and under saturated conditions (however, note that those experiments do not reach a steady-state friction condition and thus the validity of the results on velocity dependence are unclear). For a given normal stress and sliding velocity, (*a-b*) does not vary over a range of shear strain from ~3 to 14 [*Saffer and Marone*, 2003].

In contrast, illite shale exhibits only velocity-strengthening behavior over a wide range of normal stresses (5–150 MPa) and sliding velocities (0.1–200 μm/s) (fig. 12.10). Values of (*a-b*) range from ~0.0015 to 0.0035, and unlike smectite, do not vary appreciably with normal stress (fig. 12.8a) or velocity. For illite gouge, values of (*a-b*) are also generally independent of shear strain at normal stresses of 10–50 MPa [*Saffer and Marone*, 2003]. The values of (*a-b*) (0.0015–0.004) reviewed here are consistent with a value of (*a-b*) = 0.005 reported by *Morrow et al.* [1992] for illite shale under normal stresses of 100 and 300 MPa (however, note that those experiments do not reach a steady-state friction condition, and thus the validity of the results on velocity dependence are unclear).

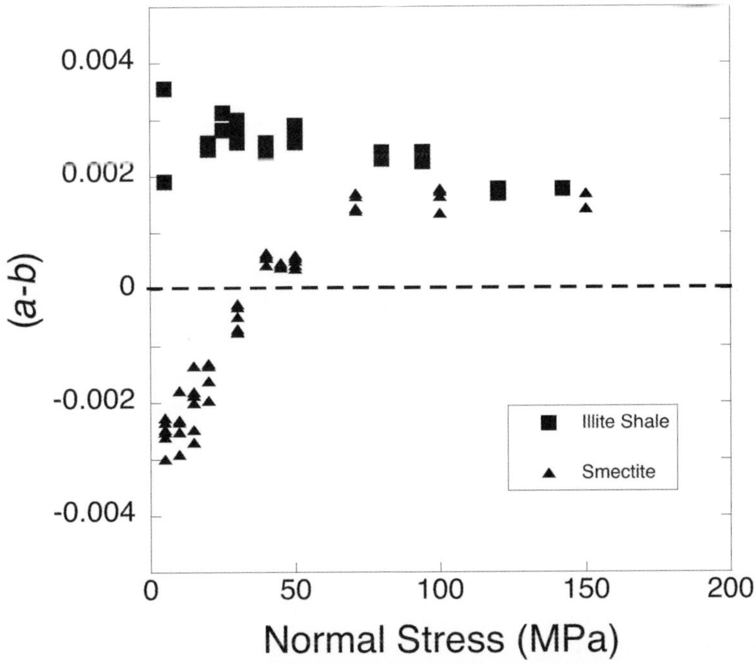

Figure 12.10 Friction-rate parameter for illite-shale and smectite clay. Smectite shows a transition from velocity-weakening frictional behavior at low normal stress to velocity-strengthening behavior at higher normal stresses. Data from *Saffer and Marone* [2003]

The Dieterich rate and state friction law provides good fits to the clay gouge data [*Saffer and Marone*, 2003]. Visual comparisons of data and model are very close, allowing for some misfit associated with data noise. For illite shale gouge the friction parameter a exceeds the evolution parameter b throughout the range of velocities and normal stresses studied, consistent with the results showing steady-state velocity strengthening (figs. 12.9 and 12.10). Both friction parameters for illite show a minimum at normal stresses of 10–20 MPa and maximum values at 5 MPa [*Saffer and Marone*, 2003]. The critical slip distance D_c is 10–30 μm for 3–4 mm thick layers of illite shale, independent of normal stress.

In order to quantify volume changes in gouge layers associated with velocity stepping, layer thickness changes are recorded continuously during shear. The tendency of a gouge layer to dilate or contract with changes in sliding velocity is important for understanding velocity or strain rate dependence of strength and fabric development [e.g., *Mair and Marone*, 1999; *Beeler*, this volume]. The effects are also important for rupture models that integrate friction laws and pore-fluid pressure changes driven by gouge layer volume changes [e.g., *Segall and Rice*, 1995]. We have documented a measure of gouge layer dilatancy, $\alpha = \Delta h/\Delta \log V$, where h is gouge layer thickness and V is sliding velocity [e.g., *Marone and Kilgore*, 1993]. To calculate α, we first detrend the measured layer thickness to account for monotonic thinning owing to simple shear.

For both smectite and illite shale, α is generally positive (indicating dilation with increased sliding velocity), and we observe little variability in α with velocity or shear strain [*Saffer and Marone*, 2003]. For smectite, α ranges from −0.2 to 2.5 μm; values for illite shale range from 0.65 to 3.2 μm. Smectite exhibits some variability in α with normal stress, with higher values at low normal stress, whereas for illite shale values of α are scattered but exhibit no systematic variation over the range of normal stresses investigated.

Existing studies have not investigated the role of saturation or controlled pore pressure within gouge layers. The frictional strength of pure smectite has been shown to decrease with saturation, by as much as a factor of 2 [*Morrow et al.*, 1992, 2000]. In contrast, *Morrow et al.* [1992] showed that the coefficient of sliding friction for illite shale is not appreciably affected by saturation, consistent with the fact that illite is not a hydrous clay mineral. For both smectite and illite the effects of saturation on constitutive parameters, including frictional velocity dependence, are not well characterized, although *Morrow et al.* [1992, 2000] showed that illite shale exhibits velocity-strengthening behavior when saturated, albeit under an extremely limited set of experimental conditions. These observations suggest that saturation state does not affect illite frictional properties significantly, although additional experiments are clearly needed. We also note that α is positive for illite shale, indicating dilatancy with increased sliding velocity. This volume increase should result in further velocity strengthening in saturated gouges, by decreasing pore pressure during slip (dilation), rather than potentially causing weakening by transient pore-pressure increase.

Figure 12.11 shows the friction-rate parameter (*a-b*) for smectite and illite shale [from *Saffer and Marone,* 2003] as a function of depth, assuming an effective stress gradient of 10 MPa/km. The dependence of (*a-b*) on shear strain for quartz [from *Mair and Marone,* 1999] is shown for comparison. For the quartz gouge, shear strain is converted to depth *z*, using a linear relationship between depth and shear strain, under the assumption that because fault displacement along subduction megathrusts increases with depth and distance from the trench, shear strain should increase systematically with depth. In addition, shearing-induced localization and gouge consolidation in laboratory experiments are reasonable proxies for consolidated behavior caused by greater pressures and temperatures at depth. Although the transition depth from velocity strengthening to velocity weakening in figure 12.11 is arbitrary, the form of the curve mimics the thermally driven transition of (*a-b*) documented in several studies [e.g., *Blanpied et al.,* 1995, 1998].

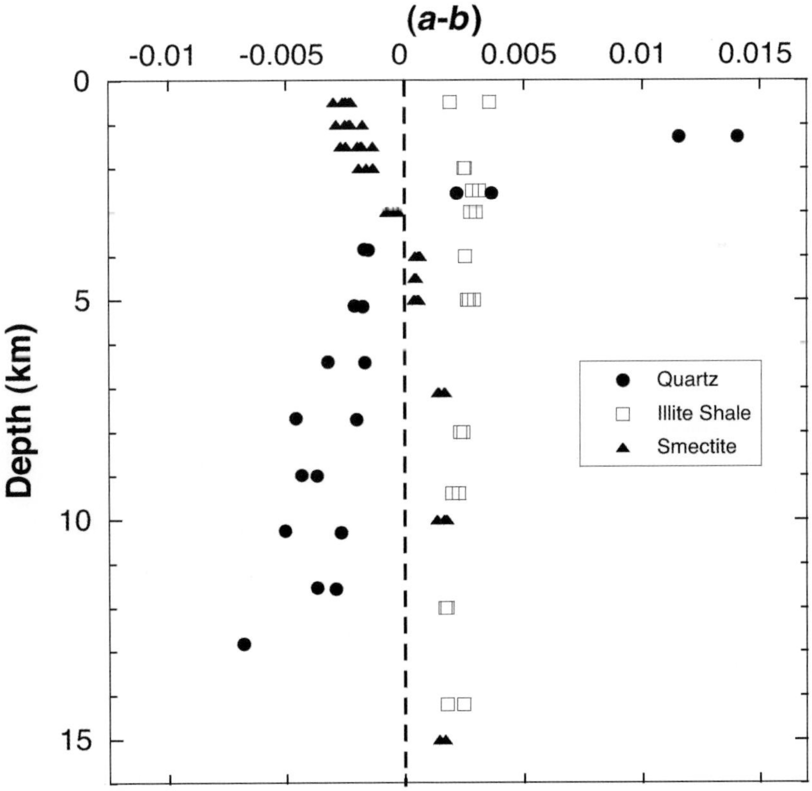

Figure 12.11 Variation of the friction rate parameter with depth for quartz and clays. Shear strain is converted to depth *z*, assuming that because fault displacement along subduction megathrusts increases with depth and distance from the trench, shear strain should increase systematically with depth. Data from *Saffer and Marone* [2003] and *Mair and Marone* [1999].

Conclusions

Contrary to the hypothesized velocity weakening of illite invoked to explain the onset of seismogenic slip at subduction zones, illite shale exhibits only velocity-strengthening behavior over the range of experimental conditions that have been explored to date. Although we cannot rule out the possibility of velocity-weakening behavior at higher temperatures or larger shear strains, these results over a wide range of experimental conditions (normal stress, shear strain, and sliding velocity) are inconsistent with the hypothesis that mineral transformation of stably sliding smectite to illite acts as a sole control on the updip limit of the seismogenic zone.

The results of laboratory experiments on relevant materials to date indicate that the transformation of smectite is unlikely to control the shallow transition from stable sliding to unstable slip along subduction megathrusts. Although it is clear that additional work is needed to rigorously evaluate the clay mineral hypothesis under a wider range of experimental conditions, such as elevated temperature, saturation and controlled pore pressure, and large shear strains, additional evidence from drilling also suggests that smectite transformation is not a universal explanation for the updip limit of seismicity. First, many convergent margins are characterized by little or no smectite in incoming sediments; yet these subduction zones still exhibit a clear updip limit of earthquakes [e.g., *Moore and Saffer,* 2001]. Second, in some cases, such as the southwest Nankai margin, smectite within the incoming sediments is transformed to illite outboard of the trench [e.g., *Steuerer and Underwood,* 2003]; yet the updip extent of microseismicity occurs tens of kilometers landward.

Taken together, the existing evidence suggests that other processes act to control the location of the upper aseismic-seismic transition [e.g., *Moore and Saffer,* 2001; *Saffer and Marone,* 2003]. These processes include the following: increased shear strain with depth along the plate interface and associated increased consolidation state of gouge and sediments (resulting in a transition from distributed to localized shear) [e.g., *Marone and Scholz,* 1988; *Davis et. al,* 1994], quartz cementation and pressure solution that may be associated with clay transformation [*Moore and Saffer,* 2001], and increasing effective stress, which may be mediated by pore pressure variations [*Moore and Saffer,* 2001; *Scholz,* 1998].

Acknowledgments

We thank the organizers of the SEIZE workshop and theoretical institute and the Margins community for feedback and insight on the many factors that influence the seismogenic zone. We are grateful for a thorough and insightful review by Nick Beeler and for the editorial work of Tim Dixon. This work is a result of partial support under grants OCE-0196462 and EAR-0196570, which is gratefully acknowledged. The authors also thank Jennifer Anthony for help with laboratory work.

References

Archuleta, R. J. (1984), A faulting model for the 1979 Imperial Valley earthquake, *J. Geophys. Res.*, *89*, 4559–4585.

Beeler, N. M. (2007), Laboratory-observed faulting in intrinsically and apparently weak materials: Strength, seismic coupling, dilatancy, and pore fluid pressure, this volume.

Beeler, N. M., T. E. Tullis, and J. D.Weeks (1996), Frictional behavior of large displacement experimental faults, *J. Geophys. Res.*, *101*, 8697–8715.

Blanpied, M. L., D. A. Lockner, and J. D. Byerlee (1995), Frictional slip of granite at hydrothermal conditions, *J. Geophys. Res.*, *100*, 13,045–13064.

Blanpied, M. L., C. Marone, D. A. Lockner, J. D. Byerlee, and D. P. King (1998), Quantitative measure of the variation in fault rheology due to fluid-rock interactions, *J. Geophys. Res.*, *103*, 9691–9712.

Bos, B., and C. J. Spiers (2000), Effect of phyllosilicates-on fluid-assisted healing behaviour of gouge-bearing faults, *Earth Planet. Sci. Lett.*, *184*, 199–210.

Bos, B., and C. J. Spiers, (2002), Frictional-viscous flow of phyllosilicate-bearing fault rock: Microphysical model and implications for crustal strength profiles, *J. Geophys. Res.*, *107I*(B2), 2028, 10.1029/2001JB000301.

Bowden, F. P., and D. Tabor. (1950), *The Friction and Lubrication of Solids. Part I.* Clarendon, Oxford, U.K.

Brace, W. F., and D. L. Kohlstedt (1980), Limits on lithospheric stress imposed by laboratory experiments, *J. Geophys. Res.*, *85*, 6248–6252.

Bridgeman, P. W. (1936), Shearing phenomena at high pressure of possible importance to geology, *J. Geol.*, *44*, 653–669.

Davis, D. M., et al. (1994), Porosity loss in the evolution of accretionary wedges; some mechanical and seismic implications, in *Proceedings of the Workshop LXIII, USGS Red-Book Conference on the Mechanical Involvement of Fluids in Faulting, U.S. Geol. Surv. Open File Rep., 94-0228*, 460–465.

Deng, X., and M. B. Underwood (2001), Abundance of smectite and the location of a plate-boundary fault, Barbados accretionary prism, *Geol. Soc. Am. Bull.*, *113*, 495-507.

Dieterich, J. H. (1979), Modeling of rock friction: 1. Experimental results and constitutive equations, *J. Geophys. Res.*, *84*, 2161–2168.

Dieterich, J. H. (1981), Constitutive properties of faults with simulated gouge, in *Mechanical Behavior of Crustal Rocks, Geophys. Monogr. Ser.*, vol. 24, edited by N. L. Carter et al., pp. 103–120, AGU, Washington, D. C.

Dieterich, J. H. (1986), A model for the nucleation of earthquake slip, in *Earthquake Source Mechanics, Geophys. Monogr. Ser.*, vol. 37, edited by S. Das, J. Boatwright, and C. H. Scholz, pp. 37–47, AGU, Washington, D. C.

Doser, D. I., and H. Kanamori (1986), Depth of seismicity in the Imperial Valley region 1977–1983 and its relationship to heat flow, crustal structure, and the October 15, 1979 earthquake, *J. Geophys. Res.*, *91*, 675–688.

Dragert, H., K. Wang, and T. S. James (2001), A silent slip event on the deeper Cascadia subduction interface, *Geophys. Res. Lett.*, *292*, 1526–1528.

Fuis, G. S., W. D. Mooney, J. H. Healy, G. A. McMechan, and W. J. Lutter (1982), Crustal structure of the Imperial Valley region, the Imperial Valley, California, earthquake of October 15, 1979, *U.S. Geol. Surv. Prof. Pap.*, *1254*, 25–49.

Hyndman, R. D., M. Yamano, and D. A. Oleskevich (1997), The seismogenic zone of subduction thrust faults, *Island Arc*, *6*, 244–260.

Ida, Y. (1972), Cohesive force across the tip of a longitudinal-shear crack and Griffith's specific surface energy, *J. Geophys. Res.*, *77*, 3796–3805.

Logan, J. M., and K. A. Rauenzahn (1987), Frictional dependence of gouge mixtures of quartz and montmorillonite on velocity, composition, and fabric, *Tectonophysics*, *144* , 87–108.

Lupini, J. F., A. E. Skinner, and P. R. Vaughan (1981), The drained residual strength of cohesive soils, *Geotechnique*, *31*, 181–213.

Mair, K. and C. Marone (1999), Friction of simulated fault gouge for a wide range of velocities and normal stresses, *J. Geophys. Res. (B), 104*, 28899-28914.

Marone, C. (1998), Laboratory-derived friction constitutive laws and their application to seismic faulting, *Annu. Rev. Earth Planet. Sci., 26*, 643–696.

Marone, C., and B. Kilgore (1993), Scaling of the critical slip distance for seismic faulting with shear strain in fault zones, *Nature, 362*, 618–621.

Marone, C. , and C. H. Scholz (1988), The depth of seismic faulting and the upper transition from stable to unstable slip regimes. *Geophys. Res. Lett., 15*, 621–624.

Marone, C., C. B. Raleigh, and C. H. Scholz (1990), Frictional behavior and constitutive modeling of simulated fault gouge, *J. Geophys. Res., 95*, 7007–7025.

Marone, C., C. H. Scholz, and R. Bilham (1991), On the mechanics of earthquake afterslip, *J. Geophys. Res., 96*, 8441–8452.

Marone, C., B. E. Hobbs, and A. Ord (1992), Coulomb constitutive laws for friction: Contrasts in frictional behavior for distributed and localized shear, *Pure Appl. Geophys., 139*, 195–214.

McGuire, J. J., and P. Segall (2003), Imaging of aseismic fault slip transients recorded by dense geodetic networks, *Geophys. J. Int., 155*, 778–788.

Miller, M. M., T. Melbourne, D. J. Johnson, and W. Q. Sumner (2002), Periodic slow earthquakes from the Cascadia subduction zone, *Science, 295*, 2423–2424.

Moore, G. F., et al. (2001), New insights into deformation and fluid flow processes in the Nankai Trough accretionary prism: Results of Ocean Drilling Program Leg 190, *Geochem. Geophys. Geosyst.*, 2(10), doi:10.1029/2001GC000166.

Moore, J. C., and D. M. Saffer (2001), Updip limit of the seismogenic zone beneath the accretionary prism of southwest Japan: An effect of diagenetic to low-grade metamorphic processes and increasing effective stress, *Geology, 29*(2), 183–186.

Morrow, C. A., L. Q. Shi, and J. D. Byerlee (1982), Strain hardening and strength of clay-rich fault gouges, *J. Geophys. Res., 87*, 6771–6780.

Morrow, C. A., B. Radney, and J. D. Byerlee (1992), Frictional strength and the effective pressure law of montmorillonite and illite clays, in: B. Evans, Ed., *Fault Mechanics and Transport Properties of Rocks*, edited by B. Evans, pp. 69–88, CRC Press, Boca Raton, Fla.

Morrow, C. A., D. E. Moore, and D. A. Lockner (2000), The effect of mineral bond strength and absorbed water on fault gouge frictional strength, *Geophys. Res. Lett., 27*,815–818.

Oleskevich, D. A., R. D. Hyndman, and K. Wang (1999), The updip and downdip limits to great subduction earthquakes: Thermal and structural models of Cascadia, South Alaska, SW Japan, and Chile, *J. Geophys. Res., 104*, 14,965–14,991.

Palmer, A. C. and J. R. Rice, The growth of slip surfaces in the progressive failure of over- consolidated clay, *Proc. Roy. Soc. Lond. A332*, 527-548, 1972.

Peacock, S. M., and R. D. Hyndman (1999), Hydrous minerals in the mantle wedge and the maximum depth of subduction thrust earthquakes. *Geophys. Res. Lett., 26*, 2517–2520.

Perfettini, H., J. Schmittbuhl, J. R. Rice, and M. Cocco (2001), Frictional response induced by time-dependent fluctuations of the normal load, *J. Geophys. Res., 106*, 13,455–13,472.

Pytte, A. M., and R. C. Reynolds (1988), The thermal transformation of smectite to illite, in *Thermal Histories of Sedimentary Basins*, edited by T. H. McCulloh and N. D. Naeser, pp. 133–140, Springer, New York.

Quin, H. (1990), Dynamic stress drop and rupture dynamics of the October 15, 1979 Imperial Valley earthquake, *Tectonophysics, 175*, 93–117.

Rabinowicz, E. (1951), The nature of static and kinetic coefficients of friction, *J. Appl. Phys., 22*, 1373–1379.

Rabinowicz, E. (1956), Stick and slip, *Sci. Am., 194*, 109–118.

Reinen, L. A., J. D. Weeks, and T. E. Tullis (1994), The frictional behavior of lizardite and antigorite serpentinites: Experiments, constitutive models and implications for natural faults, *Pure Appl. Geophys., 143*, 317–358.

Rice, J. R. (1983), Constitutive relations for fault slip and earthquake instabilities, *Pure Appl. Geophys., 121*, 443–475.

Rice, J. R. (1993), Spatio-temporal complexity of slip on a fault, *J. Geophys. Res., 98*, 9885-9907.
Rice, J. R., and A. L. Ruina (1983), Stability of steady frictional slipping, *J. Appl. Mech., 50*, 343–349.
Richardson, E., and T. H. Jordan (2002), Seismicity in deep gold mines of South Africa: Implications for tectonic earthquakes, *Bull. Seismol. Soc. Am., 92* , 1766–1782.
Roy, M., and C. Marone (1996), Earthquake nucleation on models faults with rate and state dependent friction: The effects of inertia, *J. Geophys. Res., 101*, 13,919–13,932.
Ruina, A. (1983), Slip instability and state variable friction laws, *J. Geophys. Res., 88*, 10, 359–10,370.
Saffer, D. M., and C. Marone (2003), Comparison of smectite- and illite-rich gouge frictional properties: Implications for the updip limit of the seismogenic zone along subduction megathrusts, *Earth Planet. Sci. Lett., 215*, 219–235.
Saffer, D. M., K. M. Frye, C. Marone, and K. Mair (2001), Laboratory results indicating complex and potentially unstable frictional behavior of smectite clay, *Geophys. Res. Lett., 28*, 2297–2300.
Satake, K. (1994), Mechanism of the 1992 Nicaragua tsunami earthquake, *Geophys. Res. Lett., 21*, 2519–2522.
Scholz, C. H. (1988), The critical slip distance for seismic faulting, *Nature, 336*, 761–763.
Scholz, C. H. (2002), *The Mechanics of Earthquakes and Faulting*, 2nd ed., Cambridge Univ. Press, New York.
Segall, P., and J. R. Rice (1995), Dilatancy, compaction, and slip instability of a fluid infiltrated fault, J. *Geophys. Res., 100*, 22,155–22,173.
Sibson, R. H. (1982), Fault zone models, heat flow, and the depth distribution of earthquakes in the continental crust of the United States, *Bull. Seismol. Soc. Am., 72*, 151–163.
Sleep, N. H., and M. L. Blanpied (1992), Creep, compaction and the weak rheology of major faults, *Nature, 359*, 687–692.
Sleep, N. H., and M. L. Blanpied (1994), Ductile creep and compaction: A mechanism for transiently increasing fluid pressure in mostly sealed fault zones, *Pure Appl. Geophys., 143*, 9–40.
Steurer, J. F. and M. B. Underwood (2003), Clay Mineralogy of Mudstones from the Nankai Trough Reference Sites 1173 and 1177 and frontal accretionary prism site 11741, edited by H. Mikada, G. F. Moore, A. Taira, K. Becker, J. C. Moore, and A. Klaus, in *Proc. Ocean Drill. Program Sci. Results 190/196*, 22–33.
Tobin, H., P. Vannucchi, and M. Meschede (2001), Structure, inferred mechanical properties, and implications for fluid transport in the décollement zone, Costa Rica convergent margin: *Geology, 29, 10,* 907–910.
Tullis, T. E. (1988), Rock friction constitutive behavior from laboratory experiments and its implications for an earthquake prediction field monitoring program, *Pure Appl. Geophys., 126*, 555–588.
Underwood, M. B., et al. (1993), Sediment geochemistry, clay mineralogy, and diagenesis: A synthesis of data from leg 131, Nankai Trough, *Proc. Ocean Drill. Program Sci. Results, 131*, 343–363.
Vrolijk, P. (1990), On the mechanical role of smectite in subduction zones, *Geology, 18*(8), 703–707.
Vrolijk, P., and B. A. Van der Pluijm (1999), Clay gouge, *J. Struct. Geol., 21* 1039–1048.

Wald, D. J., and T. H. Heaton (1994), Spatial and temporal distribution of slip for the 1992 Landers, California, earthquake, *Bull. Seis. Soc. America, 84*, 668-691.
Zhang, J., D. Davis, and T-F. Wong (1993), The brittle-ductile transition in porous sedimentary rocks: Geological implications for accretionary wedge aseismicity, *J. Struct. Geol., 15*, 819–830.

13 Laboratory-Observed Faulting in Intrinsically and Apparently Weak Materials

Strength, Seismic Coupling, Dilatancy, and Pore-Fluid Pressure

N. M. Beeler

Abstract

Under some circumstances, subduction thrust faulting produces the Earth's largest and most hazardous earthquakes and earthquake-induced tsunamis, but in other localities, large-scale aseismic slip is common and portions of some subduction zones are nearly completely aseismic. Understanding the particular conditions that lead to seismic rather than aseismic fault slip is critical for improving earthquake and tsunami hazard estimates in subduction zones and is a major objective of the NSF-MARGINS initiative. This paper reviews laboratory rock mechanics observations of rock fracture and fault slip relevant to seismic shallow crustal faulting, emphasizing differences between behavior of intrinsically weak and strong faults and identifying the underlying physical processes and conditions that influence fault strength. These laboratory rock friction and rock failure observations, made primarily on strong faults and intact rock, largely suggest that earthquake occurrence is a result of dilatancy. As many subduction zones appear to be profoundly weak, the conditions for seismicity are considered in the context of intrinsic fault weakness, and under circumstances of apparent weakness owing to high pore pressure. The principal conclusion of this review is that in the absence of chemical effects of fluids on fault strength, both intrinsic weakness and apparent weakness resulting from high pore pressure tend to encourage aseismic over seismic slip. However, the sliding stability of intrinsically and apparently weak rocks and minerals is not well studied. The effect of high pore pressure on sliding stability has only been considered for an intrinsically strong, highly dilatant, apparently weak fault. Since many weak phases found in subduction zones have low dilatancy, the effect of high pore pressure on sliding stability is presently unknown. In addition, chemical effects on fault strength are not well studied to date. Many weak phases have strengths that are sensitive to the chemical environment; in particular, clays and other sheet silicate minerals have strengths that are influenced by the presence of water. The role of such effects in seismogenesis is not known at present. Given the present state of knowledge, sliding stability and fault strength cannot be unequivocally related.

Introduction

The San Andreas fault between San Juan Batista and Parkfield and sections of the Hayward-Calaveras fault south of San Francisco Bay are of the few well-characterized locations in the world where slip on large displacement through-going crustal-scale continental faults is predominately aseismic [*Scholz*, 1990]. However, in subduction zones, large-scale aseismic slip is very common, and portions of some subduction zones are nearly completely aseismic [*Peterson and Seno*, 1984]. At the same time, subduction thrust faulting produces the Earth's largest and most hazardous earthquakes [*Kelleher et al.*, 1974] and the Earth's largest earthquake-induced tsunamis [*Furumoto*, 1991]. Understanding the particular conditions that lead to seismic rather than aseismic response is critical for improving earthquake and tsunami hazard estimates in subduction zones and is a major objective of the NSF-MARGINS initiative. At present, seismogenesis in subduction zones is not well understood for any particular subduction thrust. With the exception of temperature [*Goetze and Evans*, 1979; *Brace and Kohlstedt*, 1980; *Kirby*, 1980], seismicity in subduction zones has not been clearly related to specific observables such as strain rate, strain, intrinsic material strength, mineralogy, porosity, or fluid pressure. Though qualitative arguments suggest nearly cohesionless weak accretionary prism sediments are aseismic and that stronger cemented and lithified materials are associated with the onset of seismicity [*Byrne et al.*, 1988; *Moore and Saffer*, 2001], some great subduction zones earthquakes occur on faults that are weaker than the intrisinic strength of the weakest geomaterials. *Wang and He* [1999] estimate a spatially averaged apparent friction $\mu' = 0.05–0.09$ for the Nankai and Cascadia subduction zones; here, $\mu' = \mu(1 - p/\sigma_L)$, μ is friction, the ratio of shear to normal stress and, p is fluid pressure, σ_L is lithostatic pressure (see Background). Weak phases in subduction environments, illite, and chrysotile serpentine, have friction coefficients between 0.2 and 0.45 [*Morrow et al.*, 1992, 2000; *Moore et al.*, 1996; *Moore and Lockner*, 2004]. Saturated montmorillonite can have friction as low as 0.08 at low effective stress [*Logan and Rauenzahn*, 1987], but since montmorillonite is chemically unstable above 125°C, the weakness of the seismic Nankai and Cascadia subduction zones is apparently not due to intrinsic material properties alone. Furthermore, saturated smectite, illite, and serpentine, though pressure dependent, are apparently rheologically ductile in the shallow crust (see Strength and Rate Dependence of Subduction Zone Minerals section below); at least they do not permit seismic stress drops in laboratory tests at natural low loading rates. Thus, on the basis of laboratory observations to date, the Nankai and Cascadia subduction zones are both weaker and more seismic than expected mineral composition permits.

This paper reviews laboratory rock mechanics observations of rock fracture and fault slip relevant to seismic and aseismic shallow crustal faulting, emphasizing differences between behavior of intrinsically weak and strong faults. The physical processes that influence fault strength in the laboratory under shallow crustal conditions are also identified. Representative constitutive equations for rock failure and for slip on preexisting faults are used to illustrate implications of the laboratory observations. These relations are used to develop the standard laboratory-derived conditions for seismic fault slip (instability). The conditions for seismicity are considered in the context of intrinsic fault weakness and under circumstances of apparent weakness due to high pore pressure. A conclusion of this review is that in the absence of chemical effects of fluids on fault strength, both intrinsic weakness and apparent weakness resulting from high pore pressure encourage aseismic over seismic slip. Thus absolute fault strength, apparent strength, and the stability of faulting may correlate in some cases, which might explain or help to explain the spatial and temporal distribution of seismicity in weak-faulting environments.

However, existing theories of frictional instability, based on frictional sliding on preexisting faults [*Dieterich*, 1978b, 1979] and on analysis of the resulting constitutive equations [*Ruina*, 1983, *Rice*, 1983; *Rice and Ruina*, 1983], may not be comprehensive. The experiments that established the condition for unstable slip were done dry at room temperature on artificially well-localized faults in quartzofeldspathic rocks. Inferences of the role of fluid pressure in instability are also based on analysis of these intrinsically strong and highly dilatant rocks. Such laboratory faults are much stronger than many seismic subduction zones and many weak materials have material properties (e.g., molecular bond strengths [*Morrow et al.*, 2000; *Moore and Lockner*, 2004], material anisotropy [e.g., *Mares and Kronenburg*, 1993], and dilatancy [*Escartin et al.*, 1997] that are quite different. In particular, many weak phases found in subduction thrusts have strengths that are significantly influenced by the presence of water and that deform by contact-scale processes that are distinctly different from those occurring in quartzofeldspathic materials. The final section of this paper reviews faulting experiments in intrinsically weak saturated fault zones and the implications of these results for seismicity.

Background: Pressure Dependence

Knowledge of crustal stress, earthquake occurrence and related crustal processes have been guided by laboratory studies in rock mechanics. To begin to put laboratory observations in context, let's briefly illustrate the specific observations and terminology used to define rock failure, fault slip, pressure dependence, and the strength losses of brittle failure (fig. 13.1a) using stress-strain data from a series of experiments at different confining pressure conducted

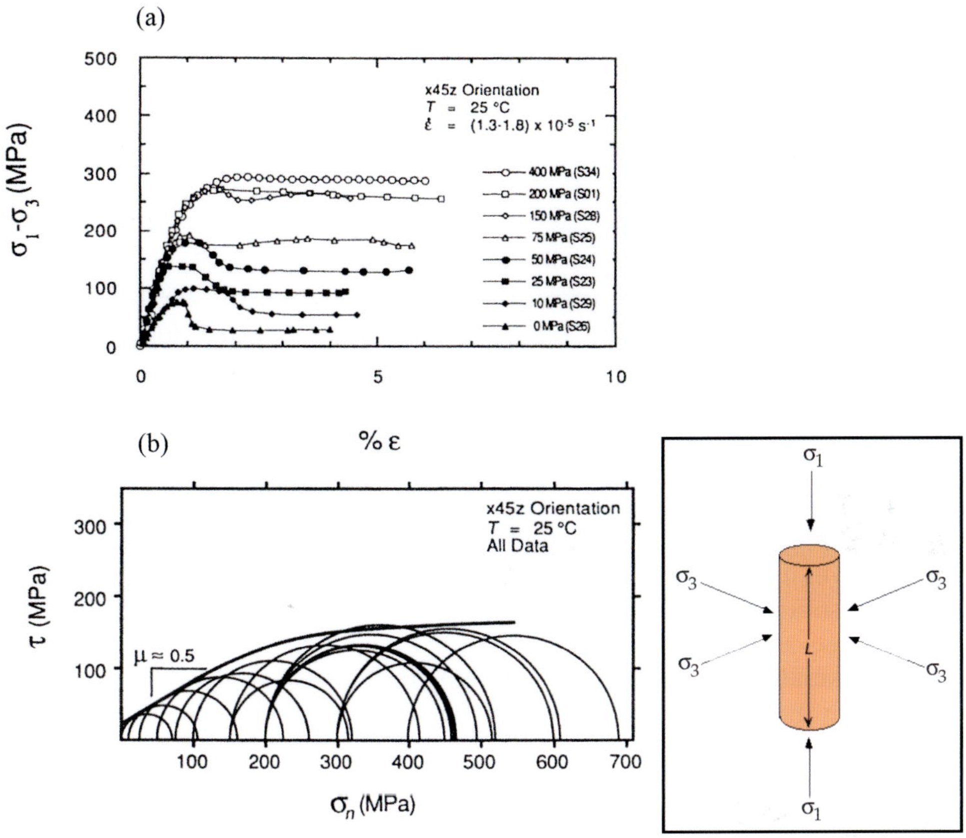

Figure 13.1 Observations of rock failure from *Shea and Kronenberg* [1992]. (a) Differential stress versus axial strain (percent) data from a series of experiments at different confining pressure conducted on initially intact mica schist at room temperature. Tests use a cylindrical sample and two independently controlled applied stresses (inset). The differential stress is $\sigma_\Delta = \sigma_1 - \sigma_3$. Axial strain ε_z in percent is $100 \times (L - L_0)/L_0$, where L is sample length and L_0 is the starting length. The failure condition is defined by the peak stress on the stress strain curve, which in the case of faulting can be related to the shear τ and normal stresses σ_n on the resulting fault (see text). The measured shear and normal stress at peak are used to define the brittle failure envelope from the tests at different confining pressures; the envelope is constructed by collecting the failure strengths from the different tests on a single shear versus normal stress plot (fig. 13.1b). (b). The failure envelope (heavy line) for the tests shown in figure 13.1a. Also shown are the Mohr stress circles, at peak stress in each test.

on initially intact mica schist at room temperature from *Shea and Kronenburg* [1992]. Failure tests are conducted using a cylindrical sample and two independently controlled applied stresses, an axial stress σ_1 provided by a mechanically driven hardened metal piston, and a constant confining pressure σ_3 provided to the jacketed sample by pressurized fluid (fig. 13.1, inset). The differential stress $\sigma_\Delta = \sigma_1 - \sigma_3$ when raised to a sufficient level causes failure, either a throughghoing fracture at angle ϕ referenced to σ_1, or from deformation distributed throughout the sample. The failure condition is usually defined by the peak stress (fig. 13.1), which in the case of faulting can be related to the shear τ and normal stresses σ_n on the fault by the relationships $\tau = 0.5\sigma_\Delta \sin 2\phi$, and $\sigma_n = \sigma_3 + 0.5\sigma_\Delta(1 - \cos 2\phi)$.

It is these so measured shear and normal stress associated with the peak strength that are used to define the brittle failure condition from a sequence of tests at different confining pressures as shown in figure 13.1a; these failure shear and normal stresses are collected on a single plot, defining the failure envelope shown in figure 13.1b. The failure envelope is curved with pressure dependence $\mu_f = \partial\tau/\partial\sigma_e$ that is high at low stress and decreases with increasing stress. The individual stress strain data (fig. 13.1a) also show a postpeak strain weakening associated with failure. The strain weakening is transient, and strength drops to a steady level defining the residual sliding strength. The residual strength usually has pressure dependence quite similar to the failure strength over most of the brittle regime (fig. 13.2). The residual sliding strength and peak strength define the shear strength loss. The strength loss does not occur in tests at high pressure; for example in figure 13.1a the strength loss is becoming smaller with increasing confining pressure.

Though the examples in figure 13.2 show a nonlinear dependence of the shear strength at failure on confining or fault normal stress, laboratory observations of rock failure strength and fault strength at low to modest confining pressure, corresponding to shallow crustal conditions, can be described for illustrative purposes by the linear empirical Coulomb failure relation between shear stress in the direction of slip τ and the effective normal stresses σ_e across the fracture plane or fault at the time of failure

$$|\tau| = S_o + \mu_f \sigma_e \tag{13.1a}$$

Here S_o, the strength at zero effective normal stress, is cohesion. And μ_f is a coefficient of friction $\mu_f = \partial\tau/\partial\sigma_e$, the normal pressure dependence of fault strength, referred to in the case of rock failure as the internal friction coefficient and in the case of slip on a preexisting fault as the coefficient of friction.

Pore-fluid pressure affects rock failure and fault strength by decreasing normal stress throughout the rock mass according to the effective pressure law [*Terzaghi*, 1923; *Hubbert and Rubey*, 1959], $\sigma_e = \sigma_n - \alpha p$, where σ_n is the applied normal stress, p is pore-fluid pressure, and α is a constant related to pore geometry. For the onset of frictional sliding and for the fracture strength of rock with well-connected porosity, $\alpha = 1$ and $\sigma_e = \sigma_n - p$. The normal hydraulic head for well-connected porous material $p = \rho_{H_20} g z$ will offset the overburden pressure $\sigma_v = \rho g z$ by approximately one third owing to the density difference between water and rock. Here z is depth from the surface, g is the acceleration due to gravity 9.8 m/s^2, ρ is rock density, and ρ_{H_20} is the density of water. How-

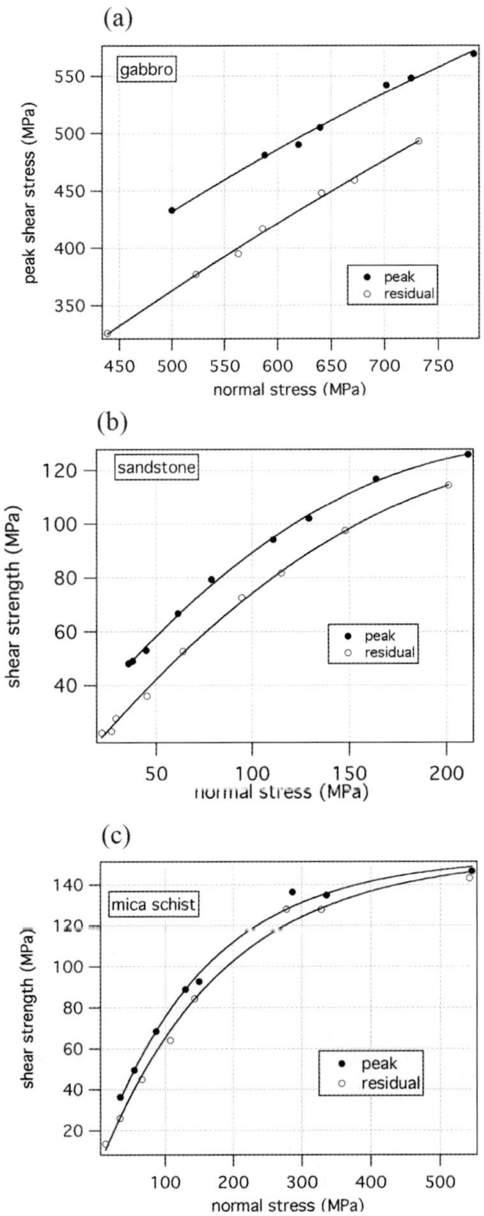

Figure 13.2 Peak and residual shear strengths defining the strength loss during failure for three rock types. Tests were conducted at constant confining pressure, so failure produces a reduction in both shear strength and normal stress across the failure plane. Shear and normal stress are estimated from the measured confining pressure and differential stress, and the orientation of the fault as measured from the recovered sample (see equations in the text). (a) A strong gabbro [*Wong*, 1986]. (b) Intermediate strength sandstone [*Bernabe and Brace*, 1990]. (c) A weak mica schist [*Shea and Kronenberg*, 1993].

ever, for isolated porosity at depth, especially under conditions of rapid burial or underthrusting, consolidation and dewatering processes can compress pore fluids to exceed hydrostatic pressure. Under some circumstances, burial and tectonic forcing may lead to repeated episodes of hydraulic fracture. Overpressure is of considerable interest and presumed importance in subduction fault-zone mechanics owing to known high volume fluid sources (buried porous sediments, dehydration reactions) [e.g., *Saffer and Bekins*, 2002], the apparent weakness of subduction faults [*Wang and He,* 1999], and indirect evidence of elevated pressure [e.g., *Moore and von Huene*, 1980; *Westbrook and Smith*, 1983].

Intact crystalline rock, particularly quartzofeldspathic rock, has significant cohesion, where as preexisting faults more often do not. In this latter case, the failure strength (equation (13.1a)) has the simpler form

$$|\tau| = \mu_f \sigma_e \quad (13.1b)$$

known as Amontons' law. The strength quantity most easily measured and most often used in studies of faulting is the simple ratio of shear to normal stress. Throughout this paper, this ratio μ, is referred to as friction or frictional resistance, and for equation (13.1a) is related to the friction coefficient μ_f, by

$$\mu = \frac{|\tau|}{\sigma_e} = \frac{S_0}{\sigma_e} + \mu_f \quad (13.1c)$$

The laboratory observations indicate that the failure strength of crustal rock increases with normal stress; since the ambient level of stress in the crust is controlled by the overburden stress $\sigma_v = (\rho - \rho_{H_20})gz$, effective normal stress increases linearly to a limiting depth. Throughout this paper, rocks or faults are regarded as pressure dependent, as in equation (13.1), meaning that their failure strength or sliding strength increases with effective pressure.

Temperature also increases with depth in the Earth, eventually reaching the high temperatures and pressures where bulk material strength is controlled by deformation mechanisms other than those underlying the behavior described by equation (13.1). Examples of high temperature/pressure mechanisms are dislocation glide, dislocation creep, and pressure solution. All of these processes are ductile; that is, they accommodate large permanent deformation without loss of strength. For pressure solution and dislocation creep, the ductile flow strength, equal to the differential stress σ_Δ during yielding, rapidly decreases with temperature as follows from the creep law of the form

$$\dot{\varepsilon} = C\sigma_d^n \exp\frac{-E}{RT} \quad (13.2)$$

where R is the gas constant, E is the activation energy, n is a dimensionless constant with values in the range 1–4 depending on the particular mechanism, and C is a constant ($\mathrm{MPa}^{-n}/\mathrm{s}$).

Laboratory observations suggest earthquakes should be limited to conditions where the differential stress does not exceed the ductile flow strength of rocks, because ductile deformation precludes dynamic stress drop. Thus, using equations (13.1) and (13.2) as representative, the majority of crustal seismicity occurs at conditions where strength equation (13.1) is lower than equation (13.2). Diagrams, originated by *Goetze and Evans* [1979], such as figure 13.3a, where quartz strength is assumed representative of the continental crust and olivine of oceanic crust, can crudely explain depth-dependent features of natural seismicity (figures 13.3b and 13.3c). However, uncertainties in the form of the flow law [*Brace and Kohlstedt*, 1980; *Kirby*, 1980] and uncertainty of the Earth's temperature gradient, depth variation of lithology, and the strain rate dependence of equation (13.2) do not allow rock strength or other material properties to be well constrained by natural seismicity.

Furthermore, while the intersection of equations (13.1) and (13.2) is often referred to as the brittle-ductile transition, when considering particular distributions of seismicity, using brittle to describe this transition can be completely misleading. Rheologically, brittle describes materials whose ability to resist deformation decreases with increasing permanent deformation, indicat-

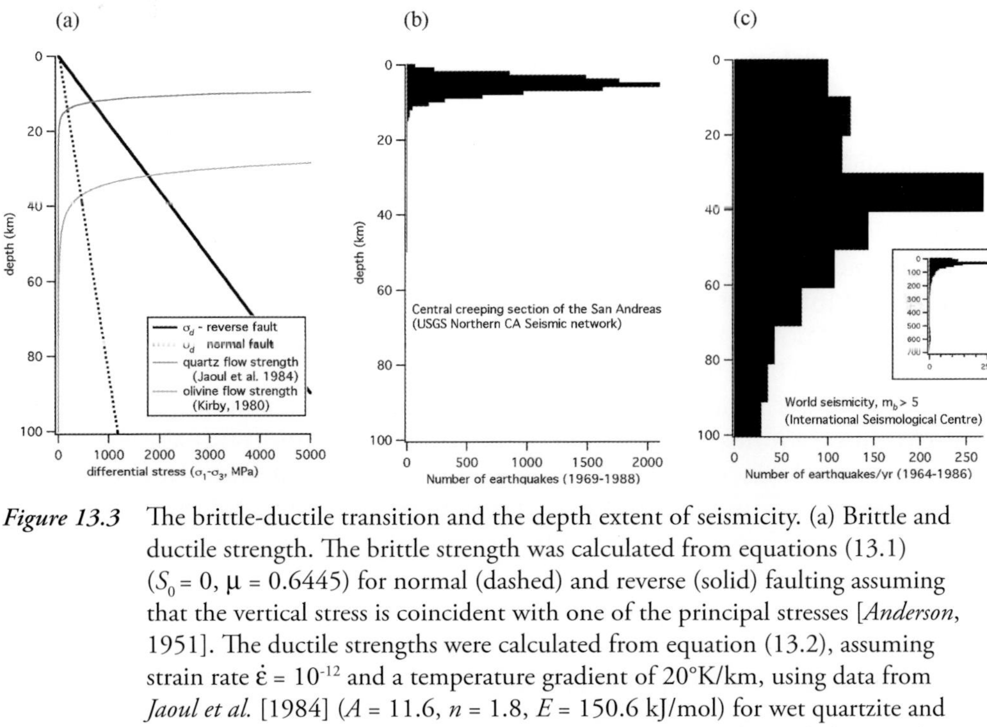

Figure 13.3 The brittle-ductile transition and the depth extent of seismicity. (a) Brittle and ductile strength. The brittle strength was calculated from equations (13.1) ($S_0 = 0$, $\mu = 0.6445$) for normal (dashed) and reverse (solid) faulting assuming that the vertical stress is coincident with one of the principal stresses [*Anderson*, 1951]. The ductile strengths were calculated from equation (13.2), assuming strain rate $\dot{\varepsilon} = 10^{-12}$ and a temperature gradient of 20°K/km, using data from *Jaoul et al.* [1984] ($A = 11.6$, $n = 1.8$, $E = 150.6$ kJ/mol) for wet quartzite and data from *Kirby* [1980] ($A = 1.2 \times 10^{10}$, $n = 4.8$, $E = 502$ kJ/mol) for wet olivine g. (b) Seismicity from the central creeping section of the San Andreas fault, recorded by the U.S. Geological Survey northern California seismic network from 1969 to 1988 [from *Blanpied et al.*, 1991]. (c) World seismicity for events with body wave magnitude $m_b > 5$ as recorded in the International Seismological Centre catalog from January 1964 to February 1986. The inset shows the complete data set [from *Frolich*, 1989].

ing strength loss with strain. Instead of rheologically brittle, many pressure-dependent materials of interest to subduction zone mechanics are rheologically ductile, do not allow material strength loss with strain, and therefore do not permit seismic stress drop. For such materials, there is brittle-ductile transition from a brittle pressure-dependent regime with associated faulting, to a rheologically ductile but pressure-dependent semibrittle regime (transitional or cataclastic flow) (e.g., review by *Evans et al.* [1990]). The semibrittle regime is not associated with strength loss and may not be associated with well-localized deformation. So there may be two transitions with increasing depth: a brittle to semibrittle and a semibrittle to fully plastic; the former is the brittle-ductile transition. The latter is a transition with increasing depth from a pressure-sensitive, approximately temperature and strain rate insensitive rheology such as equation (13.1) to a pressure insensitive, strongly temperature and strain-rate dependent regime (13.2). As this review considers seismic and aseismic faulting in the shallow crust, in the remainder of this paper, only brittle and semibrittle pressure-sensitive behavior of rock failure and fault slip are considered.

Framework for Addressing Rock Failure and Fault Strength

While sliding friction and rock failure can, in some instances, be described adequately by equation (13.1) with constant pressure dependence, more usually, rock failure and friction have a nonlinear pressure dependence. Furthermore, without additional information, equations such as equation (13.1) provide little insight into why particular rock types have high strength or high internal friction μ_f. Another drawback is that simple threshold failure relations do not predict failure time well, particularly at natural low loading rates. In addition, sliding stability is not addressed by relations such as equation (13.1). Seismic stress drop requires a loss of strength sufficiently rapid with shear strain to radiate seismic energy. If instead, the failure relation accounts for the size of the strength loss, the rate and time dependence of failure and the loading/unloading characteristics of the fault system, the fault properties necessary for seismic slip can be established.

To describe low-temperature rock failure and fault slip a simple physical model, similar to *Mogi* [1974] and *Savage et al.* [1996] can be used. Assume that rock failure and fault slip involve only two kinematic actions: deformation of intact rock and relative motion between two rigid rock surfaces or fragments. The deformation of intact rock might involve some combination of fracturing and ductile flow, and it may occur on any scale, e.g., tensile fracture of a single particle within a fault gouge, or generating a throughgoing shear fracture in large intact mass. Similarly, the motion between rigid rock surfaces might represent sliding on a discontinuity at large or small scales, or slip between particles in a fault gouge, between large contacting asperities across a fault, or rolling contact between particles in a fault gouge. At any particular instant during a failure test or fault slip experiment, the total fault or incipient fault area A is divided into two distinct areas: intact material with area A_i and strength τ_i that must be broken or otherwise inelastically strained to accommodate a further increment of fault creation or fault slip, and broken area. Within the broken area is area of asperity contact A_f across fractures or between fragments within the fractured rock, with strength τ_f. The remaining fractured area A_d is the area crossing open pore space. Because noncontacting area has no strength, the shear resistance is the sum of the intact resistance T_i and the contacting resistance T_f, the total shear resisting force is

$$T = T_i + T_f = A_i \tau_i + A_f \tau_f \qquad (13.3a)$$

and the total resisting shear strength is

$$\tau = \frac{A_i \tau_i}{A} + \frac{A_f \tau_f}{A} \qquad (13.3b)$$

Changes in shear resistance, for example due to changes in normal stress (fig. 13.2) or displacement (strain) (fig. 13.1a), are attributable to changes in one or more of A_i, τ_i, A_f and τ_f, and the general description of changes in shear resistance is

$$d\tau = \frac{\partial \tau}{\partial A_i} dA_i + \frac{\partial \tau}{\partial \tau_i} d\tau_i + \frac{\partial \tau}{\partial A_f} dA_f + \frac{\partial \tau}{\partial \tau_f} d\tau_f \qquad (13.3c)$$

Depending on the conditions and material type, different terms in equation (13.3c) dominate. The subsequent discussion of laboratory observations of rock failure and rock friction, and constitutive equations will refer to the specific terms of equation (13.3c) thought to underlie changes in fault strength with displacement, time, and slip.

Rock Failure

This section summarizes processes underlying pressure dependence of rock failure strength and the cause of strain rate and time dependence of strength. The required conditions for failure with stress drop and those that lead to seismicity are illustrated using a simplified constitutive description.

Dilatancy

The source of pressure dependence of intact rock failure strength and the pressure dependence of fault sliding strength can be illustrated using equation (13.3c) [*Mogi*, 1974; *Savage et al.*, 1996]. If the shear strength of intact regions is a material constant τ_s (the ultimate strength from single crystal deformation experiments [*Savage et al.*, 1996]) and the shear strength of contacting regions is a material constant τ_f, then changes in fault strength are due entirely to changes in contacting and intact area. For example, the pressure dependence

$$\frac{d\tau}{d\sigma_e} = \frac{\tau_i}{A}\frac{dA_i}{d\sigma_e} + \frac{\tau_f}{A}\frac{dA_f}{d\sigma_e} \tag{13.3d}$$

is due to increases with normal stress of either the intact area or the area of asperity contact within the broken regions. Both increases in intact area and contact area can only occur at the expense of dilatant volume.

Consistent with the qualitative representation (equation (13.3d)), pressure dependence of the failure strength (fig. 13.2) in rock failure tests is thought to be due to inelastic dilatancy, the generation of open (mode I) microcracks, fracture networks, and granular comminuted rock. Dilatancy reduces the contacting area within the highly stressed rock volume, ultimately across the incipient fault plane. Dilatational strain requires work done against the normal stress across the incipient fault or against confining stress, e.g., $\sigma_3 \int d\varepsilon_v$. Thus work necessary to weaken the fault to the point of failure increases with confining pressure, and the failure strength is confining pressure dependent. In most cases, because pressure dependence decreases with increasing confining pressure (figs. 13.2b and 13.2c), the "ease of dilatancy" decreases with increasing pressure (also see Chemical Effects of Pore Fluid section).

Some aspects of the dilatant volume associated with failure are seen in laboratory tests where volume strain is measured as differential stress is increased to failure at constant confining pressure (fig.13.4) [*Brace et al.*, 1966]. Initially, the volume decreases with increasing stress, reflecting elastic compression. For highly porous rock (not shown), there is additional inelastic compaction due to pore collapse [e.g., *Wong*, 1990]. Eventually, at high stress, open pores, contrasts in elastic properties of adjacent minerals and small amounts of slip at weak grain boundaries act as stress concentrators [*Peng and Johnson*, 1972; *Tapponnier and Brace*, 1976], producing large local stresses [*Peng and Johnson*, 1972; *Horii and Nemat-Nasser*, 1986; *Sammis and Ashby*, 1986; *Ashby and Sammis*, 1990; *Lockner et al.*, 1992]. The local stresses contain a component of tension that induces dilatant crack propagation. The onset of measurable dilatancy is shown as C′ in figure 13.4. As crack density increases, cracks begin to interact and a protofault nucleates: a shear fracture connecting en echelon arrays of the dilatant microcracks [*Peng and Johnson*, 1972; *Horii and Nemat-Nasser*, 1986; *Sammis and Ashby*, 1986; *Costin*, 1987; *Kemeny and Cook*, 1987; *Ashby and Sammis*, 1990; *Du and Aydin*, 1991; *Lockner et al.*, 1992; *Reches and Lockner*, 1994]. At this point, labeled C in figure 13.4, a macroscopic fault propagates in-plane owing to its own stress field usually with continuing dilatancy [*Lockner et al.*, 1991, 1992]. The amount of dilatancy prior to failure increases with porosity [*Brace*, 1978].

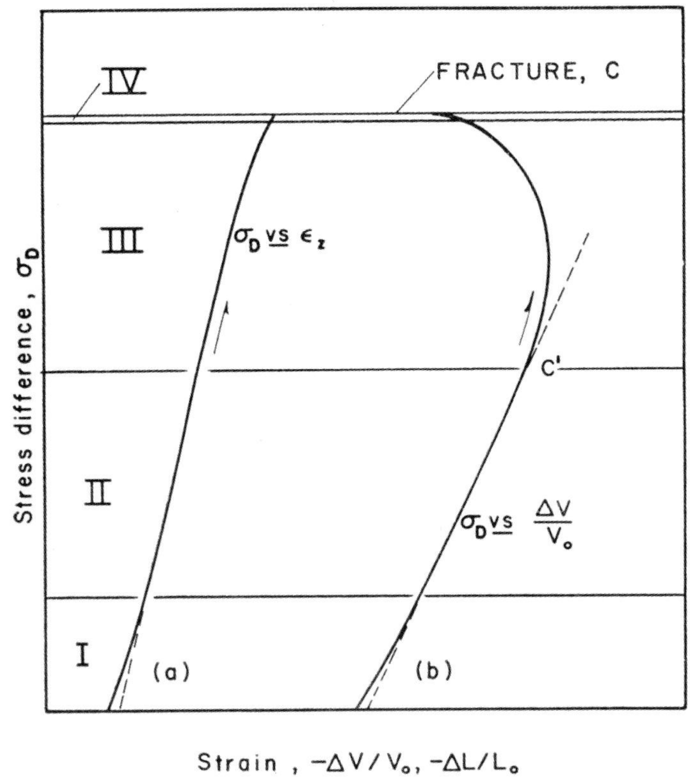

Figure 13.4 Strain and dilatancy during brittle failure [from *Brace et al.*, 1966]. Differential stress (stress difference) versus two strain measures; at left (a) is axial strain and at right (b) is the volumetric strain. Axial strain and differential stress are defined in the caption to figure 13.1. Volumetric strain is the sum of the axial strain, and twice the radial strain of the cylindrical sample; volumetric strain is also defined as $(V - V_0)/V_0$, where V is sample volume and V_0 is the initial volume. The horizontal lines define different parts of stress strain behavior. Region I is a nonlinear elastic region associated with closing open fractures. Region II is linear elastic with compressive volumetric and axial strain. Region III is the inelastic region associated with dilatant cracking and large volume increase. C′ marks the onset of region III. Region IV is that of macroscopic fracture and failure occurs at C.

In weak phyllosilicate rocks, weak pressure dependence reflects low dilatancy [*Escartin et al.*, 1997]. For example, anitigorite and lizardite serpentinites are markedly weaker than nonhydrous silicate rocks such as granite and dunite ($\mu_f = 0.73–0.43$), and serpentinites have lower pressure dependence $\mu_f = 0.35$. The difference in pressure dependence indicates low bulk dilatancy (fig. 13.5, compare with fig. 13.4). As serpentine has weak [001] cleavage, shear is accommodated preferentially along these cleavage surfaces resulting in crys-

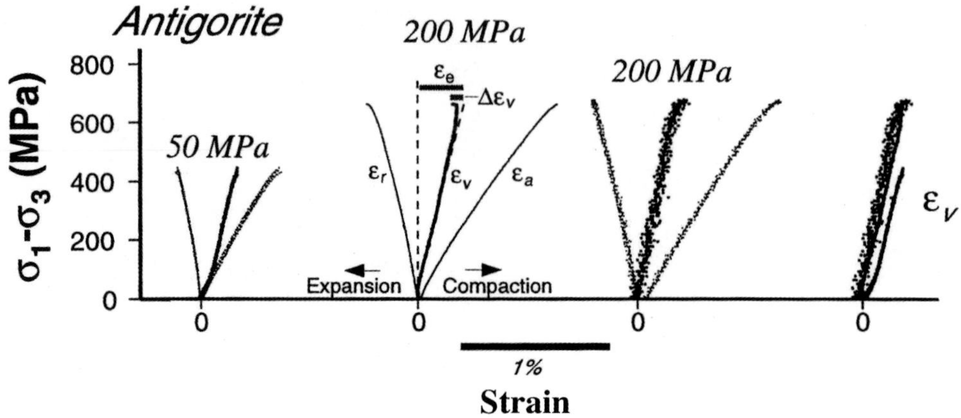

Figure 13.5 Strain and dilatancy during failure of weak serpentine [from *Escartin et al.*,1997]. Differential stress versus axial, radial, and volumetric strain measures for three experiments, one at 50 MPa confining pressure and two at 200 MPa. The volumetric strains for all three cases are superimposed at right. Axial strain (here ε_a) and differential stress are defined in the caption to figure 13.1. Volumetric strain (ε_v) is defined in the caption to figure 13.4. Radial strain is $(r - r_0)/r_0$, where r is cylinder radius and r_0 is the initial radius. Volumetric strain is constructed from the measured values of ε_a and ε_r as described in the caption to figure 13.4. The net volumetric strain at failure (coinciding with the peak stress) is compaction, as opposed to the strong material shown in figure 13.4 that shows net dilatancy.

tallographically controlled (nonopening) mode II shear fractures. Deformed serpentinite has scarce mode I cracks that produce the strong pressure dependence in quartzofeldspathic rock. Dilatancy that does occur in serpentinite is restricted to rigid motion owing to roughness along the shear planes [*Escartin et al.*, 1997] and in dilatant jogs within regions of shear step-over.

Anisotropy and Failure Strength

The comparison between serpentines [*Escartin et al.*, 1997] and quartzofeldspathic rocks [*Brace et al.*, 1966] suggests that weak pressure dependence and intrinsic material weakness are associated with low dilatancy and strong pressure dependence with high dilatancy. This causal relationship is consistent with the failure strengths of other rocks and minerals with strongly anisotropic material properties at low temperature and low stress, e.g., foliated metamorphic rocks, schists, and gneisses [*Shea and Kronenberg*, 1993], shale ($\mu_f = 0.33$ [*Donath*, 1961], $\mu_f = 0.3$ [*Ibanez and Kronenberg*, 1993]), and muscovite single crystals ($\mu_f = 0.4$ [*Mares and Kronenberg*, 1993]). The experiments on muscovite single crystals illustrate that preferred orientation (p.o.) of anisotropy and the geometric relationship between the p.o. and the principal stress directions also are involved in pressure dependence and weakness (fig. 13.6). If the musco-

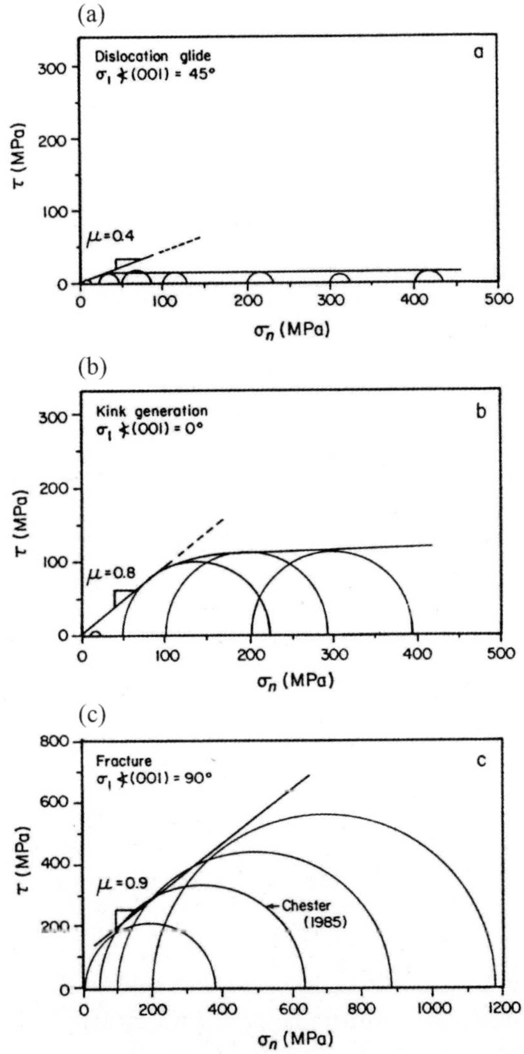

Figure 13.6 Effect of stress orientation on strength of "weak" material. Strength of muscovite single crystals in three different orientations. Reprinted from Journal of Structural Geology, 15, *Mares and Kronenberg*, Experimental deformation of muscovite, 1061–1075, Copyright (1993), with permission from Elsevier. Schematics at left show the orientation of the basal cleavage [001] with respect to a vertical greatest principal stress σ_1. At right are the Mohr diagrams corresponding to deformation tests at different confining pressures in these orientations. (a) σ_1 at 45° to [001]. Sample is weak in a relative sense with low-pressure dependence, and absolutely weak in the sense that the transition to the plastic regime occurs at 50 MPa. (b) σ_1 at 0° to [001]. Sample is stronger in the relative and absolute senses; pressure dependence is twice that in figure 13.6a, and the transition to the plastic regime occurs at twice the normal stress. (c) σ_1 at 90° to [001]. Sample is extremely strong with pressure dependence larger than quartzofeldspathic rocks, and the transition to the plastic regime occurs at higher than the maximum testing strength (>400 MPa normal stress).

vite cleavage direction [001] is aligned 45 to greatest principal stress, then the pressure dependence is weak (μ_f = 0.4) and the transition to the pressure-independent regime, here controlled by dislocation glide, occurs at <50 MPa. If instead the cleavage is parallel to σ_1, the pressure dependence is twice as large (μ_f = 0.8) and the transition to the pressure-independent regime occurs at between 100 and 150 MPa. Furthermore, if the cleavage is perpendicular to σ_1, then muscovite is extremely strong, has very high pressure dependence (μ_f = 0.9), and the transition to true ductile shearing occurs at stresses higher than attained in the highest stress tests >750 MPa. In this severely misoriented regime the sample fails by brittle fracture accompanied by significant stress drop. Similar anisotropy of failure strength is seen in shale [*Donath*, 1961], schists, and phyllities [*Shea and Kronenberg*, 1993]. So even intrinsically weak minerals can be very strong, and strongly dilatant in polycrystalline form if there is no p.o. or if there is a p.o. but the fabric is misoriented with respect to the stress field. Observable intrinsic weakness in rocks therefore requires anisotropy, preferred orientation of the anisotropy, and that the fabric is aligned optimally for failure.

Failure With Strength Loss

Under certain pressure and temperature conditions for many materials, faulting of intact rock is associated with strength loss. Other materials can fail at the same conditions with no stress drop at all; these different responses reflect differences in microscopic behavior. Strength loss during failure of intact rock is expected if the fault sliding strength is lower than the incipient fault strength (e.g., fig. 13.2) or if the fault sliding strength is slip rate dependent. Here, the focus is on the former; the latter case is discussed below. If the processes associated with fault formation produce a fault that is weaker than the incipient fault, then a peak strength is followed by strain weakening. At low pressure the strength of preexisting faults (see below) is usually lower than the peak strength in an intact failure test; thus strength loss is expected at low confining stress for most materials.

The physical process underlying the strength loss can be rationalized by using equation (13.3). Again, taking the intact and contact strengths as material constants, the difference between the failure strength τ^p and the residual sliding strength τ^r of a failure test involves differences in the intact and asperity contact areas

$$\Delta\tau = \tau^p - \tau^r = \frac{\tau_i}{A}(A_i^p - A_i^r) + \frac{\tau_f}{A}(A_f^p - A_f^r) \qquad (13.4)$$

where the superscripts p and r refer to the peak and residual values. For equation (13.4) both the peak and residual strengths involve resistance due to contributions from intact rock and sliding contact [*Mogi*, 1974; *Savage et al.*, 1996]. The contribution of intact resistance to residual strength is justified by laboratory results where wear debris and gouge are always observed as by-products of fault slip on preexisting faults and initially clean joint surfaces. Nonzero strength loss requires net reduction of intact and contacting area, which cannot occur unless dilatant volume is generated during the strength loss. So strength loss, like the pressure dependence of failure strength, is due to dilatancy, reflecting weakening due to shear-induced reduction of intact and contacting area across the fault surface.

To illustrate the tacit correlation between the strength loss, here measured as change in friction (see caption), and the pressure dependence of failure strength are data from selected rock failure tests on schist [*Shea and Kronenberg*, 1993], Berea sandstone [*Bernabe and Brace*, 1990] and gabbro [*Wong*, 1986] (fig. 13.7). For a particular rock type, the strength loss is a fraction of the pressure dependence, and as the pressure dependence decreases, the strength loss decreases. The implication is that low pressure dependence results from low dilatancy and any associated strength losses during faulting are small. The clearest expression of this expected relation between pressure dependence and strength loss for any particular rock is at the transition to truly ductile behavior where both the pressure dependence and the strength loss must vanish.

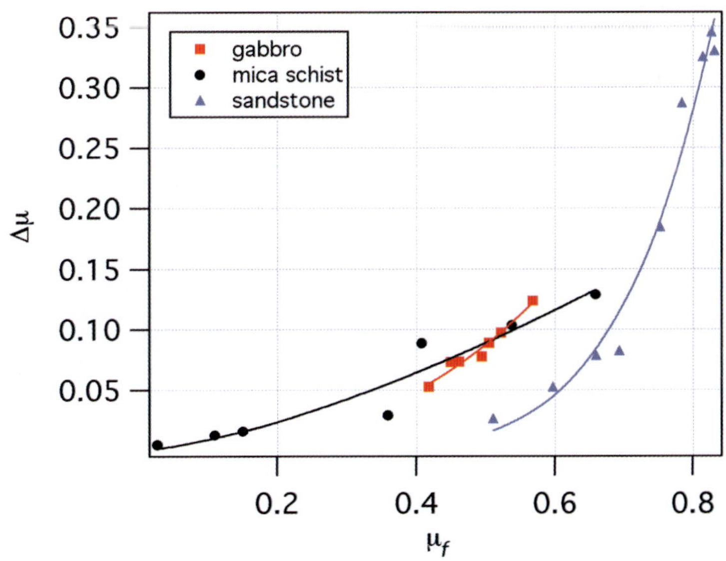

Figure 13.7 Relations between strength drop $\Delta\mu$ and pressure dependence $\mu_f = \partial\tau/\partial\sigma_e$ from failure tests for three rock types. These are the same experiments shown in figure 13.2. The strength drops are expressed as changes in friction between the peak strength and the residual strength, accounting for the change in normal stress that accompanies the strength loss. The pressure dependence was estimated from the failure envelope (fig. 13.2).

The transition with increasing confining pressure from the brittle strain weakening response to ductile deformation can be a transition to semibrittle deformation [e.g., *Escartin et al.*, 1997] or to a pressure-independent regime [e.g., *Shea and Kronenberg*, 1993]. For quartzofeldspathic rocks, strength loss can persist to high confining stresses [e.g., *Byerlee*, 1968]. In anisotropic material with preferred orientation the transition occurs at much lower pressure (e.g., fig. 13.6a). In any case, the transition from the strain-weakening regime is associated with reduction in the pressure dependence or elimination of the pressure dependence. Mechanistic and stress-strain behavior spanning the brittle-ductile transition are summarized in figure 13.8, and readers are directed to the review paper by *Evans et al.* [1990] for additional details.

In characterizing fault strength during failure it is important to recognize that stress threshold models of failure such as equation (13.1) do not relate well to the processes that determine the timing and occurrence brittle failure. In particular, microcrack initiation and growth produces the accelerating inelastic dilatancy that precedes failure (fig.13.4). There is measurable precursory

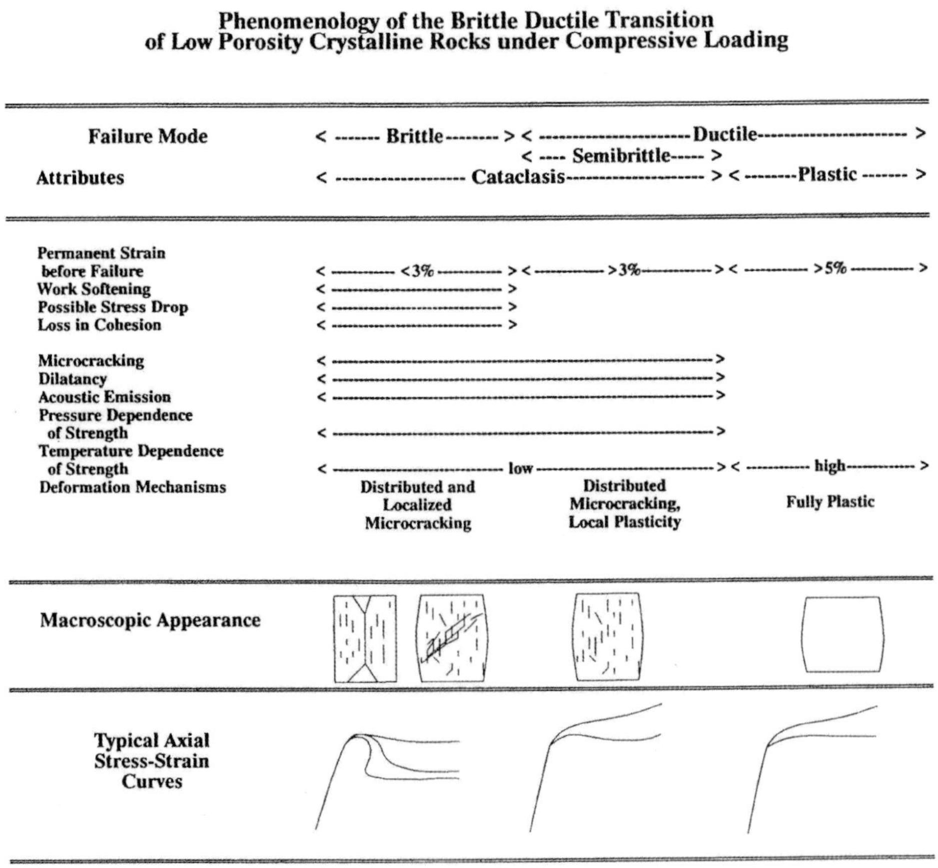

Figure 13.8 Rheological, mechanistic, and microstructural characteristics of the brittle-ductile transition. Figure from the comprehensive review of *Evans et al.* [1990].

inelastic creep prior to the peak strength and significant displacement associated with the postpeak loss of strength (fig. 13.4). Thus the loss of strength is not instantaneous at a strain or displacement threshold, and the weakening associated with fault formation is gradual, involving a displacement or strain length scale. To determine the seismic potential of faulting, the displacement rate of weakening is key.

Second-Order Effects on Strength

That rock failure is not abrupt in strain (fig. 13.4) requires it also to be gradual in time. As shown in figure 13.9, failure is delayed. There is lag time between

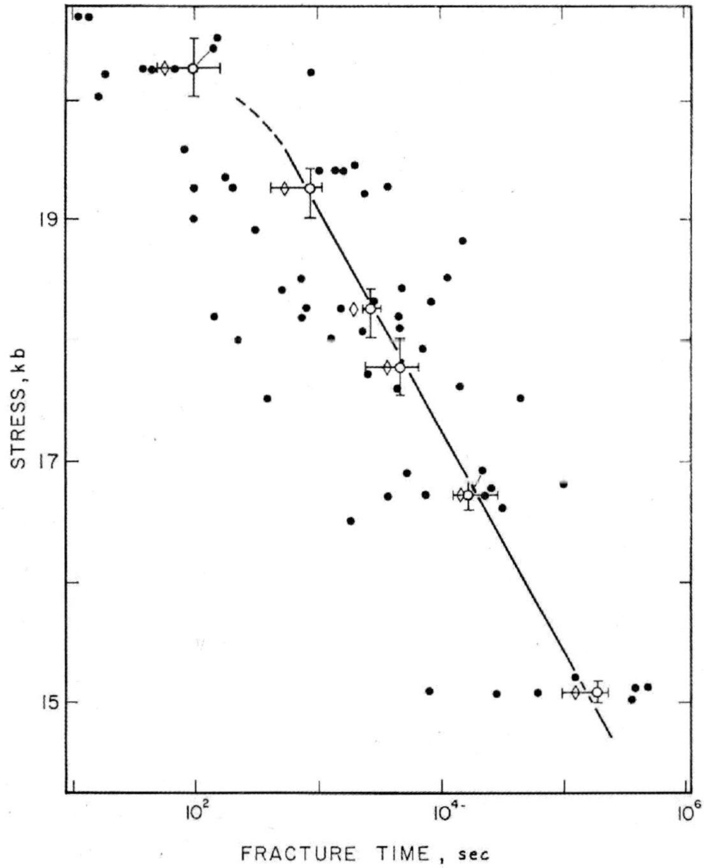

Figure 13.9 Dependence of failure time on stress from experiments on quartz single crystals in water at room temperature. Figure from *Scholz* [1972]. In these tests at ambient conditions a uniaxial stress sufficient to induce failure is applied and held constant until failure occurs. This is the phenomenon known as static fatigue. Solid symbols are individual failure tests. The open symbols are the mean failure times with error bars, and the open triangles are median failure times. The line is a fit to the means.

the application of a critical stress and the time of failure. This delay originates from a small viscous component to rock strength. At constant loading rate and low temperature, rock failure strength and creep strength depend weakly and logarithmically on the loading strain rate approximately following the empirical relationship [e.g., *Rutter*, 1974; *Lockner*, 1998]

$$\dot{\varepsilon} = \dot{\varepsilon}_o \exp(\tau - \tau_o/\mathbf{A}) \tag{13.5a}$$

reflecting an underlying strain rate dependent deformation mechanism. The empirical constant **A** scales the size of the rate dependence. Because of the role of microcracking in rock failure, many studies, dating to *Scholz* [1968c] have related the rate dependence (equation (13.5a)) to the growth rate of single tensile cracks in double-cantilever experiments on glass [e.g., *Lawn*, 1993] and single mineral crystals [*Atkinson*, 1984]. Figure 13.10 shows examples of this "subcritical" behavior, rate data from single cracks propagating in fused silica glass, [*Lawn*, 1993], synthetic quartz, and Westerly granite [*Atkinson and Meredith*, 1987]. In figure 13.10 the horizontal axis is crack propagation velocity and the

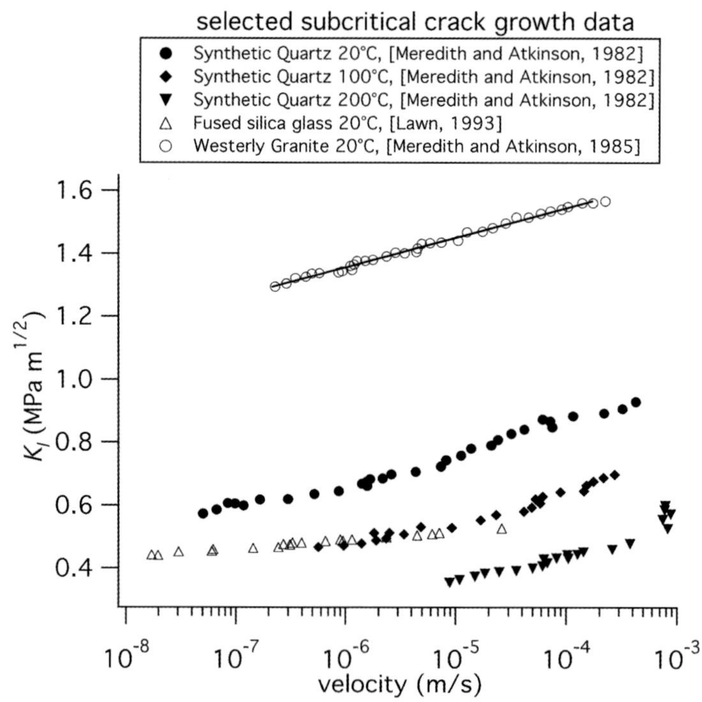

Figure 13.10 Subcritical crack growth. Crack stress intensity K_I versus velocity for fused silica glass [*Lawn*, 1993], synthetic quartz [*Atkinson and Meredith*, 1987a], and Westerly granite [*Atkinson and Meredith*, 1987a]. The stress intensity is proportional to the remotely applied stress; thus the vertical axis reflects the driving stress for subcritical crack extension. Also shown is a fit to the granite data using $K_I = K_0 + \xi \ln V$, with $\xi = 0.041$ MPa m$^{1/2}$.

vertical axis is the stress intensity K_I; stress intensity is related to the remotely applied driving stress σ_r and to the crack length c as $K_I \propto \sigma_r \sqrt{c}$. Subcritical fracture growth has a nonlinear viscous rheology; as the stress on the sample increases, crack velocity increases, or alternatively, as the crack propagation velocity increases, the crack strength increases. There is temperature dependence indicating that a thermally activated mechanism controls growth. Arguably, subcritical crack growth occurs in all instances of natural faulting and in laboratory rock fracture as indicated by damage such as pervasive dilatant and shear microcracks, macroscopic fracture, and wear material (gouge).

For well-studied minerals and for other brittle materials, propagation rate is determined by the rate of chemical reaction at the crack tip (the rate at which chemical bonds are broken) [*Charles and Hillig*, 1962; *Atkinson and Meredith*, 1987a] following [*Charles and Hillig*, 1962]

$$v = v_o \exp\left(\frac{-E + \Omega\sigma_t - \Omega_m \Gamma/\rho}{RT}\right) \qquad (13.5b)$$

where T is temperature in degrees Kelvin, E is the stress free activation energy, Ω is the activation volume, Ω_m is molar volume, Γ is the interfacial energy, R is the gas constant (8.3144 J/mol °K), and ρ is the radius of curvature at the crack tip. The dependence of rate on stress has the same form as the empirical relationship (equation (13.5a)), and precursory creep and volumetric strain observed in rock failure tests [*Scholz*, 1968a; *Lockner*, 1998] are expected from the collective growth of many subcritical fractures.

Precursory Creep and Static Fatigue

Accounting for a rate-dependent strength, equation (13.5) produces much better simulations of laboratory observations than threshold models such as equation (13.1). Observations such as slow, controlled strength loss (fig. 13.1) or rapid seismic strength loss (fig. 13.11) can be characterized using an equation for fault strength (constitutive relation) and one for elastic interaction with the surroundings. The fault strength relation accounts for the reduction of contact area by progressive dilatancy and for the second-order strain-rate dependence,

$$\tau = \tau^r + \frac{\Delta\tau(d_* - \delta)}{d_*} + \mathbf{A}\ln\frac{V}{V_L} \qquad \delta < d_*$$
$$\tau = \tau^r + \mathbf{A}\ln\frac{V}{V_L} \qquad \delta \ge d_* \qquad (13.6a)$$

The first term on the right side is the residual resistance of the fault sliding at the loading rate V_L. For slip less than d_*, the second term is the fault strength

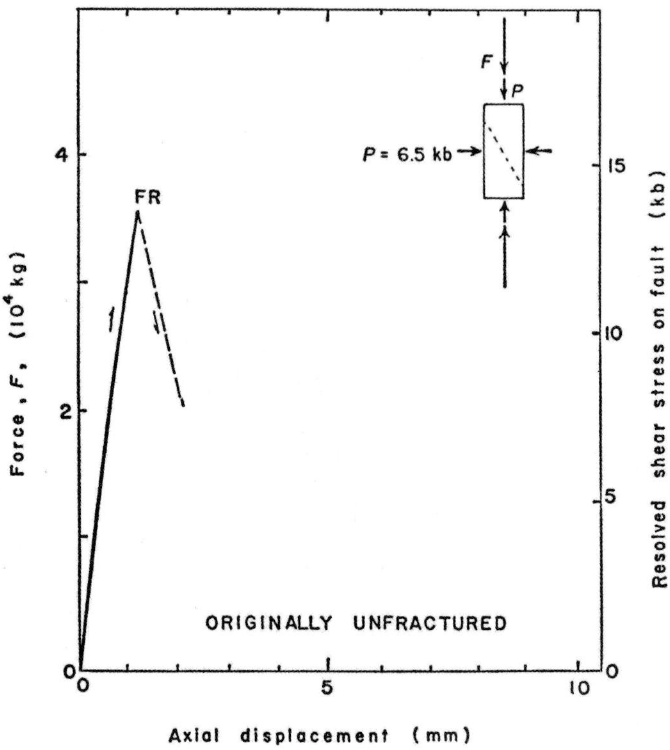

Figure 13.11 Rock failure with rapid slip and seismic radiation. Reprinted with permission from Science, 153, *Brace and Byerlee*, Stick-slip as a mechanism for earthquakes, 990–992, Copyright [1966] AAAS. This is the force displacement record during failure of granite at 650 MPa confining pressure. Stress drop occurs at the point labeled FR; subsequent motion of the fault and the loading system during the stress drop is uncontrolled, and the stress drop occurs effectively instantaneously. The dashed line connects the measurements before and after the rapid slip event.

loss due to dilatancy, represented by slip weakening, which operates over the characteristic displacement d_*. The weakening relationship is linear, as most often assumed in representations of material and earthquake failure [*Ida*, 1972; *Palmer and Rice*, 1973], dynamic rupture modeling [*Andrews*, 1976], and in some laboratory-derived friction representations [*Dieterich*, 1992]. $\Delta\tau$ is the strength loss associated with fault formation. The other term represents the viscous effect having same form as observed in subcritical crack growth experiments [*Atkinson*, 1984] (fig. 13.10) and in rock failure tests (equation (13.5a)) [*Scholz*, 1968c; *Lockner*, 1998]. The strain rate dependence of rock strength is a second-order effect, e.g., having a value $d\tau/\sigma_e d\ln V$ of 0.008 to 0.022 for granite [*Lockner*, 1998] compared to the failure strength τ/σ_e = 1.5 to 0.6 between 300 and 2000 MPa normal stress [see *Savage et al.*, 1996]. Equation (13.6a) is intended to be illustrative, but it may be useful for quantitative analysis as well.

An elastic element is used to describe the elastic loading and unloading behavior of the fault surroundings [*Dieterich*, 1978a, b]. Elastic coupling of the fault to the surroundings is through a stiffness k. The stress $\boldsymbol{\tau}$ supplied to the fault is specified by assuming a constant loading rate V_L and inertia is ignored:

$$\boldsymbol{\tau} = k(\delta_L - \delta) \tag{13.6b}$$

δ_L is the loading displacement $\delta_L = tV_L$, and δ is fault slip. In laboratory experiments, stiffness is the combined stiffness of the intact portion of the sample and the testing apparatus, whereas in the Earth the stiffness is related to the elastic modulus of the rock and for simple representations is inversely proportional to the characteristic dimension of the portion of fault that is slipping [e.g., *Byerlee*, 1970; *Walsh*, 1971; *Dieterich*, 1978a; *Dieterich*, 1979].

Equating the fault strength τ (equation (13.6a)) with the fault stress $\boldsymbol{\tau}$ (equation (13.6b)) well illustrates the range of behavior seen in laboratory faulting experiments, for example fault generation under conditions of strain neutral creep, faulting with controlled strength loss, and seismic faulting (fig. 13.12). These example simulations are conducted at identical conditions using different values of the strength loss. In the case of faulting with no observed stress drop the entire strength loss $\Delta\tau$ is accomplished during the prepeak creep while the fault slip rate is simultaneously accelerating. In this case the $\mathbf{A}\ln V$ term in equation (13.6a) entirely masks the strength loss. For stable stress drop the strength loss is complete before the slip rate reaches seismic slip rates.

Precursory creep and delayed failure in laboratory failure tests result in a related phenomenon known in mechanical engineering, material science, and rock mechanics as static fatigue (fig. 13.9), an inherent time delay between the application of a stress increase and the occurrence of rock failure induced by the stress change [e.g., *Griggs*, 1936; *Scholz*, 1972; *Kranz*, 1980]. Static fatigue behavior is predicted by (equation (13.6a)). In a fatigue test, stress is raised to a particular level τ_s at time $t = 0$ and then held constant; the sliding rate evolves from the starting velocity V_s at $t = 0$ according to

$$V = \frac{V_s \mathbf{A} d_*}{\mathbf{A} d_* - \Delta\tau t V_s} \tag{13.7a}$$

where $V_s = V_* \exp[(\tau_s - \tau^r - \Delta\tau)/\mathbf{A}]$ and V_* is an arbitrary slip velocity (fig. 13.13). The delay time t_d between the application of the stress that induces failure and actual time of failure is

$$t_d = \frac{\mathbf{A} d_*}{\Delta\tau V_*} \exp[(\tau_s - \tau^r - \Delta\tau)/\mathbf{A}] \tag{13.7b}$$

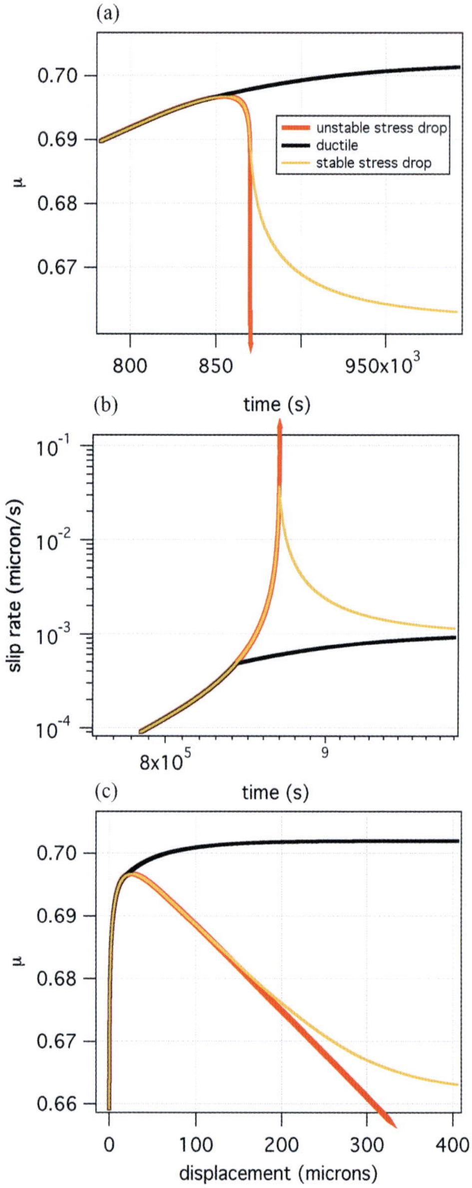

Figure 13.12 Characteristics of simulated laboratory failure with the failure relationship equation (13.6). Three cases corresponding to ductile (black), semibrittle (yellow), and brittle (red) are calculated with σ_n = 100 MPa, k = 0.01 MPa/μm, V_L = 0.001 μm/s, τ_0 = 70 MPa, A = 0.8 MPa, $\Delta\tau/d_*$ = 0.06 MPa/μm, and an initial slip rate of 1×10^{-10} μm/s. For the ductile case the strength loss is $\Delta\tau$ = 1 MPa, and stress drop is progressively larger for the semibrittle ($\Delta\tau$ = 5 MPa) and brittle ($\Delta\tau$ =30 MPa) cases. (a) Friction (τ/σ_n) versus time. As the parameters are the same except for the slip weakening term in equation (13.6), the calculations follow the same stress-strain path until the net slip becomes significant and the paths diverge. For the brittle case the stress drops rapidly in an uncontrolled fashion. For the semibrittle case, strength drops rapidly initially but is damped. The ductile response is monotonic. (b) Slip velocity versus time. (c) Friction versus displacement.

Longer delay times result from smaller stress changes. If earthquake failure is controlled by the same processes as in laboratory rock failure, the failure time of an earthquake triggered by a particular stress change depends on the size of stress change and on how close the fault is to failure prior to the stress change (proximity to failure). That rock failure time that is related to stress change in this way has been known since *Griggs* [1936]. Figure 13.13 shows static fatigue of granite [*Kranz*, 1980].

As the relationship between stress change and failure time is nonlinear, when one considers populations of faults with an initially random distribution

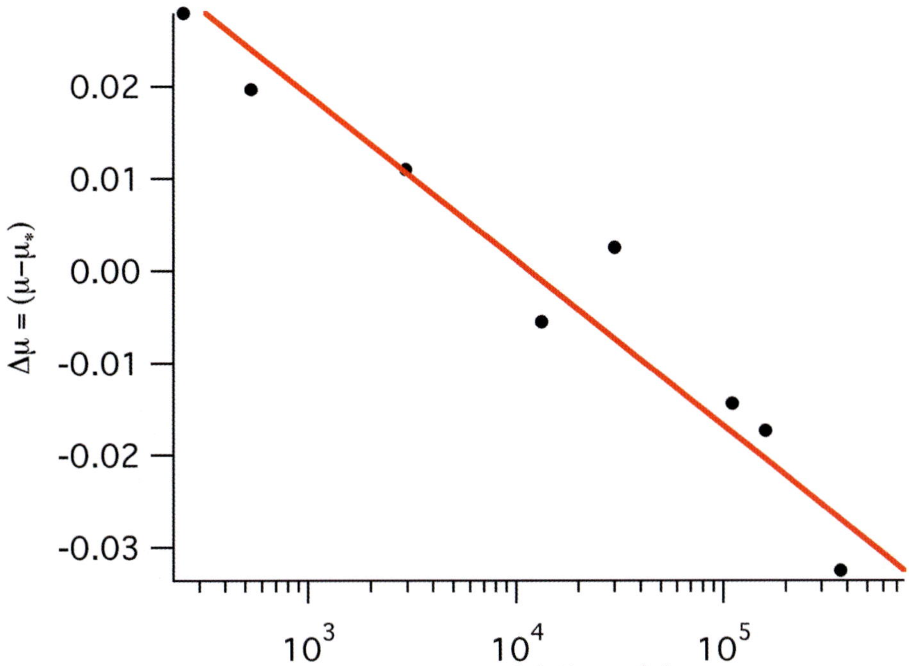

Figure 13.13 Model fit to static fatigue data. Experimental static fatigue data from rock fracture of granite at room temperature and 53 MPa confining pressure [*Kranz*, 1980]. Failure stress was converted to friction assuming a 30° angle between the greatest principal stress and the incipient failure plane. $\Delta\mu$ was calculated using μ at 10^4 s as the arbitrary reference μ_*. Shown is a fit using equation (13.7b), which produces a value of $\mathbf{A}/\sigma_n = 0.0078$.

of proximity to failure, a stress change produces a seismicity rate due to triggered events that is initially very high and decreases with time following the stress change. This is the behavior seen in induced acoustic emissions, microseismicity, and Omori aftershock sequences, as was noted by *Mogi* [1962], and subsequently by many researchers [e.g., *Scholz*, 1968b; *Knopoff*, 1972; *Das and Scholz*, 1981; *Yamashita and Knopoff*, 1987; *Hirata*, 1987; *Reuschle*, 1990; *Marcellini*, 1995, 1997]. Many of these authors have developed fracture growth based models of aftershocks.

Conditions for Seismic Failure

Accelerating deformation rate is only possible if strength drops so that stored elastic energy from the surrounding rock can be used to drive further deformation. Accelerating slip rate depends on inequality between strength and the stress available to drive shear deformation, which is supplied by the loading system. In the laboratory the test sample, piston, and load frame store elastic energy just as the brittle crust stores energy around a fault during tectonic loading. When the yield strength is reached, deformation will commence and stored elastic energy is converted to frictional heat or surface energy as the fault zone shears. If an increment of deformation reduces the strength faster than the driving stress is reduced, an unstable condition develops in which deformation can accelerate into an earthquake and some of the stored elastic energy is radiated as seismic waves [*Byerlee*, 1970; *Dieterich*, 1979].

The conditions that distinguish faulting with seismic strength loss from aseismic slip can be deduced directly from equation (13.6) by equating equations (13.6a) to (13.6b). The rate of stress change with displacement from equation (13.6b) is $d\tau/d\delta = k(V_L/V - 1)$. So, if the fault is sliding faster than V_L, the stress is decreasing; if sliding is slower than V_L, stress increases. The displacement rate of strength change at constant sliding velocity from equation (13.6a) is $\partial\tau/\partial\delta = -\Delta\tau/d_*$. Equating the rate of strength change with the rate of stress change gives

$$k = \frac{\dot{\tau}}{V} + \frac{\Delta\tau}{d_*} \tag{13.8a}$$

where the stressing rate $\dot{\tau} = kV_L$. Remember that constant slip rate is assumed in equation (13.8a), so equation (13.8a) is the stiffness that allows constant slip rate given a particular stressing rate and displacement rate of strength loss [see also *Dieterich*, 1992]. As all terms of equation (13.6c) are positive, if $\Delta\tau/d_* > k$, then the equality (13.8a) cannot possibly be met. Under this circumstance the slip rate is not constant and increases to increase the fault strength through the $\mathbf{A}\ln V$ term in equation (13.6a). However, because the elastic reduction in stress with slip k is smaller than the strength loss $\Delta\tau/d_*$, each increment of slip requires a further increase in slip rate to maintain stress (equation (13.6b)) equivalent to strength (equation (13.6a)). This is the recipe for accelerating slip rate, possibly to seismic rates. Thus

$$\Delta\tau/d_* > k \tag{13.8b}$$

is a necessary condition for seismic stress drop. In the rock friction literature an unbounded increase in slip rate with slip is referred to as a slip instability [e.g., *Rice*, 1983; *Rice and Ruina*, 1983; *Ruina*, 1983] and for linearly and approximately linear slip weakening relations, equation (13.8b) is considered the necessary condition for earthquake nucleation. This condition for instability can also be illustrated graphically (see fig. 13.14).

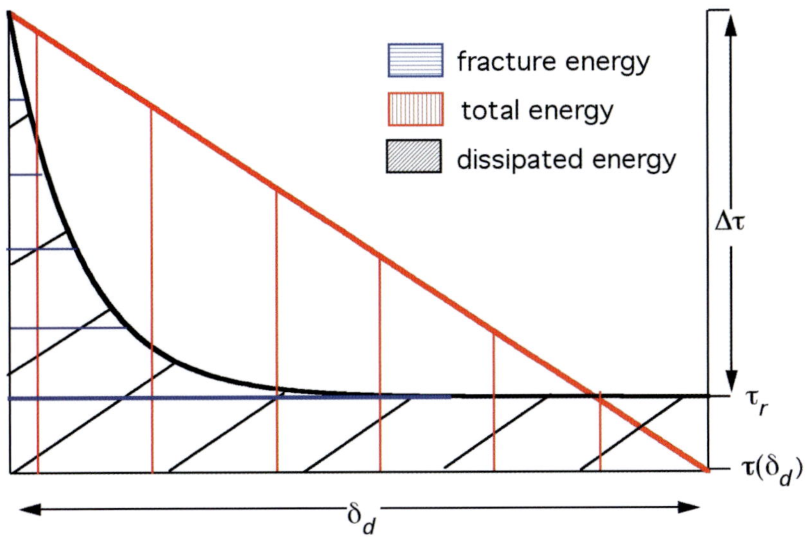

Figure 13.14 Schematic diagram showing the energies per unit fault area (e.g., Joules/m^2) released and dissipated during strength loss in a laboratory experiment. Energy per fault area can be expressed equivalently as stress times displacement; stress and displacement are the axes of the figures, and the energy/fault area corresponds to an area on the plot (e.g., the blue hatched region which corresponds to fracture energy). The displacement rate of stress drop is fixed by the machine stiffness (heavy red line); the area under this stress line is the total energy released per unit fault area. The strength loss (heavy black line), which can occur more rapidly with displacement than the stress drop, defines the energy dissipated by fault slip, and the effective shear fracture energy G_e (blue horizontal hatching), The effective shear fracture energy is the portion of dissipated energy associated with the strength loss. The requirement for earthquake nucleation, equation (13.9d), is illustrated here graphically. For nucleation to occur the total energy must exceed the dissipated energy; in other words, the red area exceeds the black area.

Effective Shear Fracture Energy In instances where the strength loss with displacement is not linear, a more general condition can be derived from an energy balance. For slip acceleration the total work W accomplished in dropping the stress in the source region must exceed that absorbed or dissipated in the fault zone E_f, $W - E_f > 0$. The energy difference $W - E_f$ is the energy available to accelerate the slip rate and to be radiated as seismic energy. Ignoring small amounts of dilatancy, the total work (energy per unit fault area) is

$$W = \int_{\delta_o}^{\delta_d} \tau d\delta \tag{13.9a}$$

This stress drop is accomplished in some displacement δ_d. The total work can be divided into that associated with the stress drop and the energy $\tau(\delta_d)\delta_d$. The partitioning is shown in the stress versus displacement diagram (fig. 13.14). The total energy released (in Joules/m^2 = Pa m), is the area under the heavy red line. That associated with the stress drop is the area $k\delta_d^2/2$.

The energy absorbed in the fault zone consists of shear-generated heat and that of inelastic processes such as ductile yielding and creation of new surface area by fracturing. These dissipative energy changes per unit area within the fault zone can be represented by changes in the shear resistance integrated over the total displacement

$$E_f = \int_{\delta_o}^{\delta_d} \tau d\delta \tag{13.9b}$$

If the collective energy E_f is divided into the energy associated with the strength loss G_e and the energy τ_r associated with sliding the residual strength, then G_e is defined as [*Rice*, 1980]

$$G_e = \int_{\delta_o}^{\delta_d} \tau d\delta - \tau_r \delta_d \tag{13.9c}$$

With reference to figure 13.14, the effective shear fracture energy G_e is the hatched region; the energy expended in dropping the fault strength from a failure strength τ_f to the residual strength τ_r. The energy difference is $W - E_f = k\delta_d^2/2 + \tau(\delta_d)\delta_d - \tau_r\delta_d - G_e$, and the necessary requirement for earthquake nucleation is

$$k\delta_d^2/2 + \tau(\delta_d)\delta_d > \tau_r\delta_d + G_e \tag{13.9d}$$

Equation (13.9d) can be rewritten without explicit reference to the residual strength and residual stress as

$$k > 2\left(\Delta\tau/\delta_d - G_e/\delta_d^2\right) \tag{13.9e}$$

For linear slip weakening, this expression is equivalent to equation (13.8b).

The fracture energy concept in laboratory rock mechanics originates from the Griffith energy balance for tensile fracture [*Griffith*, 1920]. For a fracture to propagate, the energy release rate, the total energy per unit area of crack advance, must exceed the energy adsorbed by the creation of new surface area as the crack advances. The critical energy release rate G_c, so defined when the energy released equals the energy adsorbed, is thus the energy associated with inelastic breakdown processes of the material at the crack tip. In the case of purely brittle behavior, G_c can be equated with the specific surface energy (the surface energy per unit area) Γ of the fractured material, $G_c = 2\Gamma/\Delta A$, where ΔA is the increment of surface area that is created as the crack advances. This shear fracture energy is often defined, as in equation (13.8), using a simple linear slip-weakening model where the fault strength drops from the failure strength to the sliding strength over a characteristic displacement d_*, defining a strength loss $\Delta\tau_d$ [*Ida*, 1972; *Palmer and Rice*, 1973]. So long as the spatial size of the breakdown region is small with respect to the crack dimensions, the energy per unit area of new surface created is $G_c = \Delta\tau_d d_*/2$ [*Palmer and Rice*, 1973], equivalent to G_e in figure 13.14.

However, despite its name, laboratory experiments and theory suggest that the effective fracture energy relevant to rock failure and the onset of unstable slip between rock surfaces results primarily from changes in fault strength with displacement, rather than energy dissipated by the creation of new surface area. If a rupture is thought of as a single ideal surface, the slip-weakening distance corresponds to the interatomic spacing [*Ida*, 1973] and is many orders of magnitude smaller than the d_* observed in laboratory shear fracture experiments. Furthermore, such small values of d_* will not give reasonable seismological predictions [*Ida*, 1973; *Andrews*, 1976]. By carefully accounting for the number and size of stress-induced microcracks during rock failure, *Wong* [1982] found that the total energy over the area of the fault A inferred from measured fault strength and inelastic strain during the breakdown, that is integrating to obtain the total energy associated with the shaded region in figure 13.14 ~ $\Delta\tau_d d_* A/2$, is 10–100 times greater than the energy stored in new fracture surface area $2\Gamma S_v w$, where S_v is the surface area per unit volume that is created as the crack advances, and w is the fault-zone width. In contrast, a similar comparison between a stress-strain estimate of fracture energy and an accounting for increased microcrack surface area during tensile fracture shows complete agreement [*Friedman et al.*, 1972]. Comparing the tensile case of *Friedman et al.* [1972] with *Wong* [1982] suggests that the difference between effective and true shear fracture energy seen by *Wong* [1982] is due to frictional heating associated with the breakdown, as shear heating should be much smaller during tensile fracture. For breaking intact rock, *Wong* [1986] found G_e to increase with increasing normal stress, as *Okubo and Dieterich* [1981] found for dynamic slip of a preexisting fault, confirming that the effective shear fracture energy is a pressure-dependent quantity.

The idea that effective shear fracture energy is due mostly to displacement-dependent changes in shear resistance, rather than the creation of fracture surface area, implies that models of laboratory-observed strength breakdown during fault slip are appropriate for use in simulating earthquake rupture nucleation.

Slip Rate For rock failure tests, the fault strength loss is bounded (fig. 13.1) and the acceleration can be limited to subseismic slip rates if the strength loss is not large enough. If the rate of strength reduction is fixed $d\tau/d\delta$, the maximum slip velocity associated with equations (13.6a) and (13.6b) depends on the strength loss as

$$V = \frac{V_s + \dot{\tau}/\gamma[1 - \exp(-\gamma\delta/\mathbf{A})]}{\exp(-\gamma\delta/\mathbf{A})} \tag{13.9f}$$

where $\gamma = \Delta\tau/d_* - k$. Slip rate depends nonlinearly on the strength loss and to achieve seismic slip velocities during rock failure a sufficient loss is needed.

In summary, earthquake nucleation involving the failure of intact rock requires strength loss more rapid with slip than the unloading stiffness of the surroundings (equation (13.8b)), and strength loss of sufficient size (equation (13.9f)) to achieve seismic velocities. Materials with low pressure dependence have low dilatancy and will tend to have small strength losses associated with faulting. Weakly pressure-dependent materials such as shale and serpentinite will therefore tend to not meet the conditions of high displacement rate of strength loss (equation (13.8b)) and sufficiently large strength loss (equation (13.9f)) and tend to aseismic slip more so than strongly dilatant rocks.

Rock Friction

Laboratory fault surfaces are rock that is machined flat and roughened with grinding compound. The applied stresses are supported by a small number of contacts and the fault is dilatant with pore volume fixed by the character of the small-scale roughness (fig. 13.15). Force applied normal to the plane F_n is concentrated at the asperity contacts. Assuming that resistance to shear comes from the sliding strength of asperity contacts, the shear resistance from equation (13.3) is

$$\tau \approx \frac{A_f \tau_f}{A} \tag{13.10a}$$

Further assuming that the contact shear resistance is a material constant, the pressure dependence of strength is

$$\frac{d\tau}{d\sigma_e} \approx \frac{\tau_f}{A}\frac{dA_f}{d\sigma_e} \tag{13.10b}$$

If the pressure dependence is constant and there is low cohesion, as is often observed in laboratory tests for dry weak materials [e.g., *Scruggs and*

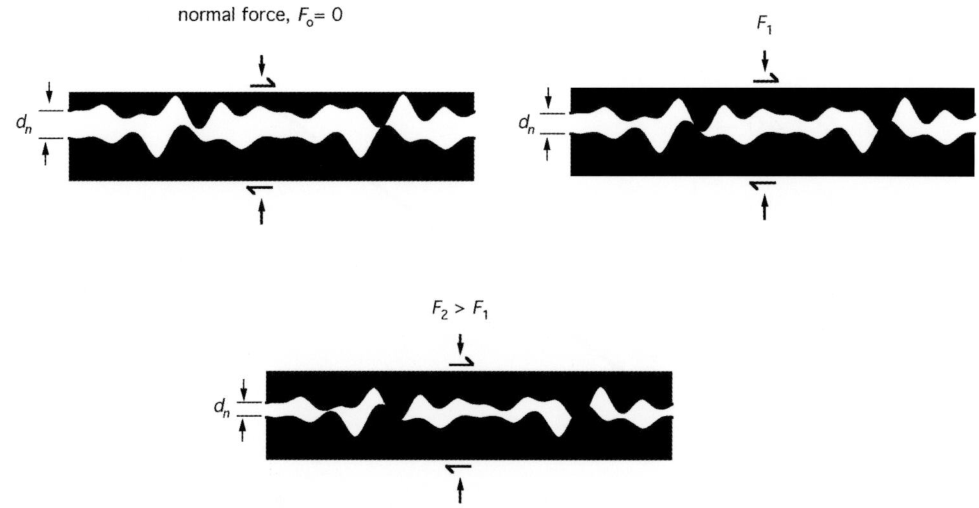

Figure 13.15 Schematic diagram of two rough surfaces in contact as normal stress is increased, clockwise from top left. Direct observations of rough surfaces in contact show that most of the surface area is not in contact, that applied normal forces F_n are concentrated at few asperity contacts across the discontinuity, and the fault contains open pore space [e.g., *Dieterich and Kilgore*, 1994]. As the normal force is increased, the asperity contact area increases through elastic or inelastic processes, and the dilatant volume is reduced.

Tullis, 1998] (fig. 13.16), then sliding friction adheres to Amonton's law (equation (13.1b)).

For constant pressure dependence, $dA_f/d\sigma_e$ in equation (13.10b) must be constant such that contact area increases in proportion to the applied normal stress. An additional requirement of constant pressure dependence follows from the relationship between contact area and macroscopic normal stress, $A_f = A\sigma_e/\sigma_f$, which produces constant $dA_f/d\sigma_e$ only when the asperity contact normal stress σ_f is constant. Thus the observation of approximately constant pressure dependence and the assumption of constant contact shear resistance requires that the contact normal stress be independent of the macroscopic normal stress ($d\sigma_f/d\sigma_e = 0$).

This section describes laboratory observations of sliding friction in the context of classic and rock-specific (rate and state) theories of friction, paying particular attention to physical processes thought to be underlie the observations, emphasizing the contrasts between weak and strong materials, and considering the implications of these contrasts for seismicity. An objective of theories of sliding friction is to explain the linear pressure dependence that is observed for virtually all materials [*Bowden and Tabor*, 1950; *Greenwood and Williamson*, 1966] including dry rock [*Dieterich and Kilgore*, 1996]. As friction theories assume nominally flat surfaces that are rough only at small scales

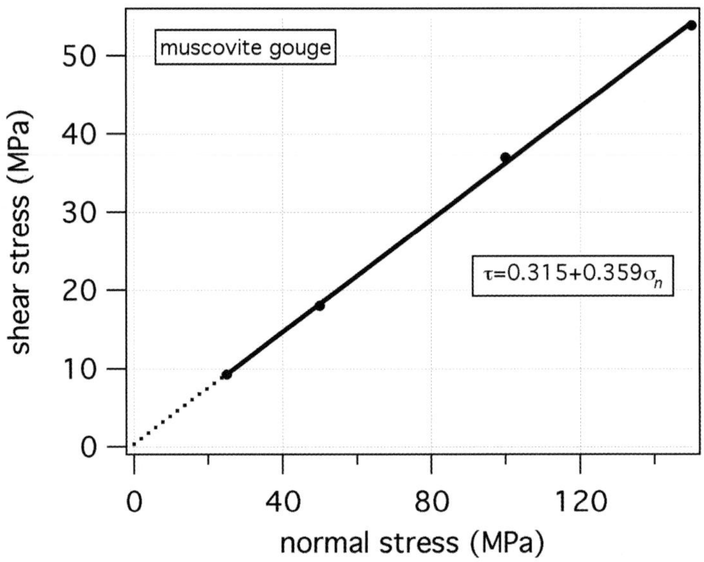

Figure 13.16 Linear pressure dependence of weak fault zone, a 1-mm layer of muscovite gouge. Friction is constant over the normal stress range, and cohesion (0.32 MPa) is low as expected for fault gouge at room temperature [*Scruggs and Tullis*, 1998].

[e.g., *Bowden and Tabor*, 1950; *Greenwood and Williamson*, 1966], these theories can be used to interpret laboratory rock friction results.

Friction Due to Inelastic Asperity Contact

One easy way to have the asperity contact normal stress be independent of the applied macroscopic normal stress is to assume that contact stress is limited by a material property. Classic friction theory [*Bowden and Tabor*, 1950] assumes that the stress at the asperity contacts is limited by the material yielding strength $\sigma_f = \sigma_y$, usually equated with the indentation hardness. At yield the average contact area is $\bar{A}_f = F_n / n\sigma_y$, where n is the number of contacts. If the normal stress $\sigma_e = F_n / A$ is increased, the contact area $\bar{A}_f = A\sigma_e / \sigma_y$ increases in proportion owing to the fixed yielding strength. If shear load F_s is applied to the surface sufficient to cause shear displacement, the contact shear yield strength is $\tau_f = F_s / n\bar{A}_f$. The shear stress is $\tau = F_s / A$. Thus the shear $\tau = \tau_f \bar{A}_f / A$ and normal stresses can be written without specific reference to the forces. So long as τ_f and σ_y are material constants, the ratio of shear stress to normal stress is a constant

$$\mu = \frac{\tau}{\sigma_e} = \frac{\tau_f}{\sigma_y} \tag{13.11a}$$

the coefficient of friction [*Bowden and Tabor*, 1950]. This theory of friction was developed to explain friction of metals that deform at highly stressed contacts by dislocation creep. This particular yielding mechanism is not possible for most silicates at room temperature, but there is considerable evidence that yielding is possible at room temperature for common minerals [e.g., *Scholz and Engelder*, 1976; *Dieterich and Kilgore*, 1994] (see Second-Order Effects section below).

Friction Due to Elastic Asperity Contacts

For rough surfaces, increasing normal stress produces larger area at individual contacts and an increasing number of contacts as the rough surfaces are brought closer together. An increase in the number of contacts tends to reduce the average contacting area somewhat. Experimental determination of the normal stress dependence of average contact dimension and contact stress of rough surfaces in contact [*Dieterich and Kilgore*, 1996] suggest that for some materials the increase in area of individual contacts is nearly exactly counterbalanced by an increase in the number of contacts such that the average contact dimension is essentially independent of normal stress (J. Dieterich, personal communication 2003). At the same time, the average contact stress remains nearly constant and the total contact area increases linearly with normal stress. While this result is not inconsistent with yielding at asperity contacts, it suggests that the observed friction is determined to a large degree by the statistics of the surface roughness. If the surface roughness is key, then linear pressure dependence can arise even if the contact-scale interactions are elastic rather than inelastic [*Greenwood and Williamson*, 1966]. The distribution of asperity heights on man-made and some natural surfaces is such that increments of increased normal stress bring the appropriate number of new asperities into contact to maintain constant average contact stress [*Greenwood and Williamson*, 1966; *Walsh and Grosenbaugh*, 1979; *Brown and Scholz*, 1985]. *Greenwood and Williamson* [1966] showed that for surfaces with roughness of laboratory test surfaces, with asperities assumed elastic, the total area of asperity contact increases essentially linearly over 4 orders of magnitude of increased load. Though they assume a nonlinear elastic contact compliance, the resulting linear pressure dependence is entirely independent of the assumed contact-scale elastic interaction. Instead, the pressure dependence of contact area is determined by the statistics of the surface topography. In this case the contact normal stress σ_f is constant but not a material constant and the friction coefficient

$$\mu = \frac{\tau}{\sigma_e} = \frac{\tau_f}{\sigma_f} \qquad (13.11b)$$

is constant.

Laboratory rock friction appears to have both elastic [e.g., *Scholz*, 1988] and inelastic [*Dieterich*, 1972] components. This is discussed in more detail in the Relative Roles of Elastic and Inelastic Processes section below. The important point is that the pressure-dependent behavior of dry sliding strength is due to dilatancy. As in equation (13.10a), the pressure dependent reflects changing of contacting area with normal stress that accompanies changes in dilatant volume. In this regard, slip of preexisting faults and failure of intact rock is similar. The cause of intrinsic weakness of preexisting faults is also the same as for intact rock; materials with weak pressure dependence are weakly dilatant and strongly anisotropic, with a preferred orientation in the shearing direction [e.g., *Morrow et al.*, 2000, *Moore and Lockner*, 2004]. The weakest materials, talc, graphite, wet smectite, and chrysotile serpentinite, slide between cleavage surfaces that produce little rigid dilatancy and minimal tensile fracturing. Thus, in general for preexisting faults, dry pressure dependence directly reflects the dilatancy associated with shear.

Second-Order Effects

Time Dependence While simple inelastic theories of friction imply instantaneous deformation at asperity contacts, real materials instead show time-dependent yielding. In static contact, time-dependent yielding results in both increasing contact area and fault normal closure. Figure 13.17 shows changes in asperity contact area with time of static contact in transparent samples [*Dieterich and Kilgore*, 1994]. Contact area increases approximately linearly with the log of contact time. Similarly, in laboratory slide-hold-slide tests [*Dieterich*, 1972; *Johnson*, 1981] (fig. 13.18a) fault normal convergence occurs at constant normal stress when sliding velocity is approximately zero [*Beeler and Tullis*, 1997]. In hold tests the fault surface slides initially at a steady-state velocity and then the loading velocity is set to zero and held for some duration, the hold time t_h. During the hold the fault surfaces are held in a condition approaching stationary contact, and there is measurable closure of the fault surfaces (fig. 13.18c), consistent with creep normal to the fault surface at asperity contacts, increase of contact area, and reduction of the dilatant volume of the fault. The closure is linear or nearly linear in log time, consistent with the area change results. This fault normal closure is also associated with an increase in frictional resistance (fig. 13.18b) [*Dieterich*, 1972; *Nakatani and Mozichi*, 1995, *Karner and Marone*, 1998], which has been previously observed for other materials [*Dokos*, 1946; *Rabinowicz*, 1958]. The increase in resistance is measured somewhat indirectly. Following the period of approximately stationary contact, the loading velocity is returned to its original value resulting in progressive increase in shear stress. The shear stress increases, reaches a peak τ_{peak}, and subsequently returns to its previous steady-state value. The amount of strengthening that occurs during the hold period is given by the difference between τ_{peak} and the steady-state strength. *Dieterich* [1972] found $\mu_{peak} = \tau_{peak}\sigma_e$ to increase linearly with log of

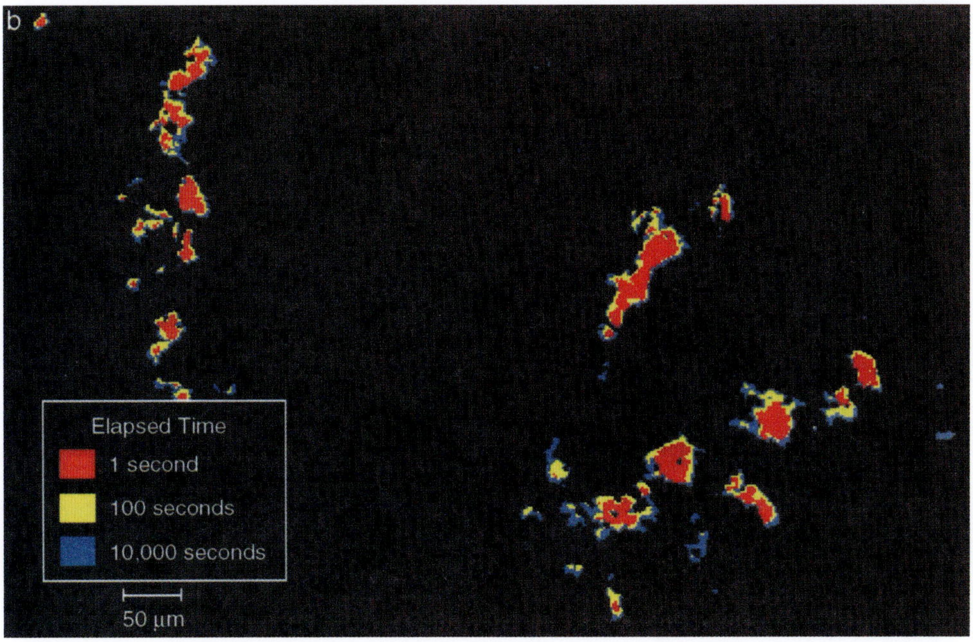

Figure 13.17 Three superimposed transmitted light photomicrographs of asperity contact area on an analogue laboratory fault surface. Reprinted from Pure and Applied Geophysics, 143, *Dieterich and Kilgore*, Direct observation of frictional contacts: New insights for state-dependent properties, 283–302, Copyright (1994), Birkhäuser Verlag, Switzerland. The pictures were taken at 1, 100, and 10,000 s contact time and the contacting areas are colored to illustrate the time-dependent growth of contact area by yielding.

hold time $b\ln t_h$, but it is a second-order effect, for example a strengthening coefficient for granite of 0.01 relative to the frictional strength of 0.7.

Shear-Induced Weakening, Shear-Induced Dilatancy and Age Dependence On the basis of the hold and static contact tests, it is tempting to represent contact normal strength as time dependent, but that is incomplete. While there may be time-dependent effects ongoing during sliding at appreciable rates, friction at constant sliding rates is not an explicit function of time. Whatever yielding process occurs during approximately static contact is modulated by other effects at appreciable slip rates.

Because asperity contacts are very small, they have limited time of contact while the fault is sliding at a significant rate. Therefore any time-dependent creep process operating at an asperity has a limited duration, and there will be ongoing reduction in the contact area due to shearing. After slip equivalent to the contact dimension, the asperity will no longer be in contact. So during slip, there are two competing processes, time, which tends to increase contact area, and slip, which tends to replace yielding contacts and bring undeformed asperi-

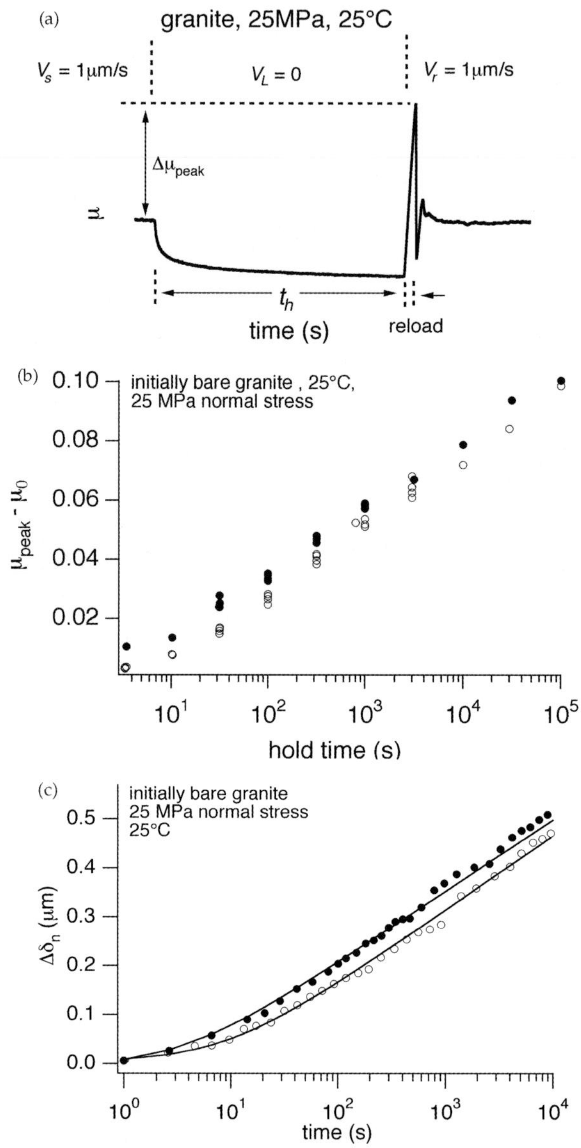

Figure 13.18 Slide-hold-tests. (a) Slide-hold-slide test procedure in friction versus time. The sample slides initially at a steady-state starting/reload velocity $V^{s/r}$. At some time the loading velocity is set to zero, and the measured strength as the fault creeps a bit due to the nonzero shear stress, relaxing the elastic distortion of the testing machine. After some length of time the sample is reloaded by resetting the load point velocity to $V^{s/r}$. Strength increases, reaches a peak μ_{peak} and subsequently returns to its previous steady-state value. (b) The amount of strengthening ($\mu_{peak} - \mu_0$) from slide-hold-slide tests on initially bare Westerly granite, as a function of hold time [after *Dieterich*, 1972]. (c) Time dependence of fault thickness measured during the hold portion of a slide-hold-slide test for initially bare Westerly granite. Solid lines are fits to the data using an empirical equation $\delta_n = \delta_{n0} + \alpha \ln (t + c_3)$, where δ_{n0} is δ_n at the start of the hold and α and c_3 are constants.

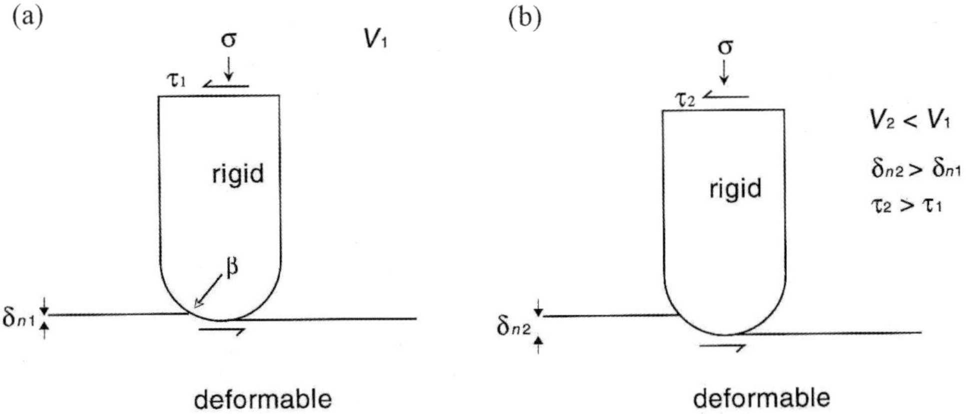

Figure 13.19 Schematic of rigid asperity model. Here, β is the radius of the spherical asperity tip. (a) High sliding velocity, with asperity riding high. (b) Low sliding velocity, with asperity sinking deeper [after *Scholz and Engelder*, 1976].

ties into contact. This concept of competition between slip and time can be represented by the average age θ of the asperity contacts. For constant slip rate *V*, average age takes the value $\theta = d_c/2V$, where d_c is a representative dimension of asperity contacts on the fault surface; conversely, during static contact $d\theta = dt$.

Interactions between time-dependent and displacement-dependent effects have implications for volume change that can be considered using an asperity model consistent with the experiments of *Engelder and Scholz* [1976] (fig. 13.19). Consider a rigid asperity at constant normal force, indenting a deformable material. Assume that the rheology of the deformable material in the absence of slip requires contact area to increase with time, consistent with the hold data (fig. 13.18) and the size variation of indentations in quartz with time [*Scholz and Engelder*, 1976] (see fig. 13.26). At steady-state sliding velocity the depth of penetration of the asperity δ_n decreases with sliding velocity, owing to the asperity shape and the reduced residence time for penetration at higher velocity. Therefore, at high sliding velocity the contact area and the contact's resistance to slip (fig. 13.19a) are small. When the velocity is decreased, there will be a transition from one steady-state depth to another (i.e., the change from that of figs. 13.19a to 13.19b), over a distance related to the asperity dimension. Fault normal strain (as measured by δ_n), contact area *A*, and shear resistance will vary similarly with displacement as the sliding velocity is varied.

For this model, at steady-state prior to an imposed step change in velocity, there is no net dilation or compaction. However, there is a dynamic balance between ongoing time-dependent compaction and ongoing shear-induced dilatancy. Using the appropriate requirement at steady-state $d\varepsilon_n = 0$, the balance can be written as

$$\left[\left(\frac{\partial \delta_n}{\partial \delta_s}\right)_t\right]_{ss} = -\frac{1}{V}\left[\left(\frac{\partial \delta_n}{\partial t}\right)_{\delta_s}\right]_{ss} \qquad (13.12)$$

If the velocity is increased, the fault will dilate. The amount of time-dependent interpenetration of the substrate by the asperity depends on the residence time of the asperity over a particular spot, and the residence time depends on the sliding velocity; thus fault-zone thickness is inherently velocity dependent. That time-dependent yielding is modulated by slip-induced dilatancy during sliding is evident in experiments where fault normal motion is monitored during changes in sliding velocity (rate stepping tests) [*Morrow and Byerlee*, 1989; *Marone et al.*, 1990; *Wang and Scholz*, 1994]. Changes in sliding velocity produce an immediate displacement-dependent change in fault normal closure (fig. 13.20).

These observations have been incorporated into rock friction theory by allowing the asperity contact strength to be an explicit function of age [e.g., *Dieterich*, 1979]. Using the Amonton form equation (13.11), the normal strength is

$$\sigma_f = \sigma_o + f(\theta) \tag{13.13a}$$

where σ_0 is a reference value of contact stress and f is some function of age that produces the log time-dependent effect seen in the area, closure, and strength data during static contact and the implied changes in contact area with slip (figs. 13.18 and 13.20).

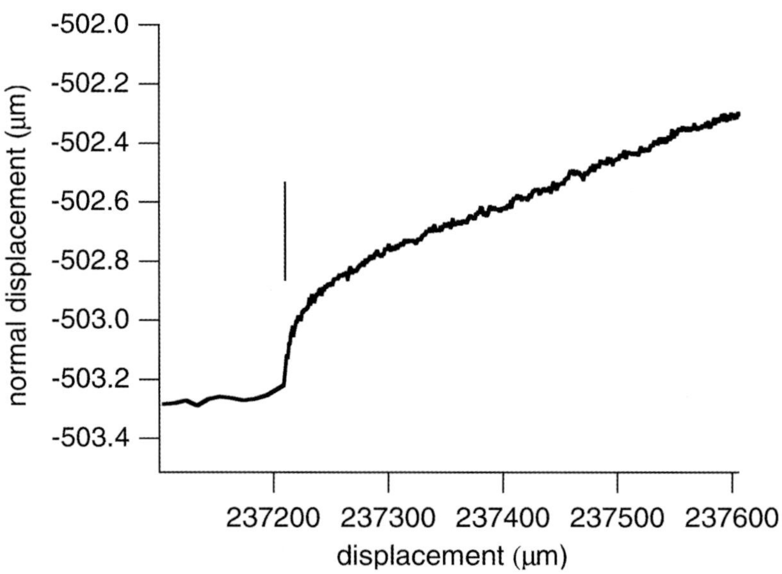

Figure 13.20 Response of fault thickness to changes in sliding rate for a 1-mm-thick granite gouge layer at 25 MPa normal stress and room temperature. An increase in sliding from 1 to 10 μm/s is imposed at displacement marked by the vertical line. The fault zone dilates in response to the change of deformation rate [after *Morrow and Byerlee*, 1989; *Marone et al.*, 1990].

Rate Dependence Sets of slide-hold-slide tests at different loading/reloading rates [*Kato et al.*, 1992; *Marone*, 1998] show more apparent restrengthening during the hold at higher loading velocity (fig. 13.21). This suggests that fault strength depends on slip rate or loading rate in addition to age. A slip-rate effect is also seen in rate stepping tests where slip velocity is changed. An increase in sliding velocity produces the age-dependent effect discussed in the previous section: a displacement-dependent reduction in fault strength and dilation. Superimposed on this is an immediate change in fault strength of the opposite sense (fig. 13.22). Thus, in both hold and rate stepping tests, there is a positive rate-dependent effect. This viscous effect has been incorporated into friction theory by allowing the shear strength of asperity contacts to depend on sliding rate [*Dieterich*, 1979]

$$\tau_f = \tau_o + g(V) \tag{13.13b}$$

The observations show that the rate dependence has the same logarithmic form as seen in rock fracture tests (equation (13.5a)) and that the effect is second order (for example, a rate-dependent coefficient a for granite of 0.008 relative to the frictional strength of 0.7).

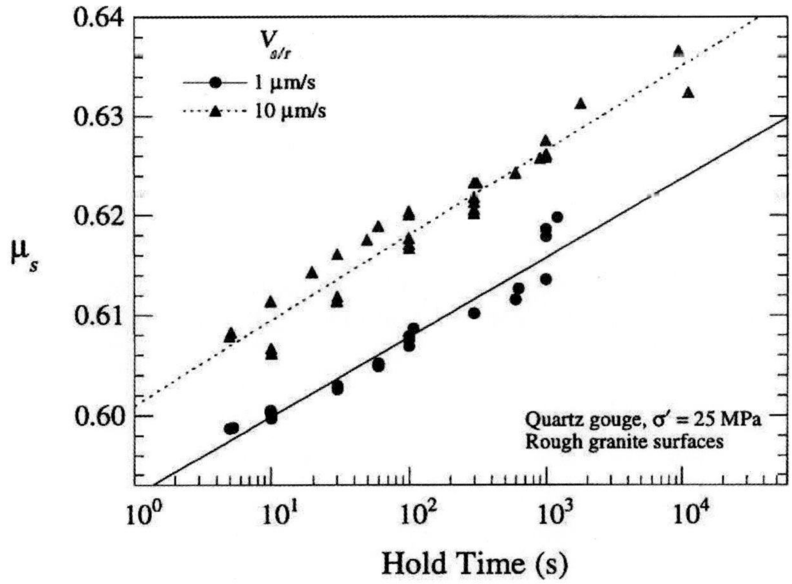

Figure 13.21 Direct measure of the instantaneous rate dependence of rock friction [*Marone*, 1998]. Reprinted, with permission, from the Annual Review of Earth and Planetary Sciences, Volume 26 (c) 1998 by Annual Reviews, www.annualreviews.org. Peak strengths from slide-hold-slide tests conducted at different load/reloading rates. The peak strengths depend systematically on the sliding rate.

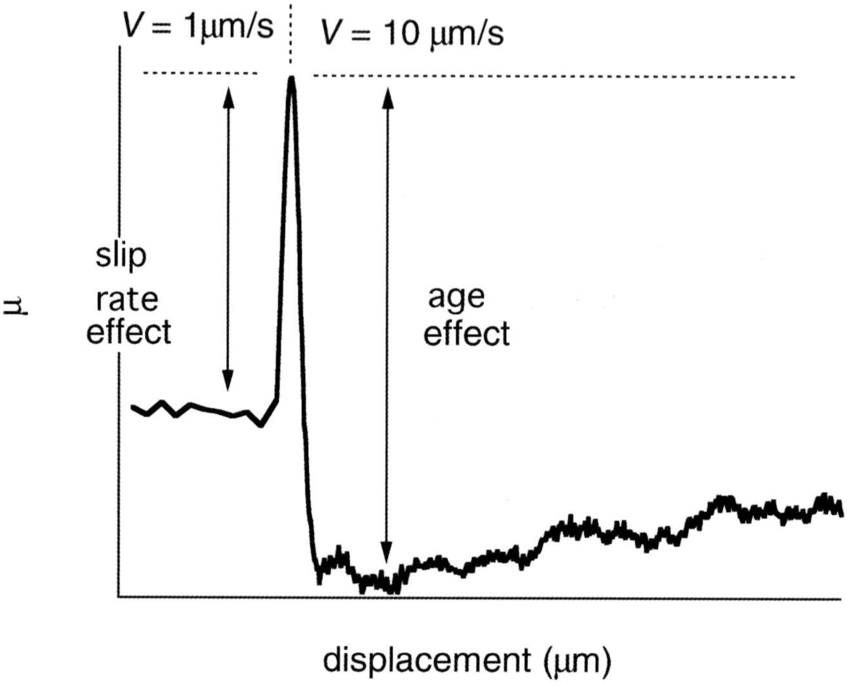

Figure 13.22 Fault strength for granite at 25 MPa normal stress and room temperature in response to an imposed change in sliding velocity. The loading velocity is increased by an order of magnitude at displacement corresponding to the vertical dashed line. The net rate dependence $d\mu_{ss}/d\ln V$ is the sum of two opposing effects, the instantaneous rate effect and an age dependence that is reset by constant slip rate.

Constitutive Relations for Friction

The second-order dependencies of fault strength on age and slip rate (equations (13.13a) and (13.13b)) can be combined with equation (13.10) into a friction relation

$$\mu = \frac{\tau_f}{\sigma_f} = \frac{[\tau_o + g(V)][\sigma_o - f(\theta)]}{[\sigma_o + f(\theta)][\sigma_o - f(\theta)]} \tag{13.13c}$$

Discarding the product of second-order terms leads to the form of friction relation developed by *Dieterich* [1979], which is the basis of *Ruina* [1983] and *Rice and Ruina's* [1983] rate and state friction equations:

$$\mu = \frac{\tau_o}{\sigma_o} + \frac{g(V)}{\sigma_o} - \frac{\tau_o f(\theta)}{\sigma_o^{\,2}} \tag{13.13d}$$

On the right side of equation (13.13d), the first term is first order, the second term is the second-order positive direct dependence of contact shear resistance on sliding velocity, and the third term is the second-order negative dependence of contact normal strength on contact age. Experimental observations constrain the rate and age dependencies: $g(V) = \mathbf{A}\ln(V/V_0)$ (fig. 13.22) and $g(\theta) = \beta\ln(\theta/\theta_0)$ (fig. 13.18). Expressing the first-order term as $\tau_0/\sigma_0 = \mu_0$ gives

$$\mu = \mu_o + \frac{\mathbf{A}\ln(V/V_0)}{\sigma_o} - \frac{\mu_o\beta\ln(\theta/\theta_0)}{\sigma_o} \tag{13.14}$$

Some implications of equation (13.14) are discussed subsequently in the Some Physical Implications section below. Equation (13.14) can be expressed using more conventional notation as

$$\mu = \mu_o + a\ln(V/V_0) + b\ln(V_0\theta/d_c) \tag{13.15a}$$

where the standard coefficients are $a = A/\sigma_0$ and $b = -\mu_0\beta/\sigma_0$ and $\theta_0 = d_c/V_0$. An expression for the slip and time dependencies that underlie dependence of friction on age is derived from a partial derivative expansion of $\theta = h(\delta, t)$ and the requirements that age increases linearly with time in the absence of slip (fig. 13.18) and decreases exponentially with slip, $\left(\frac{\partial\theta}{\partial t}\right)_\delta = 1$ and $\left(\frac{\partial\theta}{\partial d}\right)_t = -\frac{\theta}{d_c}$, respectively. This leads to [*Ruina*, 1983; *Linker and Dieterich*, 1992]

$$\frac{d\theta}{dt} = 1 - \frac{\theta V}{d_c} \tag{13.15b}$$

The steady-state value of age is $\theta_{ss} = dc/V$ and the steady-state value of friction is $\mu_{ss} = \mu_0 + (a\text{-}b)\ln(V/V_0)$.

Simulations of Aseismic Laboratory Observations

Equations (13.15) reproduce the observations of time, slip, and rate dependence seen in aseismic laboratory hold and rate stepping tests and simulations with equations (13.15) illustrate how constitutive parameters a and b can be inferred from experimental measurements. The log time dependence of the peak stress in hold tests is reproduced and the change in peak stress with log of hold time is the constitutive parameter b (fig. 13.23a). In hold tests conducted at different loading reloading rates, the difference in hold peak strengths from the different loading reloading rates is $a\ln(V_2/V_1)$. The spac-

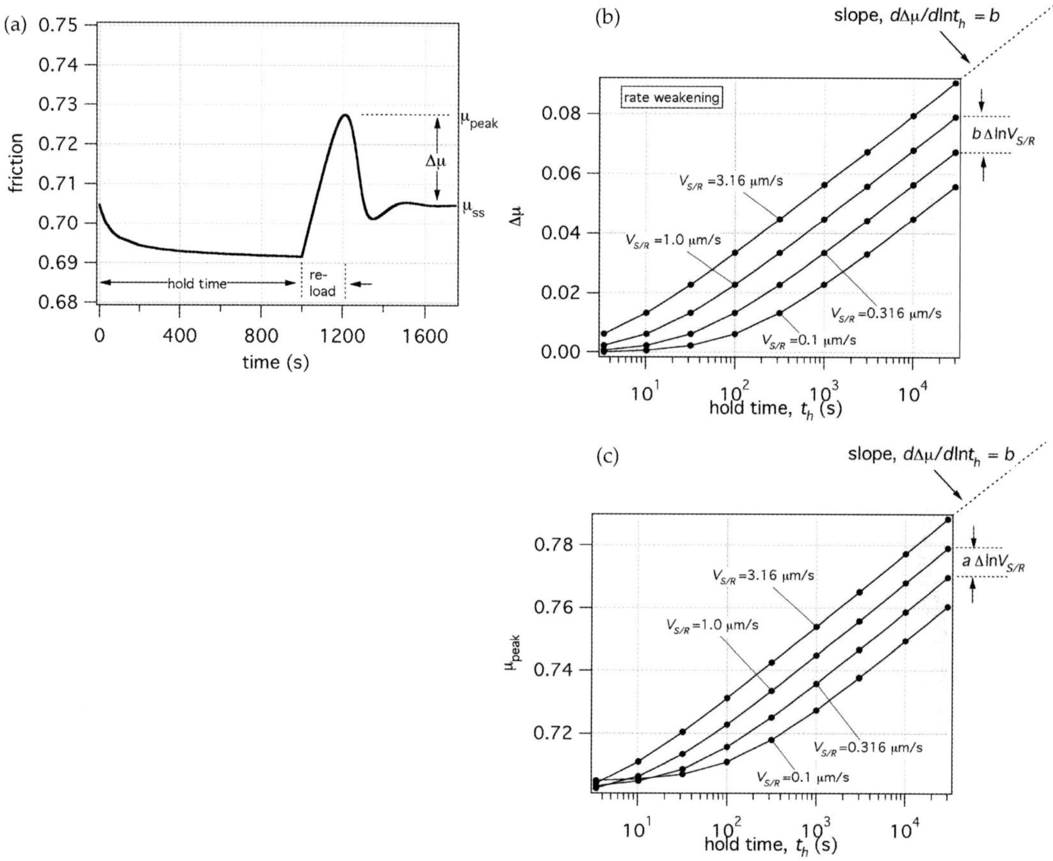

Figure 13.23 Numerical simulations of slide-hold-slide tests with rate and state constitutive equations (13.15). For all simulations the starting and reloading velocities are equal, $a = 0.008$, $b = 0.010$, $d_c = 5$ μm, $\mu_0 = 0.7$, and $V_0 = 0.1$ μm/s. The machine stiffness normalized by normal stress k/σ_n =0.0021 1/μm. This is a compliant testing machine; however, the individual constitutive parameters a and b are directly measured in the simulated tests. (a) Simulation of a single 1000-s slide-hold-slide test. (b) Slide-hold-slide delta friction versus log hold time for 36 different hold time and loading/reloading velocities. Test at the same starting/reloading velocity are connected by lines. The spacing between the sets and the slope of the lines reflect the rate of aging. (c) Peak strength versus log hold time for 36 different hold time and loading/reloading velocities. Tests at the same starting/reloading velocity are connected as in figure 13.23b. The spacing between the sets reflects the instantaneous rate dependence, and the slope reflects the rate of aging.

ing between the changes in friction between the peak and steady-state Δμ for the different sliding velocities is $b\ln V_2/V_1$ for rate-weakening faults (fig. 13.23b); for rate-strengthening faults the spacing is $(2a\text{-}b)\ln V_2/V_1$. Rate stepping test simulations produce the typical response of a direct rate dependence

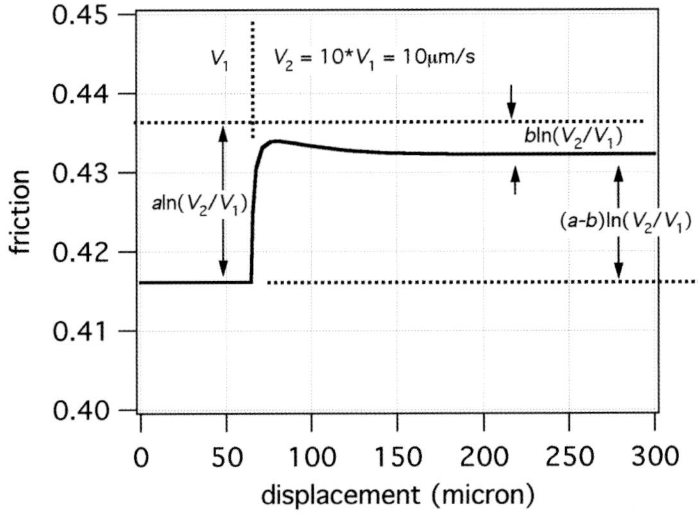

Figure 13.24 Numerical simulation of a rate step test with rate and state constitutive relation (13.15), $a = 0.008$, $b = 0.001$, $d_c = 20$ μm, and $\mu_0 = 0.4$. The machine stiffness normalized by normal stress is $k/\sigma_n = 0.0021$ 1/μm. The steady-state rate dependence $d\mu/d\ln V = (a\text{-}b)$ is measured directly in this test; however, owing to the finite compliance of the system, the individual parameters *a* and *b* are not directly apparent from the simulation. This indicates that, unlike for hold tests (fig. 13.23), the individual parameters *a* and *b* cannot be directly inferred from rate stepping tests and must be retrieved from data using an indirect procedure such as numerical inversion.

followed by the slip-dependent change of age. The net change of friction is $(a\text{-}b)\ln V_2/V_1$ (fig. 13.24).

Stick-Slip and the Condition for Frictional Instability

Seismic slip rather than aseismic sliding is also observed in some laboratory experiments (fig. 13.25). Stick-slip is periodic or quasi-periodic seismic stress drop resulting from constant loading. Simulations using equations (13.15) produce stick-slip at the appropriate loading conditions. Stick-slip simulations with equation (13.15) and direct observations show that the steady-state rate dependence controls aspects of the cycle duration and amplitude [*Gu and Wong*, 1991; *Wong and Zhao*, 1990] Peak strength goes as $d\mu_{\text{peak}}/d\ln V_L = (b\text{-}a)$ [*Gu and Wong*, 1991]. The amplitude of the static stress drop also varies with loading rate V_L approximately as $d\mu_{\text{static}}/d\ln V_L \approx (b\text{-}a)(1 + \zeta)$, where ζ is the fractional overshoot. If earthquakes correspond to slip on preexisting fault surfaces, as opposed to failure of intact rock, then the conditions leading to

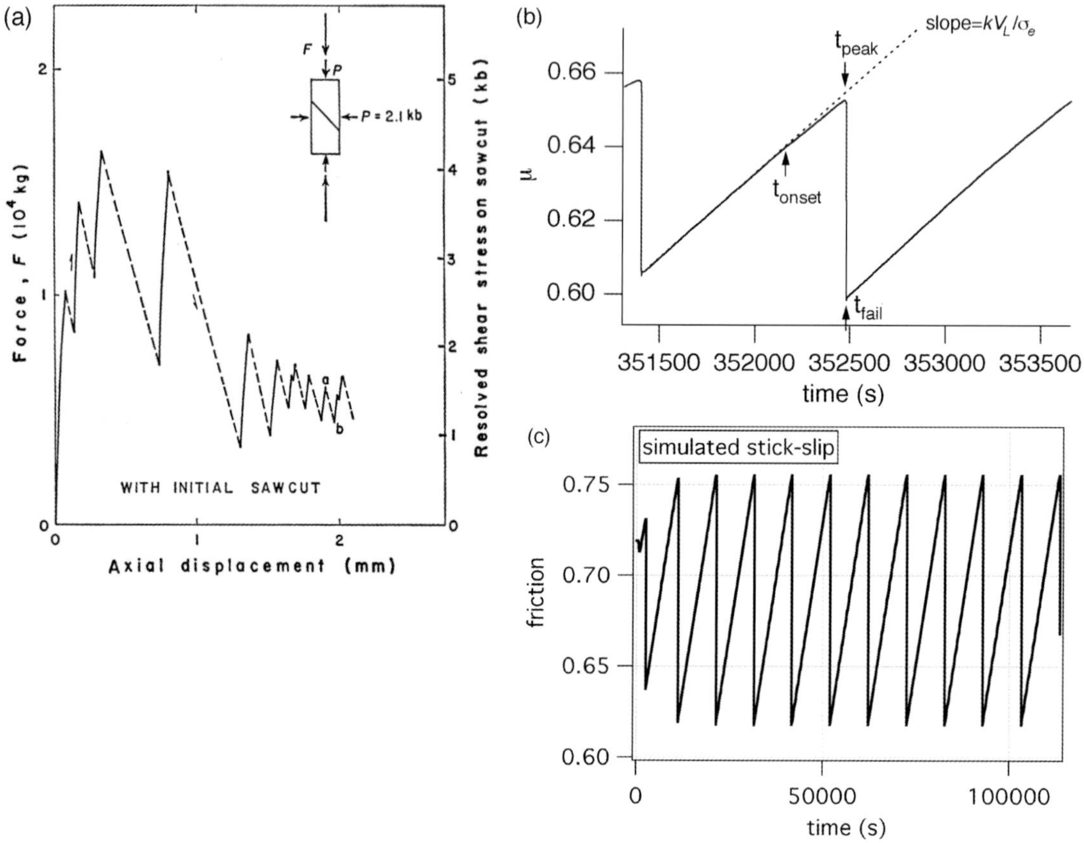

Figure 13.25 Stick-slip sliding. (a) Original stick-slip sliding on sawcut granite fault at 210 MPa confining pressure and room temperature. Reprinted with permission from Science, 153, *Brace and Byerlee*, Stick-slip as a mechanism for earthquakes, 990–992, Copyright [1966] AAAS. (b) A laboratory experiment on a granite sawcut at 50 MPa confining pressure showing evidence of precursory slip; the loading deviates from the linear elastic load indicating on-going slip prior to peak strength. (c) Simulation of laboratory stick-slip using rate and state variable equation (13.15), $a = 0.008$, $b = 0.012$, $d_c = 0.8$ μm, $V_0 =$ 1 μm/s, and $\mu_0 = 0.7$. The machine stiffness normalized by normal stress is $k/\sigma_n = 0.00137$ 1/μm. This is dynamic slip of single degree of freedom slider block and simulates the effects of radiated energy using the radiation damping procedure of *Rice* [1993]. Normal stress is 21.2 MPa, shear wave velocity is 3000 m/s, and the shear modulus is 30 GPa. The calculation is started from steady-state sliding at 0.009 μm/s. The system is perturbed by increasing the loading rate to 0.01 μm/s, inducing the unstable slip.

stick-slip faulting in friction experiments are relevant to studies of earthquake nucleation. Thus stick-slip sliding is considered to be the laboratory equivalent of repeating earthquake cycles [*Brace and Byerlee,* 1966], and the onset of slip instability is analogous to earthquake nucleation.

If the elastic stiffness of the loading system is expressed as $d = d\tau/d\delta$, where δ is fault slip, then the requirement for slip instability at constant normal stress can be expressed in terms of a critical stiffness [*Dieterich*, 1979]

$$k_c = -\frac{d\mu}{d\delta}\sigma_e \tag{13.16a}$$

where $d\mu$ is the drop in friction. A slip instability can occur for $k < k_c$. This relation has been verified experimentally for faults with artificial but well-characterized displacement weakening [*Lockner and Byerlee*, 1990]. However, for the constitutive equation (13.15) in addition to the intrinsic displacement dependence of strength, there are time and rate dependencies that must be considered. Instead of $d\mu/d\delta$, for equation (13.15) the sign of the steady-state rate dependence $d\mu_{ss}/d\ln V$ is key in determining whether faulting is potentially seismic. When the dependence of steady-state frictional strength on the logarithm of sliding velocity is negative ($d\mu_{ss}/d\ln V < 0$), any perturbation that causes the fault to accelerate will lead to a loss of strength and will encourage further slip acceleration [*Rice and Ruina*, 1983]. In this case, slip instability is possible. Conversely, for neutral or positive rate dependence ($d\mu_{ss}/d\ln V \geq 0$), increasing slip rate produces no strength decrease so sliding will be stable. Linearized perturbation analysis of equation (13.15) [*Rice and Ruina*, 1983; *Ruina*, 1983] results in an expression for critical stiffness analogous to equation (13.16a)

$$k_c = -\frac{d\tau_{ss}}{d\ln V}\frac{1}{d_c} = \frac{(b-a)\sigma_e}{d_c} \tag{13.16b}$$

where τ_{ss} is the steady-state frictional resistance and d_c is the characteristic displacement of the age-dependent term. Accordingly, instability is favored by more negative values of $d\tau_{ss}/d\ln V$ and by low values of k [*Rice and Ruina*, 1983; *Gu et al.*, 1984].

Laboratory measurements of (*a-b*) have been conducted using initially bare rock surfaces (ground, flat, parallel rock roughened with abrasive) and simulated gouge layers (powdered rock flour sheared between rigid blocks). Thick quartzofeldspathic gouge layers have positive rate dependence unless shear displacement becomes localized within the layer [*Dieterich*, 1981; *Marone et al.*, 1990; *Marone and Kilgore*, 1993] (also see review of *Marone* [1998]). While gouge layers show variability in the sign of (*a-b*), initially bare rock surfaces of quartzofeldspathic rock [*Dieterich*, 1978b; *Tullis and Weeks*, 1986] are rate weakening (*b-a* > 0). Slip between bare rock surfaces remains localized even at large displacement and such faults show little tendency for fault-zone width growth after an initial period of rapid wear [*Biegel et al.*, 1989] because they are flat at long wavelengths (see *Power et al.* [1988] and as discussed below).

In contrast to experiments, a continually expanding fault zone is expected for natural faults [*Robertson*, 1983]. Increased thickness of natural fault zones may result from geometrical and dynamic effects during shear. Natural rock surfaces such as fracture and joint surfaces have a fractal roughness [*Brown and Scholz*, 1985], As a result, increasing displacement of an initially matched shear fracture (protofault) leads to an increasing volume of mismatched surface irregularity and therefore increasing wear material (gouge) with displacement [*Power et al.*, 1988]. During earthquake slip, large dynamic stresses that develop in front of an advancing shear rupture during propagation may be higher on off-fault planes than on the fault plane itself [*Rice*, 1980]. These high stresses may induce branching or off-fault deformation that may lead to nonplanar ruptures, fault surface roughening and gradual growth of the damage zone with successive earthquakes. Thus natural large displacement fault zones are wide, for example, the San Andreas fault zone, with total offset of hundreds of kilometers, has a thickness of as much as a kilometer in places [*Sibson*, 2002]. The experimental observations thus require that slip during earthquake nucleation is localized so that $d\mu_{ss}/d\ln V < 0$, even within very wide fault zones. At least in experiments to date, fault slip in simulated gouge layers at the largest displacements measured does not localize to the degree that the rate dependence is negative [*Beeler et al.*, 1996], though localization does occur and persist at short displacements in some experimental geometries [*Marone*, 1998]. It is not known exactly how such localization arises in experiments or in mature natural fault zones. The remainder of this paper focuses on the physical implications of slip between initially bare rock surfaces where shear is localized and which are better understood and less complicated than measurements made on simulated gouge.

Some Physical Implications

Constraints on Contact-Scale Mechanism, Mostly for Strong Materials

Laboratory measurements have not determined the physical processes responsible for the rate or time-dependent mechanisms. A significant limitation to equation (13.15) in its application to earthquake phenomenon is this uncertain physical basis. So long as rate and state effects are not attributed to thermally activated deformation mechanisms, it will remain difficult to extrapolate the observations in pressure and temperature to conditions other than those of the particular experiments, to establish the dependence of the effects on environmental factors such as the presence or absence of fluids and fluid chemistry, and to determine how the specific rate-dependent effects compare with others that might compete at significant depth in the Earth. However, some inferences about the mechanisms underlying rock friction can be made from existing laboratory data. For instance, the logarithmic dependencies of friction on slip rate and on contact time are consistent with low temperature deformation studies and chemical reaction rates, as follows. At modest temperatures and pressures, deformation mechanisms at highly stressed asperity contacts such as subcritical crack growth [*Atkinson*, 1984] and dislocation glide [*Evans*, 1984] have deformation rates that depend logarithmically on pressure or stress, e.g.,

$$v = v_* \exp(-E/RT) \tag{13.17a}$$

where the activation energy $E = E_* - \sigma_c \Omega$ is the difference between a nominal value and the product of stress at the crack tip σ_c and the activation volume Ω [e.g., *Rice et al.*, 2001, *Nakatani*, 2001]. Equation (13.17a) is the Arrhenius equation that is characteristic of processes controlled by the rate of chemical reaction.

Rate Dependence If the expression (13.17a) is solved for stress as a function of rate, stress is

$$\sigma_c = \frac{E_*}{\Omega} + \frac{RT}{\Omega} \ln \frac{v}{v_*} \tag{13.17b}$$

That equation (13.17a) underlies second-order dependencies of friction and rock failure on time and on slip rate is a common argument in the rock mechanics literature. As previously discussed in the paper, many studies since *Scholz* [1968c] have related the rate dependence of rock fracture strength (equation (13.3a)) to the rate dependence of single tensile crack growth rate in double-cantilever experiments on glass [*Lawn*, 1993] and single mineral crystals [*Atkinson*, 1984]. More recently, it was found that both friction and intact failure tests yield values of the *a* coefficient in the range 0.003–0.009 for granite at low temperature [*Lockner*, 1998]. Similar agreement between *a* for friction and intact rock was found by *Tullis and Weeks* [1987] for carbonate rocks; they argued that a common mechanism was responsible. In the case of granite, the similarity between rock failure and rock friction rate dependence lead *Lockner* [1998] to suggest that subcritical crack growth is the mechanism underlying this effect in both friction and rock failure of granite.

Direct comparison between the rate dependence of subcritical crack growth and that of rock friction can be made using a normalization scheme that follows from one proposed by *Tullis and Weeks* [1987] for comparing results of rock friction to intact failure tests. For crack growth, rather than equation (13.5b), which is specific to water-assisted crack growth (stress corrosion), a general form is

$$v = v_* \exp\left(\frac{-E_* + \varsigma K_i}{RT}\right) \tag{13.17c}$$

where ζ is an empirical constant and E_* is an activation energy [*Wiederhorn and Bolz*, 1970. K_i is the stress intensity factor that relates the applied macroscopic stress σ, crack dimension c, and the specific experimental geometry to the crack tip stress as $K_i \propto \sigma\sqrt{c}$. In crack growth experiments, K_i is independent of crack dimensions and $K_i \propto \sigma$ [*Atkinson and Meredith*, 1987b]. Expressing equation (13.17c) with crack velocity v as the independent variable leads to

$$\sigma \propto \left[\frac{E_*}{\varsigma} + \frac{RT}{\varsigma}\ln\frac{v}{v_*}\right] \quad (13.17d)$$

Normalizing this expression by the resistance σ_* resulting from slip at the reference slip rate v_* is

$$\frac{\sigma}{\sigma_*} = 1 + \frac{RT}{E_*}\ln\frac{v}{v_*} \quad (13.17e)$$

Here, the normalized rate dependent coefficient (RT/E_*) is the actual rate dependence (RT/ζ) divided by the nominal stress (E_*/ζ), and can be interpreted as the rate dependence of the remotely applied stress driving crack propagation, divided by the nominal value of that stress. The unknown constant of proportionality is eliminated by the normalization.

An analogous expression for friction can be derived from the standard rate and state equation (13.15) by following the same procedure of normalizing by an absolute friction μ_0 resulting from the arbitrary slip rate V_0

$$\frac{\mu}{\mu_0} = 1 + \frac{\mathbf{A}}{\tau_0}\ln\frac{V}{V_0} + \frac{\beta}{\sigma_0}\ln\frac{\theta}{\theta_0} \quad (13.17f)$$

Note with reference to equation (13.15a), the coefficient of the rate-dependent term ($\mathbf{A}/\tau_0$) is the actual contact shear strength coefficient normalized by the nominal contact shear strength and has the same interpretation as the coefficient in equation (13.17e). Similarly, the coefficient of the age-dependent term in equation (13.17f) (β/σ_0) is the actual coefficient normalized by the nominal contact normal strength.

To compare the rate dependence of crack growth experiments to friction data, the rate of crack propagation v must be equated with the fault slip rate V or fault-zone strain rate. The difference between the fault slip rates and crack growth rates in the raw experimental data is very large; rate stepping test velocity for friction is in the 1–10 μm/s range while crack growth rates are up to 1×10^{-2} m/s. If v is equated with V, implying that frictional slip produces crack growth at the same rate, and the reference velocity is 1.8 μm/s, then for marble the ratio $RT/E_* = 0.055$ from equation (13.17e) does not compare favorably with the analogous friction parameter $a/\mu_0 = 0.011$. With this assumption of $v = V$ a discrepancy of similar size results for quartz and granite. While this discrepancy might be due to very large differences in the rate of crack growth associated with a particular fault sliding velocity, the sense of the difference in velocity necessary to account for the difference in normalized rate dependence is unexpected. To produce a normalized rate dependence similar to that seen in the friction experiments, the reference velocity for crack growth must greatly exceed the compressive wave speed. Obviously, frictional slip cannot produce subcritical crack growth velocity approaching the elastic wave speed.

In the following estimates of the normalized rate dependence of crack growth in geologic materials the crack growth during laboratory friction experiments is arbitrarily assumed to occur at subcritical rates of 1 m/s (v_* in equation (13.17e) is 1 m/s). This is ~1 order of magnitude higher than the largest rates in subcritical crack growth experiments. The normalized coefficient for subcritical crack growth in Westerly granite at room temperature from *Atkinson and Meredith* [1987a] is 0.021, while the analogous normalized rate-dependent coefficient from Westerly rock friction tests [e.g., *Beeler et al.*, 1996] is 0.011, roughly ½ the crack growth value. The normalized rate-dependent coefficient from room temperature experiments on synthetic quartz and natural quartzite [*Atkinson and Meredith* [1987a, 1987b] is 0.033–0.035, while the coefficient from quartzite friction tests is 0.009, less than ⅓ the crack growth value. Essentially identical results are found for Carrara marble; for crack growth the room temperature normalized rate dependence is 0.032, while the normalized friction data of *Tullis and Weeks* [1987] has rate dependence of 0.011, about ⅓ the crack growth value.

The conclusion is that quartzofeldspathic rocks and marble show rate dependencies that are not consistent with crack growth rate dependencies. While there is great uncertainty in the proper reference crack velocity to use in comparisons of friction and crack growth data, the difference cannot be reconciled by adjustments in the rates chosen as the reference. A possible exception is the rock analogue, silica glass whose crack growth normalized rate dependence is 0.018 [*Lawn*, 1993]; similar normalized values were found by *Weeks et al.* [1991]; however, the quality of the friction data is poor.

On the basis of a comparison between the size of the rate-dependent coefficient at low-strain rate for faulting of serpentinite and rock failure of biotite at room temperature, *Reinen et al.* [1994] argued that dislocation glide controls the rate dependence of serpentinite at low slip rate. This argument, which ignores temperature dependence of the rate-dependent process, was discounted somewhat by *Escartin et al.* [1997], who showed that dislocation does not occur in rock failure tests at room temperature in serpentinite. Similarly, dislocation is not expected in quartzofeldspathic materials at room temperature. However, the *Reinen et al.* [1994] argument probably does apply at asperity contacts on fault surfaces of other weak phases that deform via dislocation glide at room temperature, for example biotite, muscovite, and illite (see Some Other Aspects of Rock Friction section below), and applies to strong materials known to deform via dislocation motion at room temperature such as olivine [*Evans and Goetze*, 1979].

Time-Dependent Strengthening The time-dependent increase in static friction [*Dieterich*, 1972] (fig. 13.18) is also seen in analogue materials, soda lime glass, and lucite plastic where it has been shown to result directly from ductile flow or distributed fracturing at highly stressed asperity contacts on the fault surface [*Dieterich and Kilgore*, 1994]. Similar behavior can also be seen directly in single asperity hardness tests, where a loaded indenter of diamond or sapphire is placed in contact with material surfaces. Surfaces of quartz or olivine show a time-dependent change in indent area [*Scholz and Engelder*, 1976] (fig. 13.26), as previously shown in metals and oxides [*Westbrook and Jorgenson*, 1965], implying time-dependent increase in the shear strength of the contact. Indentation tests showing the same kind of effect have been conducted on the glass and lucite analogue materials [*Dieteich and Kilgore*, 1994]. In this case, the indentation tests were compared directly with the tests measuring time-dependent frictional strengthening in which contact area was monitored directly. There is good agreement between area changes with time in these two types of tests on analogues. Furthermore, if an Arrhenius flow law (equation (13.17a)) is used to model indentation tests, under certain circumstances the contact area increase follows the logarithmic increase seen in most hold tests and that is approximately seen indentation tests [*Berthoud et al.*, 1999]. This strongly suggests that the Arrhenius representation is also appropriate for representing time-dependent processes on friction surfaces.

However, the physical processes responsible for time-dependent yielding during hold tests or during indentation tests on analogue materials are unknown and have not been clearly related to the similarly unknown processes operating in minerals. Direct comparison between rates of time-dependent strengthening and indentation rates in minerals does not yield consistent re-

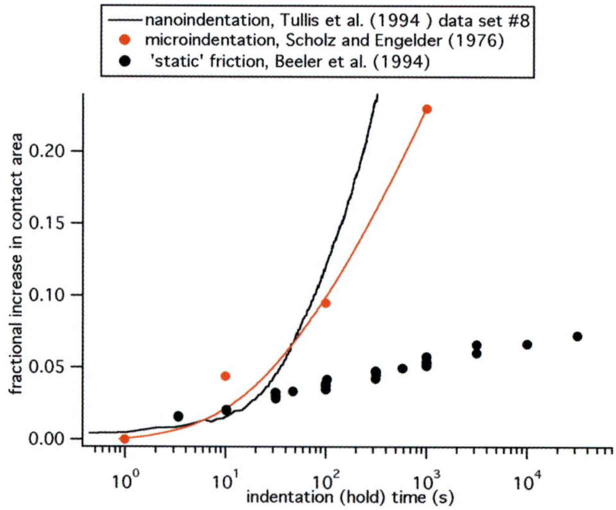

Figure 13.26 Comparison between quartz indentation data and change in quartzite friction with hold time. The data are from microindentation [*Scholz and Engelder*, 1976], nanoindentation [*Tullis et al.*, 1994], and slide-hold-slide tests. The red line is a fit to the microindentation data using an equation for static indentation based on the Arrhenuis rheology (equation (13.17a)).

sults (fig. 13.26) [*Goldsby et al.*, 2004]. The indentation rates are significantly faster, and at least for quartz, there is no known mechanism of ductile flow at these conditions [*Westbrook*, 1958; *Evans*, 1984]. It is possible that the mechanisms of contact-scale deformation are different in indentation and friction sliding tests because, even if the contact normal stress were the same in both kinds of tests, the frictional contacts support large shear stresses that are absent in the indentation tests. *Sharp et al.* [1993] estimate that the normal force necessary to induce fracture under a sliding contact is roughly one tenth that for a static contact, and they estimate that materials deforming plastically during static indentation are markedly more brittle during sliding contact. Thus differences in the deformation mechanism in these two kinds of tests are expected. However, contrary to what is actually observed, the difference in shear stress should produce more rapid rates of strengthening in hold tests.

More recent nanoindentation tests on quartz indicate plastic deformation [*Masuda et al.*, 2000] and recovered indents, examined both with atomic force microscopy and transmission electron microscopy (TEM), showed no evidence of dislocation motion or fracture. The deformation seems to have been accommodated by a previously unrecognized lattice misorientation mechanism they term "lattice creasing," which may involve generation of small amorphous regions or lattice distortion [*Masuda et al.*, 2000]. *Kailer et al.* [1999] present more detailed microstructural observations that demonstrate that indentation leads to amorphization of quartz. Possibly, similar deformation is associated with frictional sliding; *Yund et al.* [1990] report amorphous material results from frictional slip, though it is unclear whether the amorphous material is a high-pressure phase or highly comminuted granular material. So, it is plausible, even likely, that the deformation mechanisms operating in indentation tests and frictional sliding tests in minerals do not involve conventional dislocation or fracture growth mechanisms.

The most significant constraint on mechanism of time-dependent strengthening observed in laboratory friction tests comes from the study of *Dieterich and Conrad* [1984]. Hold tests were conducted on bare surfaces of quartzite in dry argon and at controlled and room humidity in air. The time-dependent strengthening effect vanished in the dry tests indicating that it is a water assisted effect. More recent tests on quartz gouge by *Frye and Marone* [2003] find a likely related correlation between time-dependent strengthening and water. For these gouges, ~5% relative humidity produces negligible aging, whereas >10% relative humidity leads to healing observed at ambient conditions. While these studies have not allowed the specific mechanism to be identified the prevailing theory of Dieterich [*Dieterich and Conrad*, 1984] is that the effect of water seen in these friction tests is the same effect seen in microindentation tests in nonmetallic solids where extremely dry conditions inhibit yield at highly stressed contacts [*Westbrook and Jorgenson*, 1965].

Determining the mechanisms responsible for frictional strengthening in laboratory experiments is not possible without direct microstructural examination of the deformed material. Such examinations are rarely performed. A preliminary study to match mechanical data from static and sliding indentation tests with detailed optical, scanning electron microscopy (SEM) and TEM observation of the deformed mineral surfaces was undertaken by B. Evans and D. Kohlstedt [*Viswanathan et al.*, 1995; *Viswanathan and Kohlstedt*, 1996; *Eckhardt et al.*, 1996]. In experiments on calcite and olivine, deformation consisted of cooperative fracture, dislocation motion, and sliding contact; on the basis of those preliminary results, static indentation and frictional sliding of rock surfaces involve mixed brittle and ductile processes. It is hard to imagine a more fundamentally important study in rock friction than one that establishes the physics of contact-scale deformation, as Evans and Kohlstedt intended. However, owing to lack of funding, that study was not completed.

Temperature Dependence While there is little agreement on mechanisms associated with rate and state effects it is accepted that whatever the mechanism, the temperature dependence of second-order friction effects is consistent with the Arrhenius relation, which requires a linear scaling (equation (13.17b)) RT/Ω of the rate-dependent coefficient with temperature. Linear temperature dependence was first suggested by *Stetsky* [1978]. Later *Chester* [1994] developed a temperature-dependent friction formulation based on laboratory quartz friction data at hydrothermal conditions that incorporated the Arrhenius dependence of the direct rate and aging effects. Using his formulation, Chester successfully modeled rate step data from hydrothermal experiments on quartz to determine the individual temperature dependence of the rate and age dependencies, finding that both effects have similar temperature dependence with an average activation energy 90 kJ/mol. This is consistent with the activation energy of crack growth in quartz [*Atkinson and Meredith*, 1987b]. Similarly, *Blanpied et al.* [1998] analyzed hydrothermal rate stepping tests in granite to determine the temperature dependence of the individual parameters a and b. Though they did not determine them, activation energies can be inferred from their analysis and tabulated data.

The most compelling observations of this temperature dependence are *Nakatani*'s [2001], whose experimental technique isolated the direct rate dependent coefficient a at temperatures between 25° and 800°C in a feldspar gouge layer (fig. 13.27). While the low-temperature activation energy for the process underlying a again is in the range of 90 kJ/mol, suggestive of subcritical crack growth, Nakatani argues against crack growth and implies that ductile flow of some form is responsible. Again, there is no consensus on mechanism, only on the form of the temperature dependence. Comparable direct determinations of the temperature dependence of b have not been made.

Relative Roles of Elastic and Inelastic Processes

While either inelastic or elastic friction theories predict a constant friction coefficient [*Bowden and Tabor*, 1950; *Greenwood and Williamson*, 1966], laboratory tests can easily distinguish between these two end-member models. Tests by

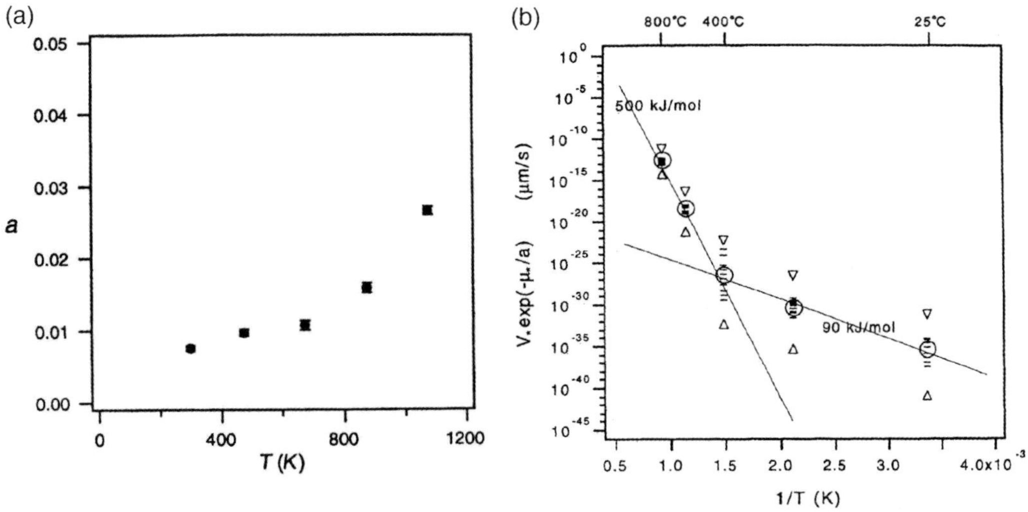

Figure 13.27 Data on the temperature dependence and activation energy of the instantaneous rate dependence of friction of sodium feldspar gouge [from *Nakatani*, 2001]. (a) Approximately linear dependence of *a* on temperature, consistent with expectations from the Arrhenius rheology (e.g., equation (13.17). (b) Inferred activation energy from the same experiments as in figure 13.27a, assuming an equation of the form of equation (13.17).

Linker and Dieterich [1992] provide the current state of knowledge on the roles of elastic and inelastic processes in rock friction, albeit for granite and room temperature. Both reversible (elastic) and displacement-dependent (inelastic) changes in shear resistance occur when the normal stress is changed [*Linker and Dieterich*, 1992] (fig. 13.28a). The reversible changes in shear stress with normal stress probably reflect the increase in the number of contacts and increase in area of existing contacts as a function of normal stress due to elastic closure as in the work of *Greenwood and Williamson* [1966]. The inelastic changes with associated displacement dependence are of the same sense as the elastic changes and probably reflect changes in dilatant volume due to time-dependent ductile flow at asperity contacts, modulated by slip. In other words, the inelastic effects are due to the same kind of age dependence as seen in constant normal stress experiments [*Linker and Dieterich*, 1992]. The reversible changes are as large as or larger than the inelastic changes so a significant portion of the pressure dependence of fault strength in experiments is probably due to elastic changes in contact area and dilatant volume.

Normal stress effects are well modeled by modification of the rate and state equation (13.15) [*Linker and Dieterich*, 1992]; the following formulation is equivalent to *Linker and Dieterich* [1992] but is slightly modified to more clearly reflect the partitioning between elastic and inelastic changes in contact area associated with normal stress change. For step changes in normal stress at con-

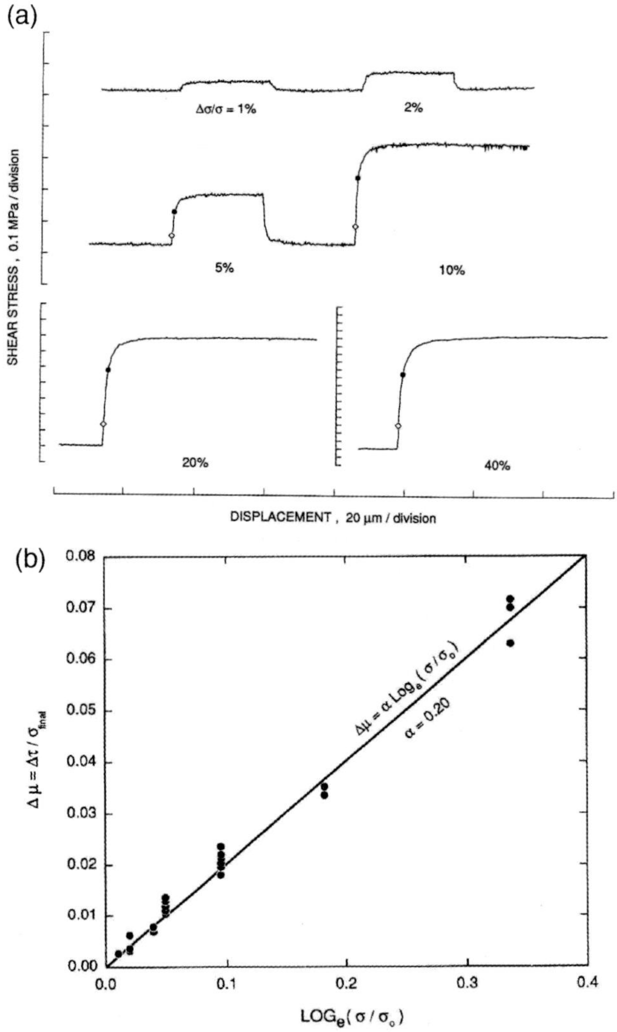

Figure 13.28 Effect of changes in normal stress on shear stress for bare granite at room temperature and 5 MPa normal stress [*Linker and Dieterich*, 1992]. (a) Shear stress versus displacement for changes in normal stress of 1, 2, 5, 10, 20, and 40% of the ambient value. Note that the 40% change is shown at a compressed vertical scale compared to the other tests. The response has three components; there is an instantaneous change marked by the open circle. This is a machine effect due to Poisson expansion of the fault surface as the fault normal stress is changed. There is an additional instantaneous change from the open circle to the solid circle; this is interpreted as elastic increase in contact area due to the increased normal stress. Subsequently, with displacement, there is a further increase in shear resistance that is interpreted as inelastic, time-dependent yielding of asperity contacts. (b) Summary of tests of the type shown in figure 13.28a as a function of the size of the normal stress change. Here σ is the final normal stress, σ_0 is the initial normal stress, and $\Delta\mu$ is the inelastic component of the change in shear stress, normalized by the final normal stress.

stant slip rate, the inelastic change in shear stress depends logarithmically on the size of the shear stress change as

$$\frac{\Delta\tau}{\sigma_e} = \alpha \ln \frac{\sigma_e}{\sigma_0} \tag{13.18a}$$

(fig. 13.28b), where σ_0 is the starting normal stress [*Linker and Dieterich*, 1992]. If these shear stress changes are equated with changes in steady-state contact age $\Delta\tau/\sigma_e = b \ln(\theta_{ss}/\theta_0)$, state changes with normal stress as

$$\theta_{ss} = \theta_0 \ln \left(\frac{\sigma_e}{\sigma_0}\right)^{\alpha/b} \tag{13.18b}$$

Linker and Dieterich [1992] assume that at the time of the normal stress change, the contact age is instantaneously offset by the amount given by equation (13.18b). However, changing contact age without significant slip or time passage is confusing. In the present treatment it's assumed there is no instantaneous change in age, rather there is reversible instantaneous change in shear resistance that represents the elastic change in contacting area. A normal stress change subsequently induces the asperity contact population to evolve to a new steady-state age that represents the inelastic change in contacting area at the new normal stress. A modification of the standard evolution law (equation (13.15b)), which has the required steady-state value and that behaves otherwise identically to equation (13.15b), is

$$\frac{d\theta}{dt} = \left(\frac{\sigma_e}{\sigma_0}\right)^{\alpha/b} - \frac{V\theta}{dc} \tag{13.19a}$$

σ_0 can be considered a constant reference value, and the steady-state value of age is $\theta_{ss} = (d_c/V)(\sigma_e/\sigma_0)^{\alpha/b}$. To maintain steady-state friction independent of normal stress, equation (13.19a) is used with the modified relation for friction

$$\mu = \mu_o - \alpha \ln(\sigma_e/\sigma_0) + a \ln(V/V_0) + b \ln(V_0\theta/d_c) \tag{13.19b}$$

Here, steady-state friction is assumed independent of normal stress, and the total pressure dependence of strength is $d\tau/d\sigma_e = \mu_{ss}$. For this relation (13.19) the total pressure dependence is the sum of the elastic pressure dependence $\mu_{ss} - \alpha$, and the inelastic pressure dependence α. The ratio of inelastic to total pressure dependence is α/μ_{ss}, and α is some measure of the role of inelastic fault normal creep in rock friction. The value α is bounded $0 \le \alpha \le \mu_{ss}$ [*Dieterich and Linker*, 1992]. For materials corresponding to purely inelastic friction [*Bowden and Tabor*, 1950] $\alpha = \mu_{ss}$; for purely elastic friction [*Greenwood and Williamson*, 1966] $\alpha = 0$. For Westerly granite $\mu_{ss} = 0.7$ and $\alpha = 0.2$ and roughly 30% of the pressure dependence results from inelastic creep.

Stability Condition for Normal Stress Changes As described immediately above, information about contact-scale process associated with the second-order friction effects can be derived from normal stress stepping tests in the laboratory; unfortunately, such tests are rarely implemented. The stability of slip following normal stress change is important in natural circumstances of nonplanar faults, such as in most subduction zones, and where faults intersect the Earth's surface. Stability of an expression equivalent to equation (13.19) was analyzed using an inclined slider-block geometry with spring angle ϕ by *Dieterich and Linker* [1992]. The value ϕ is the angle between the loading direction and the fault. Ignoring second-order effects for the moment, the inclusion of coupling between slip and normal stress can have a stabilizing or destabilizing effect. Since $\tau = \mu\sigma_e$, if normal stress goes down with slip, the shear resistance goes down. In this way, decreasing normal stress reduces the resistance to slip rate acceleration, a potentially unstable situation. The second-order effect on normal stress is a tendency for the normal stress to initially change by a bit less than $\mu\Delta\sigma_e$ so that, in the potentially unstable case of decreasing normal stress, the $\alpha\ln\sigma$ term stabilizes slip [*Dieterich and Linker*, 1992]. These characteristics are well reflected by the *Dieterich and Linker* [1992] stability criteria for equations (13.19a) and (13.19b):

$$k_c = \frac{(b-a)\sigma_e}{d_c[1+(\mu_{ss}-\alpha)\tan\phi]} \qquad (13.19c)$$

The critical stiffness increases for negative friction angle (normal stress decreases with slip), encouraging instability. However, this tendency is counteracted somewhat by nonzero α.

Dilatancy and Slip Instability

In applying the stability criteria (13.16a) or (13.19c) to models of crustal seismicity to date [e.g., *Tse and Rice*, 1986; *Stuart*, 1988] it is always assumed that the steady-state rate dependence is independent of the ambient fault strength μ_0, as the empirical parameters (*b-a*) and α are assumed independent of nominal frictional strength. Were this true aseismic faulting is no less likely in intrinsically weak materials than in strong materials, despite some observational evidence to the contrary in weak transform environments [*Boettcher and Jordan*, 2004] and in subduction zones [*Byrne et al.*, 1988; *Moore and Saffer*, 2001]. However, close examination of the underpinning of rate and state equations suggests the steady-state rate dependence and the nominal strength are not independent. In equation (13.14) the constitutive parameter b, which controls the size of the age dependence is $b = -\mu_0\beta/\sigma_0$, recalling that μ_0 is the nominal strength, σ_0 is the contact normal stress and β is the intrinsic time dependence of the contact normal stress. So b depends directly on the nominal frictional strength and the a parameter is not so tied to μ_0; all other factors being equivalent, intrinsically weak materials should tend to have smaller aging, more positive rate dependence and therefore more aseismic response than strong materials.

This relationship between b and μ_0 results from an interpretation of empirical constitutive equations. However, it is consistent with the role of dilatancy in controlling seismic potential, as follows. Recall from the earlier discussion of pressure dependence of intact rock failure and friction that pressure dependence is a consequence of reduction of dilatant volume with increasing normal stress. In this sense, pressure dependence measures "ease of dilatancy"; large pressure dependence associated with large changes in dilatant volume. Thus weak materials are less easily dilated than strong materials. Recall also from the discussion of second-order effects that at constant fault normal stress in the absence of slip, contact area increases with time (fig. 13.17), fault strength increases (fig.13.18), and dilatant volume is reduced. This time-dependent reduction in dilatant volume is counterbalanced by slip-induced dilatancy (fig. 13.20), and the aging effect is directly related to dilatancy. The parameters μ_0 and b are expected to be strongly correlated. Equation (13.15) reflects this connection naturally. As b is a small fraction of total pressure dependence, e.g., 1–2 % for strong quartzofeldspathic material with $\mu = 0.7$ $b = 0.01$, it is reasonable to expect that for weak materials with low friction say $\mu = 0.2$ that b is proportionally smaller at 0.003. However, this is a speculative and untested hypothesis.

Note also that the normal stress effect on friction (equation (13.19)) is dilatancy related: α is the inelastic pressure dependence of strength having value equivalent to up to 30% of the total pressure dependence for granite. Though there is no data for α for any weak material, the inelastic dilatancy is expected to be a similar percentage of the total. So for a material with $\mu = 0.2$, α should be proportionally small at 0.06. This is also an untested prediction.

Some Other Aspects of Rock Friction

Other aspects of rock friction and the character of frictional sliding on monomineralic weak faults are illustrated using experiments of *Scruggs* [1997] conducted on gouge layers of muscovite between rings of porous ceramic in a high-pressure, computer-interfaced rotary shear apparatus [*Tullis and Weeks*, 1986; *Reinen et al.*, 1994]. Muscovite was chosen so that results could be compared with single-crystal deformation experiments of *Mares and Kronenberg* [1993]. Additionally, a number of surface physics and surface forces measurements have been made on muscovite cleavage fragments in the presence of fluids [*Israelachvili*, 1992], making it an ideal material for initial studies of the physics of high-pressure mineral friction. Muscovite has similar crystal structure and friction as illite, and therefore is an appropriate analogue for studying aspects of subduction zone faulting within the illite stability field.

The gouges were prepared by crushing and passing the fragments through standard sieves at room temperature and humidity; the size fraction smaller than 74 μm is used. The amount of gouge used resulted in an ~1.0-mm-thick layer when compacted under load. Surface preparation of the ceramic forcing blocks was performed at room temperature and humidity. The rock rings were ground to height with a surface grinder and then abraded with #100 SiC grit on a glass plate. At the start of an experiment, confining pressure was raised to the appropriate value. Axial load was then raised to bring the normal stress to final load. Both shear stress (torque) and axial load were measured with a combined load and torque cell located within the pressure vessel. Constant rate loading was provided by an electrohydraulic stepping motor. Fault displacement is measured directly by a high-resolution resolver [*Weeks and Tullis*, 1992]. Changes in gouge thickness are determined from the relative length of the sample column recorded by a linear variable differential transformer (LVDT). Data were recorded digitally at fixed time intervals, with a maximum frequency of 10 samples/s. Measurements were obtained simultaneously for gas pressure, axial load, torque, room temperature, displacement (rotation angle), and the length of the loading column.

Sliding tests on muscovite were conducted at 25, 50, 100, and 150 MPa, and the shear stresses are linear with increasing normal stress defining friction $\mu = \mu_f = 0.364$ [*Scruggs and Tullis*, 1998] (fig. 13.16). Similarly, the steady-state rate dependence of shear strength is linear with increasing normal stress (fig. 13.29a). The rate dependence is positive and rate stepping tests show a very small age-dependent effect, roughly ⅙ of the direct rate-dependent effect (fig. 13.29b). Measured in this way, there appears to be very little ongoing time-dependent creep and shear-induced dilatancy relative to the direct rate dependence of shearing. This conclusion is borne out by slide-hold-slide tests; these show a small component of time-dependent strengthening $b = 0.0007$ (fig. 13.29c). The coefficient b inferred from these tests is also insensitive to normal stress. These results are consistent with the assumptions leading to equation (13.14), suggesting that at all pressures in this range, the nominal and second-order effects are independent of normal stress.

Degree of Localization

Characteristic Displacement These tests illustrate two other aspects of rock friction observed in laboratory tests. In rate stepping tests, the age-dependent effect has a characteristic length $d_c \approx 125$ μm more than 25 times larger than observed in step tests on bare surfaces of granite or quartz. It seems unlikely that d_c this large is controlled by asperity contact dimension alone and slip to return friction to its residual level ($\sim 4 \times d_c$) is of the order of the gouge thickness. As suggested first by *Marone and Kilgore* [1993], deformation in gouge can be distributed rather than localized, and if it is distributed, the observed characteristic sliding dimension must be longer and related to both to the asperity contact dimension and the width of the deforming zone, as follows. For fully localized slip, the steady-state value of fault age can be related to the contact strain $d\delta/d_{c,}$ where d_c is the representative contact dimension. The contact shear

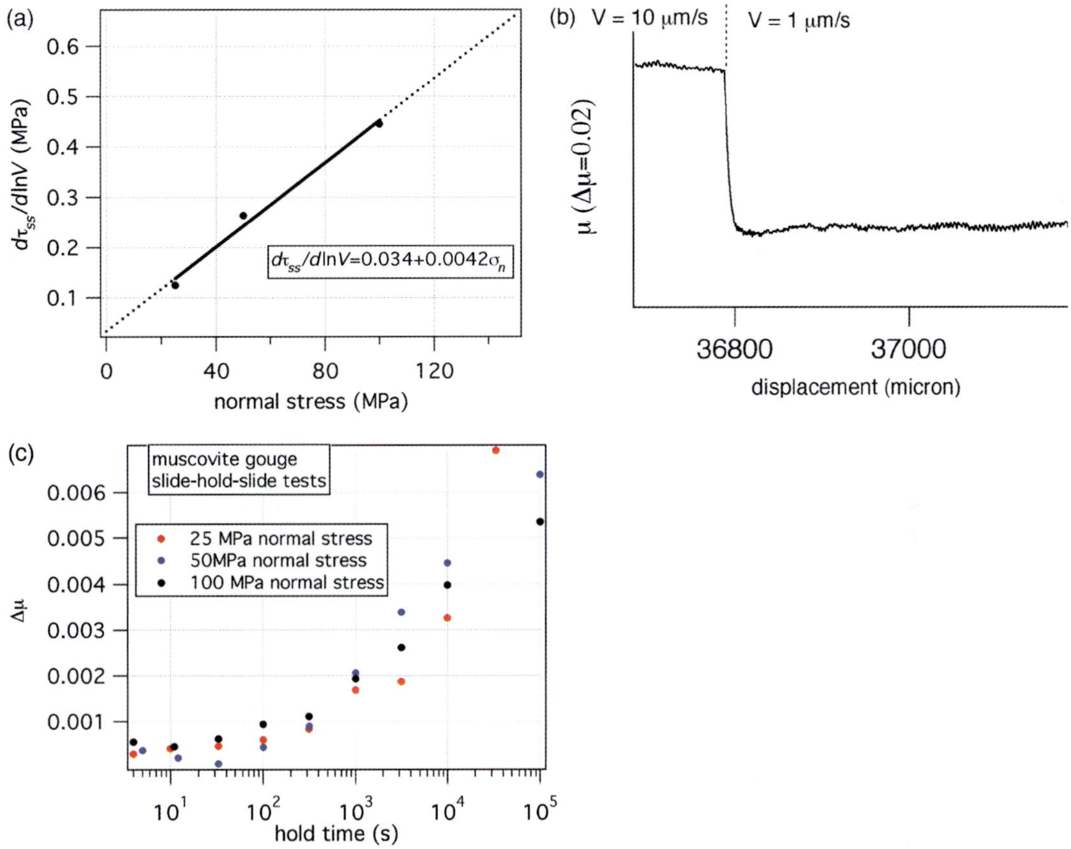

Figure 13.29 Muscovite gouge rate dependence and time dependence from *Scruggs*, 1998]. (a) Steady-state rate dependence from rate stepping tests at three normal stresses. The rate steps were between 1 and 10 μm/s. The points represent the average of 52, 62, and 50 measurements made at 25, 50, and 100 MPa, respectively. The standard deviation is between 13 and 8% for these measurements. There is no evidence of significant changes in the rate dependence with normal stress in this range. A fit to the data gives a slope of 0.0042, which is *a-b* from equation (13.15). (b) Typical rate step from 10 to 1 mm/s, showing the relatively large instantaneous rate dependence and the small age dependence. (c) Results of slide hold slide tests at the same three normal stresses shown in figure 13.29a. There is no evidence of significant changes in the rate of time dependent strengthening with normal stress. The rate of strengthening is very small, compare with typical quartzofeldspathic rates (fig 13.18).

strain rate then is $\gamma_c = V/d_c = 1/\theta_{ss}$. If instead strain is distributed within a layer, slip occurs on many slip surfaces within the layer. The number of slip surfaces is related to the average grain size D and the thickness of the deforming layer w, $N_c \propto w/D$. The characteristic length is the average contact dimension d_c times the number of slip surfaces $D_c = d_c N_c$. So for gouge the characteristic displacement is expected to be larger than for bare surfaces by an amount related to the thickness of the deforming layer $D_c \propto d_c w/D$. For gouge the contact strain rate is $\gamma_c = 1/\theta_{ss} = V/D_c \propto VD/wd_c$.

Effective Cutoff Time The characteristic length also influences aspects of slide hold tests results via the effective cutoff time [*Nakatani*, 2001]. Time-dependent strengthening in hold tests can only be measured at hold times longer than an effective characteristic time. The effective characteristic time is approximately the age at the start of the hold. For example, if a hold test could be conducted where fault slip ceased immediately during the hold period, then age (equation (13.15b)) would change with time during the hold as $\theta = \theta_o + t$ and age would contribute to fault strengthening as $b\ln(\theta_o + t)$. The parameter θ_0 is the starting value of state and the effective cutoff time of *Nakatani* [2001]. As hold tests start at steady state, the cutoff time is $\theta_{ss} = \theta_o = D_c/V$ and will scale directly with the characteristic slip distance. The muscovite experiment illustrates the long cutoff time for gouge (fig. 13.29c) relative to bare surfaces of granite or quartzite.

Shear Localization and Rate Dependence Because of the longer displacement length scale for gouge layers, gouges can be more stable than bare surfaces (e.g., equation (13.16b)). In addition, the actual rate dependence of gouge layers is more positive than for bare surfaces of the same composition [*Byerlee and Summers*, 1976; *Marone et al.*, 1990]. Although explanations of this difference in rate dependence have been proposed, most notably that the difference is related to the shear dilatancy rate [*Marone et al.*, 1990], the source of the difference in rate dependence is not known [*Marone*, 1998].

A convincing explanation notwithstanding, the degree of localization in gouge should directly reflect the rate dependence. Velocity-dependent fault strength relations lead directly to a strain localization/delocalization criteria derived by requiring that work is minimized. Ignoring dilation, mechanical work is normally written as $W = \int \tau d\gamma$ but when $d\tau/d\gamma$ does not change sign, W can be equivalently expressed as $W = \int \gamma d\tau$. Then $dW = \gamma d\tau$, and an increment of work is $dW/d\delta = \gamma d\tau/d\delta$. Reduction of work requires $dW/d\delta < 0$ and so shear stress must decrease with displacement $0 > d\tau/d\delta$. For a velocity-dependent material, $\tau = f(\dot{\gamma})$ and

$$\frac{d\tau}{dw} = \frac{\partial \tau}{\partial \dot{\gamma}} \frac{d\dot{\gamma}}{dw} \tag{13.20a}$$

Here, the engineering shear strain is $\gamma = \delta/w$ and w is the thickness of the deforming zone. The shear strain rate is $\dot{\gamma} = V/w$. Taking derivative of strain rate with respect thickness leads to

$$d\tau = -\frac{V}{w^2} \frac{\partial \tau}{\partial \dot{\gamma}} dw \tag{13.20b}$$

equivalently
$$\frac{d\tau}{d\delta} = -\frac{V}{w^2}\frac{\partial\tau}{\partial\dot{\gamma}}\frac{dw}{d\delta} \tag{13.20c}$$

Using the requirement of decreasing shear strength with shear strain yields

$$0 > -\frac{V}{w^2}\frac{\partial\tau}{\partial\dot{\gamma}}\frac{dw}{d\delta} \tag{13.20d}$$

$\partial\tau / \partial\dot{\gamma}$ is the rate dependence of fault strength, which can be either negative or positive and because V and w are both positive, equation (13.20) defines the localization criteria. If the velocity dependence is positive, then $dw/d\delta$ is positive and the shear zone widens with displacement. *Mair and Marone* [1999] found that D_c increased with slip velocity for a case in which shear is distributed, consistent with delocalization. Conversely, if the velocity dependence is negative $dw/d\delta$ is negative and the shear zone localizes. In rate-strengthening gouges such as the muscovite example, the deformation will tend to be distributed leading to the large characteristic slip distance.

Constraints on Contact-Scale Deformation Mechanism

The experimentally inferred coefficients a and b from the muscovite experiments can be compared for consistency of mechanism with the rate dependence of intact strength measured at room temperature in single crystals of muscovite by *Mares and Kronenberg* [1993]. The normalization scheme developed in the Constraints on Contact-Scale Mechanism, Mostly for Strong Materials section is applied to correct for the unknown contact stresses in friction tests. The dislocation glide flow law [e.g., *Mares and Kronenberg*, 1993], has the Arrhenius form (equations (13.17) and (13.5a)), where the appropriate stress measure in this case is the differential stress $\sigma_\Delta = \sigma_1 - \sigma_3$, $\dot{\varepsilon} = \dot{\varepsilon}_* \exp[-(E_* - \sigma_\Delta\Omega)/kT]$. The reference strain rate $\dot{\varepsilon}_*$ is not arbitrary. To use an arbitrary reference, strain rate proportional to $\dot{\varepsilon}_*$, define $\dot{\varepsilon}_0 = c\dot{\varepsilon}_*$. Doing so makes the nominal strength $\sigma_{\Delta 0} = E_*/\Omega + \mathbf{A}\ln c$. The rate-dependent coefficient is $\mathbf{A} = kT/\Omega$. The normalized equation analogous to equation (13.17e) is

$$\frac{\sigma_\Delta}{\sigma_{\Delta 0}} = 1 + \frac{\mathbf{A}}{\sigma_{\Delta 0}}\ln\frac{\dot{\varepsilon}}{\dot{\varepsilon}_0} \tag{13.21}$$

for comparison with equation (13.17f). Geometric differences between measured differential stress σ_Δ in rock deformation tests and shear stress τ in friction tests, $\tau = 0.5\sigma_\Delta \sin 2\beta$, where β is the angle between the fault normal and the greatest principal stress, are eliminated by the normalization. To compare the instantaneous rate-dependent coefficients, the slip velocity in equation (13.17f) must be converted to strain rate and the nominal strength should be associated with the same reference strain rate $\dot{\varepsilon}_0$ for both equations (13.17f) and (13.21). Taking d_c from the friction relation to be ½ the average contact dimension, the strain rate for friction is $\dot{\varepsilon} = V/2d_c$.

Microstructural analysis of muscovite single crystals deformed at pressure between 50 and 400 MPa and temperatures between 20° and 400°C with σ_1 at 45° to the 001 plane, indicates deformation predominately by dislocation glide [*Mares and Kronenberg*, 1993]. Analysis of the three 400°C rate stepping experiments of *Mares and Kronenberg* [1993] (fig. 13.30), taking $\dot{\varepsilon}_0 = 1 \times 10^{-1}$, produces a normalized coefficient of rate dependence of dislocation glide $\mathbf{A}/\sigma_{\Delta 0} = 0.034$. To compare with room temperature friction data, the difference in temperature (673K to 295K) introduces factor of 2.3 reduction in **A** of the single-crystal data. The *Mares and Kronenberg* [1993] single-crystal data extrapolated to room temperature is then $\mathbf{A}/\sigma_{\Delta 0} = 0.015$.

To estimate the normalized coefficient of rate dependence for room temperature friction of muscovite, $\dot{\varepsilon} = V/2d_c$ is assumed. There is ambiguity in the strain rate due to shear being possibly distributed in a 1-mm-thick layer. If the deformation is distributed within the layer, the characteristic distance d_c is related both to the average contact dimension and the thickness of the actively deforming layer [*Marone and Kilgore*, 1993]. The value $d_c = 125$ μm inferred from rate stepping tests is larger than bare surface experiments, consistent with more distributed deformation. Fortunately, the normalization is insensitive to differences in strain rate of this magnitude; using a d_c of 9–125 μm, μ_0 varies from 0.359 to 0.372 and $\mathbf{A}/\tau_0 = 0.017$ for either choice. This normalized

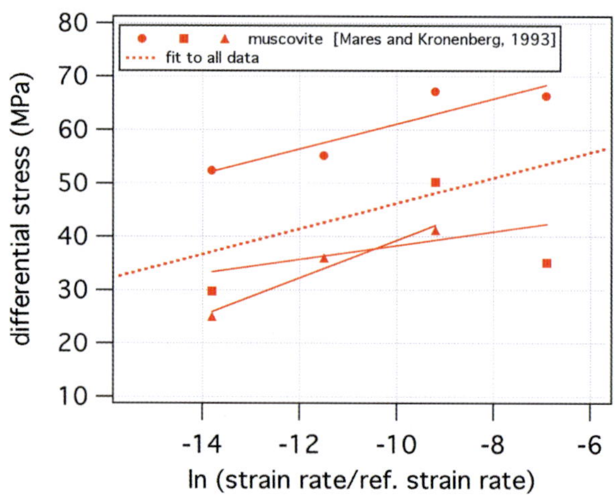

Figure 13.30 Rate dependence of low-temperature plasticity in muscovite at 400°C and 200 MPa confining pressure [*Mares and Kronenberg*, 1993]. These are rate stepping test data from three experiments. The dashed line represents the average strength and rate dependence. The lower axis has been normalized by a reference strain rate of 1×10^{-1}/s.

rate dependence is nearly identical to the single-crystal glide value, suggesting that the instantaneous rate dependence in muscovite could be related to dislocation motion. If glide does control the instantaneous rate dependence of friction, a would scale linearly with temperature, consistent with recent process based rock friction theories [e.g., *Rice et al.*, 2001; *Nakatani*, 2001] and direct measurements made by *Nakatani* [2001].

In low-velocity friction tests on serpentinite, *Reinen et al.* [1994] observed a transition to a more strongly rate-dependent process. Similar low-velocity rate stepping and hold tests on muscovite do not produce such a transition and the rate dependence in the velocity range 0.001–10 μm/s is approximately constant. A preliminary conclusion is that dislocation glide is possibly responsible for the direct rate dependent effect in muscovite. As the age-dependent coefficient $b/\sigma_0 = 0.0028$ is small, it seems to be unrelated to conventional low-temperature crystal flow.

Bond Strength in Sheet Silicates Consistent with the implied role of dislocation in muscovite rate dependence are recent results on the frictional strength of truly dry muscovite and other weak sheet silicate minerals (micas, clays) where that base level friction is controlled by the interlayer bond strengths [*Morrow et al.*, 2000; *Moore and Lockner*, 2004] (fig. 13.31). Experiments on 6-sheet silicate minerals by *Morrow et al.* [2000] and 10-sheet silicate minerals by *Moore and Lockner* [2004], carefully vacuum dried and deformed under dry conditions, show a systematic relationship between measured friction and the estimated bond strength [*Giese*, 1978, 1980; *Bish and Giese*, 1981]. Bond strength in this case is the separation energy, the energy necessary to separate individual sheets along the [001] basal cleavage. Separation energy was calculated by Giese and coauthors by considering energy change of electrostatic bonding with distance of separation. This correlation between measured fault strength and calculated bond strength extends from the weakest (talc) to lizardite serpentine, gibbsite, and kaolinite whose dry friction is equivalent to Byerlee's law. The electrostatic bond energy of these minerals exceed ~70 kcal/mol and the brittle mica margarite, which has separation energy nearly 3 times these more weakly bonded sheet silicates has the same frictional strength (fig. 13.31). These observations suggest a threshold bond strength below which interlayer bond energy determines friction and above which other mineralogy independent processes determine the nominal frictional strength. At least for the sheet silicates with separation energies below 70 kcal/mol, nominal dry frictional strength is determined by breaking interlayer bonds rather than pure sliding between previously broken cleavage fragments [*Moore and Lockner*, 2004]. This inference is consistent with direct measurement of the shear strength of previously cleaved mica sheets and observations that indicate the adhesive strength of cleaved surfaces is comparable to the interlayer bond strength [*Bailey and Courtney-Pratt*, 1955; *Israelachvili and Adams*, 1976]. Asperity contact-scale processes determining the strength of weak sheet silicate minerals are likely to

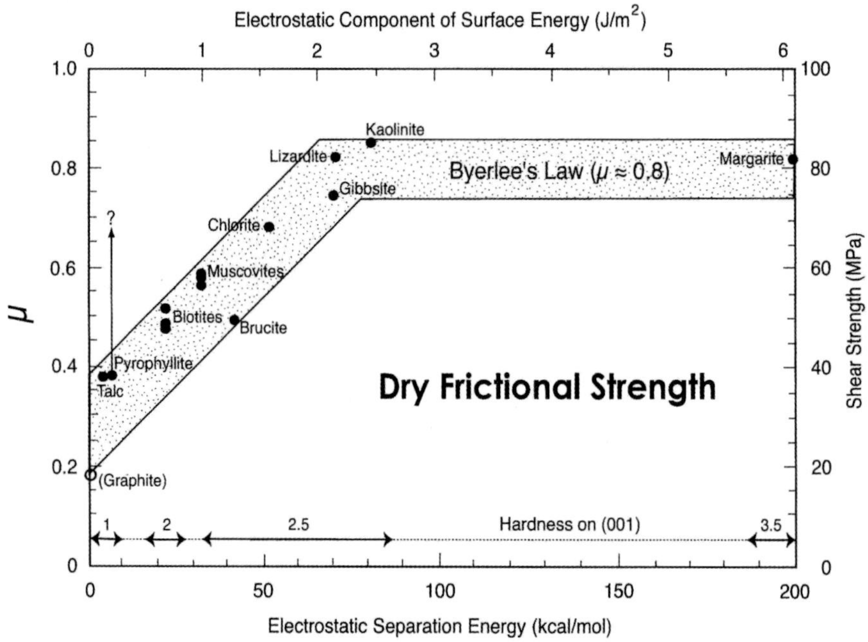

Figure 13.31 Variation of dry peak friction (left) versus estimated electrostatic separation energy for sheet silicate minerals at room temperature and 100 MPa confining pressure. The samples were oven and vacuum dried, and deformed dry. Hardness measured on the basal plane is shown as an alternative axis at bottom. Figure from *Moore et al.* [2004].

be fundamentally different than those of stronger phases whose strengths are well characterized by Byerlee's law [*Byerlee*, 1978].

Effects of Pore-Fluid Pressure on Fault Strength and Stability

The presence for water can have a large effect on fault strength and stability. Discussion of chemical effects of water on strength is deferred to the next section. Absolute fault strength depends to first order on pore pressure through the effective pressure relation (see the Rock Failure section). Pore pressure affects fault stability in two ways. First, effective normal stress reduces the seismic potential by reducing the critical stiffness for unstable slip (equation (13.16b)). As the effective normal stress goes down, the displacement rate of strength loss $(b-a)\sigma_e/d_c$ decreases. The other way that pore pressure affects fault stability is through dilatancy hardening [*Frank*, 1965; *Rice*, 1975; *Rudnicki and Chen*, 1988]. Dilatancy hardening is an increase in the effective pressure caused by decreasing pore pressure and increasing porosity owing to shear-induced dilatancy. As all faults appear to dilate prior to seismic slip, if the fault zone is not fully drained, the pore pressure goes down as failure is approached and the effective normal stress goes up causing the fault to become incrementally stronger.

The implications of dilatancy hardening on sliding stability are an extension of the constant normal stress equation (13.14b) [*Segall and Rice*, 1995]. Dilatancy associated with changes in sliding velocity (fig. 13.20) [*Morrow and Byerlee*, 1989; *Marone et al.*, 1990] can be characterized by assuming a steady-state porosity associated with each slip rate and that porosity evolves exponentially with slip,

$$\frac{d\phi}{dt} = -\frac{V}{d_c}(\phi - \phi_{ss}) \tag{13.22a}$$

Figure 13.32 shows comparison by *Segall and Rice* [1995] of a simulation with equation (13.22a) to data of *Marone et al.* [1990]. Steady-state porosity is assumed purely rate dependent

$$\phi_{ss} = \phi_o + \varepsilon \ln\left(\frac{V}{V_o}\right) \tag{13.22b}$$

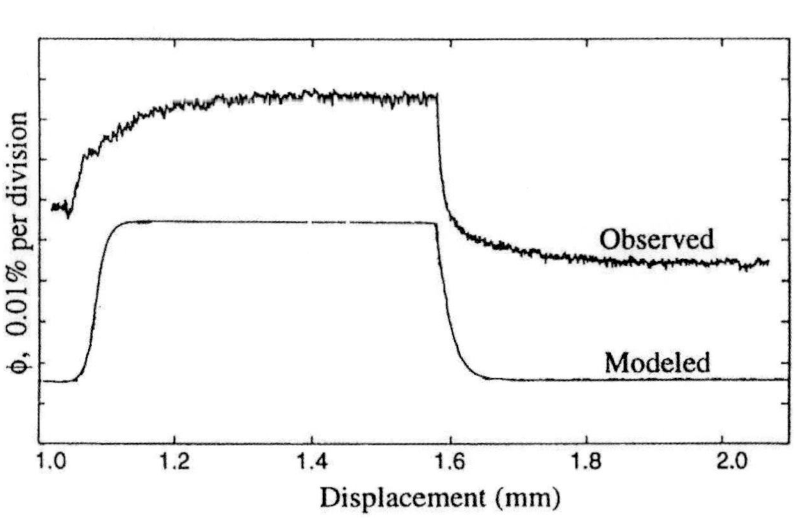

Figure 13.32 Observed and modeled changes in porosity due to a change in slip velocity from 1 to 10 μm/s for a 3-mm quartz gouge layer at room temperature and 150 MPa normal stress. The data are from the study of *Marone et al.* [1990], the model is calculated with equations (13.22a) and (13.22b) by *Segall and Rice* [1995]. The figure is from *Segall and Rice* [1995].

At constant normal stress and undrained conditions (constant fluid mass) the rate of pore pressure change is fixed by the rate of inelastic porosity change

$$\frac{dp}{dt} = -\frac{d\phi_{in}}{dt}\frac{1}{\phi(\beta_f + \beta_\phi)} \tag{13.22c}$$

where β_f and β_ϕ are the fluid and pore compressibilities respectively. Equations (13.22) and (13.15) describe slip within a saturated fault zone. Using the effective pressure relation $\tau = \mu(\sigma_n - p)$ and assuming that the undrained steady-state rate dependence results from the sum of intrinsic rate dependence and a contribution due to dilatant pore pressure change leads to

$$\left.\frac{d\tau_{ss}}{d\ln V}\right|_m = \left.\frac{d\tau_{ss}}{d\ln V}\right|_p - \mu\left.\frac{dp}{d\ln V}\right|_m \tag{13.23a}$$

The first term on the right is the drained rate dependence $(a - b)(\sigma_n - p)$. The second right-side term is found by *Segall and Rice* [1995] from equation (13.22) to be

$$\left.\frac{dp}{d\ln V}\right|_m = -\frac{\varepsilon}{\phi(\beta_f + \beta_\phi)} \tag{13.23b}$$

giving

$$\left.\frac{d\tau_{ss}}{d\ln V}\right|_m = (a - b)(\sigma_n - p) + \frac{\mu_{ss}\varepsilon}{\phi(\beta_f + \beta_\phi)} \tag{13.23c}$$

Substituting this result into equation (13.16b) gives the undrained critical stiffness [*Segall and Rice*, 1995]

$$k_c = \frac{1}{d_c}\left[(b - a)(\sigma_n - p) - \frac{\mu_{ss}\varepsilon}{\phi(\beta_f + \beta_\phi)}\right] \tag{13.23d}$$

This expression includes both effects of pore pressure on stability: the first term on the right-hand side describes the sensitivity to effective stress where high pore pressure (low effective stress) decreases the critical stiffness. This is a stabilizing effect of high pore pressure. The second right-side term is dilatancy hardening; as the ratio of the dilatancy coefficient to porosity ε/ϕ increases, the critical stiffness goes down.

As shown by the derivation of equation (13.23d) and by simulations and analysis conducted by *Segall and Rice* [1995], the presence of undrained pore pressure tends to stabilize faults and high pore pressure significantly discourages seismicity. *Segall and Rice* [1995] concluded for laboratory-measured values of μ_{ss}, ε, *a-b*, and d_c for quartz that it is not possible to have seismic slip on faults that are apparently weak enough to satisfy the San Andreas heat-flow constraint $\mu < 0.2$). Partly, this result is due to the high values of ε and μ_{ss} from the limited existing laboratory data from quartz gouge layers. These constraints lead to the conclusion that apparently weak fault zones tend to be more aseismic than strong ones.

However, the implications of equation (13.23) for intrinsically weak, low dilatancy materials such as clays and phyllosilicates, as may exist in subduction zones, are unexplored. By using data from quartz, dilatancy hardening and intrinsic strength assumed by *Segall and Rice* [1995] are at their respective maximums, making the second right-side term large, favoring stability. Intrinsically weak, low dilatancy materials will have much smaller contributions from the second term. On the other hand, as argued in the Dilatancy and Slip Instability section above, intrinsically weak materials tend to have small *b* coefficients, making them less likely to be intrinsically unstable, keeping in mind that the relation between intrinsic stability and intrinsic strength proposed in the Dilatancy and Slip Instability section is speculative. Because some seismic subduction zones may be intrinsically weak [e.g., *Moore and Saffer*, 2001] and may have pore pressure elevated above hydrostatic [e.g., *Moore and von Huene*, 1980], given the present state of knowledge, it would be unwise to unequivocally relate sliding stability and apparent fault strength.

Chemical Effects of Pore Fluid

To this point, the pressure dependence of strength has been explicitly assumed to arise from increased area of contact (simplifications of equation (13.3c)). An implication of this assumption is that with increasing confining pressure as the fault-zone material becomes denser, larger increments of elastic or inelastic strain energy are required to further compact the fault and increase the contact area. In the limit, as the contact area approaches the total fault area, the amount of strain energy required to further increase the contact area becomes infinite. Thus it is expected and observed that the pressure dependence decreases with increasing confining pressure, both for friction [e.g., *Raleigh and Paterson*, 1965] and for intact failure (e.g., fig. 13.2). However, there are exceptional instances where pressure dependence increases with normal stress or confining pressure. These cases are associated with fault weakness. The best-documented examples of this behavior are for wet clay [*Morrow et al.*, 1992] and wet chrysotile serpentine [*Moore and Lockner*, 2004] (fig. 13.33). Both are weak when dry and weaker when saturated. In general, sheet silicates are weakened by the presence of water; apparently, the strength reduction measures water affinity [*Morrow et al.*, 1992; *Moore and Lockner*, 2004]. Cleavage surfaces are highly

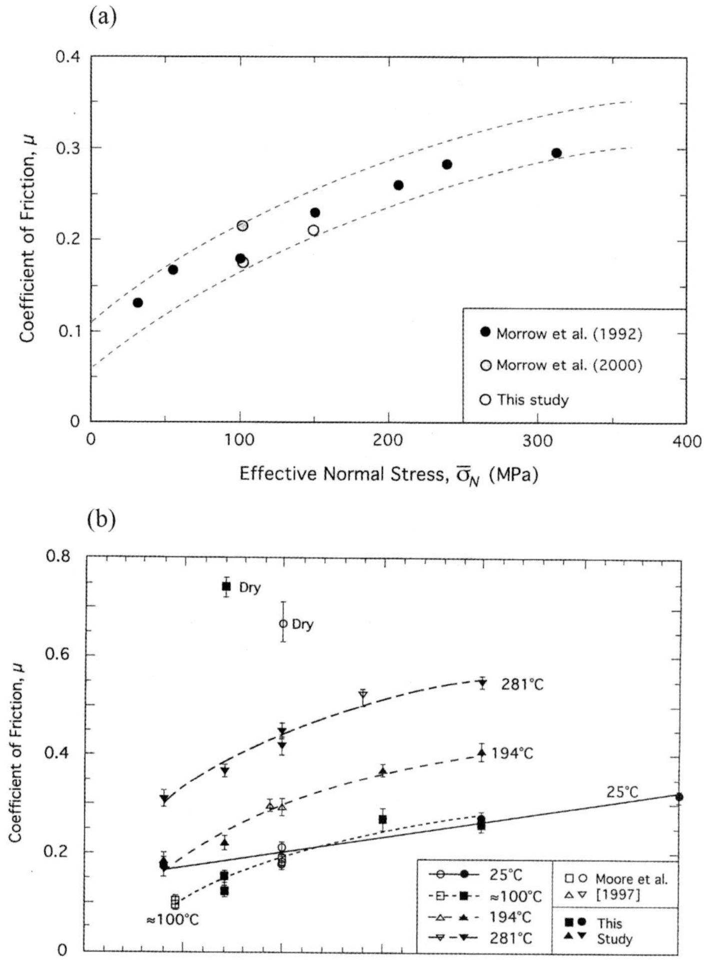

Figure 13.33 Examples of "reverse" pressure dependence where the coefficient of friction $\mu_f = \partial\tau/\partial\sigma_e$ increases with normal stress. Both examples are for drained, water saturated gouge layers. (a) Montmorillonite from *Morrow et al.* [1992, 2000] and *Moore and Lockner* [2004]. Figure is from *Moore et al.* [2004]. (b) Chrysotile serpentinite, data from *Moore et al* [1997], *Moore and Lockner* [2004], figure from *Moore and Lockner* [2004].

charged and water is attracted and bonded to surfaces in a structured film [*Israelachvili et al.*, 1988]. The bonding of structured water influences macroscopic shear strength. The increasing pressure dependence seen for montmorillonite and chrysotile is probably due to the reduction in thickness of the bound intergranular layer with increasing normal stress, as shown by *Rutter* [1983] in statically loaded kaolinite layers. The presence of a water layer with pressure-dependent thickness is a case that violates the equation (13.10b) and instead the pressure dependence results from the sum of changes in contacting area and changes in shear strength of the water layer

$$\frac{d\tau}{d\sigma_e} = \frac{\partial\tau}{\partial A_f}\frac{dA_f}{d\sigma_e} + \frac{\partial\tau}{\partial\tau_f}\frac{d\tau_f}{d\sigma_e} \tag{13.24}$$

As water dramatically reduces the strength of all sheet silicates [*Morrow* et al., 2000; *Moore and Lockner*, 2004], it is likely that friction at water-saturated conditions of all these minerals increases to some degree with effective pressure.

Rate Dependence

One largely unexplored implication of equation (13.24) is that rate and displacement-dependent changes of strength are possible without dilatancy and changes in area of contact. Fluid layers might be expected to be viscous and therefore rate strengthening; however, structured fluids may have quite anomalous properties [e.g., *Thompson and Robbins*, 1990], and it is difficult to anticipate even the sign of the steady-state rate dependence without some experimental observation. As yet there are no comprehensive studies of the rate dependence of saturated weak faults.

Strength and Rate Dependence of Subduction Zone Minerals

Many measurements of rate dependence of individual mineral constituents found in subduction zones have been made. Localized slip between strong $\mu = 0.65–0.8$) quartzofeldspathic mineral and rock surfaces, for example quartz [*Dieterich and Conrad*, 1984], feldspar (V. J. Scruggs and T. E. Tullis, unpublished data 1997), and granite [*Dieterich*, 1978] are rate weakening and allow for earthquake nucleation.

Basic rocks and minerals as represented by gabbro [*Cox*, 1990; *Tsutsumi and Shimamoto*, 1997] are rate weakening and strong $\mu = 0.7–0.8$).

The data for localized slip between carbonate mineral and rock surfaces are not extensive and are somewhat ambiguous. *Weeks and Tullis* [1985] report relatively high strength ($\mu = 0.5$) rate strengthening for a dolomite. The same laboratory finds that calcite is similarly rate strengthening and somewhat stronger ($\mu = 0.73$) [*Tullis and Weeks*, 1987]. However, *Spetzler et al.* [1991] reports both stable and unstable slip for limestone surfaces. As carbonate is a common cement in accretionary and subduction environments, additional experiments are warranted.

Serpentine has more complicated friction than more approximately isotropic phases. Serpentine can be either weak or strong, rate strengthening or rate weakening, depending on temperature, pressure, and sliding velocity [*Reinen et al.* 1994; *Moore et al.*, 1997]. Chrysotile can be weak (μ = 0.2) at low temperature and pressure except in very dry environments. The other serpentine varieties antigorite and lizardite are stronger (μ > 0.4.) at all conditions, but relatively weak compared to quartzofeldspathic minerals. At low sliding rates, chrysotile serpentines are rate weakening only at high temperatures (>150°C) and therefore could produce earthquake nucleation near the base of the seismogenic zone [*Moore et al.*, 1997]. However, at rates slip rates of around 1×10^{-3} μm/s (31 mm/yr) all serpentines studied are rate neutral or rate strengthening and at corresponding plate motion rates would therefore be a barrier to earthquake nucleation [*Reinen et al.*, 1994; *Moore et al.*, 1997]. There is virtually no low slip rate data for lizardite and antigorite at crustal conditions however.

Montmorillonite clay may also have complicated strength and rate dependence [e.g., *Saffer et al.*, 2001]. Under saturated conditions, montmorillonite can be weak (μ = 0.08–0.15) at low confining pressure [*Logan and Rauenzahn*, 1987; *Morrow et al.*, 1992; *Brown et al.* 2003], but under increasing effective stress becomes considerably stronger (μ = 0.35). While *Logan and Rauenzahn* [1987] and *Saffer et al.* [2001] suggest that rate weakening may occur at very low sliding rates ($\sim 1 \times 10^{-3}$ μm/s), at all saturated conditions tested to date, montmorillonite is rate strengthening [*Morrow et al.*, 1992; *Brown et al.*, 2003].

The sources of illite in laboratory samples are illite-rich shales; these are not pure illite (>90%). Saturated illite /illite-rich shale is somewhat stronger than montmorillonite (μ = 0.2–0.45) [*Morrow et al.*, 1992, *Brown et al.*, 2003] but is weak relative to quartzofeldspathic rocks. At all saturated conditions tested to date, illite is rate strengthening at temperatures below 400°C [*Moore et al.*, 1989; *Morrow et al.*, 1992; *Brown et al.*, 2003].

Friction in Polyphase Fault Zones

That accretionary sediments, often involved in shallow faulting and underthrusting in subduction zones, have mineral components with widely varying frictional strength (e.g., quartz and smectite) requires that the friction of polyphase aggregates be considered. Continuum rheology of polyphase components is well studied [e.g., *Tullis et al.*, 1991, and references therein], however, because friction is controlled by processes operating at contacts between grains, mixing relationships from standard rheological studies are not directly applicable. For friction that is controlled in part by intergranular water films, the polyphase boundary may have strength that is different from the single-phase boundaries of either component.

Basic aspects of polyphase friction can be considered using a single roughly planar surface between two arrays of grains as the locus of all deformation (fig. 13.34a). There are two phases present. The resistance to shearing is the sum of the resistance of each contact to shear. There are three types of contacts, two different single-phase boundaries and interphase boundaries. The macroscopic shear strength is

$$\tau = P_{11}\tau^{11} + P_{22}\tau^{22} + P_{12}\tau^{12} \qquad (13.25a)$$

where τ^{11}, τ^{22}, and τ^{12} are the strengths of a single-phase boundary between phase 1, a single-phase boundary between phase 2, and an interphase boundary, respectively. The coefficients P_{11}, P_{22}, and P_{12} are the corresponding fractions of each boundary type present. For a fully localized shear boundary whose composition is the same of the entire gouge layer, the fractions are known as a function of composition; they are the probabilities

$$P_{11} = \frac{\binom{K}{0}\binom{M-K}{2}}{\binom{M}{2}}$$
$$P_{22} = \frac{\binom{K}{2}\binom{M-K}{0}}{\binom{M}{2}} \qquad (13.25b)$$
$$P_{12} = \frac{\binom{K}{2}\binom{M-K}{0}}{\binom{M}{2}}$$

where M is the number of grains in the boundary, K is the number of grains of phase 1, and the probability notation used is defined by

$$\binom{i}{j} = \frac{i!}{(i-j)!\,j!} \qquad (13.25c)$$

Figure 13.34b shows the coefficients P as a function of composition. At a composition of 50%, half of the grain boundaries are interphase boundaries. It is possible to have enhanced weakening owing to a weak interphase boundary, for example $\tau^{11} = 0.7$, $\tau^{22} = 0.2$, and $\tau^{12} = 0.2$ (fig. 13.34c). Similar behavior is seen in analog experiments where wet interphase boundaries induce dissolution effects not present at single-phase boundaries [*Bos and Spiers*, 2001]. Mixing also implies mixed rate dependence, which can be analyzed using a similar approach to that used for strength above. The very limited work on mixtures to date by *Logan and Rauenzahn* [1987], *Brown et al.* [2003], and *Saffer and Marone* [2003] suggest that interphase boundaries between quartz and clay have intermediate strength (fig. 13.34d). It is also possible that mixed clays have anomalous strengths, either stronger or weaker than linear combinations

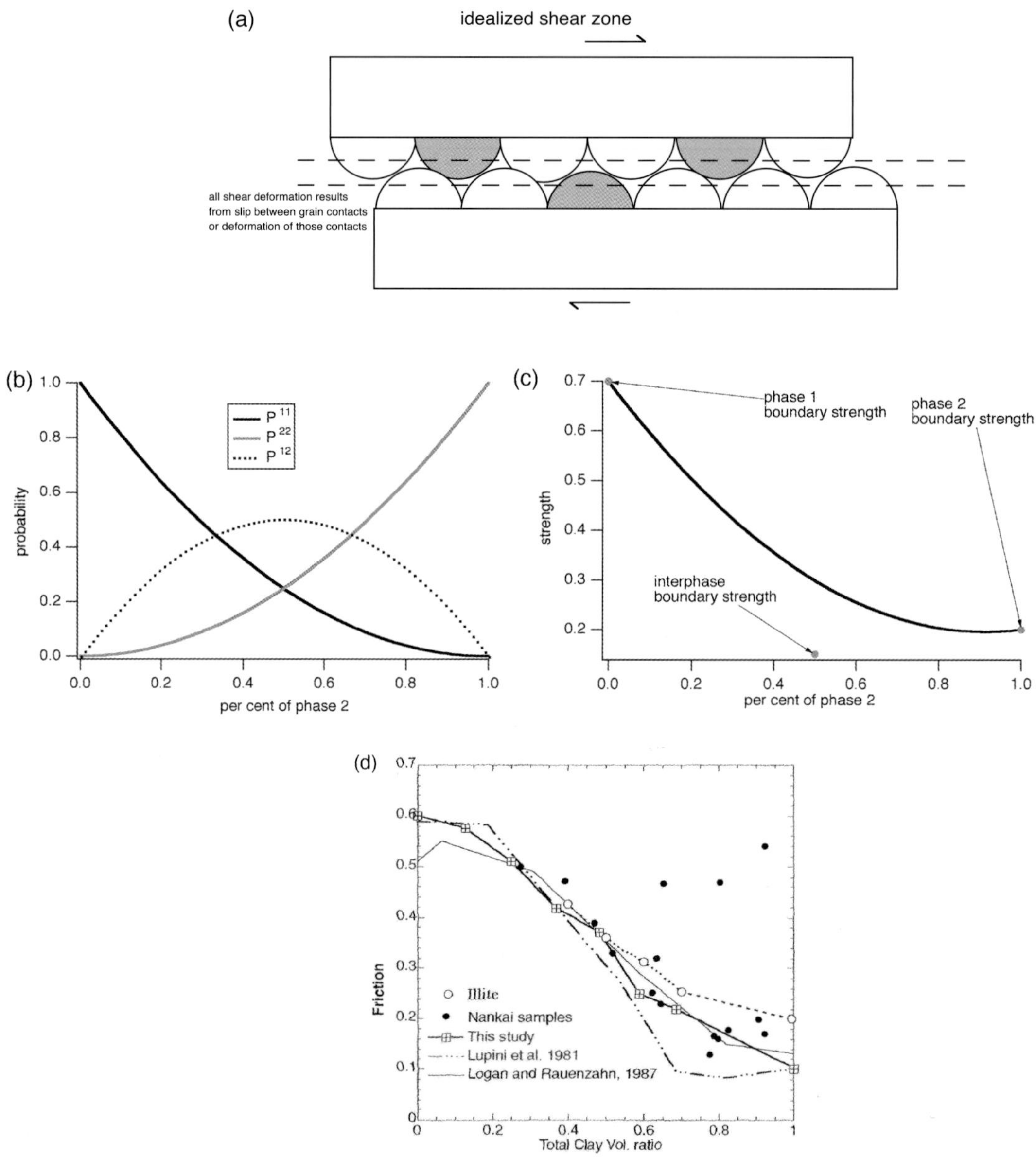

Figure 13.34 Simple model of strength of two-phase granular material. (a) The model is a fully localized slip on a surface composed of particles whose relative numbers reflect the overall composition. (b) The number of phase 1, phase 2, and interphase contacts as a function of percent phase 2. (c) Example strength with variation in percent phase 2. Here the phase 1 contact strength is $\mu = 0.7$ (quartz), the phase 2 contact strength is $\mu = 0.2$ (smectite), and the interphase boundary is anomalously weak $\mu = 0.15$. (d) Variation of mixed clay–quartz samples with clay content. Reprinted from Earth and Planetary Science Letters, 214, *Brown, Kopf, Underwood, and Weinberger*, Compositional and fluid pressure controls on the state of stress on the Nankai subduction thrust, 589–603, Copyright (2003), with permission from Elsevier.

of the constituents, and as strengths are related to water adhesion, water pH, and ionic properties may come into play.

Summary

Strength loss during failure of dry intact rock occurs when deformation is accompanied by dilatancy and an associated loss of contacting area across the incipient fault. Dilatancy is also the source of the pressure dependence of rock failure strength, the observation that as confining pressure is increased the failure strength increases. As confining pressure is increased, dilatancy is suppressed so more shear stress is required to dilate the rock to the point of failure. The primary differences between intrinsically strong and intrinsically weak rocks are that intrinsically weak rocks have low dilatancy and have small strength losses associated with fault formation. Intrinsic weakness results from the presence of weak, anisotropic minerals that slip on cleavage surfaces and a preferential orientation of the weak minerals within the stress field.

The basic observations stated above also apply to dry slip on preexisting faults: failure strength is pressure dependent, rapid slip and strength loss are accompanied by dilatancy and loss of contacting area across the fault, and intrinsic weakness is due to the presence and preferential alignment of weak minerals. Because strength loss during rapid slip is a result of dilatancy, it is expected that small strength losses are associated with the rapid slip of preexisting intrinsically weak faults. This expected correlation between failure strength and strength loss implies that for laboratory-based rate and state friction relationships there is proportionality between the coefficient of the rate weakening term (b) and the nominal strength (μ_0).

As applied to earthquake occurrence, the theory of slip instability requires that fault strength loss ($\Delta\tau$) with slip (δ) is larger than the elastic stiffness (k) of the surroundings, defining a critical stiffness (k_c), such that for instability $\Delta\tau/\delta > k_c$. As dry intrinsically weak rocks and faults have low strength loss, instability is less likely in weak materials. The presence of undrained pore fluid at pressure p affects sliding stability in two ways. First, the effective normal stress is lower, making the strength loss proportionally lower, which favors aseismic over seismic slip. Second, shear induces dilatancy, which reduces the local pore pressure making the fault strong. This dilatancy hardening prevents slip instability, again promoting aseismic over seismic slip. The principal conclusion of this review of laboratory observations of dry and saturated faulting is that in the absence of chemical effects of fluids on fault strength, both intrinsic weakness and apparent weakness resulting from high pore pressure tend to encourage aseismic over seismic slip.

However, the sliding stability of intrinsically weak rocks and minerals is not well studied. The effect of high pore pressure on sliding stability has only been considered for an intrinsically strong, highly dilatant, apparently weak fault. Since many weak phases found in subduction zones have low dilatancy, the effect of high pore pressure on sliding stability is presently unknown. In addition, chemical effects on fault strength are not well studied to date. Many weak phases have strengths that are sensitive to the chemical environment, in particular clays and other sheet silicate minerals have strengths that are influenced by the presence of water. The role of such effects in seismogenesis is not known at present. Given the present state of knowledge, sliding stability and fault strength cannot be unequivocally related.

Acknowledgments

Some results presented here are from unpublished collaborations with Terry Tullis, David Lockner, T.-F. Wong, and Joan Gomberg. Long-standing discussion and correspondence with D. Lockner, T. Tullis, J. D. Weeks, T.-F. Wong, M. Nakatani, G. Hirth, D. Goldsby, J. Dieterich, C. Marone, L. Reinen, and M. Blanpied are gratefully acknowledged. A. Kronenberg provided unpublished data from studies of talc single crystals and shale. I was fortunate to be associated with D. Moore, L. Reinen, C. Morrow, and G. Hirth during periods when each were considering low temperature deformation of weak minerals. Masao Nakatani reviewed and made corrections to the analysis of contact-scale deformation leading to comparisons between the rate dependencies of friction, crack growth and dislocation glide. Thanks to Harold Tobin and Casey Moore for the opportunity to contribute to the SEIZE and MARGINS initiatives, and to T. Dixon, G. Hirth, S. Karner and C. Marone for comments on the manuscript.

This review was supported by the Southern California Earthquake Center. SCEC is funded by NSF Cooperative Agreement EAR–0106924 and USGS Cooperative Agreement 02HQAG0008. The SCEC contribution number for this paper is 1021.

References

Anderson, E. M. (1951), *The Dynamics of Faulting and Dyke Formation With Application to Britain*, 206 pp., Oliver and Boyd, Edinburgh, U.K.

Andrews, D. J. (1976), Rupture velocity of plane strain shear cracks, *J. Geophys. Res., 81*, 5679–5687.

Ashby, M. F., and C. G. Sammis (1990), The damage mechanics of brittle solids in compression, *Pure Appl. Geophys., 133*(3), 489–521.

Atkinson, B. K. (1984), Subcritical crack growth in geologic materials, *J. Geophys. Res., 89*, 4077–4114.

Atkinson, B. K., and P. G. Meredith (1987a), The theory of subcritical crack growth with applications to minerals and rocks, in *Fracture Mechanics of Rock, Geol. Ser.*, edited by B. K. Atkinson, pp. 111–166, Elsevier, New York.

Atkinson, B. K., and P. G. Meredith (1987b), Experimental fracture mechanics data for rocks and minerals, in *Fracture Mechanics of Rock, Geol. Ser.*, edited by B. K. Atkinson, pp. 477–525, Elsevier, New York.

Bailey, A. I., and J. S. Courtney-Pratt (1955), The area of real contact and the shear strength of monomolecular layers of a boundary lubricant, *Proc. R. Soc., Ser. A, 227*, 500–515.

Beeler, N. M., T. E. Tullis, and J. D. Weeks (1994), The roles of time and displacement in the evolution effect in rock friction, *Geophys. Res. Lett., 21* 1987–1990.

Beeler, N. M., and T. E. Tullis (1997), The roles of displacement in velocity dependent volumetric strain of fault zones, *J. Geophys. Res., 102*, 22,595–22,609.

Beeler, N. M., T. E. Tullis, M. L. Blanpied, and J. D. Weeks (1996), Frictional behavior of large displacement experimental faults, *J. Geophys. Res.*, *101*, 8697–8715.
Bernabe, Y. and W. F. Brace (1990), Deformation and fracture of berea sandstone, in *The Brittle-Ductile Transition in Rocks, Geophys. Monogr. Ser.*, vol. 56, edited by A. G. Duba et al., pp. 91–101, AGU, Washington, D. C.
Berthoud, P., T. Baumberger, C. G'Seil, and J.-M. Hiver (1999), Physical analysis of the state- and rate-dependent friction law: Static friction, *Phys. Rev. B.*, *59*, 14,313–14,327.
Biegel, R. L., C. G. Sammis, and J. H. Dieterich (1989), The frictional properties of a simulated gouge having a fractal particle distribution, *J. Struct. Geol.*, *11*, 827–846.
Bish, D. L., and R. F. Giese (1981), Interlayer boding in IIb chlorite, *Am. Mineral.*, *66*, 1216–1220.
Blanpied, M. L., D. A. Lockner, and J. D. Byerlee (1991), Fault stability inferred from granite sliding experiments at hydrothermal conditions, *Geophys. Res. Lett.*, *18*(4), 609–612.
Blanpied, M. L., C. J. Marone, D. A. Lockner, J. D. Byerlee, and D. P. King (1998), Quantitative measure of the variation in fault rheology due to fluid-rock interactions, *J. Geophys. Res.*, *103*, 9691–9712.
Boettcher, M. S., and T. H. Jordan (2004), Earthquake scaling relations for mid-ocean ridge transform faults, *J. Geophys. Res.*, *109*, B12302, doi:10.1029/2005JB003753.
Bos, B., and C. J. Spiers (2001), Experimental investigation into the microstructural and mechanical evolution of phyllosilicate-bearing fault rock under conditions favoring pressure solution, *J. Struct. Geol.*, *23*, 1187–1202.
Bowden, F. P., and D. Tabor (1950), *The Friction and Lubrication of Solids*, 374 pp., Oxford Univ. Press, New York.
Brace, W. F., and J. D. Byerlee (1966), Stick-slip as a mechanism for earthquakes, *Science*, *153*, 990–992.
Brace, W. F. (1978), Volume changes during fracture and frictional sliding, *Pure Appl. Geophys.*, *116*, 603–614.
Brace, W. F., and D. L. Kohlstedt (1980), Limits on lithospheric stress imposed by laboratory experiments, *J. Geophys. Res.*, *85*, 6248–6252.
Brace, W. F., B. W. Paulding, and C. H. Scholz (1966), Dilatancy in the fracture of crystalline rock, *J. Geophys. Res.*, *71*, 3939–3953.
Brown, K. M., A. Kopf, M. B. Underwood, and J. L. Weinberger (2003), Compositional and fluid pressure controls on the state of stress on the Nankai subduction thrust, *Earth Planet. Sci. Lett.*, *214*, 589–603.
Brown, S. R., and C. H. Scholz (1985), The closure of random elastic surfaces in contact, *J. Geophys. Res.*, *90*, 5531–5545.
Byerlee, J. D. (1967). Frictional characteristics of granite under high confining pressure, *J. Geophys. Res.*, *72*, 3639–3648.
Byerlee, J. D. (1968), Brittle-ductile transition in rocks, *J. Geophys. Res.*, *73*, 4741–4750.
Byerlee, J. D., (1970), The mechanics of stick-slip, *Tectonophysics*, *9*, 475–486.
Byerlee, J. D. (1978), Friction of rocks, *Pure Appl. Geophys.*, *116*, 615–626.
Byerlee, J. D., and R. Summers (1976), A note on the effect of fault gouge thickness on fault stability, *Int. J. Rock Mech. Min. Sci.*, *13*, 35–36.
Byrne, D. E., D. M. Davis, and L. R. Sykes (1988), Loci and maximum size of thrust earthquakes and the mechanics of the shallow region of subduction zones, *Tectonophysics*, *7*, 833–857.
Charles, R. J., and W. B. Hillig (1962), The kinetics of glass failure by stress corrosion, in *Symposium sur la Resistance due Verre et les Moyens de l'ameliorer, Union Scientific Continentale du Verr, Charleroi, Belgium*, pp. 511–527, Union Sciences Continentale du Verre, Charleroi, Belgium.
Chester, F. M. (1994), Effects of temperature on friction: Constitutive equations and experiments with quartz gouge, *J. Geophys. Res.*, *99*, 7247–7262.
Costin, L. S. (1987), Deformation and failure, in *Fracture Mechanics of Rock*, edited by B. K. Atkinson, pp. 167–215, Elsevier, New York.
Cox, S. J. D. (1990), Velocity-dependent friction in a large direct shear experiment on gabbro, in *Deformation Mechanisms, Rheology and Tectonics, Spec. Publ. Geol. Soc.*, *54*, 63–70.

Das, S., and C. H. Scholz (1981), Theory of time-dependent rupture in the Earth, *J. Geophys. Res., 86*, 6039–6051.
Dieterich, J. H. (1972), Time-dependent friction in rocks, *J. Geophys. Res., 77*, 3690–3697.
Dieterich, J. H. (1978a), Mechanics of earthquake faulting, in *Summaries of Technical Reports, U.S. Geological Survey National Earthquake Hazards Reduction Program*, edited by J. Evernden, pp. 328–329, U.S. Geol. Surv., Boulder, Colo.
Dieterich, J. H. (1978b), Time-dependent friction and the mechanics of stick slip, *Pure Appl. Geophys., 116*, 790–806.
Dieterich, J. H. (1979), Modeling of rock friction: 1. Experimental results and constitutive equations, *J. Geophys. Res., 84*, 2161–2168.
Dieterich, J. H. (1981), Constitutive properties of faults with simulated gouge, in mechanical Behavior of Crustal Rocks, edited by N. L. Carter, M. Friedman, J. M. Logan, and D. W. Stearns, *Geophys. Monor. Ser., 24*, AGU, Washington, D. C. 103–120.
Dieterich, J. H. (1992). Earthquake nucleation on faults with rate- and state-dependent strength, in *Earthquake Source Physics and Earthquake Precursors*, edited by T. Mikumo et al., pp. 115–134, Elsevier, New York.
Dieterich, J. H., and G. A., Conrad (1984). Effect of humidity on time- and velocity-dependent friction in rocks, *J. Geophys. Res., 89*, 4196–4202.
Dieterich, J. H., and B. D. Kilgore (1994), Direct observation of frictional contacts: New insights for state-dependent properties, *Pure and Applied Geophys., 143*, 283–302.
Dieterich, J. H., and B. Kilgore (1996), Implications of fault constitutive properties for earthquake prediction, *Proc. Natl. Acad., Sci. U. S. A., 93*, 3787–3794.
Dieterich, J. H., and B. D. Kilgore (1994), Direct observation of frictional contacts: New insights for state-dependent properties, *Pure Appl. Geophys., 143*, 283–302.
Dieterich, J. H., and M. F. Linker (1992), Fault stability under conditions of variable normal stress, *Geophys. Res. Lett., 19*, 1691–1694.
Dokos, S. J. (1946), Sliding friction under extreme pressures, *J. Appl. Mech., 13A*, 148–156.
Donath, F. A. (1961), Experimental study of shear failure in anisotropic rocks, *Bull. Geol. Soc. Am., 72*, 985–900.
Du, Y., and A. Aydin (1991), Interaction of multiple cracks and formation of echelon crack arrays, *Int. J. Numer. Anal. Methods Geomech., 15*, 205–218.
Eckhardt, M. M, B. Evans, and S. Viswanathan (1996), Microstructural damage due to frictional sliding in calcite, *Eos, Trans. AGU, 77*, Fall Meet. Suppl., F696.
Engelder, J. T., and C. H. Scholz (1976), The role of asperity indentation and ploughing in rock friction, II, Influence of relative hardness and normal load, *Int. J. Rock Mech. Sci. Geomech. Abstr. 13*, 155–163.
Escartin, J., G. Hirth, and B. Evans (1997), Nondilatant brittle deformation of serpentinites; implications for Mohr-Coulomb theory and the strength of faults, *J. Geophys. Res., 102*, 2897–2913.
Evans, B. (1984), The effect of temperature and impurity content on indentation hardness of quartz, *J. Geophys. Res., 89*, 4213–4222.
Evans, B., and C. Goetze (1979), The temperature variation of hardness of olivine and its implication for polycrystalline yield stress, *J. Geophys. Res., 84*, 5505–5524.
Evans, B., J. T. Fredrich, and T.-F. Wong (1990), The brittle-ductile transition in rocks: Recent experimental and theoretical progress, in *The Brittle-Ductile Transition in Rocks, Geophys. Monogr. Ser.*, vol. 56, edited by A. G. Duba et al., pp. 1–21, AGU, Washington, D. C.
Frank, F. C. (1965), On dilatancy in relation to seismic sources, *Rev. Geophys., 3*, 485–503.
Friedman, M., J. Handin, and G. Alani (1972), Fracture surface energy of rocks, *Int. J. Rock Mech. Min. Sci., 9*, 757–766.
Frolich, C. (1989), The nature of deep-focus earthquakes, *Annu. Rev. Earth Planet. Sci., 17*, 227–254.
Frye, K. M., and C. Marone (2002), Effect of humidity on granular friction at room temperature, *J. Geophys. Res., 107*(B11), 2309, doi:10.1029/2001JB000654.
Furumoto, A. S. (1991), Source parameters of destructive tsunamis, *Sci. Tsunami Hazards, 9*, 95–113.

Giese, R. F. (1978), The electrostatic interlayer forces of layer structure minerals, *Clays Clay Miner., 26*, 51–57.
Giese, R. F. (1980), Hydroxyl orientations and interlayer bonding in amesite, *Clays Clay Miner., 28*, 81–86.
Goldsby, D. L., A. Rar, G. M. Pharr, and T. E. Tullis (2004), Nanoindentation creep of quartz, with implications for rate-and state-variable friction laws relevant to earthquake mechanics, *J. Mater. Res., 19*, 357–365.
Goetze, C., and B. Evans (1979), Stress and temperature in the bending lithosphere as constrained by experimental rock mechanics, *Geophys. J. R. Astron. Soc., 59*, 463–478.
Greenwood, J. A., and J. Williamson (1966), Contact of nominally flat surfaces, *Proc. R. Soc. London, Ser. A, 295*, 300–319.
Griffith, A. A., (1920), The phenomena of rupture and flow in solids, *Philos., Trans. R. Soc. London, Ser. A 221*, 163–198.
Griggs, D. (1936), Deformation of rocks under high confining pressure, *J. Geol., 44*, 541–577.
Gu, J., J. R. Rice, A. L. Ruina, and S. Tse (1984), Stability of frictional slip for a single degree of freedom elastic system with non-linear rate and state dependent friction, *J. Mech. Phys. Solids, 32*, 167–196.
Gu, Y., and T.-F. Wong (1991), Effects of loading velocity, stiffness and inertia on the dynamics of a single degree of freedom spring-slider system, *J. Geophys. Res., 96*, 21,677–21,691.
Hirata, T. (1987), Omori's power law aftershock sequences of microfracturing in rock fracture experiment, *J. Geophys. Res.*, 6215–6221.
Horii, H., and S. Nemat-Nasser (1986), Brittle failure in compression: splitting, faulting and brittle-ductile transition, *Philos. Trans. R. Soc. London, Ser. A, 319*, 337–374.
Hubbert, M. K., and W. W. Rubey (1959), Role of fluid pressure in mechanics of overthrust faulting, *Bull. Geol. Soc. Am., 70*, 115–166.
Ibanez, W. D., and A. K. Kronenberg (1993), Experimental deformation of shale: Mechanical properties and microstructural indicators of mechanisms, *Int. J. Rock Mech. Min. Sci., 30*, 723–734.
Ida, Y. (1972), Cohesive force across the tip of a longitudinal shear crack and Griffith's specific surface energy, *J. Geophys. Res., 77*, 3796–3805.
Ida, Y., (1973), The maximum acceleration of seismic ground motion, *Bull. Seismol. Soc. Am. , 63*, 959–968.
Israelachvili, J. N. (1992), Adhesion forces between surfaces in liquids and condensable vapours, *Surface Sci. Rep., 14*, 109–160.
Israelachvili, J. N., and G. E. Adams (1976), Direct measurement of long range forces between two mica surfaces in aqueous KNO_3 solutions, *Nature, 262*, 774–776.
Israelachvili, J. N., P. M. McGuiggan, and A. M. Homola (1988), Dynamic properties of molecularly thin liquid films, *Science, 240*, 189–191.
Jaoul, O., J. Tullis, and A. Kronenberg (1984), The effect of varying water contents on the creep behavior of Heavitree quartzite, *J. Geophys. Res., 89*, 4298–4312.
Johnson, T. (1981), Time dependent friction of granite: Implications for precursory slip on faults, *J. Geophys. Res., 86*, 6017–6028.
Kailer, A., K. G. Nickel, and Y. G. Gogotsi (1999), Raman microspectroscopy of nanocrystalline and amorphous phases in hardness indentations, *J. Raman Spectrosc., 30*, 939–946.
Karner, S. L., and C. J. Marone (1998), The effect of shear load on frictional healing in simulated fault gouge, *Geophys. Res. Lett., 25*, 4561–4564.
Kato, N., K. Yamamoto, H. Yamamoto, and T. Hirasawa (1992), Strain-rate effect on frictional strength and the slip nucleation process, *Tectonophysics, 211*, 269–282.
Kelleher, J., J. Savino, H. Rowlett, and W. McCann (1974), Why and where great thrust earthquakes occur along island arcs, *J. Geophys. Res., 79*, 4889–4899.
Kemeny, J. M., and N. G. W. Cook (1987), Crack models for the failure of rocks in compression, in *Constitutive Laws for Engineering Materials: Theory and Applications*, edited by C. S. Desai, pp. 878–887, Elsevier, New York.
Kirby, S. H. (1980), Tectonic stresses in the lithosphere: Constraints provided by the experimental deformation of rocks, *J. Geophys. Res., 85*, 6353–6363.

Knopoff, L. (1972), Model of aftershock occurrence, in *Flow and Fracture of Rocks, Geophys. Monogr. Ser.*, vol. 16, edited by H. C. Heard et al., pp. 259–263, AGU, Washington, D. C.
Kranz, R. L. (1980), The effects of confining pressure and stress difference on static fatigue of granite, *J. Geophys. Res., 85.*, 1854–1866.
Kronenberg, A. K., S. H. Kirby, and J. C. Pinkston (1990), Basal slip and mechanical anisotropy of biotite, *J. Geophys., Res., 95*, 19,257–19,278.
Lawn, B. (1993), *Fracture of Brittle Solids*, 378 pp., Cambridge Univ. Press, New York.
Linker, M. F., and J. H. Dieterich (1992), Effects of variable normal stress on rock friction: Observations and constitutive equations, *J. Geophys. Res., 97*, 4923–4940.
Lockner, D. A. (1998), A generalized law for brittle deformation of Westerly granite, *J. Geophys. Res., 103*, 5107–5123.
Lockner, D. A., and J. D. Byerlee (1990), An example of slip instability resulting from displacement-varying strength, *Pure Appl. Geophys., 133*, 398–410.
Lockner, D. A., J. D. Byerlee, V. Kuksenko, A. Ponomarev, and A. Sidorin (1991), Quasi-static fault growth and shear fracture energy in granite, *Nature, 350*, 39–42.
Lockner, D. A., D. E. Moore, and Z. Reches (1992), Microcrack interaction leading to shear fracture, in *33rd U.S. Rock Mechanics Symposium*, edited by J. R. Tillerson and W. R. Wawersik, pp. 807–816, A. A. Balkema, Brookfield, VT.
Logan, J. M., and K. A. Rauenzahn (1987), Frictional dependence of gouge mixtures of quartz and montmorillonite on velocity, composition and fabric, *Tectonophysics, 144*, 87–108.
Mair, K., and C. Marone (1999), Friction of simulated fault gouge for a wide range of velocities and normal stresses, *J. Geophys. Res., 137*, 28,899–28,914.
Marcellini, A. (1995), Arrhenius behavior of aftershock sequences, *J. Geophys. Res., 100*, 6463–6468.
Marcellini, A. (1997), Physical model of aftershock temporal behavior, *Tectonophysics, 277*, 137–146.
Mares, V. M., and A. K. Kronenberg (1993), Experimental deformation of muscovite, *J. Struct. Geol., 15*, 1061–1075.
Marone, C. J. (1998), Laboratory-derived friction laws and their application to seismic faulting, *Annu. Rev. Earth Planet. Sci., 26*, 643–696.
Marone, C., and B. Kilgore (1993), Scaling of the critical slip distance for seismic faulting with shear strain in fault zones, *Nature, 362*, 618–621.
Marone, C., C. B. Raleigh, and C. H. Scholz (1990), Frictional behavior and constitutive modeling of simulated fault gouge, *J. Geophys. Res., 95*, 7007–7025.
Masuda, T., T. Hiraga, H. Ikei, H. Kanda, Y. Kugimiya, and M. Akizuki (2000), Plastic deformation of quartz at room temperature: A Vickers nano-indentation test, *Geophys. Res. Lett., 27*, 2773–2776.
Mogi, K. (1962), Study of elastic shocks caused by the fracture of heterogeneous materials and their relation to earthquake phenomenon, *Bull. Earthquake Res. Inst. Univ. Tokyo, 40*, 1438.
Mogi, K. (1974), On the pressure dependence of strength of rocks and the Coulomb fracture criterion, *Tectonophysics, 21*, 273–285.
Moore, D., and D. A. Lockner (2004), Crystallographic controls on the frictional behavior of dry and water-saturated sheet structure minerals, *J. Geophys. Res., 109*, B03401, doi:10.1029/2003JB002582.
Moore, D. E., and D. A. Lockner (2007), Friction of the smectite clay montmorillonite: A review and interpretation of data, this volume.
Moore, D. E., R. Summers, and J. D. Byerlee (1989), Sliding behavior and deformation textures of heated illite gouge, *J. Struct. Geol., 11*, 329–342.
Moore, D. E., D. A. Lockner, R. Summers, M. Shengli, and J. D. Byerlee (1996), Strength of chrysotile-serpentinite gouge under hydrothermal conditions: Can it explain a weak San Andreas fault?, *Geology, 24*(11), 1041–1044.
Moore, D. E., D. A. Lockner, S. Ma, R. Summers, and J. D. Byerlee (1997), Strengths of serpentinite gouges at elevated temperatures, *J. Geophys. Res., 102*, 14,787–14,801.
Moore, J. C., and R. von Huene (1980), Abnormal pore pressure and hole instability in forearc regions, preliminary report, p. 29, Ocean Margin Drill. Proj., Menlo Park, Calif.

Moore, J. C., and D. Saffer (2001), Updip limit of the seismogenic zone beneath the accretionary prism of southwest Japan: An effect of diagenetic to low-grade metamorphic processes and increasing effective stress, *Geology, 29*(2), 183–186.

Morrow, C., and J. Byerlee (1989), Experimental studies of compaction and dilatancy during frictional sliding on faults containing gouge, *J. Struct. Geol., 11*, 815–825.

Morrow, C. A., Radney, B, and J. D. Byerlee (1992), Frictional strength and the effective pressure law of montmorillonite and illite clays, in *Fault Mechanics and Transport Properties of Rocks*, pp. 69–88, Elsevier, New York.

Morrow, C. A., D. E. Moore, and D. A. Lockner (2000), The effect of mineral bond strength and adsorbed water on fault gouge frictional strength, *Geophys. Res. Lett, 27*, 815–818.

Nakatani, M. (2001), Conceptual and physical clarification of rate and state friction: Frictional sliding as a thermally activated rheology, *J. Geophys. Res., 106*, 13,347–13,380.

Nakatani, M., and H. Mochizuki (1996), Effects of shear stress applied to surfaces in stationary contact on rock friction, *Geophys. Res. Lett., 23*, 869–872.

Okubo, P. G., and J. H. Dieterich (1981), Fracture energy of stick-slip events in a large scale biaxial experiment, *Geophys. Res. Lett., 8*, 887–890.

Palmer, A. C., and J. R. Rice (1973), The growth of slip surfaces in the progressive failure of over-consolidated clay, *Proc. R. Soc. London, Ser. A, 332*, 527–548.

Peng, S., and A. M. Johnson (1972), Crack growth and faulting in cylindrical specimens of Chelmsford granite, *Int. J. Rock Mech. Min. Sci. Geomech. Abstr., 9*, 37–86.

Peterson, E. T., and T. Seno (1984), Factors affecting seismic moment release rates in subduction zones, *J. Geophys. Res, 89*, 10,233–10,248.

Power, W. L., T. E. Tullis, and J. D. Weeks (1988), Roughness and wear during brittle faulting, *J. Geophys. Res., 93*, 15,268–15,278.

Rabinowicz, E. (1958), The intrinsic variables affecting the stick-slip process, *Proc. Phys. Soc. London, 71*, 668–675.

Raleigh, C. B., and M. S. Paterson (1965), Experimental deformation of serpentinite and its tectonic implications, *J. Geophys. Res., 70*, 3965–3985.

Reches, Z., and D. A. Lockner (1994), Nucleation and growth of faults in brittle rocks, *J. Geophys. Res., 99*, 18,159–18,173.

Reuschle, T. (1990), Slow crack growth and aftershock sequences, *Geophys. Res. Lett., 17*, 1525–1528.

Rice, J. R. (1975), On the stability of dilatant hardening for saturated rock masses, *J. Geophys. Res., 80*, 1531–1536.

Rice, J. R. (1980), The mechanics of earthquake rupture, in *Physics of the Earth's Interior*, edited by A. M. Dziewonski and E. Boschi, pp. 555–649, Elsevier, New York.

Rice, J. R. (1983), Constitutive relations for fault slip and earthquake instabilities, *Pure Appl. Geophys., 121*, 443–475.

Rice, J. R. (1993), Spatio-temporal complexity of slip on a fault, *J. Geophys. Res., 98*, 9885–9907.

Rice, J. R., and A. L. Ruina (1983), Stability of steady frictional slipping, *J. Appl. Mech., 50*, 343–349.

Rice, J. R., N. Lapusta, and K. Ranjith (2001), Rate and state dependent friction an the stability of sliding between elastically deformable solids, *J. Mech. Phys. Sol., 49*, 1865–1898.

Reinen, L. A., J. D. Weeks, and T. E. Tullis (1994), The frictional behavior of lizardite and antigorite serpentinites: Experiments, constitutive models, and implications for natural faults, *Pure Appl. Geophys., 143*, 317–358.

Robertson, E. C. (1983), Relationship of fault displacement to gouge and breccia thickness, *Trans. Am. Inst. Min. Metall. Pet. Eng., 35*, 1426–1432.

Rudnicki, J. W., and C.-H. Chen (1988), Stabilization of rapid frictional slip on a weakening fault by dilatant hardening, *J. Geophys. Res., 93*, 4745–4757.

Ruina, A. L. (1983), Slip instability and state variable friction laws, *J. Geophys. Res., 88*, 10,359–10,370.

Rutter, E. H. (1974), The influence of temperature, strain rate and interstitial water in the experimental deformation of calcite rocks, *Tectonophysics, 22*, 311–334.

Rutter, E. H. (1983), Pressure solution in nature, theory, and experiment, *J. Geol. Soc. London, I40*, 725–740.

Saffer, D. M., and B. A. Bekins (2002), Hydrologic controls on the morphology and mechanics of accretionary wedges, *Geology, 30*(3), 271–274.
Saffer, D. M., and C. Marone (2003), Comparison of smectite- and illite-rich gouge frictional properties: Application to the updip limit of the seismogenic zone along subduction zone megathrusts, *Earth Planet. Sci. Let., 215,* 219–235.
Saffer, D. M., K. M. Frye, C. Marone, and K. Mair (2001), Laboratory results indicating complex and potentially unstable frictional behavior of smectite clay, *Geophys. Res. Lett., 28,* 2297–2300.
Sammis, C. G., and M. F. Ashby (1986), The failure of brittle porous solids under compressive stress states, *Acta Metall., 34,* 511–526.
Savage, J. C., J. D. Byerlee, and D. A. Lockner (1996), Is internal friction, friction?, *Geophys. Res. Lett., 23,* 487–490.
Scholz, C. (1968a), The frequency magnitude relation of microfracturing in rock and its relation to earthquakes, *Bull. Seismol. Soc. Am., 58,* 399–415.
Scholz, C. (1968b), Microfractures, aftershocks and seismicity, *Bull. Seismol. Soc. Am., 58,* 117–130.
Scholz, C. H. (1968c), Mechanism of creep in brittle rock, *J. Geophys. Res., 73,* 3295–3302.
Scholz, C. H. (1972), Static fatigue in quartz, *J. Geophys. Res., 77,* 2104–2114.
Scholz, C. H. (1988), The critical slip distance for seismic faulting, *Nature, 336,* 761–763.
Scholz, C. H. (1990), *The Mechanics of Earthquakes and Faulting,* 439 pp., Cambridge Univ. Press, New York.
Scholz, C. H., and J. T. Engelder (1976), The role of asperity indentation and ploughing in rock friction, I, Asperity creep and stick-slip, *Int. J. Rock Mech. Sci. Geomech. Abstr., 13,* 149–154.
Scruggs, V. J. (1997), Frictional constitutive properties and related microstructures of albite, muscovite, biotite and talc, Ph.D. dissertation, Brown Univ., Providence, R. I.
Scruggs, V. J., and T. E. Tullis (1998), Correlation between velocity dependence of friction and strain localization in large displacement experiments on feldspar, muscovite and biotite gouge, *Tectonophysics, 295,* 15–40.
Segall, P., and J. R. Rice (1995), Dilatancy, compaction, and slip instability of a fluid infiltrated fault, *J. Geophys. Res., 100,* 22,155–22,173.
Sharp, S. J., M. F. Ashby, and N. A. Fleck (1993), Material response under static and sliding indentation loads, *Acta Metall. Mater., 41,* 685–692.
Shea, W. T., and A. K. Kronenberg (1992), Rheology and deformation mechanisms of an isotropic mica schist, *J. Geophys. Res., 97,* 15,201–15,237.
Shea, W. T., and A. K. Kronenberg (1993), Strength and anisotropy of foliated rocks with varied mica content, *J. Struct. Geol., 15,* 1097–1121.
Sibson, R. H. (2002), Geology of the crustal earthquake source, in *International Handbook of Earthquake and Engineering Seismology,* edited by W. H. K. Lee, H. Kanamori, and P. C. Jennings, Elsevier, New York, 455–473.
Spetzler, H., G. Sobolev, A. Koltsov, A. Zang, and I. C. Getting (1991), Some properties of unstable slip on rough surfaces, *Pure Appl. Geophys. 137,* 95–112.
Stesky, R. M. (1978), Mechanisms of high temperature frictional sliding in Westerly granite, *Can. J. Earth Sci., 15,* 361–375.
Stuart, W. D. (1988), Forecast model for great earthquakes at the Nankai trough subduction zone, *Pure Appl. Geophys. 126,* 619–641.
Tapponnier, P., and W.F. Brace (1976), Development of stress-induced microcracks in Westerly granite, *Int. J. Rock Mech. Min. Sci. Geomech. Abstr., 13,* 103–112.
Terzaghi, K. V. (1923), Die berechnung der durchlassigkeitsziffer des tones asu dem verlauf der hydrodynamischen spannumgserscheinungen, *Sber. Akad. Wiss. Wien, 132,* 105.
Thomas, P. A., and M. O. Robbins (1990), Origin of stick-slip motion in boundary lubrication, *Science, 250,* 792–794.
Tse, S. T., and J. R. Rice (1986), Crustal earthquake instability in relation to the depth variation for frictional slip properties, *J. Geophys. Res., 91,* 9452–9472.
Tsutsumi, A., and T. Shimamoto (1997), High-velocity frictional properties of gabbro, *Geophys. Res. Lett., 24,* 699–702.

Tullis, T. E., and J. D. Weeks (1986), Constitutive behavior and stability of frictional sliding of granite, *Pure Appl. Geophys., 124,* 384–414.
Tullis, T. E., and J. D. Weeks (1987), Micromechanics of frictional resistance of calcite, *Eos Trans. AGU, 68,* 405.
Tullis, T. E., F. G. Horowitz, and J. A. Tullis (1991), Flow laws of polyphase aggregates from end-member flow laws, *J. Geophys. Res., 96,* 8081–8096.
Tullis, T. E., Cooper, C., Sanford, D., and A. Lochhead (1994), The influence of fluid chemistry on nanoindentation of quartz: relevance for the friction evolution effect, *Eos Trans. AGU, 75, Fall Meeting Abs. Supp.,* 443.
Viswanathan, S., and D. Kohlstedt (1996), Evolution of friction coefficient during sliding of a single asperity on semi-brittle mineral surfaces, *Eos Trans. AGU, 77*(46), Fall Meet. Suppl., F696.
Viswanathan, S., D. Kohlstedt, G. Eckhardt, and B. Evans (1995), Partitioning of strain between brittle and ductile deformation in microscratches on calcite, *Eos Trans. AGU, 76*(46), Fall Meet. Suppl., F583.
Walsh, J. B. (1971), Stiffness in faulting and friction experiments, *J. Geophys. Res., 76,* 8597–8598.
Walsh, J. B., and M. A. Grosenbaugh (1979), A new model for analyzing the effect of fractures on compressibility, *J. Geophys. Res., 84,* 3532–3536.
Wang, K., and J. He (1999), Mechanics of low-stress forearcs: Nankai and Cascadia, *J. Geophys. Res., 104,* 15,191–15,205.
Wang, W., and C. H. Scholz (1994), Micromechanics of the velocity and normal stress dependence of rock friction, *Pure Appl. Geophys., 143,* 303–316.
Weeks, J. D., and T. E. Tullis (1985), Frictional sliding of dolomite: A variation in constitutive behavior, *J. Geophys. Res., 90,* 7821–7826.
Weeks, J. D., and T. E. Tullis (1992), High-resolution measurement of displacement in rock friction experiments, *Eos Trans. AGU, 73,* 565.
Weeks, J. D., N. M. Beeler, and T. E. Tullis (1991), Glass is like a rock, *Eos Trans. AGU, 72,* 457–458.
Westbrook, G. K., and M. J. Smith (1983), Long decollements and mud volcanoes: Evidence from the Barbados Ridge Complex for the role of high pore-fluid pressure in the development of an accretionary complex, *Geology, 11*(5), 279–283.
Westbrook, J. H. (1958), Temperature dependence of strength and brittleness of some quartz structures, *J. Am Ceram. Soc., 41,* 433–440.
Westbrook, J. H., and P. J. Jorgensen (1965), Indentation creep of solids, *Trans. Am. Inst. Min. Metall. Pet. Eng., 233,* 425–428.
Westbrook, J. H., and P. J. Jorgensen (1968), Effects of water desorption on indentation microhardness anisotropy in minerals, *Am. Mineral., 53,* 1899–1909.
Wiederhorn, S. M., and L. H. Bolz (1970), Stress corrosion and static fatigue of glass, *J. Am. Ceram. Soc., 53,* 543–548.
Wong, T.-F. (1982), Shear fracture energy of Westerly granite from post-failure behavior, *J. Geophys. Res., 87,* 990–1000.
Wong, T.-F. (1986), On the normal stress dependence of the shear fracture energy, in *Earthquake Source Mechanics, Geophys. Monogr. Ser.,* vol. 37, edited by S. Das et al., pp. 1–11, AGU, Washington, D. C.
Wong, T.-F. (1990), Mechanical compaction and the brittle-ductile transition in porous sandstones, in *Deformation Mechanisms, Rheology and Tectonics, Spec. Publ. Geol. Soc., 54,* 111–122.
Wong, T.-F., and Y. Zhao (1990), Effects of load point velocity on frictional instability behavior, *Tectonophysics, 175,* 177–195.
Yamashita, Y., and L. Knopoff (1987), Models of aftershock occurrence, *Geophys. J. R. Astron. Soc., 91,* 13–26.
Yund, R. A., M. L. Blanpied, T. E. Tullis, and J. D. Weeks (1990), Amorphous material in high strain experimental fault gouges, *J. Geophys. Res., 95,* 15,589–15,602.

PART V

Seismic and Geodetic Studies

14

Asperities and Quasi-Static Slips on the Subducting Plate Boundary East of Tohoku, Northeast Japan

Akira Hasegawa, Naoki Uchida, Toshihiro Igarashi, Toru Matsuzawa, Tomomi Okada, Satoshi Miura, and Yoko Suwa

Abstract

We review recent studies on interplate coupling and recurrence of earthquakes along the subducting plate boundary offshore east of northeastern Japan. Waveform inversions of a few successive earthquakes that occurred in common source areas have revealed that large slip areas of successive ruptures are in the same place (asperities) on the plate boundary. Many small earthquake sequences having similar waveforms, including an M 4.8 ± 0.1 earthquake sequence regularly occurring with a recurrence interval of 5.35 ± 0.53 years, have been found on the plate-boundary zone. They are interpreted as caused by repeated ruptures of small asperity patches surrounded by stably sliding areas. Space-time distribution of quasi-static slip on the plate boundary was estimated from activity of these repeating earthquake sequences, which is approximately consistent with that estimated from backslip inversions of GPS data on land. These observations strongly suggest that the asperity model is applicable to the process of seismic and aseismic slip occurring on the plate boundary east of NE Japan. Three large asperities are left as unruptured patches on the plate boundary. One asperity south of Tokachi ruptured on 26 September 2003, the M 8.0 Tokachi-oki earthquake. In the vicinity of another asperity east of Miyagi prefecture, an intermediate-depth M 7.1 event occurred in the subducted slab on 26 May 2003, and a shallow M 6.4 event occurred in the overriding plate on 26 July 2003. These events could be triggered by strong coupling at the asperity and quasi-static slip in the surrounding areas.

Introduction

Beneath Tohoku, northeastern Japan, the Pacific plate is subducting downward into the mantle at a convergence rate of ~8 cm/yr [*Demets et al.*, 1994]. It subducts at a very low angle of <10° for its initial descent to a distance ~100 km away from the trench. Then it changes dip angle and descends with a steeper angle of ~30° beneath the land area [*Hasegawa et al.*, 1994; *Umino et al.*, 1995; *Hino et al.*, 1996; *Zhao et al.*, 1997].

The seismic coupling coefficient is estimated to be ~25% for the plate boundary offshore Sanriku (northern Tohoku) and ~10% for the plate boundary off Fukushima (southern Tohoku) [e.g., *Pacheco et al.*, 1993; *Peterson and Seno*, 1984]. This suggests that much of the interplate motion is accommodated by aseismic slip in these regions. Nonetheless, many shallow earthquakes including great earthquakes occur beneath the Pacific Ocean along the plate boundary or in its vicinity associated with the plate subduction. Figure 14.1 shows the epicentral distribution of shallow (<60 km) earthquakes for each 5-year period from 1981 to 2000. An uneven distribution pattern seems to be persistent, regardless of time.

Recent studies based on data obtained by seismic observations and GPS observations on land have made an important contribution to understanding how aseismic and seismic slip proceeds along the plate boundary for these regions and their relation with each other. Moreover, waveform inversion analyses for some interplate earthquakes have shown that the asperity model [*Lay and Kanamori*, 1981; *Lay et al.*, 1982] is applicable to the plate boundary in these regions.

In this paper, we review recent studies and discuss the present state of interplate coupling and its relation to recent seismic activity in the northeastern Japan convergent plate boundary zone.

Distribution of Large Slip Areas by Successive Earthquakes

Two end-member models have been used to describe interplate coupling and the recurrence of great earthquakes at convergent plate-boundary zones. One is the asperity model, in which the plate boundary is divided into areas of strong frictional coupling dominated by stick-slip behavior (asperities), and intervening areas of weak frictional coupling dominated by stable sliding. The other is the uniform coupling model, in which frictional coupling along the plate boundary is uniform.

In the asperity model, asperities are persistent features. Large slip areas of successive ruptures will be in the same place (asperities) for a given section of the plate boundary. Earthquakes may rupture one or more adjacent asperities; earthquake magnitudes will be large if the total size of the ruptured asperities is large. In contrast, in the uniform coupling model, both the recurrence of

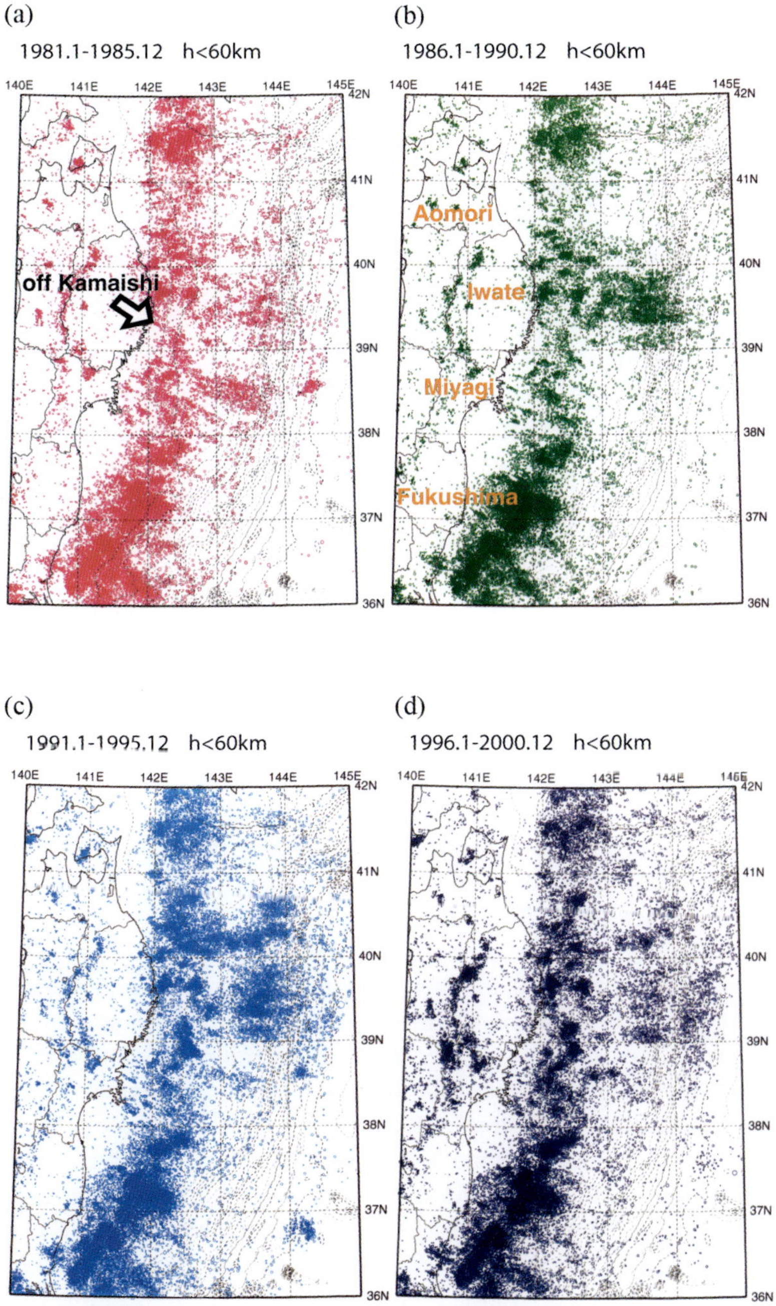

Figure 14.1 Epicentral distribution of shallow earthquakes located in northeastern Japan by the seismic network of Tohoku University for the 5-year periods of (a) 1981–1985, (b) 1986–1990, (c) 1991–1995, and (d) 1996–2000, respectively. Earthquakes shallower than 60 km are plotted. An arrow in figure 14.1a shows the location of the repeating earthquake sequence off Kamaishi.

great earthquakes and their slip distributions vary with time. A large slip area for one great earthquake will be a small slip area in the next great earthquake in a given section of the plate boundary. It has been difficult, however, to know from direct observations which model is applicable to the process actually occurring on the plate boundary, since there have been few examples of well-documented recurrent large earthquakes.

Recently, some large earthquakes in subduction zones observed with modern seismographs reruptured the same portions of the plate boundary that failed in previous large earthquakes, providing the opportunity to compare the slip distribution for successive events. *Schwartz* [1999] investigated such earthquake pairs (or triplets) rupturing overlapping regions. They are (1) the 1968 (M 7.9) Tokachi-oki and 1994 (M 7.6) Sanriku-oki earthquakes in NE Japan; (2) the 1971 (M_w 8.0) and 1995 (M_w 7.7) Solomon Islands earthquakes; (3) the 1963 (M_w 8.5) and 1995 (M_w 7.9) Kuril Islands earthquakes; and (4) the 1957 (M_w 8.6), 1986 (M_w 8.0) and 1996 (M_w 7.9) Aleutian Islands earthquakes. *Schwartz* [1999] systematically studied the spatial distribution of moment release of these earthquake pairs or triplets based on inversion of teleseismic broadband body waves and long-period surface waves. Comparison of the spatial distribution of moment release for these successive earthquake ruptures shows two types of recurrent fault slip pattern. The NE Japan earthquakes and the Solomon Islands earthquakes are not consistent with the asperity model: the 1994 Sanriku-oki and 1995 Solomon Islands ruptures filled in areas of slip deficit left by their preceding earthquakes. In contrast, the Kuril and Aleutian earthquakes are consistent with the asperity model: the 1995 Kuril Islands and the 1996 Aleutian Islands earthquakes both re-ruptured portions of the asperity distribution defined by the preceding events. However, the amount of slip at the same asperities varied for the sequential earthquake ruptures.

Nagai et al. [2001] reinvestigated the slip distributions of the 1968 M 7.9 Tokachi-oki earthquake and the 1994 M 7.6 Sanriku-oki earthquake and obtained a result that supports the asperity model. They estimated detailed slip distributions of the two earthquakes by waveform inversions of both strong motion records at a regional network and teleseismic body wave records at global networks. The results show that the 1968 event has more than two large slip areas (asperities) and the 1994 event has one large slip area as shown in figure 14.2.

Comparison of the slip distributions between the two events shows that the large slip area of the 1994 event is located in the same place as one of the two large slip areas of the 1968 event. This indicates that one of the two asperities in the south ruptured by the 1968 event was ruptured again by the 1994 event. *Nagai et al.* [2001] estimated that the seismic coupling is almost 100% in this common asperity. This observation suggests that asperities are persistent features and that the asperity model better explains the recurrence of large plate-boundary earthquakes in this region. They also reexamined aftershock hypocenters by using the same station corrections for the two events. Relocated aftershocks fringe the estimated asperities and are not distributed within the asperities for both events.

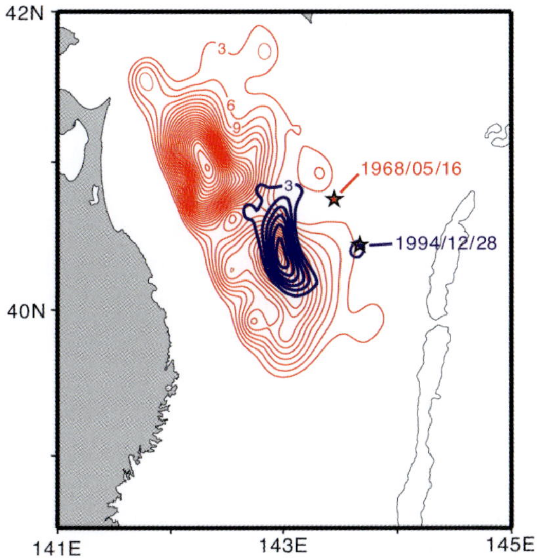

Figure 14.2 Moment release distribution of the 1968 *M* 7.9 Tokachi-oki earthquake and that of the 1994 *M* 7.6 Sanriku-oki earthquake [*Nagai et al.*, 2001]. Contour interval is 3×10^{16} Nm/km^2. Red and blue stars show the hypocenters of the 1968 and 1994 events, respectively.

The study by *Nagai et al.* [2001] updated the previous result for the two large earthquakes in NE Japan reported by *Schwartz* [1999]. Use of regional strong-motion network data together with teleseismic body wave data enabled estimation of slip distributions with higher resolution. Thus they could detect the second largest asperity in the south for the 1968 Tokachi-oki earthquake, which was not detected in the previous studies using teleseismic body and surface wave data and tsunami waveform data [*Hartog and Schwartz*, 1996; *Tanioka et al.*, 1996]. Consequently, recurrent fault slip for three of the four plate-boundary segments investigated by *Schwartz* [1999] proved to be consistent with the asperity model.

Yamanaka and Kikuchi [2004] have systematically studied slip distributions of large interplate earthquakes that occurred offshore east of NE Japan for the last 70 years by inverting waveforms recorded by strong motion seismometers of Japan Meteorological Agency. They found a clear tendency that large slip areas of successive large earthquakes were in the same places on the plate boundary, which again supports the asperity model. Moreover, they found that the typical size of individual asperities in NE Japan is *M* 7 class and that *M* 8 class great earthquakes are caused when more than one asperities are ruptured simultaneously.

Aseismic Slip and Interplate Coupling Estimated from GPS Observations

Asperities slip dynamically as earthquakes, while the surrounding stably sliding areas (or conditionally stable sliding areas) slip quasi-statically during the interseismic period. If so, it is very important to know when the quasi-static slip actually occurs in the stably sliding areas of the plate boundary. Recently, remarkable progress has been made on this point for the plate boundary east off NE Japan. *Nishimura* [2000] and *Heki et al.* [1997] detected large postseismic slip after the 1994 *M* 7.6 Sanriku-oki earthquake, which was comparable to the seismic moment of the main shock. Similarly, large postseismic slip was also observed after the 1989 and 1992 Sanriku-oki earthquakes of *M* 7.1 and *M* 6.9, respectively [*Kawasaki et al.*, 2001; *Miura et al.*, 1993]. These are based on observations of crustal deformation by GPS, strainmeters and tiltmeters. GPS data also provide important information on the spatial distribution of interplate coupling. *Nishimura et al.* [2000] estimated spatial and temporal distribution of seismic and aseismic slips along the plate boundary offshore east of NE Japan based on back-slip inversions of GPS data on land. Back-slip distribution estimated for the period of 1996 – 1999 showed a strong interplate coupling in the area off Miyagi prefecture and large forward slip in the downward extension of the focal area of the 1994 *M* 7.6 Sanriku-oki earthquake.

Recently, *Suwa et al.* [2003] estimated back-slip distribution on the plate boundary east of Tohoku and south of Hokkaido by the inversion method of *Yabuki and Matsuura* [1992]. Vertical components of the displacement velocity field estimated from GPS observations, in addition to horizontal components, were used for the back-slip inversions. Addition of the vertical component provided more stable results for estimation of the back-slip distribution. Estimated back-slip distribution for the 5-year period 1997–2002 is shown by contours in figure 14.3. It clearly shows two strongly coupled areas (areas with large

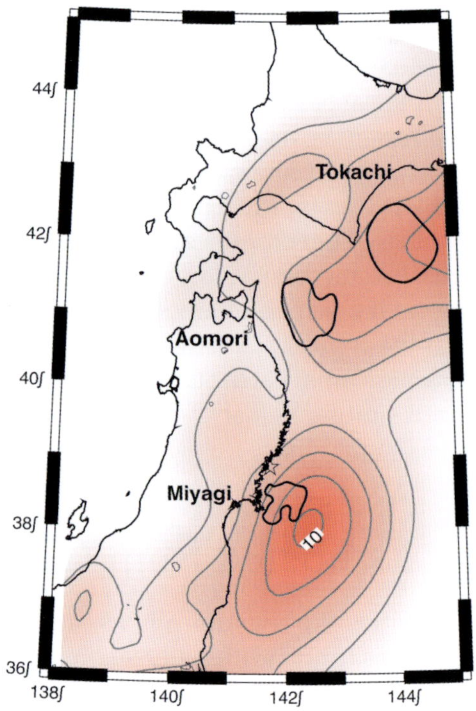

Figure 14.3 Back-slip distribution along the plate boundary east off NE Japan estimated from GPS data for the 5-year period 1997–2002 [*Suwa et al.*, 2003]. Amount of back slip, which expresses the degree of interplate coupling, is shown by contours with an interval of 2 cm/yr. Estimated asperities left unruptured are shown by solid lines.

back-slip amounts) on the plate boundary offshore; one east of Miyagi prefecture and the other an area east of Aomori prefecture through south of Tokachi. In those areas the estimated seismic coupling for the 5 years is ~75%–100%.

Back slip in a small area east of Miyagi prefecture slightly exceeds the plate conversion rate, but this is probably because of the estimation error. Figure 14.3 also shows that the back-slip area is not confined to the area beneath the Pacific Ocean but also to the area beneath the land. This means that the interplate coupling extends to a depth of ~100 km almost beneath the volcanic front, although the amount of back slip is small (seismic coupling of ~25%). This is contradictory to our knowledge of the seismogenic depth along the plate boundary obtained from seismic observations. Interplate earthquakes usually occur only at depths shallower than 50–60 km in NE Japan [*Hasegawa et al.*, 1991, 1994; *Igarashi et al.*, 2001]. If the interplate coupling really extends down to ~100 km depth, the coupled portion of the plate interface deeper than 50–60 km must slip episodically at sometime during the interseismic period. Aseismic slip after large interplate earthquakes (postseismic slip) is a possible candidate for this. In fact, large postseismic slip amounting to 25–70 cm was detected in the deep plate interface down to ~100-km depth on the downdip extension of the mainshock fault after the previous *M* 7.4 Miyagi-oki earthquake of 1978 [*Ueda et al.*, 2001]. Similar postseismic slip on the deep portion of the plate interface has been reported for the 1938 *M* 7.7 Fukushima-oki earthquake [*Ueda*, 2001] and for the 1994 *M* 7.6 Sanriku-oki earthquake [*Nishimura*, 2000].

Alternatively, *Heki* [2004] interpreted that the subsidence in the fore-arc region of NE Japan observed by recent GPS measurements is caused by basal subduction erosion, which does not require the interplate coupling on the deep portion of the plate interface beneath the land. He estimated the rate of material loss at the plate interface (the rate of the erosion of the upper plate by the underthrusting plate) as 15 mm/yr in the depth range of 5–90 km, using vertical velocity data by GPS observations for 4 years of 1996–2000. However, the estimated rate of 15 mm/yr is significantly faster than that geologically estimated [*von Huene and Lallemand*, 1990]. Moreover, his interpretation leaves a problem that basal erosion with such a high rate must have occurred for the 4 years in regions where the plate interface has been almost completely coupled (offshore east of Miyagi prefecture and south of Tokachi), while the rate must have been very low in regions where the plate interface has been slipping quasi-statically (a region between 39°N and 40°N). Accumulation of crustal deformation data for longer periods is needed to solve this problem.

Similar Earthquakes Off Kamaishi Recurring at a Regular Interval

Similar earthquakes of *M* 4.8 ± 0.1 that regularly occur at the same location have been found on the plate boundary offshore Kamaishi, east of northern

Tohoku [*Matsuzawa et al.*, 1999, 2002] (hereafter Kamaishi sequence). The location of this peculiar earthquake sequence is shown by an arrow on a map of epicentral distribution of shallow earthquakes in figure 14.1a. These similar earthquakes recur at a regular interval of 5.35 ± 0.53 years (fig. 14.4). They have identical low-angle thrust fault type focal mechanisms, typical of interplate events, and nearly the same waveforms. Dimensions of rupture areas of the recent two events of this earthquake sequence were estimated from corner frequencies determined from acceleration seismograms recorded at a nearby station [*Matsuzawa et al.*, 2002]. Results show source lengths of ~1 km and source areas of 0.5–1 km^2. Slip amounts for the two events were estimated

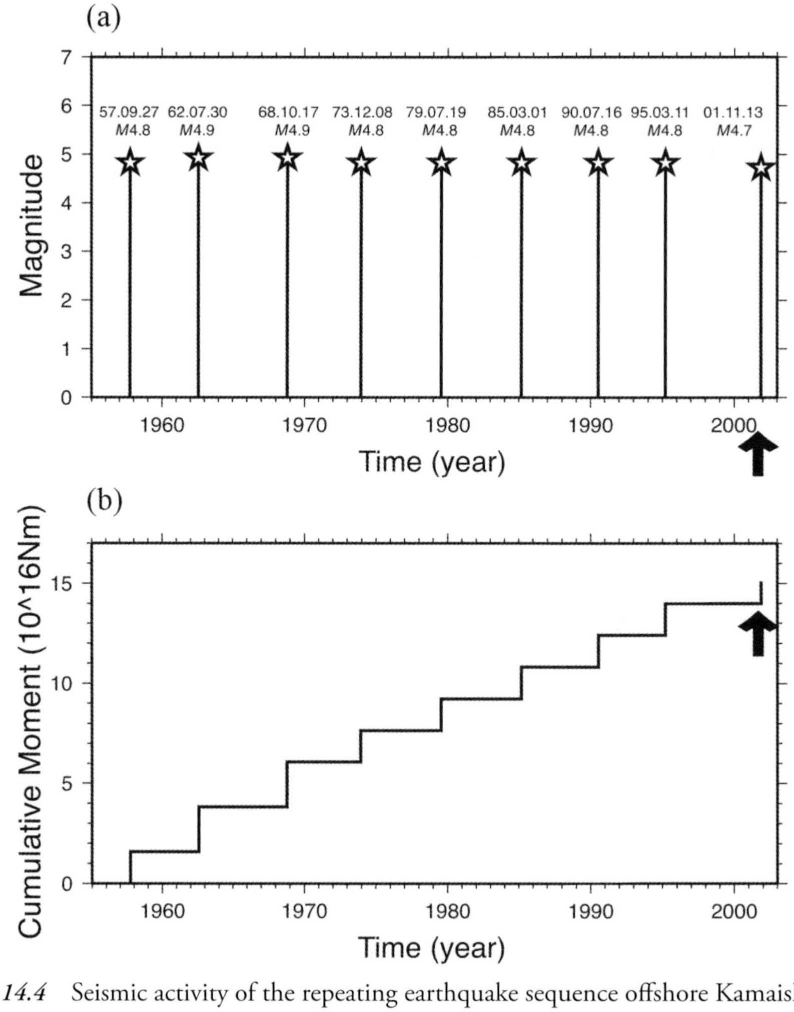

Figure 14.4 Seismic activity of the repeating earthquake sequence offshore Kamaishi [*Matsuzawa et al.*, 2002]: (a) magnitude-time plot and (b) cumulative seismic moment estimated from *Mj* (JMA magnitude) using Aki's relation [*Aki*, 1972].

to be 23–46 cm. Hypocentral relocations of the recent four events in this sequence using the homogeneous station method [*Ansell and Smith*, 1975] show that their hypocenters, i.e., rupture initiation points, are confined to an area with a diameter of ~0.8 km (fig. 14.6) [*Igarashi et al.*, 2000].

This earthquake sequence is interpreted to be generated by repeated ruptures of the same asperity patch (frictionally locked patch) with a size of ~1 km surrounded by stably sliding areas on the plate boundary. The seismic coupling ratio on this asperity amounts to almost 100%. There are no historical records of earthquakes with magnitudes larger than 6 in the area surrounding this earthquake sequence, although microearthquake activity is very high here (fig. 14.1). This suggests that the plate boundary in this area is creeping similar to the creeping segment of the San Andreas fault. The regular occurrence of M 4.8 ± 0.1 earthquakes is perhaps due to the repeating slip of the isolated asperity patch surrounded by the creeping area slipping at nearly a constant rate. On the basis of this interpretation, the next event was expected to occur by the end of November 2001 with 99% probability [*Matsuzawa et al.*, 1999]. An earthquake with M 4.7 actually occurred on 13 November 2001 at the same location on the plate boundary as expected [*Matsuzawa et al.*, 2002].

Rupture area of this 2001 event was compared with that of the last 1995 event in this sequence [*Okada et al.*, 2003]. Hypocenters of the two events are precisely relocated by carefully picking P- and S-wave arrival times at surrounding stations and by using the homogeneous station method. Waveforms of the two events observed by a regional broad-band seismograph network were inverted to estimate moment release distribution on the fault plane by using the method of *Hartzell and Heaton* [1983]. The results show that the spatial extent of both rupture areas is almost the same with a length of 1.0–1.5 km. The rupture areas of the two events almost completely overlap with each other, although their ruptures were initiated from different points ~200 m apart (fig. 14.5). This clearly shows that the two events result from repeated slip of the same asperity patch on the plate boundary and again strongly supports the hypothesis of persistent asperities.

We do not have strong evidence to understand formation of asperities (frictionally locked patches) on the plate boundary; however, the hypocentral distribution of the four recent events of the Kamaishi sequence provides useful information. Relative locations of hypocenters for these events are precisely located by the method of *Nakamura et al.* [2002] using differences of S-P times between a master event and the other three events. Relocated hypocenters are plotted on EW vertical cross section and are shown in figure 14.6 [*Igarashi et al.*, 2000]. Hypocenters of these M 4.8 ± 0.1 events are distributed on an inclined plane with a dimension of ~0.8 km. The dip angle of this plane is ~40°, slightly steeper than that of the subducting plate boundary in this region (~28°). This locally steeper dip angle at the asperity suggests a possibility that the asperity of the M 4.8 ± 0.1 events are generated by a small protuberance existing on the top of the subducting Pacific plate with a radius of ~1 km, as schematically

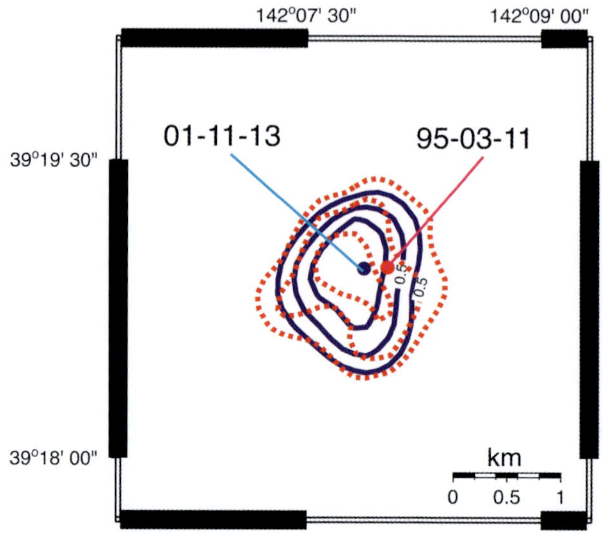

Figure 14.5 Moment release distribution of the 1995 *M* 4.8 event and the 2001 *M* 4.7 event of the Kamaishi repeating earthquake sequence [*Okada et al.*, 2003]. The red dashed contour line denotes the distribution of the 1995 event, and the blue contour denotes the 2001 event. Each moment release distribution is normalized by the peak value. Contour interval is 0.25. The absolute peak values are 2.2×10^{16} and 2.5×10^{16} Nm/km^2 for the 1995 and 2001 events, respectively.

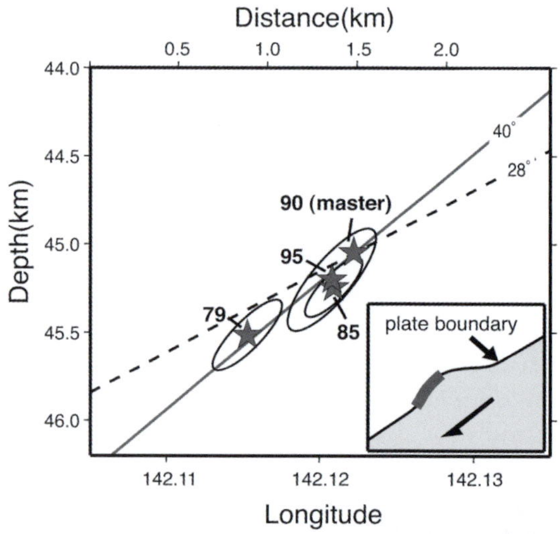

Figure 14.6 EW vertical cross section of hypocenters of the Kamaishi earthquake sequence [*Igarashi et al.*, 2000]. Hypocenters of the 1979, 1985, 1990, and 1995 events are shown by stars. Ellipses denote errors in hypocentral locations relative to that of the 1990 event. Solid line shows a plane on which the hypocenters are distributed. Dashed line shows a plane with a dip angle of 30°, which corresponds the plate boundary in this area.

shown in the inset of figure 14.6, although the difference in dip angle is not very significant if we take estimation errors of hypocenters into consideration. Focal mechanisms of the five recent events determined by *P*-wave polarity data have the average dip angle of 33.4° ± 4.3°. This value is again slightly steeper than the local dip of the plate boundary, which seems to support the above suggestion.

Repeating Earthquakes and Interplate Quasi-Static Slip

Many similar earthquakes (earthquakes with similar waveforms) were found along the plate-boundary zone east of NE Japan [*Igarashi et al.*, 2003]. Most of these events were not found within the subducting Pacific plate or within the continental North American/Okhotsk plate with a few exceptions (fig. 14.7). We interpret these similar earthquakes to be caused by repeating ruptures of small asperity patches with dimensions of 0.1 – 1 km surrounded by stably sliding areas along the plate boundary, similar to the *M* 4.8 ± 0.1 Kamaishi earthquake sequence already described.

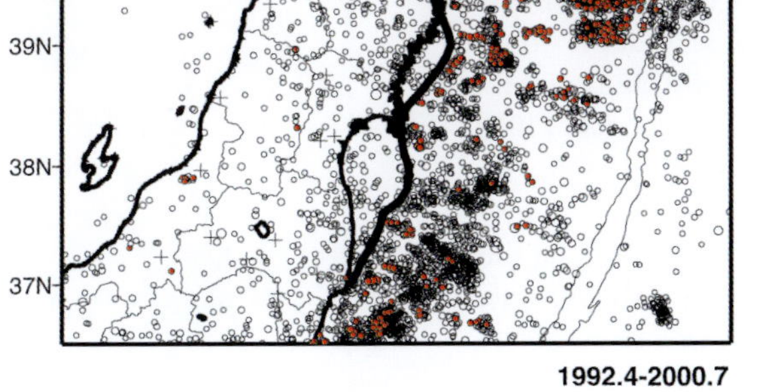

Figure 14.7 Epicentral distribution of repeating earthquakes with *M* 3 or larger that occurred for the period from April 1992 to July 2000 [*Igarashi et al.*, 2003]. Red circles show repeating earthquakes, and open circles show other earthquakes. Pluses show locations of stations used for waveform cross-correlation analyses. Thick line shows the aseismic front, i.e., the western limit of the interplate coupling area [*Igarashi et al.*, 2001].

Figure 14.8 shows the spatial distribution of the proportion of similar earthquakes to ordinary earthquakes along the plate boundary and its vicinity east of NE Japan. Back-slip distribution for the 5-year period shown in figure 14.3 is also shown by contours in this figure. Comparison between the two indicates that areas with back slip larger than ~8 cm/yr, i.e., 100% coupling for this period, have no similar earthquakes. On the contrary, the proportions of similar earthquakes are relatively high for areas with small back slips. This observation seems to support the above interpretation that the similar earthquakes are caused by repeating ruptures of small asperity patches surrounded by freely slipping regions (repeating earthquakes) [*Igarashi et al.*, 2003].

If seismic coupling within these small asperities is 100%, we can estimate the cumulative amount of aseismic slip in the area surrounding each asperity on the plate boundary [*Nadeau and McEvilly*, 1999]. *Nadeau and Johnson* [1998] derived a scaling relation between seismic moment and seismic slip for repeating earthquakes in Parkfield and Stone Canyon, California, by assuming the average slip rate on the plate boundary based on the relative plate motion velocity and 100% seismic coupling at each asperity. We investigated whether or not this scaling relation, derived from repeating earthquakes on the transform-fault type plate boundary in California, is applicable to those on the subduction zone plate boundary in NE Japan [*Igarashi et al.*, 2003]. The repeating earthquake sequence off Kamaishi with M 4.8 ± 0.1 described above, a subcluster sequence with M ~3.5 next to it, and earthquakes off Miyagi pre-

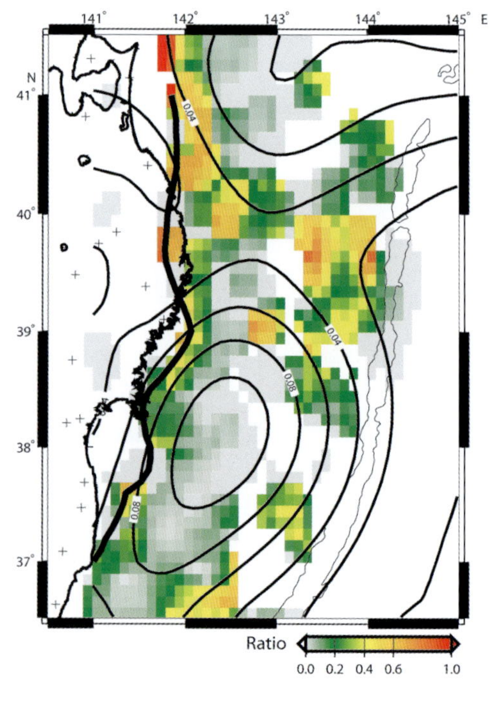

Figure 14.8 Distribution of proportion of repeating earthquakes to ordinary earthquakes [*Uchida et al.*, 2002]. Proportion of repeating events in each area of 0.1° × 0.1° is shown by the color scale at the bottom. Back-slip distribution estimated by *Suwa et al.* [2003] is also shown by contours with an interval of 2 cm/yr. Thick line nearly along the Pacific coast is the aseismic front [*Igarashi et al.*, 2001].

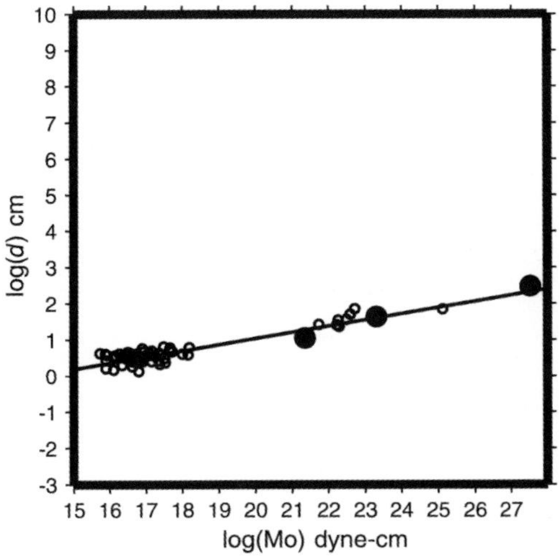

Figure 14.9 Relation between seismic slip and scalar moment [*Igarashi et al.*, 2003]. Open circles and the regression line are the result of *Nadeau and Johnson* [1998]. Solid circles represent the *M* 7.5 off Miyagi earthquake sequence, the *M* 4.8 off Kamaishi earthquake sequence, and *M* 3.5 subcluster of the off Kamaishi sequence, respectively.

fecture with *M* ~7.5 occurring repeatedly with a recurrence interval of ~37 years [*Earthquake Research Committee of the Headquarters for Earthquake Research Promotion*, 2001] are used for this analysis. Slip rates on the plate boundary surrounding these earthquake sequences are assumed to be 8 cm/yr since those surrounding areas are supposed to be creeping. The results show that these earthquake sequences also satisfy the scaling relation of *Nadeau and Johnson* [1998] (fig. 14.9). This indicates that the cumulative amount of aseismic slips on the plate boundary surrounding repeating earthquake sequences can be estimated for the NE Japan subduction zone using this scaling relation between the moment and slip.

Spatial distribution of the slip rate along the plate boundary east of NE Japan was estimated by applying the scaling relation of *Nadeau and Johnson* [1998] to repeating earthquakes occurring there [*Igarashi et al.*, 2003]. The estimated distribution of slip rate is almost consistent with the back-slip distribution estimated from GPS data [*Nishimura et al.*, 2000; *Suwa et al.*, 2003]. Large forward-slip areas estimated from the back-slip inversions also have a large slip amount estimated from the repeating earthquake analysis. This suggests that repeating earthquakes can be used for estimating temporal and spatial distribution of the interplate aseismic slip.

Recently, many more repeating earthquakes occurring offshore along the plate boundary east of NE Japan were identified based on systematic analyses of waveform cross-correlation calculations for earthquakes with magnitudes 2.5 or more for the last 17 years [*Uchida et al.*, 2003]. Cross-correlation coefficients are calculated for ~36,000,000 waveform pairs. Detailed space-time distributions of quasi-static slip are estimated using the repeating earthquakes as creepmeters embedded in the subducting plate boundary. Figure 14.10a shows the spatial distribution of the number of repeating earthquake sequences (number of groups). We can see that many repeating earthquakes occur just east of

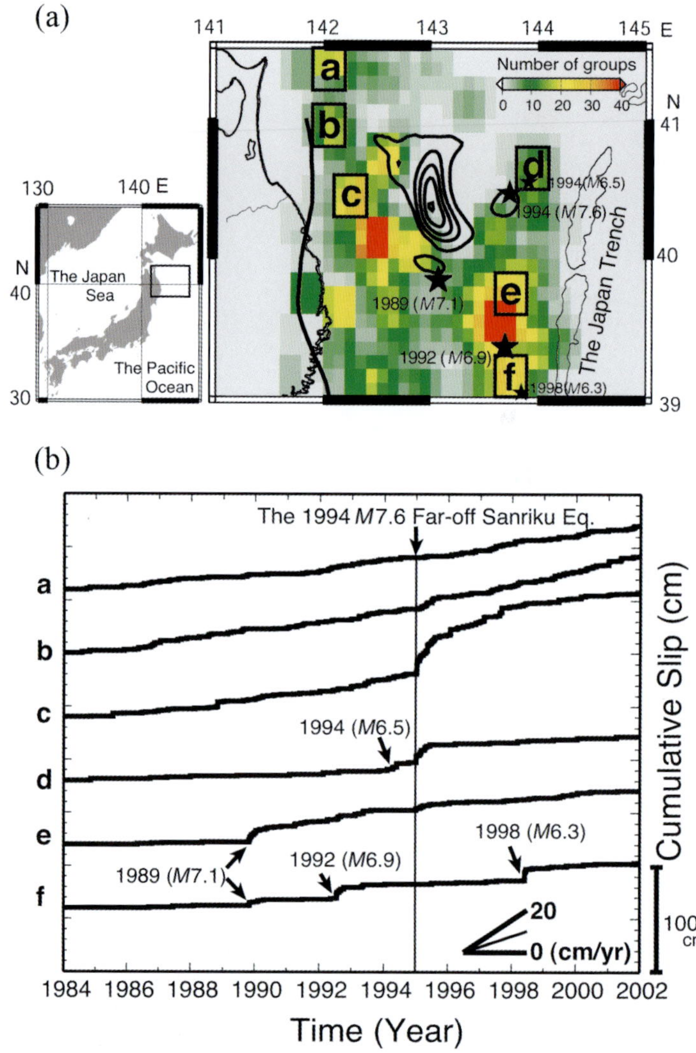

Figure 14.10 (a) Distribution of repeating earthquake sequences in the area east of northern Tohoku shown in the inserted map. The number of repeating earthquake sequences (the number of groups) in each rectangle is shown by the color scale on the top. Solid line shows the aseismic front [*Igarashi et al.*, 2001]. Stars denote epicenters of the 1994 Sanriku-oki earthquake and other major events indicated in figure 14.2b. Contours show moment release distribution of the 1994 Sanriku-oki earthquake [*Nagai et al.*, 2001]. (b) Averaged slip histories of the plate boundary in the six areas shown in figure 14.10a. The occurrence time of the five major earthquakes are indicated by arrows [after *Uchida et al.*, 2003].

the aseismic front shown by a black line in the figure, the western limit of low-angle thrust-fault type earthquakes, which is thought to be a transition zone from the coupled to decoupled plate boundary [*Hasegawa et al.*, 1994; *Igarashi et al.*, 2001]. Many repeating events also occur close to the Japan Trench. Very

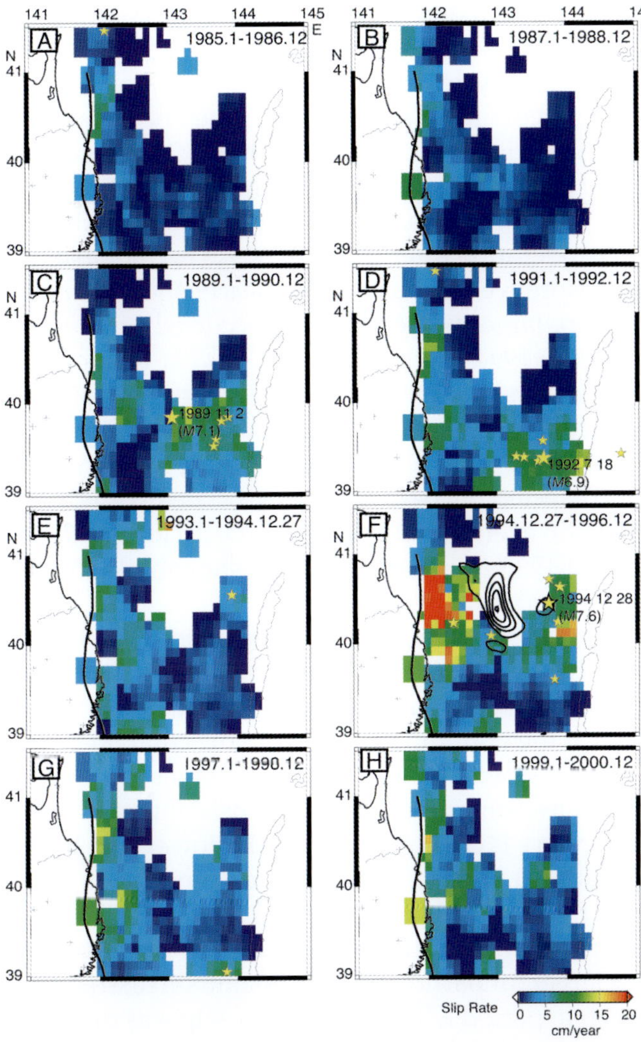

Figure 14.11 Slip-rate distribution for every 2 years from 1985 to 2001 [*Uchida et al.*, 2003]. Slip rate for each rectangle in which three or more sequences are included is shown by the color scale at the bottom. The end of the period E is set just before the 1994 Sanriku-oki earthquake and the start of the period F is set just after it. Stars show epicenters of earthquakes with magnitudes 6 or larger. Contours in the period F show the moment release distribution of the 1994 Sanriku-oki earthquake.

few repeating earthquake sequences are located inside the large slip area of the 1994 M 7.6 Sanriku-oki earthquake shown by contours [*Nagai et al.*, 2001], which perhaps corresponds to the location of a large asperity as described in the Distribution of Large Slip Areas by Successive Earthquakes section.

Figure 14.10b shows averaged cumulative slip for six rectangles of $0.3^\circ \times 0.3^\circ$ in Figure 14.10a. Each of these slip histories was obtained by averaging (stacking) cumulative slips for many repeating earthquake sequences located within each rectangle. These cumulative slips, *a* through *f*, are interpreted as quasi-static slips taking place on the plate boundary in the six rectangular areas, respectively. Interplate quasi-static slip with an almost constant rate of ~4 cm/yr is occurring in rectangle *a*, located near the aseismic front and far from the large asperity that caused the 1994 *M* 7.6 Sanriku-oki earthquake. Figure 14.10b also shows that quasi-static slip accelerated after the 1994 event in areas *b*, *c*, and *d*, located around the large asperity of the 1994 event. This quasi-static slip acceleration after the 1994 *M* 7.6 earthquake probably reflects postseismic slip. Moreover, Figure 14.10b clearly shows the characteristics of the postseismic slip: large postseismic slip occurred on the landward side of the source area of the 1994 event and lasted ~1 year (rectangle *c*). Postseismic slip on the trenchward side of the source area (rectangle *d*) is smaller and terminated earlier. Clear postseismic slip is also seen in Figure 14.10b after the 1989 *M* 7.1, 1992 *M* 6.9, 1994 *M* 6.5, and 1998 *M* 6.3 events. Analyses of GPS and strainmeter data detected postseismic slip associated with the 1989 *M* 7.1, 1992 *M* 6.9, and 1994 *M* 7.6 events [*Kawasaki et al.*, 2001; *Miura et al.*, 1993] but not for the 1994 *M* 6.5 and 1993 *M* 6.3 earthquakes, showing higher detectability by the repeating earthquake analysis.

Figure 14.11 shows the space-time distribution of quasi-static slip on the plate boundary offshore east of northern Tohoku [*Uchida et al.*, 2003]. Slip rate for every 2 years estimated from repeating earthquake sequences is indicated by the color scale at the bottom. This figure shows that quasi-static slip occurs almost stably in the deeper and shallower parts of the plate boundary between the trench and the aseismic front. This is consistent with the understanding that the plate boundary is decoupled in the shallower part because of the existence of unconsolidated sediments and in the deeper parts because of thermal conditions [e.g., *Hyndman and Wang*, 1993; *Oleskevich et al.*, 1999]. Figure 14.11 also shows the spatial distribution of postseismic slip associated with earthquakes *M* 6.3 or larger. In particular, postseismic slip associated with the 1994 *M* 7.6 event is seen in period *F*. The postseismic slip was distributed not only to the west of the asperity causing the 1994 event but also to the east of it, which was not observed by GPS data analyses [*Nishimura et al.*, 2000].

Present State of Interplate Coupling and Recent Seismic Activity

As described in the previous sections, recent studies based on seismic and GPS data have revealed that the asperity model is applicable to the process of seismic and aseismic slip occurring on the convergent plate boundary of NE Japan. Figure 14.12 is a simplified cartoon schematically showing a spatial distribution of asperities on the subducting plate boundary offshore Sanriku. Scattered on the plate boundary are small and large asperities, surrounded by stably sliding areas. Asperities are locked during the interseismic period, and their subsequent failure and repeated rupture represent the source of earthquakes. Each large asperity ruptures repeatedly as large earthquakes at long time intervals, while each small asperity rupture as small earthquakes at short time intervals. The location of each asperity is unchangeable and is a property of the plate boundary. This seems to explain at least partly why shallow earthquakes beneath the Pacific Ocean are distributed unevenly along the plate boundary and why their general distribution pattern does not change with time as can be seen in Figures 14.1a–14.1d.

Repeating ruptures of small asperity patches generate almost identical waveforms to each other and are identified by waveform cross-correlation analyses. *Igarashi et al.* [2003] and *Uchida et al.* [2003] found many small asperities with dimensions of ~0.1–1 km as repeating earthquakes with magnitudes of 2.5–5. If small asperities are sparsely distributed and are surrounded by stably sliding areas, repeated ruptures of those small asperities are governed by the aseismic slip rate in the surrounding areas. This means that quasi-static slip on the plate boundary can be estimated from the spatiotemporal distribution of repeating earthquake activity [*Nadeau and Johnson*, 1998]. Actually, the estimated slip distribution from repeating earthquakes agreed well with the distribution of back slip and forward slip estimated from GPS observations [*Igarashi et al.*, 2003; *Uchida et al.*, 2003].

Larger asperities (large coseismic slip areas) can be detected at the time of their ruptures by waveform inversions. *Yamanaka and Kikuchi* [2004] estimated the locations of many large asperities by inverting strong motion seismograms for large earthquakes with magnitudes ~7 or more that occurred over the last 70 years. Some of the large asperities estimated by them recently ruptured in the 1989 *M* 7.1, 1991 *M* 6.9, and 1994 *M* 7.6 Sanriku-oki earthquakes, whereas other large asperities remain as unruptured patches on the plate boundary. They are located at the areas east off Miyagi prefecture, east of Aomori prefecture, and

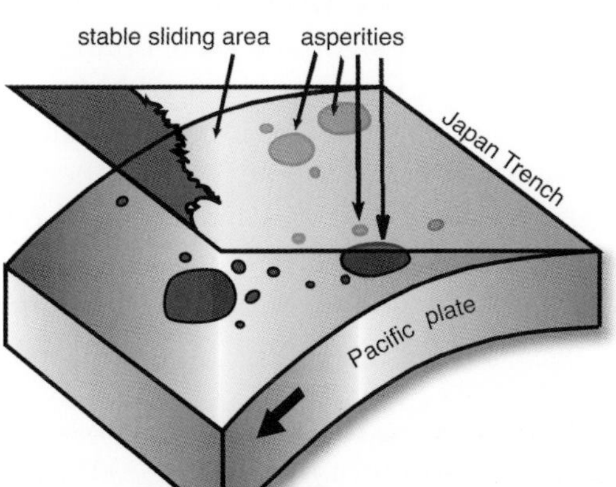

Figure 14.12 Schematic diagram showing scattered asperities and surrounding stable sliding area on the plate boundary east of NE Japan.

south of Tokachi and are shown as solid lines on the back-slip distribution map of figure 14.3. The offshore asperities east of Miyagi prefecture and south of Tokachi were ruptured previously by the 1978 *M* 7.4 Miyagi-oki earthquake and by the 1952 *M* 8.2 Tokachi-oki earthquake. The offshore asperity east of Aomori prefecture is one of the two major asperities that caused the 1968 *M* 7.9 Tokachi-oki earthquake. The other asperity in the south ruptured again recently during the 1994 *M* 7.6 Sanriku-oki earthquake as described above. Comparison of these large asperities left unruptured with the back slip distribution in the last 5 years (fig. 14.3) shows that these large asperities coincide with the areas of large back slip amount (almost 100% coupling for the asperities east off Miyagi prefecture and south off Tokachi). Moreover, the large back-slip areas extend over wide areas surrounding these asperities. This suggests that strong interplate coupling on these large asperities causes the large back-slip amounts for those wide areas.

Very recently, remarkable seismic activity has occurred within the subducted Pacific slab, in the upper crust of the overriding continental plate and along the plate boundary in NE Japan. It seems that these earthquakes are closely related to the present stage of interplate coupling described above, although quantitative modeling of stress interactions are needed to better understand these observations. An intermediate-depth earthquake with *M* 7.1 occurred on 26 May 2003 at a depth of 68 km beneath in the northern part of Miyagi prefecture. It is located on the upper seismic plane of the double seismic zone, which is clearly seen within the subducted Pacific slab in NE Japan [*Hasegawa et al.*, 1978a, 1978b]. This event caused slight damage. The focal mechanism of this event reveals downdip compression, typical of events in the upper plane of the double seismic zone. Hypocenters of this event and its aftershocks, located by the homogeneous station method, are shown in figures 14.13 and 14.14a [*Okada et al.*, 2003]. The focal area is close to the large asperity that ruptured in the last Miyagi-oki earthquake of 1978. The hypocentral distribution and main shock focal mechanism delineate the fault plane of this event steeply dipping to the WNW.

Before the occurrence of the 26 May *M* 7.1 event, seismic activity in the area surrounding the asperity of the Miyagi-oki earthquake on the plate boundary increased. Several events occurred including the 12 October *M* 5.6, 03 November *M* 6.1, 05 December *M* 5.2, 2002, and 03 March *M* 5.8, 2003, event. Particularly, remarkable postseismic slip occurred associated with the 3 November *M* 6.1 event [*Miura et al.*, 2003] (figs. 14.13 and 14.14a). The focal area of the *M* 7.1 event is located right below the plate boundary at the deeper extension of this postseismic slip area. This suggests that the 3 November *M* 6.1 event and its postseismic slip on the plate boundary affected the occurrence of the 26 May *M* 7.1 event in the slab.

Two months after the *M* 7.1 event, a shallow earthquake with *M* 6.4 occurred again in the northern part of Miyagi prefecture on 26 July 2003 and

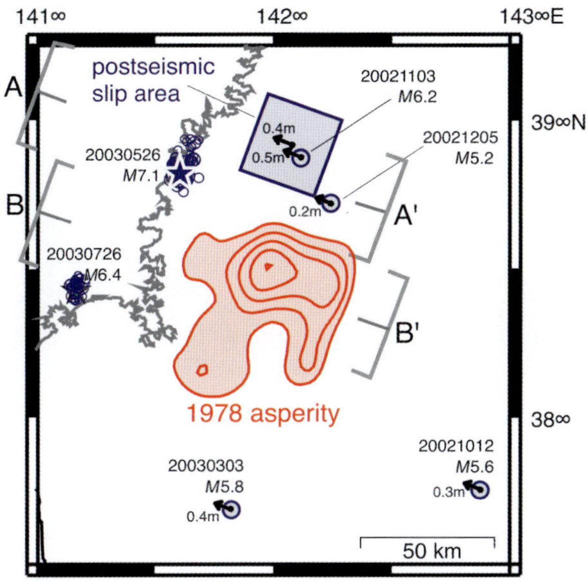

Figure 14.13 Asperity east off Miyagi prefecture and recent seismic activity surrounding it. The asperity ruptured in the 1978 *M* 7.4 Miyagi-oki earthquake [*Yamanaka and Kikuchi*, 2004] is contoured by red lines. Epicenters of earthquakes with *M* 5 or larger that occurred in 2002 and 2003 are indicated by blue open circles. The postseismic slip area associated with the 2002 *M* 6.1 event (for 3 months) is outlined by a blue rectangle [*Miura et al.*, 2003]. Numeral attached to the arrow shows estimated slip amount of each event. The epicenters of the main shocks and aftershocks of May 26 2003 *M* 7.1 event and July 26 2003 *M* 6.4 event are denoted by asterisks and circles, respectively.

caused severe damage to the area right above its fault plane; 1273 houses were completely destroyed. The aftershock distribution and focal mechanism reveal a thrust fault, dipping to the west at an angle of ~50° in the depth range from 2 to 12 km [*Umino et al.*, 2003]. The focal area of this event is located to the west of the large asperity of the Miyagi-oki earthquake, which is completely locked at present (figs. 14.13 and 14.14b). This almost 100% coupling at the asperity causes compressional stress in the direction of the plate convergence, i.e., EW direction. The compressional stress within the seismogenic upper crust of the overriding plate reaches its maximum in the area surrounding the focal area of the 26 July *M* 6.4 event since the coupling area on the plate boundary is located just to the east of the focal area of the *M* 6.4 event (fig. 14.3) [*Suwa et al.*, 2003]. This observation suggests that the present state of the interplate coupling at the asperity east off of Miyagi prefecture contributed to building up the stress and attaining the level of the final rupture on the fault plane of the *M* 6.4 event.

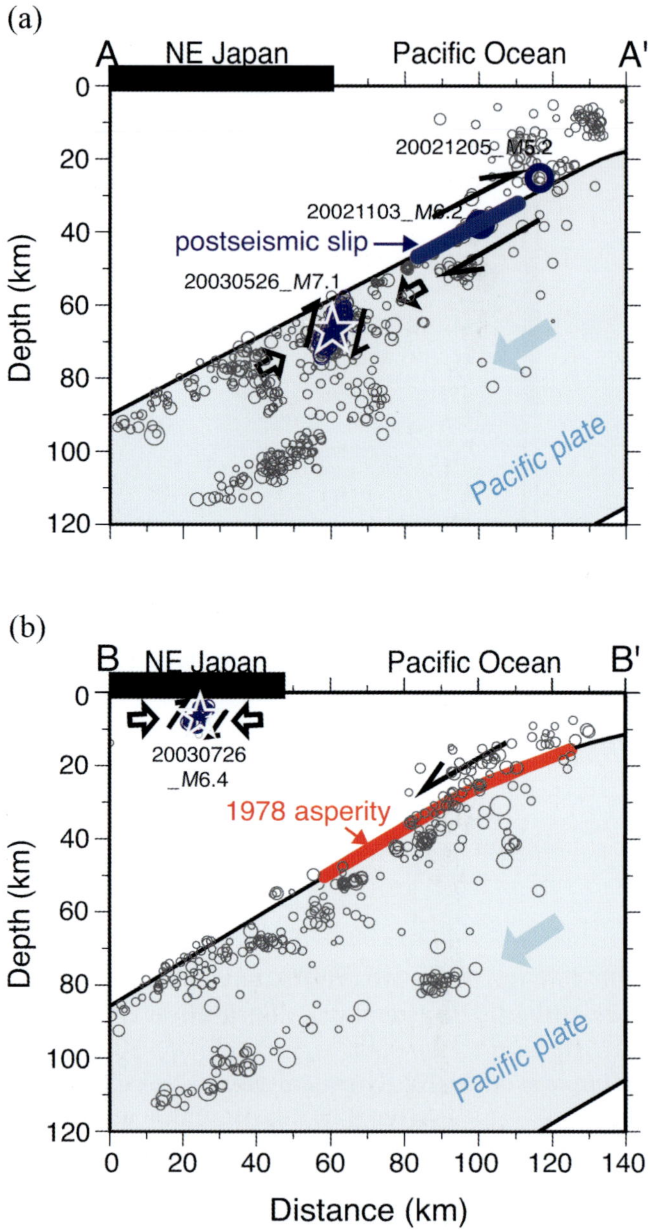

Figure 14.14 Vertical cross sections of earthquakes along lines (a) AA′ and (b) BB′ in figure 14.13. Circles show earthquakes located by the seismic network of Tohoku University. The postseismic slip area (for 3 months) is indicated by a blue line in figure 14.14a, and the asperity is indicated by a red line in figure 14.14b. The hypocenters of the mainshocks and aftershocks of May 26 *M* 7.1 and July 26 *M* 6.4 events are indicated by asterisks and blue circles, respectively.

A large shallow earthquake with *M* 8.0 occurred on 26 September 2003 along the plate boundary south of Tokachi. This caused extensive damage particularly in the eastern part of Hokkaido. Two persons were missing associated with the tsunami, and 60 houses were completely destroyed. The focal mechanism of this event is a low-angle thrust fault typical of interplate earthquakes. Aftershocks spread in a wide area of ~150 km × 150 km along the plate boundary. Its location and extent are approximately consistent with those of the aftershock area of the 1952 *M* 8.2 Tokachi-oki earthquake, which suggests that the large asperity south off Tokachi that caused the last 1952 *M* 8.2 event ruptured again as the present event. Preliminary results of comparative studies of waveform inversions of the 1952 *M* 8.2 and 2003 *M* 8.0 earthquakes show nearly identical locations of large slip areas by the two events with each other, although the large slip area by the present *M* 8.0 event is slightly smaller than that by the last *M* 8.2 event. [*Yamanaka and Kikuchi*, 2003]. This observation again supports the hypothesis of persistent asperities.

Earthquake Research Committee of the Headquarters for Earthquake Research Promotion [2003] has been systematically evaluating the long-term probability of earthquake occurrence in each area of Japan based on the history of past earthquakes. Estimated probability of occurrence in the coming 30 years was ~60% for the offshore area south of Tokachi, which was the third highest in Japan. The highest probability is east of Miyagi, with ~99% probability in the next 30 years. The second highest is east of Aomori, with ~90% probability. As described before, these three areas coincide with large back-slip areas estimated from GPS data (fig. 14.3).

Conclusion

We have reviewed recent studies of interplate coupling along the subducting plate boundary offshore and east of NE Japan. These studies involve waveform inversions of large to medium size earthquakes, small repeating earthquake analyses, and back-slip inversion of GPS data. Integration of the results of these studies strongly suggests that the asperity model is applicable to the plate boundary in this region. Large to small asperities are scattered on the plate boundary surrounded by stable sliding areas. Large asperities are locked for the interseismic period. The location and extent of each asperity is almost unchangeable. Large asperities rupture as large earthquakes, and small asperities as small earthquakes. Sometimes two or more asperities rupture simultaneously, which causes larger magnitude earthquakes.

The present state of interplate coupling is discussed based on the above interpretation. There exist three large asperities left unruptured on the plate boundary east of NE Japan. Back slips are large, and the activity of small repeating earthquakes is very low in those asperities, consistent with strong interplate coupling. One asperity, south of Tokachi, ruptured on 26 September 2003 as the 2003 *M* 8.0 Tokachi-oki earthquake. There occurred remarkable earthquakes in the vicinity of another asperity east of Miyagi prefecture; one in the subducted slab on 26 May and the other in the overriding plate on 26 July this year. Strong coupling at this asperity and quasi-static slip in its surrounding area perhaps contributed to stress buildup within the overriding plate and within the subducted slab, respectively, and are likely to trigger these disastrous earthquakes. These observations strongly suggest high probability of the next earthquake occurrences in the two large asperities left unruptured.

Acknowledgments

We are grateful to all the members of Research Center for Prediction of Earthquakes and Volcanic Eruptions, Tohoku University, for valuable discussions. We appreciate the critical reviews of Timothy Dixon, Susan Y. Schwartz, and an anonymous reviewer, who helped to improve this manuscript. This research was partially supported by a grant from the Ministry of Education, Culture, Sports, Science and Technology.

References

Aki, K. (1972), Scaling law of earthquake source time function, *Geophys. J. R. Astron. Soc., 31,* 3–25.

Ansell, J. H., and E. G. C. Smith (1975), Detailed structure of a mantle seismic zone using the homogeneous station method, *Nature, 253,* 518–520.

DeMets, C., R. G. Gordon, D. F. Argus, and S. Stein (1994), Effect of recent revisions to the geomagnetic reversal time scale on estimates of current plate motions, *Geophys. Res. Lett., 21,* 2191–2194.

Earthquake Research Committee of the Headquarters for Earthquake Research Promotion (2001), *Long-Term Evaluation of Earthquakes in the Sea Off Miyagi Prefecture* (in Japanese), pp. 601–618, Minist. of Educ., Culture, Sports, Sci. and Technol., Govt. of Jpn., Tokyo.

Hartog, R., and S. Y. Schwartz (1996), Directivity analysis of the December 28, 1994 Sanriku-Oki earthquake (Mw = 7.7), Japan, *Geophys. Res. Lett., 23,* 1037–2040.

Hartzell, S., and T. Heaton (1983), Inversion of strong ground motion and teleseismic waveform data for the fault rupture history of the 1979 Imperial Valley, California, earthquake, *Bull. Seismol. Soc. Am., 73,* 1553–1583.

Hasegawa, A., N. Umino, and A. Takagi (1978), Double-planed structure of the deep seismic zone in the northeastern Japan arc, *Tectonophysics, 47,* 43–58.

Hasegawa, A., S. Zhao, S. Hori, A. Yamamoto, and S. Horiuchi (1991), Deep structure of the northeastern Japan arc and its relationship to seismic and volcanic activity, *Nature, 352,* 683–689.

Hasegawa, A., S. Horiuchi, and N. Umino (1994), Seismic structure of the northeastern Japan convergent margin: A synthesis, *J. Geophys. Res., 99,* 22,295–22,311.

Heki, K. (2004), Space geodetic observation of deep basal subduction erosion in northeastern Japan, *Earth Planet. Sci. Lett.,* 219, 13–20.

Heki, K., S. Miyazaki, and H. Tsuji (1997), Silent fault slip following an interplate thrust earthquake at the Japan Trench, *Nature, 386,* 595–598.

Hino, R., T. Kanazawa, and A. Hasegawa (1996), Intraplate seismic activity near the northern Japan Trench deduced from ocean bottom and land-based seismic observations, *Phys. Earth Planet. Inter., 93,* 37–52.

Hyndman, R. D., and K. Wang (1993), Thermal constraints on the zone of major thrust earthquake failure: The Cascadia subduction zone, *J. Geophys. Res., 98,* 2039–2060.

Igarashi, T., T. Matsuzawa, and A. Hasegawa (2000), Characteristics of earthquake clusters in northern Japan arc extracted from waveform similarity analysis, paper presented at Western Pacific Geophysics Meeting, AGU, Tokyo.

Igarashi, T., T. Matsuzawa, N. Umino, and A. Hasegawa (2001), Spatial distribution of focal mechanisms for interplate and intraplate earthquake associated with the subducting Pacific plate beneath the northeastern Japan arc: A triple-planed deep seismic zone, *J. Geophys. Res., 106,* 2177–2191.

Igarashi, T., T. Matsuzawa, and A. Hasegawa (2003), Repeating earthquakes and interplate aseismic slip in the northeastern Japan subduction zone, *J. Geophys. Res., 108*(B5), 2249, doi:10.1029/2002JB001920.

Kawasaki, I., Y. Asai, and Y. Tamura (2001), Space-time distribution of interplate moment release including slow earthquakes and the seismo-geodetic coupling in the Sanriku-oki region along the Japan trench, *Tectonophysics, 330,* 267–283.

Lay, T., and H. Kanamori (1981), An international review, in *Earthquake Prediction,* edited by. M. Ewing, D. W. Simpson, and P. G. Richards, pp. 579–592, AGU, Washington, D. C.

Lay, T., H. Kanamori, and L. Ruff (1982), The asperity model and the nature of large subduction zone earthquakes, *Earthquake Predict. Res., 1,* 3–71.

Matsuzawa, T., T. Igarashi, and A. Hasegawa (1999), Characteristic small earthquake sequence off Sanriku, Japan, *Eos Trans. AGU, 80*(46), Fall Meet. Suppl., F724.

Miura, S., K. Tachibana, T. Sato, K. Hashimoto, M. Mishina, N. Kato, and T. Hirasawa (1993), Postseismic slip events following interplate thrust earthquakes occurring in subduction zone, in *Proceedings of the Eighth International Symposium on Recent Crustal Movements (CRCM'93),* published as a special issue of the *J. Geodetic Soc. Jpn.,* Local Organizing Committee for the CRCM'93, Kyoto, pp. 83–84.

Miura, S., T. Sato, K. Tachibana, Y. Satake, and A. Hasegawa (2002), Strain accumulation in and around Ou Backbone Range, northeastern Japan as observed by a dense GPS network, *Earth Planets Space, 54,* 1071–1076.

Miura, S., Y. Suwa, T. Sato, K. Tachibana, and A. Hasegawa (2003), Slow slip events on the plate boundary off Miyagi and Fukushima detected by GPS and strain measurements, paper presented at Japan Earth and Planetary Science Joint Meeting, Chiba.

Nadeau, R. M., and L. R. Johnson (1998), Seismological studies at Parkfield VI: Moment release rates and estimates of source parameters for small repeating earthquakes, *Bull. Seismol. Soc. Am., 88,* 790–814.

Nadeau, R. M., and T. V. McEvilly (1999), Fault slip rates at depth from recurrence intervals of repeating microearthquakes, *Science, 285,* 718–721.

Nagai, R., M. Kikuchi, and Y. Yamanaka (2001), Comparative study on the source processes of recurrent large earthquakes in Sanriku-oki region: The 1968 Tokachi-oki earthquake and the 1994 Sanriku-oki earthquake, *J. Seismol. Soc. Jpn., 54,* 267–280.

Nakamura, A., A. Hasegawa, N. Hirata, T. Iwasaki, and H. Hamaguchi (2002), Temporal variations of seismic wave velocity associated with 1998 M6.1 Shizukuishi earthquake, *Pure Appl. Geophys., 159,* 1183–1204.

Nishimura, T. (2000), Spatio-temporal pattern of plate coupling estimated from GPS observations in northeastern Japan, thesis, Tohoku Univ., Tohoku, Jpn.

Nishimura, T., et al. (2000), Distribution of seismic coupling on the subducting plate boundary in northeastern Japan inferred from GPS observations, *Tectonophysics, 323,* 217–238.

Okada, T., and A. Hasegawa (2003), The M7.1 May 26, 2003 off-shore Miyagi Prefecture Earthquake in northeast Japan: Source process and aftershock distribution of an intra-slab event, NE Japan, *Earth Planets Space, 55,* 731–739.

Okada, T., T. Matsuzawa, and A. Hasegawa (2003), Comparison of source areas of M 4.8 ± 0.1 earthquakes off Kamaishi, NE Japan – Are asperities persistent feature?, *Earth Planet. Sci. Lett., 213,* 361–374.

Oleskevich, D. A., R. D. Hyndman, and K. Wang (1999), The updip and downdip limits to great subduction earthquakes: Thermal and structural models of Cascadia, south Alaska, SW Japan, and Chile, *J. Geophys. Res., 104,* 14,965–14,991.

Pacheco, J. F., L. R. Sykes, and C. H. Scholz (1993), Nature of seismic coupling along simple plate boundaries of the subduction type, *J. Geophys. Res., 98,* 14,133–14,159.

Peterson, E. T., and T. Seno (1984), Factors affecting seismic moment release rates in subduction zones, *J. Geophys. Res., 89,* 10,233–10,248.

Schwartz, S. Y. (1999), Noncharacteristic behavior and complex recurrence of large subduction zone earthquakes, *J. Geophys. Res., 104,* 23,111–23,125.

Suwa, Y., S. Miura, A. Hasegawa, T. Sato, and K. Tachibana (003), Inter-plate coupling beneath the NE Japan arc Inferred from 3 dimensional crustal deformation, IUGG 2003 Scientific Program and Abstracts, p.B10, JSG01/08P/D-002.

Tanioka, Y., and L. Ruff, and K. Satake (1996), The large Sanriku-oki Japan, earthquake of December 28, 1994: Rupture of an asperity unbroken in a previous earthquake, *Geophys. Res. Lett., 23,* 1465–1468.

Uchida, N., T. Igarashi, T. Matsuzawa, and A. Hasegawa (2002), Spatio-temporal distribution of interplate quasi-static slip in the northeastern Japan subduction zone, Estimated from repeating earthquake analyses, *Eos Trans. AGU, 83*(47), Fall Meet. Suppl., Abstract S52B-1109.

Uchida, N., T. Matsuzawa, T. Igarashi, and A. Hasegawa (2003), Interplate quasistatic slip off Sanriku, NE Japan, estimated from repeating earthquakes, *Geophys. Res. Lett., 30*(15), 1801, doi:10.1029/2003GL017452.

Ueda, H. (2001), Postseismic crustal deformation associated with large earthquakes and its generation mechanism in the NE Japan arc, thesis, Tohoku Univ., Tohoku, Jpn.

Ueda, H., M. Ohtake, and H. Sato (2001), Afterslip of the plate interface following the 1978 Miyagi-Oki, Japan, earthquake, as revealed from geodetic measurement data, *Tectonophysics, 338,* 45–57.

Umino, N., A. Hasegawa, and T. Matsuzawa (1995), sP depth phase at small epicentral distances and estimated subducting plate boundary, *Geophys. J. Int., 120,* 356–366.

Umino, N., et al. (2003), Hypocenter and focal mechanism distribution of aftershocks of July 26 2003 *M*6.4 northern Miyagi, NE Japan, earthquake revealed by temporary seismic observation, *Earth Planets Space, 55,* 719–730.

von Huene, R., and S. Lallemand (1990), Tectonic erosion along the Japan and Peru convergent margins, *Geol. Soc. Am. Bull., 102,* 704–720.

Yabuki, T., and M. Matsuura (1992), Geodetic data inversion using a Bayesian information criterion for spatial distribution of fault slip, *Geophys. J. Int., 109,* 363–375.

Yagi, Y. (2001), Complementary relationship between seismic and aseismic slip in Hyuga-nada and Sanriku-oki regions, thesis, 57 pp., Univ. of Tokyo, Tokyo, Jpn.

Yamanaka, Y., and M. Kikuchi (2003), 2003/09/26 Tokachi-oki earthquake, *Seismol. Note 139,* Earthquake Res. Inst., Tokyo, Jpn. (Available at http://www.eri.utokyo.ac.jp/topics/TOKACHIOKI03/index.html)

Yamanaka, Y., and M. Kikuchi (2004), Asperity map along the subduction zone in northeastern Japan inferred from regional seismic data, *J. Geophys. Res.*, *109*, B07307, doi:10.1029/2003JB002683.

Zhao, D., T. Matsuzawa, and A. Hasegawa (1997), Morphology of the subducting slab boundary in the northeastern Japan arc, *Phys. Earth Planet. Inter.*, *102*, 89–104.

15

Anomalous Earthquake Ruptures at Shallow Depths on Subduction Zone Megathrusts

Thorne Lay and Susan Bilek

Abstract

Interplate thrust faults in subduction zones host the largest earthquakes and majority of seismic energy release on Earth. The seismically coupled portion of these megathrusts typically extends 100 ± 50 km across the depth range from 5–10 to 25–55 km, with convergent motions between underthrusting and overriding plates being accommodated by a mixture of earthquake slip, post-seismic deformation, and interseismic creep. There are varying proportions of seismic and aseismic slip distributed with depth and varying seismic coupling coefficients in different subduction zones. Rupture of the shallow portion of megathrusts is the focus of this review; both large and small thrust earthquakes in the upper 10 km of the seismogenic zone sometimes display unusual seismic energy release attributes, with low rupture velocity and/or low stress drop. The largest of these events can be exceptionally tsunamigenic, with seismic slip extending updip of the apparent seismic front defined by smaller events. The transition from aseismic stable sliding on the shallow megathrust contact with any sedimentary wedge to the unstable slip regime of the seismically coupled zone at greater depth is variable from region to region and may involve zones of conditional stability that display bimodal deformation behavior as strain rates fluctuate. Systematic variations of rupture process with depth across the upper 10 km of the seismogenic regime have been observed, suggesting depth dependence of mechanical properties of the megathrust that may provide key information about sediment induration, hydrothermal processes, and frictional behavior of the megathrust. The acute seismic faulting heterogeneity observed in the shallow megathrust complicates generalization about the environment based on isolated spatial sampling of its properties, but there is great need for in situ ground truth determinations of structures and properties in the shallow megathrust.

Introduction

Subduction of oceanic lithosphere occurs along massive interplate thrust faults that are the contact surfaces between overriding and underthrusting plates in convergent margins. These megathrusts accommodate the convergent motions by varying portions of seismic and aseismic slip, with significant variations in geometry and maximum earthquake size from region to region [e.g., *Kanamori*, 1977; *Uyeda and Kanamori*, 1979; *Ruff and Kanamori*, 1980; *Pacheco et al.*, 1993; *Scholz and Campos*, 1995]. About 90% of the seismic moment released by global earthquakes occurs near subduction zones, with most events involving slip on megathrusts, including the largest recorded events, the 1960 Chile and 1964 Alaska earthquakes [e.g., *Pacheco and Sykes*, 1992]. The important tectonic role played by megathrusts has prompted many efforts to characterize their seismic and aseismic properties, including the depth extent of interplate faulting, variations in maximum earthquake size, and the underlying causes of faulting complexity observed for individual ruptures.

The updip and downdip bounds on interplate seismic slip (fig. 15.1) are generally located near depths of ~5–10 and 25–55 km below the seafloor, respectively [e.g., *Byrne et al.*, 1988; *Hyndman et al.*, 1995, 1997; *Oleskevich et al.*, 1999; *Tichelaar and Ruff*, 1991, 1993; *Zhang and Schwartz*, 1992; *Pacheco et al.*, 1993; *Suárez and Sánchez*, 1996; *Deshon et al.*, 2003]. The updip limit usually involves a transition from stable-sliding to unstable stick-slip behavior [e.g., *Marone and Saffer*, this volume], commonly below a shallow tapering sedimentary wedge, at a lateral distance of 20–100+ km from the trench called the "seismic front" [e.g., *Yoshii*, 1979; *Marone and Scholz*, 1988; *Pacheco et al.*, 1993], although earthquake ruptures may extend right to the trench when little sediment is present [e.g., *Polet and Kanamori*, 2000]. Temperatures at the updip limit are estimated as around 100°–150°C [e.g., *Hyndman et al.*, 1997]. A variety of mechanisms likely control the downdip edge of the seismically coupled zone, which is typically at temperatures from ~350°–450°C. Thermal control on the frictional mechanism at the downdip edge in the case of subduction of relatively young oceanic crust, or transition in the fore-arc mantle to weak, serpentinized wedge material for island arcs and for subduction of older oceanic crust, has been proposed by *Tichelaar and Ruff* [1993], *Hyndman and Wang* [1993], *Hyndman et al.* [1997], and *Peacock and Hyndman* [1999]. The overall seismogenic depth extent and dip angle define the maximum width available for the largest earthquake ruptures that can occur in each subduction zone, with widths ranging from 50 to 150 km [*Pacheco et al.*, 1993]. A few regions with flattened megathrusts tend to have broader fault widths and larger megathrust earthquakes [e.g., *Gutscher and Peacock*, 2003].

Studies of the largest earthquakes along megathrusts in different subduction zones have suggested some correlations with convergence rate, lithospheric age, sediment supply, bathymetric features, and back-arc spreading [e.g., *Ruff and Kanamori*, 1980, 1983b; *Peterson and Seno*, 1984; *Jarrard*, 1986; *Ruff*,

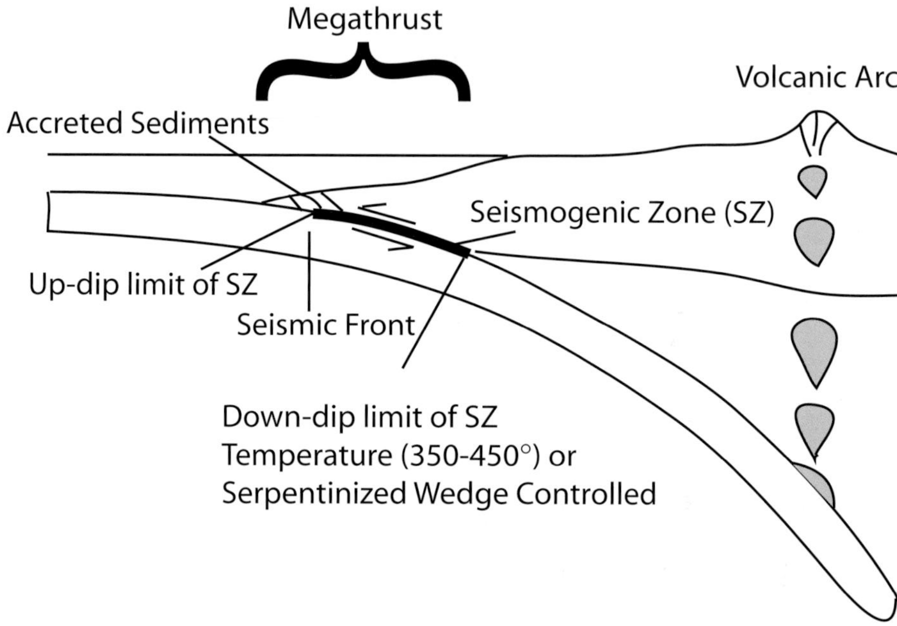

Figure 15.1 Schematic diagram illustrating the megathrust fault in subduction zones at approximately correct scale. The thick black line highlights the seismogenic range of the megathrust on which large thrust earthquakes occur. This has an updip limit and a downdip limit controlled by a variety of processes. Below the downdip limit earthquakes occur only within the subducting slab as intraplate faulting. The nature of the upper plate, the amount of accreted sediments, and the geometry of the megathrust vary from region to region as described in the text. The shallow portion of the seismically coupled megathrust is the focus of this review.

1989; *van der Hilst and Seno*, 1993; *Pacheco et al.*, 1993; *Scholz and Campos*, 1995]. While gross physical attributes of subduction zone environments may affect the width of the seismogenic interface and maximum earthquake size, the statistics are not particularly compelling with regard to any overall model [e.g., *Pacheco et al.*, 1993]. One of the more prominent trends is that when back-arc spreading is pronounced, the largest megathrust events tend to be smaller [e.g., *Kanamori*, 1977; *Uyeda and Kanamori*, 1979; *Scholz and Campos*, 1995].

The substantial variations between earthquake ruptures within an individual subduction zone and from event to event for repeated failures of a given region of a megathrust have prompted numerous studies. Notions of variable fault zone failure strength, and resultant seismic slip heterogeneity, such as the asperity model [e.g., *Lay and Kanamori*, 1981; *Lay et al.*, 1982; *Ruff and Kanamori*, 1983b; *Kanamori*, 1986; *Schwartz and Ruff*, 1987; *Ruff*, 1992; *Ruff and Miller*, 1994], emerged from studies of subevents of large ruptures. *Scholz* [1990, 1998] discusses frictional perspectives of such rupture heterogeneity, arguing that it is more useful to invoke notions of stable and unstable sliding, with observed rupture complexity arising from variations in state variables such as effective normal stress, temperature from frictional heating, and material properties such as rock and sediment composition, permeability, and roughness.

It is generally recognized that significant interplate slip must occur aseismically, even within the seismogenic depth range, to account for total plate convergence in most subduction zones [e.g., *Ruff and Kanamori*, 1983a; *Peterson and Seno*, 1984; *Pacheco et al.*, 1993; *Scholz and Campos*, 1995]. Precise determination of seismic coupling is statistically difficult given the short historic record available [*McCaffrey*, 1997], but the current evidence favors a full range of seismic coupling values from 0 (purely aseismic subduction) to 1 (purely seismic subduction). Regions such as Kamchatka, Alaska, South Chile, East Aleutians, West Aleutians, and Nankai have high seismic coupling coefficients, while the Marianas, Kermadec, South Tonga, Izu Bonin, Java, North Peru, Vanuatu, and Central America have low seismic coupling [e.g., *Scholz and Campos*, 1995].

The depth distribution of interplate thrusting along the megathrust has been explored in an effort to understand this seismic coupling variability [e.g., *Zhang and Schwartz*, 1992; *Pacheco et al.*, 1993]. Most subduction zones exhibit peaked distributions of earthquake numbers and moment release as functions of depth, tapering toward both updip and downdip limits of the seismogenic zone. The data suggest that a relatively narrow portion of the seismogenic zone tends to have the highest proportion of seismic coupling, and even this may be laterally "patchy" [e.g., *Pacheco et al.*, 1993]. Localized asperity regions may undergo purely seismic slip, corresponding to total seismic coupling, thereby controlling overall failure process of the megathrust [e.g., *Lay et al.*, 1982; *Thatcher*, 1989]. Seamounts, horst and graben structures, and other bathymetric features on the underthrusting plate may account for locally enhanced frictional contacts [e.g., *Kelleher and McCann*, 1976; *Cloos*, 1992; *Pacheco et al.*, 1993; *Cloos and Shreve*, 1996; *Scholtz and Small*, 1997; *Tanioka et al.*, 1997; *von Huene et al.*, 1999; *Abercrombie et al.*, 2001; *Husen et al.*, 2002; *Bilek*, this volume], or these features may provide barriers that delimit rupture [e.g., *Kodaira et al.*, 2000; *Hirata et al.*, 2003], but as yet there is not a good understanding of the nature of the localized high-slip regions for interplate events or of the persistence of such features over time [e.g., *Scholz*, 1990; *Thatcher*, 1990].

The apparent complexity of sliding properties of the megathrust is even more pronounced in the shallowest region, <15 km below the ocean bottom. This region generally has a higher ratio of aseismic to seismic slip than the deeper interface and is likely to have acute variations in physical properties due to the presence of indurating sediments, expelled fluids, rough bathymetric structure on the subducting plate, complex structure of the overriding wedge, and vulnerability to transient high strain rates upon failure of the deeper interface. This review will focus on the seismological observations of earthquakes at shallow depths on the megathrust, considering first large, then smaller events, to assess possible constraints on this complex environment.

Large Earthquakes on the Shallow Megathrust

Systematic investigations of the rupture processes of interplate thrust events have identified high-slip regions within the rupture zones for most of the very large events that have occurred during the past 50 years or so [e.g., *Schwartz and Ruff*, 1987; *Kikuchi and Fukao*, 1987; *Beck and Ruff*, 1989; *Ruff and Miller*, 1994]. In almost all cases the primary slip occurs in the depth range 15–40 km, with relatively few documented instances of large slip extending to shallower depths, even after allowing for substantial uncertainties in subevent locations. Studies of the depth distribution of underthrusting events larger than magnitude 5 or so during the past 25 years show that relatively few events tend to occur in the shallowest part of the seismogenic zone [e.g., *Zhang and Schwartz*, 1992; *Pacheco et al.*, 1993]. These observations are generally attributed to the shallow portion of the megathrust undergoing a transition from stable sliding to unstable friction as depth increases, possibly with an intervening zone of conditional stability [*Scholz*, 1998]. Subduction of low permeability, water-saturated clays could increase fluid pressures in the shallow fault zone, lowering the effective normal stress and allowing stable sliding to depths of 5–10 km. Anelastic deformation of weak, unconsolidated fore-arc sediments in contact with the megathrust at very shallow depth could also play a role [e.g., *McCaffrey*, 1993]. Given ongoing motion of the underthrusting plate due to combined forces from slab-pull and ridge-push, large, elastic shear strains would not be expected to accumulate along the shallow megathrust as its frictional properties accommodate the deformation by stable sliding. High strain-rate loading caused by sudden failure of the deeper unstable sliding regime would not always induce large slip in any conditionally stable region at shallow depth and the shallowest region of any velocity-strengthening material would quickly terminate any slip event. These general considerations are at odds with the observation that in some locations the shallow portion of the megathrust actually has failed in large earthquakes, with very large slip, both with and without failure of the deeper portion of the rupture zone. When this occurs, resulting displacements of the seafloor are large, generating strong tsunamis, in a special class of earthquakes called tsunami earthquakes.

The 1896 and 1946 Tsunami Earthquakes

Tsunami earthquakes were identified by *Kanamori* [1972] as a distinct class of seismic events for which the tsunami excitation was larger than expected given the measured seismic magnitude (usually, the 20-s surface wave magnitude M_s). He considered two underthrusting events notable for their great tsunamis: The 1896 Sanriku (M_s 7.2) and 1946 Aleutian (M_s 7.4) events (fig. 15.2). The global tsunami magnitudes M_t estimated for these events by *Abe* [1979] are 8.6 and 9.3, respectively, so there are huge differences between M_s and M_t,

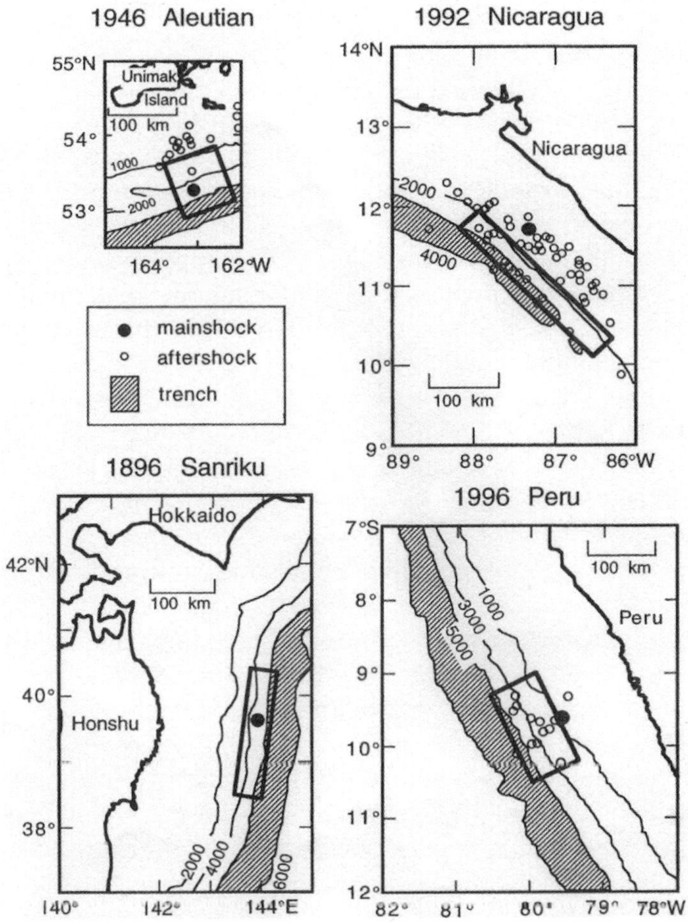

Figure 15.2 Fault models for four tsunami earthquakes: the 1946 Aleutian [*Johnson and Satake*, 1997], 1896 Sanriku [*Tanioka and Satake*, 1996a], 1992 Nicaragua [*Satake*, 1994] and 1996 Peru [*Satake and Tanioka*, 1999] events. Bathymetry is shown with contour lines. Boxes outline fault models used by the various authors. [From *Satake and Tanioka*, 1999].

distinguishing these from typical events. The limited seismic data available for these events yield effective seismic moment determinations that increase greatly with increasing period, indicative of source process time constants of ~100 s [*Kanamori*, 1972]. At long periods the discrepancy between seismic source strength and tsunami magnitude is thus greatly reduced, suggesting that faulting displacements may be able to account for the strong tsunami excitation without need for additional effects such as submarine slumping. Both events have locations relatively close to their respective trenches. *Kanamori* [1972] suggested that the long rupture durations of these events are related to weakness in the source zones, possibly associated with reduction of strength in the thrust zone due to nearby normal faulting in the oceanic plate.

Subsequent investigations of the tsunami data for the 1896 event confirm that the source area is very close to the trench, 210 km long, and ~50 km in width (fig. 15.2) [*Tanioka and Satake*, 1996a]. Slip of 5.7 m (for a dip angle of 20°) near the trench was estimated, a surprisingly large amount of slip to occur at very shallow depth along the Honshu coast. Rupture in low-rigidity sediments was suggested as a mechanism to give low rupture velocity (which can account for the long source process time) and resultant weak excitation of 20-s surface waves. Using a relatively low rigidity of 20 GPa based on observations of sediments being subducted (at greater depth around the plate interface it is common to assume a "hard rock" rigidity of 50 GPa), *Tanioka and Satake* [1996a] estimated a moment magnitude, M_w, of 8.0; still lower than the global tsunami magnitude of 8.6 or even the local tsunami magnitude of 8.2 [*Abe*, 1981]. Using a more realistic dip angle of ~10° requires a slip of 10.4 m, increasing the estimated M_w to 8.2 [*Tanioka and Seno*, 2001], and reducing, but not eliminating, the $M_w - M_t$ discrepancy without violating any seismic constraints. Even lower dip angles can be reconciled with regional microearthquakes [*Umino et al.*, 1995], so some of the difference between seismic and tsunami magnitudes can be accounted for. Very shallow dip angles for thrust faults lead to poor excitation of long-period seismic waves, but tsunami excitation is not as dependent on source depth or dip angle [*Ward*, 1982], so reducing dip of the shallow megathrust goes in the right direction to account for the relative tsunami excitation. *Tanioka and Seno* [2001] sought to further reconcile the seismic and tsunami strength estimates by accounting for possible lateral plowing and uplift of the very small sedimentary wedge seaward of the slip zone [e.g., *Seno*, 2000], an effect that can enhance tsunami excitation by ~50% because of broadened ocean bottom uplift that results. This reduces the slip needed at depth for a 10° dipping fault, lowering the M_w back to 8.0–8.1, but predicting some of the contrast to M_t.

The 1946 event produced a truly remarkable tsunami, which caused a 30-m local run-up on Unimak Island and 16-m run-up in Hawaii. The tide gauge recordings for this event were studied in detail by *Johnson and Satake* [1997]. Using a 100-km-long rupture based on the aftershock distribution [*Sykes*, 1971], an estimate of 9.6–9.8 m of slip was obtained to match the tsunami records, with strong slip extending up to the trench, seaward of the aftershock zone (fig. 15.2). A limited set of body waves analyzed by *Pelayo* [1990] yielded a 90-s duration source time function, suggesting a very slow rupture velocity of <1 km/s. For an assumed rigidity of 20 GPa, an M_w of 8.2–8.3 was estimated, still well below the M_t of 9.3, but *Johnson and Satake* [1997] argue that the latter value may be biased upward by harbor resonances. *Pelayo* [1990] estimated a very large seismic moment from surface wave spectra that gives an $M_w = 8.6$. This predicts a 27-m slip using a standard rigidity, but such a large slip overpredicts the early (pre-Harbor resonance) tsunami amplitudes. *Kanamori*

[1985] suggested that the focal mechanism for long period signals of the 1946 event may be best represented by a single force, implying that a landslide contributed to the tsunami excitation, as is preferred for the 1929 Grand Banks event [*Hasegawa and Kanamori*, 1987]; however, the model of *Johnson and Satake* [1997] indicates that the pre-Harbor resonance tsunami observations can be well explained by a megathrust faulting source with slow slip and rupture extending to the trench.

Tsunami Earthquakes Between 1960 and 1992

Following the initial study of the 1896 and 1946 tsunami earthquakes by *Kanamori* [1972], several smaller events with anomalous tsunami excitation were identified in the shallow megathrusts along the Kurile Islands and Peru. In particular, the 20 October 1963 (M_s 6.9) and 10 June 1975 (M_s 7.0) events in the Kurile subduction zone were classified as tsunami earthquakes [e.g., *Fukao*, 1979; *Ward*, 1982; *Pelayo and Wiens*, 1992], and the 20 November 1960 (M_s 6.75) tsunami event occurred along the Peru margin [*Pelayo and Wiens*, 1990]. The two events in the Kuriles locate seaward (updip) of large slip regions in the rupture zones of larger interplate thrusts along the megathrust [*Schwartz and Ruff*, 1987]. The aftershock zones overlap those of earlier great thrusts at depth but extend all the way to the trench. *Fukao* [1979] suggested that the Kurile events are very shallow, and their source process times are <100 s (unusually long, but insufficient to account for the discrepancy between M_s and tsunami strength). On the basis of some indications of change in focal mechanisms during rupture, he proposed that relatively steeply dipping splay faulting into the sedimentary wedge might account for strong ocean-bottom deformation needed to explain the tsunamis. Low rigidity μ of the sediments (13 GPa) was assumed, requiring ~ 4 times more slip at the source to give the same seismic moment, M_0. The seafloor deformation scales as $M_0/\mu = Ad$, where A is the fault area and d is the average displacement, thus large displacement and shallow depth would combine to give strong tsunami excitation [see *Okal*, 1988]. Splay fault branching into the accretionary wedge has also been suggested to occur along the Nankai subduction zone [*Cummins and Kaneda*, 2000; *Park et al.*, 2002].

Pelayo and Wiens [1992] modeled P and SH waveforms for the Kurile events, favoring source depths <10 km with a smooth, broad, 90-s source function for the 20 October 1963 event and a 60-s duration pulse for the 10 June 1975 event, with both events having weak slip in the first 20 s or so. Shallow dips of 7°–8°, with no change in mechanism, were found. The 20 November 1960 Peru event was modeled as a shallow (5–25 km) thrust event with very shallow dip angle (6°) and a rupture duration of 110 s [*Pelayo and Wiens*, 1990]. For these three events the long source durations and moderate aftershock zone dimensions suggest low rupture velocities, but this was not directly determined because point-source models were used. *Pelayo and Wiens* [1992] conclude that all three events involve slow rupture along the basal décollement of the accretionary prism, with low rigidity around the shallow megathrust being responsible for the low rupture velocities. The long source process time reduces M_s relative to M_o, leading to discrepancy between the tsunami size and M_s, as in the model of *Kanamori* [1972]. The very shallow source depths and shallow dip angles further reduce the seismic wave excitation relative to the tsunami excitation.

Tsunami Earthquakes of the Broadband Seismology Era

Several tsunami earthquakes have occurred subsequent to deployment of global networks of broadband seismic equipment; this has provided the opportunity for improved seismological characterization of their rupture processes. By far the most important of these events was the 1992 Nicaragua earthquake, which struck on 2 September in a seismic gap that had been identified by *Harlow et al.* [1981]. This event has an M_t = 7.9–8.0, and caused a 10-m tsunami run-up at the coastal village of El Transito and 2–6 m run-up along the entire Nicaraguan coast [*Satake et al.*, 1993; *Ide et al.*, 1993; *Imamura et al.*, 1993], despite weak ground shaking (≤ Modified Mercalli III). The aftershock zone is 200 km long and ~100 km wide, an unusually large size for an M_s 7.2 event. A modest number of aftershocks locate close to the trench. *Ihmlé* [1994] relocated the aftershocks, finding that most events concentrate in a 40-km-wide zone ~250 km long, with an updip region ~50 km wide being almost devoid of aftershocks (fig. 15.2). The aftershock distribution has the appearance of a conventional seismic front, but as discussed below, it appears that the primary mainshock slip was updip of this region. *Ide et al.* [1993] and *Kanamori and Kikuchi* [1993] found that the surface wave radiation patterns for the Nicaragua event are consistent with a shallow underthrusting event rather than with a single force, so a landslide source is not favored. *Kanamori and Kikuchi* [1993] note that there is no sedimentary wedge off of Nicaragua, and they suggest that rupture propagated all the way to the trench with the megathrust frictional properties being affected by the subducted sediments.

The 256-s period surface wave moment magnitude for the Nicaragua event is M_w = 7.6 [*Ide et al.*, 1993; *Kanamori and Kikuchi*, 1993], substantially larger than the M_s; this partly closes the gap with M_t. The rupture centroid time shift is ~50 s, suggesting a source duration of ~100 s [*Ide et al.*, 1993; *Kanamori and Kikuchi*, 1993; *Velasco et al.*, 1994]. The source process time estimated using a procedure similar to that which *Kanamori* [1972] applied to the 1896 and 1946 tsunami events, is 40% as long as for those earlier tsunami events [*Ide et al.*, 1993]. *Kanamori and Kikuchi* [1993] find that the Nicaragua event did not have an increase in apparent seismic moment for periods longer than 200 s, which is also consistent with a shorter source process time than for the 1896 and 1946 events. This event thus appears to be similar to the 1960 Peru, 1963 Kurile Island, and 1975 Kurile Island events, but it generated far superior seismic data because of the increased bandwidth of global seismic recording stations.

Several investigations of the source time function for the 1992 Nicaragua event were undertaken using the broadband digital seismic data, including body and surface wave analyses [e.g., *Ide et al.*, 1993; *Kanamori and Kikuchi*, 1993; *Satake et al.*, 1993; *Velasco et al.*, 1994; *Kikuchi and Kanamori*, 1995; *Ihmlé* 1996a, 1996b]. The data are abundant enough and have sufficient bandwidth to directly determine some aspects of the source rupture finiteness from directivity effects in the seismic waves, the first time this was possible for a tsunami earthquake. This is important, given that the aftershocks may not illuminate the updip region of the rupture zone if it has unusual frictional properties. The Nicaragua event commenced with a 10-s interval of low energy release, then ruptured updip and asymmetric bilaterally, with a rupture velocity of ~1.0–2.2 km/s [e.g., *Ide et al.*, 1993; *Velasco et al.*, 1994; *Kikuchi and Kanamori*, 1995; *Ihmlé*, 1996b], with slip distributed along a 160–200 km long fault length. The low-frequency centroid locations estimated for the overall energy release [*Dziewonski et al.*, 1993; *Velasco et al.*, 1994; *Ihmlé*, 1996b] are updip of the relocated aftershocks, suggesting that significant slip occurred in a region with no aftershocks [*Ihmlé*, 1996b]. The published slip models differ in slip distribution over the fault, apparently due to the smooth nature of the source time functions and intrinsic trade-offs between changes in rupture area expansion rate and slip. The early stage of slow energy release during the rupture complicates location of the regions of strong slip. The model parameterization (such as use of constant or variable rupture velocity) also has strong effects on the inferred slip distribution. There are also problems in recovering the body wave ground motion baselines associated with near-field terms visible in the unusually smooth teleseismic records [e.g., *Kanamori*, 1993; *Kikuchi and Kanamori*, 1995]. The slip models are consistent in having strong slip just northwest of the epicenter that dominates the early portion of the radiation, with slow southeastward expansion of the slip dominating the latter portion of the radiation. The seismic waveform data provide little direct constraint on the actual width of the rupture zone, but shallow centroid depths of 10 km or less are favored by *Dziewonski et al.* [1993] and *Velasco et al.* [1994].

Ide et al. [1993] estimated the static stress drop as 0.26 MPa for a rupture zone width of 100 km, a rigidity of 30 GPa, and average dislocation of 0.5 m. *Kikuchi and Kanamori* [1995] assumed a 50-km-wide rupture zone and obtained a static stress drop of 1.1 MPa with an average dislocation of 1.4 m for the same rigidity, along with estimating the average rise time of the dislocation to be ~10 s. The average dislocation velocity is thus estimated to be 0.14 m/s, significantly slower than for other subduction zone events. The seismic wave energy E_s to moment, M_o ratio was estimated as $E_s/M_o = 6.2 \times 10^{-7}$, about an order of magnitude lower than typical for earthquakes [e.g., *Choy and Boatwright*, 1995; *Ruff*, 1999]. The apparent stress, defined by $\sigma_a = 2\mu\, E_s/M_o = 0.037$ MPa, is much lower than the static stress drop. Using source time functions inferred from surface waves and the centroid-moment tensor moment of *Dziewonski et al.* [1993], *Velasco et al.* [1994] estimated the apparent stress as 0.03 MPa, so this result is fairly stable for relatively uniform slip models. *Ihmlé* [1996b] found a more heterogeneous slip model, with two patches having average slips of 3.5–4.5 m (for assumed rigidity of 22 GPa). This gives higher localized static stress drop values of 3.0–7.0 MPa. Rise times of 10–18 s were preferred in the models of *Ihmlé* [1996b], so dislocation velocities of 0.2–0.7 m/s are estimated.

The tsunami observations for the Nicaragua event have been modeled by *Imamura et al.* [1993], *Titov and Synolakis* [1993], *Satake* [1994, 1995], *Piatanesi et al.* [1996], and *Geist and Bilek* [2001]. *Imamura et al.* [1993] used linear long-wave theory to study the tsunami propagation, with an assumed run-up ratio of 2. They found a uniform 3.75-m slip for assumed 200×100 km^2 fault area and rigidity of 30 GPa, giving a seismic moment one order of magnitude greater than observed. A similar result was found by *Titov and Synolakis* [1993], who included a one-dimensional nonlinear run up computation. The discrepancy with the seismic model was reduced to a factor of 5.6 by *Imamura et al.* [1993] by taking a smaller fault area of 150×75 km^2, although the larger fault area was slightly preferred by the authors. *Satake* [1994, 1995] modeled tide-gauge waveforms and run-up observations using linear shallow water equations and nonlinear equations including bottom friction and showed that the required fault width is actually only 40 km, much narrower than the aftershock area and extending only as deep as 10 km. Uniform slip over a 250-km-long fault was preferred, although a 200-km fault with larger slip at the southeastern end of the fault can match the data well (fig. 15.2). The slip needed for this model is 3 m, which can be reconciled with the seismic moment by using a rigidity of 10 GPa. *Piatanesi et al.* [1996] modeled the tsunami run-up data with a variable slip model, using a nonlinear shallow-water theory. A slip function intermediate between a uniform slip model and the variable slip model of *Ihmlé* [1996b] was preferred, with two predominant slip patches having slip of 3.5–4.5 m near the epicenter and at the southeastern end of the rupture zone, and slip elsewhere along the fault being 1–2 m. A rigidity of 10 GPa was assumed, with a 50×250 km long fault. *Geist and Bilek* [2001] further reconciled the seismic and tsunami models by using the slip model of *Ihmlé* [1996b] with a depth-dependent shear modulus [*Bilek and Lay*, 1998, 1999b] increasing from 3.6 to 30 GPa across the rupture plane, finding good agreement with the tide-gauge waveform at Corinto.

These studies, by far the most comprehensive for any tsunami earthquake, demonstrate that the best overall model for the 1992 Nicaraguan event involves shallow thrusting on a 40–50 km wide fault extending from the trench down to depths of no more than 10 km or so (fig. 15.2). Few aftershocks fall within the area of large slip (~3 m). There is no accretionary prism in the trench; thus it is assumed that sediments have been subducted. The overall rupture took ~110 s, commencing with 10 s of weak slip and then bilateral expansion over the fault zone at low slip velocities. The average source volume rigidity is ~10 GPa.

There has been one additional recent tsunami earthquake of note; the 21 February 1996 Peruvian earthquake (M_s = 6.6, M_w = 7.5, M_t = 7.8) [*Satake and*

Tanioka, 1999]. *Bourgeois et al.* [1999] find that the 1996 event is similar to the 1960 Peru event analyzed by *Pelayo and Wiens* [1990], with these two tsunami earthquakes being the only large subduction zone earthquakes in northern Peru since 1619, in what is inferred to be a subduction zone with a large component of aseismic convergence. The event occurred where the Mendaña Fracture zone is being subducted under the Peru margin (fig. 15.2). The seismic observations for this event were modeled by *Ihmlé et al.* [1998] and tsunami data were modeled by *Heinrich et al.* [1998] using the seismic slip model as a constraint. A 40-km-wide, 110-km-long fault model was used in the seismic modeling. The event ruptured bilaterally, updip of the epicenter with rupture velocity varying from 1.0 to 2.0 km/s and a total rupture duration of ~50 s. This is a relatively long duration for the moment magnitude, but not dramatically so, as for most of the events discussed above [e.g., *Polet and Kanamori*, 2000]. *Heinrich et al.* [1998] adopted this slip model to predict run-up heights and tide-gauge recordings along the Peruvian coast using a finite difference method for long wave equations. They found that a rigidity of 20 GPa reconciles the seismic and tsunami source models. Tsunami modeling by *Bourgeois et al.* [1999] using a 60-km-wide fault that is 120 km long with 4-m slip suggests a rigidity of 7.5 GPa. *Satake and Tanioka* [1999] modeled the tide-gauge records using either 40- or 80-km-wide faults extending up to the trench with slip of 4.5 and 2.2 m, respectively. Using a low rigidity of 10 GPa allows these slip models to be consistent with the seismic moment magnitude.

The tsunami earthquakes discussed above have common characteristics of shallow dipping thrust motion at shallow depths in material with very low rigidity of 5–10 GPa [*Satake and Tanioka*, 1999; *Synolakis et al.*, 1997]. Figure 15.2 summarizes the narrow rupture zones at shallow depths favored for the 1896, 1946, 1992, and 1996 events [*Satake and Tanioka*, 1999]. All of these events are documented to have relatively long source process times, and in some cases, directly measured low rupture velocities. Figure 15.3 shows a schematic suggesting the distinct nature of rupture at very shallow depth in low rigidity material around the megathrust and resulting enhanced seafloor uplift documented for several of the large tsunami earthquakes. Note that an accretionary prism need not be present for this to occur, as in the case of the 1992 Nicaragua event, as long as sediments modify the frictional properties of the shallow portion of the megathrust.

There have been several other events in the broadband seismic data era that have generated unusually large tsunamis by other mechanisms. The 1992 Flores Islands (M_w 7.9) back-arc thrusting earthquake did not have slow rupture [e.g., *Imamura and Kikuchi*, 1994; *Beckers and Lay*, 1995; *Polet and Kanamori*, 2000], and most of the local tsunami is attributable to the seismic slip model, except for a 26.2-m run-up at Riangaroko, which is attributed to a local landslide effect by *Imamura et al.* [1995], *Tsuji et al.* [1995b], and *Hidayat et al.* [1995]. The small tsunami generated by the 1989 Loma Prieta earthquake was attributed to slumping in the Monterey Bay by *Ma et al.* [1991]. The 17 July 1998

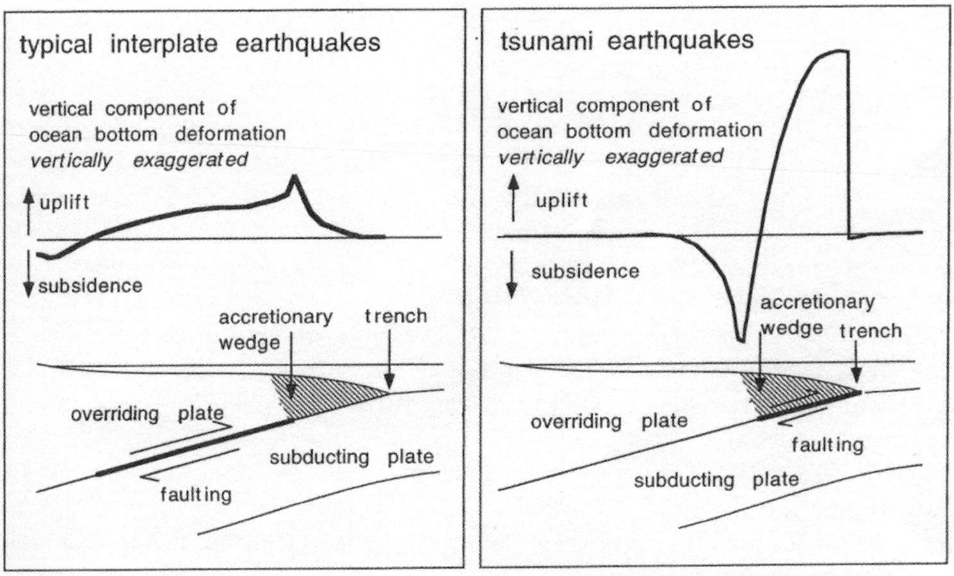

Figure 15.3 Schematic of slip at varying depth on the megathrust and resulting surface (ocean bottom) vertical displacement. Large slip at very shallow depth in low rigidity sediments causes large ocean bottom displacement relative to comparable seismic moment events with less slip in higher rigidity material at greater depth along the megathrust. This accounts for the strong tsunami excitation of some tsunami events; in some places, there is no accretionary wedge, and sediments subducted along the megathrust may play the same role. [From *Satake and Tanioka*, 1999].

Papua New Guinea (M_w 7.0) event produced a maximum tsunami run-up of 10–15 m on the coast in the vicinity of Sissano Lagoon. The seismic magnitudes for this event do not characterize it as having a particularly long rupture process or slow rupture velocity [e.g., *Newman and Okal*, 1998b; *Polet and Kanamori*, 2000]. Tsunami generation for this event is controversial, but it appears that this event is at least not a conventional interplate thrusting tsunami earthquake [*Geist* 1998, 2000; *Kawata et al.*, 1999; *Matsuyama et al.*, 1999; *Tanioka*, 1999; *Tappin et al.*, 1999; *Synolakis et al.*, 2002]. It is unclear whether the seismic models are able to account for the local tsunami run-up. *Kikuchi et al.* [1999] and *Tanioka* [1999] favor a steeply dipping fault plane rather than a shallowly dipping thrust plane as both the seismic and teleseismic tsunami source, although *Satake and Tanioka* [2003] suggest a combination of low angle faulting and a submarine landslide as the most likely source.

The 2 June 1994 East Java underthrusting earthquake (M_w 7.9) occurred in a region normally characterized as weakly seismically coupled. It produced local tsunami run ups that ranged from 1 to 14 m [*Tsuji et al.*, 1995a], and *New-*

man and Okal [1998a] and *Polet and Kanamori* [2000] suggest that this event has low energy release, slow rupture, and location near the trench, which are consistent with it being a tsunami earthquake. *Abercrombie et al.* [2001] studied the seismic waves, finding a solution with shallow interplate thrusting ~12–20 km depth with an initial weak updip radiation over ~12 s. They found no evidence for anomalously slow, shallow slip, although it is notable that all of the larger aftershocks of this event involve normal faulting rather than thrusting. It was proposed that rupture of a seamount surrounded by stably sliding interface explains this event. The stress drop was estimated as 0.3~3 MPa, a low value comparable to estimates for the 1992 Nicaragua event, but the event was deemed to be different from typical tsunami events. *Tanioka and Satake* [1996b] model the tsunami using a shallow slip event near the trench, constraining the model by the seismic moment and using a rigidity of 40 GPa. This model underpredicts the tsunami by 30–50%, even allowing for horizontal movement of the ocean bottom, which enhances tsunami excitation in this case. Use of lower rigidity or more accurate run-up model could account for this, but *Abercrombie et al.* [2001] argue that the tsunami arrival times rule out a slip model as far seaward as the trench. *Polet and Thio* [2003] suggest slow particle velocity rather than slow rupture velocity might account for the weak short period radiation from the Java event relative to that for nearby aftershocks.

Large Events Deeper on the Megathrust Triggering Shallow Slip

The occurrence of significant seismic slip on the shallow portion of the megathrust, in some cases extending right up to the trench or to the margin of any sedimentary wedge, raises the question of whether this region fails during the rupture of great events at larger depth in the megathrust [*Geist*, 1998]. Inferences of earthquake rupture zones from aftershock distributions may be misleading, as in the case of the 1992 Nicaragua event, so it would seem plausible, and even likely, that during a large interplate rupture, slip may extend into the shallow megathrust where few, if any, aftershocks occur. Combined assessments of seismic, geodetic, and tsunami observations for great megathrust earthquakes have actually revealed that this is sometimes the case. *Hirata et al.* [2003] show that slip extended to shallow depth along a limited portion of the 1952 Tokachi-Oki earthquake, with rupture toward the southwest possibly being delimited by a subducted seamount. *Johnson and Satake* [1999] found that there was minor slip at shallow depth along a portion of the 1952 Kamchtaka rupture zone. *Satake* [1993] and *Tanioka and Satake* [2001a, 2001b] demonstrated that significant slip occurred at shallow depths along the Nankai region in the 1994 Tonankai and 1946 Nankai events. *Ichinose et al.* [2003] also find very shallow slip near the trench along the Kii peninsula for the 1994 Tonankai earthquake. *Johnson and Satake* [1996] estimated 1–3 m of slip along the shallow portion of the 1965 Rat Island earthquake. *Johnson et al.* [1996] found shallow slip for the 1964 Alaska earthquake as well. The 9 October 1995 (M_w 8) Colima-Jalisco earthquake ruptured to shallow depths near the trench [*Mendoza and Hartzell*, 1999; *Pacheco et al.*, 1997]. *Polet and Kanamori* [2000] note that some large events near Guerrero and Oaxaca, Mexico, have centroid locations displaced toward the trench as well. The occurrence of shallow slip may well have been more common in earlier events but simply was not recognized in the seismic data.

For some events it appears that slow slip on the updip portion of the megathrust precedes the mainshock at greater depth. This occurred for the 1968 Tokachi Oki earthquake [*Schwartz and Ruff*, 1985; *Satake*, 1989], for which a 30-s duration of low energy release preceded the mainshock, and for the 28 December 1994 Sanriku-oki earthquake (M_w 7.7), which initiated with a 20-s updip precursory slip followed by downdip rupture propagation [e.g., *Sato et al.*, 1996; *Hartog and Schwartz*, 1996; *Tanioka et al.*, 1996]. *Nakayama and Takeo* [1997] demonstrate that the precursory slip had a lower rupture velocity than the deeper primary slip. GPS observations revealed large aseismic afterslip following this event on both short term [*Heki and Tamura*, 1997] and long term [*Heki et al.*, 1997], equaling the seismic moment of the mainshock, but with a time constant of over a year and with slip possibly overlapping the coseismic slip zone. Subsequent inversions of the GPS data indicate that postseismic slip occurred both updip and downdip of the coseismic slip, without significant spatial overlap with the co-seismic rupture area [*Uchida et al.*, 2003; *Yagi et al.*, 2003]. The aftershocks of the event extend far seaward of the main slip zone [e.g., *Matsuzawa et al.*, 1995; *Hino et al.*, 1996], overlapping the updip portion of the 1968 Tokachi-oki rupture zone. *Kawasaki et al.* [1995] found geodetic evidence for a very slow slip event in 1992 with a time constant of ~1 day to the south of this event. *Seno* [2002] summarizes the mainshock asperity zones and updip aftershock zones for the 1994, 1992, and 1989 events in this region, all of which have shallow aftershock zones overlapping the region of the 1896 tsunami event.

Moderate Size Earthquakes on the Shallow Megathrust

The unusual attributes of tsunami events and their destructive effects have focused attention on very large shallow megathrust earthquakes. While it is possible that these events have some uniqueness associated with their large slip and large total seismic moment, the interpretation of unusual frictional or mechanical properties at the source raises the question of whether smaller events rupturing the same area as a tsunami event have anomalous rupture properties, such as unusually low stress drop, slow rupture velocity, or slow particle velocities. This can be addressed by systematic analysis of moderate size earthquakes that are large enough to be well-recorded teleseismically, which simplifies the analysis of the rupture process. In a series of papers, *Bilek and Lay* [1998, 1999a, 1999b, 2000] address this topic using data for 417 megathrust events. They find that after earthquake source time functions (determined by removing propagation effects from teleseismic *P* waves) are scaled to eliminate predictable differences due to total seismic moment, there is a systematic occurrence of unusually broad source time functions in the upper 10–15 km of many circum-Pacific megathrusts (fig. 15.4). Comparisons of moment-scaled source durations of these moderate size earthquakes with moment-scaled duration estimates from the known tsunami earthquakes suggests similar behavior (fig. 15.5), indicating that the frictional properties are not dependent on earthquake size. In addition, the general shape of the time functions for the shallow earthquakes is similar to time functions for several of the tsunami earthquakes [*Bilek and Lay*, 2002], with long tails of moment release and several subevents. Efforts to detect any correlation of the depth pattern in individual subduction zones with macroscopic attributes of each region (age of plate, bathymetric structures, etc.) have not yielded any strong correlations [*Bilek*, 2001].

The anomalously long source durations for the moderate size events cannot be unambiguously associated with slow rupture velocities, given that

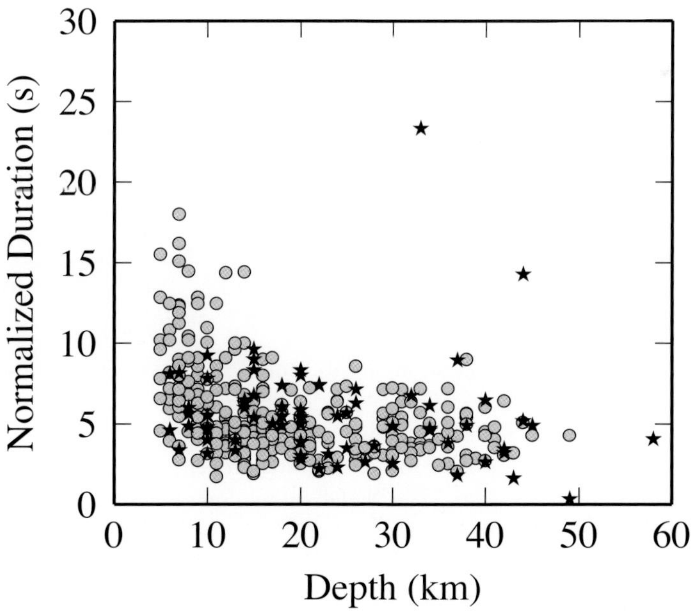

Figure 15.4 Earthquake rupture durations as a function of source depth for circum-Pacific subduction zone thrust events. The events are scaled to be normalized to $M_w = 6.0$. The circles are from the studies of *Bilek and Lay* [1999b, 2000] and *Bilek et al.* [2004], while the stars are from *Tanioka and Ruff* [1997] and updates in the University of Michigan source time function catalog.

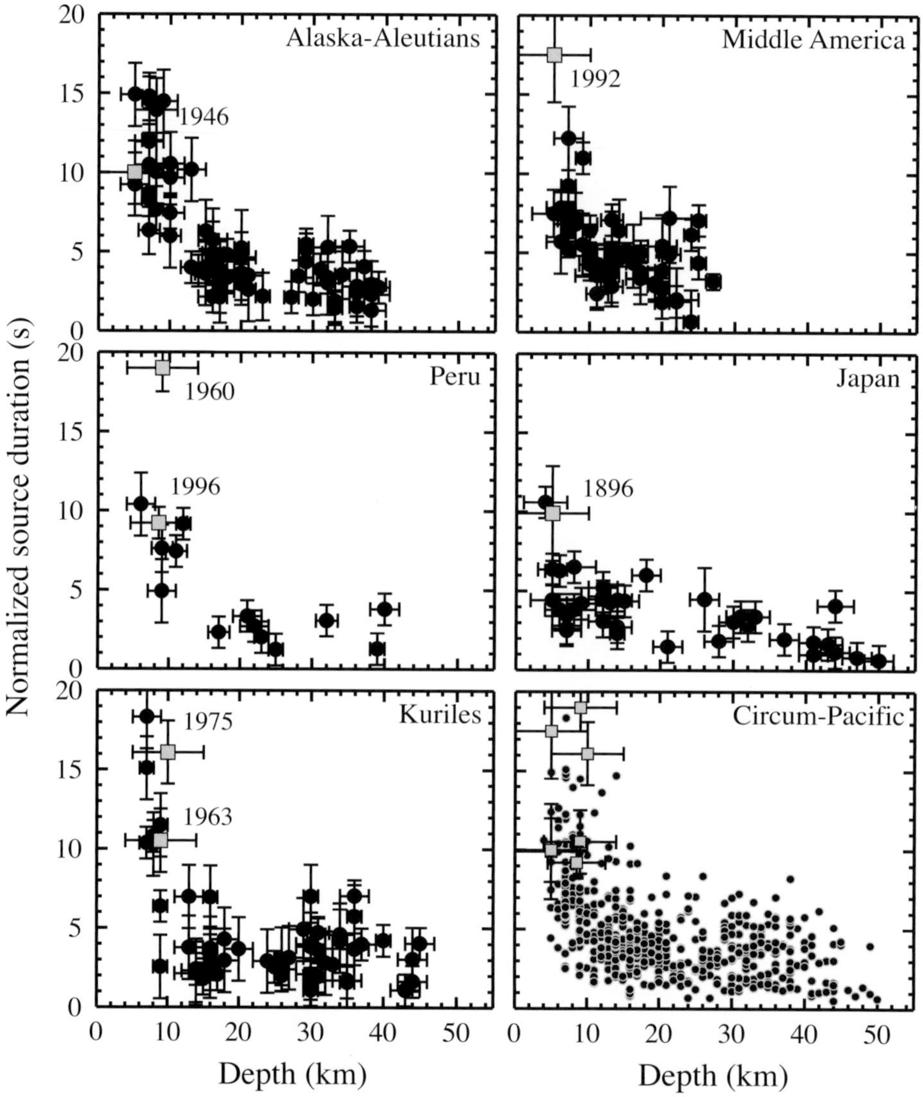

Figure 15.5 Earthquake rupture durations versus source depth for individual regions for the data from *Bilek and Lay* [1999b, 2000] and *Bilek et al.* [2004] along with moment-scaled source durations from large tsunami earthquakes in each of the five separate regions. All data are combined together at the lower right. Note that the tsunami events generally fall on or along the trend defined by the shallow events of intermediate size. [From *Bilek and Lay*, 2002].

point-source models were used to determine the source-time functions and no spatial finiteness is resolved. This leads to a complete trade-off between end-member models of events with low rupture velocity but normal (scaled) rupture area, or low stress drop events with normal (scaled) slip, but anomalously large rupture area [e.g., *Bilek and Lay*, 1999b, 2000]. If it is assumed that rupture velocity is anomalously low (as demonstrated unambiguously to be the case for the 1992 Nicaragua event) and that rigidity is low, giving low shear velocity at the source and corresponding low rupture velocity, the source durations can be used to infer source rigidity variations as a function of depth (fig. 15.6a). Alternatively, one can map the durations into fault areas (assuming a given "normal" shear velocity), and then static stress drop variations as a function of depth can be estimated (fig. 15.6b). While it is likely that rupture velocity variations and stress drop variations are actually coupled, the data suggest an overall pattern of depth-dependent variations of rigidity and/or stress drop across the upper 10 km of the seismogenic zone. There is some indication of depth dependence of the ratio of seismic energy, *Es*, to seismic moment [*Bilek et al.*, 2004] for individual regions, and all of the lowest values of this ratio (including the values for large tsunami earthquakes) are for events in the upper 10 km of the seismogenic zone. If future finite-source modeling of some of the long duration events is able to reveal the actual rupture velocities involved, it would allow the trade-off between rupture velocity and stress drop to be resolved, which would be a valuable contribution.

Additional studies have indicated anomalous rupture properties for moderate size earthquakes in the shallow subduction zone environment. *Houston* [2001] uses the source time function catalog of *Tanioka and Ruff* [1997] to examine variations in the shape and duration of time functions for a variety of tectonic settings. For a modest number (69) of megathrust events, she finds a general decrease in duration of about a factor of 1.5 from the shallowest (5–10 km) depth earthquakes to events at 30–40 km; a trend somewhat less than that apparent in the much larger data set of *Bilek and Lay* [2000]. Some differences in duration estimates for specific events are pointed out by *Houston* [2001], but she fails to note that these are associated with differences in depth estimates in most cases. *Ruff* [1999] uses the same catalog to explore dynamic stress drop variations along subduction zone megathrusts. In particular, *Ruff* [1999] highlights several earthquakes along the Oaxaca segment of the Mexican subduction zone that have significantly lower stress drops at the updip edge than earthquakes occurring at the downdip edge of seismicity. Other studies support a systematic depth variation in earthquake rupture properties along the Mexican subduction zone. *Shapiro et al.* [1998] and *Iglesias et al.* [2003] identify several near-trench shallow earthquakes that have low peak accelerations and are depleted in high-frequency radiation, similar to the depleted high-frequency radiation for tsunami earthquakes.

Okal and Newman [2001] use historical data to examine possible rupture anomalies for moderate and large earthquakes in regions of known tsunami earthquake occurrence, specifically Central America, Java, and Peru. They do not find a significant correlation between their slowness parameter (used to discriminate tsunami earthquakes) and depth, although they do question the values used for the historical earthquake depths. They do find regional

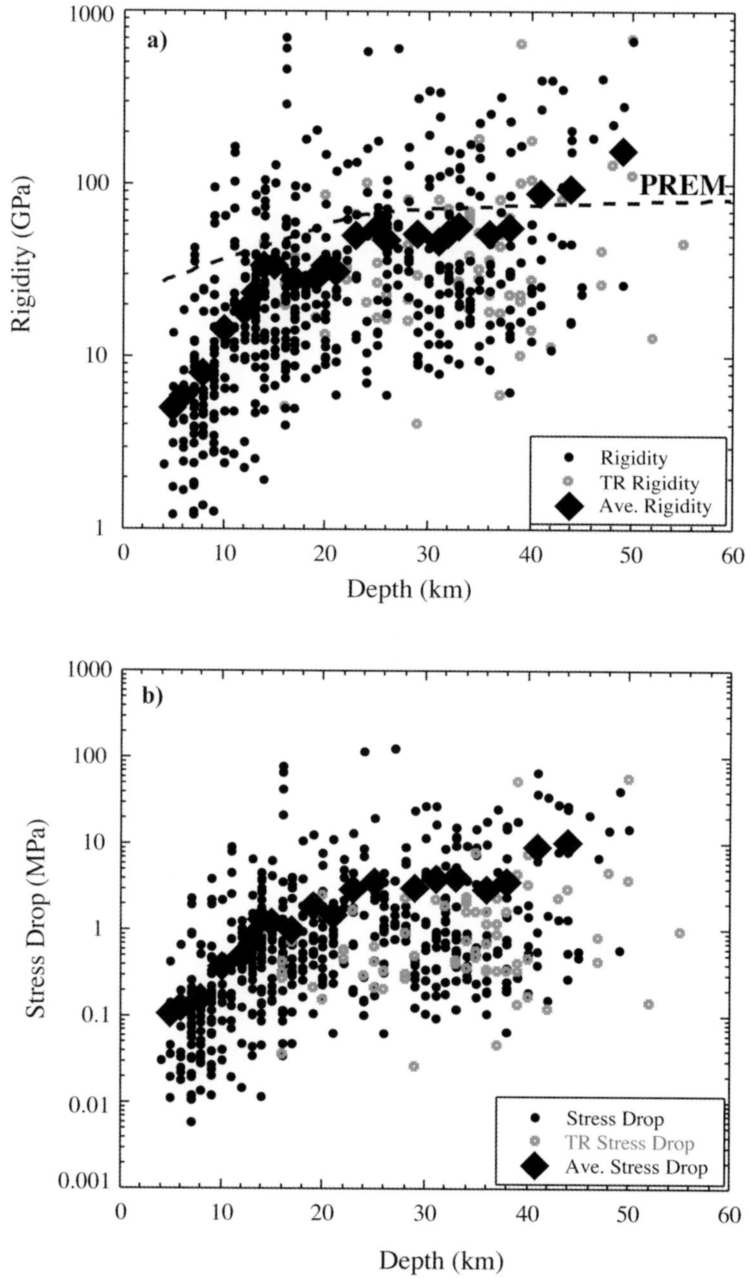

Figure 15.6 Estimates of source region rigidity (under assumption of constant stress drop) and source static stress drop (under the assumption of constant rupture velocity) for the source duration data of figure 15.4, plotted as functions of source depth. Open circles indicate data from *Tanioka and Ruff* [1997] and the solid circles are from the studies of *Bilek and Lay* [1999b, 2000] and *Bilek et al.* [2004]. Solid diamonds indicate mean values in depth bins. The values of rigidity from model PREM [*Dziewonski and Anderson*, 1981] are shown for reference.

correlations in that other slow earthquakes may have occurred in the regions of the 1992 Nicaragua and the 1960 and 1996 Peru tsunami earthquakes, and they suggest a local control on source slowness. The inclusion of events off the megathrust makes it difficult to assess this work.

The many observations of earthquake duration, stress drop, and source slowness variations along the shallow subduction megathrust suggest depth dependence of earthquake rupture process, even though it is unclear whether stress drop or rupture velocity increases with depth are dominant. This trend is similar to that which can be inferred on the basis of large earthquakes alone, although it appears to be more gradational than bimodal, as might be inferred just given large events. The systematic behavior observed for moderate and large events underlies the strategies being pursued for rapid tsunami hazard assessment [e.g., *Schindele et al.*, 1995; *Newman and Okal*, 1998a].

Small Earthquakes on the Shallow Megathrust

Given the trends in large and moderate size earthquake rupture process with source depth, it is obviously important to examine yet smaller events to assess whether there is a scale dependence. Determining rupture characteristics for smaller earthquakes is actually a harder problem due to a lack of good quality data for most subduction zones. As the small event seismicity in the shallow megathrust is usually far offshore, adequate recording of the small earthquake signals usually requires ocean bottom seismometer (OBS) arrays.

Some early work on small magnitude earthquakes does suggest the possibility of anomalous behavior of the shallowest events. *Fukao and Kanjo* [1980] note a near-trench zone of shallow earthquakes extending along most of northern Japan that appear to generate low to very low frequency radiation. Their study of over 600 shallow earthquakes shows a distinct change in the frequency content of the earthquakes, with many high-frequency events occurring close to shore and presumably deeper along the megathrust. Examination of seismicity in the Kamchatka subduction zone suggests a gradient in source properties with depth, as the earthquakes located downdip from the trench radiate more high-frequency energy than those earthquakes occurring near the trench [*Zobin*, 1999].

OBS networks have been placed near several subduction zone margins in recent years, providing high-quality records for microseismicity in the megathrust zone. Earthquake distributions in northern Japan shows very little activity at the shallowest subduction zone depths beneath the margin wedge, with most activity occurring at depths of ~15 km even near the 1896 tsunami earthquake zone, which appears to have ruptured at shallow depth [*Nishizawa et al.*, 1990; *Umino et al.*, 1995; *Hino et al.*, 1996; *Takahashi et al.*, 2000]. Earthquake clustering in the shallow megathrust off Japan is common, suggesting lateral variations in the properties along the trench strike. To the south, around the

Nankai Trough, seismicity levels are very low; however, using an OBS network, *Obana et al.* [2003] note some plate interface microseismicity that suggests an updip limit offshore Cape Muroto, consistent with the 150° isotherm computed in regional thermal models. Rupture properties of the small events along Japan have not been systematically determined.

Several examinations of small earthquakes off Sanriku, Japan, suggest repeating earthquakes at small asperities. *Matsuzawa et al.* [2002] and *Igarashi et al.* [2003] find several clusters of microseismicity that are not collocated with regions of large moment release in previous earthquakes nor in areas identified as strongly coupled, and they estimate asperity size on the order of 0.1–1 km. These observations are used to describe a model for the NE Japan subduction zone including large asperities near the trench to support large earthquake occurrence, with smaller, sparsely distributed asperities and adjacent stable sliding zones deeper on the plate boundary. These observations have more recently been used to estimate quasi-static slip at this boundary. Assuming that the small asperities on the fault break in a characteristic manner, the time interval between each small earthquake is governed by the amount of aseismic slip in the adjacent regions [*Uchida et al.*, 2003]. By estimating the slip released in these small earthquakes, *Uchida et al.* [2003] find regions of quasi-static slip in both the shallow and deepest portion of the plate boundary.

A combination of land and offshore seismic networks has been deployed in other subduction zones as well. *Husen et al.* [1999] use land stations and OBHs to relocate aftershocks of the 1995 M_w = 8.0 Antofagasta underthrust earthquake, defining a minimum seismogenic depth of 29 km. The upper limit of seismicity in northern Costa Rica appears to be transitional along the Nicoya Peninsula, shallowing from 20 km in the north to 10 km in the southern portion of the peninsula [*Newman et al.*, 2002]. *Newman et al.* [2002] suggest a thermal cause for this transition with colder subducting crust in the north and warmer underthrusting crust in the south. In central Costa Rica, near the Osa Peninsula, *DeShon et al.* [2003] suggest that disrupted subducting crust influences the limits of seismogenesis, with an updip edge at a depth of 10 km. At this time, most effort has been placed on accurate event locations and defining the seismogenic limits in these margins, and comparing the seismic distribution with geodetic inferences of slipping and locked regions. Additional work will be needed for systematic exploration of possible rupture velocity or stress drop variations in these microseismicity data sets.

Nature of Friction on the Shallow Megathrust

The preceding overview of seismic failure characteristics for the shallow megathrust environment reveals substantial complexity. It is possible for large strains and/or slip deficits to accumulate in the shallow megathrust environment, for deeper slip events to induce slip at shallow depths, for slow shallow slip to precede deeper thrust faulting, and for slip to occur in large earthquakes at very shallow depth with almost no small events rupturing the same region. Clearly, the frictional properties of this region are not simply governed by pressure or temperature variations alone, with it being likely that bathymetric structures, sediments, pore-water pressures, and strain rates all play roles in the seismic behavior.

Estimates of the temperatures near the updip limit beneath sedimentary wedges are around 100°–150°C [e.g., *Hyndman et al.*, 1997; *Oleskevich et al.*, 1999], corresponding to the temperatures at which smectite undergoes most of its transformation to illite/chlorite [e.g., *Pytte and Reynolds*, 1989] along with occurrence of other diagenetic to low-grade metamorphic and consolidation phenomena that can affect fluid pressures in the megathrust (fig. 15.7; some of these processes are discussed elsewhere in this volume). The progressive dehydration of smectite along with consolidation and lithification of the prism sediments have thus been proposed as important mechanisms in the change in frictional properties associated with the onset of faulting [e.g., *Vrolijk*, 1990; *Morrow et al.*, 1992; *Moore and Saffer*, 2001]. Recent experimental work suggests that illite actually does not exhibit the velocity-weakening behavior necessary for seismic slip, making it unlikely that the smectite to illite transition is directly responsible for the updip limit of seismicity in subduction zones [*Saffer and Marone*, 2003]. Smectite itself may exhibit unstable sliding as well [*Saffer et al.*, 2001]. Thus more work is needed in order to understand other possible mechanisms and their role in controlling the updip seismicity limit.

Shallow megathrust faulting is commonly associated with low rupture velocity, for both large and small events. This observation raises the question of whether there is any effect of rupture in proximity to the free surface that influences faulting. The mechanics of dip-slip rupture are likely to be influenced if rupture extends to the surface, where there will essentially be a zero stress intensity condition. This will skew the dip-directed slip distribution in the updip direction [e.g., *Dmowska and Kostrov*, 1973; *Tse and Rice*, 1986; *Rudnicki and Wu*, 1995], with the large water depth near the trench enhancing the potential energy for excitation of tsunami gravity waves, as noted by *Geist and Bilek* [2001]. Even if seismic slip ends at the seismic front, anelastic seaward shoving of the shallow sedimentary wedge could effectively extend rupture to the surface, as in the idea of *Tanioka and Seno* [2001].

In the absence of specific knowledge about materials controlling friction in the shallow megathrust, it is still possible to characterize the overall behavior. The transition from stable sliding to unstable sliding appears to involve an intermediate behavior of conditional stability [e.g., *Scholz*, 1990, 1998]. Conditionally stable materials will typically undergo steady aseismic sliding but may fail coseismically if loaded at high strain rate. Earthquakes only nucleate in unstable zones, with their slip propagating into conditionally stable zones if they involve a strong enough velocity jump [*Scholz*, 1998]. Variation of the stability regime with depth and laterally with varying state variables gives

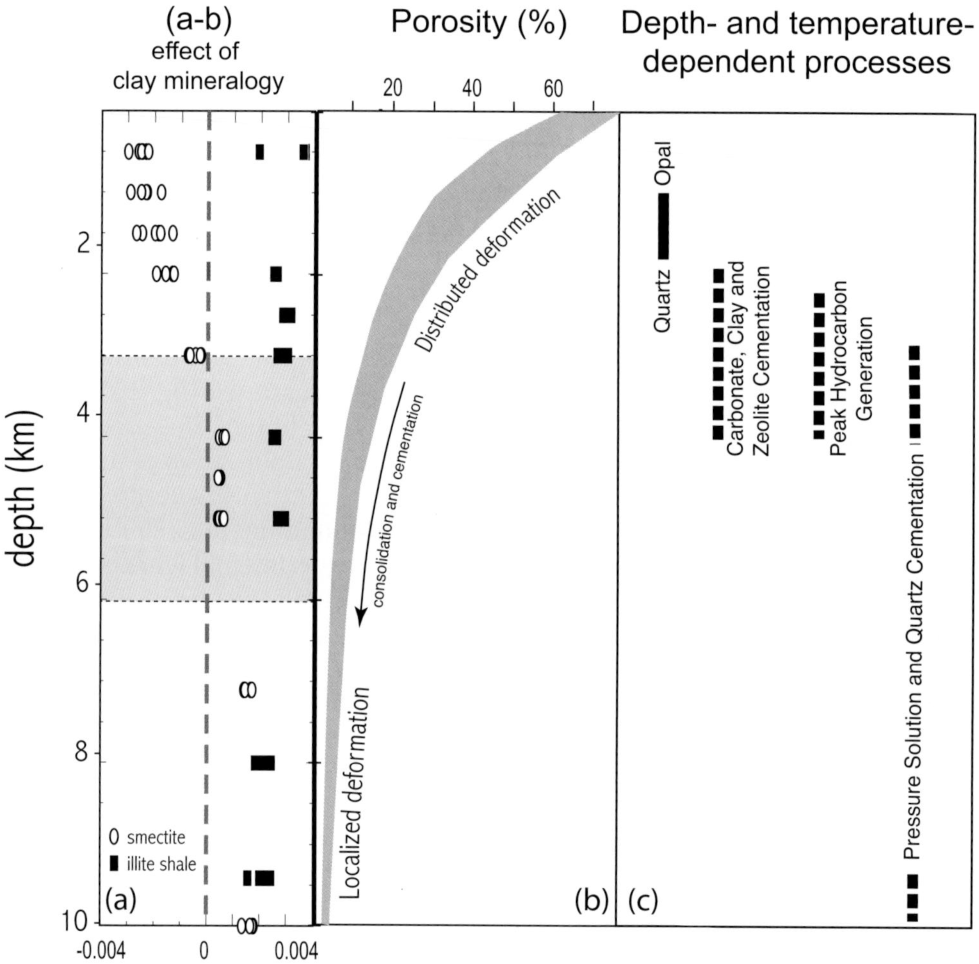

Figure 15.7 (a) Frictional velocity dependence for smectite (open circles) and illite shale (black squares) for up-step velocities of 20 mm/s, plotted with increasing depth assuming that effective stress increases by 10 MPa/km. Grey shaded region shows depth range for smectitie transformation. Dashed line denotes velocity-neutral behavior. Note that both materials exhibit only velocity-strengthening behavior at normal stresses above those corresponding to 4-km depth, and illite shale is velocity strengthening at all depths. (b) Porosity loss and associated transition from distributed deformation to localized slip hypothesized to explain updip limit of seismicity in accretionary wedges [e.g., *Moore and Saffer*, 2001]. (c) A suite of diagenetic and dehydration processes, including quartz phase transformation, cementation, fluid release from hydrous clays, and hydrocarbon generation and pressure. Courtesy of D. Saffer (2004)

rise to one-dimensional and two-dimensional characterizations of friction on the megathrust as shown in figure 15.8. *Scholz* [1998] presents a simple one-dimensional depth-varying model (fig. 15.8a), recognizing that there are going to be many lateral variations in properties. This suggests that the shallow

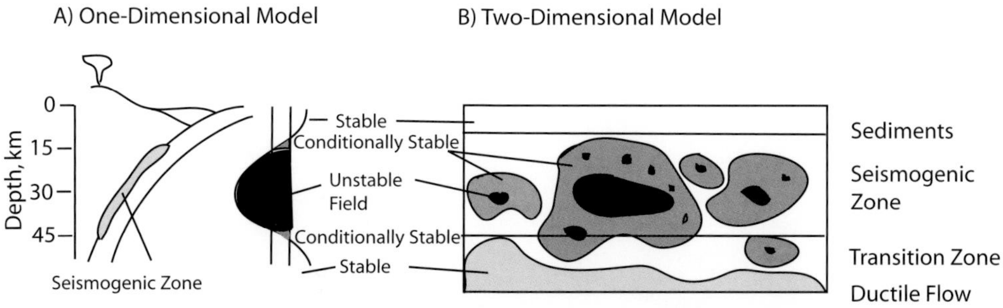

Figure 15.8 Conceptual models of frictional stability zones in subduction zones. A one-dimensional cartoon is shown on the left [modified from *Scholz*, 1999], emphasizing the transitions from stable sliding to unstable sliding involving zones of conditional stability at the updip and downdip edges of the seismogenic zone. A two-dimensional cartoon is shown on the right [modified from *Pacheco et al.*, 1993], emphasizing lateral variations along strike of the megathrust, with patches of unstable regime surrounded by conditionally stable and stable environments.

seismic zone just downdip from the aseismic front is likely to be conditionally stable. Shallow slip triggered by deeper strong faulting, as observed in many regions, would readily be accounted for. Tsunami earthquakes would involve slip within the conditionally stable regime in this case, presumably triggered by onset of slip at depth in the unstable regime. This is compatible with the updip rupture observed for several of the recent tsunami earthquakes such as the 1992 Nicaragua event [*Polet and Kanamori*, 2000]. *Pacheco et al.* [1993] envision the conditionally stable regime as surrounding patchy areas of unstable sliding regime throughout the seismogenic zone (fig. 15.8b). This is motivated by the need to accommodate large amounts of aseismic convergence in almost all regions. This is essentially a frictional stability regime perspective of the asperity model that has been used to discuss many earthquake ruptures.

A puzzle remains as to why very large strains/slip deficits accumulate in the conditionally stable zone to be released in large tsunami earthquakes. Given continuous convergence (with both slab-pull and ridge-push forces serving as driving forces), some type of buttressing of the conditional stability zone is needed to prevent slip from simply occurring at the plate motion rate. Downdip locked zones that are accumulating strain could allow some updip slip deficit to develop, with updip slip then occurring during large events at greater depth or possibly separately, triggered by minor slip of the downdip region. Localized regions of shallow frictional instability within the shallowest portion of the megathrust, associated with subducted bathymetric features such as seamounts or horst and graben structures (fig. 15.9) could also account for accumulation of strain [e.g., *Tanioka et al.*, 1997; *Polet and Kanamori*, 2000].

It is also not evident why rupture in conditionally stable material will proceed at lower rupture velocity than in the unstable regime. The seismic and tsunami observations for large shallow tsunami events favor low source

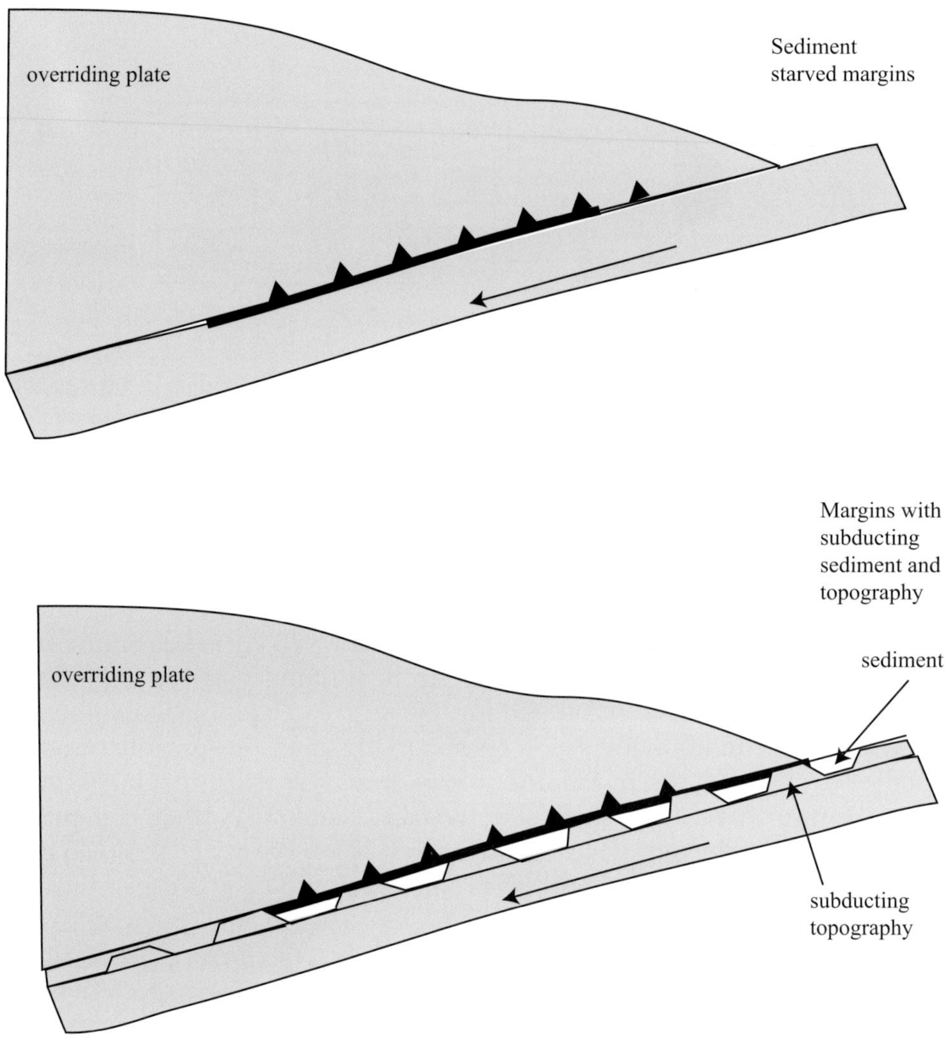

Figure 15.9 Bathymetric roughness in the form of seamounts or horst/graben structures, combined with sediment supply may have strong affects on the shallow megathrust environment. Sediment-starved margins may have no accretionary wedge, and if the bathymetry is relatively smooth, there may be uniform coupling of the interface with some sediments on the interface; rupture in these regions could extend to the trench. Regions with rough bathymetry and substantial sediment supply could have localized regions of unstable conditions within the conditionally stable zone.

zone rigidity, and this is compatible with the observations of long source process time and the few direct determinations of low rupture velocity. For smaller events the rupture durations are compatible with low source region rigidity as well, although low stress drop cannot be ruled out at this time, and it is possible to lower the particle slip velocity rather than the rupture velocity to account for long processes times as well [e.g., *Polet and Thio*, 2003]. It is reasonable to associate the very low rigidities with the presence of sediments in the shallow megathrust, as these materials are expected to have low rigidities prior to dehydration and induration [e.g., *Bray and Karig*, 1985; *Schreve and Cloos*, 1986; *von Huene and Scholl*, 1991; *Kanamori and Kikuchi*, 1993; *Polet and Kanamori*, 2000]. The rigidity of sediments is dependent on sediment composition, water saturation, porosity, and pressure, with order of magnitude variations being observed in the laboratory [e.g., *Guerin and Goldberg*, 1996; *Mavko et al.*, 1998]. The role of sediments may be twofold: such materials may tend to develop conditionally stable frictional behavior progressively with depth, and their low rigidities may control the overall strain release behavior by delimiting rupture velocity. Thus major aspects of anomalous faulting in the shallow megathrust may be explained. It remains difficult to assess subducted sediment budgets and to relate them to megathrust faulting [*Ruff*, 1989; *von Huene and Scholl*, 1991; *Rea and Ruff*, 1996; *Polet and Kanamori*, 2000]; however, the evidence for large slip in low rigidity material at shallow depth is strong.

Sediments may play the additional important role of controlling fluid content in the shallow megathrust. Many studies of shallow sedimentary conditions suggest that fluid pressures in the accretionary prism may be high and laterally variable [e.g., *Moore and Vrolijk*, 1992; *Tobin et al.*, 1994; *Saffer and Bekins*, 1998; *Bekins and Screaton*, this volume; *Saffer*, this volume]. Presumably, heterogeneity in the permeability and pore pressures extends to variable depths in the megathrust, and *Pacheco et al.* [1993] invoke permeability variations as a primary state variable influencing heterogeneity in the frictional stability regime in their models. Fluids and weak properties of the fault zone are likely to play important roles in dynamic faulting processes [e.g., *Kanamori and Heaton*, 1999; *Brodsky and Kanamori*, 2001], which may provide an additional mechanism for distinctive rupture properties of very shallow megathrust events relative to deeper interplate thrusting. *Seno* [2002] suggests that increases of the area of zones of elevated fluid pressure may cause transition in the frictional sliding properties of subducted unconsolidated sediments. His idea is that transient changes in frictional stability occur owing to sealing and pressure buildup of the pore fluids, reducing effective normal stress. If pore pressures approach 90% of the lithostatic normal stress and the tectonic stress is not large, smectite could enter the unstable frictional regime [*Saffer et al.*, 2001]. In this perspective the updip rupture extent for larger interplate thrusts may be influenced by the fluid state of the accretionary contact, with rupture either terminating far from the trench or diverting into splay faults or with rupture extending toward the trench resulting in strong tsunami excitation. Fluctuations in fluid pressure may thus help to account for the lack of shallow slip in the conditionally stable zone for many large megathrust ruptures and for subsequent rupture of those zones in tsunami earthquakes.

Figure 15.10, adapted from *Bilek and Lay* [2002], summarizes the frictional attributes of the shallow megathrust in a simplified form. This should be viewed as a snapshot in a steadily converging system loaded by slab pull and ridge-push, with frictional interactions across the megathrust being responsible for variations in slip on the interface. When there are off-scraped sediments in an accretionary wedge, there will typically be a shallow zone of aseismic stable sliding, with slip keeping pace with the plate convergence. Even large slip events at shallow depth will not rupture through this region as it will serve as a strain energy sink, although anelastic plowing of the material may occur. As pressure and temperature increase, the shallow megathrust enters the conditionally stable regime, possibly with localized patches of unstable frictional zones associated with bathymetric features such as seamounts, horst/graben structures, and variations in sediment permeability and pore pressure. The conditionally stable regime is heterogeneous laterally, and large tsunami earthquakes may induce large slip in this region that varies along strike (as for the 1992 Nicaragua event). Failure may initiate at the downdip edge of this region, as for the 1946 Aleutian and 1992 Nicaraguan events, or it may initiate on a patch of unstable friction such as an asperity, as in the 1994 Java event. In other regions this zone may slip continuously and aseismically. At greater

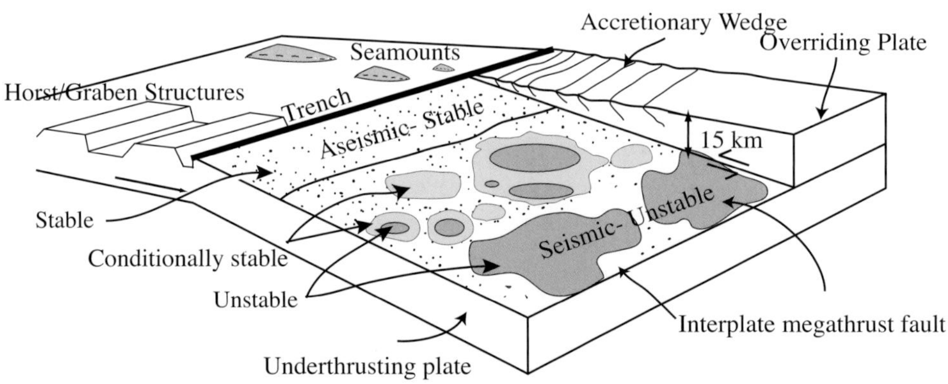

Figure 15.10 Cartoon of frictional properties on the shallow megathrust combining downdip and along-strike properties of the cartoons in figure 15.8. The shallowest region of the megathrust may be aseismic, with stable sliding if there are sediments accumulating in a sedimentary wedge of any significance. The upper 10 km or so of the megathrust seismogenic zone is dominated by stable and conditionally stable conditions, possibly with intermittent patches of unstable friction associated with variations in bathymetry, sediment, or fluid conditions. At greater depth the megathrust has an increasing percentage of unstable frictional conditions, although the percentage varies greatly from region to region. The unstable zones appear to be patchy, surrounded by conditionally stable or stable environments, giving rise to asperity type behavior of the large earthquake ruptures. Modified from *Bilek and Lay* [2002].

depth a larger percentage of the fault zone is in the unstable frictional regime, although there is still substantial aseismic slip in this domain in many subduction zones. Progressing from this cartoon notion will require improved constraints on the frictional properties of sediments, improved characterization of the fluid pressures at depth in the megathrust, and, ideally, in situ constraints on the detailed properties of the shallow megathrust.

Conclusions

The shallow portion of the megathrust in subduction zones exhibits complex seismogenic properties. While generally having low levels of seismicity, large slip events can occur at very shallow depth, and the resulting large ocean bottom displacement generates large tsunamis. The effective source region rigidity for these events is 2–4 times lower than typical hard rock values, suggesting that low rigidity sediments are on the fault plane. This is consistent with unusually long source process times and low rupture velocities documented for tsunami earthquakes. Slip in the tsunami earthquakes may extend updip of the aftershock zones and of the seismic front as defined by smaller events. Moderate size events on the megathrust tend to show some depth-dependent variations in rupture as well, with longer rupture durations found for events in the upper 10 km of the seismogenic zone. This is likely to reflect low rigidity of the shallow fault zone, increasing with depth due to sediment induration, fluid expulsion, and phase transformations in the sediments. The shallow megathrust appears to have a variable distribution of unstable and stable friction, with extensive zones of conditionally stable friction. Temporal fluctuations may occur due to variations in fluid properties, complicating the interpretation of this region.

Acknowledgments

We made extensive use of the Harvard CMT catalog. GMT software [*Wessel and Smith*, 1991] was used for some figure preparation. Kelin Wang provided a thorough review of the manuscript. All data from the IRIS Global Seismic Network were obtained from the IRIS DMS. This work was supported by NSF EAR 0125595. This is contribution number 474 of the Center for the Study of Imaging and Dynamics of the Earth, IGPP, University of California, Santa Cruz.

References

Abe, K. (1979), Size of great earthquakes of 1873–1974 inferred from tsunami data, *J. Geophys. Res.*, *84*, 1561–1568.

Abe, K. (1981), Physical size of tsunamigenic earthquakes of the northwestern Pacific, *Phys. Earth Planet Inter.*, *27*, 194–205.

Abercrombie, R. E., M. Antolik, K. Felzer, and G. Ekström (2001), The 1994 Java tsunami earthquake: Slip over a subducting seamount, *J. Geophys. Res., 106*, 6595–6607.

Beck, S., and L. Ruff (1989), Great earthquakes and subduction along the Peru trench, *Phys. Earth Planet. Int., 57*, 199–224.

Beckers, J., and T. Lay (1995), Very broadband seismic analysis of the 1992 Flores, Indonesia, earthquake (M_w = 7.9), *J. Geophys. Res., 100*, 18,179–18,193.

Bekins, B. A., and E. J. Screaton (2007), Pore pressure and fluid flow in the northern Barbados accretionary complex: A synthesis, this volume.

Bilek, S. L. (2001), Earthquake rupture processes in circum-Pacific subduction zones. Ph.D. thesis, 180 pp., Univ. Calif., Santa Cruz.

Bilek, S. L. (2007), Influence of subducting topography on earthquake rupture, this volume.

Bilek, S. L., and T. Lay (1998), Variation of interplate fault zone properties with depth in the Japan subduction zone, *Science, 281*, 1175–1178.

Bilek, S. L., and T. Lay (1999a), Comparison of depth dependent fault zone properties in the Japan trench and Middle America trench, *Pure Appl. Geophys., 154*, 433–456.

Bilek, S. L., and T. Lay (1999b), Rigidity variations with depth along interplate megathrust faults in subduction zones, *Nature, 400*, 443–446.

Bilek, S. L., and T. Lay (2000), Depth dependent rupture properties in circum-Pacific subduction zones, in *Geocomplexity and the Physics of Earthquakes, Geophys. Monogr. Ser.*, vol. 120, edited by J. B. Rundle, D. L. Turcotte, and W. Klein, pp. 165–186. AGU, Washington, D. C.

Bilek, S. L., and T. Lay (2002), Tsunami earthquakes possibly widespread manifestations of frictional conditional stability, *Geophys. Res. Lett., 2*(14), 1673, doi:10.1029/2002GL015215.

Bilek, S. L., T. Lay, and L. J. Ruff (2004), Radiated seismic energy and earthquake source duration variations from teleseismic source time functions for shallow subduction zone thrust earthquakes, *J. Geophys. Res., 109*, B09308, doi:10.1029/2004JB003039.

Bourgeois, J., C. Petroff, H. Yeh, V. Titov, C. E. Synolakis, B. Benson, J. Kuroiwa, J. Lander, and E. Norabuena (1999), Geologic setting, field survey and modeling of the Chimbote, northern Peru, tsunami of 21 February 1996, *Pure Appl. Geophys., 154*, 513–540.

Bray, C., and D. Karig (1985), Porosity of sediments in accretionary prisms and some implications for dewatering processes, *J. Geophys. Res., 90*, 768–778.

Brodsky, E. E., and H. Kanamori (2001), Elastohydrodynamic lubrication of faults, *J. Geophys. Res., 106*, 16,357–16,374.

Byrne, T., D. Davis, and L. Sykes (1988), Loci and maximum size of thrust earthquakes and the mechanics of the shallow region of subduction zones, *Tectonics, 7*, 833–857.

Choy, G., and J. Boatwright (1995), Global patterns of radiated seismic energy and apparent stress, *J. Geophys. Res., 100*, 18,205–18,228.

Cloos, M. (1992), Thrust-type subduction zone earthquakes and seamount asperities: A physical model for seismic rupture, *Geology, 20*(7), 601–604.

Cloos, M., and R. L. Shreve (1996), Shear-zone thickness and the seismicity of Chilean- and Marianas-type subduction zones, *Geology, 24*(2), 107–110.

Cummins, P. R., and Y. Kaneda (2000), Possible splay fault slip during the 1946 Nankai earthquake, *Geophys. Res. Lett., 27*, 2725–2728.

DeShon, H. R., S. Y. Schwartz, S. L. Bilek, L. M. Dorman, V. Gonzalez, J. M. Protti, E. R. Flueh, and T. H. Dixon (2003), Seismogenic zone structure of the southern Middle America trench, Costa Rica, *J. Geophys. Res., 108*(B10), 2491, doi:10.1029/ 2002JB002294.

Dmowska, R., and B. V. Kostrov (1973), A shearing crack in a semi-space under plane strain conditions, *Arch. Mech., 25*, 421–440.

Dziewonski, A., and D. L. Anderson (1981), Preliminary reference earth model, *Phys. Earth Planet. Int., 25*, 297–356.

Dziewonski, A. M., G. Ekström, and M. P. Salganik (1993), Centroid-moment tensor solutions for July-September 1992, *Phys. Earth Planet Int., 79*, 287–297.

Fukao, Y. (1979), Tsunami earthquakes and subduction processes near deep-sea trenches, *J. Geophys. Res., 84,* 2303–2314.

Fukao, Y., and K. Kanjo (1980), A zone of low-frequency earthquakes beneath the inner wall of the Japan Trench, *Tectonophysics, 67,* 153–162.

Geist, E. L. (1998), Local tsunamis and earthquake source parameters, *Adv. Geophys., 39,* 117–209.

Geist, E. L. (2000), Origin of the 17 July 1998 Papua New Guinea tsunami: Earthquake or landslide? *Seismol. Res. Lett., 71,* 344–350.

Geist, E. L., and S. Bilek (2001), Effect of depth-dependent shear modulus on tsunami generation along subduction zones, *Geophys. Res. Lett., 28,* 1315–1318.

Guerin, G., and D. Goldberg (1996), Acoustic and elastic properties of calcareous sediments across a siliceous diagenetic front on the eastern U.S. continental slope, *Geophys. Res. Lett., 23,* 2697–2700.

Gutscher, M.-A., and S. M. Peacock (2003), Thermal models of flat subduction and the rupture zone of great subduction earthquakes, *J. Geophys. Res., 108*(B1), 2009, doi:10.1029/2001JB000787.

Harlow, D. H., R. A. White, I. L. Cifuentes, and A. Aburto (1981), Quiet zone within a seismic gap near western Nicaragua: Possible location of a future large earthquake, *Science, 213,* 648–651.

Hartog, R., and S. Schwartz (1996), Directivity analysis of the December 28, 1994 Sanriku-oki earthquake ($M_w = 7.7$), Japan, *Geophys. Res. Lett., 23,* 2037–2040.

Hasegawa, H. S., and H. Kanamori (1987), Source mechanism of the magnitude 7.2 Grand Banks earthquake of November 1927: Double couple or submarine landslide? *Bull. Seismol. Soc. Am., 77,* 1984–2004.

Heinrich, P., F. Schindele, S. Guibourg, and P. F. Ihmlé (1998), Modeling of the February 1996 Peruvian tsunami, *Geophys. Res. Lett., 25,* 2687–2690.

Heki, K., and Y. Tamura (1997), Short term afterslip in the 1994 Sanriki-Haruka-Oki earthquake, *Geophys. Res. Lett., 24,* 3285–3288.

Heki, K., S. Miyazaki, and H. Tsuji (1997), Silent fault slip following an interplate thrust earthquake at the Japan trench, *Nature, 386,* 595–598.

Hidayat, D., J. S. Barker, and K. Satake (1995), Modeling the seismic source and tsunami generation of the December 12, 1992, Flores island, Indonesia, earthquake, *Pure Appl. Geophys., 144,* 537–554.

Hino, R., T. Kanazawa, and A. Hasegawa (1996), Interplate seismic activity near the northern Japan Trench deduced from ocean bottom and land-based seismic observations, *Phys. Earth Planet. Inter., 93,* 37–52.

Hirata, K., E. Geist, K. Satake, Y. Tanioka, and S. Yamaki (2003), Slip distribution of the 1952 Tokachi-Oki earthquake (M 8.1) along the Kuril Trench deduced from tsunami waveform inversion, *J. Geophys. Res., 108*(B4), 2196, doi:10.1029/2002JB001976.

Houston, H. (2001), Influence of depth, focal mechanism, and tectonic setting on the shape and duration of earthquake source time functions, *J. Geophys. Res., 106,* 11,137–11,150.

Husen, S., E. Kissling, E. Flueh, and G. Asch (1999), Accurate hypocentre determination in the seismogenic zone of the subducting Nazca Plate in northern Chile using a combined on-/offshore network, *Geophys. J. Int., 138,* 687–701.

Husen, S., E. Kissling, and R. Quintero (2002), Tomographic evidence for a subducted seamount beneath the Gulf of Nicoya, Costa Rica: The cause of the 1990 $M_w = 7.0$ Gulf of Nicoya earthquake, *Geophys. Res. Lett., 29*(8), 1238, doi:10.1029/2001GL014045.

Hyndman, R. D., and K. Wang (1993), Thermal constraints on the zone of major thrust earthquake failure: The Cascadia subduction zone, *J. Geophys. Res., 98,* 2039–2060.

Hyndman, R. D., K. Wang, and M. Yamano (1995), Thermal constraints on the seismogenic portion of the southwestern Japan subduction thrust, *J. Geophys. Res., 100,* 15,373–15,392.

Hyndman, R. D., M. Yamano, and D. A. Oleskevich (1997), The seismogenic zone of subduction thrust faults, *Island Arc, 6*, 244–260.

Ichinose, G. A., H.-K. Thio, P. G. Somerville, T. Sato, T., and T. Ishii (2003), Rupture process of the 1944 Tonankai earthquake (M_s 8.1) from the inversion of teleseismic and regional seismograms, *J. Geophys. Res. 108*(B10), 2497, doi:10.1029/2003JB002393.

Ide, S., F. Imamura, Y. Yoshida, and K. Abe (1993), Source characteristics of the Nicaraguan tsunami earthquake of September 2, 1992, *Geophys. Res. Lett., 20*, 863–866.

Igarashi, T., T. Matsuzawa, and A. Hasegawa (2003), Repeating earthquakes and interplate aseismic slip in the northeastern Japan subduction zone, *J. Geophys. Res., 108*(B5), 2249, doi:10.1029/2002JB001920.

Iglesias, A., S. K. Singh, J. F. Pacheco, L. Alcántara, M. Ortiz, and M. Ordaz (2003), Near-trench Mexican earthquakes have anomalously low peak accelerations, *Bull. Seismol. Soc. Am., 93*, 953–959.

Ihmlé, P. F. (1994), Teleseismic study of rupture processes with long duration, Ph.D. thesis, Mass. Inst. of Technol., Cambridge.

Ihmlé, P. F. (1996a), Frequency-dependent relocation of the 1992 Nicaragua slow earthquake: An empirical Green's function approach, *Geophys. J. Int., 127*, 75–85.

Ihmlé, P. F. (1996b), Monte Carlo slip inversion in the frequency domain: Application to the 1992 Nicaragua slow earthquake, *Geophys. Res. Lett., 23*, 913–916.

Ihmlé, P. F., J.-M. Gomez, P. Heinrich, and S. Guibourg (1998), The 1996 Peru tsunamigenic earthquake: Broadband source process, *Geophys. Res. Lett., 25*, 2691–2694.

Imamura, F. M., and Kikuchi (1994), Moment release of the 1992 Flores Island earthquake inferred from tsunami and teleseismic data, *Sci. Tsunami Hazards, 12*, 67–76.

Imamura, F., E. Gica, T. Takahashi, T., and N. Shuto (1995), Numerical simulation of the 1992 Flores tsunami: Interpretation of tsunami phenomena in northeastern Flores Island and damage at Babi Island, *Pure Appl. Geophys., 144*, 555–568.

Imamura, F., N. Shuto, S. Ide, Y. Yoshida, and K. Abe (1993), Estimate of the tsunami source of the 1992 Nicaraguan earthquake from tsunami data, *Geophys. Res. Lett., 20*, 1515–1518.

Jarrard, R. (1986), Relation among subduction parameters, *Rev. Geophys., 24*, 217–284.

Johnson, J. M., and K. Satake (1996), The 1965 Rat Islands earthquake: A critical comparison of seismic and tsunami wave inversions, *Bull. Seismol. Soc. Am., 86*, 1229–1237.

Johnson, J. M., and K. Satake (1997), Estimation of seismic moment and slip distribution of the April 1, 1946, Aleutian tsunami earthquake, *J. Geophys. Res., 102*, 11,765–11,779.

Johnson, J. M., and K. Satake (1999), Asperity distribution of the 1952 great Kamchatka earthquake and its relation to future earthquake potential in Kamchatka, *Pure Appl. Geophys., 154*, 541–553.

Johnson, J. M., K. Satake, S. R. Holdahl, and J. Sauber (1996), The 1964 Prince William Sound earthquake: Joint inversion of tsunami and geodetic data, *J. Geophys. Res., 101*, 523–532.

Kanamori, H. (1972), Mechanism of tsunami earthquakes, *Phys. Earth Planet. Int., 6*, 246–259.

Kanamori, H. (1977), Seismic and aseismic slip along subduction zones and their tectonic implications, in *Island Arcs, Deep Sea Trenches and Back-Arc Basins, Maurice Ewing Series,* vol. 1, edited by M. Talwani and W. C. Pittman, pp. 163–174. AGU, Washington, D. C.

Kanamori, H. (1985), Non double-couple seismic source, in XXIIIrd General Assembly, vol. 2, p. 425, Abstract S7-56, Int. Assoc. of Seismol. and Phys. of Earth Inter., Tokyo, Japan.

Kanamori, H. (1986), Rupture process of subduction-zone earthquakes, *Annu. Rev. Earth Planet. Sci., 14*, 293–322.

Kanamori, H. (1993), W phase, *Geophys Res. Lett., 20*, 1691–1694.

Kanamori, H., and T. H. Heaton (1999), Microscopic and macroscopic physics of earthquakes, in *Geocomplexity and the Physics of Earthquakes, Geophys. Monogr. Ser.,* vol. 120, edited by J. B. Rundle, D. L. Turcotte, and W. Klein, pp. 147–163, AGU, Washington, D. C.

Kanamori, H., and M. Kikuchi (1993), The 1992 Nicaragua earthquake: A slow tsunami earthquake associated with subducted sediments, *Nature, 361*, 714–716.

Kawasaki, I., Y. Asai, Y. Tamura, T. Sagiya, N. Mikami, Y. Okada, M. Sakata, and M. Kasahara (1995), The 1992 Sanriku-Oki, Japan, ultra-slow earthquake, *J. Phys. Earth, 43*, 105–116.
Kawata, Y., B. Benson, J. Borrero, H. Davies, W. de Lange, F. Imamura, H. Letz, J. Nott, and C. Synolakis (1999), Tsunami in Papua New Guinea was as intense as first thought, *Eos Trans. AGU, 80*, 101–105.
Kelleher, J., and W. McCann (1976), Buoyant zones, great earthquakes and unstable boundaries of subduction, *J. Geophys. Res., 81*, 4885–4896.
Kikuchi, M., and Y. Fukao (1987), Inversion of long-period P waves from great earthquakes along subduction zones, *Tectonophysics, 144*, 231–247.
Kikuchi, M., and H. Kanamori (1995), Source characteristics of the 1992 Nicaragua tsunami earthquake inferred from teleseismic body waves, *Pure Appl. Geophys., 144*, 441–453.
Kikuchi, M., Y. Yamanaka, K. Abe, and Y. Morita (1999), Source rupture process of the Papua New Guinea earthquake of 17 July 1998 inferred from teleseismic body waves, *Earth Planets Space, 51*, 1319–1324.
Kodaira, S., N. Takahashi, A. Nakanishi, S. Miura, and Y. Kaneda (2000), Subducted seamount imaged in the rupture zone of the 1946 Nankaido earthquake, *Science, 289*, 104–106.
Lay, T., and H. Kanamori (1981), An asperity model of great earthquake sequences, in *Earthquake Prediction: An International Review, Maurice Ewing Ser.*, vol. 4., edited by D. Simpson and P. Richards, pp. 579–592. AGU, Washington, D. C.
Lay, T., H. Kanamori, and L. Ruff (1982), The asperity model and the nature of large subduction zone earthquakes, *Earthquake Pred. Res., 1*, 3–71.
Ma, K.-F., K. Satake, and H. Kanamori (1991), The origins of the tsunami excited by the 1989 Loma Prieta earthquake--faulting or slumping, *Geophys. Res. Lett., 18*, 637–640.
Marone, C., and D. Saffer (2007), Fault friction and the upper transition from seismic to aseismic faulting, this volume.
Marone, C., and C. H. Scholz (1988), The depth of seismic faulting and the upper transition from stable to unstable slip regimes, *Geophys. Res. Lett., 15*, 621–624.
Matsuyama, M., J. P. Walsh, and H. Yeh (1999), The effect of bathymetry on tsunami characteristics at Sissano Lagoon, Papua New Guinea, *Geophys. Res. Lett., 26*, 3513–3516.
Matsuzawa, T., T. Igarashi, and A. Hasegawa (2002), Characteristic small-earthquake sequence off Sanriku, northeastern Honshu, Japan, *Geophys. Res. Lett., 29*(11), 1543, doi:10.1029/2001GL014632.
Matsuzawa, T., N. Umino, A. Hasegawa, M. Kosuga, K. Tanaka, K., and M. Kasahara (1995), Aftershock activity of the 1994 far off Sanriku earthquake observed by a microearthquake observation network (in Japanese), Rep. 06306019, Sci. Res. Fund by the Minist. of Educ. Sci. and Culture of Japan.
Mavko, G., T. Mukerji, T., and J. Dvorkin (1998), *The Rock Physics Handbook: Tools for Seismic Analysis in Porous Media*, Cambridge Univ. Press, New York.
McCaffrey, R. (1993), On the role of the upper plate in great subduction zone earthquakes, *J. Geophys. Res., 98*, 11,953–11,966.
McCaffrey, R. (1997), Statistical significance of the seismic coupling coefficient, *Bull. Seismol. Soc. Am., 87*, 1069–1073.
Mendoza, C., and S. Hartzell (1999), Fault-slip distribution of the 1995 Colima-Jalisco, Mexico, earthquake, *Bull. Seismol. Soc. Am., 89*, 1338–1344.
Moore, J. C., and D. Saffer (2001), The updip limit of the seismogenic zone beneath the accretionary prism of southwest Japan: An effect of diagenetic to low-grade metamorphic processes and increasing effective stress, *Geology, 29*(2), 183–186.
Moore, J. C., and P. Vrolijk (1992), Fluids in accretionary prisms, *Rev. Geophys., 30*, 113–135.
Morrow, C. A., B. Radney, and J. Byerlee (1992), Frictional strength and the effective pressure law of montmorillonite and illite clays, in *Fault Mechanics and Transport Properties of Rocks*, edited by B. Evans, pp. 69–88, Elsevier, New York.
Nakayama, W., and M. Takeo (1997), Slip history of the 1994 Sanriku-Haruka-Oki, Japan earthquake deduced from strong-motion data, *Bull. Seismol. Soc. Am., 87*, 918–931.

Newman, A. V., and E. A. Okal (1998a), Teleseismic estimates of radiated seismic energy: The E/Mo discriminant for tsunami earthquakes, *J. Geophys. Res., 103*, 26, 885–26,898.
Newman, A. V., and E. A. Okal (1998b), Moderately slow character of the 17 July 1998 Sandaun earthquake as studied by teleseismic energy estimates, *Eos Trans. AGU*, 79, Fall Meet. Suppl., F564.
Newman, A. V., S. Y. Schwartz, V. Gonzalez, H. R. DeShon, J. M. Protti, and L. M. Dorman (2002), Along-strike variability in the seismogenic zone below Nicoya Peninsula, Costa Rica, *Geophys. Res. Lett., 29*(20), 1977, doi:10.1029/2002GL015409.
Nishizawa, A., T. Kono, A. Hasegawa, T. Hirasawa, T. Kanazawa, and T. Iwasaki (1990), Spatial distribution of earthquake off Sanriku, northeaster Japan, in 1989 determined by ocean-bottom and land-based observation, *J. Phys. Earth, 38*, 347–360.
Obana, K., S. Kodaira, Y. Kaneda, K. Mochizuki, M. Shinohara, and K. Suyehiro (2003), Microseismicity at the seaward updip limit of the western Nankai Trough seismogenic zone, *J. Geophys. Res., 108*(B10), 2459, doi:10.1029/2002JB002370.
Okal, E. A. (1988), Seismic parameters controlling far-field tsunami amplitudes: A review, *Nat. Hazards, 1*, 67–96.
Okal, E. A., and A. V. Newman (2001), Tsunami earthquakes: the quest for a regional signal, *Phys. Earth Planet. Inter., 124*, 45–70.
Oleskevich, D. A., R. D. Hyndman, and K. Wang (1999), The updip and downdip limits to great subduction earthquakes: Thermal and structural models of Cascadia, south Alaska, SW Japan, and Chile, *J. Geophys. Res., 104*, 14,965–14,991.
Pacheco, J. F., and L. R, Sykes (1992), Seismic moment catalog for large shallow earthquakes from 1900 to 1989, *Bull. Seismol. Soc., 82*, 1306–1349.
Pacheco, J. F., L. R. Sykes, and C. H. Scholz (1993), Nature of seismic coupling along simple plate boundaries of the subduction type, *J. Geophys. Res., 98*, 14,133–14,159.
Pacheco, J., et al. (1997), The October 9, 1995 Colima-Jalisco, Mexico earthquake (M_w 8): An aftershock study and a comparison of this earthquake with those of 1932, *Geophys. Res. Lett., 24*, 2223–2226.
Park, J. O., T. Tsuru, S. Kodaira, P. R. Cummins, and Y. Kaneda (2002), Splay fault branching along the Nankai subduction zone, *Science, 297*, 1157–1160.
Peacock, S. M., and R. D. Hyndman (1999), Hydrous minerals in the mantle wedge and the maximum depth of subduction thrust earthquakes, *Geophys. Res. Lett., 26*, 2517–2520.
Pelayo, A. M. (1990), Earthquake source parameter inversion using body and surface waves: Applications to tsunami earthquakes and Scotia Sea seismotectonics, Ph.D. thesis, Washington Univ., St. Louis, Mo.
Pelayo, A. M., and D. A. Wiens (1990), The November 20, 1960 Peru tsunami earthquake: Source mechanism of a slow earthquake, *Geophys. Res. Lett., 17*, 661–664.
Pelayo, A. M., and D. A. Wiens (1992), Tsunami earthquakes: Slow thrust-faulting events in the accretionary wedge, *J. Geophys. Res., 97*, 15,321–15,337.
Peterson, E., and T. Seno (1984), Factors affecting seismic moment release rates in subduction zones, *J. Geophys. Res., 89*, 10,233–10,248.
Piatanesi, A., S. Tinti, and I. Gavagni (1996), The slip distribution of the 1992 Nicaragua earthquake from tsunami run-up data, *Geophys. Res. Lett., 23*, 37–40.
Polet, J., and H. Kanamori (2000), Shallow subduction zone earthquakes and their tsunamigenic potential, *Geophys. J. Int., 142*, 684–702.
Polet, J., and H. K. Thio (2003), The 1994 Java tsunami earthquake and its "normal" aftershocks, *Geophys. Res. Lett., 30*(9), 1474, doi:10.1029/2002Gl016806.
Pytte, A M., and R. C. Reynolds (1989), The thermal transformation of smectite to illite, in *Thermal History of Sedimentary Basins: Methods and Case Histories*, edited by Naeser, N. D. and T. H. McCulloh, pp. 33–40, Springer, New York.
Rea, D., and L. J. Ruff (1996), Composition and mass flux of sediment entering the world's subduction zones: Implications for global sediment budgets, great earthquakes and volcanism, *Earth and Planet. Sci. Lett., 140*, 1–12.

Rudnicki, J. W., and M. Wu (1995), Mechanics of up-slip faulting in an elastic half-space, *J. Geophys. Res., 100*, 22,173–22,186.
Ruff, L. J. (1989), Do trench sediments affect great earthquake occurrence in subduction zones? *Pure Appl. Geophys., 129*, 263–282.
Ruff, L. J. (1992), Asperity distributions and large earthquake occurrence in subduction zones, *Tectonophysics, 211*, 61–83.
Ruff, L. J. (1999), Dynamic stress drop of recent earthquakes: Variations within subduction zones, *Pure Appl. Geophys., 154*, 409–431.
Ruff, L. J., and H. Kanamori (1980), Seismicity and the subduction process, *Phys. Earth Planet. Inter., 23*, 240–252.
Ruff, L. J., and H. Kanamori (1983a), Seismic coupling and uncoupling at subduction zones, *Tectonophysics, 99*, 99–117.
Ruff, L. J., and H. Kanamori (1983b), The rupture process and asperity distribution of three great earthquakes from long period diffracted *P*-waves, *Phys. Earth Planet. Int., 31*, 202–230.
Ruff, L. J., and A. D. Miller (1994), Rupture process of large earthquakes in the northern Mexico subduction zone, *Pure Appl. Geophys., 142*, 101–171.
Saffer, D. (2007), Pore pressure within underthrust sediments in subduction zones, this volume.
Saffer, D. M., and B. A. Bekins (1998), Episodic fluid flow in the Nankai accretionary complex: Timescale, geochemistry, flow rates, and fluid budget, *J. Geophys. Res., 103*, 30,351–30,370.
Saffer, D. M., and C. Marone (2003), Comparison of smectite- and illite-rich gouge frictional properties: Application to the updip limit of the seismogenic zone along subduction megathrusts, *Earth Planet. Sci. Lett., 215*, 219–235.
Saffer, D. M., K. M. Frye, C. Marone, and K. Mir (2001), Laboratory results indicating complex and potentially unstable frictional behavior of smectite clay, *Geophys. Res. Lett., 28*, 2297–2300.
Satake, K. (1989), Inversion of tsunami waveforms for the estimation of heterogeneous fault motion of large submarine earthquakes: The 1968 Tokachi-oki and the 1983 Japan Sea earthquakes, *J. Geophys. Res., 94*, 5627–5636.
Satake, K. (1993), Depth distribution of coseismic slip along the Nankai Trough, Japan, from joint inversion of geodetic and tsunami data, *J. Geophys. Res., 98*, 4553–4565.
Satake, K. (1994), Mechanism of the 1992 Nicaragua tsunami earthquake, *Geophys. Res. Lett., 21*, 2519–2522.
Satake, K. (1995), Linear and nonlinear computations of the 1992 Nicaragua earthquake tsunami, *Pure Appl. Geophys., 144*, 455–470.
Satake, K., and Y. Tanioka (1999), Sources of tsunami and tsunamigenic earthquakes in subduction zones, *Pure Appl. Geophys., 154*, 467–483.
Satake, K., and Y. Tanioka (2003), The July 1998 Papua New Guinea earthquake: Mechanism and quantification of unusual tsunami generation, *Pure Appl. Geophys. 160*, 2087–2118.
Satake, K., J. Bourgeois, J., K. Abe, K. Abe, Y. Tsuji, F. Imamura, Y. Iio, H. Katao, E. Noguera, and F. Estrada, (1993), Tsunami field survey of the 1992 Nicaragua earthquake, *Eos Trans. AGU, 74*, 145, 156–157.
Sato, T., K. Imanishi, and M. Kosuga (1996), Three-stage rupture process of the 28 December 1994 Sanriku-oki earthquake, *Geophys. Res. Lett., 23*, 33–36.
Schindele, F., D. Reymond, E. Caucher, and E. Okal (1995), Analysis and automatic processing in near-field of eight 1992–1994 tsunamigenic earthquakes: Improvement towards real-time tsunami warning, *Pure Appl. Geophys. 144*, 381–408.
Scholz, C. (1990), *The Mechanics of Earthquakes and Faulting*, Cambridge Univ. Press, New York.
Scholz, C. (1998), Earthquakes and friction laws, *Nature, 391*, 37–42.
Scholz, C., and J. Campos (1995), On the mechanism of seismic decoupling and back arc spreading at subduction zones, *J. Geophys. Res., 100*, 22,103–22,115.

Scholz, C., and C. Small (1997), The effect of seamount subduction on seismic coupling, *Geology, 25*(6), 487–490.
Schwartz, S. Y., and L. J. Ruff (1985), The 1968 Tokachi-Oki and the 1969 Kurile Islands earthquakes: Variability in the rupture process, *J. Geophys. Res., 90*, 8613–8626.
Schwartz, S. Y., and L. J. Ruff (1987), Asperity distribution and earthquake occurrence in the southern Kurile Islands Arc, *Phys. Earth Planet. Int., 49*, 54–77.
Seno, T. (2000), The 21 September, 1999 Chi-Chi earthquake in Taiwan: Implications for tsunami earthquakes, *Terr. Atmos. Ocean Sci., 11*, 701–708.
Seno, T. (2002), Tsunami earthquakes as a transient phenomena, *Geophys. Res. Lett., 29*(10), 10.1029/2002Gl014868.
Shapiro, N. M., S. K. Singh, and J. Pacheco (1998), A fast and simple diagnostic method for identifying tsunamigenic earthquakes, *Geophys. Res. Lett., 25*, 3911–3914.
Shreve, R. L., and M. Cloos (1986), Dynamics of sediment subduction, melange, formation, and prism accretion, *J. Geophys. Res., 91*, 10, 229–10,245.
Suárez, G., and O. Sánchez (1996), Shallow depth of seismogenic coupling in southern Mexico: Implications for the maximum size of earthquakes in the subduction zone, *Phys. Earth Planet. Inter., 93*, 53–61.
Sykes, L. (1971), Aftershock zones of great earthquakes, seismicity gaps, and earthquake prediction for Alaska and the Aleutians, *J. Geophys. Res., 76*, 8021–8041.
Synolakis, C., P. Liu, G. Carrier, and Y. Yeh (1997), Tsunamigenic seafloor deformations, *Science, 278*, 598–600.
Synolakis, C. E., J. P. Bardet, J. C. Borrero, H. L. Davies, E. A. Okal, E. A. Silver, S. Sweet, and D. R. Tappin (2002), The slump origin of the 1998 Papua New Guinea tsunami, *Proc. R. Soc. London, 458*, 763–789.
Takahasi, N., S. Kodaira, T. Tsuru, J. Park, Y. Kaneda, H. Kinoshita, S. Abe, M. Nishino, and R. Hino (2000), Detailed plate boundary structure off northeast Japan coast, *Geophys. Res. Lett., 27*, 1977–1980.
Tanioka, Y. (1999), Analysis of the far-field tsunamis generated by the 1998 Papua New Guinea earthquake, *Geophys. Res. Lett., 26*, 3393–3396.
Tanioka, Y., and L. J. Ruff (1997), Source time functions, *Seismol. Res. Lett., 68*, 386–400.
Tanioka, Y., and K. Satake (1996a), Fault parameters of the 1896 Sanriku tsunami earthquake estimated from tsunami numerical modeling, *Geophys. Res. Lett., 23*, 1549–1552.
Tanioka, Y., and K. Satake (1996b), Tsunami generation by horizontal displacement of ocean bottom, *Geophys. Res. Lett., 23*, 861–864.
Tanioka, Y., and K. Satake (2001a), Coseismic slip distribution of the 1946 Nankai earthquake and aseismic slips caused by the earthquake, *Earth Planets Space, 53*, 235–241.
Tanioka, Y., and K. Satake (2001b), Detailed coseismic slip distribution of the 1944 Tonankai earthquake estimated from tsunami waveforms, *Geophys. Res. Lett., 28*, 1075–1078.
Tanioka, Y., and T. Seno (2001), The sediment effect on tsunami generation of the 1896 Sanriku tsunami earthquake, *Geophys. Res. Lett., 28*, 3389–3392.
Tanioka, Y., L. Ruff, and K. Satake (1996), The Sanriku-Oki, Japan, earthquake of December 28, 1994 (M_w 7.7): Rupture of a different asperity from a previous earthquake, *Geophys. Res. Lett., 23*, 1465–1468.
Tanioka, Y., L. Ruff, and K. Satake (1997), What controls the lateral variation of large earthquake occurrence along the Japan trench? *Island Arc, 6*, 261–266.
Tappin, D. R., et al. (1999), Sediment slump likely caused 1998 Papua New Guinea tsunami, *Eos Trans. AGU, 80*, 329–340.
Thatcher, W. (1989), Earthquake recurrence and risk assessment in circum-Pacific seismic gaps, *Nature, 341*, 432–434.
Thatcher, W. (1990), Order and diversity in the modes of circum-Pacific earthquake recurrence, *J. Geophys. Res., 95*, 2609–2623.

Tichelaar, B. W., and L. J. Ruff (1991), Seismic coupling along the Chilean subduction zone, *J. Geophys. Res., 96,* 11,997–12,022.

Tichelaar, B. W., and L. J. Ruff (1993), Depth of seismic coupling along subduction zones, *J. Geophys. Res., 98,* 2017–2037.

Titov, V. V., and C. E. Synolakis (1993), A numerical study of wave runup of the September 2, 1992 Nicaraguan tsunami, in *Proceedings of the IUGG/IOC International Tsunami Symposium,* Int. Union Geodesy and Geophysics, Intergov. Oceanographic Commission, Boulder, CO, pp. 627–635.

Tobin, H. J., J. C. Moore, and G. F. Moore (1994), Fluid pressure in the frontal thrust of the Oregon accretionary prism: Experimental constraints, *Geology, 22*(11), 979–982.

Tse, S. T., and J. R. Rice (1986), Crustal earthquake instability in relation to the depth variation of frictional slip properties, *J. Geophys. Res., 91,* 9452–9472.

Tsuji, Y., F. Imamura, H. Matsutomi, C. E. Synolakis, P. T. Nanang Jamadi, S. Harada, S. Han, K. Arai, and B. Cook (1995a), Field survey of the East Java earthquake and tsunami of June 3, 1994, *Pure Appl. Geophys., 144,* 839–854.

Tsuji, Y., H. Matsutomi, F. Imammura, M. Takeo, Y. Kawata, M. Matsuyama, T. Takahashi, and P. Sunarjo Harjadi (1995b), Damage to coastal villages due to the 1992 Flores island earthquake tsunami, *Pure Appl. Geophys., 144,* 481–524.

Uchida, N., T. Matsuzawa, A. Hasegawa, and T. Igarashi (2003), Interplate quasi-static slip off Sanriku, NE Japan, estimated from repeating earthquakes, *Geophys. Res. Lett., 30*(15), 1801, doi:10.1029/2003GL017452.

Umino, N., A. Hasegawa, and T. Matsuzawa (1995), sP depth phase at small epicentral distances and estimated subducting plate boundary, *Geophys. J. Int., 120,* 356–366.

Uyeda, S., and H. Kanamori (1979), Back-arc opening and the mode of subduction, *J. Geophys. Res., 84,* 1049–1061.

van der Hilst, R., and T. Seno (1993), Effects of relative plate motion on the deep structure and penetration depth of slabs below the Izu-Bonin and Mariana island arcs, *Earth Planet. Sci. Lett., 120,* 395–407.

Velasco, A., C. Ammon, T. Lay, and J. Zhang (1994), Imaging a slow bilateral rupture with broadband seismic waves: The September 2, 1992 Nicaraguan tsunami earthquake, *Geophys. Res. Lett., 21,* 2629–2632.

von Huene, R., and D. W. Scholl (1991), Observations at convergent margins concerning sediment subduction, subduction erosion, and the growth of continental crust, *Rev. Geophys., 29,* 279–316.

von Huene, R., D. Klaeschen, and J. Fruehn (1999), Relation between the subducting plate and seismicity associated with the great 1964 Alaska earthquake, *Pure Appl. Geophys., 154,* 575–591.

Vrolijk, P. (1990), On the mechanical role of smectite in subduction zones, *Geology, 18*(8), 703–707.

Ward, S. (1982), Earthquake mechanisms and tsunami generation: The Kurile Islands event of 13 October, 1963, *Bull. Seismol. Soc. Am., 72,* 759–777.

Wessel, P., and W. H. F. Smith (1991), Free software helps map and display data, *Eos Trans. AGU, 72,* 441–445.

Yagi, Y., M. Kikuchi, and T. Nishimura (2003), Co-seismic slip, post-seismic slip, and largest aftershock associated with the 1994 Sanriku-haruka-oki, Japan, earthquake, *Geophys. Res. Lett., 30*(22), 2177, doi:10.1029/2003GL018189.

Yoshii, T. (1979), A detailed cross-section of the deep seismic zone beneath northeastern Honshu, Japan, *Tectonophysics, 55,* 349–360.

Zhang, Z., and S. Y. Schwartz (1992), Depth distribution of moment release in underthrusting earthquakes at subduction zones, *J. Geophys. Res., 97,* 537–544.

Zobin, V. M. (1999), Changes in earthquake source properties across a shallow subduction zone: Kamchatka peninsula, *Pure Appl. Geophys., 154,* 457–466.

16

Secular, Transient, and Seasonal Crustal Movements in Japan from a Dense GPS Array

Implication for Plate Dynamics in Convergent Boundaries

Kosuke Heki

Abstract

The Japanese nationwide dense array of Global Positioning System (GPS) receivers has been yielding large amounts of crustal movement data in Japan over the last decade. In this paper, I review its contribution to our knowledge of crustal dynamics in a subduction zone. The secular velocity field provides information on rigid plate motion and coupling on the plate interface and allows us to discuss seismic moment budgets and recurrence intervals of large interplate earthquakes. The Japanese Islands are divided into two plates colliding with each other in the central Japan, and this causes complex seismicity along the Nankai-Suruga Trough, especially in the Tokai area. Three-dimensional velocity profiles across northeastern Japan show unexpected secular subsidence of the fore-arc region. This might be a manifestation of deep basal subduction erosion of the upper plate by the subducting slab. The GPS array has revealed many coseismic and postseismic fault slips, some of which were slow and aseismic. In northeastern Japan, most interplate earthquakes are followed by after slip over the regions surrounding the ruptured asperities. Acceleration and deceleration of creeping movement of plate interface can also be inferred from activities of small repeating earthquakes, which represent ruptures of small isolated asperities within the region of stable or conditionally stable frictional properties. In southwestern Japan, slow slip tends to occur as independent silent earthquakes. Some of them are quasi-periodic and are often accompanied by nonvolcanic deep tremors. Seasonal crustal movements are caused by changes of various kinds of surface loads, among which snow loading is by far the largest. Understanding seasonal movements is important in order to isolate subtle signals of transient phenomena from the raw data, and to investigate the possible causal relationship between surface loads and seasonal changes in seismic activities.

Introduction

Space geodetic techniques such as very long baseline interferometry (VLBI), satellite laser ranging (SLR), Global Positioning System (GPS), and interferometric synthetic aperture radar (InSAR) have been the main tool to study crustal movements for nearly 20 years. The benefits of GPS among such techniques are (1) the relatively inexpensive ground stations (receiver and antenna), (2) a modest amount of raw data, and (3) a dense sampling in time. Point 1 makes GPS suitable for dense deployment of receiving stations in tectonically active regions, while point 2 enables the exchange of data through Internet or public telephone line for rapid data analysis in a central station. InSAR provides data much denser in space, but owing to the need for almost identical orbits over studied areas, its temporal sampling is sparse. In contrast, point 3 enables GPS receivers to be used as long period seismometers, recording displacements associated with passage of seismic waves [*Larson et al.*, 2003]. In this respect, GPS and InSAR are complementary; i.e., the two techniques are especially powerful when combined.

The dense GPS array, GEONET (GPS Earth Observation Network), deployed over the Japanese Islands by Geographical Survey Institute (GSI) started operation in 1993. The number of sites has been increasing since then and attained ~1200 sites in 2004 (fig.16.1). The RINEX (Receiver Independent Exchange format) raw data, as well as official solutions obtained by the Bernese software with the analysis strategy described by *Hatanaka* [2003b], can be obtained freely from GSI's Web site (http://www.gsi.go.jp).

There is a standard framework to interpret crustal deformation in convergent plate margins. Secular crustal movement from tectonic plate motion is overprinted with interseismic crustal deformation due to elastic loading by the subducting slab. A substantial part of the interseismic displacement accumulated over an earthquake cycle is considered to be canceled instantaneously by coseismic movement of the next earthquake, i.e., coseismic and interseismic movements are approximately opposite in direction and equal in magnitude. Recent geodetic observations have shown that this simple model is valid as a first approximation. However, there are examples that do not match this model. These include transient crustal deformation associated with slow fault slip, seasonal crustal deformation, subsidence not canceled over an earthquake cycle, and so on. In this paper I review crustal dynamics in Japan revealed by GEONET emphasizing such nonstandard phenomena and their geophysical implications.

Secular Crustal Movements

The Japanese Islands used to be thought to reside on the Eurasian Plate (EU), with two oceanic plates, namely the Pacific (PA) and the Philippine Sea (PH)

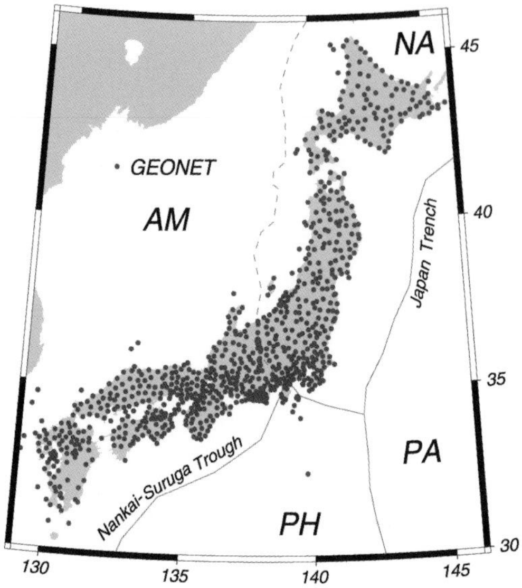

Figure 16.1 Distribution of the GEONET GPS stations and plate tectonic setting of the Japanese Islands. Abbreviations are the North American (NA), Pacific (PA), Amurian (AM), and the Philippine Sea (PH) plates.

plates, subducting at the Japan Trench and the Nankai Trough, respectively. A new plate boundary was proposed along the eastern margin of the Japan Sea and central Japan, which divides the country into two plates colliding with each other [*Nakamura*, 1983; *Kobayashi*, 1983] (fig. 16.1). This was originally considered to be the boundary between EU and the North American plate (NA). Later, *Seno et al.* [1996] suggested that northeastern Japan (NEJ) is a part of the newly proposed Okhotsk plate (OK). GPS observations in Korea, Russia, China, and southwestern Japan (SWJ) showed significant eastward components relative to the stable interior of EU. *Heki et al.* [1999] attributed this to the eastward movement of the Amurian plate (AM), covering SWJ, Korean Peninsula, northeastern China, Russian Far East, at a rate of 9–10 mm/yr with respect to EU. They also suggested that NEJ is a part of NA rather than OK and collides with SWJ at a rate of ~20 mm/yr in central Japan. Later, *Sagiya et al.* [2000] identified the concentration zone of compressional strain running from Niigata to Kobe from velocity fields obtained by GEONET and named it the Niigata-Kobe Tectonic Zone (fig. 16.2). AM has offered a good platform to model the interseismic velocity field in SWJ caused by the subduction of PH at the Nankai Trough [*Miyazaki and Heki*, 2001].

Recurrence of Interplate Earthquakes Along the Nankai/Suruga Trough

The fact that Japan consists of two plates colliding with each other has significant implications for seismotectonics in Japan, especially for interplate earthquakes along the Nankai-Suruga Trough. There, *M* 8 class interplate thrust events have occurred repeatedly during the last 1500 years with an average

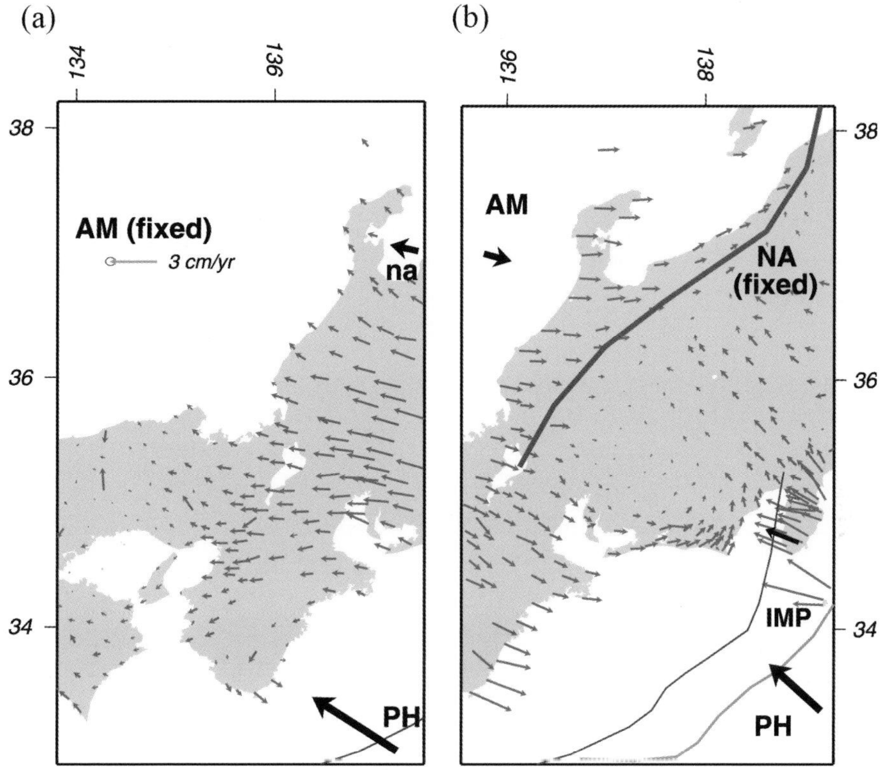

Figure 16.2 Long-term velocity fields of central Japan in the frames fixed to (a) AM and (b) NA. Thin gray arrows and thick black arrows show observed velocities at GPS sites and those predicted by Euler vectors. The northern most tip of PH is suggested to be detached from the rest as the Izu Microplate (IMP). Niigata-Kobe Tectonic Zone [*Sagiya et al.*, 2000] is shown by a gray broad line. The typical error ellipse size is indicated at the arrowhead of the legend vector.

recurrence interval of 120 years [*Sangawa*, 1993]. The next large event is anticipated in the Tokai district (the easternmost segment of the trough) since the last major events, the 1944/1946 Tonankai/Nankaido earthquakes, ruptured the rest of the trough [*Ishibashi*, 1981]. This seismic gap hypothesis, however, assumes that both the landward and oceanic sides of the trough are single plates. Transition between AM and NA somewhere in the landward plate implies along-trough differences in plate convergence rates. *Sagiya* [1999] pointed out that the incipient plate convergence along the Zenisu Ridge and detachment of the northern tip of PH as the Izu Microplate causes further complexities in convergence rates.

Heki and Miyazaki [2001] suggested that the convergence rate at the Tokai district is reduced to only ~⅓ of the rate at the western segments by two factors, (1) the landward plate belongs to NA rather than AM, and (2) the Izu

Microplate subducts there instead of PH. A smaller convergence rate makes the recurrence interval longer. This can be confirmed in past records (fig. 16.3, bottom), where the Tokai segment often fails to rupture just like it did in the last cycle. *Sagiya* [1999], by inversion of GPS data, inferred the slip deficit to be 30–40 mm/yr in the Sea of Enshu and 25–30 mm/yr in Suruga Bay, smaller than the PH-AM convergence rate, consistent with *Heki and Miyazaki* [2001] (fig.16.3, top).

Using a larger data set of GPS observations in Siberia, *Kogan et al.* [2000] suggested that a newly determined Euler pole of EU can account for velocity vectors in Europe and in eastern Asia simultaneously and dismissed the exis-

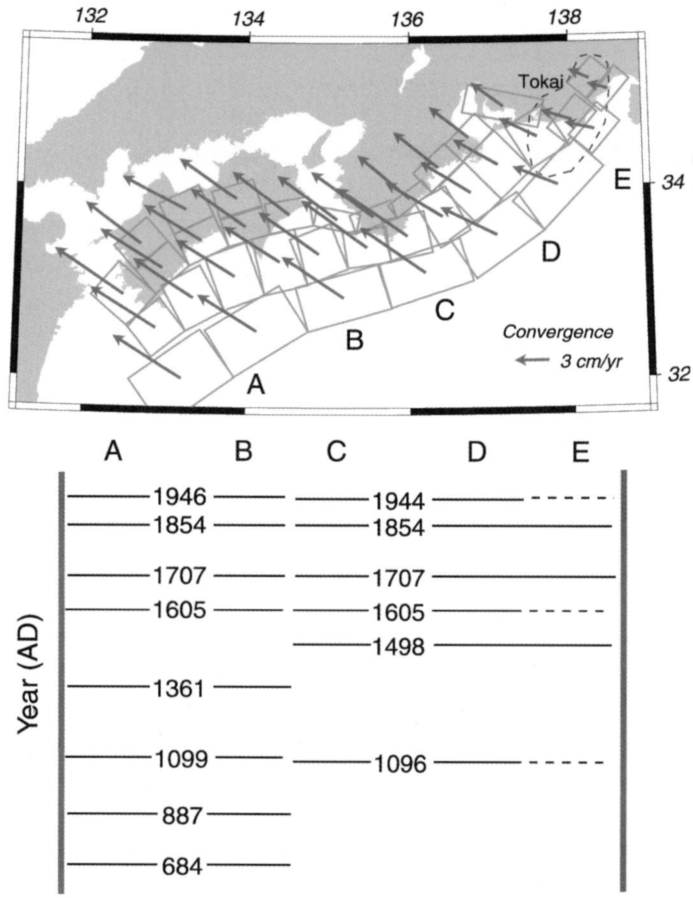

Figure 16.3 Plate convergence vectors at the centers of the fault segments along the Nankai-Suruga Trough. Convergence vectors are significantly shorter in the expected focal region of the future Tokai earthquake shown by the dashed line. The lower part shows years (AD) of occurrences of historic interplate thrust events for five segments along the trough, A, B, C, D, and E [*Sangawa*, 1993]. The Tokai region corresponds to segment E.

tence of AM. Additional velocity data between Europe and the Far East would help to solve this problem: should the Euler vector for EU be revised or is it similar to the NUVEL-1 result [*DeMets et al.*, 1990] implying several smaller plates in eastern Asia? Such an uncertainty in the large-scale plate tectonic setting has little effect on geodetic interpretation in the Japanese Islands, i.e., they are composed of two plates colliding with each other in central Japan, and convergence rates vary along the Nankai-Suruga Trough.

Interplate Coupling and Vertical Velocities

Interseismic crustal deformation due to a locked plate interfaces in subduction zones is often modeled as the sum of steady-state reverse slip and normal slip at the locked surface [*Savage*, 1983]. We often disregard crustal deformation by the former, and model interseismic deformation assuming only the latter hypothetical slip, referred to as back slip or slip deficit. The interseismic velocity field provides information on the distribution of slip deficits (i.e., strength of plate coupling) at seismogenic depths of the plate interface. Crustal strain observed by GPS suggests a rather simple view that the plate interface is almost fully coupled over the depths 5–50 km in NEJ and 5–25 km in SWJ. Coupling gradually decreases and disappears at the depths of 65 km in NEJ and 35 km in SWJ [*Mazzotti et al.*, 2000]. However, *Nishimura et al.* [2000] and *Suwa et al.* [2003] suggested nonuniform distribution of slip deficits in NEJ, i.e., stronger coupling in the regions where relatively large asperities exist [*Yamanaka and Kikuchi*, 2004].

The majority of past studies have been based only on horizontal crustal movements or strains because of the much lower signal-to-noise ratio of vertical signals. However, increasing observation periods are making it possible to discuss secular uplift or subsidence rates. *Aoki and Scholz* [2003] confirmed that the observed vertical crustal velocity field in SWJ is consistent with the back-slip model predictions, i.e., the area closest to the trough subsides, and the subsidence changes to uplift across the "hinge line." Uplift decays as we go farther inland (see the velocity profile discussed in the next section). The largest uplift occurs approximately at the surface projection of the lower limit of the coupled zone. At the same time, *Aoki and Scholz* [2003] did not find such a pattern in NEJ; i.e., sites along the Pacific coast subsides secularly by a few millimeters per year, decaying toward the west. They did not discuss this issue in much detail and attributed the absence of the uplift zone to the remoteness of the shoreline from the trench (i.e., uplifted zone lies offshore).

Tide gauge data at Ayukawa (see fig. 16.6 for location) provide an example of the peculiar situation of NEJ (fig. 16.4). The data span >40 years and record the last interplate event, the 1978 Miyagi-Oki (*M* 7.3) earthquake. According to the basic concept of interseismic crustal deformation, cumulative displacement during the recurrence interval is to be canceled substantially by the next coseismic displacement. The Ayukawa tide gauge data violate this concept;

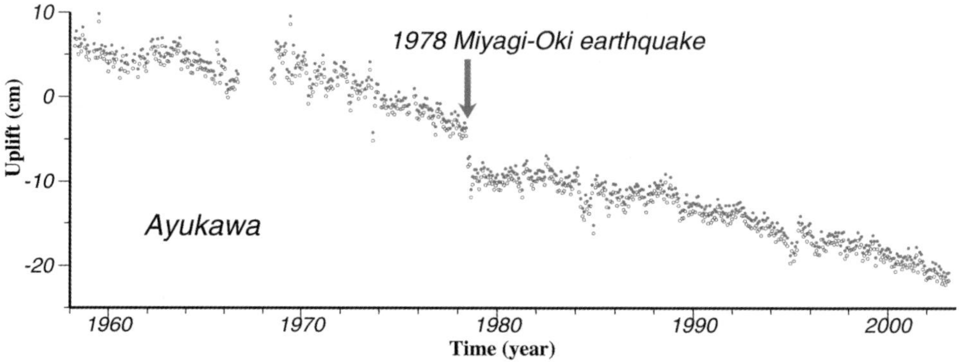

Figure 16.4 Ayukawa (fig. 16.6) tide gauge data converted to show the land uplift. Secular sea-level change due to eustatic sea-level rise of ~1.9 mm/yr (fig. 16.6) was removed from the time series. The data show coseismic subsidence in the 1978 event as well as interseismic slow subsidence. This example violates the classical concept that the coseismic jump cancels the cumulative interseismic displacement.

the site slowly subsides during the interseismic period and coseismic subsidence is added when an earthquake occurs. Figure 16.5 shows secular vertical velocities derived from the daily coordinates of GEONET 1996.2–2000.5 by modeling the time series with linear and seasonal (annual and semiannual) components. As shown by *Aoki and Scholtz* [2003], GPS points along the Pacific coast of NEJ show secular subsidence. In the next section, I will confirm the accuracy of vertical velocity results obtained by GPS by comparing them with sea-level rise data.

Secular Subsidence of the NEJ Fore Arc

Figure 16.6 compares long-term sea-level change rates at 76 tide gauges [*Geographical Survey Institute* 2002] obtained after removing oceanographic signals using *Kato*'s [1983] method, and vertical velocities of nearby GPS points. Absolute vertical velocities (those relative to the Earth's center of gravity) of GPS points were obtained by fixing the Tsukuba velocity to the value given in the International Terrestrial Reference Frame (ITRF) 2000 [*Altamimi et al.*, 2002]. They are negatively correlated with an offset (sea-level rise is larger) of ~1.9 mm/yr, which agrees with the eustatic sea-level rise expected in this region [*Mitrovica et al.*, 2001]. The large scatter about the regression line may reflect (1) different time spans represented by the two techniques, (2) local differences of vertical velocities between tide gauges and GPS sites, and (3) underestimation of vertical rate errors from tide gauges.

We now compare horizontal (normal to trench strike) and vertical velocity profiles of NEJ (fig. 16.7) and SWJ (fig. 16.8) with the velocities predicted by the slip deficit model [*Savage*, 1983] assuming the depth-dependent cou-

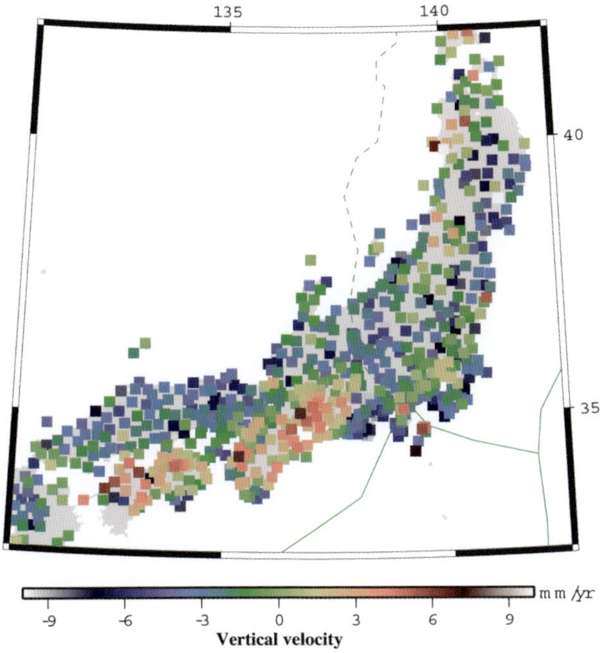

Figure 16.5 Vertical velocities of the GEONET GPS points. The vertical velocity of the reference station (Tsukuba) was fixed to the value given in ITRF2000 [*Altamimi et al.*, 2002], and these velocities are hence relative to the geocenter.

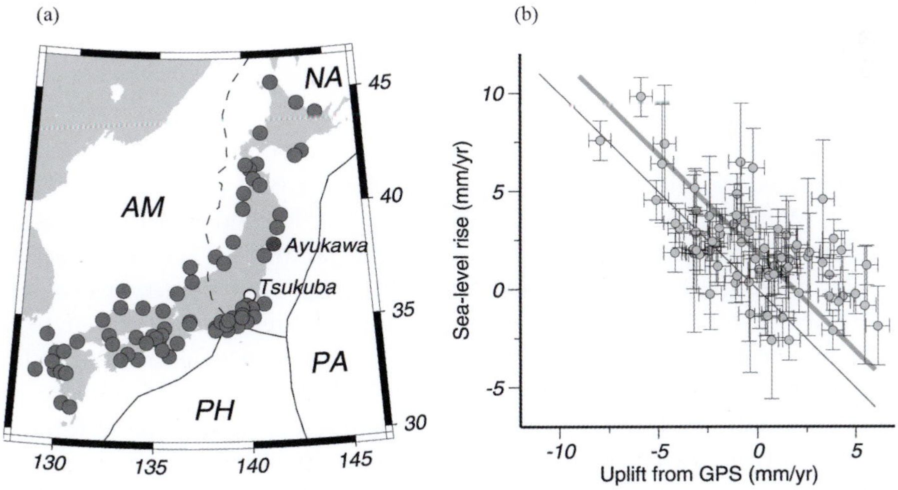

Figure 16.6 Comparison of vertical velocities from GPS and tide gauges. (a) Dots denote tide gauge stations. Sea-level rise at tide gauges and vertical velocities at nearest GPS sites are negatively correlated with a small excess sea-level rise of 1.9 mm/yr (offset between the best-fit thick gray line and the thin line) corresponding to the (b) eustatic sea-level rise in Japan. Errors for GPS velocities are 1σ (error of Tsukuba velocity in ITRF2000 [*Altamimi et al.*, 2002] is considered), and those for tide gauges are ±1, ±2, ±3, and ±4 mm/yr for sites observed for >40, >30, >20, and <20 years, respectively.

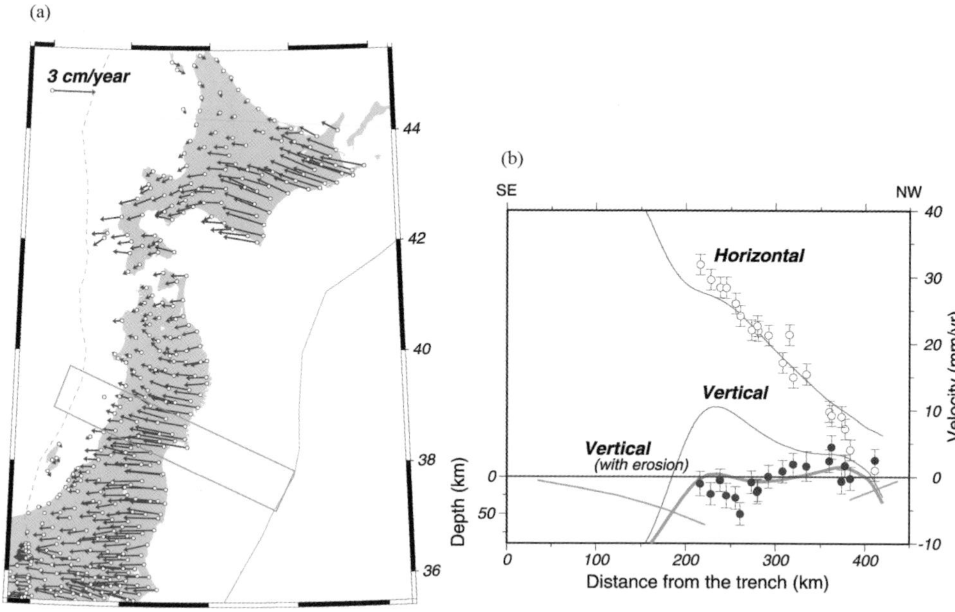

Figure 16.7 Comparison of observed and calculated velocities in NEJ. (a) Trench-normal horizontal (open circles), and vertical (solid circles) velocities in green rectangles plotted against (b) the distance from the Japan Trench. Curves are the velocity profiles predicted by the slip deficits at the plate interface, whose cross section is shown with gray curves. Error bars show 3σ. Another heavier gray curve, indicated as "with erosion," is the vertical profile obtained taking account of tectonic erosion of 15 mm/yr over the depths 45–90 km.

pling strength after *Mazzotti et al.* [2000]. Euler vectors derived by space geodesy were used to calculate convergence vectors at NEJ [*Altamimi et al.*, 2002] and SWJ [*Miyazaki and Heki*, 2001], assuming NA-PA and AM-PH plate pairs, respectively. Along the eastern margin of the Japan Sea, we considered additional convergence between AM and NEJ, whose convergence vector was taken from *Heki et al.* [1999]. Slab geometry is assumed to be similar to fault planes of past earthquakes along this zone, e.g., the 1983 Central Sea of Japan earthquake [*Earthquake Research Committee*, 1998]. The vertical velocity reference is reasonably defined by the geocenter, but selecting horizontal references is not straightforward. For SWJ, I fixed Kamitsushima (fig. 16.8), which is >500 km distant from the Nankai Trough. In NEJ, because subduction occurs at both sides of Japan, a stable reference cannot be found within the arc. There an artificial translation is given to the horizontal velocity profile to let them match the prediction.

In SWJ, both horizontal and vertical crustal movements are consistent with the model predictions (fig. 16.8) as suggested by *Aoki and Scholtz* [2003]. In NEJ, however, vertical velocities show subsidence relative to the model (fig. 16.7).

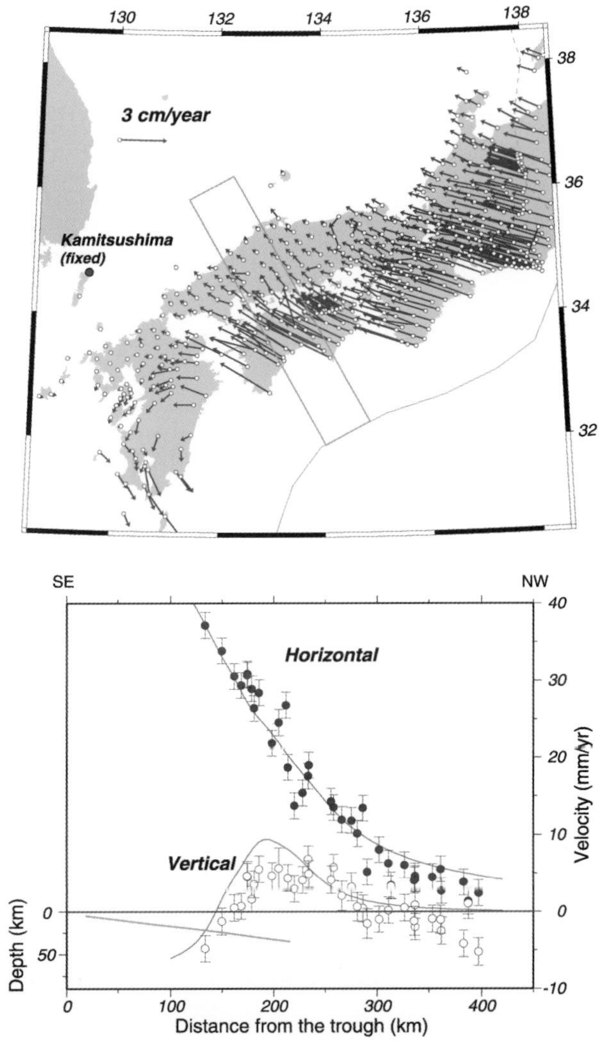

Figure 16.8 Comparison of observed and calculated velocities in SWJ. Unlike NEJ (fig. 16.7), both horizontal and vertical velocity profiles are consistent with the model curves.

This anomaly might indicate a viscous effect; i.e., crustal velocity may vary during an earthquake cycle due to viscous relaxation in an asthenosphere. *Cohen* [1994] calculated such cyclic components for several cases. He showed that viscous effects modulate interseismic displacement rates to some extent but rarely change directions; i.e., the uplift/subsidence pattern remains similar throughout the cycle. Tide gauge data in Ayukawa (fig. 16.4) provide additional insight. Although there is some difference between the rates immediately before and after the 1978 event, the site subsides persistently throughout the cycle. It is hence unlikely for viscous effects to cause the fore-arc subsidence observed in NEJ. Another possibility is that we applied an incorrect coupling strength in the model. We could make the fore arc subside, by extending the coupled (i.e., seismogenic) zone to the depth of 100–120 km [*Suwa et al.*, 2003]. However, this is inconsistent with our knowledge that interplate thrust earthquakes occur at depths shallower than 50–60 km in NEJ [*Hasegawa et al.*, 1991] and most other regions. *Heki* [2004a] proposed a new interpretation for the NEJ fore-arc subsidence, which I will briefly introduce in the next section.

Deep Basal Erosion and Secular Subsidence of NEJ Fore Arc

Subduction (tectonic) erosion is the detachment of upper plate rock and sediment and the downward transport of this material along with the subducting lower plate. *Von Huene and Scholl* [1991] proposed two basic erosion mechanisms. At a trench, material collapsed from the landward slope is trapped in horst-graben structures of the subducting plate (frontal erosion). Material at the base of the upper plate may also be detached and move toward the mantle with the subducting slab (basal erosion). These two processes carry down upper plate material, make the trench retreat landward, and cause fore-arc subsidence. *Heki* [2004a] evaluated the amount of basal erosion using the formulation of *Okada* [1992] assuming compressional dislocation (fault-normal shortening) on the plate boundary. Figure 16.7 shows the model prediction curve where 15 mm/yr shortening is assumed to occur over the depth range 45–90 km (updip extent of the erosion is not well constrained, see *Heki* [2004a]).

According to *von Huene and Lallemand* [1990], the Deep Sea Drilling Project site off the Sanriku coast was subaerial at 16 Ma. This site is only 30 km distant from the trench location today. *Heki* [2004a] inferred the trench to have retreated 170 km in 16 myr, with an average rate of 10 mm/yr, by assuming a similar subduction zone geometry to the present (i.e., the dip angle and land-trench distance of ~200 km have been constant). If the erosion takes place uniformly over the slab dipping 30°, the average deep basal erosion rate would be 5 mm/yr. The current rate of 15 mm/yr from GPS is three times as large, i.e., erosion at this segment of NEJ might be temporarily accelerated by a subducting seamount acting as an asperity [*Wells et al.*, 2003].

Water is considered to play an essential role in basal subduction erosion [*von Huene and Scholl*, 1991]. A significant amount of water is released at various depths during subduction in NEJ [*Iwamori*, 1998]. At shallow locations, loose oceanic sediments release water as they consolidate. At greater depth (~50 km in NEJ), owing to high temperature and pressure, hydrous and clay minerals in the oceanic crust release water. This water is accommodated by the upper plate as olivine transforms to serpentine and chlorite. Serpentinized mantle is detached from the wedge mantle and is dragged down by the slab.

It eventually releases water beneath the volcanic front (~150 km in NEJ) encouraging magma genesis and arc volcanism. Geodetic data may provide a new quantitative way to explore such deep processes. A comparative study of world subduction zones, with or without subduction erosion, using vertical velocity profiles, would be meaningful in clarifying the roles of subduction erosion in convergent plate boundaries.

Transient Movements

Shortly after the start of operation of GEONET in 1994, coseismic crustal movements of two earthquakes were detected, namely the 1994 Hokkaido-Toho-Oki (Shikotan) earthquake ($M_w = 8.3$) [*Tsuji et al.*, 1995], and the 1994 Sanriku-Haruka-Oki (far-off Sanriku) earthquake ($M_w = 7.7$) [*Tanioka et al.*, 1996]. After the 1994 earthquake, GEONET detected slow crustal movement with time scales of about a year (fig. 16.9) [*Heki et al.*, 1997]. This first showed that slow fault slip is able to release seismic moment comparative to coseismic moment. Similar discoveries continued in NEJ, where seismic moment released by slow fault slip is now considered as important a process as coseismic releases. In SWJ, GEONET confirmed the existence of slow slip events not preceded by normal-speed ruptures, and some of those are found to recur fairly regularly. Here, past examples of such transient crustal movements are reviewed.

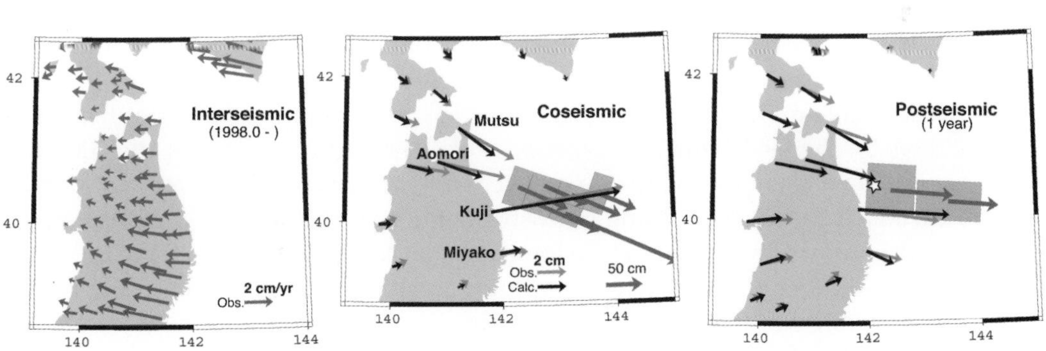

Figure 16.9 Distribution of (left) interseismic, (center) coseismic, and (right) postseismic displacement of GPS stations for the 1994 Sanriku earthquake. Since the number of GPS sites was much smaller in 1994, coseismic and postseismic movements are recorded at fewer GPS sites than interseismic movements. Postseismic movements are relative to Tsukuba over a 1-year period following the earthquake. Interseismic movements were derived using time series after 1998.0 to avoid influences of the postseismic movements (fig. 16.10). An open star in the right panel indicates the position of the small repeating earthquake [*Uchida et al.*, 2003] whose cumulative slip is shown in figure 16.10.

After Slip

Along the Japan Trench, NEJ, the distribution of past interplate earthquakes is fairly nonuniform, and cumulative slip amounts fall short of expected plate convergence rates at the trench. This suggests the existence of aseismic ("silent") plate convergences there. Slow fault slip was first detected by quartz-tube extensometers at the Esashi Earth Tides Station, National Astronomical Observatory after the 1992 Sanriku earthquake ($M_w \sim 6.9$) [*Kawasaki et al.*, 1995]. It recorded a slow change in strain with time scale of about a day immediately after the step-like coseismic change.

After the 1994 Sanriku earthquake, GEONET detected a slow crustal movement with a time scale of about a year (fig. 16.9) [*Heki et al.*, 1997]. This was the first observation of slow fault slip in NEJ by multiple space geodetic stations, and it was found that seismic moment comparable to the coseismic moment was released during the 1-year period following the earthquake. Past earthquakes were then examined using tide gauge records, and *Ueda et al.* [2001] recognized the existence of large after slip associated with the 1978 Miyagi-Oki earthquake, similar to the 1994 Sanriku earthquake. Densification of GEONET enabled the detection of slow slip after smaller earthquakes. There was a medium-sized ($M_w \sim 6.4$) earthquake offshore east of Aomori in August 2001, for which *Sato et al.* [2004] detected a subtle (few millimeters) after-slip signal. Recent analysis by the *Geographical Survey Institute* [2003] also detected small after-slip signals following the 2002 November earthquake off-Miyagi (MJMA ~6.1). After the September 2003 Tokachi-Oki earthquake (MJMA ~8.0), fairly large after slip was observed to start in the region surrounding the main rupture [*Takahashi et al.*, 2004], whose total amount is still unknown.

It seems that after slip is a common phenomenon following interplate earthquakes at the Japan Trench. All interplate earthquakes during the last decade have manifested after slip. The time scale of after slip varies from 1 or 2 days for the 1992 Sanriku [*Kawasaki et al.*, 1995], to 2 months for the 2001 Aomori [*Sato et al.*, in press], to one half year for the 1994 Sanriku [*Heki et al.*, 1997]. *Heki and Tamura* [1997], using both strainmeter and GPS data, demonstrated a smooth transition from short- to long-term after slip of the 1994 Sanriku earthquake (fig. 16.10). These two techniques are sensitive to different time scales, and it is difficult to discuss after-slip time scales when only one kind of data are available. For example, the 1992 Sanriku event may have shown a yearlong after slip if a dense GPS array were available then (this was found to be the case by analysis of small repeating earthquakes, see the next section). The 2001 Aomori event may have shown large after slip with a time scale of a few days if a good strainmeter were available near the focal region.

Crustal deformation studies following inland earthquakes have been done by dense deployment of GPS receivers for campaign observations in and around the focal region. After slip for these earthquakes, e.g., the 1995 Kobe earthquake [*Nakano and Hirahara*, 1997] and the 2000 Tottori earthquake [*Hashimoto et al.*, 2003], was found to be much smaller than for interplate thrust events.

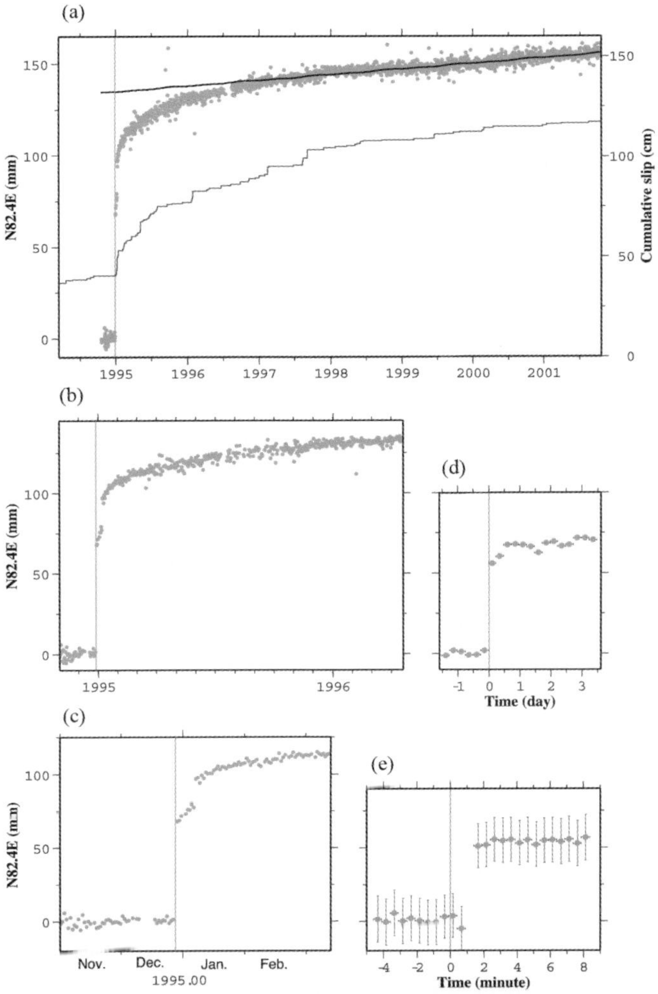

Figure 16.10 Movement of the Kuji station in the N82.4E direction viewed from the Miyako station (see fig. 16.9 for locations). They are daily solutions over (a) 6 years, (b) 1 year, (c) 4 months, and (d) quarter-daily solutions over the 5 days and epoch-wise (i.e., every 30 s) solution over (e) the ~12 min period. In figure 16.10a the cumulative slip of a small repeating earthquake in the afterslip area (see fig. 16.9 for position) is drawn after *Uchida et al.* [2003].

Asperities and Small Repeating Earthquakes

In addition to geodesy, seismological studies of small repeating earthquakes provide independent evidence for slow fault slip. As seen with typical repeating earthquakes, such as the *M* 4.8 earthquake sequence off Kamaishi [*Matsuzawa et al.*, 2002], they recur with fairly uniform intervals and similar waveforms. They are considered to be the ruptures of the same asperities that are small in size and isolated from larger asperities [*Okada et al.*, 2003]. A transient decrease of their recurrence interval may indicate an accelerated creeping rate of that part of the plate interface. They provide useful information on after slip in several cases, e.g., GPS does not have enough sensitivity for slip occurring far off the coastline, and for earthquakes before the GPS era. For example, *Igarashi et al.* [2003] found a sudden increase in the occurrence rate of one such repeating earthquake within the 1994 Sanriku earthquake after-slip area (fig. 16.9) shortly after the earthquake (fig. 16.10). Detailed studies of such repeaters in the same region by *Uchida et al.* [2003] clarified similar accelerations after the 1992 Sanriku earthquake as well and showed propagation of the after-slip area by comparing different repeaters around the focal region.

Positions, shapes, and behaviors of larger asperities, which have caused *M* 7~8 class earthquakes in NEJ in the past, have been investigated by *Yamanaka and Kikuchi* [2004]. According to them, the 1994 Sanriku earthquake is caused by one of the two relatively large asperities that ruptured during the 1968 Tokachi-Oki earthquake. Figure 16.9 compares distribution of interseismic, coseismic, and postseismic crustal deformation of the 1994 Sanriku earthquake. The difference between coseismic and postseismic displacement vectors indicates that coseismic slip occurred in a rather small area corresponding to this asperity while after slip occurred over a larger area [*Heki et al.*, 1997]. Recent seismological studies of source mechanisms and aftershock distribution and a reexamination of geodetic data for this earthquake suggested that the regions of coseismic slip and after slip do not overlap; i.e., after slip occurred in the region surrounding the asperity [*Yagi et al.*, 2003]. Figure 16.10 shows the change of the ENE component of the coordinate of the Kuji GPS station relative to the Miyako station in various time scales (modified from *Heki et al.* [1997] and *Heki and Tamura* [1997]). The transition from coseismic to short-term postseismic movements is not seamless but distinctive, suggesting that the coseismic rupture and the after slip are processes occurring at different parts of the fault with different frictional properties. *Heki and Tamura* [1997] compared strain patterns and found that even the earliest part of after slip has different slip distribution from coseismic slip but similar to that of the yearlong after slip.

Silent Events and Their Recurrences

A totally silent earthquake (i.e., without normal-speed fault rupture) was first detected with GEONET in 1996, first, beneath the Boso Peninsula, central Japan [*Sagiya*, 1997, 2004], and second, in the Bungo Channel, SWJ, in 1997 [*Hirose et al.*, 1999]. Similar slow fault slip events found in the Cascadia subduction zone [*Dragert et al.*, 2001] seem to recur regularly with average intervals of ~14 months [*Miller et al.*, 2002] and are associated with low-frequency tremor activities [*Rogers and Dragert*, 2003]. Such low-frequency tremors are also discovered in Japan along the 30-km depth line of the Philippine Sea plate slab [*Obara*, 2002]. However, slow crustal movements synchronized with the tremor activities have not been detected yet with GEONET.

Recently, a dike intruded along a line connecting two of the Izu Islands, Miyake, and Kozu, over a 3-month period in summer 2000, causing crustal movements of several tens of centimeters in the Kanto-Tokai region [*Kaidzu et al.*, 2000]. This was immediately followed by a slow crustal movement toward southeast in the Tokai area beginning in autumn 2000 [*Ozawa et al.*, 2002]. Its initial stage resembled to the phenomenon suggested to be the 1944 Tonankai earthquake preslip [*Mogi*, 1985], but the 2000 event turned out to be the onset of an episodic slow convergence event (silent earthquake) that occurred at the downdip extension of the Tokai segment of the Nankai-Suruga Trough. The movement still goes on, and it is difficult to tell whether it is decaying or not because the signal is overprinted with seasonal components with possible interannual amplitude variations (see the last section and figure 16.14).

An important recent finding concerning silent earthquakes is their fairly regular recurrences, as revealed by GPS data in the Cascadia subduction zone [*Miller et al.*, 2002]. Exploration for such periodicity is done backward in time beyond the GPS age in Japan. Past observations of conventional ground survey suggested that silent fault slips similar to the 2000 Tokai silent event may have occurred several times, with the last one being around 1988–1990 [*Kimata and Yamauchi*, 1998]. This is also supported by past records of swarm earthquake activity and characteristic signals in records of a tiltmeter above the slipping region [*Yamamoto et al.*, 2005]. A silent earthquake similar to the 1996 Boso event occurred again in 2002, and old strainmeter records suggested that similar events have occurred in 1983 and 1991 [*Ozawa et al.*, 2003].

Another important finding is that deep nonvolcanic tremor activities and slow slip events occur simultaneously, as first pointed out in the Cascadia subduction zone [*Rogers and Dragert*, 2003]. In Japan, activities of such deep tremors were known to have certain periodicities [*Obara*, 2002]; e.g., it becomes active approximately every half year beneath the western Shikoku region. Using the nationwide array of borehole tiltmeters, *Hirose and Obara* [2005] revealed that slow changes of crustal tilts of ~0.1 μrad occur together with the enhancements of tremor activities. They showed that such tilt changes indicate slow reverse slip events on what we call transition zone of the plate interface (depth 25–35 km in SWJ) where full coupling at its upper edge decays to zero at its lower edge. Surface displacements by such slow fault slips are not more than a few millimeters in amplitude, which would be difficult to detect with GEONET.

Apart from such periodicity, some of the silent earthquakes seem to have been triggered by the occurrence of nearby normal-speed earthquakes. For example, the 1997 Bungo Channel silent earthquake was preceded by two normal earthquakes in 1996 in Hyuga-nada [*Yagi et al.*, 2001; *Yagi and Kikuchi*, 2003]. The 2000 Tokai slow event may also have been triggered by the Miyake-Kozu dike intrusion event that should have increased the Coulomb failure stress at the Tokai event fault surface. Silent events in SWJ and Cascadia seem to occur on the transition zone of the plate interface. In NEJ, neither such periodic silent fault slips nor deep tremor activities have been found. This might reflect observational difficulties coming from the larger depth of the transition zone in NEJ (50–65 km).

Seasonal Crustal Movements

Seasonal crustal deformation signals in the GEONET GPS site coordinate time series were studied first by *Murakami and Miyazaki* [2001]. They were highly systematic in space and suggested origins other than orbital errors that would cause signals with simpler spatial patterns. On a global scale, comprehensive studies of potential seasonal displacement signal sources were done by *Dong et al.* [2002]. Seasonal signals in northeastern Honshu were found to be due largely to snow loads along the western flanks of the backbone range by *Heki* [2001]. *Hatanaka* [2003a] pointed out the existence of seasonal signals unrelated to crustal deformation in GEONET, namely seasonal-scale changes of unknown origin and spurious displacement signals due to atmospheric gradients.

Although such periodic signals are not of tectonic origin, it is important to understand them for crustal deformation studies in active plate-boundary regions. For example, moment release history of the 2000 Tokai slow event is discussed using transient components inferred by eliminating seasonal signals [*Ozawa et al.*, 2002]. A millimeter-level after-slip signature of the 2001 off-Aomori event [*Sato et al.*, 2004] would not have been detected unless seasonal changes had been adequately removed. In these studies, elimination of seasonal components is done expecting that the same seasonal pattern repeats every year. Such interannual regularity, however, has not been well studied. In order to explore a realistic method to model seasonal components, *Heki* [2004b] evaluated various kinds of seasonally changing loads and compared the modeled variations with observed signals. Here, I briefly summarize its essential parts.

Seasonal Changes in Surface Loads

Heki [2004b] improved the approach of *Heki* [2001] by considering (1) factors other than snow, (2) the whole of Japan rather than northeastern Honshu, and (3) waveforms of seasonal load changes. *Heki* [2004b] divided the Japanese Islands into 130 blocks as large as 0.5° latitude and 0.67° longitude and estimated loads of various factors at monthly epochs for individual blocks. Then, the displacement of each GPS station was calculated as the sum of the Green's function [*Farrell*, 1972] describing the elastic response of the Earth to a point load.

Snow Load

Snow depth data are obtained daily at the Automated Meteorological Data Acquisition System (AMeDAS) sites, run by Japan Meteorological Agency.

Figure 16.11 shows the snow depth curves at three points and averages of the maximum snow depths of the last seven winters. Despite their dense deployment, these AMeDAS sites exist along inhabited valleys with altitudes lower than the regional averages. Since snow depths are highly dependent on the elevation, *Heki* [2004b] performed "altitude correction" by (1) obtaining linear relationships between snow depths and site altitudes for individual snowy prefectures and (2) performing a correction to bring the AMeDAS snow depth data to the values at the prefectural average heights. Then, the altitude-corrected snow depth data were compiled to calculate average seasonal change of snow depths within the blocks.

The average density of the snow cover increases toward the end of winter because of compaction. Using snow density data collected in situ in Japan, *Heki* [2004b] modeled the snow density to be 0.25 g/cm^3 on 1 December and to increase linearly with time reaching 0.4 g/cm^3 on 31 March. This time-dependent model density was used to convert snow depths to loads (fig. 16.12). Large snow loads are distributed along the northwestern flanks of the back-

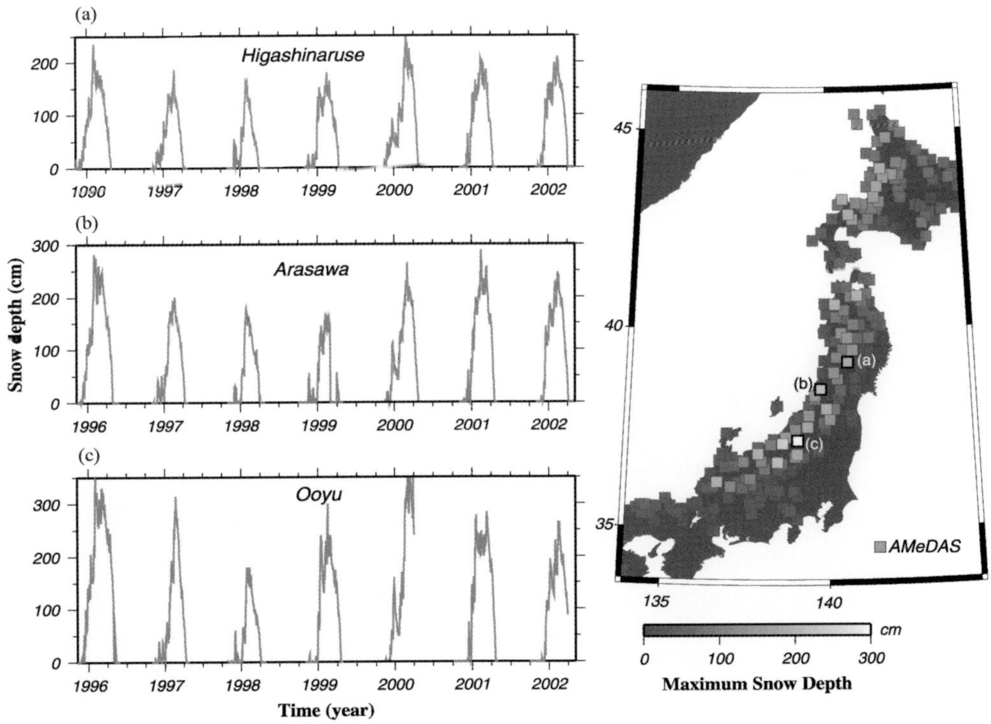

Figure 16.11 Daily snow depth measurements at three AMeDAS sites, (a) Higashinaruse (Akita Prefecture), (b) Arasawa (Yamagata Prefecture), and (c) Ooyu (Niigata Prefecture), indicated as squares with bold frames in the map, from north to south, over the (left) seven winter seasons 1996–2002 (left graphs), and (right) distribution of the average of the maximum snow depths of the seven winters at the AMeDAS sites equipped with snow depth meters.

bone range, while they are almost absent along the Pacific coast and in SWJ. As shown in the insets of figure 16.12, snow load changes are characterized by strong peak in winter and relatively flat period during the rest of the seasons.

Other Loads The second largest seasonal changing load is the atmosphere. Unlike snow, atmospheric loading exists both on the land and ocean, but at relatively long periods (e.g., annual frequency), pressure changes are exerted only on the land area (inverted barometer). *Heki* [2004b] calculated monthly mean pressures for individual blocks, using meteorological data at 80 sites from *National Astronomical Observatory* [2002]. In general, pressure tends to be higher in winter in Japan, and the amplitude of seasonal variation increases toward the west. The peak-to-peak changes amount to 1.5 kPa (~15-cm water column) in Kyushu but are <1 kPa in NEJ. They are smaller than snow loads but are responsible for the main seasonal load in snow-free parts of Japan.

Heki [2004b] also modeled seasonal variation of soil moisture throughout Japan, using a simple model based on the potential evapotranspiration (PET) [*Thornthwaite*, 1948], precipitation, and the water-holding capacity of soil [*Milly*, 1994]. Simulation of seasonal change in soil water performed with monthly time steps at various points in Japan showed that the load changes are smaller than atmosphere in most regions. Their peak-to-peak changes are largest in the inland area of southwestern Japan and Hokkaido but do not exceed 1 kPa. *Heki* [2004b] also compiled capacities and coordinates of 546 dams that can accommodate more than five million tons of water and calculated sea-

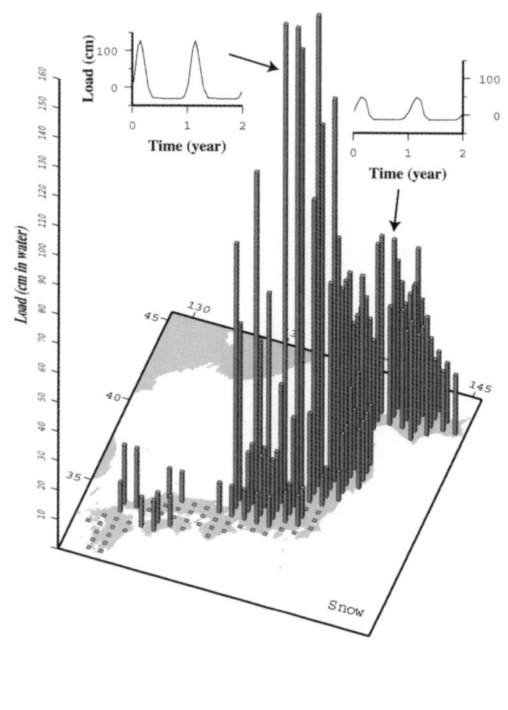

Figure 16.12 Distribution of the maximum snow load in Japan divided into 130 blocks as large as 0.5° in latitude and 0.67° in longitude, obtained as the average of the last seven winters. Insets are the seasonal snow load changes at the two selected blocks indicated by arrows. Since the annual averages are adjusted to zero there, loads take negative values during summer.

sonal variations of the load assuming the typical water-level change patterns of Japanese dams. The peak-to-peak amplitudes of the dam load are the largest in the mountain area of central Japan. However, they are ~1 kPa at most and are much less than the two major contributors, i.e., snow and atmosphere.

Sea surface height (SSH) in September is known to be higher than in March by 20 cm or more along the Pacific coast of Japan [*Okada*, 1982; *Kato*, 1983]. This includes thermal expansion irrelevant to load changes, and *Sato et al.* [2001] removed it assuming uniform thermal steric coefficient of 6.0 mm/deg. After applying such a correction, SSH is found to be almost stationary throughout a year in the Japan Sea. However, a large summer increase in SSH still remains in the Pacific Ocean. This seasonal ocean load change gives rise to a seasonal translation of the whole Japanese Islands by ~1 mm, but crustal deformation within the Japanese Islands does not much exceed those by soil moisture and dam. Apart from loads, *Munekane et al.* [2003] revealed that inflation and deflation of soil by seasonal extraction of groundwater for agricultural purposes cause vertical movements of GPS sites with amplitudes exceeding 10 mm at certain regions including Tsukuba. This can be confirmed by collocating water level meters of deep wells with GPS receivers [*Ohtani et al.*, 2000], and the areal extent of such regions should be clarified in future by using inSAR technique.

Apparent Seasonal Crustal Movement In addition to real seasonal crustal deformation due to surface load, *Hatanaka* [2003a] found that the entire GPS network manifests annual cycles of uniform expansion in summer and contraction in winter (seasonal scale change) of ~5–6 part per billion (ppb). Isotropic contraction causes substantial shortening of arc-parallel baselines, which lets us discriminate scale changes easily from load changes characterized by arc-normal contractions [*Heki*, 2004b]. Seasonal degree-one deformation [*Blewitt et al.*, 2001] causes an apparent scale change in Japan but its amplitude is <1 ppb. *Hatanaka* [2003a] considered the scale change as an artifact irrelevant to deformation of the real Earth, e.g., scale change caused by an error in absolute atmospheric delay [*Beutler et al.*, 1988]. *Heki* [2004b] showed that the seasonal-scale change patterns are similar from year to year and made an empirical model to enable a seasonal-scale correction.

In addition to seasonal-scale changes, *Hatanaka* [2003a] pointed out that atmospheric delay gradients [*Miyazaki et al.*, 2003] are significant in the north-south direction at stations along the Pacific coast of SWJ. GSI does not estimate atmospheric gradients in their official GEONET solutions [*Hatanaka*, 2003b]. Atmospheric delay gradients tend to cause apparent northward movement of the Pacific side stations in winter relative to the Japan Sea side. This contributes to the arc-normal winter contraction in SWJ and enhances the signals of seasonally changing loads. By comparing solutions with and without estimation of atmospheric gradient parameters, *Hatanaka* [2003a] found that this effect may contribute to seasonal signals by up to a few millimeters.

Comparison With GPS Observations

Heki [2004b] calculated the effect of the sum of crustal deformation due to seasonal changing loads including snow, atmosphere, soil moisture, dam, and SSH and contributions from the seasonal-scale changes on baseline lengths. Figure 16.13 compares modeled seasonal signals with those observed by GEONET (lengths of five arc-normal baselines) from the newest solution based on the improved analysis strategy [*Hatanaka*, 2003b]. Snow-induced signals are characterized by sharp negative peaks in winter and a relatively flat period from spring to autumn. Other components, mainly scale and atmospheric load, bring about a broad positive peak in summer making the total waveform a mild one. In total, arc-normal contraction of up to 4–5 mm across the backbone range is expected in winter in NEJ. This feature seems to be well reproduced in the real data. The southernmost baseline (baseline E, fig. 16.13) show somewhat larger seasonal variations than predicted. Part of this discrepancy may come from the north-south atmospheric gradient at southern stations. Arc-parallel baselines are more affected by scale changes, although snow load has significant contributions in NEJ. Unlike baseline length changes that mainly reflect horizontal coordinates, vertical components are less affected

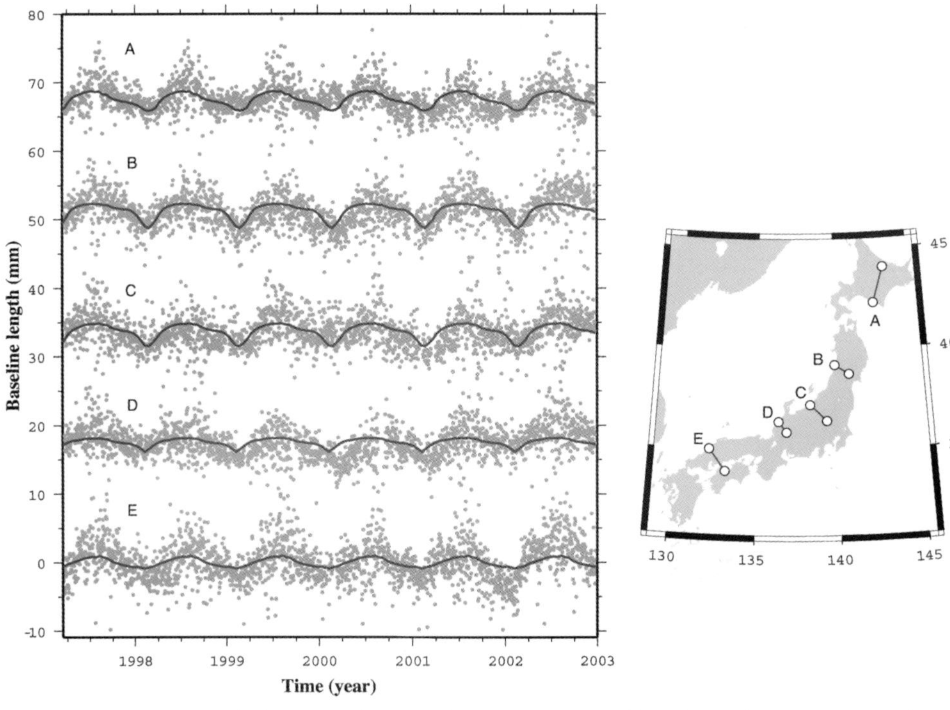

Figure 16.13 Seasonal change in lengths for five arc-normal baselines shown in the right map. Dots are daily solution of GPS (secular trends are removed beforehand), while curves denote model seasonal variation curves calculated by combining contributions from scale, snow, atmosphere, soil moisture, dam, and ocean.

by scale changes and respond mainly to change in load. *Heki* [2004b] showed that seasonal vertical movements along the snowiest zone show subsidence in winter of ~10 mm. Some sites, such as Tsukuba, show vertical seasonal movements larger than model expectations, which may reflect seasonal change in groundwater levels.

Relationship with Earthquake Occurrence Seasonality

Seismic activity in some regions in Japan is known to change seasonally [*Mogi*, 1969; *Okada*, 1982]. Using the newest historical earthquake catalog, *Heki* [2003] confirmed the tendency that inland earthquakes $M > 7.0$ in snow-covered regions in Japan occur more in spring and summer, as originally suggested by *Okada* [1982], and hypothesized that they may have been triggered by removal of snow load by spring thaw. As explained in the previous section, snow is by far the largest seasonally changing load in Japan. This justifies the approach by *Heki* [2003], in which only snow effects are considered. To verify their causal relationship, however, we need to solve several problems, e.g., ambiguous seasonality for smaller ($M < 7.0$) earthquakes, insufficient statistical significance due to a small number of historical large inland earthquakes, and so on.

Interplate thrust events along the Nankai, Suruga, and Sagami Troughs are known to occur more in autumn and winter [*Ohtake and Nakahara*, 1999]. This is qualitatively consistent with the seasonal change of nontidal oceanic loads; the SSH shows maximum in August/September [*Okada*, 1982], and the highest seismicity in December coincide with the period of the highest rate of load decrease (¼ cycle after the load peak). However, such seasonal concentration is not so clear for interplate earthquakes at the Japan Trench, NEJ; they seem to have a small peak in March in addition to the broad peak in autumn/winter similar to SWJ. Deployment of ocean bottom pressure gauges on the landward slope of the Japan Trench is in progress so that we can monitor in situ water mass changes above the sensor [e.g., *Fujimoto et al.*, 2003]. Such data will help us conclude if there is causal relationship between change in ocean loads and interplate earthquake occurrences in the future.

Conclusions

About 10 years of the GEONET GPS data have clarified various new aspects of crustal dynamics and earthquake physics in subduction zones. The large-scale plate tectonic setting of the Japanese Islands is still controversial, but there is a general agreement that Japan is divided into two plates colliding with each other somewhere in central Japan. This has a certain influence on the discussion on the seismic moment budget along the Nankai-Suruga Trough. Interseismic crustal movements are generally consistent with those predicted by slip deficits at the seismogenic parts of the plate boundaries. Figure 16.14

shows the movement in a certain horizontal direction of Hamamatsu relative to Oogata in which all of secular, transient, and periodic crustal deformation signals discussed in this paper are present. There the secular component corresponds to the movement of the Hamamatsu station owing to interseismic elastic straining by the subducting PH slab.

Vertical velocities in the fore arc of NEJ are found to deviate negatively from the model, suggesting the existence of a certain geological process responsible for its secular subsidence. This might indicate the thinning of the upper plate at depths 45–90 km due to basal subduction erosion, which may be a manifestation of the process at depth that mantle material serpentinized by water released from the slab is detached from the upper plate and transported downward.

One of the most important findings by GPS is the discovery of aseismic slow faulting. In NEJ, they often occur immediately after normal-speed earthquakes over areas surrounding asperities (after slips). Their time constants seem to range from days to a year. Recent studies of small repeating earthquakes have given independent support to this view. In central and southwestern Japan, slow slip tends to occur as single events (silent earthquakes). Many of those have been found to occur repeatedly with recurrence inter-

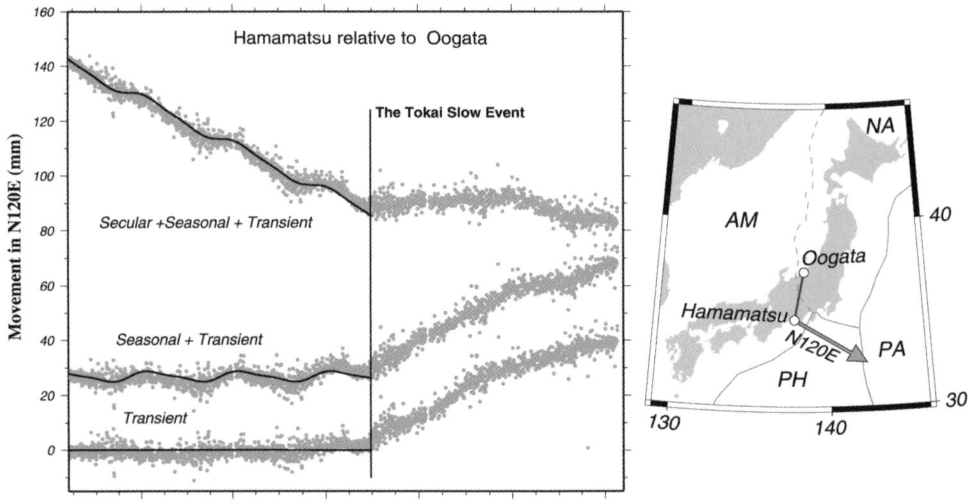

Figure 16.14 Movement of the horizontal coordinate of the Hamamatsu GPS station, in the Tokai area, with respect to the Oogata station (see the right map for location). (top) The raw data includes secular, transient, and seasonal components. In the middle curve, the secular trend is removed. In the lower curve, seasonal changes are also removed. Seasonal and secular movements are inferred using the time series before the onset of the Tokai slow event in the 2000 autumn [*Ozawa et al.*, 2002]. Solid lines indicate modeled secular and seasonal changes.

vals ranging from half year (western Shikoku), 13 months (Cascadia), 6 years (Boso Peninsula), to >10 years (Tokai). In western Shikoku and Cascadia they are accompanied by deep nonvolcanic tremor activities, which suggests a causal relationship between water released from the subducting slab and the occurrences of slow slip. In figure 16.14 the sudden departure from the normal trend in the middle of 2000 shows the onset of the 2000 Tokai slow event recorded at the Hamamatsu station.

Mechanisms of seasonal crustal deformation due to changes of surface loads are being understood, but mechanisms responsible for seasonal-scale changes of a few ppb in Japan are yet to be investigated. The top curve of figure 16.14 includes seasonal components whose large portion comes from snow load changes near Oogata. The seasonal signals are removed in the bottom curve of figure 16.14 by modeling them with annual and semiannual components using time series before the onset of the 2000 Tokai event. It is important to understand seasonal signals and their interannual variations in order to study subtle transient signals, e.g., slow fault slips at depth. For the seasonal changes of seismic activities, a century-long issue of Japanese seismology, quantitative studies of influence of seasonal load changes on seismicity are in progress.

Acknowledgments

I thank Tim Dixon (University of Miami) and Seth Stein (Northwestern University) for inviting me to present a review at the Snowbird meeting on which this paper is based. Naoki Uchida (Tohoku University) sent me data on the repeating earthquakes. I also thank Jeff Freymueller (University of Alaska) and an anonymous referee for critical reviews of the manuscript.

References

Altamimi, Z., P. Sillard, and C. Boucher (2002), ITRF2000: A new release of the International Terrestrial Reference Frame for earth science applications, *J. Geophys. Res., 107*(B10), 2214, doi:10.1029/2001JB000561.

Aoki, Y., and C. H. Scholz (2003), Vertical deformation of the Japanese islands, 1996–1999, *J. Geophys. Res., 108*(B5), 2257, doi:10.1029/2002JB002129.

Beutler, G., I. Bauersima, W. Gurtner, M. Rothacher, T. Schildknecht, and A. Geiger (1988), Atmospheric refraction and other important biases in GPS carrier phase observations, in *Atmospheric Effects on Geodetic Space Measurements, Monogr. Ser.*, vol. 12, edited by F. K. Brunner, p. 26, Sch. of Surv., Univ. New South Wales, Kensington.

Blewitt, G., D. Lavallée, P. Clarke, and K. Nurutdinov (2001), A new global mode of earth deformation: Seasonal cycle detected. *Science, 294*, 2342–2345.

Cohen, S. C. (1994), Evaluation of the importance of model features for cyclic eformation due to dip-slip faulting, *Geophys. J. Int., 119*, 831–841.

DeMets, C., D. Argus, S. Stein, and R. Gordon (1990), Current plate motions, *Geophys. J. Int., 101*, 425–478.

Dong, D., P. Fang, Y. Bock, M. K. Cheng, and S. Miyazaki (2002), Anatomy of apparent seasonal variations from GPS-derived site position time series, *J. Geophys. Res., 107*(B4), 2075, doi:10.1029/2001JB000573.

Dragert, H., K. Wang, and T.S. James (2001), A silent slip event on the deeper Cascadia subduction interface, *Science, 292*, 1525–1528.

Earthquake Research Committee (1998), Seismic activity in Japan–Regional perspectives on the characteristics of destructive earthquakes (excerpt), report, 222 pp., Headquarters of the Earthquake Res. Promotion, Prime Minist. Off., Gov. of Jpn., Tokyo.

Farrell, W. E. (1972), Deformation of the Earth by surface loads. *Rev. Geophys., 10*, 761–797.

Fujimoto, H., M. Mochizuki, K. Mitsuzawa, T. Tamaki, and T. Sato (2003), Ocean bottom pressure variations in the southeastern Pacific following the 1997–98 El Niño event, *Geophys. Res. Lett., 30*, 1456, doi:10.1029/2002GL016677.

Geographical Survey Institute (2002), Crustal deformation of entire Japan, in *Report of the Coordinating Committee for Earthquake Prediction,* vol. 68, pp.504–533, Tsukuba, Japan.

Geographical Survey Institute (2003), Crustal movements in the Tohoku District, in *Report of the Coordinating Committee for Earthquake Prediction,* vol. 70, pp. 45–68, Tsukuba, Japan.

Hasegawa, A., D. Zhao, S. Hori, A. Yamamoto, and S. Horiuchi (1991), Deep structure of the northeastern Japan arc and its relationship to seismic and volcanic activity, *Nature, 352*, 683–689.

Hashimoto, M., T. Ito, T. Tabei, and T. Sagiya (2003), Postseismic deformation following the 2000 Western Tottori earthquake, southwest Japan, using GEONET and urgent campaign GPS data, paper presented at the 23rd IUGG General Assembly, Sapporo, Japan, 7 July.

Hatanaka, Y. (2003a), Seasonal variation of scale of GEONET network and ZTD biases, paper presented at the 23rd IUGG General Assembly, Sapporo, Japan, 7 July.

Hatanaka, Y. (2003b), Improvement of the analysis strategy of GEONET, *Bull. Geogr. Surv. Inst., 49*, 11–37.

Heki, K. (2001), Seasonal modulation of interseismic stress buildup in northeastern Japan driven by snow loads, *Science, 293*, 89–92.

Heki, K. (2003), Snow load and seasonal variation of earthquake occurrence in Japan, *Earth Planet. Sci. Lett., 207*, 159–164.

Heki, K. (2004a), Space geodetic observation of deep basal subduction erosion in northeastern Japan, *Earth Planet. Sci. Lett., 219*, 13–20.

Heki, K. (2004b), Dense GPS array as a new sensor of seasonal changes of surface loads, in *The State of the Planet: Frontiers and Challenges in Geophysics, Geophys. Monogr. Ser.*, vol. 150, edited by R. S. J. Sparks and C. J. Hawkesworth, pp. 177–196, AGU, Washington, D. C.

Heki, K., and S. Miyazaki (2001), Plate convergence and long-term crustal deformation in central Japan, *Geophys. Res. Lett., 28*, 2313–2316.

Heki, K., and Y. Tamura (1997), Short term afterslip in the 1994 Sanriku-Haruka-Oki earthquake, *Geophys. Res. Lett., 24*, 3285–3288.

Heki, K., H. Tsuji, and S. Miyazaki (1997), Silent fault slip following an interplate thrust earthquake at the Japan Trench, *Nature, 386*, 595–597.

Heki, K., S. Miyazaki, H. Takahashi, M. Kasahara, F. Kimata, S. Miura, N. Vasilenco, A. Ivashchenco, and K. An (1999), The Amurian Plate motion and current plate kinematics in eastern Asia, *J. Geophys. Res., 104*, 29,147–29,155.

Hirose, H. and K. Obara (2005), Repeating short- and long-term slow slip events with deep tremor activity around the Bungo channel region, southwest Japan, *Earth Planets Space, 57*, 961–972.

Hirose, H., K. Hirahara, F. Kimata, N. Fujii, and S. Miyazaki (1999), A slow thrust slip event following the two 1996 Hyuganada earthquakes beneath the Bungo Channel, southwest Japan, *Geophys. Res. Lett., 26*, 3237–3240.

Igarashi, T., T. Matsuzawa, and A. Hasegawa (2003), Repeating earthquakes and interplate aseismic slip in the northeastern Japan subduction zone, *J. Geophys.Res., 108*(B5), 2249, doi:10.1029/2002JB001920.
Ishibashi, K. (1981), Specification of a soon-to-occur seismic faulting in the Tokai district, central Japan, based upon seismotectonics, in *Earthquake Prediction: An International Review, Maurice Ewing Ser.*, vol. 4, D. W. Simpson and P. G. Richards, pp. 297–332, AGU, Washington, D. C.
Iwamori, H. (1998), Transportation of H_2O and melting in subduction zones, *Earth Planet. Sci. Lett., 160*, 65–80.
Kaidzu, M., T. Nishimura, M. Murakami, S. Ozawa, T. Sagiya, H. Yarai, and T. Imakiire (2000), Crustal deformation associated with crustal activities in the Northern Izu Islands area during the summer, 2000, *Earth Planets Space, 52*, ix–xviii.
Kato, T. (1983), Secular and earthquake-related vertical crustal movements in Japan as deduced from tidal records [1951–1981]. *Tectonophysics, 97*, 183–200.
Kawasaki, I., Y. Asai, Y. Tamura, T. Sagiya, N. Mikami, Y. Okada, M. Sakata, and M. Kasahara (1995), The 1992 Sanriku-Oki, Japan, ultra-slow earthquake, *J. Phys. Earth, 43*, 105–116.
Kimata, F., and T. Yamauchi (1998), Time series of line lengths on the baselines in the Tokai region detected by EDM in the period of 1978 to 1997 (in Japanese with English abstract), *J. Seismol. Soc. Jpn., 51(35)*, 229–232.
Kobayashi, Y. (1983), On the initiation of subduction of plates (in Japanese), *Earth Monthly, 5*, 510–514.
Kogan, M., G. M. Steblov, R. W. King, T. A. Herring, D. I. Frolov, S. G. Egorov, V. Ye. Levin, A. Lerner-Lam, and A. Jones (2000), Geodetic constraints on the rigidity and relative motion of Eurasia and North America, *Geophys. Res. Lett., 27*, 2041–2044.
Larson, K. M., P. Bodin, and J. Gomberg (2003) Using 1-Hz GPS data to measure deformations caused by the Denali Fault earthquake, *Science, 300*, 1421–1424.
Matsuzawa, T., T. Igarashi, and A. Hasegawa (2002), Characteristic small-earthquake sequence off Sanriku, northeastern Honshu, Japan, *Geophys. Res. Lett., 29*(11), 1543, doi:10.1029/2001GL014632.
Mazzotti, S., X. LePichon, P. Henry, and S. Miyazaki (2000), Full interseismic locking of the Nankai and Japan-West Kurile subduction zones: An analysis of uniform elastic strain accumulation in Japan constrained by permanent GPS, *J. Geophys. Res., 105*, 13,159–13,177.
Miller, M. M., T. Melbourne, D. J. Johnson, and W. Q. Sumner (2002), Periodic slow earthquakes from the Cascadia subduction zone, *Science, 295*, 2423–2423.
Milly, P. C. D. (1994), Climate, soil water storage, and the average annual water balance. *Water Resour. Res., 30*, 2143–2156.
Mitrovica, J. X., M. E. Tamisiea, J. L. Davis, and G. A. Milne (2001), Recent mass balance of polar ice sheets inferred from patterns of global sea-level change, *Nature, 409*, 1026–1029.
Miyazaki, S., and K. Heki (2001), Crustal velocity field of southwest Japan: Subduction and arc-arc collision, *J. Geophys. Res., 106*, 4305–4326.
Miyazaki, S., T. Iwabuchi, K. Heki, and I. Naito (2003), An impact of estimating tropospheric gradient on precise positioning in summer using the Japanese nationwide GPS array, *J. Geophys. Res., 108*(B7), 2335, doi:10.1029/2000JB000113.
Mogi, K. (1969), Monthly distribution of large earthquakes in Japan (in Japanese with English abstract), *Bull. Earthquake Res. Inst. Univ. Tokyo, 47*, 419–427.
Mogi, K. (1985), Temporal variation of crustal deformation during the days preceding a thrust-type great earthquake – the 1944 Tonankai earthquake of magnitude 8.1. *Pure Appl. Geophys., 122*, 765–780.
Munekane, H., M. Tobita, K. Takashima, S. Matsuzaka, Y. Kuroishi, and Y. Masaki (2003), Groundwater-driven vertical movement in Tsukuba detected by GPS, paper presented at the Fall Meeting, AGU, San Francisco, Calif., 12 Dec.

Murakami, M., and S. Miyazaki (2001), Periodicity of strain accumulation detected by permanent GPS array: Possible relationship to seasonality of major earthquakes' occurrence, *Geophys. Res. Lett.*, *28*, 2983–2986.

Nakamura, K. (1983), Possible nascent trench along the eastern Japan Sea as the convergent boundary between Eurasian and North American plates (in Japanese with English abstract), *Bull. Earthq. Res. Inst. Univ. Tokyo*, *58*, 711–722.

Nakano, T., and K. Hirahara (1997), GPS observation of postseismic deformation for the 1995 Hyogo-ken Nanbu earthquake, Japan, *Geophys. Res. Lett.*, *24*, 503–506.

National Astronomical Observatory (2002), *Chronological Scientific Tables*, vol. 76, 942 pp., Maruzen, Tokyo.

Nishimura, T., et al. (2000), Distribution of seismic coupling on the subducting plate boundary in northeastern Japan inferred from GPS observations, *Tectonophysics*, *323*, 217–238.

Obara, K. (2002), Nonvolcanic deep tremor associated with subduction in southwest Japan, *Science*, *296*, 1579–1681.

Ohtake, M., and H. Nakahara (1999), Seasonality of great earthquake occurrence at the northwestern margin of the Philippine Sea plate, *Pure Appl. Geophys.*, *155*, 689–700.

Ohtani, R., N. Koizumi, N. Matsumoto, and E. Tsukuda (2000), Preliminary results from permanent GPS array by the Geological Survey of Japan in conjunction with groundwater-level observations, *Earth Planets Space*, *52*, 663–668.

Okada, M. (1982), Seasonal variation in the occurrence rate of large earthquakes in and near Japan and its regional differences (in Japanese with English abstract), *J. Seismol. Soc. Jpn.*, *35*, 53–64

Okada, Y. (1992), Internal deformation due to shear and tensile faults in a half-space, *Bull. Seismol. Soc. Am.*, *82*, 1018–1040.

Okada, T., T. Matsuzawa, and A. Hasegawa (2003), Comparison of source areas of $M4.8 \pm 0.1$ repeating earthquakes off Kamaishi, NE Japan: Are asperities persistent features? *Earth Planet. Sci. Lett.*, *213*, 361–374.

Ozawa, S., M. Murakami, M. Kaidzu, T. Tada, T. Sagiya, Y. Hatanaka, H. Yarai, and T. Nishimura (2002), Detection and monitoring of ongoing aseismic slip in the Tokai Region, Central Japan, *Science*, *298*, 1009–1012.

Ozawa, S, S. Miyazaki, Y. Hatanaka, T. Imakiire, M. Kaidzu, and M. Murakami (2003), Characteristic silent earthquakes in the eastern part of the Boso Peninsula, central Japan, *Geophys. Res. Lett.*, *30*(6), 1230, doi:10.1029/2002GL016665.

Rogers, G., and H. Dragert (2003), Episodic tremor and slip on the Cascadia subduction zone: the chatter of silent slip, *Science*, *300*, 1942–1943.

Sagiya, T. (1997), Boso peninsula silent earthquake of May 1996, *Eos Trans. AGU*, *78*(46). Fall Meet. Suppl., Abstract G42A-07.

Sagiya, T. (1999), Interplate coupling in the Tokai District, central Japan, deduced from continuous GPS data, *Geophys. Res. Lett.*, *26*, 2315–2318.

Sagiya, T. (2004) Interplate coupling in the Kanto District, central Japan, and the Boso Peninsula silent earthquake in May 1996, *Pure Appl. Geophys.*, *161*, 2327–2342.

Sagiya, T., S. Miyazaki, and T. Tada (2000), Continuous GPS array and present-day crustal deformation of Japan, *Pure Appl. Geophys.*, *157*, 2303–2322.

Sangawa, A. (1993), The paleo-earthquake study using traces of the liquefaction (in Japanese), *Quat. Res.*, *32*, 249–255.

Sato, T., Y. Fukuda, Y. Aoyama, H. McQueen, K. Shibuya, Y. Tamura, K. Asari, and M. Ooe (2001), On the observed annual gravity variation and the effect of sea surface height variations, *Phys. Earth Planet. Inter.*, *123*, 45–63.

Sato, T., K. Imanishi, N. Kato, and T. Sagiya (2004), Detection of a slow slip event from small signal in GPS data, *Geophys. Res. Lett.*, *31*, L05606, doi:10.1029/2004GL019514.

Savage, J. C. (1983), A dislocation model of strain accumulation and release at a subduction zone, *J. Geophys. Res.*, *88*, 4984–4996.

Seno, T., T. Sakurai, and S. Stein (1996), Can the Okhotsk plate be discriminated from the North American plate? *J. Geophys. Res., 101*,11,305–11,315.

Suwa, Y., S. Miura, A. Hasegawa, T. Sato, and K. Tachibana (2003), Inter-plate coupling beneath the NE Japan arc inferred from 3 dimensional crustal deformation, paper presented at the 23rd IUGG General Assembly, AGU, Sapporo, Japan, 8 July.

Takahashi, H., et al. (2004), GPS observation of the first month of postseismic crustal deformation associated with the 2003 Tokachi-Oki earthquake [MJMA8.0], off southeastern Hokkaido, Japan, *Earth Planets Space, 56*, 377–382.

Tanioka, Y., L. Ruff, and K. Satake (1996), The Sanriku-oki, Japan, earthquake of December 28, 1994 [*Mw* 7.7]: Rupture of a different asperity from a previous earthquake, *Geophys. Res. Lett. 23*:1465-1468.

Thornthwaite, C. W. (1948), An approach toward a rational classification of climate, *Geogr. Rev., 38*, 55–94.

Tsuji, H., Y. Hatanaka, T. Sagiya, and M. Hashimoto (1995), Coseismic crustal deformation from the 1994 Hokkaido-Toho-Oki earthquake monitored by a nationwide continuous GPS array in Japan, *Geophys. Res., Lett., 22*, 1669–1672.

Uchida, N., T. Matsuzawa, A. Hasegawa, and T. Igarashi (2003), Interplate quasi-static slip off Sanriku, NE Japan, estimated from repeating earthquakes, *Geophys. Res. Lett., 30*(15), 1801, doi:10.1029/2003GL017452.

Ueda, H., M. Ohtake, and H. Sato (2001), Afterslip of the plate interface following the 1978 Miyagi-Oki, Japan, earthquake, as revealed from geodetic measurement data, *Tectonophysics, 338*, 45–57.

von Huene, R., and D. Scholl (1991), Observations at convergent margins concerning sediment subduction, subduction erosion, and the growth of continental crust, *Rev. Geophys., 29*, 279–316.

von Huene, R., and S. Lallemand (1990), Tectonic erosion along the Japan and Peru convergent margins, *Geol. Soc. Am. Bull., 102*, 704–720.

Wells, R. E., R. J Blakely, Y. Sugiyama, D. W. Scholl, and P. A. Dinterman (2003), Basin-centered asperities in great subduction zone earthquakes: A link between slip, ubsidence, and subduction erosion? *J. Geophys. Res., 108*(B5), 2507,doi:10.1029/2002JB002072.

Yagi, Y., and M. Kikuchi (2003), Partitioning between seismogenic and aseismic slip as highlighted from slow slip events in Hyuga-nada, Japan, *Geophys. Res. Lett., 30*(2), 1087, doi:10.1029/2002GL015664.

Yagi, Y., M. Kikuchi, and T. Sagiya (2001), Co-seismic slip, post-seismic slip, and aftershocks associated with two large earthquakes in 1996 Hyuga-nada, Japan, *Earth Planets Space, 53*, 793–803.

Yagi, Y., M. Kikuchi, and T. Nishimura (2003), Co-seismic slip, post-seismic slip, and largest aftershock associated with the 1994 Sanriku-haruka-oki, Japan, earthquake, *Geophys. Res. Lett., 30*(22), 2177, doi:10.1029/2003GL018189.

Yamamoto, E., S. Matsumura, and T. Ohkubo (2005), A slow slip event in the Tokai area detected by tilt and seismic observation and its possible recurrence, *Earth Planets Space, 57*, 917–923.

Yamanaka, K., and M. Kikuchi (2004), Asperity map along the subduction zone in northeastern Japan inferred from regional seismic data, *J. Geophys. Res., 109*, B07307, doi:10.1029/2003JB002683.

17 Elastic and Viscoelastic Models of Crustal Deformation in Subduction Earthquake Cycles

Kelin Wang

Abstract

An ideal model of subduction earthquake cycles that includes tectonic loading, rate-dependent friction, and viscoelastic stress relaxation is not yet available. For comparison with geodetic observations, the most widely used models treat fault motion in a purely kinematical fashion; that is, the fault slip (or state of locking) is estimated from surface observations regardless of the loading mechanism and frictional properties. These include forward and inverse elastic dislocation and viscoelastic models. Without addressing the loading mechanism, extra care should be taken to ensure that the assigned or estimated fault motion is physically valid. Time dependence is an important aspect of interseismic deformation, although it is often difficult to distinguish between contributions from time-dependent fault motion and stress relaxation of the upper mantle. A brief review of modeling work for the Cascadia margin is provided to demonstrate the advantages and limitations of the elastic and viscoelastic models and to explain a number of practical issues. A finite element viscoelastic model is used to model the evolution of fore-arc crustal deformation since the 1700 great earthquake. Elastic models can also fit most of the Cascadia geodetic observations but by assuming that all deformation is due to current fault motion. The fault motion thus determined effectively includes a contribution from stress relaxation.

Introduction

A "subduction earthquake cycle" includes a great earthquake and subsequent strain accumulation that leads to the next event. Here, the use of the word "cycle" by no means implies periodicity: neither the size of the earthquakes nor the duration of the interseismic interval between two events need be a constant. Ideally, a model of earthquake cycles should account for tectonic stress loading of the system, stress relaxation of the rock medium in response to previous earthquakes and ongoing loading, and fault rupture as a frictional instability. Such a comprehensive model is not yet available.

The loading mechanism is rarely addressed in earthquake-cycle models because many fundamental issues regarding driving forces, strength of plate boundary faults, and the coupling of long-term tectonic processes and short-term earthquake cycles are not known or poorly understood. Assigning a slip rate on one part of the fault to drive seismic and aseismic slip of another part does not really represent loading by tectonic forces (we often know the velocity boundary conditions, but seldom know the stress boundary conditions). Models based on rate- and state-dependent friction laws are useful in demonstrating how seismic fault slips may start and stop as a result of the interplay between fault frictional behavior and system rigidity [*Stuart*, 1988; *Marone and Saffer*, this volume; *Beeler*, 2005], although the parameters to use when applying these friction laws to subduction faults are not well known. For comparison with geodetic observations, however, the most widely used models treat the fault motion in a purely kinematical fashion; that is, surface deformation is predicted from given fault slip (or state of locking) regardless of the loading mechanism and friction properties. These are the models to be discussed in the present paper.

In this type of model the motion of the subduction fault is usually decomposed into a steady subduction component and an earthquake perturbation component [*Savage*, 1983]. It is assumed that multiple earthquake cycles cause no net deformation, and any long-term net deformation is attributed to steady subduction. There are also models that mix earthquake cycles and steady subduction [e.g., *Matsu'ura and Sato*, 1989; *Cohen*, 1994]. Unless necessary to explore nonlinear coupling between the two processes, long-term deformation associated with steady subduction can be studied with separate deformation models. With steady subduction subtracted, a fully locked segment of the fault can be equivalently described as backward slip at the plate convergence rate, and the slip deficit becomes "back slip" (fig. 17.1a). Therefore a sudden forward slip, representing an earthquake, and subsequent slow back slip, representing fault locking, constitute the basic earthquake cycle.

Slip of the seismogenic portion of the fault as a function of time is schematically illustrated in figure 17.1b, including (top panel) and excluding (bottom panel) steady subduction. Each vertical line segment, i.e., an instantaneous slip, represents an earthquake. The commonly assumed, idealized staircase function (thick line in the top panel) depicts a pure stick-slip process. It is possible that aseismic slip occurs before, between, or after earthquakes, as illustrated by points 1, 2, and 3 in figure 17.1b, respectively, but these scenarios are usually not considered in earthquake-cycle models. The simplest way to account for some aseismic slip between earthquakes is to assume that the slip

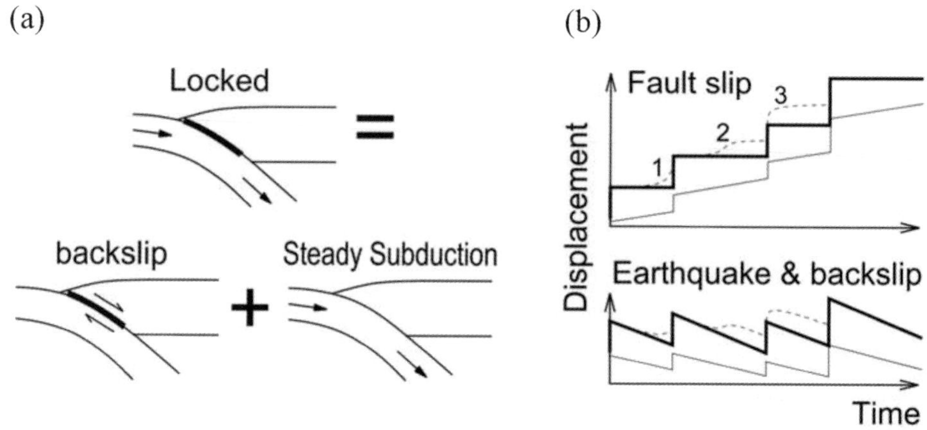

Figure 17.1 (a) Decomposition of an earthquake cycle problem into a perturbation component and steady subduction component. (b) Schematic illustration of the slip history of a subduction fault (top) including and (bottom) excluding steady subduction. The thick solid line shows the commonly assumed slip history in earthquake-cycle models. The thin solid line is a modification of this history, with the assumption of some constant-rate aseismic slip between earthquakes. Numbered dashed lines represent possible complications of fault slip.

is at a constant slow rate and to remove it together with steady subduction (thin solid line in fig. 17.1b) [e.g., *Zheng et al.*, 1996]. The effect is to reduce the amplitude of the sawtooth slip function.

This review paper consists of two parts. First, a critical review of the subject is provided. Attention is paid to the clarification of various concepts. No effort is made to compile an exhaustive reference list. Second, a brief review of recent modeling work for the Cascadia subduction zone is provided. The examples provide a perspective for understanding model results when the same data can be explained using either elastic or viscoelastic models. Comparison is then made with other subduction zones that have had recent great earthquakes.

Deformation Models with Prescribed Fault Motion

Kinematics Versus Driving Force

Driving force, or tectonic loading, is often studied in subduction zone stress models [*Wang and He*, 1999] or long-term deformation models such as for the accretionary prism [*Davis et al.*, 1983]. In earthquake-cycle models with prescribed fault motion, the driving force is not considered. In this section, I wish to make the point that while we usually deal with the kinematics of fault motion, we should pay attention to what the prescribed motion implies about the forces that drive the motion. For simplicity, only purely elastic models are used as examples.

Figure 17.2 shows horizontal velocities predicted by three simple two-dimensional (2-D) models with different assumptions on the behavior of the most updip segment of the subduction fault. In one model (green), the most

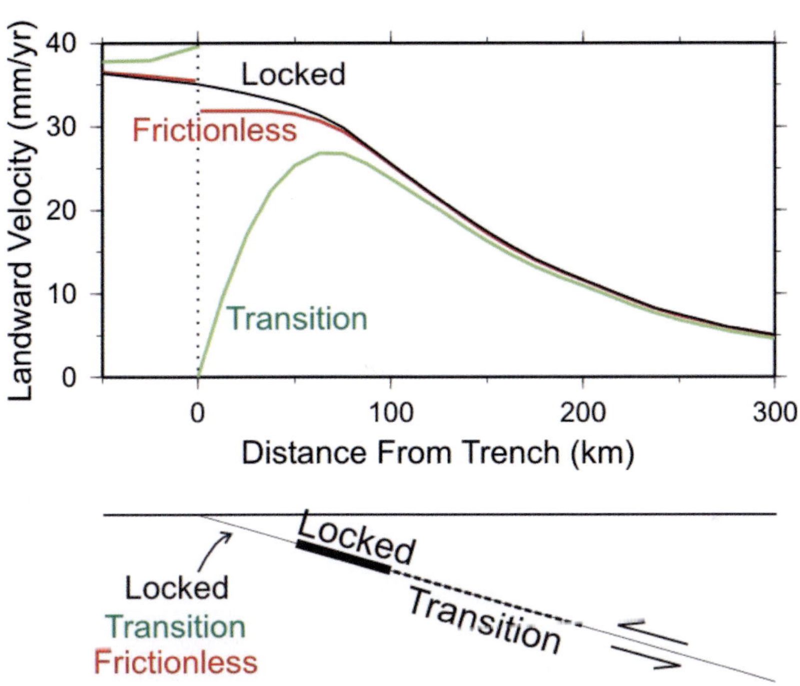

Figure 17.2 Horizontal surface velocities predicted by three simple two-dimensional (2-D) examples with different assumptions (color-coded) on the behavior of the segment updip of the locked zone. Velocities are relative to very remote regions of the upper plate that are not affected by earthquake deformation. The plate convergence rate is 40 mm/yr. Even if the updip segment is frictionless (end member of a very weak fault), it has little interseismic slip because of the presence of the locked zone immediately downdip. The predicted surface velocity (red) is similar to that predicted by assuming the segment to be locked (black). The green curve is the velocity profile if the updip segment is assigned a slip rate that decreases from the plate convergence rate at the trench to zero at the locked zone. Fixing the updip end of the fault at the plate convergence rate (i.e., zero back slip) is a common but possibly incorrect practice in geodetic inversion. Land-based GPS observations usually cannot distinguish between the three models. Because analytical dislocation solutions cannot handle a frictionless fault, these models have been developed using the finite element method. The "locked" and "transition" models reproduce the dislocation solutions almost exactly.

updip end of the fault is assigned the plate convergence rate (i.e., zero back slip) while the downdip zone is locked. The steady slip of the updip end in this model is questionable, because there is no driving force for it [*Wang and Dixon*, 2004]. In the presence of a locked zone it is very unlikely that any part of the updip segment can slip at this rate in a sustained fashion. If the updip segment is weak because of unconsolidated sediments or high pore-fluid pressure [*Moore et al.*, 2005], it may not have much shear stress and is not locked in a mechanical sense. This is represented by the model with a frictionless updip segment (red). The predicted surface velocities are similar to that predicted by assuming the fault to be locked all the way to the trench (black). Because of the presence of the downdip locked zone, this updip segment has little slip and, kinematically, it is better to consider it part of the "locked" zone. This reasoning is consistent with the lack of resolvable slip of the shallowest part of the subduction fault off Peru [*Gagnon et al.*, 2005].

If the updip segment has a stable frictional behavior, it may be pushed to slip when the locked zone ruptures in the next earthquake. If the updip segment does not slip during the earthquake or slips more slowly when the locked segment ruptures, compressive stress is increased in the upper plate above the two segments. The relief of this incremental stress (release of strain energy) can cause the updip segment to slip in a transient fashion while the downdip segment is locked, as observed after the M_w 8.7 2005 Nias-Simeulue, Sumatra, earthquake [*Hsu et al.*, 2006]. However, it cannot slip if the stress is relieved. At a given point in time we may see an updip segment slip faster than a downdip segment. However, slipping at the plate convergence rate is unlikely to last for a long time. The behavior of the velocity-strengthening updip segment in earthquake cycles is discussed in *Wang and Y. Hu* [2006] in the context of critical and stable Coulomb wedges.

Many papers claim that inversion of interseismic land-based geodetic, mainly GPS, data have shown the updip edge of the subduction fault (offshore) to be slipping continuously at or near the plate convergence rate. However, the continuous slip was not necessarily "resolved" by the geodetic data but may have been assigned zero back slip as a boundary condition in the inversion. GPS networks are usually too far from the trench to adequately resolve this motion, as shown in figure 17.2.

A few other problematic back-slip patterns are schematically shown in figure 17.3 (left panels). Similar patterns are frequently seen in geodetic inversion results, and the patches with higher back-slip rates are very often called "asperities." For each pattern the actual slip distribution it represents is shown in the right panel. The slip vector of steady subduction is indicated with an arrow between the two panels. In figure 17.3a the back-slip rate is faster than plate convergence, such that when steady subduction is added back, the fault actually slips backward. This pattern is surprisingly popular in the literature, but it represents an unlikely scenario in which a segment of a subduction thrust fault becomes a (aseismically slipping) normal fault. In

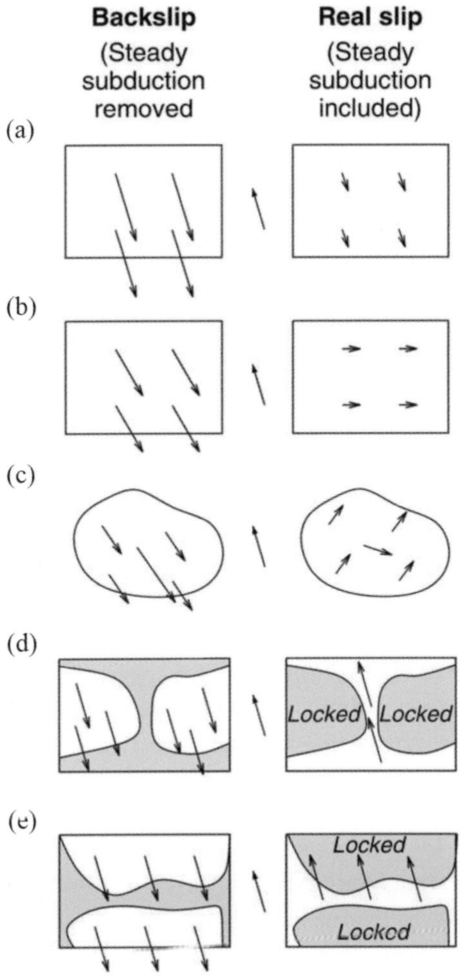

Figure 17.3 Examples of problematic (left) back-slip patterns often seen in the literature. The arrows represent slip of the upper plate relative to the lower plate along the plate interface. The arrow in the middle is the steady plate convergence vector. Adding the plate convergence vector to the (left) back-slip vectors gives the (right) real slip vectors.

figure 17.3b the back-slip vector is at a significant angle with the direction of plate convergence. With steady plate motion added back, the actual slip of the fault segment is nearly perpendicular to plate convergence. Figure 17.3c is similar to figure 17.3b but is a more common version of an "asperity," with the back-slip rate attaining a maximum value at the center of the patch. The back-slip distribution appears reasonable, but it represents a very complex, if not unlikely, slip pattern if steady subduction is added back: the center of the asperity slips at a different direction from its surrounding areas. It is difficult to accept that the back-slip patterns in figures 17.3a, 17.3b, or 17.3c represent actual fault locking. Figures 17.3d and 3e show that a small fault patch between two locked patches slips by itself at the plate convergence rate (zero back-slip rate). In reality, the slipping corridor in each of these two cases, similar to the updip segment in figure 17.2, is unlikely to slip in a sustained fashion when its neighboring areas are locked.

Except perhaps for the first one, the scenarios in figure 17.3 may be possible as transient phenomena under special circumstances. One part of the subduction fault may occasionally slip aseismically in an odd fashion, especially in response to the rupture of neighboring fault patches. These scenarios require special explanations about their driving forces and do not represent general interseismic behavior. The least we can do (or the editors can request) is to avoid directly interpreting back-slip patterns like the left panels of figure 17.3. Adding back the removed steady-subduction motion as in the right panels of figure 17.3 often immediately reveals problems.

Time-Independent Versus Time-Dependent Deformation

Numerous geological and geodetic observations have been made to constrain coseismic and interseismic crustal deformation in response to the rupture and locking of subduction faults. It is generally not difficult to fit most of the data by adjusting parameters or performing an inversion. The challenge is to establish a first-order deformation pattern from the data that allows us to explore the underlying physical processes.

When using a purely elastic model to study interseismic deformation and analyze hazard potential for future earthquakes, there is often a tacit assumption that interseismic deformation rate does not change with time. Some key observations tell us that the deformation rate is strongly time dependent. Geodetic observations made at a given time, and to some extent the elastic models based on these observations, represent a "snapshot" of an evolving deformation field. Important characteristics of subduction earthquake-cycle deformation are schematically summarized in figure 17.4, where the velocities and displacements are relative to distant regions of the upper plate that are not affected by the earthquake cycle.

1. At the coseismic stage, most crustal deformation occurs near the rupture area, although deformation caused by very large events such as the M_w ~ 9.2 2004 Sumatra earthquake can be detected at rather large epicentral distances [*Chlieh et al.*, 2006]. It is well known that the coseismic deformation can be modeled using an elastic earth model. According to these models, maximum subsidence occurs roughly above the fault area in which the rupture zone tapers to zero in the downdip direction (point *a* in fig. 17.4). For active continental margins, if the coastline is around this area, geological evidence for past coseismic subsidence can often be observed [*Plafker*, 1972; *Atwater*, 1987; *Atwater and Hemphill-Haley*, 1997].

2. Within days to months following a subduction earthquake, deformation near the coseismic rupture zone appears to be much faster than decades or centuries after. The fast postseismic deformation was apparent in elevation changes inferred from tide gauge records after of the 1944/1946 Nankai earthquakes [*Thatcher and Rundle*, 1984]. Similar postseismic transients have been

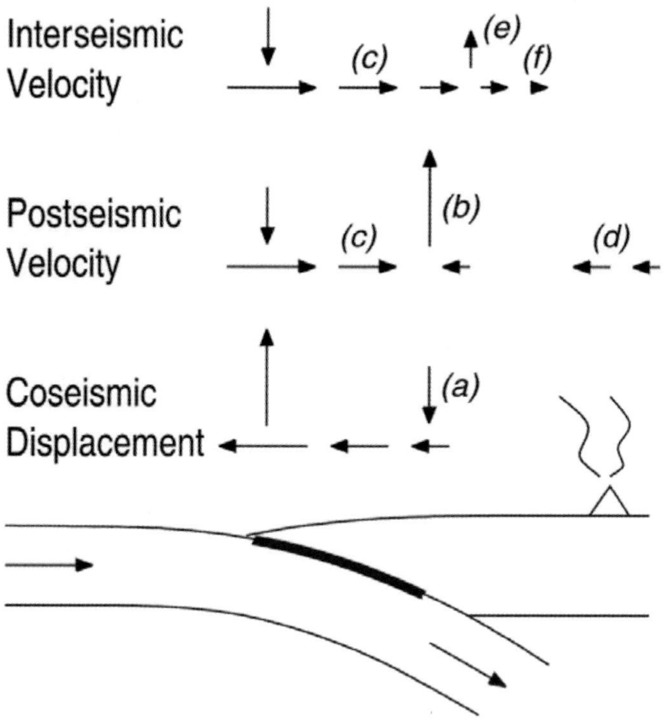

Figure 17.4 Summary of important features of observed earthquake cycle deformation. For simplicity, co- and postseismic deformation near the trench are not included in this illustration. Here postseismic indicates a few years to a few decades after a great earthquake, and interseismic is a few decades to centuries after the earthquake. See the Time-Independent Versus Time-Dependent Deformation section in text for discussion of alphabetically labeled points.

detected by continuously monitoring GPS networks in other subduction zones [*Bürgmann et al.*, 2001; *Melbourne et al.*, 2002; *Yagi et al.*, 2003].

3. Geological and geodetic observations made within a few years after great subduction earthquakes indicate that the region of maximum coseismic subsidence quickly bounces back to become a region of uplift (point *b* in fig. 17.4). The observations include repeat leveling data before and after the 1944/1946 great Nankai, southwest Japan, earthquakes [*Thatcher*, 1984] and comparison of tide gauge records after the 1964 great Alaska earthquake with *Plafker*'s [1971] coseismic deformation observations [*Cohen and Freymueller*, 2001].

4. With rare exceptions, GPS measurements at subduction zones show landward motion of coastal regions (point *c* in fig. 17.4), understood to reflect the interseismic locking of the subduction fault [e.g., *Dixon*, 1993; *Norabuena et al.*, 1998]. However, at least for great earthquakes that rupture a very long segment of the margin such as the 1960 $M_w = 9.5$ Chile earthquake and the 1964 $M_w = 9.2$ Alaska earthquake, some inland areas 200–400 km from the trench still move slowly seaward a few decades after the event (point *d* in fig. 17.4), as

if to slowly catch up with the coseismic motion [*Savage et al.*, 1999; *Freymueller et al.*, 2000; *Klotz et al.*, 2001; *Khazaradze et al.*, 2002].

5. The only place where geodetic observations have spanned a nearly complete subduction earthquake cycle is the Nankai margin. Repeat leveling observations [*Thatcher*, 1984] show that vertical crustal deformation in response to the 1944/1946 great earthquakes gradually spreads out from the rupture area, with the amplitude decreasing with time (point *e* in fig. 17.4). The pattern has been shown to be consistent with stress relaxation of the upper mantle after the earthquake [*Thatcher and Rundle*, 1984; *Miyashita*, 1987]. Horizontal strain observations that were made a few decades ago did not have sufficient accuracy to define the deformation history clearly, but by inference the evolution of horizontal deformation should be similar to that of the vertical deformation (point *f* in fig. 17.4).

6. One phenomenon not illustrated in figure 17.4 is the recently discovered episodic silent slip events, presumably occurring along the subduction zone plate interface, especially downdip from the seismogenic zone. GPS networks have recently detected such slips in a number of places, the best examples including a few centimeters of slip over a couple of weeks at northern Cascadia [*Dragert et al.*, 2001] and tens of centimeters of slip over a couple of years at eastern Nankai [*Ozawa et al.*, 2002]. The Cascadia slip events show intriguing recurrence regularity and are accompanied by low-frequency nonvolcanic seismic tremors [*Rogers and Dragert*, 2003]. The physical mechanism of these silent events and their implications to earthquake-cycle models are currently under intense investigation.

In addition to the above main characteristics, there are significant differences between different subduction zones or different segments of the same subduction zone. The differences may be related to the age of the subducting plate, convergence rate, sediment thickness and type on the incoming plate, and subduction of special features such as seamounts or fossil and active spreading ridges, etc. A great earthquake may rupture only one segment of a subduction zone, and therefore different segments along the same margin may be at different stages of strain accumulation towards future earthquakes.

Linear Versus Nonlinear Mantle Rheology

In the Maxwell viscoelastic rheology most frequently employed in geodynamic modeling, the rock material exhibits an instantaneous elastic behavior and longer-term viscous behavior. In response to fast loading such as an earthquake or rapid deglaciation, the deformation is initially elastic. The material gradually becomes more fluid like and eventually obeys a "steady-state" viscous flow law such as the following one:

$$\dot{\varepsilon} = A\sigma^n \exp\left(-\frac{Q}{RT}\right) \qquad (17.1)$$

where ε is strain rate, σ is stress, T is absolute temperature, A is a constant that may depend on grain size, Q is the activation energy, and R is the universal gas constant. The exponent n for mantle rocks depends on the microscopic deformation mechanism: $n = 1$ for diffusion creep and around 3 for dislocation creep. Conditions controlling the deformation mechanism have been discussed by *Karato and Wu* [1993]. Numerous deformation mechanisms at the microscopic scale result in various macroscopic flow laws [*Blenkinsop*, 2000], but the above power law is the most widely used in geodynamics. It is customary to quote flow laws for uniaxial shortening under axisymmetric compression, with ε being the shortening rate and σ the difference between the axial stress and the uniform stress normal to the axis.

The above flow law is a steady-state one because it is applicable to rock specimens that deform at a constant strain rate under a given loading stress in laboratories. Deformation behavior between the initial elastic response and the eventual steady-state flow is described by the transient rheology. During the transient phase, strain rate decreases from a large value to that of the steady state. For temperatures appropriate for the lower crust and mantle, strain in the transient phase may follow t^m, with $1/3 < m < 2/3$ [e.g., *Weertman and Weertman*, 1975; *Ranalli and Schloessin*, 1989] or other forms. The transient deformation may also be described using the linearly Kelvin viscoelasticity [e.g., *Pollitz*, 2003; *Pollitz et al.*, 2006] in which the strain follows $[1 - \exp(-t/\tau_K)]$, where τ_K is a time constant. The transient rheology is probably very important for postseismic deformation, but parameters involved are poorly known. The following discussions are based on the untested assumption that the transient rheology can be ignored.

The above steady-state flow law can be cast into a more general expression of the following form [e.g., *Melosh*, 1980; *Wu*, 1992]:

$$\dot{\varepsilon} = B\sigma_{II}^{n-1}\sigma \tag{17.2}$$

where ε and σ are components of strain rate and stress tensors, respectively, σ_{II} is the second stress invariant, and parameter B combines the contributions of A and T. An effective viscosity $\eta(n)$ can be defined as

$$\eta(n) = \frac{\sigma}{2\dot{\varepsilon}} = \frac{1}{2B\sigma_{II}^{n-1}} \tag{17.3}$$

Although we focus on the stress dependence of the effective viscosity in this discussion, it should be mentioned that the effective viscosity is affected by many other conditions. Above all, it depends strongly on temperature and hence depth, although nonlinear coupling between temperature and flow can usually be ignored for earthquake-cycle models, so that a constant background geothermal field can be used. Addition of water significantly decreases viscosity [e.g., *Karato and Wu*, 1993], and fluids from dehydrating subducting slab may weaken materials along the plate interface. Water released from the slab may serpentinize the fore-arc mantle wedge [*Hyndman and Peacock*, 2003]. The rheology of serpentinites is relatively poorly known.

For linear Maxwell viscoelasticity, assumed in most viscoelastic earthquake-cycle and postglacial rebound models, $n = 1$, and $\eta(1)$ is the Newtonian viscosity. Given suddenly imposed and fixed strain, the induced elastic stress relaxes with time. The time scale of the stress relaxation is characterized by a constant Maxwell time $\tau(1)$

$$\tau(1) = \frac{\eta(1)}{\mu} = \frac{1}{2B\mu} \tag{17.4}$$

where μ is the shear modulus. Within time $\tau(1)$ the solid exhibits mostly elastic behavior. For the power-law rheology, there is a similar Maxwell time definition. *Melosh and Raefsky* [1983] showed that for post-seismic deformation, the power-law Maxwell time can be defined as

$$\tau(n) = \frac{1}{2B\mu\sigma_o^{n-1}} \tag{17.5}$$

where σ_o is some characteristic value of the stress perturbation due to the earthquake, and a convenient value to use is the average stress drop in the earthquake.

For the nonlinear rheology the Maxwell time depends on the initial stress perturbation. Stress relaxation is faster if the earthquake has a greater stress drop. Because of the power-law relation, the effective viscosity in the highly stressed area such as around the deep end of the rupture is locally decreased, allowing fast shear deformation. However, as the stress is relaxed and hence the effective viscosity increased, the deformation drastically slows down. Compared with the linear rheology, the power-law rheology is characterized by very fast initial deformation and much slower later deformation. *Melosh and Raefsky* [1983] showed that the patterns of surface deformation following a single dip slip earthquake are virtually the same for Newtonian and power-law mantle rheologies, except for their very different time dependence: Solution for a power-law system can be obtained from that of a Newtonian system by replacing normalized time $t/\tau(1)$ in the latter solution with $[t/\tau(n)]^{1/n}$.

The effective viscosity for power-law flow depends on the total stress (or strain rate). It can be shown [*Melosh*, 1980] that if the stress perturbation is much smaller than the background stress associated with mantle convection, its contribution to the effective viscosity can be neglected. The effective viscosity then depends only on the background stress and can be regarded as Newtonian when used to evaluate deformation associated with the stress perturbation. This reasoning may provide some justification for using a Newtonian mantle for postglacial rebound analysis. An earthquake can generate shear stresses greater than the background stress around the rupture zone, and Newtonian viscosity, at least in the highly stressed region, may be a rather poor approximation. A few fault dimensions away from the rupture zone and/or many $\tau(n)$ after the earthquake, the stress perturbation may be sufficiently small to justify the use of a Newtonian viscosity.

Until recently, there were few postseismic observations to help distinguish between the Newtonian and power-law mantle rheology. Rapid development of space geodesy for crustal deformation monitoring has now allowed us to make accurate, high-density, and continuous near-fault observations. Consideration of the nonlinear rheology is therefore becoming necessary and practical. For example, geodetically observed time-varying crustal deformation rates after recent earthquakes in California led *Freed and Bürgmann* [2004] to propose that power-law rheology applies to the upper mantle in that region. They showed that if the postseismic deformation is modeled using a Newtonian mantle, a much lower viscosity is required to explain observations within the first couple of months than that required to explain later observations. However, a linear transient rheology may also be used to explain these observations [*Pollitz*, 2003].

Fault Slip Versus System Stress Relaxation

Variable crustal deformation rates in earthquake cycles have two primary sources. One is temporal variation of fault slip, and the other is viscoelastic stress relaxation in the rock medium, particularly in the upper mantle. Time-dependent fault slip is an intrinsic feature of the rate- and state-dependent friction laws [*Scholz*, 2002]. Viscoelastic mantle rheology best explains postglacial rebound observations and has also been widely applied to earthquake deformation analyses.

Other processes that may become important under certain conditions include deformation due to pore-fluid pressure evolution of poroelastic rock systems in response to earthquakes. The effect of poroelasticity on postseismic deformation has been reported for continental earthquakes [e.g., *Peltzer et al.*, 1998; *Fialko*, 2004]. Limited modeling results for subduction earthquakes indicate that the effect may be important around the center of the rupture area [*Masterlark et al.*, 2001]. Unless the rupture extends significantly landward of the coast, the effect is difficult to detect using land-based geodetic observations. Poroelasticity will not be further addressed here. Some aspects of its application to earthquake problems are discussed in a separate review paper [*Wang*, 2004].

The two primary sources of time-dependent deformation rates are not mutually exclusive, but models developed for them usually are. Models that use rate-dependent friction laws usually employ a purely elastic medium, and viscoelastic models usually ignore the friction laws. Besides numerical difficulties, one argument for this separation is time scale. It is thought that transient fault slip may be important at a time scale of months to years, but viscoelastic stress relaxation is important at a time scale of decades to centuries. The "after slip" of a downdip extension of the rupture zone of the 1994 Sanriku-oki earthquake off northeast Japan lasted for no more than 2 years [*Uchida et al.*, 2003] or even within 100 days after the earthquake [*Yagi et al.*, 2003]. Postseismic transients of duration of hours to months in other subduction zones have been explained using the after-slip model [*Thatcher and Rundle*, 1984; *Bürgmann et al.*, 2001; *Melbourne et al.*, 2002]. For short time scales such as seconds to months, we usually assume a purely elastic medium. However, there is no obvious reason why transient fault slip cannot last longer than several years. For example, the after-slip model has been used to explain crustal deformation at the Chile and Alaska margins a few decades following great earthquakes [*Barrientos et al.*, 1992; *Zweck et al.*, 2002]. Conversely, if nonlinear rheology is considered, as discussed in the Linear Versus Nonlinear Mantle Rheology section, viscoelastic stress relaxation may also explain some of the short-term postseismic transients [*Freed et al.*, 2006], as the power-law Maxwell time $\tau(n)$ can be as short as days and weeks. Application of transient rheology without invoking after slip or power law can also fit observations of short-term postseismic deformation [*Pollitz et al.*, 2006]. In a kinematical model, however, one may choose to allow arbitrary fault-slip patterns to fit surface deformation data, as is done in most geodetic inversion work.

Most of the characteristics of time-dependent deformation in figure 17.4 can be modeled either with time-variable fault slip or with viscoelastic stress relaxation, or a combination of the two. That is, the same deformation history may be explained in different ways. The nonuniqueness cannot always be removed by making more observations. The acceptance of a model is often based on other knowledge. If the mantle viscosity is higher than 10^{20} Pa s ($\tau(1)$ > 50 years), a purely elastic model is not a bad approximation for earthquake recurrence intervals of a couple of hundred years. However, there is consensus that mantle viscosity for active margins is around 10^{19} Pa s (table 17.1), and for such a low viscosity stress relaxation should be considered. In some cases, the source of the deformation is too shallow to be mantle stress relaxation, so that fault slip becomes a better explanation. For example, continuous near-trench GPS observations following the 2005 Nias-Simeulue earthquake can only be explained as indicating after slip of the fault segment updip of the coseismic rupture zone [*Hsu et al.*, 2006]. In other cases, the fault-slip model is not favored because it requires the slip to occur at very large depth where the temperature and pressure condition makes distributed viscous deformation more likely. However, nonlinear rheology tends to localize deformation, and the distinction between aseismic slip of the fault plane and rapid power-law

TABLE 17.1 **Newtonian Upper Mantle or Mantle Wedge Viscosities in Viscoelastic Subduction Zone Models**

Subduction Zone	Viscosity, 10^{19} Pa s	Method*	Reference
Aleutian/Alaska	1.2–2.2	F-2	*Wahr and Wyss* [1980]
Alaska	~2	F-2	*Zheng et al.* [1996]
Alaska	1–5	A-3	*Piersanti et al.* [1997]
Cascadia	0.1–1	F-2	*Wang et al.* [1994]
Cascadia	1	F-3	*Wang et al.* [2001]
Chile	8–10	A-3	*Piersanti* [1999]
Chile	3	F-3	*Khazaradze et al.* [2002]
Chile	2.5	F-3	*Hu et al.* [2004]
Nankai	~0.5	A-2	*Thatcher and Rundle* [1984]
Nankai	~2	F-2	*Miyashita* [1987]
Nankai	0.5	A-2	*Sato and Matsu'ura* [1992]
Nankai	0.8	A-3	*Pollitz and Sacks* [1997]
Nankai	1	F-3	*Yoshioka and Suzuki* [1999]
Kanto (Japan)	~0.5	A-2	*Matsu'ura and Iwasaki* [1983]
NE Japan	~1	A-2	*Thatcher et al.* [1980]
NE Japan	~1	F-2	*Cohen* [1984]
NE Japan	0.93	F-3	*Suito and Hirahara* [1999]
NE Japan	0.7–1.3	A-2	*Rydelek and Sacks* [1990]
NE Japan	0.4	F-3	*Ueda et al.* [2003]
NE Japan	0.93	F-3	*Hyodo and Hirahara* [2003]

*F: Finite element; A: Analytical solution; 2: two-dimensional; 3: three-dimensional

creep of a shear band along the fault zone may become obscure downdip of the rupture zone. Exactly how much of the surface deformation is due to fault slip and how much is due to viscoelastic stress relaxation will remain a subject of debate for a long time to come.

Coseismic Versus Interseismic Fault Slip

Purely stick-slip motion of the seismogenic zone of subduction faults is commonly assumed (thick line in fig. 17.1b). For example, for hazard analysis the size of a future earthquake is often estimated by multiplying the plate convergence rate with the average recurrence interval of past earthquakes. Rather deterministic relationships between the size of earthquake slip and the length of interseismic intervals have been proposed, "time predictable" or "slip predictable," depending on whether the length of an interseismic interval is assumed proportional to the slip of the preceding or ensuing earthquake [*Shimazaki and Nakada*, 1980]. Although simple and seemingly plausible, these relationships have not yet been substantiated by reliable observations [e.g., *Murray and Segall*, 2002].

It is not clear if the seismogenic zone is indeed purely stick-slip or, in other words, if coseismic slips in many earthquake cycles should sum up to the total amount of plate convergence over a long period of time. Total seismic slip estimated from moment release of subduction earthquakes appears to be less than the total plate convergence for many subduction zones [*Pacheco et al.*, 1993], although there are significant uncertainties in estimating slip from seismic moment, especially in the assumed downdip width of the rupture zone [*Hyndman*, this volume]. To balance the plate convergence budget, insufficient coseismic slip has to be compensated by interseismic creep of the seismogenic portion of the fault. The frequently reported very slow interseismic slip of subduction faults (usually called "partial coupling") from fitting coastal GPS observations may pertain to this issue.

At present, we do not know how the seismogenic zone actually moves during the time between great earthquakes. The seismogenic zone is mostly offshore and difficult to monitor using land-based GPS networks. Recent seismological findings from the northeast Japan subduction zone may shed some light [*Hasegawa et al.*, 2005]. *Matsuzawa et al.* [2002] and *Igarashi et al.* [2003] reported recordings of earthquakes that repeatedly rupture the same fault patches of ~0.1–1 km dimension with variable recurrence intervals of weeks to a few years. Assuming these patches to be little stick-slip islands (asperities) surrounded by a sea of aseismic slip, *Igarashi et al.* [2003] and *Uchida et al.* [2003] were able to use the cumulative coseismic slips of repeating earthquakes at individual asperities to represent the total aseismic slip of their surrounding area in a given time window. They envisioned that at a given time the subduction fault is a mosaic of interspersed large areas of little or no slip and regions of mostly aseismic slip that include the small repeating earthquakes. It is reasonable to speculate that the mosaic distribution is present at all scales (fig. 17.5). A group of closely spaced small asperities may form a large asperity, and a collection of these groups constitutes an even larger asperity, and so on. More widely spaced small asperities or groups of asperities surrounded by slipping regions may move in a stick-slip mode to produce repeating earthquakes. The distribution pattern may evolve with time, as influenced by earthquake rupture, pore-fluid pressure, and fault-zone materials, etc.

When developing an earthquake-cycle model for a specific subduction zone, we usually have to assume a coseismic slip without knowing how much interseismic slip actually takes place. In other words, we are not too sure which of the two sawtooth functions in figure 17.1b should be used. Most multiple-cycle models with prescribed fault slips suffer from this uncertainty. It is important to remember this uncertainty when using model results.

Forward Modeling Versus Inversion

Tarantola and Valette [1982, p. 219] succinctly explained the nonuniqueness of inverse problems: "There are two reasons for nonuniqueness. In some problems the nonuniqueness comes from the fact that the data are discrete; if the

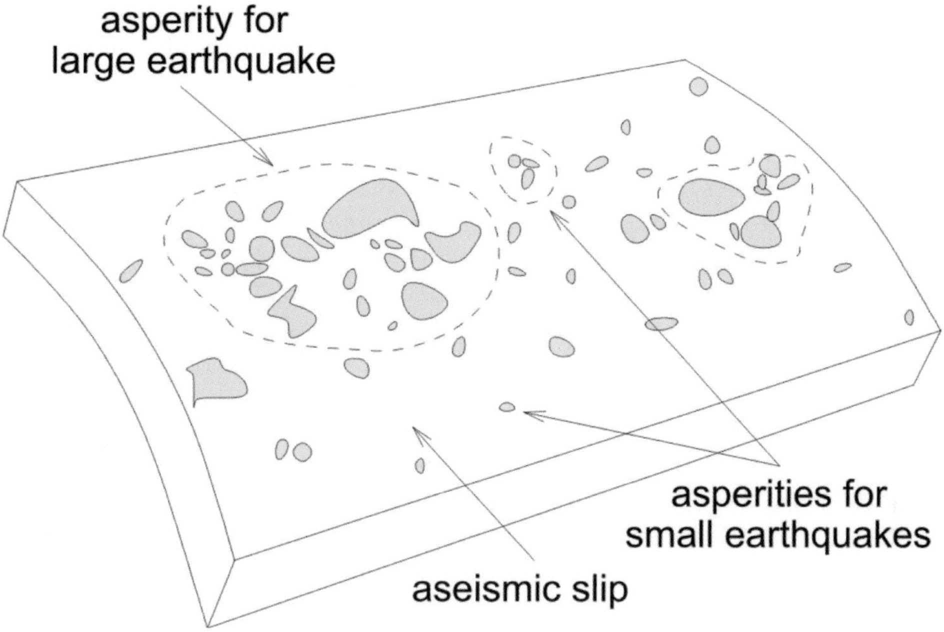

Figure 17.5 A modification of the conceptual asperity model of *Igarashi et al.* [2003] and *Uchida et al.* [2003]. This is similar to the asperity models of *Lay and Kanamori* [1981] and *Lay and Bilek* [this volume], but the fault is envisioned to be a mosaic of stronger and weaker patches at all scales. Closely spaced stronger patches (asperities or groups of asperities) may form a larger asperity. Sparsely spaced asperities or groups of asperities may cause repeating earthquakes.

data are dense, the solution would be unique. . . . In other problems, nonuniqueness may be deeper, as, for example, in the inverse problem of obtaining the density structure of a region of the earth from the measurements of the local gravitational field: Gauss' theorem states that an infinity of different density configurations give identical gravitational fields."

The problem of determining fault slip from surface deformation data suffers from both sources of nonuniqueness. An inversion method can allow additional information to be incorporated in a mathematically rigorous manner. The additional information can be added explicitly as constraints such as a priori values, bounds, smoothness criteria, and "regularization" factors used to cure solution instability, or it can be added implicitly when an inverse problem is "parameterized." A simple example of parameterization is to divide a fault into a few patches of uniform slip, which reduces the number of unknown slip vectors from infinity to the number of fault patches. The fewer patches one uses, the "better resolved" (smaller error bars) are the slip vectors, simply because more information about the slip distribution has been assumed in the inversion. The error introduced in the parameterization is usually not formulated into the error bars of the resolved parameters.

Additional information is the key to the solution of any inverse problem. For this reason, inversion results are always influenced by the researcher's own understanding of the problem. Therefore it is normal to see different researchers obtain different results by using different inversion schemes or even by using the same scheme. This is not to say that inversion is the same as trial-and-error based forward modeling. Given all the information, a properly performed inversion does yield the most probable estimate of the parameter values, with an error bar representing the range of other possible, but less probable, solutions allowed by the same data. A more difficult question is: Is this best mathematical solution the best geophysical solution?

To perform an inversion, we assume that the system can be described by a set of physical laws, expressed in the form of mathematical equations. When the equations accurately describe the real physics, generally believed to be the case for static coseismic deformation and seismic wave propagation, inversion may indeed yield the best geophysical solution. However, for problems like postseismic and interseismic deformation, the processes are much less well understood. We probably have to deal with poorly known processes more than just poorly known parameters. Deciding what processes to include and to neglect is the real challenge. The importance of reasoning, judgment, and intuition often outweighs that of statistical measures of data fit. In this case, forward modeling allows more flexibility in testing various processes and parameters and evaluating competing conceptual models. The usefulness of inverse methods will increase with the maturity level of the subject.

Cascadia Models

Geodetic Observations

At the Cascadia subduction zone the Juan de Fuca (JDF) plate subducts beneath the North America (NA) plate. The last great subduction earthquake at this margin occurred in 1700 with an estimated moment magnitude of 9 [*Satake et al.*, 2003]. The southern Cascadia fore arc, primarily the relatively mafic Siletzia terrane in southern Washington and Oregon and the Klamath Mountains in northern California, moves northward toward the Canadian coast mountains (fig. 17.6) [*Wells et al.*, 1998; *Wells and Simpson*, 2001]. Its motion relative to the rest of North America can be described as a rigid fore-arc sliver rotating around an Euler pole. The position of this Oregon Coast–North America (OC-NA) pole as defined by *Wells and Simpson* [2001] is shown in figure 17.6. There are questions regarding the presence of an eastern boundary of the fore-arc block [*McCaffrey*, 2002], but the questions do not concern the deformation of the coastal area in Cascadia great earthquake cycles.

Strain rates and GPS velocities along the Cascadia fore arc are summarized in figures 17.7a and 17.7b, respectively. The velocities are relative to ref-

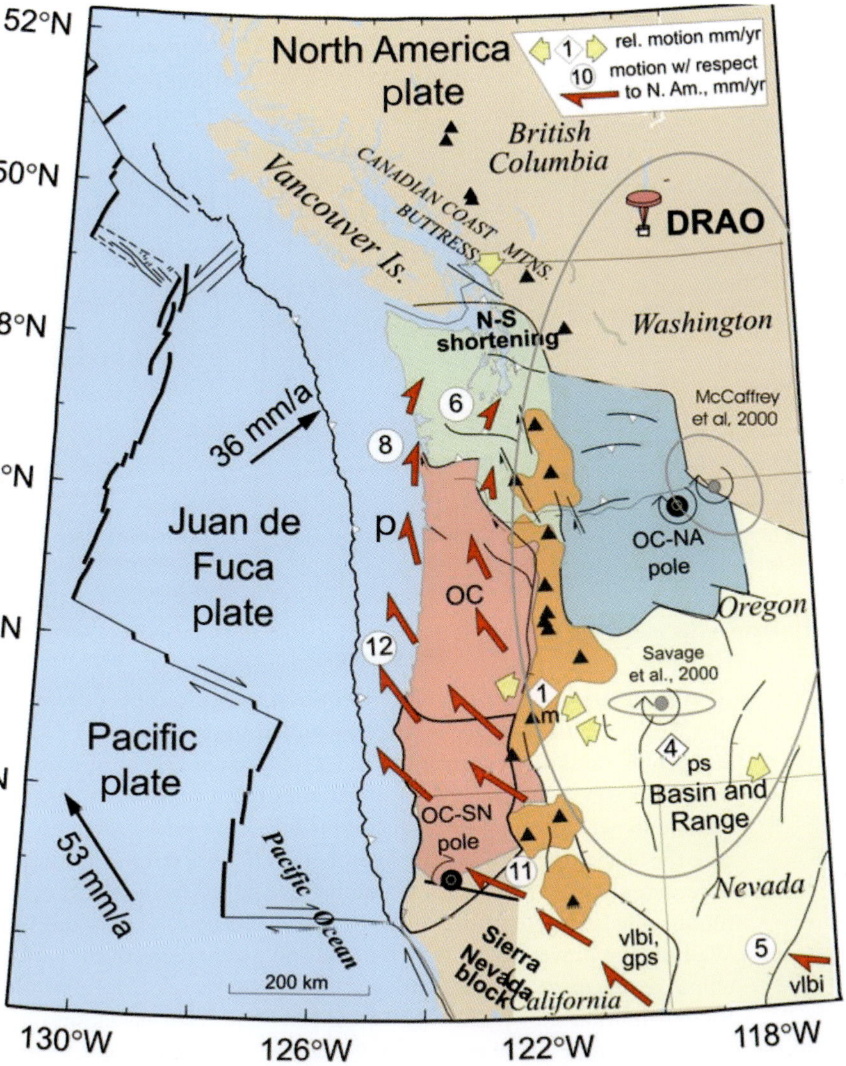

Figure 17.6 Fore-arc motion model of *Wells et al.* [1998] and *Wells and Simpson* [2001]. Fore-arc motion rates (red half arrows) relative to North America (NA) are defined by the OC-NA Euler pole. Rates of motion in millimeters per year are given in circles. Oregon block (OC) rotating at Neogene paleomagnetic rate is linked by the OC-SN Euler pole to the Sierra Nevada block (SN), itself rotating about a distant pole at a rate constrained by very long baseline interferometry (VLBI) and GPS. Poles marked "M" and "S" are OC-NA poles determined using GPS by *McCaffrey et al.* [2000] and *Savage et al.* [2000]. Pairs of yellow arrows indicate relative motion with rates in millimeters per year given in diamonds. Where necessary, types of evidence used to constrain motion rates are labeled as follows: ps, paleoseismic; m, paleomagnetic; vlbi, very long baseline interferometry; gps, global positioning system. The figure is modified from *Wells and Simpson* [2001]. Courtesy of R. Wells.

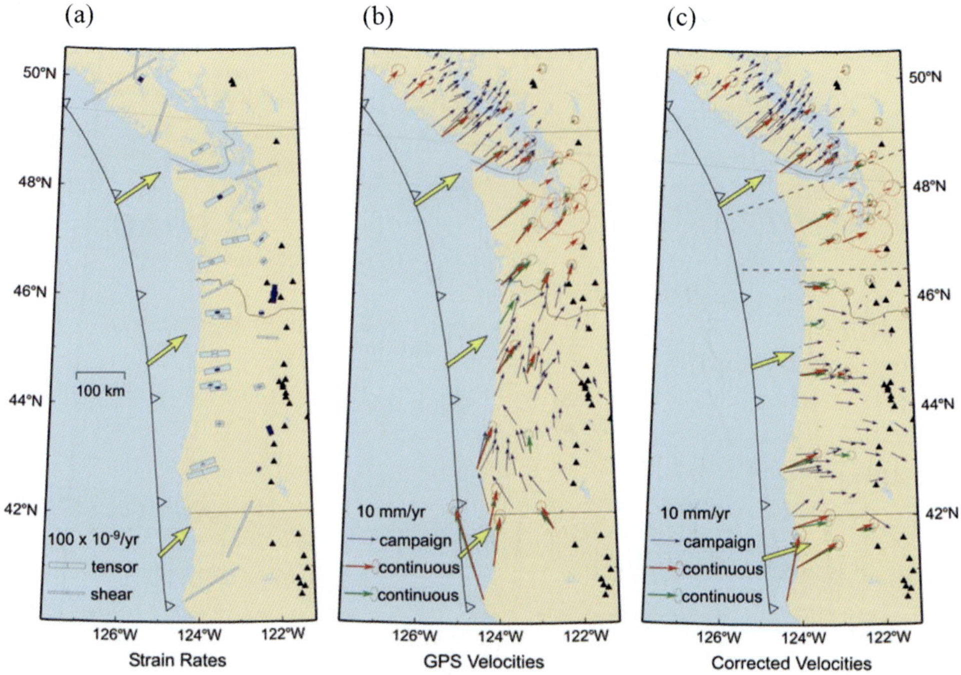

Figure 17.7 (a) A summary of geodetic strain rate measurements compiled from triangulation, laser ranging, and GPS observations. For strain-rate tensor estimates, an open bar indicates contraction, and a solid bar indicates extension. Where only shear strain rates were determined, maximum contraction direction and rate are shown assuming uniaxial contraction. Each value represents an average over the area of the strain network used. (b) GPS velocities relative to reference station DRAO in British Columbia (see fig. 17.6). Values at continuously monitoring sites determined at the Pacific Geoscicence Centre and Central Washington University are shown in red and green, respectively, with the error ellipse representing one standard deviation. (c) GPS velocities after the secular fore-arc motion shown as red half arrows in figure 17.6 are subtracted from sites south of the two dashed lines. No correction is made for velocities north of the dashed lines, and correction applied to sites between the dashed lines is a linear transition. Yellow vectors offshore show the direction of Juan de Fuca plate motion relative to North America in figures 17.7a and 17.7b but relative to Cascadia fore arc in figure 17.7c.

erence GPS site DRAO in British Columbia (shown in fig. 17.6), Some of the strain rates in figure 17.7a (blue crosses) were derived from the GPS data of figure 17.7b. Sources for most of the strain-rate data were reviewed by *Wang et al.* [2003], but figure 17.7a displays an additional eight strain-rate values in Oregon and southern Washington derived by *McCaffrey* [2002] from a larger campaign GPS data set. Campaign GPS velocities in Canada were reported by *Henton* [2000] and *Mazzotti et al.* [2003], and those in the United States by

McCaffrey et al. [2000] and *Savage et al.* [2000]. Velocities at continuously monitoring GPS stations were determined at the Pacific Geoscience Centre [*Mazzotti et al.*, 2003] and at the Central Washington University (M. Miller, personal communication, 2003). The strain rates in figure 17.7a represent interseismic elastic strain buildup due to the locking of the subduction fault. There are two contributions to the GPS velocities, namely interseismic deformation and long-term fore-arc motion shown in figure 17.6. For example, as coastal Oregon rigidly translates northward (fig. 17.7b), it is being elastically shortened in the east-west direction at the same time (fig. 17.7a).

For fore-arc deformation in earthquake cycles, it is the JDF-fore arc not JDF-NA convergence that should be considered. By subtracting the OC-NA motion of figure 17.6, we obtain the "corrected" GPS velocities in figure 17.7c that represent crustal deformation in response to earthquake cycles. The northern fore arc is not affected by the OC motion [*Wells et al.*, 1998], and therefore the JDF-fore-arc convergence in this region is just the JDF-NA convergence. The transition from the JDF-OC convergence in southern Cascadia to JDF-NA convergence in northern Cascadia is assumed to take place linearly over the area between the two dashed lines in figure 17.7c [*Mazzotti et al.*, 2002].

There is a great deal of uncertainty in the long-term fore-arc motion. However, the removal of the rigid-body motion does not affect strain rates (fig. 17.7a) that were derived locally over small networks. The strain-rate data and the northern Cascadia GPS data that are not affected by the sliver correction provide primary constraints for the contemporary interseismic deformation of the Cascadia fore arc. Vertical deformation data such as leveling, gravity, and tide gauge measurements provide ground tilt information [*Hyndman and Wang*, 1995; *Flück et al.*, 1997], but they are best regarded as corroborative information because of a number of unresolved interpretation and observation issues [*Wang et al.*, 2003; *Wolynec*, 2004].

Elastic Half-Space Dislocation Models

Two-dimensional elastic dislocation models were used to model pre-GPS geodetic data for a few margin-normal profiles [*Savage et al.*, 1991, 2000; *Hyndman and Wang*, 1993, 1995; *Dragert et al.*, 1994]. The objective was to define the currently locked patch of the subduction fault. How the locked zone is terminated at its downdip end is not known, but we know the termination cannot be abrupt. So a zone of transition from full locking (modeled as back slip at plate convergence rate) to slipping at plate convergence rate (modeled as zero back slip) was assumed. For a 2-D model (zero along-strike strain) assuming a uniform elastic half space, the results are independent of elastic moduli. For a 3-D half space the results depend only on the Poisson's ratio, which is usually taken to be 0.25.

Geodetic data could allow a wider locked zone with a narrower transition zone or a narrower locked zone with a wider transition zone. *Hyndman and*

Wang [1993, 1995] argued that the updip and downdip ends of the locked zone could be marked by temperatures of about 125° and 350°C, respectively, and developed thermal models to determine these temperatures along the subduction fault. They defined the downdip limit using the thermal results but found that the Cascadia subduction fault was too warm to apply the updip limit. The 3-D dislocation model of *Flück et al.* [1997] was an extension of the 2-D model of *Hyndman and Wang* [1995] but included a more realistic curved subduction fault. Variations of this 3-D model were applied to smaller regions along the Cascadia margin [*Khazaradze et al.*, 1999; *Henton*, 2000].

A more recent 3-D dislocation model, called CAS3D-2, was published by *Wang et al.* [2003]. The model fault geometry and results are shown in figures 17.8a and 17.9, respectively. The model was constrained primarily by strain-rate data and northern Cascadia GPS data. Other GPS data and vertical deformation data were used to validate the model. In addition to using substantially different observational constraints and removing the secular fore-arc motion (fig. 17.6), *Wang et al.* [2003] reevaluated the concept of the transition zone. As discussed in the Time-Independent Versus Time-Dependent Deformation section, crustal deformation in earthquake cycles has a strong time dependence due to transient and episodic fault slip and viscoelastic stress relaxation. As explained in the Slip Versus System Stress Relaxation section, it is often possible to find an equivalent fault-slip distribution to model the effect of stress relaxation. Since elastic models can only deal with fault slip, such an equivalent approach becomes necessary.

The transition zone in CAS3D-2 and in many other elastic dislocation models serves to account for the viscoelastic effects that cannot be modeled with elastic models and for this reason is called the "effective transition zone" [*Wang et al.*, 2003]. Like the "effective elastic thickness" of the lithosphere, the effective transition zone characterizes integrated effects of many controlling factors using a single quantity. The effective transition zone in the CAS3D-2 model has a back-slip distribution that decreases in the downdip direction following an exponential function. Assuming a thermally controlled locked zone as in the work of *Hyndman and Wang* [1995], *Wang et al.* [2003] adjusted the width of the effective transition zone and a parameter that controls the exponential shape of its back-slip distribution to fit geodetic observations. The model applies to Cascadia crustal deformation 300 years after a great earthquake. If a dislocation model were to be developed for deformation earlier in the earthquake cycle, a much narrower effective transition zone would be needed.

There are other types of elastic models for Cascadia interseismic deformation. *Williams and McCaffrey* [2001] used a 2-D elastic plate model to represent the upper plate. A basal traction is used to resist the landward push by the subducting plate across the locked subduction fault. The role of the traction force is similar to that of the effective transition zone and can be adjusted to account for the effect of stress relaxation in the underlying mantle. Three-dimensional inverse models were developed by *McCaffrey et al.* [2000] and *McCaffrey* [2002] to determine back-slip distribution and fore-arc block rotation at the same

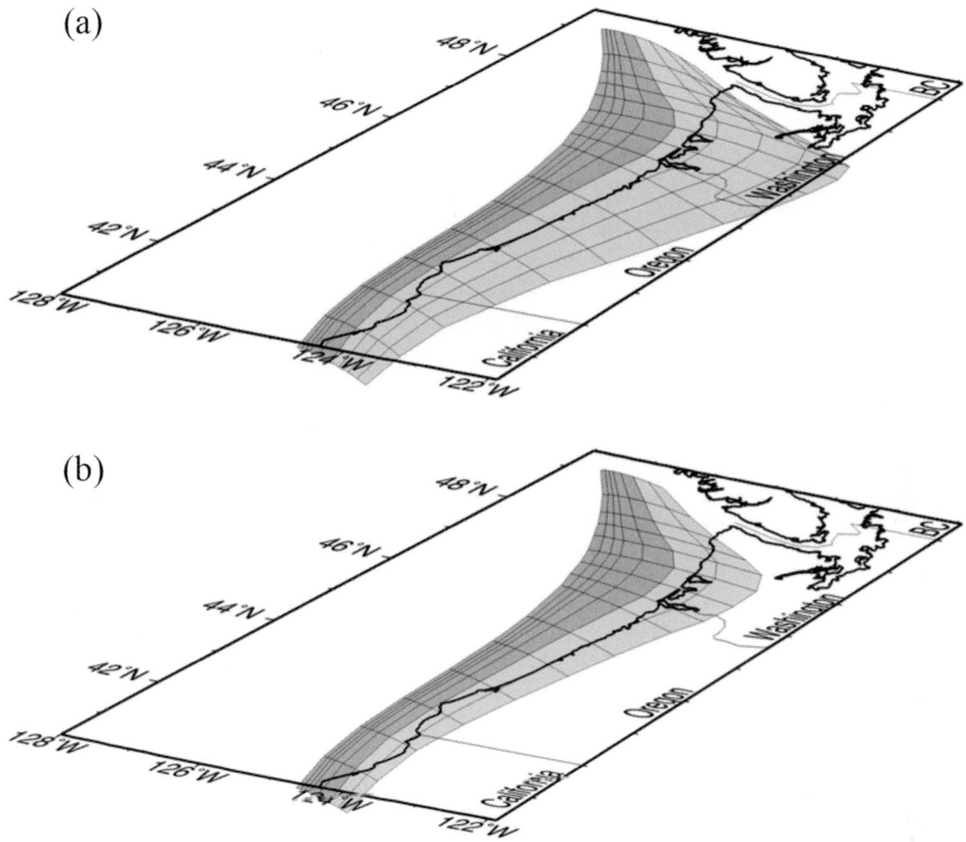

Figure 17.8 Locked (dark shading) and transition (light shading) zones of the Cascadia margin seismogenic zone used in the (a) elastic dislocation model and (b) viscoelastic finite element model.

time. *Yoshioka et al.* [2005] applied a 3-D inversion scheme to northern Cascadia GPS data. Inversion results tend to give greater variations in back-slip distribution and some slow slip of the locked zone. As discussed in the Coseismic Versus Interseismic Fault Slip section, differences between different models are due mostly to how the models are constrained and parameterized. For example, *McCaffrey* [2002] allowed significant ongoing slow slip of the seismogenic zone off Oregon to fit the GPS data, but strain rates (not affected by block rotation) derived from the same data can be well explained by CAS3D-2 that assumes full locking (fig. 17.9a).

Viscoelastic Models

The reason for pursuing viscoelastic models at Cascadia is to establish a connection between the 1700 great earthquake and contemporary geodetic deformation. Many parameters in viscoelastic deformation of the Earth are poorly

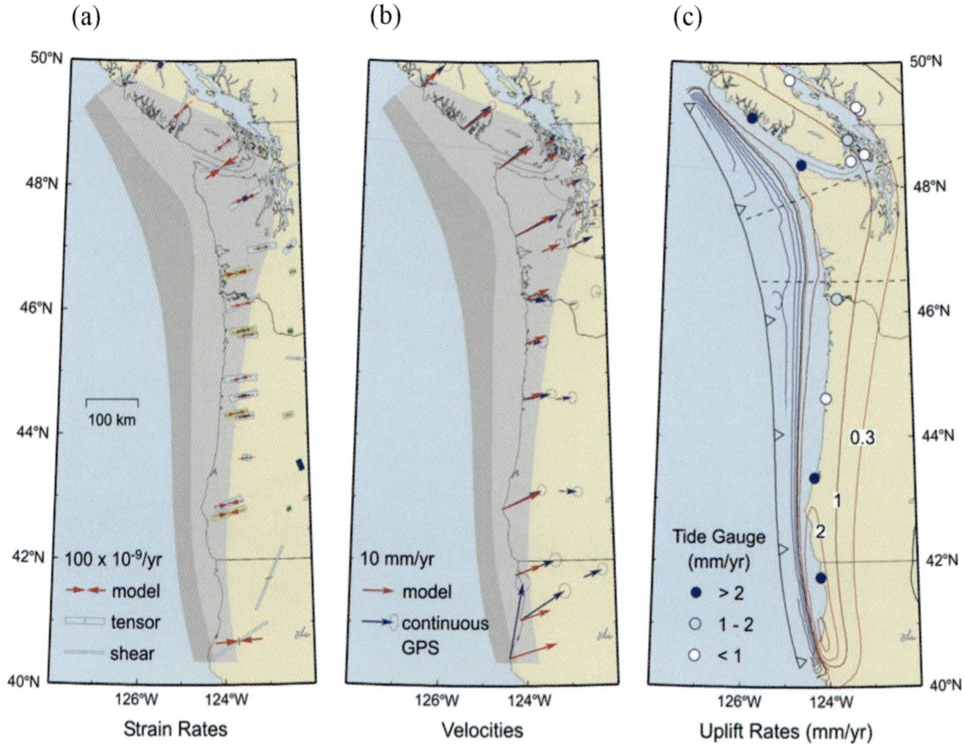

Figure 17.9 Model results of elastic dislocation model CAS3D-2 compared with geodetic observations. (a) Model and observed strain rates. The "tensor" strain rates are the best geodetic data constraints for an interseismic deformation model. Strain rates reported by McCaffrey [2002] (green symbols) were not available as model constraints when CAS3D-2 was developed. (b) Model velocities and continuous GPS velocities. GPS data for central and southern Cascadia have been corrected for secular fore-arc motion (fig. 17.6). (c) Model uplift rates (contour lines) and uplift rates derived from tide gauge records. Modified from Wang et al. [2003].

known, but we wish to know whether the available models can explain what we know about the 1700 earthquake and current crustal deformation and, if not, what additional data and/or assumptions are needed. The finite element method has been used for this type of model at Cascadia because of its flexibility in dealing with geological structure such as a dipping elastic slab. The presence or absence of a slab makes a difference to the stress relaxation process [*Miyashita*, 1987]. A 2-D viscoelastic model was developed for northern Cascadia [*Wang et al.*, 1994]. 3-D Cartesian models were developed for the entire subduction zone for different plate convergence scenarios [*Wang et al.*, 2001]. In this paper, we present some preliminary results of an ongoing modeling effort using spherical earth geometry and 27-node isoparametric triquadratic finite elements. The effect of gravity is simulated using a prestress advection

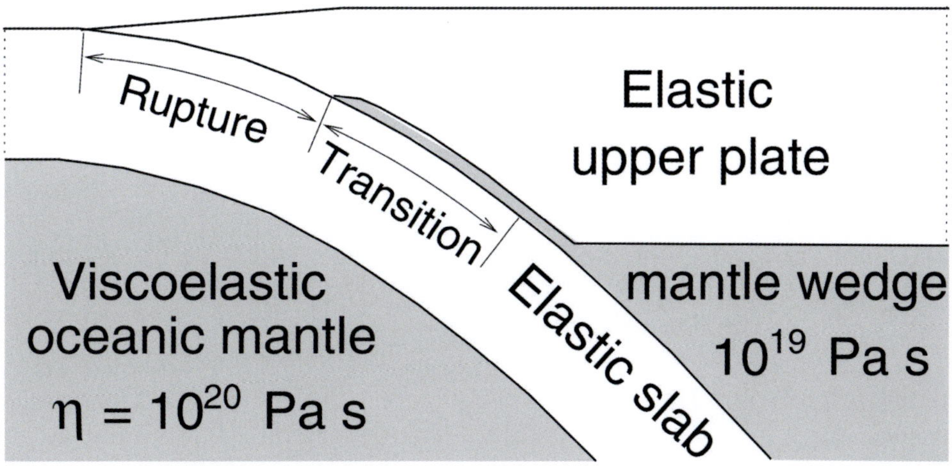

Figure 17.10 Schematic illustration of the fault structure in the viscoelastic model. The rupture zone is assigned a forward slip to simulate earthquake rupture or a back-slip rate to simulate interseismic fault locking. The prescribed slip or back-slip rate taper to zero over the transition zone. The behavior of the transition zone is also controlled by the thin viscoelastic layer along it.

term [*Wang et al.*, 2001]. More detailed descriptions of the modeling technique and model results will be presented elsewhere.

Structure and Fault Slip The basic structure of the model is schematically illustrated in figure 17.10. Material properties for the preferred model are given in table 17.2. Newtonian viscosities are assumed, and the values will be discussed separately below. The finite element mesh of the new model is shown in figure 17.11.

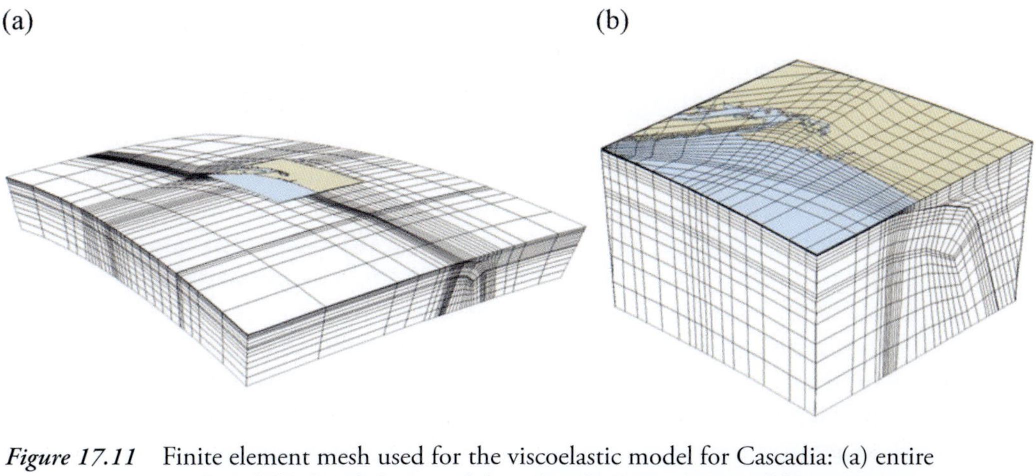

Figure 17.11 Finite element mesh used for the viscoelastic model for Cascadia: (a) entire mesh and (b) detailed view of the colored center portion of figure 17.11a.

TABLE 17.2 **Material Properties for Cascadia Viscoelastic Model***

Structural Unit	Viscosity, Pa s	Young's Modulus, GPa
Continental plate	∞	120
Oceanic plate	∞	120
Continental mantle	10^{19}	160
Oceanic mantle	10^{20}	160

*Poisson's ratio = 0.25, rock density = 3300 kg/m^3, g = 10 m/s^2

We use thicknesses of 40 and 30 km for the overriding and subducting plates, respectively. The plate geometry is the same as in the 3-D Cartesian viscoelastic model of *Wang et al.* [2001]. The shallowest part (<50 km depth) of the plate interface is also the same as in the elastic CAS3D-2 model. The construction of this type of models involves many simplifying assumptions. A very complex model would not reduce the number of assumptions but would introduce many more parameters and make it very difficult to evaluate the results.

For the rupture zone the forward slip is calculated as the JDF-fore-arc convergence rate (discussed in the Geodetic Observations section) times 500 years, a rough average of the recurrence interval of Cascadia subduction earthquakes. The prescribed slip linearly decreases to zero over a zone of transition that is much narrower than that of CAS3D-2 (fig. 17.8b). Although the seismic slip takes place in a single time step in the model, the linear transition is meant to account for some rapid postseismic after slip. Overlapping with the transition zone and farther downdip is a viscoelastic layer of 1-km thickness that allows for continued slow slip after the earthquake. Details of the short-term postseismic transients have little impact on the results 300 years after the earthquake. The JDF-fore-arc convergence rate and direction vary along strike (fig. 17.7c), and the forward slip thus calculated varies accordingly. The along-strike average of the coseismic slip is ~19 m. After the earthquake step, the rupture zone becomes the locked zone and is assigned a back slip at the plate convergence rate and direction to represent complete locking.

Results for the Preferred Model The coseismic slip distribution described above is exactly the same as in the half-space elastic dislocation model used to provide initial seafloor deformation for a trans-Pacific tsunami model for the 1700 earthquake [*Satake et al.*, 2003]. Predicted coseismic deformation, not displayed here, is nearly identical to that shown by *Sakate et al.* [2003]. Variable elastic moduli in the finite element model only result in very small differences in the predicted surface deformation. *Satake et al.* [2003] has compared the model coseismic deformation and resultant tsunami heights with geological observations and Japanese historical tsunami descriptions. New insights from the viscoelastic model are in the time dependence of crustal deformation.

Model deformation 50 and 300 years after the earthquake are shown in figures 17.12 and 17.13, respectively, with the 300-year results compared with modern geodetic observations. Velocities are shown only at continuous GPS stations. Results within 100 km of the northern and southern ends of the subduction zone should be interpreted with caution, because the model is not designed to model deformation around triple junctions. The model reproduces the first-order characteristics of time-dependent crustal deformation in subduction zone earthquake cycles summarizes in figure 17.4 and fits the modern geodetic data reasonably well. The model predicts that strain rates and uplift rates are expected to decrease with time. It also shows that at an early stage of interseismic deformation, such as 50 years, although coastal sites all move landward as a result of fault locking, the inland area is still moving seaward in the direction of coseismic slip (fig. 17.12b).

Viscosities The most critical yet uncertain parameter controlling the time-dependent deformation is the viscosity of the upper mantle. Postglacial rebound models and subduction zone deformation models usually assume a linearly

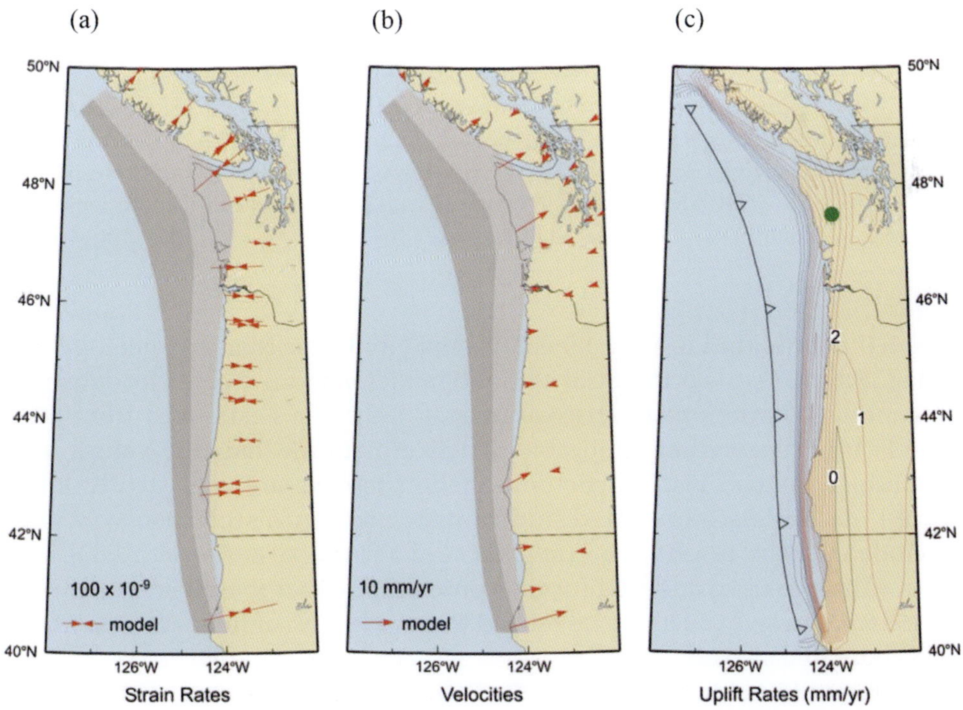

Figure 17.12 Results of viscoelastic Cascadia model at 50 years after the great earthquake. (a) Strain rates, evaluated at observations points. (b) Velocities shown only at continuous GPS sites. (c) Uplift rates, contoured at 1 mm/yr interval. Green spot shows the location where uplift history is displayed in figure 17.14.

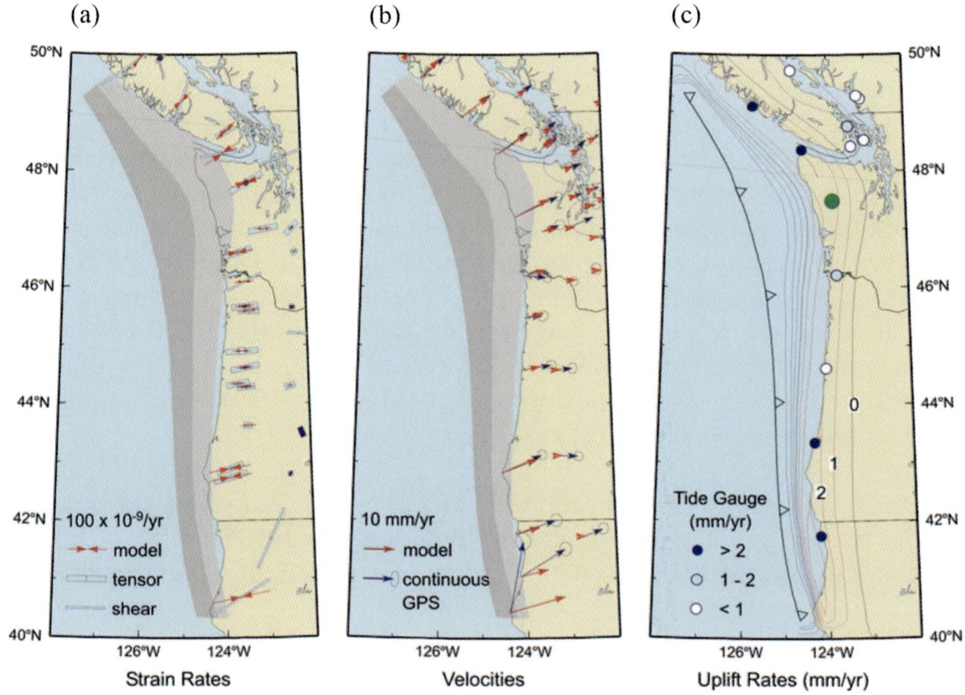

Figure 17.13 Results of viscoelastic Cascadia model at 300 years after the great earthquake (present) and comparison with geodetic observations. See figure 17.7 for explanation of geodetic observations. (a) Strain rates, evaluated at observation points. (b) Velocities, shown only at continuous GPS sites. (c) Uplift rates, contoured at 1 mm/yr interval. Green spot shows the location where uplift history will be displayed in figure 17.14. Green spot shows the location where uplift history is displayed in figure 17.14.

Maxwell mantle (but see the Linear Versus Nonlinear Mantle Rheology section). Continent-scale postglacial rebound models are constrained by data from the continental interior, and viscosities of $10^{21}-10^{22}$ Pa s estimated from these models are probably not applicable to active margins. Viscosity values used in subduction zone viscoelastic deformation models are consistently lower (table 17.1). Postglacial rebound studies at the northern Cascadia margin yield a mantle viscosity of ~10^{19} Pa s [*James et al.*, 2000; *Clague and James*, 2002]. This is the value we have adopted for the continental upper mantle in our model. The relatively low value may be related to the addition of fluids released from the subducting slab.

Different from our previous model [*Wang et al.*, 2001], a value of 10^{20} Pa s is used for the oceanic mantle in the new model. We found that a smaller value would not produce the third characteristic of the time-dependent deformation discussed in the Time-Independent Versus Time-Dependent Deformation section, that is, the region of maximum coseismic subsidence quickly becomes a region of fastest uplift. This effect is illustrated in figure 17.14, which shows

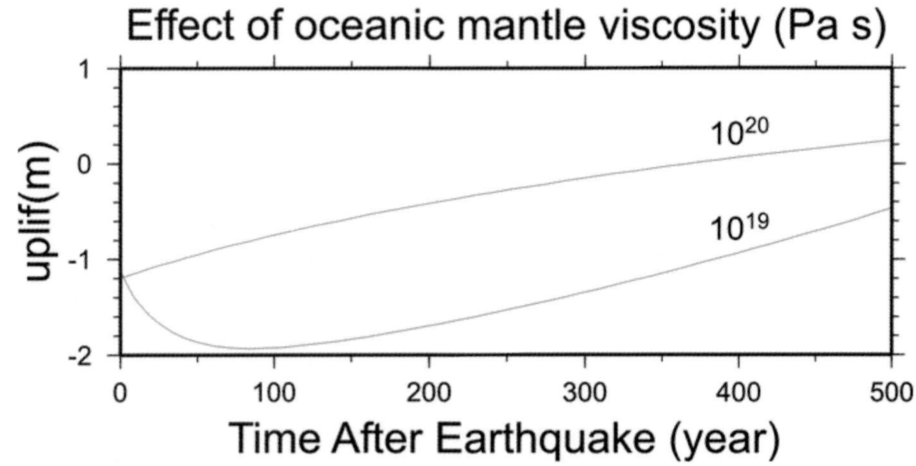

Figure 17.14 Uplift histories at the location indicated in figures 17.12c and 17.13c by a green spot for models with different oceanic mantle viscosities. The value of 10^{20} Pa s is preferred.

uplift histories at a location near the Washington coast (indicated by a green dot on the uplift-rate maps of figs. 17.12c and 17.13c) predicted by two models with different oceanic mantle viscosities. Coseismic subsidence (at time zero) is not affected by the viscosity value. With a viscosity of 10^{19} Pa s for the oceanic mantle, the site will continue to subside after the earthquake for several decades. With a value of 10^{20} Pa s the site reverses the sense of vertical motion immediately after the earthquake. That the oceanic mantle has a higher viscosity than the continental mantle wedge may reflect the lack of additional fluids due to slab dehydration.

Instead of a higher viscosity for the oceanic mantle, we may produce the immediate postseismic uplift in the same area by assuming aseismic fault slip at depths >30 km. We have already lumped the possible short-term after slip into the coseismic transition zone, but a deeper after slip that lasts longer in time cannot be ruled out. No relevant data are available at Cascadia to constrain the short-term postseismic deformation.

Comparison with the Chile and Nankai Margins The most important aspect of viscoelastic deformation in subduction earthquake cycles is the role of stress relaxation in the continental mantle wedge. The coseismic fault slip stretches the upper plate fore arc in the seaward direction. At this time the entire system behaves elastically, and elastic shear stress is generated in the mantle wedge to resist the seaward stretch. This is why coseismic crustal deformation is limited to the proximity of the rupture area. The stress relaxes after the earthquake. A lower viscosity leads to a faster relaxation. The stress relaxation allows the deformation in the upper plate to spread out, as is obvious in the evolution of the uplift rate pattern in figures 17.12c and 17.13c. For earthquakes that rupture very long segments of the subduction fault, the stress relaxation also allows the inland area to move seaward slowly, to catch up with the seaward coseismic motion of the fore arc (point d in fig. 17.4).

Seaward motion of inland sites a few decades after a great earthquake has indeed been observed at the Chile and Alaska margins, as discussed in the Time-Independent Versus Time-Dependent Deformation section. A 3-D viscoelastic model [*Hu et al.*, 2004] similar to the Cascadia model explains the GPS observations 35 years after the 1960 great Chile earthquake very well. The successful application of the viscoelastic stress relaxation model to the Chile data lends support to the Cascadia model. For Alaska, *Zweck et al.* [2002] explained the present-day seaward motion of inland GPS sites by proposing an after slip that lasted at least for several decades in a purely elastic Earth model. However, *Sato et al.* [2003] showed that some viscoelastic stress relaxation is also required to explain these observations.

The Cascadia and the Nankai, southwest Japan, margins are very similar in many ways [*Wang*, 2000], but modern strain rates derived from GPS observations at the Nankai fore arc [e.g., *Kato et al.*, 1998] are larger than those at Cascadia. Part of the reason is that the Cascadia locked zone is generally farther offshore, but the contrast between figures 17.12a and 17.13a adds another perspective. The most recent great earthquakes at Nankai occurred in 1944 ($M_w = 8.2$) and 1946 ($M_w = 8.2$), much more recent than the 1700 Cascadia event. The viscoelastic model indicates that the strain rates several decades after the earthquake should be larger than those 300 years after. The strain rate decreases with time owing to stress relaxation is partially responsible for the difference in the strain-rate size between Nankai and Cascadia. If stress relaxation is not considered, this difference will have to be explained entirely using different fault locking scenarios (i.e., back-slip distributions).

The along-strike rupture lengths of the 1944 and 1946 Nankai events are ~100 and 200 km, respectively, in contrast to the ~1000 km long ruptures of the 1960 Chile, 1964 Alaska, and 1700 Cascadia earthquakes. The shorter rupture perhaps explains why seaward motion similar to the Chile and Alaska margins is presently not observed at Nankai. Numerical experiments by *Hu et al.* [2004] show that such a short rupture will not result in significant seaward motion of the inland area a few decades after the earthquake, because it induces significant coseismic shear stress in a much smaller region of the upper mantle. Shorter ruptures also tend to have smaller coseismic slip and therefore, even near the rupture zone, cause less stress perturbation.

Summary

This paper has focused on subduction zone earthquake-cycle models with fault motion kinematically assigned (in forwarding modeling) or determined (in inverse modeling). For deformation of very short time scales, the earth material exhibits elastic behavior. Elastic models, even with the assumption of a uniform half space, generally describe coseismic deformation and short-term postseismic and interseismic deformation transients very well. For decadal and longer time scales, such as interseismic deformation and postglacial rebound, the earth material, especially that of the upper mantle, exhibits viscoelastic behavior. The process of interseismic crustal deformation is poorly understood. Linear Maxwell viscoelasticity with a viscosity value of around 10^{19} Pa s is commonly used for subduction earthquake-cycle models, but there are large uncertainties in the assumed rheology.

A critical review of earthquake-cycle modeling is provided in the first half of this paper. The main points of emphasis are summarized as follows.

1. Because the driving force for fault slip is not considered in the back-slip-type modeling, it is easy to come up with interseismic fault-slip patterns that are unphysical or unlikely. The physical meaning of the back-slip patterns used to fit geodetic data, especially those determined by inversion, should be carefully examined before inferences on fault properties and loading processes are made.

2. Crustal deformation in subduction earthquake cycles is strongly time dependent, especially in the first few decades after a great earthquake. Coseismic deformation is limited near the rupture zone, with maximum subsidence occurring above the downdip end of the rupture, often around the coast. Rapid deformation within days to months of an earthquake (postseismic transients) occurs in the proximity of the coseismic rupture, but later interseismic deformation involves a wider area in the trench-normal direction. The zone of coseismic subsidence quickly becomes the zone of fast uplift after the earthquake. For earthquakes with large rupture lengths along strike, the inland area a few hundred kilometers from the trench may continue to move seaward decades after the earthquake.

3. Viscous deformation following an earthquake may follow the power law and/or be affected by the transient rheology. Linear (stress-independent) viscosity in viscoelastic earthquake-cycle models is arguably a reasonable approximation for areas away from the fault rupture and sufficiently long after the earthquake. Much effort is needed to investigate the role of nonlinear and transient rock rheology.

4. The time scale of postseismic fault slip around the rupture zone, particularly the downdip segment, appears to be short. The same time-dependent surface deformation can often be explained by either time-dependent fault slip or viscoelastic stress relaxation. Additional knowledge is required to alleviate the nonuniqueness.

5. The commonly assumed purely stick-slip motion of the seismogenic zone may not be valid for many subduction zones. How the seismogenic zone moves between great earthquakes deserves great future attention.

6. The purpose of inversion is to determine parameters. Inverse methods are best applied to processes are relatively well understood, such as coseismic deformation and seismic wave propagation. Forward modeling is

needed to test conceptual models for poorly known processes such as interseismic deformation.

The second part of this paper presents a brief review of the earthquake-cycle modeling work conducted for the Cascadia subduction zone. Various models can be used to fit modern geodetic observations. Elastic models can be viewed to some degree as snapshots of time-dependent deformation. The slip distribution along the fault effectively accounts for the effect of viscoelastic stress relaxation. A viscoelastic model that assumes a Newtonian viscosity value of 10^{19} Pa s for the continental mantle and 10^{20} Pa s for the oceanic mantle yields a deformation history since the 1700 great earthquake that reproduces all the primary features of interseismic deformation observed at different subduction zones. The model is also in general agreement with modern strain rate, GPS, and tide gauge observations in the Cascadia fore arc. Although the model is specifically for the Cascadia subduction zone, the results are useful for understanding deformation patterns in other subduction zones. For example, the higher strain rates currently observed at the Nankai fore arc and the seaward motion of inland GPS stations at the Chile and Alaska margins are consistent with model-predicted deformation patterns several decades after a great earthquake.

Acknowledgments

J. He developed the computer program and carried out most of the simulation for the 3-D viscoelastic finite element model. Comments from B. Atwater, T. Dixon, R. D. Hyndman, W. Thatcher, and an anonymous reviewer substantially improved the paper. Geological Survey of Canada contribution 2004098.

References

Atwater, B. F. (1987), Evidence for great Holocene earthquakes along the outer coast of Washington State, *Science, 236*, 942–944.

Atwater, B. F., and E. Hemphill-Haley (1997), Recurrence intervals for great Earthquakes of the past 3500 years at northeastern Willapa Bay, Washington, *Prof. Pap. U.S. Geol. Surv.*, 1576, 108 pp.

Barrientos, S. E., G. Plafker, and E. Lorca (1992), Postseismic coastal uplift in southern Chile, *Geophys. Res. Lett., 19*, 701–704.

Beeler, N. M. (2007), Laboratory-observed faulting in intrinsically and apparently weak materials: Strength, seismic coupling, dilatancy and pore-fluid pressure, this volume.

Blenkinsop, T. (2000), *Deformation Microstructures and Mechanisms in Minerals and Rocks*, 150 pp., Springer, New York.

Bostock, M. G., R. D. Hyndman, S. Rondenay, and S. M. Peacock (2002), An inverted continental Moho and serpentinization of the forearc mantle, *Nature, 417*, 536–538.

Bürgmann, R., M. G. Kogan, V. E. Levin, C. H. Scholz, R. W. King, and G. M. Steblov (2001), Rapid aseismic moment release following the 5 December, 1997 Kronosky, Kamchatka, earthquake, *Geophys. Res. Lett., 28*, 1331–1334.

Chlieh, M., J.-P. Avouac, V. Hjorleifsdottir, T.-R. A. Song, C. Ji, K. Sieh, A. Sladen, et al. (2006), Coseismic slip and afterslip of the great (Mw9.15) Sumatra-Andaman earthquake of 2004, Bull. Seismol. Soc. Am., in press.

Clague, J., and T. S. James (2002), History and isostatic effects of the last ice sheet in southern British Columbia, *Quat. Sci. Rev., 21*, 71–87.

Cohen, S. C. (1984), Postseismic deformation due to subcrustal viscoelastic relaxation following dip-slip earthquakes, *J. Geophys. Res., 89*, 4538–4544.

Cohen, S. C. (1994), Evaluation of the importance of model features for cyclic deformation due to dip-slip faulting, *Geophys. J. Int., 119*, 831–841.

Cohen, S. C., and J. T. Freymueller (2001), Crustal uplift in the south central Alaska subduction zone: New analysis and interpretation of tide gauge observations, *J. Geophys. Res., 106*, 653–668.

Davis, D. M., J. Suppe, and F. A. Dahlen (1983), Mechanics of fold-and-thrust belts and accretionary wedges, *J. Geophys. Res., 88*, 1153–1172.

Dixon, T. H. (1993), GPS measurement of relative motion of the Cocos and Caribbean plates and strain accumulation across the Middle America trench, *Geophys. Res. Letters, 20*, 2167–2170.

Dragert, H., R. D. Hyndman, G. C. Rogers, and K. Wang (1994), Current deformation and the width of the seismogenic zone of the northern Cascadia subduction thrust, *J. Geophys. Res., 99*, 653–668.

Dragert, H., K. Wang, and T. S. James (2001), A silent slip event on the deeper Cascadia subduction interface, *Science, 292*, 1525–1528.

Fialko, Y. (2004), Evidence of fluid-filled upper crust from observations of postseismic deformation due to the 1992 *Mw* 7.3 Landers earthquake, *J. Geophys. Res., 109*, B08401, doi:10.1029/2004JB002985.

Flück, P., R. D. Hyndman, and K. Wang (1997), Three-dimensional dislocation model for great earthquakes of the Cascadia subduction zone, *J. Geophys. Res., 102*, 20, 539–20, 550.

Freed, A. M., and R. Bürgmann (2004), Evidence of power-law flow in the Mojave desert mantle, *Nature, 430*, 548–551.

Freed, A. M., R. Bürgmann, E. Calais, and J. Freymueller (2006), Stress-dependent power-law flow in the upper mantle following the 2002 Denali, Alaska, earthquake. *Earth Planet. Sci. Lett., 252*, 481–489.

Freymueller, J. T., and J. Beavan (1999), Absence of strain accumulation in the western Shumagin segment of the Alaska subduction zone, *Geophys. Res. Lett., 21*, 3233–3236.

Freymueller, J. T., S. C. Cohen, and H. J. Fletcher (2000), Spatial variations in continuously monitoring GPS stations about 100 to 250 km from present-day deformation, Kenai Peninsula, Alaska, and their implications, *J. Geophys. Res., 105*, 8079–8101.

Gagnon, K., C. D. Chadwell, and E. Norabuena (2005), Measuring the onset of locking in the Peru–Chile trench with GPS and acoustic measurements, *Nature, 434*, 205–208.

Hasegawa, A., N. Uchida, T. Igarashi, T. Matsuzawa, T. Okada, S. Miura, and Y. Suwa (2007), Asperities and quasi-static slip on the subducting plate boundary east off Tohoko, northeast Japan, this volume.

Henton, J. A. (2000), GPS Studies of crustal deformation in the northern Cascadia subduction zone, Ph.D. thesis, Univ. of Victoria, B. C., Canada.

Hsu, Y.-J., M. Simons, J.-P. Avouac, J. Galetzka, K. Sieh, M. Chlieh, D. Natawidjaja, L. Prawirodirdjo, and Y. Bock (2006), Frictional afterslip following the 2005 Nias-Simeulue earthquake, Sumatra, *Science, 312*, 1921–1926.

Hyndman, R. D. (2007), The seismogenic zone of subduction thrust faults: What we know and what we don't know, this volume.

Hyndman, R. D., and S. M. Peacock (2003), Serpentinization of the forearc mantle, *Earth Planet. Sci. Lett.*, *212*, 417–432.

Hyndman, R. D., and K. Wang (1993), Thermal constraints on the zone of major thrust earthquake failure: The Cascadia subduction zone, *J. Geophys. Res.*, *98*, 2039–2060.

Hyndman, R. D., and K. Wang (1995), Current deformation and thermal constraints on the zone of potential great earthquakes on the Cascadia subduction thrust, *J. Geophys. Res.*, *100*, 22,133–22,154.

Hyodo, M., and K. Hirahara (2003), A viscoelastic model of interseismic strain accumulation in Niigata-Kobe tectonic zone of central Japan, *Earth Planets Space*, *55*, 667–675.

Hu, Y., K. Wang, J. He, J. Klotz, and G. Khazaradze (2004), Three-dimensional viscoelastic finite element model for post-seismic deformation of the great 1960 Chile earthquake, *J. Geophys. Res.*, *109*, B12403, doi:10.1029/2004JB003163.

Igarashi, T., T. Matsuzawa, and A. Hasegawa (2003), Repeating earthquakes and interplate aseismic slip in the northeastern Japan subduction zone, *J. Geophys.Res.*, *108*(B5), 2249, doi:10.1029/2002JB001920.

James, T. S., J. J. Clague, K. Wang, and I. I. Hutchinson (2000), Postglacial rebound at the northern Cascadia subduction zone, *Quart. Sci. Rev.*, *19*, 1527–1541.

Karato, S., and P. Wu (1993), Rheology of the upper mantle: A synthesis, *Science*, *260*, 771–778.

Kato, T., G. S. El-Fiky, and E. N. Oware (1998), Crustal strains in the Japanese islands as deduced from dense GPS array, *Geophys. Res. Lett.*, *25*, 3445–3448.

Khazaradze, G., A. Qamar, and H. Dragert (1999), Tectonic deformation in western Washington from continuous GPS measurements, *Geophys. Res. Lett.*, *26*, 3153–3156.

Khazaradze, G., K. Wang, J. Klotz, Y. Hu, and J. He (2002), Prolonged post-seismic deformation of the 1960 great Chile earthquake and implications for mantle rheology, *Geophys. Res. Lett.*, *29*(22), 2050, doi:10.1029/2002GL015986.

Klotz, J., G. Khazaradze, D. Angermann, C. Reigber, R. Perdomo, and O. Cifuentes (2001), Earthquake cycle dominates contemporary crustal deformation in central and southern Andes, *Earth Planet. Sci. Lett.*, *193*, 437–446.

Lay, T., and S. Bilek (2007), Anomalous earthquake ruptures at shallow depths on subduction zone megathrusts, this volume.

Lay, T., and H. Kanamori (1981), An asperity model of great earthquake sequences, in *Earthquake Prediction: An International Review, Maurice Ewing Ser.*, vol. 4, edited by D. Simpson and P. Richards, pp. 579–592, AGU, Washington, D. C.

Marone, C., and D. Saffer (2007), Fault friction and the upper transition from seismic to aseismic faulting, this volume.

Masterlark, T., C. Demets, H. F. Wang, O. Sánchez, and J. Stock (2001), Homogeneous vs heterogeneous subduction zone models: Coseismic and postseismic deformation, *Geophys. Res. Lett.*, *28*, 4047–4050.

Matsu'ura, M., and T. Iwasaki (1983), Study on coseismic and post-seismic crustal movements associated with the 1923 Kanto earthquake, *Tectonophysics*, *97*, 201–215.

Matsu'ura, M., and T. Sato (1989), A dislocation model for the earthquake cycle at convergent plate boundaries, *Geophys. J. Int.*, *96*, 23–32.

Matsuzawa, T., T. Igarashi, and A. Hasegawa (2002), Characteristic small-earthquake sequence off Sanriku, northeastern Honshu, Japan, *Geophys. Res. Lett.*, *29*(11), 1543, doi:10.1029/2001GL014632.

Mazzotti, S., H. Dragert, R. D. Hyndman, M. M. Miller, and J. A. Henton (2002), GPS deformation in a region of high crustal seismicity: N. Cascadia forearc, *Earth Planet. Sci. Lett.*, *198*, 41–48.

Mazzotti, S., H. Dragert, J. Henton, M. Schmidt, R. Hyndman, T. James, Y. Lu, and M. Craymer (2003), Current tectonics of northern Cascadia from a decade of GPS measurements, *J. Geophys. Res.*, *108*(B12), 2554, doi:10.1029/2003JB002653.

McCaffrey, R. (2002), Crustal block rotations and plate coupling, in *Plate Boundary Zones*, Geodyn. Ser., vol. 30, edited by S. Stein and J. Freymueller, pp. 101–122, Washington, D. C.

McCaffrey, R., M. D. Long, C. Goldfinger, P. C. Zwick, J. L. Nabelek, C. K. Johnson, and C. Smith (2000), Rotation and plate coupling along the southern Cascadia subduction zone, *Geophys. Res. Lett, 27*, 3117–3120.

Melbourne, T. I., F. H. Webb, J. M. Stock, and C. Reigber (2002), Rapid postseismic transients in subduction zones from continuous GPS, *J. Geophys. Res., 107*(B10), 2241, doi:10.1029/2001JB000555.

Melosh, H. J. (1980), Rheology of the earth: Theory and observation, in *Physics of the Earth's Interior, Proceedings of the 1979 Enrico Fermi Summer School, Varenna, Italy*, edited by A. M. Dziewonski and E. Boschi, pp. 318–336, Elsevier, New York.

Melosh, H. J., and A. Raefsky (1983), Anelastic response of the earth to a dip slip earthquake, *J. Geophys. Res., 88*, 515–526.

Miyashita, K. (1987), A model of plate convergence in southwest Japan, inferred from levelling data associated with the 1946 Nankaido earthquake, J. *Phys. Earth, 35*, 449–467.

Moore, J. C., C. Rowe, and F. Meneghini (2007), How accretionary prisms elucidate seismogenesis in subduction zones, this volume.

Murray, J., and P. Segall (2002), Testing time-predictable earthquake recurrence by direct measurement of strain accumulation and release, *Nature, 419*, 287–290.

Norabuena, E., L. Leffler, A. Mao, T. Dixon, S. Stein, I. S. Sacks, L. Ocola, and M. Ellis (1998), Space geodetic observations of Nazca-South America convergence along the Central Andes, *Science, 279*, 358–362.

Ozawa, S., M. Murakami, M. Kaidzu, T. Tada, T. Sagiya, Y. Hatanaka, H. Yarai, and T. Nishimura (2002), Detection and monitoring of ongoing aseismic slip in the Tokai region, central Japan, *Science, 298*, 1009–1012.

Pacheco, J. F., L. R. Sykes, and C. H. Scholz (1993), Nature of seismic coupling along simple plate boundaries of the subduction type, *J. Geophys. Res., 98*, 14,133–14,159.

Peltzer, G., P. Rosen, F. Rogez, and K. Hudnut (1998), Poroelastic rebound along the Landers 1992 earthquake surface rupture, *J. Geophys. Res., 103*, 30,131–30,145.

Piersanti, A. (1999), Postseismic deformation in Chile: Constraints on the asthenospheric viscosity, *Geophys. Res. Lett., 26*, 3157–3160.

Piersanti, A., G. Spada, and R. Sabadini (1997), Global postseismic rebound of a viscoelastic Earth: Theory for finite faults and application to the 1964 Alaska earthquake, *J. Geophys. Res., 102*, 477–492.

Plafker, G. (1971), Tectonics, in *The Great Alaska Earthquake of 1964: Geology*, pp. 47–122, Natl. Acad. Sci., Washington, D. C.

Plafker, G. (1972), The Alaskan earthquake of 1964 and Chilean earthquake of 1960: Implications for arc tectonics, *J. Geophys. Res., 77*, 901–925.

Pollitz, F. F. (2003), Transient rheology of the uppermost mantle beneath the Mojave Desert, California, *Earth Planet. Sci. Lett., 215*, 89–104.

Pollitz, F. F., and I. S. Sacks (1997), The 1995 Kobe, Japan, earthquake: A long-delayed aftershock of the offshore 1944 Tonankai and 1946 Nankaido earthquakes, *Bull. Seismol. Soc. Am., 87*, 1–10.

Pollitz, F. F., R. Bürgmann, and P. Banerjee (2006), Post-seismic relaxation following the great 2004 Sumatra-Andaman earthquake on a compressible self-gravitating Earth, *Geophys. J. Int., 167*, 397–420.

Ranalli, G., and H. H. Schloessin (1989), Role of episodic creep in global mantle deformation, in *Slow Deformation and Transmission of Stress in the Earth, Geophys. Monogr. Ser.*, vol. 49, edited by S. C. Cohen and P. Vanicek, pp. 55–63, AGU, Washington, D. C.

Rogers, G. C., and H. Dragert (2003), Episodic tremor and slip on the Cascadia subduction zone: The chatter of silent slip, *Science, 300*, 1942–1943.

Rydelek, P. A., and I. S. Sacks (1990), Asthenospheric viscosity and stress diffusion: A mechanism to explain correlated earthquakes and surface deformation in NE Japan, *Geophys. J. Int., 100*, 39–58.

Satake, K., K. Wang, and B. F. Atwater (2003), Fault slip and seismic moment of the 1700 Cascadia earthquake inferred from Japanese tsunami descriptions, *J. Geophys. Res., 108*(B11), 2535, doi:10.1029/2003JB002521.

Sato, H., J. T. Freymueller, and S. C. Cohen (2003), 3-D Viscoelastic FEM modeling of postseismic deformation caused by the 1964 Alaska earthquake, southern Alaska, *Eos Trans. AGU, 84*(46), Fall Meet. Suppl., Abstract G21B-0260.

Sato, T., and M. Matsu'ura (1992), Cyclic crustal movement, steady uplift of marine terraces, and evolution of the island arc-trench system in southwest Japan, *Geophys. J. Int., 111*, 617–629.

Savage, J. C. (1983), A dislocation model of strain accumulation and release at a subduction zone, *J. Geophys. Res., 88*, 4984–4996.

Savage, J. C., M. Lisowski, and W. H. Prescott (1991), Strain accumulation in western Washington, *J. Geophys. Res., 96*, 14,493–14,507.

Savage, J. C., J. L. Svarc, and W. H. Prescott (1999), Deformation across the Alaska-Aleutian subduction zone near Kodiak, *Geophys. Res. Lett, 26*, 2117–2120.

Savage, J. C., J. L. Svarc, W. H. Prescott, and M. H. Murray (2000), Deformation across the forearc of the Cascadia subduction zone at Cape Blanco, Oregon, *J. Geophys. Res., 105*, 3095–3120.

Scholz, C. H. (2002), *The Mechanics of Earthquakes and Faulting*, 471 pp., Cambridge Univ. Press, New York.

Shimazaki, K., and T. Nakada (1980), Time-predictable recurrence model for large earthquakes, *Geophys. Res. Lett., 7*, 279–282.

Stuart, W. D. (1988), Forecast model for great earthquakes at the Nankai Trough subduction zone, *Pure Appl. Geophys., 126*, 619–641.

Suito, H., and K. Hirahara (1999), Simulation of postseismic deformations caused by the 1896 Riku-u earthquake, northeast Japan: Re-evaluation of the viscosity of the upper mantle, *Geophys. Res. Lett., 26*, 2561–2564.

Tarantola, A., and B. Vallette (1982), Generalized nonlinear inverse problem solved using the least squares criterion, *Rev. Geophys., 20*, 219–232.

Thatcher, W. (1984), The earthquake deformation cycle at the Nankai Trough, southwest Japan, *J. Geophys. Res., 89*, 3087–3101.

Thatcher, W., and J. B. Rundle (1984), A viscoelastic coupling model for the cyclic deformation due to periodically repeated earthquakes at subduction zones, *J. Geophys. Res., 89*, 7631–7640.

Thatcher, W., T. Matsuda, and J. B. Rundle (1980), Lithospheric loading by the 1896 Riku-u earthquake, northern Japan: Implications for plate flexure and asthenospheric rheology, *J. Geophys. Res., 85*, 6429–6435.

Uchida, N., T. Matsuzawa, and A. Hasegawa (2003), Interplate quasi-static slip off Sanriku, NE Japan, estimated from repeating earthquakes, *Geophys. Res. Lett., 30*(15), 1801, doi:10.1029/2003GL017452.

Ueda, H., M. Ohtake, and H. Sato (2003), Postseismic crustal deformation following the 1993 Hokkaido Naseioki earthquake, northern Japan: Evidence for a low-viscosity zone in the uppermost mantle, *J. Geophys. Res., 108*, 2151, doi:10.1029/2002JB002067.

Wahr, J., and M. Wyss (1980), Interpretation of postseismic deformation with a viscoelastic relaxation model, *J. Geophys. Res., 85*, 6471–6477.

Wang, K. (2000), Stress-strain "paradox," plate coupling, and forearc seismicity at the Cascadia and Nankai subduction zones, *Tectonophysics, 319*, 321–338.

Wang, K. (2004), Understanding processes and estimating parameters using mathematical models, in *Hydrogeology of the Oceanic Lithosphere*, edited by E. Davis and H. Elderfield, pp. 376–413, Cambridge Univ. Press, New York.

Wang, K., and T. Dixon (2004), "Coupling" semantics and science in earthquake research, *Eos Trans. AGU, 85*(18), 4 May.

Wang, K., H. Dragert, and H. J. Melosh (1994), Finite element study of uplift and strain across Vancouver Island, *Can. J. Earth Sci., 31*, 1510–1522.

Wang, K., and J. He (1999), Mechanics of low-stress forearcs: Nankai and Cascadia, *J. Geophys. Res., 104*, 15,191–15,205.

Wang, K., J. He, H. Dragert, and T. James (2001), Three-dimensional viscoelastic interseismic deformation model for the Cascadia subduction zone, *Earth Planets Space, 53*, 295–306.

Wang, K., and Y. Hu, Accretionary prisms in subduction earthquake cycles: The theory of dynamic Coulomb wedge, *J. Geophys. Res., 111*, B06410, doi:10.1029/2005JB004094.

Wang, K., R. Wells, S. Mazzotti, R. D. Hyndman, and T. Sagiya (2003), A revised dislocation model of interseismic deformation of the Cascadia subduction zone, *J. Geophys. Res., 108*(B1), 2009, doi:10.1029/2001JB001227.

Weertman, J., and J. R. Weertman (1975), High temperature creep of rock and mantle viscosity, *Annu. Rev. Earth Planet. Sci., 3*, 293–315.

Wells, R. E., and R. W. Simpson (2001), Northward migration of the Cascadia forearc in the northwestern U.S. and implications for subduction deformation, *Earth Planets Space, 53*, 275–283.

Wells, R. E., C. S. Weaver, and R. J. Blakely (1998), Forearc migration in Cascadia and its neotectonic significance, *Geology, 26*(8), 759–762.

Williams, C. A., and R. McCaffrey (2001), Stress rates in the central Cascadia subduction zone inferred from an elastic plate model, *Geophys. Res. Lett., 28*, 2125–2128.

Wolynec, L. (2004), Improved model constraints for vertical deformation across the northern Cascadia subduction zone, M.S. thesis, Univ. of Victoria, B. C., Canada.

Wu, P. (1992), Deformation of an incompressible viscoelastic flat earth with power-law creep: A finite element approach, *Geophys. J. Int., 108*, 35–51.

Yagi, Y., M. Kikuchi, and T. Nishimura (2003), Co-seismic slip, post-seismic slip, and largest aftershock associated with the 1994 Sanriku-haruka-oki, Japan, earthquake, *Geophys. Res. Lett., 30*(22), 2177, doi:10.1029/2003GL018189.

Yoshioka, S., and H. Suzuki (1999), Effects of three-dimensional inhomogeneous viscoelastic structures on postseismic surface deformations associated with the great 1946 Nankaido earthquake, *Pure Appl. Geophys., 154*, 307–328.

Yoskioka, S., K. Wang, and S. Mazzotti (2005), Interseismic locking of the plate interface in the northern Cascadia subduction zone, inferred from inversion of GPS data, *Earth Planet. Sci. Lett., 231*, 239–247.

Zheng, G., R. Dmowska, and J. R. Rice (1996), Modeling earthquake cycles in the Shumagin segment, Alaska, with seismic and geodetic constraints, *J. Geophys. Res., 101*, 8383–8392.

Zweck, C., J. T. Freymueller, and S. C. Cohen (2002), Three-dimensional elastic dislocation modeling of the postseismic response to the 1964 Alaska earth-quake, *J. Geophys. Res., 107*, 1–11.

18

Distinct Updip Limits to Geodetic Locking and Microseismicity at the Northern Costa Rica Seismogenic Zone

Evidence for Two Mechanical Transitions

Susan Y. Schwartz and Heather R. DeShon

Abstract

Results from the Costa Rica Seismogenic Zone Experiment (CRSEIZE), a joint seismic and geodetic experiment designed to elucidate seismogenic processes at the southern end of the Middle America Trench, indicate distinct patterns of geodetic strain accumulation and interplate microseismicity in northern Costa Rica. Geodetic inversions for interseismic slip on the plate interface reveal locking equal to ~75% of the plate velocity between 8 and 15 km depth below sea level, while shallow interplate microearthquakes begin deeper, between 15 and18 km below sea level, in a region that appears to be freely slipping. Interplate microseismicity terminates at ~30 km depth along the updip edge of a second patch of partial geodetic locking that represents ~30–40% of the plate velocity. The strike-parallel alternation of geodetically locked regions with no microseismicity and freely slipping regions with abundant microseismicity suggests that downdip increases in temperature and pressure may generate two distinct transitions in mechanical behavior along the plate interface. We interpret the updip limit of the shallow geodetically locked patch as the frictional stability transition from stable sliding to stick-slip behavior. The shallow transition correlates with models of the 100°C isotherm. The updip limit of interplate seismicity and the freely slipping zone most likely correspond to a change in mechanical properties along the plate interface; increased pore-fluid pressure resulting from basalt dehydration reactions and/or decreased permeability in the upper plate may lead to fault weakening. Modeled temperatures at the deeper transition to the onset of seismicity are estimated at 200°–250°C, in the range where low-grade metamorphic reactions potentially release 0.3–3 wt % water and experimental data reveal significant decreases in permeability for a variety of rock types. Our results suggest that thermally mediated diagenetic processes, low-grade metamorphic reactions, or changes in effective stress occur across several depths to produce multiple transitions in the mechanical behavior of the seismogenic zone and explain the distinct geodetic and seismic observations in northern Costa Rica.

Introduction

Subduction is a fundamental geological process generating and modifying continental crust and associated with severe natural hazards (earthquakes, volcanoes, tsunamis). Subduction zones release ~90% of the Earth's seismic energy [*Pacheco et al.*, 1993] and generate almost all of the world's great earthquakes. Most of this seismic moment reflects mechanical coupling between the underthrusting and overriding plates along a shallow (<50 km depth) portion of the dipping plate interface termed the seismogenic zone. The earthquake cycle consists of accumulation of elastic strain in the seismogenic zone during the interseismic period and rapid release during an earthquake. Factors affecting coupling and the strain accumulation/release process are important for assessing and understanding seismic and tsunami hazard and the earthquake process. Key factors include the efficiency of strain accumulation (completely locked versus partially or completely unlocked), its spatial pattern (updip and downdip limits and along-strike variability), and the relationship between strain accumulation and seismicity.

Previous studies of convergent margin seismogenic zones have found that regions accumulating strain during the interseismic cycle and releasing seismic moment in large earthquakes collocate [e.g., *Hyndman et al.*, 1997; *Oleskevich et al.*, 1999; *Zweck et al.*, 2002; *Igarashi et al.*, 2003]. It is commonly believed that the shallow transition from aseismic to seismic behavior along the subduction thrust can be defined either by the updip limit of large earthquake rupture, the geodetically defined locked zone, or the onset of microseismicity [e.g., *Newman et al.*, 2002; *Obana et al.*, 2003]. As suggested by *Scholz* [2002], the shallow transition from aseismic to seismic behavior may be dominantly temperature controlled. For several subduction zone thrusts, the updip limit of geodetic locking or seismicity has been correlated with models of the 100°–150°C isotherm [e.g., *Hyndman and Wang*, 1993; *Hyndman et al.*, 1995, 1997; *Harris and Wang*, 2002; *Currie et al.*, 2002], a temperature range that corresponds to several phase transformations from weak hydrous to stronger dehydrated minerals [*Hyndman et al.*, 1997; *Peacock and Hyndman*, 1999; *Moore and Saffer*, 2001; *Moore et al.*, this volume].

In this paper we report distinct updip limits of elastic strain accumulation or geodetic locking and microseismicity in the vicinity of the Nicoya Peninsula, Costa Rica. We suggest that the updip limit of strain accumulation reflects the aseismic to seismic transition and the updip extent of potential rupture during large underthrusting earthquakes; the transition is most likely thermally controlled. A second transition to a freely slipping plate interface and the onset of small-magnitude earthquakes suggests that additional diagenetic, metamorphic, or fluid-dominated processes occur, generating a second important transition in mechanical behavior.

Tectonic Setting

The Nicoya Peninsula directly overlies the Costa Rica seismogenic zone, making this section of the Middle America Trench (MAT) particularly well suited for land based investigations of seismogenic zone coupling (fig. 18.1). Along the Pacific coast of Costa Rica, oceanic Cocos plate formed at the East

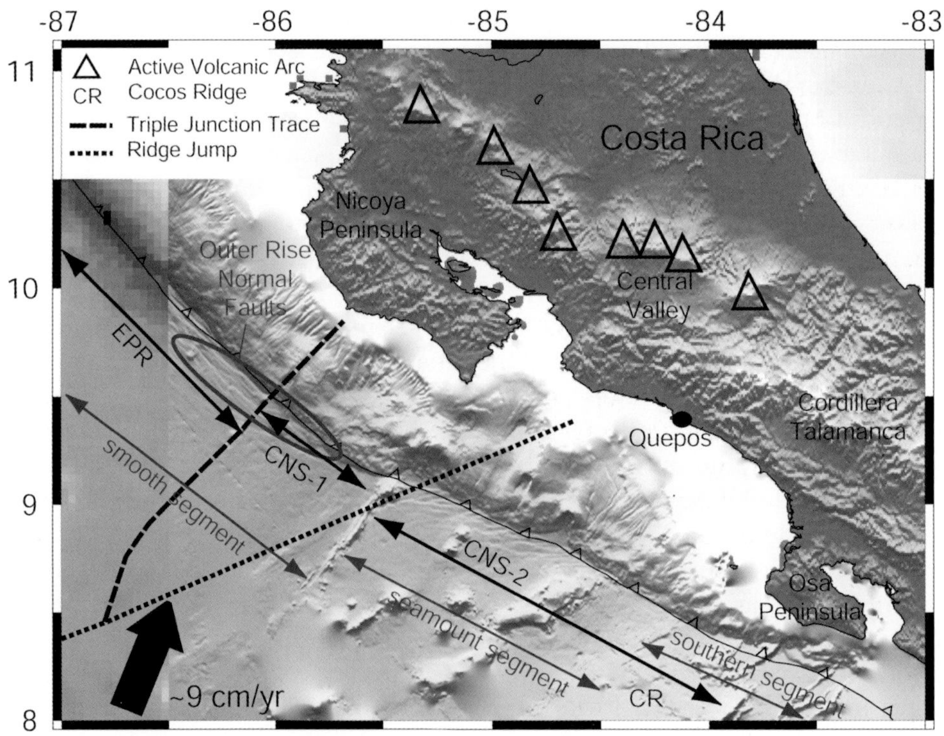

Figure 18.1 Tectonic setting of the Costa Rica subduction zone. The Cocos plate oceanic crust formed at the Cocos-Nazca Spreading Center at 22.7–19.4 Ma and at 19.5–14.5 Ma (CNS-1 and CNS-2 from *Meschede et al.* [1998] and *Barckhausen et al.* [1998, 2001]) subducts along this margin at a rate that varies between ~9 cm/yr at the Cocos Ridge and 8.3 cm/yr beneath the Nicoya Peninsula [*DeMets*, 2001]. East Pacific Rise (EPR) derived lithosphere formed at ~24 Ma subducts beneath the Nicoya Peninsula. Smooth seafloor offshore the Nicoya Peninsula abruptly transitions to seamount dominated seafloor at the trace of the Fisher seamount chain. Bathymetry is from *von Huene et al.* [2000].

Pacific Rise (EPR) and the Cocos-Nazca Spreading Center (CNS) subducts along the MAT at 8.3–9 cm/yr [*DeMets*, 2001]. The morphology, age, and formation history of the Cocos plate varies along strike [*von Huene et al.*, 1995, 2000; *Barckhausen et al.*, 2001], and the western Costa Rican margin can be divided into three morphologic and bathymetric sections: the southern Osa, the seamount-dominated central, and the smooth northern Nicoya segments (fig. 18.1) [*von Huene et al.*, 1995]. In northern Costa Rica the incoming plate exhibits variable thermal gradients across the EPR/CNS plate suture [*Langseth and Silver*, 1996; *Fisher et al.*, 2003] and is overlain by a thin sediment layer comprised of ~180 m of smectite-rich hemipelagic clays overlying ~220 m of carbonate-rich pelagic sediments [e.g., *Kimura et al.*, 1997; *Spinelli and Underwood*, 2004]. Complete sediment subduction occurs along the trench [*Kimura et al.*, 1997], and recent evidence suggests that the Costa Rica margin may be actively eroding along the subduction thrust [*Ranero and von Huene*, 2000; *Vannucchi et al.*, 2001, 2003].

Variability in lithospheric age, thermal history, and morphology of the subducting Cocos plate reflect a complex tectonic history. Oceanic plate history is divided into three stages: (1) oceanic lithosphere formation along the East Pacific Rise; (2) oceanic lithosphere formation along the Cocos-Nazca Spreading Center (CNS-1) following a reorientation of the EPR-CNS triple junction; and (3) lithosphere formation along the Cocos-Nazca Spreading Center (CNS-2) following a ridge jump along the CNS (fig. 18.1) [*Meschede et al.*, 1998; *Barckhausen et al.*, 1998, 2001]. The oldest crust, formed along the EPR at ~24 Ma, subducts offshore the northern Nicoya Peninsula and is anomalously cold for lithosphere of its age owing to strong hydrothermal circulation within the oceanic crust [*Langseth and Silver*, 1996; *Fisher et al.*, 2003]. Movement of fluids mines heat out of the crust by pulling seawater into and expelling it out through exposed basement highs, and recent bathymetric surveys show that EPR-derived oceanic crust is characterized by numerous small seamounts, exposed basement highs, and sediment-covered basement highs [see *Hutnak et al.*, this volume].

Segmentation of the Cocos plate along the southern MAT appears to strongly influence both the seismic behavior and the characteristics of arc magmas [*Carr and Stoiber*, 1990; *Protti et al.*, 1995a; *von Huene et al.*, 1995]. The volcanic arc in Costa Rica is offset westward at the transition from CNS-2 to CNS-1 oceanic crust (fig. 18.1) and shifts eastward again at the Costa Rica-Nicaragua border. Northern Costa Rica and Nicaragua volcanoes exhibit higher components of slab and subducted sediment in their isotopic compositions (e.g., Ba/La, U/Th, and ^{10}Be) than volcanoes in central Costa Rica [see *Patino et al.*, 2000 for a review]. This may reflect a larger amount of hemipelagic and carbonate-rich sediments reaching melt generation depths in the north or the dilution of oceanic sediments by upper continental crustal material derived through subduction erosion in central Costa Rica [*Morris et al.*, 2002].

Subduction erosion has been proposed along much of the Costa Rican margin [*Meschede et al.*, 1998; *Ranero and von Huene*, 2000; *Vannucchi et al.*, 1998, 2001, 2003] with two mechanisms postulated: (1) thinning of the upper plate as seamounts subduct and drag material beneath the shallow part of the fore arc; and (2) basal erosion whereby the fore arc is thinned and upper plate material is transferred to the lower plate. While evidence of fore-arc deformation caused by seamount subduction is common as furrows and scarps along the continental fore arc (fig. 18.1), evidence of deeper basal erosion, which may be unrelated to subducted topography, is more difficult to document. *Ranero and von Huene* [2000] argued for basal erosion along central Costa Rica based on an interpretation of reflection seismic images of large, fault bounded lenses of upper plate material being carried down the subduction zone. The process of subduction erosion may have important implications for the frictional properties of the plate interface because, rather than subducted sediments, the plate interface may be lined with fault gouge consisting of variably compacted, consolidated oceanic sediments and upper plate material. Additionally, large-scale normal faulting along the outer rise (fig. 18.1) may later develop into topographic basins within the subduction channel and carry gouge to magma-generating depths [*von Huene et al.*, 2004].

Protti et al. [1995a] correlated seismic behavior with subducting seafloor characteristics. The largest underthrusting earthquakes, with magnitude >7.5, have historically only occurred along the Nicoya segment where relatively smooth seafloor subducts. Intermediate magnitude interplate events, with magnitudes up to 7.4, occur in the southern Osa segment, where thickened oceanic crust of the Cocos Ridge (fig. 18.1) subducts. The central segment, where seamount dominated seafloor subducts, generates thrust events with magnitudes of only 7.0 and is the location of a general rate increase in background seismicity. More detailed examination of seismicity and its relationship to seafloor characteristics in the vicinity of the Nicoya Peninsula, made possible by CRSEIZE, revealed a shallowing of the updip limit of interplate earthquakes across the EPR/CNS-1 crustal origin boundary [*Newman et al.*, 2002]. Since EPR-derived crust is cooler than CNS-derived crust [*Langseth and Silver*, 1996; *Fisher et al.*, 2003], *Newman et al.* [2002] interpreted their observations as evidence for thermal control of the updip limit of the seismogenic zone. More recently, geodetic inversions for interseismic slip on the plate interface in the Nicoya region identified a locked patch seaward of the shallowest microearthquakes [*Norabuena et al.*, 2004]. As explored further in this paper, the geodetic inversion results suggest that the updip limit of the seismogenic zone is defined by the shallow boundary of the geodetically locked plate interface and that the updip limit of interplate seismicity may correspond to second along-dip transition in the mechanical behavior of the plate interface.

Seismic and Geodetic Observations and Results

Seismic Observations and Earthquake Locations

The Costa Rica Seismogenic Zone Experiment consisted of GPS observations and two combined land and ocean bottom seismometer (OBS) seismic deployments, one northwest of the Osa Peninsula in southern Costa Rica and the other in and around the Nicoya Peninsula in northern Costa Rica (fig. 18.2).

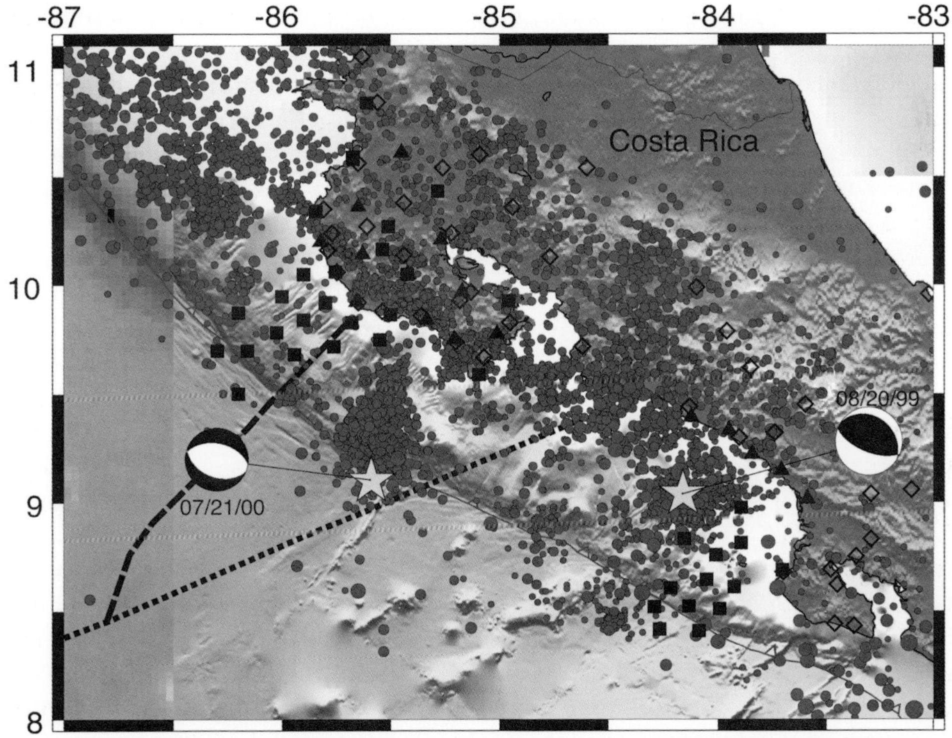

Figure 18.2 The Costa Rica Seismogenic Zone Experiment (CRSEIZE) consisted of 43 GPS stations (diamonds) and two passive on/offshore seismic experiments (squares and triangles represent broadband and short-period sensors respectively) that recorded earthquakes (circles) along the Middle America Trench. The Osa Peninsula experiment primarily recorded aftershocks of the M_w 6.9, 8/20/99 Quepos earthquake (Harvard CMT mechanism). The northern Nicoya Peninsula experiment recorded interplate events and intraplate events within the oceanic crust (e.g., the M_w 6.4, 7/21/00 outer rise event and its numerous aftershocks, shown with Harvard CMT mechanism) and within the overlying continental crust.

Experiment details, analytical techniques for earthquake data processing, location, and interpretations for the seismic component of CRSEIZE are described by *Newman et al.* [2002]; *DeShon et al.* [2003]; *DeShon* [2004]; and *DeShon et al.* [2006]. Details describing GPS data quality, processing, modeling, and interpretation can be found in the work of *Norabuena et al.* [2004].

The Nicoya seismic network recorded more than 8000 regional and local earthquakes along the Middle America subduction zone over its 18-month deployment (fig. 18.2). Shallow seismicity beneath the Nicoya Peninsula, both crustal and along the plate interface, represents ~20% of these earthquakes. Initial earthquake locations were determined using the global IASP91 velocity model (seismicity shown on fig. 18.2), and ~600 well-recorded Nicoya events were relocated using a simultaneous inversion of *P*- and *S*-wave arrival times to solve for earthquake location and three-dimensional velocity structure [*DeShon et al.* 2006]. These events have a location precision (one standard error) better than 1 km in both horizontal and vertical components. Interplate events were identified in this high-quality data set based on position relative to the plate interface as defined by the seismic velocity data, with all events located within 5 km of the plate interface assumed to be interplate earthquakes (fig. 18.3).

Two aspects of the seismicity can be discerned from figure 18.3: (1) seismicity begins at shallower depth where CNS lithosphere is subducted, and (2) the dip of the EPR slab is steeper than the CNS slab [*DeShon et al.* 2006]. The revised slab geometries for both EPR- and CNS-derived lithosphere agree well with, but extend, existing seismic refraction information [*Ye et al.*, 1996; *Christeson et al.*, 1999; *Sallarès et al.*, 1999, 2001].

Geodetic Data and Modeling

The horizontal component of GPS site velocities in northern Costa Rica relative to the Caribbean plate display a distinct north to northwest rotation of vectors relative to the plate convergence direction (fig. 18.4). This presumably reflects the influence of one or more processes in addition to strain accumulation on the plate boundary such as postseismic response to major past earthquakes and trench-parallel motion of a fore-arc block due to oblique convergence. As described by *Norabuena et al.* [2004], these are accounted for prior to estimation of locking on the plate interface. Following *Savage* [1983], interseismic strain accumulation on the plate interface is modeled with elastic dislocation theory as a superposition of continuous slip at the plate rate and "back slip" on the locked portions. Values of back slip, derived from inversions, therefore represent the part of the total plate rate presently locked and accumulating strain at particular positions along the plate boundary. Figure 18.4 shows the inferred back-slip distribution on the Nicoya portion of the plate-boundary fault for the best-fitting solution, and the corresponding observed and calculated velocity vectors. The best-fit model includes the effects of 8 mm/yr of northwest block motion and postseismic response of the 1992 Nicaragua earthquake

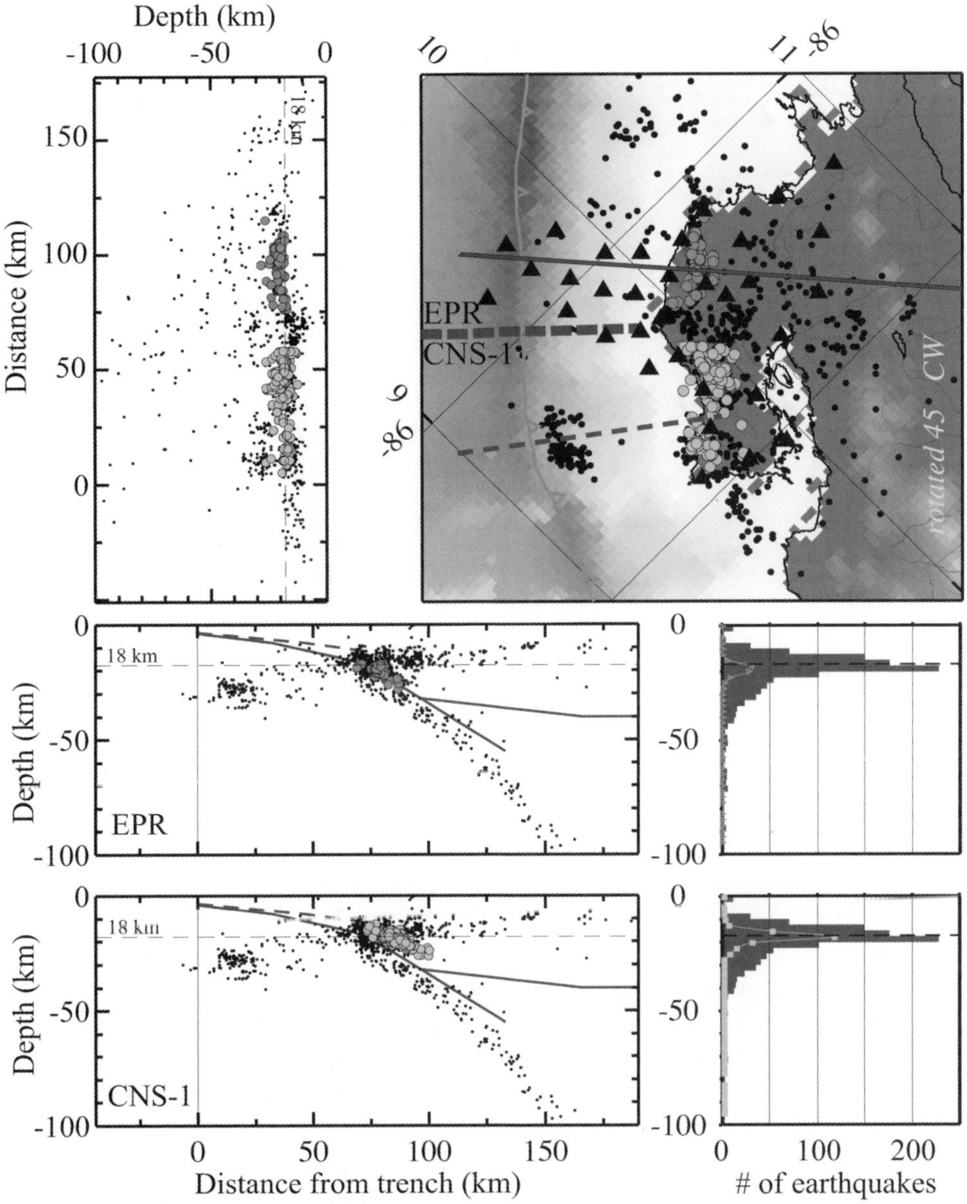

Figure 18.3 Nicoya Peninsula experiment seismicity relocated through a three-dimensional velocity model (all circles) shown in map and cross-sectional view. Larger black and gray circles represent EPR and CNS events respectively, identified as occurring on the thrust interface. Shaded histograms show the number of events in 2.5-km bins with EPR and CNS interplate event histograms overlaid. The onset of interplate events occurs at deeper depth for EPR events (18 km) compared with CNS events (15 km). The plate boundaries are from refraction information (lines on map) from *Christensen et al.* [1999] and *Sallarès et al.* [1999, 2001] for EPR (solid lines on cross sections) and from *Ye et al.* [1996] for CNS (dashed lines).

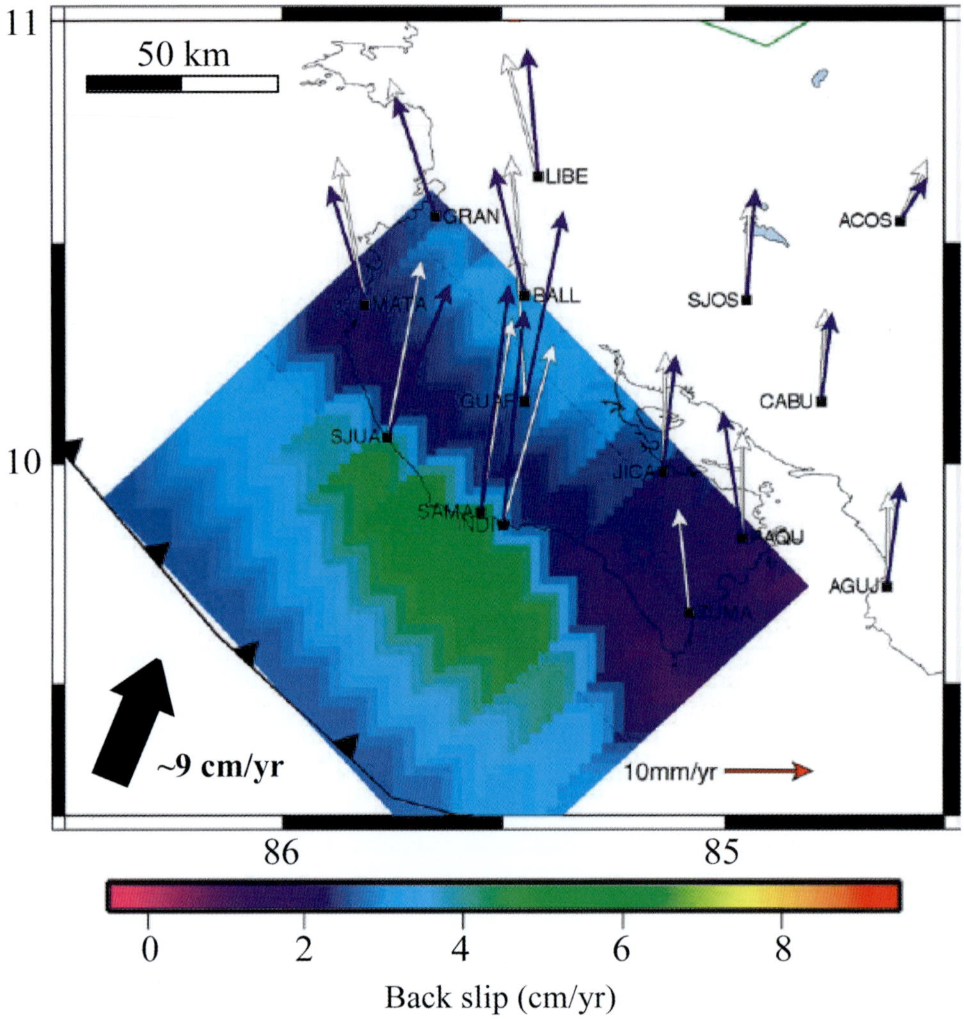

Figure 18.4 Best-fit model for locked slip on the plate boundary, from *Norabuena et al.* [2004]. Blue vectors are calculated velocities at the GPS sites, white are observed. Note elliptical patch of high locked slip (maximum, 6 cm/yr) beneath the coastal area, elongated parallel to the trench, and a second patch farther downdip (maximum, ~3 cm/yr). The black arrow is the plate convergence direction calculated using the Cocos-Caribbean angular velocity of *DeMets* [2001].

[*Norabuena et al.*, 2004]. The slip distribution shows two patches of locking, with the maximum back slip (6 cm/yr or ~70% of the plate velocity) centered at 14 ± 2 km depth, and a deeper patch (~3 cm/yr or ~35 % of the plate velocity) centered at 40 ± 5 km depth. A region of lower back slip (<2 cm/yr) centered at 25 km depth separates the two patches exhibiting highest back slip.

Checkerboard resolution tests indicate that the patches could either represent the gradational distribution illustrated, or perhaps two smaller, fully locked patches separated by a freely slipping region. Along-strike variations in back slip cannot currently be resolved beyond the Nicoya peninsula, related to station distribution. The occurrence of two patches of locked slip near the updip and downdip limits of the seismogenic zone is present in all of the low-misfit models for Nicoya, regardless of parameters used to calculate postseismic response to the 1992 Nicaragua earthquake and sliver motion [*Norabuena et al.*, 2004]. Inversions that constrain all locking to occur at depths below 15 km yield significantly degraded fits to the data and were deemed unacceptable. Therefore shallow geodetic locking between ~8 and 15 km depth is a robust result that is not sensitive to the details of the inversion parameterization.

Controls on Frictional/Mechanical Transitions

Updip Limit of the Seismogenic Zone

The updip limit of seismogenic zones has frequently been identified by the shallowest occurrence of microearthquakes [e.g., *Marone and Scholz*, 1988; *Pacheco et al.*, 1993; *Newman et al.*, 2002; *Obana et al.*, 2003] and favorably compared to other geophysical indicators of the onset of seismogenesis, such as the shallowest extent of large earthquake rupture and geodetically locked regions [e.g., *Hyndman and Wang*, 1993; *Hyndman et al.*, 1997; *Oleskevich et al.*, 1999; *Currie et al.*, 2002]. The implicit assumption has been that any of these observations or combinations therein potentially defines the shallow transition from stable sliding to stick-slip behavior. Figure 18.5 superimposes the geodetic back-slip model and interplate seismicity pattern for the Nicoya Peninsula. It clearly shows that the microseismic and geodetic definitions of the updip limit do not agree in this region over the time span sampled by the data.

While the geodetic data indicate a locked patch between ~30 and 65 km inland from the trench, microseismicity does not begin until 70–75 km from the trench. In northern Costa Rica the updip limit of the seismogenic zone corresponds to the shallowest region able to accumulate significant strain and is distinct from the updip limit of microseismicity.

Within error limits, the updip extent of the geodetically locked patch also corresponds to the updip limit of rupture in the last large earthquake in this region. Figure 18.6 plots the aftershock area of the 1950 Nicoya earthquake (M_s 7.7) determined by *Güendel* [1986] on the pattern of locked slip for the Nicoya Peninsula region. The epicentral location of this event was recomputed by *Avants et al.* [2001] relative to the well-located 1990 Gulf of Nicoya (M_w 7.0) event [*Protti et al.*, 1995b], and the entire aftershock pattern has been shifted to the position of the relocated epicenter. Although details of the coseismic slip distribution for the 1950 event are not known, preventing a comparison between coseismic slip and interseismic strain accumulation, the updip limit of the 1950 aftershock area appears to correspond reasonably well with the updip edge of the locked patch, providing confirmation that both of these adequately define the updip limit of the seismogenic zone offshore the Nicoya Peninsula, Costa Rica.

Hyndman and Wang [1993] and *Oleskevich et al.* [1999] suggested that the updip limit of the seismogenic zone corresponds to the 100°–150°C isotherm. Using the thermal model of *Spinelli and Saffer* [2004], we compared the updip locking limit in the Nicoya region to the location of their 100°C isotherm (fig. 18.5). This isotherm estimate corresponds approximately to the updip limit of the geodetically determined locked patch and the updip rupture extent of the last large earthquake (fig. 18.6). The thermal model of *Spinelli and Saffer* [2004] accounts for abrupt difference in seafloor heat flow values obtained on EPR- versus CNS-derived crust and includes 1–2 km of hydrothermal cooling of the EPR crust [*Fisher et al.*, 2003]. An offset in their 100°C isotherm, coincident with the change in oceanic crustal origin is subtly mimicked in a shallowing of the updip edge of the geodetically locked patch across the EPR-CNS boundary (figs. 18.5 and 18.6). Temperatures at the plate boundary derived from heat flux measurements and the depth to the bottom simulating reflector (BSR) (I. Grevemeyer, GEOMAR, Germany unpublished data, 2003) agree with the *Spinelli and Saffer* [2004] values where well-resolved (figs. 18.5 and 18.6).

Temperature-influenced mechanisms that have been proposed to govern the transition from stable sliding to stick-slip behavior include a suite of diagenetic and metamorphic reactions that occur near 100°–150°C and lead to a change in the frictional behavior of the plate interface or a decrease in fluid production and fluid pressure [*Moore and Saffer*, 2000]. *Vrolijk* [1990] suggested the clay-mineral transition of smectite, common in oceanic sediments, to illite controls the updip location of the aseismic to seismic transition. However, recent work on the frictional characteristics of illite under increasing normal stress, and sliding velocity has shown it to exhibit only velocity-strengthening or stable sliding behavior [*Saffer and Marone*, 2003]. *Saffer and Marone* [2003] suggested that processes such as cementation, consolidation, and localization of slip in clay minerals may exert a significant influence on the frictional state of the subduction thrust. *Moore et al.* [this volume] suggest that chemical byproducts of the smectite-illite transition, such as calcium, can facilitate carbonate cementation, which in turn may affect the frictional behavior of the thrust interface. *Spinelli and Saffer* [2004] investigated the likelihood that opal and clay dehydration reactions are responsible for the aseismic to seismic transition documented by *Newman et al.* [2002] at the Nicoya Peninsula. They found an offset in the location at which sediment dewatering is completed (between 90° and 160°C) that mimics the offset of shallow seismicity across the EPR/CNS boundary but lies ~15–20 km seaward of it. Our recent observations suggest that the shallow extent of geodetic locking, not microseismicity, may define the

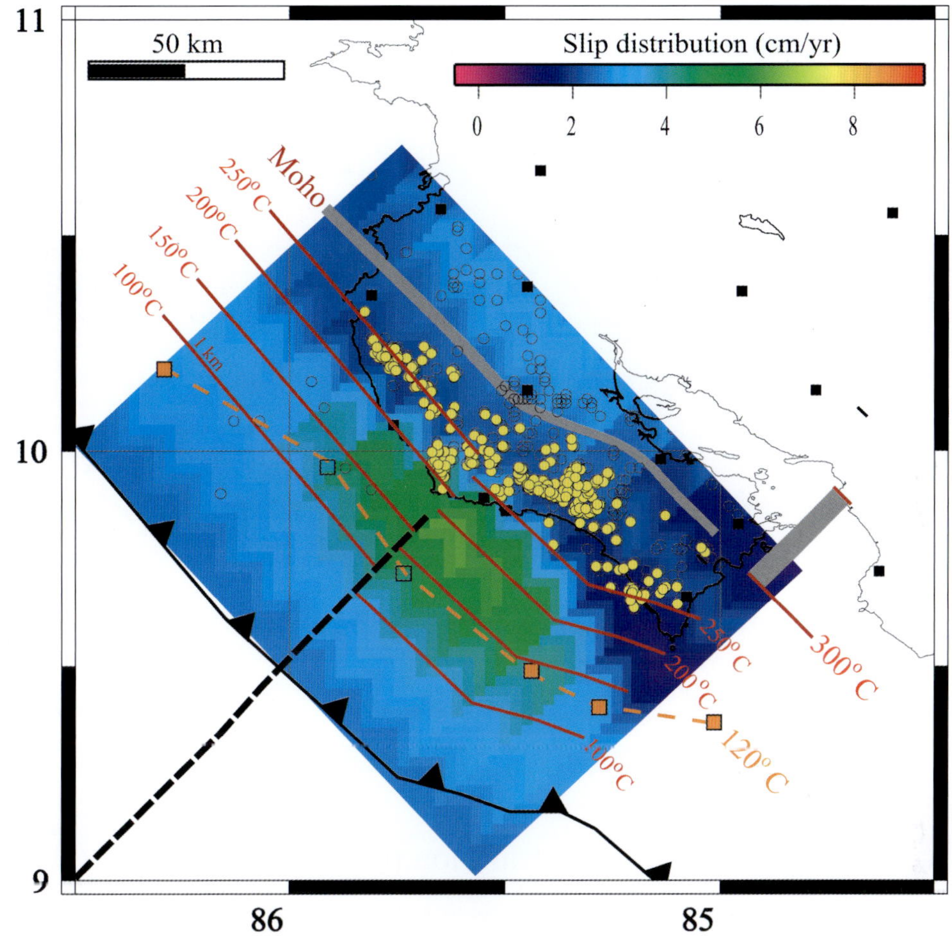

Figure 18.5 Comparison of well-located earthquakes from CRSEIZE (yellow circles are plate interface events), distribution of locked slip from inversion of the GPS data [*Norabuena et al.*, 2004], 100°–250°C isotherm estimates from *Spinelli and Saffer* [2004] calculated assuming hydrothermal cooling of the upper 1 km of the EPR crust (solid red lines), 120°C isotherm estimate from I. Grevemeyer (unpublished data, 2003) (orange dashed line; filled squares indicated well-resolved data), 300°C isotherm estimate from *Harris and Wang* [2002], and continental Moho/oceanic slab intersection from *DeShon* [2004]. Earthquakes reach an updip limit ~70 km from the trench where temperatures reach 200°–250°C. The updip extent of locked slip (~6 cm/yr) occurs ~35 km from the trench approximately coincident with the 100°–150°C isotherms that shallows across the East Pacific Rise to Cocos-Nazca Spreading Center generated crustal boundary (dashed line).

aseismic to seismic transition for the Nicoya Peninsula over the entire seismic cycle. The onset of geodetic locking agrees with the location of pore-pressure dissipation from *Spinelli and Saffer* [2004], but potential along-strike variations in locking remain unresolved.

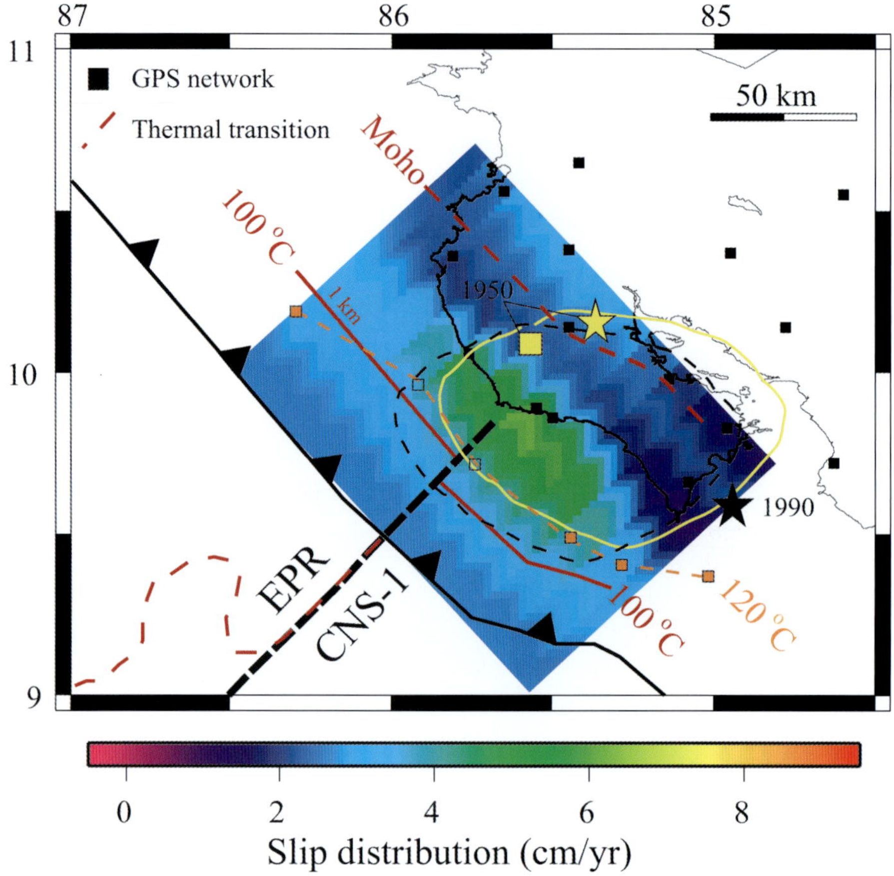

Figure 18.6 Comparison of the distribution of locked slip from inversion of GPS data and the rupture area of the $M_s = 7.7$ 1950 Nicoya earthquake, modified from *Norabuena et al.* [2004]. The mainshock epicenter (yellow box) was relocated (yellow star) by *Avants et al.* [2001] relative to the well-located $M_w = 7.0$ 1990 event (black star) and the aftershock area was shifted accordingly. Estimates of the 100° and 120°C isotherms are from *Spinelli and Saffer* [2004] and I. Grevemeyer (unpublished data, 2003) respectively and coincide with the updip limit of the rupture area and the locked patch.

Most of the progress made on understanding controls on the aseismic to seismic transition has come from the study of accretionary convergent margins where the plate interface is dominated by subducted trench sediments. In this environment, continual burial and heating transforms trench sediments in a series of progressive diagenetic/low-grade metamorphic reactions leading to the onset of seismic behavior. The association between temperature estimates of 100°–150°C and the onset of seismogenesis has been documented for erosional margins as well (e.g., Costa Rica (this study) and northeast Japan

[*Oleskevich et al.*, 1999]); however, in this environment, material eroded from the underside of the upper plate subducts and may line the plate interface [*Ranero and von Huene*, 2000; *von Huene et al.*, 2004]. This material most likely has not undergone the same progressive transformations experienced by accretionary margin sediments. The correlation of frictional stability and thermal structure at both accretionary and erosional margins suggests that fundamental processes not tied to specific mineralogies, such as smectite, operate at both types of margins to govern the transition from aseismic to seismic behavior.

Dissipation of pore-fluid pressure and its concomitant increase in effective normal stress is the common factor that has been identified as important in governing the aseismic to seismic transition at both types of margins [e.g., *Saffer and Bekins*, 2002; *Saffer*, 2003; *Spinelli and Saffer*, 2004]. *Von Huene et al.* [2004] proposed a generalized model of subduction erosion appropriate for margins with small frontal prisms such as Costa Rica. This model requires extensive hydrofracturing within the upper plate and decreased effective normal stress across the décollement due to sediment and oceanic crust dewatering. Mechanically weakened continental plate causes a vertical upward shift in the décollement, allowing inclusion of upper plate material into the subduction channel. In this model, the cessation of hydrofracturing and dissipation of pore pressures with depth terminates subduction erosion and results in an increase in effective normal stress and the beginning of seismogenesis.

Prestack depth-migrated profiles offshore the Osa Peninsula in central Costa Rica, where slab dip is significantly shallower than the Nicoya Peninsula region, image the plate boundary as high-amplitude landward dipping reflections [*Ranero and von Huene*, 2000]. High reflectivity is often associated with fluids since their presence causes a large change in impedance contrast easily detectable with seismic reflection methods. Plate-boundary reflectivity decreases beneath the continental shelf break, some 30–35 km from the trench where heat flux and depth to the BSR indicate a temperature of ~150°C [*Pecher et al.*, 2001; I. Grevemeyer, GEOMAR, Germany, unpublished data, 2003]. Assuming that the high-amplitude plate boundary reflections are caused by abundant free water, their termination at ~8 km depth most likely represents the downdip extent of significant water release along the plate boundary. The location of an M_w 6.4 underthrusting earthquake northwest of the Osa Peninsula, the shallowest earthquake to occur on the plate boundary in the south central Costa Rica region, when projected onto this reflection line, occurs where the plate-boundary reflectivity significantly decreases. This suggests that unlike the Nicoya Peninsula, the updip limit of midmagnitude seismicity in central Costa Rica coincides with the aseismic to seismic transition, but like the Nicoya region, this transition occurs where plate boundary temperatures attain 100°–150°C.

Transition from Locked with No Microseismicity to Creeping with Microseismicity

Aseismic slip along particular segments of strike-slip faults, such as the San Andreas and the North Anatolian systems, has been recognized for some time [e.g., *Steinbrugge et al.*, 1960; *Ambraseys*, 1970; *Aytun*, 1980; *Lienkaemper et al.*, 1991; *Lienkaemper and Galehouse*, 1998]. Creeping segments within these fault systems were first identified by offsets of cultural features and an abundance of microseismicity in the absence of sizeable earthquakes. More recently, creeping strike-slip fault segments have been documented and interseismic slip rates quantified through GPS geodesy. Inversion of geodetic data for slip at depth on the Hayward fault in California [*Bürgmann et al.*, 2000; *Simpson et al.*, 2001; *Malservisi et al.*, 2003] reveals that the fault zone consists of a patchwork of frictional properties with behaviors ranging from completely locked patches (no observable slip) with little microseismicity, to high-slip rate zones (unlocked and moving at the surface velocity) with abundant earthquakes. *Malservisi et al.* [2003] report the largest concentrations of microearthquakes at transition zones between locked and more freely slipping regions on the Hayward fault in California. They interpreted this to indicate accelerated strain accumulation at the transition zones. Increased strain at locked to creeping transition zones would result in deformation within the surrounding volume and might explain the abundant upper plate as well as interplate seismicity we observe near the strong gradient in locked slip (fig. 18.7).

Because much of the plate interface lies below the seafloor at subduction zones, less is known about patterns of locked and slipping segments in this environment. Our work in northern Costa Rica and recent work at convergent margins in northeast Japan, Nankai subduction zone, and Alaska have revealed distinct regions that are geodetically locked and have little to no microseismicity during the interseismic cycle and become the loci of major moment release during large earthquakes. Other regions remain unlocked with abundant microseismicity during the interseismic cycle and are sites of little to no moment release during major earthquakes. In northeastern Japan, *Igarashi et al.* [2003] found that frequently repeating small earthquakes, and adjacent freely slipping regions colocated with low-slip regions of coseismic moment release associated with two large plate-boundary earthquakes in 1968 (M_w 8.2) and 1994 (M_w 7.7). Their distribution of freely slipping regions obtained from repeating earthquakes also matched the pattern of freely slipping regions estimated from GPS. In the region of the great 1964 (M_w 9.2) Alaskan earthquake, *Zweck et al.* [2002] also found a correspondence between regions of interseismic strain accumulation or strong geodetic coupling and patches of high coseismic slip during the 1964 earthquake. However, only the Nicoya Peninsula possesses downdip alternation between geodetically locked, microseismically quiescent regions and freely slipping, microseismically active regions during an interseismic period.

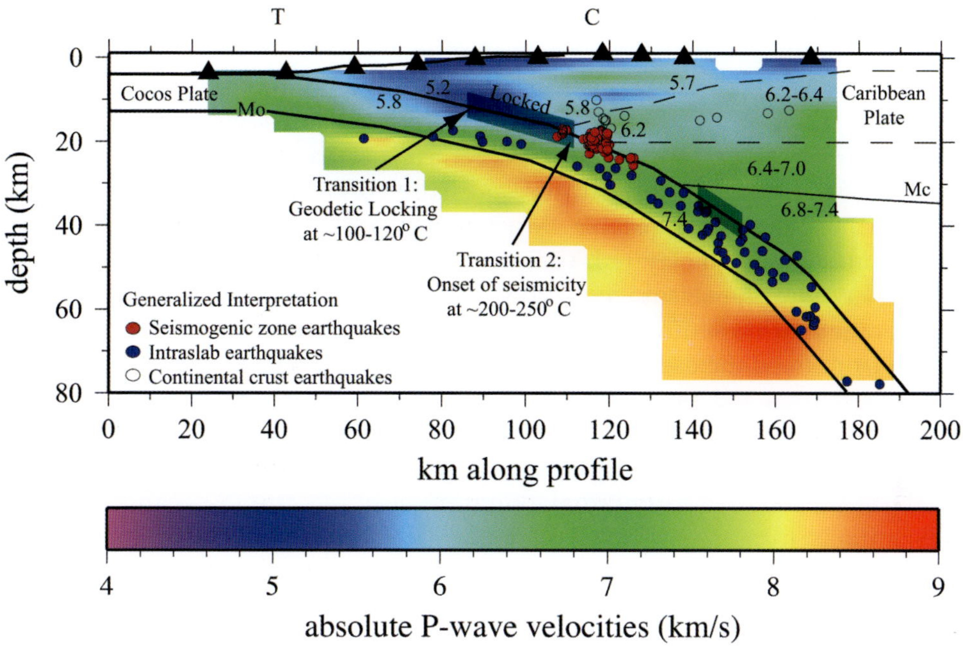

Figure 18.7 Schematic of northern Costa Rica subduction zone. Color shading represents *P*-wave velocities from *DeShon* [2004]. The aseismic to seismic transition is defined by the shallowest region presently locked (shaded) and accumulating strain [*Norabuena et al.*, 2004] and is coincident with the 100°–120°C isotherm estimates. A second transition in mechanical behavior characterized by the onset of thrust earthquakes and geodetic unlocking occurs between 200° and 250°C (modeled by *Spinelli and Saffer* [2004]), where the upper plate velocity increases velocity. A second deeper locked patch (shaded) exists below the continental Moho (Mc); however geodetic resolution at this depth is poor.

This along-dip transition occurs where temperature estimates are between 200° and 250°C (fig. 18.5), and the plate interface lies between 15 and 19 km depth or 0.5–0.6 GPa. This transition most likely reflects a weakening of the plate interface that allows it to slip freely through both creep and small earthquakes. Because this depth range is clearly in the seismogenic zone based both on historic $M_w > 7$ seismicity and the occurrence of microseismicity, the subduction megathrust is in the frictional unstable regime [e.g., *Scholtz*, 1998]. We suggest that small increases in pore-fluid pressure decrease the effective normal stress and shear strength of the fault without driving the subduction megathrust back into the frictional stability regime such as exists at the trench. In this way, strain buildup along small-scale heterogeneities, perhaps at localized regions of increased effective normal stress due to changing pore pressure or material properties, serve as initiation regions for microseismicity while the fault predominantly creeps aseismically.

An increase in pore pressure can be achieved through either or both an increase in fluid production or a reduction in permeability. Fluids in the depth range of interest (18–22 km) are most likely generated through low-grade metamorphic reactions occurring in subducted oceanic crust and/or marine sediments. Chemically bound water in oceanic crust and marine sediments will be released at different temperature-pressure conditions as hydrous silicates become progressively unstable. Most metamorphic devolatilization of siliceous limestones and clay-rich sediments occurs at temperatures greater than attained within the seismogenic zone [*Kerrick and Connolly*, 2001]. However, an average marine sediment bulk composition with an initial 7.3 wt % water (taken from the database of *Plank and Langmuir* [1998]) begins devolatilization at temperatures and pressures corresponding to the noted transition in mechanical behavior, releasing up to 0.5 wt % water [*Kerrick and Connolly*, 2001].

Although unaltered oceanic basalt typically contains <0.5 wt % water [*Dixon et al.*, 1988], hydrothermal alteration of basalt at mid-ocean ridges can substantially increase the amount of chemically bound water in the oceanic crust and lead to a highly variable water content. Estimates of the average water content of oceanic crust range from 1–2 wt % [*Peacock*, 1990] to 6 wt % [*Anderson et al.*, 1976]. Depending on the degree of hydration of the oceanic crust, dehydration will occur in association with different metamorphic facies reactions. Assuming a large degree of hydration (5–6 wt % water) and metamorphic phase equilibrium computed by *Peacock* [1993], P-T paths followed by subducted crust beneath the Nicoya Peninsula progress through zeolite (ZE) to lawsonite-chlorite (LC) to pumpellyite-actinolite (PA) facies in the depth range of 13–25 km releasing ~0.3 wt % water across the LC-PA boundary at 16 and 22 km for hotter CNS or colder EPR crust respectively (fig. 18.8). Oceanic crust containing over 8 wt % water (the maximum amount of water accommodated in zeolite facies rocks) would begin dehydration across the zeolite to lawsonite-chlorite boundary releasing 3.1 wt % water in the depth range between 14 and 17 km (fig. 18.8). The near coincidence of the onset of these basalt dehydration reactions with the updip edge of microseismicity (fig. 18.7) makes a compelling case for weakening of the plate interface due to increased pore-fluid pressure.

However, this mechanism requires hydrous oceanic crust to be subducted and thus additional information on the level of hydration is critical. *Ranero et al.* [2003] imaged pervasive bending-related normal faults in the oceanic lithosphere offshore the Nicoya Peninsula and postulated that these faults promote hydration of the surrounding crust and mantle. The faults imaged in their seismic data extend at least to 20 km into the plate and provide a mechanism for infiltration of water into the crust and mantle. *Husen et al.* [2003] suggested the existence of a shallow (<40 km) hydrous layer in the lithosphere of the downgoing Cocos plate in northern Costa Rica based on seismic tomography and petrologic modeling. Although they acknowledged that the anomaly might be due to low velocity material above the plate interface in the subduction

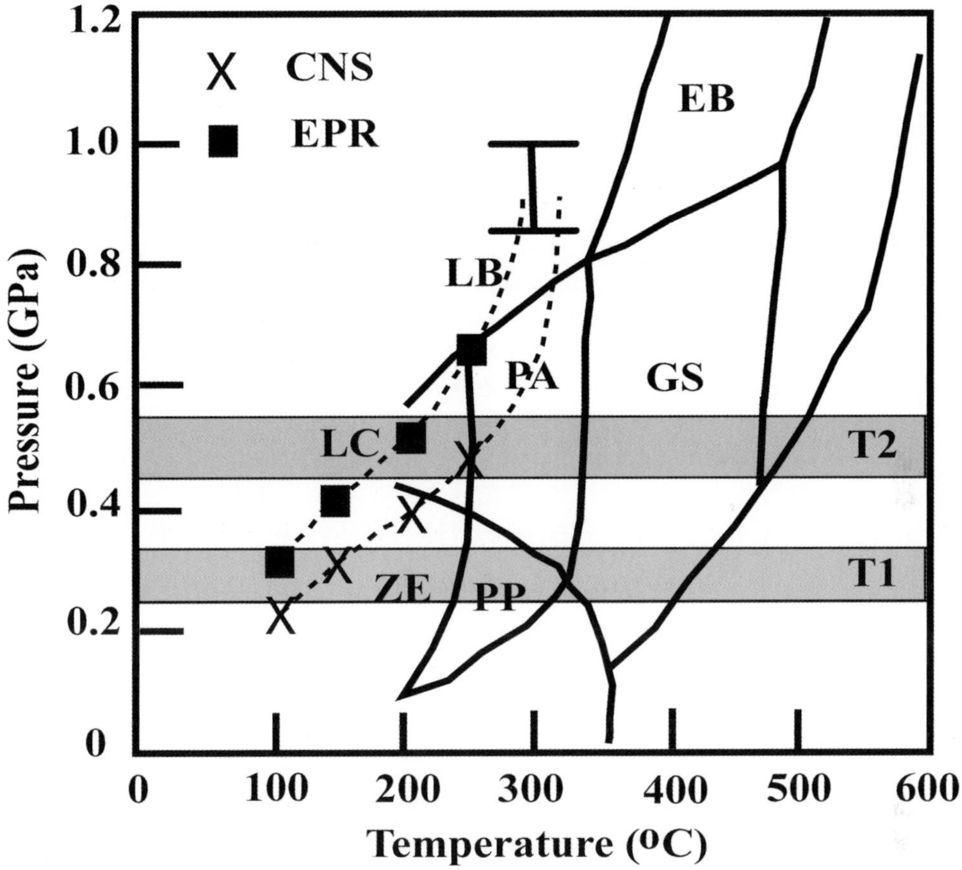

Figure 18.8 Pressure-temperature diagram of metamorphic facies modified from *Peacock* [1993] with low-temperature geotherms appropriate for EPR and CNS crust from *Spinelli and Saffer* [2004] and depth range of 300°C isotherm from *Harris and Wang* [2002]. T1 and T2 represent the depth ranges of the two mechanical transitions discussed in this paper. ZE, Zeolite (~8.5 wt % H_2O); LC, Lawsonite-Chlorite (~5.4 wt % H_2O); PP, Prenite-Pumpellyite (~5.9 wt % H_2O), PA, Prenhite-Actinolite (~5.1 wt % H_2O); LB, Lawsonite-Blueschist (5.9 wt % H_2O); GS, Greenschist (~3.4 wt % H_2O) where maximum H_2O content is indicated for a basaltic bulk composition.

channel, they favored an interpretation requiring strong hydration of the oceanic crust. Fluids released by metamorphic reactions in both the oceanic basalt and marine sediments can most efficiently increase pore-fluid pressure in the presence of reduced permeability. Recent permeability measurements on rocks from the Shimanto complex, an ancient exhumed subduction zone fault in Japan, show a decrease in permeability at temperatures above ~250°C for all rock types sampled [*Kato et al.*, 2004]. *Moore and Saffer* [2001] list a variety of diagenetic/metamorphic processes such as quartz cementation and pressure solution that reduce permeability that initiate at temperatures as low as 150°C but do not operate at full efficiency until 200°–250°C.

The estimated location of the 250°C isotherm along the plate interface [*Spinelli and Saffer*, 2004] and the onset of microseismicity occur where the upper plate material transitions to higher *P*-wave velocities (fig. 18.7). This increase in velocity may reflect a decrease in permeability achieved through fracture cementation or a fundamental change in rock type [i.e., *DeShon et al.,* 2006]. A combination of fluid production from low-grade metamorphic reactions in basaltic crust and marine sediments coupled with a decrease in permeability around 250°C has potentially led to elevated pore pressure and sufficient weakening of the thrust interface to permit failure during small-magnitude earthquakes. In contrast, the region just updip is so strongly coupled that rupture only occurs during large-magnitude earthquakes.

Transition from Creeping with Microseismicity to Locked with No Microseismicity

All acceptable geodetic models of locked slip on the plate interface produce a deeper locked patch downdip of most of the microseismicity (fig. 18.5), although it is less well constrained in magnitude and downdip extent than the shallow locked patch. The few microearthquakes that occur in this region are somewhat diffuse and have been interpreted as intraplate events within the overriding plate and subducted lithosphere (fig. 18.3) [*DeShon et al.,* 2006]. The better-resolved updip edge of this deeper locked patch at ~35 km, corresponds to the depth of the continental Moho/oceanic slab interface (figs. 18.5 and 18.6) [*Sallarès et al.,* 2001; *DeShon et al.* 2006]. Temperature estimates at this depth have been calculated to be ~300–350°C (with some degree of shear heating) [*Harris and Wang*, 2002], below those associated with stable sliding in mantle material. The significant change in upper plate composition from continental to mantle material may result in significant strengthening of the fault and account for the transition back to geodetically locked. The downdip extent of the seismogenic zone would coincide with the downdip edge of deep strain accumulation; unfortunately, our present model does not have the resolution to determine this.

Conclusion

The Nicoya segment of the Middle America seismogenic zone possesses two shallow transitions in mechanical behavior during the interseismic period. We interpret the updip aseismic to seismic frictional transition as the shallowest region presently accumulating strain (geodetically locked); this occurs between 30 and 40 km from the trench. Although we believe the frictional properties of the plate interface below this transition to be velocity weakening, no earthquakes have occurred on this portion of the plate interface during the time span of our experiment. We interpret the absence of microseismicity to reflect high strength (high shear stress) along this section of the fault. The onset of abundant interplate microseismicity, coincident with a decrease in strain accumulation (fault creep), occurs between 70 and 75 km from the trench. We interpret this as a weakening of the fault, most likely resulting from increased pore pressure. Metamorphic reactions in hydrated subducted oceanic crust coupled with low permeability are most likely responsible for generating elevated pore pressures. This deeper transition occurs where temperature estimates are between 200° and 250°C, in the range of the lawsonite-chlorite to pumpellyite-actinolite facies boundary where up to 0.3 wt % water may be released for highly hydrated crust. For a variety of rock types a significant decrease in permeability has been demonstrated at ~250°C, allowing a small influx of fluid to generate significant pore pressures.

There is already great interest in the state of hydration of subducted lithosphere due to the likely role of slab dehydration at depth in producing arc lavas and generating intermediate depth earthquakes in downgoing slabs. Slab dehydration at shallower depths and the resulting serpentinization of the fore-arc mantle wedge has also been identified as a possible mechanism for the termination of interplate earthquake rupture [e.g., *Peacock and Hyndman*, 1999]. Our observation in northern Costa Rica that the onset of interplate microseismicity is coincident with the downdip edge of a geodetically locked patch and estimates of the ~250°C isotherm (fig. 18.5) has now implicated slab dehydration in the initiation of interplate earthquakes and thus increases the importance of determining the state of hydration of subducting lithosphere.

Although we know of no other seismogenic zones that possess downdip transitions in mechanical behavior like those described here, they may exist but be beyond the present resolution of most seismic and geodetic observations. While we can generate a uniform slip solution in Nicoya with a large region of partial coupling (~40% of the plate rate), the data density and proximity to the trench allow finer resolution and indicate that the seismogenic plate interface is better represented by relatively small fully locked (or nearly so) patches, surrounded by regions that are essentially freely slipping with localized microseisms. This suggests that some subduction thrusts, previously inferred to be partially coupled based on spatially limited geodetic data, might be better represented by alternating locked and slipping regions. Spatially dense, for many subduction zones, seafloor geodetic observations will be required to resolve this issue.

Acknowledgments

Some of the ideas contained in this paper were developed while participating in a graduate seminar on the Seismogenic Zone at UC Santa Cruz. We thank the students in that seminar and especially co-instructor Casey Moore for their stimulating discussions. We also thank Tim Dixon, Nick Beeler, and Demian Saffer for their insightful comments that vastly improved this manuscript. This work was supported by NSF grant EAR-0229876 to S.Y.S.

References

Ambraseys, N. N. (1970), Some characteristic features of the Anatolian fault zone, *Tectonophysics, 9*, 143–165.

Anderson, R. N., S. Uyeda, and A. Miyashiro (1976), Geophysical and geochemical constraints at converging plate boundaries-Part I: Dehydration in the downgoing slab, *Geophys. J. R. Astron. Soc., 44*, 333–357.

Aytun, A. (1980), Creep measurements in the Ismetpasa region of the North Anatolian fault zone, in *Multidisciplinary Approach to Earthquake Prediction*, vol. 2, edited by A. M. Isakara and A. Vogel, pp. 279–292, Friedrich Vieweg, Braunschyweig/Wiesbaden, Germany.

Avants, M., S. Y. Schwartz, A. V. Newman, and H. R. DeShon (2001), Large underthrusting earthquakes beneath the Nicoya Peninsula, *Eos Trans. AGU, 82*, Fall Meet. Suppl., F1266.

Barckhausen, U., H. A. Roeser, and R. von Huene (1998), Magnetic signature of upper plate structures and subducting seamounts at the convergent margin off Costa Rica, *J. Geophys. Res., 103*, 7079–7094.

Barckhausen, U., C. R. Ranero, R. von Huene, S. C. Cande, and H. A. Roeser (2001), Revised tectonic boundaries in the Cocos Plate off Costa Rica: Implications for the segmentation of the convergent margin and for plate tectonic models, *J. Geophys. Res., 106*, 19,207–19,220.

Bürgmann, R. D. Schmidt, R. M. Nadeau, M. d'Alessio, E. Fielding, D. Manaker, T. V. McEvilly, and M. H. Murray (2000), Earthquake potential along the northern Hayward fault, California, *Science, 289*, 1178–1181.

Carr, M. J., and R. E. Stoiber (1990), Volcanism, in *The Geology of North America*, vol. H, *The Caribbean Region*, edited by G. Dengo and J. E. Case, Geol. Soc. of Am., Boulder, CO, pp. 375–391.

Christeson, G. L., K. D. McIntosh, T. H. Shipley, E. R. Flueh, and H. Goedde (1999), Structure of the Costa Rica convergent margin, offshore Nicoya Peninsula, *J. Geophys. Res., 104*, 25,443–25,468.

Currie, C. A., R. D. Hyndman, K. Wang, and V. Kostoglodov (2002), Thermal models of the Mexico subduction zone: Implications for the megathrust seismogenic zone, *J. Geophys. Res., 107*(B12), 2370, doi:10.1029/2001JB000886.

DeMets, C. (2001), A new estimate for present-day Cocos-Caribbean plate motion: Implications for slip along the Central American volcanic arc, *Geophys. Res. Lett., 28*, 4043–4046.

DeShon, H. R. (2004), Seismogenic zone structure along the Middle America subduction zone, Costa Rica. Ph.D. thesis, 330 pp., Univ. of Calif., Santa Cruz.

DeShon, H. R., S. Y. Schwartz, S. L. Bilek, L. M. Dorman, V. Gonzalez, J. M. Protti, E. R. Flueh, and T. H. Dixon (2003), Seismogenic zone structure of the southern Middle America Trench, Costa Rica, *J. Geophys. Res., 108*(B10), 2491, doi:10.1029/2002JB002294.

DeShon, H. R., S. Y. Schwartz, A. V. Newman, V. Gonzalez, M. Protti, L. M. Dorman, and T. H. Dixon et al. (2006), Seismogenic zone structure beneath the Nicoya Peninsula, Costa Rica, from three-dimensional local earthquake P- and S-wave tomography, *Geophys. J. Int., 164(1)*, 109–124.

Dixon, J. E., E. Stolper, and J. R. Delaney (1988), Infrared spectroscopic measurements of CO_2 and H_2O in Juan de Fuca Ridge basalt glasses, *Earth Planet. Sci. Lett., 90*, 87–104.

Fisher, A. T., C. A. Stein, R. N. Harris, K. Wang, E. A. Silver, M. Pfender, M. Hutnak, A. Cherkaoui, R. Bodzin, and H. Villinger (2003), Abrupt thermal transition reveals hydrothermal boundary and role of seamounts within the Cocos Plate, *Geophys. Res. Lett., 30*(11), 1550, doi:10.1029/2002GL016766.

Güendel, F. (1986), Seismotectonics of Costa Rica: An analytical view of the southern terminus of the Middle America Trench, Ph.D. thesis, 157 pp., Univ. of Calif., Santa Cruz.

Harris, R. N., and K. Wang (2002) Thermal models of the Middle America trench at the Nicoya Peninsula, Costa Rica, *Geophys. Res. Lett., 29*(21), 2010, doi:10.1029/2002GL015406.

Husen, S., R. Quintero, E. Kissling, and B. Hacker (2003), Subduction-zone structure and magmatic processes beneath Costa Rica constrained by local earthquake tomography and petrological modeling, *Geophys. J. Int., 155*, 11–32.

Hutnak, M., et al. (2007), The thermal state of 18–24 Ma upper lithosphere subducting below the Nicoya Peninsula, northern Costa Rica margin, this volume.

Hyndman, R., and S. M. Peacock (2003), Serpentinization of the fore-arc mantle, *Earth Planet. Sci. Lett., 212*, 417–432.

Hyndman, R. D., and K. Wang (1993), Thermal constraints on the zone of major thrust earthquake failure: The Cascadia subduction zone, *J. Geophys. Res., 98*, 2039–2060.

Hyndman, R. D., K. Wang, and M. Yamano (1995), Thermal constraints on the seismogenic portion of the southwestern Japan subduction thrust, *J. Geophys. Res., 100*, 15,373-15,392.

Hyndman, R. D., M. Yamano, and D. A. Oleskevich (1997), The seismogenic zone of subduction thrust faults, *Island Arc, 6*, 244–260.

Igarashi, T., T. Matsuzawa, and A. Hasegawa (2003), Repeating earthquakes and interplate aseismic slip in the northeastern Japan subduction zone, *J. Geophys. Res., 108*(B5), 2249, doi:10.1029/2002JB001920.

Kato, A., A. Sakaguchik, Y. Shingo, H. Yamaguchi, and Y. Kaneda (2004), Permeability structure around an ancient exhumed subduction-zone fault, *Geophys. Res. Lett., 31*, L06602, doi:10.1029/2003GL019183.

Kerrick, D. M., and J. A. D. Connolly (2001), Metamorphic devolatilization of subducted marine sediments and the transport of volatiles into the Earth's mantle, *Nature, 411*, 293–296.

Kimura, G., E. Silver, and P. Blum (1997), *Proceedings of the Ocean Drilling Program, Initial Reports*, vol. 170, 458 pp., Ocean Drill. Program, College Station, Tex.

Langseth, M. G., and E. A. Silver (1996), The Nicoya convergent margin-A region of exceptionally low heat flow, *Geophys. Res. Lett., 23*, 891–894.

Lienkaemper, J., and J. Galehouse (1998), New evidence doubles the seismic potential of the Hayward fault, *Seismol. Res. Lett., 69*, 519–523.

Lienkaemper, J., G. Borchart, and M. Lisowski (1991), Historic creep rate and potential for seismic slip along the Hayward fault, California, *J. Geophys. Res., 96*, 18,261–18,283.

Malservisi, R., C. Gans, and K. P. Furlong (2003), Numerical modeling of creeping faults and implications for the Hayward fault, California, *Tectonophysics, 361*, 121–137.

Marone, C., and C. H. Scholz (1988), The depth of seismic faulting and the upper transition from stable to unstable slip regimes, *Geophys. Res. Lett., 15*, 621–624.

Meschede, M., U. Barckhausen, and H.-U. Worm (1998), Extinct spreading on the Cocos Ridge, *Terra Nova, 10*, 211–216.

Moore, J. C., and D. Saffer (2001), Updip limit of the seismogenic zone beneath the accretionary prism of southwest Japan: An effect of diagenetic to low-grade metamorphic processes and increasing effective stress, *Geology, 29*(2), 183–186.

Morris, J., R. Valentine, and T. Harrison (2002), ^{10}Be imaging of sediment accretion and subduction along the Northeast Japan and Costa Rica convergent margins, *Geology, 30*(1), 59–62.

Moore, J. C., C. Rowe, and F. Meneghini (2007), How accretionary prisms elucidate seismogenesis in subduction zones, this volume.

Newman, A. V., S. Y. Schwartz, V. Gonzales, H. R. DeShon, J. M. Protti, and L. Dorman (2002), Along strike variability in the seismogenic zone below Nicoya Peninsula, Costa Rica, *Geophys. Res. Lett., 29*(20), 1977, doi:10.1029/2002GL015409.

Norabuena, E., et al. (2004), Geodetic and seismic constraints on some seismogenic zone processes in Costa Rica, *J. Geophys. Res., 109*, B11403, doi:10.1029/2003JB002931.
Obana, K., S. Kodaira, Y. Kaneda, K. Mochizuki, M. Shinohara, and K. Suyehiro (2003), Microseismicity at the seaward updip limit of the western Nankai Trough seismogenic zone, *J. Geophys. Res., 108*(B10), 2459, doi:10.1029/2002JB002370.
Oleskevich, D. A., R. D. Hyndman, and K. Wang (1999), The updip and downdip limits to great subduction earthquakes: Thermal and structural models of Cascadia, south Alaska, SW Japan and Chile, *J. Geophys. Res., 104*, 14,965–14,991.
Pacheco, J. F., L. R. Sykes, and C. H. Scholz (1993), Nature of seismic coupling along simple plate boundaries of the subduction type, *J. Geophys. Res., 98*, 14,133–14,159.
Patino, L. C., M. J. Carr, and M. D. Feigenson (2000), Local and regional variations in Central American arc lavas controlled by variations in subducted sediment input, *Contrib. Mineral. Petrol., 138*, 265–283.
Peacock, S. M. (1990), Numerical simulation of metamorphic pressure-temperature-time paths and fluid production in subducting slabs, *Tectonics, 9*, 1197–1211.
Peacock, S. M. (1993), The importance of blueschist to eclogite dehydration reactions in subducting oceanic crust, *Geol. Soc. Am. Bull., 105*, 684–694.
Peacock, S. M., and R. D. Hyndman (1999), Hydrous minerals in the mantle wedge and the maximum depth of subduction thrust earthquakes, *Geophys. Res. Lett., 26*, 2517–2520.
Pecher, I. A., N. Kukowski, C. R. Ranero, and R. von Huene (2001), Gas hydrates along the Peru and Middle America Trench systems, in *Natural Gas Hydrates: Occurrence, Distribution, and Detection, Geophys. Monogr. Ser.*, vol. 124, edited by C. K. Paull and W. P. Dillon, pp. 257–271, AGU, Washington, D. C.
Plank, T., and C. H. Langmuir (1998), The chemical composition of subducted sediments and its consequences for the crust and mantle, *Chem. Geol., 145*, 325–394.
Protti, M., F. Güendel, and K. McNally (1995a), Correlation between the age of the subducting Cocos plate and the geometry of the Wadati-Benioff zone under Nicaragua and Costa Rica, in *Geologic and Tectonic Development of the Caribbean Plate Boundary in Southern Central America*, edited by P. Mann, *Spec. Pap. Geol. Soc. Am., 295*, 309–326.
Protti, M., et al. (1995b), The March 25, 1990 ($M_w = 7.0$, $M_L = 6.8$), earthquake at the entrance of the Nicoya Gulf, Costa Rica: Its prior activity, foreshocks, aftershocks, and triggered seismicity, *J. Geophys. Res., 100*, 20,345–20,358.
Ranero, C. R., and R. von Huene (2000), Subduction erosion along the Middle America convergent margin, *Nature, 404*, 748–752.
Ranero, C. R., J. Phipps Morgan, K. McIntosh, and C. Reichert (2003), Bending-related faulting and mantle serpentinization at the Middle America trench, *Nature, 425*, 367–373.
Saffer, D. M. (2003), Pore pressure development and progressive dewatering in underthrust sediments at the Costa Rican subduction margin: Comparison with northern Barbados and Nankai, *J. Geophys. Res., 108*(B5), 2261, doi:10.1029/2002JB001787.
Saffer, D. M., and B. Bekins (2002), Hydrologic controls on the morphology and mechanics of accretionary wedges, *Geology, 30*(3), 271–274.
Saffer, D. M., and C. Marone (2003), Comparison of smectite- and illite-rich gouge frictional properties: Application to the updip limit of the seismogenic zone along subduction megathrusts, *Earth Planet. Sci. Lett., 215*, 219–235.
Sallarès, V., J. J. Dañobeitia, E. R. Flueh, and G. Leandro (1999), Seismic velocity structure across the middle American landbridge in northern Costa Rica, *J. Geodyn., 27*, 327–344.
Sallarès, V., J. J. Dañobeitia, and E. R. Flueh (2000), Seismic tomography with local earthquakes in Costa Rica, *Tectonophysics, 296*, 61–78.
Sallarès, V., J. J. Dañobeitia, and E. R. Flueh (2001), Lithospheric structure of the Costa Rican Isthmus: Effects of subduction zone magmatism on an oceanic plateau, *J. Geophys. Res., 106*, 621–643.

Savage, J. C. (1983), A dislocation model of strain accumulation and release at a subduction zone, *J. Geophys. Res., 88*, 4984–4996.

Scholz, C. H. (2002), *The Mechanics of Earthquakes and Faulting*, 2nd ed., p. 471, Cambridge Univ. Press, New York.

Simpson, R. W., J. J. Lienkaemper, and J. S. Galehouse (2001), Variations in creep rate along the Hayward fault, California, interpreted as changes in depth of creep, *Geophys. Res. Lett., 28*, 2269–2272.

Spinelli, G., and D. Saffer (2004), Along-strike variations in underthrust sediment dewatering on the Nicoya margin, Costa Rica related to the updip limit of seismicity, *Geophys. Res. Lett., 31*, L04613, doi:10.1029/2003GL018863.

Spinelli, G. A., and M. B. Underwood (2004), Characterization of sediment entering the Costa Rica subduction zone: Implications for partitioning of water along the plate interface, *Island Arc, 13(3)*, 432–451.

Steinbrugge, K. V., E. G. Zacher, D. Tocher, C. A. Whitten, and C. N. Claire (1960), Creep on the San Andreas fault, *Bull. Seismol. Soc. Am., 65*, 93–112.

Vanucchi, P., M. Meschede, and D. W. Scholl (1998), Tectonic erosion of the Pacific margin of Costa Rica: Combined implications of offshore drilling, a regional "base-of-slope-sediment-BOSS" reflector and a regional unconformity of the Nicoya and Osa Peninsulas, in *European Ocean Drilling Forum Programme and Abstracts*, edited by A. Robertson and J. Pierce, p. 102, Univ. of Edinburgh, Scotland, U.K.

Vanucchi, P., D. W. Scholl, M. Meschede, and K. McDougall-Reid (2001), Tectonic erosion and consequent collapse of the Pacific margin of Costa Rica: combined implications from ODP Leg 170, seismic offshore data, and regional geology of the Nicoya Peninsula, *Tectonics, 20*, 649–668.

Vannucchi, P., C. R. Ranero, S. Galeotti, S. M. Straub, D. W. Scholl, and K. McDougall-Ried, (2003), Fast rates of subduction erosion along the Costa Rica Pacific margin: Implications for nonsteady rates of crustal recycling at subduction zones, *J. Geophys. Res., 108*(B11), 2511, doi:10.1029/2002JB002207.

von Huene, R., et al. (1995), Morphotectonics of the Pacific convergent margin of Costa Rica, Geologic and tectonic development of the Caribbean plate boundary, in *Southern Central America*, edited by P. Mann, *Spec. Pap. Geol. Soc. Am., 295*, 291–307.

von Huene, R., C. R. Ranero, W. Weinrebe, and K. Hinz (2000), Quaternary convergent margin tectonics of Costa Rica: Segmentation of the Cocos Plate and Central American volcanism, *Tectonics, 19*, 314–334.

von Huene, R., C. R. Ranero, and P. Vannucchi (2004),. A generic model of subduction erosion, *Geology*, *32*(10), 913–916..

Vrolijk, P. (1990), On the mechanical role of smectite in subduction zones, *Geology, 18*(8), 703–707.

Ye, S., J. Bialas, E. R. Flueh, A. Stavenhagen, and R. von Huene (1996), Crustal structure of the Middle American Trench off Costa Rica from wide-angle seismic data, *Tectonics, 15*, 1006–1021.

Zweck, C., J. Freymueller, and S. C. Cohen (2002), Three-dimensional elastic dislocation modeling of the postseismic response to the 1964 Alaska earthquake, *J. Geophys. Res., 107*(B4), 2064, doi:10.1029/2001JB000409.

PART VI

Regional Scale Deformation

9

Collision Versus Subduction

From a Viewpoint of Slab Dehydration

Tetsuzo Seno

Abstract

On the basis of the observation that there is no intraslab seismicity or low-frequency tremor north of the Izu collision zone in central Honshu, I suggest that the lack of dehydration of the subducted slab may characterize collision. The plate-boundary thrust is then devoid of pore fluid and becomes strong. I show that crustal deformation in the collision zone is better described by the detachment model, whereas in the subduction zone by the back-slip model. This distinction can be explained by the nonexistence of lubrication due to insufficient pore fluid at the thrust zone of the collision zone. The complete adherence of the thrust zone in the collision zone makes the strength of the thrust zone of the order of 100–1000 MPa. This explains offscraping of the whole or upper crust (i.e., delamination of the mantle lithosphere and/or lower crust) and extensive intracontinental deformation in the hinterland. Collision zones with intermediate-depth seismicities might be a type between the typical collision and subduction zones. Fluid dehydrated from the slab will lubricate the thrust zone to some extent. This reduces the shear stress at the thrust zone and causes less deformation in the hinterland and less offscraping of the crust in the front. The collision at the arc-arc junction due to lateral migration of the fore arc caused by oblique subduction might be expected when the underthrusting fore-arc mantle wedge is not serpentinized.

Introduction

When a continent or an island arc is impinging on a subduction zone, collision takes place. It is generally believed that buoyancy of the continental or island-arc crust makes subduction difficult [e.g., *McKenzie*, 1969; *Dewey and Bird*, 1970;

Cloos, 1993] and causes delamination of the upper mantle and/or lower crust of the colliding lithosphere [*Bird*, 1978; *Sacks and Secor*, 1990; *van den Beukel*, 1992] and even stacking of the upper or whole crust [*Lyon-Caen and Molnar*, 1983; *Mattauer*, 1986].

However, there remains a question whether the buoyancy is a unique condition for generating collision. The calculation by *van den Beukel* [1992] showed that the buoyancy force of the continental crust subducted to the 60-km depth is not large enough to cause delamination of cold continental lithosphere with surface heat flow <60 mW/m^2. Furthermore, although he assumed that buoyancy operates only on the crustal part of the subducted lithosphere, it should operate on the whole lithosphere, if the crust is part of the solid lithosphere. Then this not only reduces the net buoyancy force by the amount of the excess mass of the lithospheric mantle [*Cloos*, 1993], but it also makes *van den Beukel*'s [1992] idea that the buoyancy force operates selectively on the crust to disintegrate it from the rest of the lithosphere doubtful.

Iwase and Honda [1993] showed with numerical calculations that the continental back-arc crust can be entrained into the mantle because of drag of the subducting plate unless its viscosity is low. This may also be true for the continental crust in contact with a subducting plate. *Butler* [1986] proposed that a major portion of the continental crust might have been subducted beneath the western Alps owing to eclogatization of the lower crust that is consistent with the sedimentary rocks found to be metamorphosed with a minimum pressure of ~3 GPa [*Chopin*, 1987; *Schreyer*, 1988].

These studies suggest that there may be a rheological factor that controls collision in addition to the buoyancy. The purpose of this study is to propose this factor from a viewpoint of dehydration of the subducting slab. The nonexistence of low-frequency tremors in some parts of the southwest Japan fore arc made me notice this factor. *Honda et al.* [2000] pointed out that rheological properties of plates such as low yield strength associated with dehydration of oceanic plates might be important in controlling convection patterns and subduction zone geometries. Thus the idea presented in this study may have implications not only for tectonics in collision zones but also for general geodynamics.

Low-Frequency Tremors and Dehydration from the Subducted Crust

Recently, low-frequency tremors (LFT) have been found around the Moho depth of the southwest Japan fore arc, forming a narrow zone along the strike of the arc (fig. 19.1) [*Obara*, 2002] (see also *Katsumata and Kamaya* [2003]). These are small-amplitude events, lacking sharp onsets and lasting a few minutes to a few days, with predominant frequency of 1–10 Hz, and the high-sensitivity seismograph network over Japan (Hi-net) has made it possible to detect these subtle events [*Obara*, 2002]. They likely occur in relation with fluid leaving the dehydrating slab because they are located above the 35–45 km depth surface of the subducting Philippine Sea plate. Seno and Yamasaki [2003] suggested that they are caused by hydrofracturing because they are located at the landward edge of the supposed serpentinized wedge mantle (S. Kamiya and Y. Kobayashi, manuscript in preparation, 2004) in a cross section across the easternmost Nankai Trough [see *Seno and Yamasaki*, 2003, figure 4].

LFT is absent in four areas along the southwest Japan fore arc; they are Kanto, North Izu, East Shikoku, and South Kyushu from east to west (figs. 19.1 and 19.2). The Hi-net covers these areas, and therefore they are not out of range of detection. *Seno and Yamasaki* [2003] noted that these are the places where island-arc crust is subducting (or colliding in the case of North Izu). They proposed that dehydration of the subducted crust does not occur beneath these regions, because it is composed mainly of tonalite [*Taira et al.*, 1998; *Suyehiro*

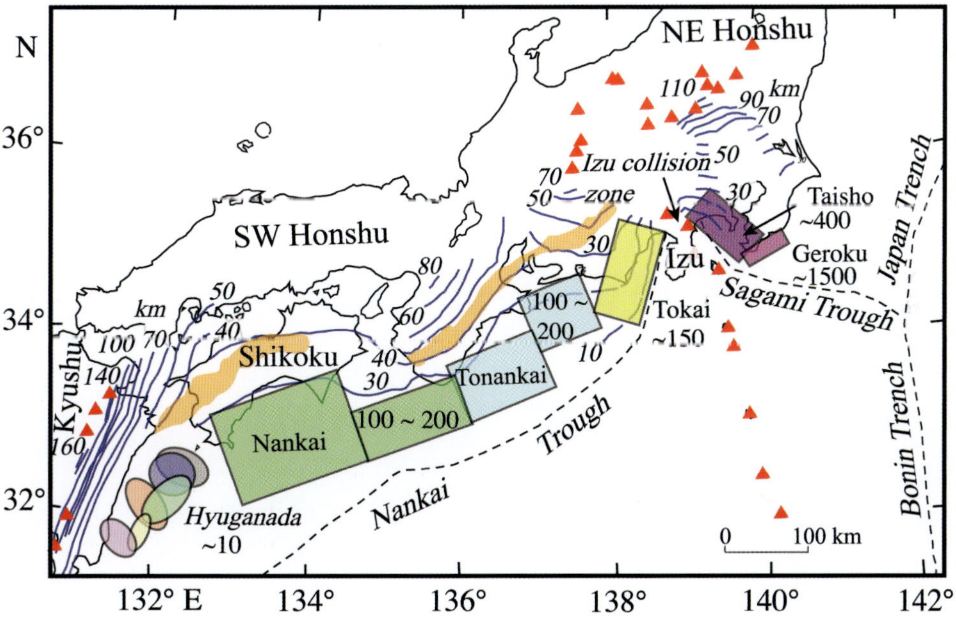

Figure 19.1 Fault planes with recurrence intervals (years) of great earthquakes, distribution of low-frequency tremors (LFT) and contours of the upper surface of intraslab seismicity along the Suruga-Sagami Troughs [*Seno and Yamasaki,* 2003]. The fault planes are those of the Taisho and Genroku Kanto earthquakes [*Kasahara et al.,* 1973], the Tokai earthquake [*Ishibashi,* 1981], and the 1944 Tonankai and 1946 Nankai earthquakes [*Ando,* 1975]. At the time of the Genroku earthquake the Taisho fault plane also ruptured. Rupture zones of large earthquakes in Hyuganada are from *Utsu* [1974]. Contours of the upper surface of the intraslab seismicity (blue lines) are from *Noguchi and Sekiguchi* [2001], *Nakamura et al.* [1997], and *Goto* [2001]. LFT (light brown) is from *Obara* [2002]. The red triangles indicate active volcanoes.

et al., 1996], lacking the quantity of hydrous minerals seen in normal subducting oceanic crust. They showed that almost no earthquake occurs within the subducted crust in these regions, consistent with the dehydration embrittlement hypothesis for intraslab seismicity [*Kirby et al.*, 1996; *Seno and Yamanaka*, 1996; *Peacock and Wang*, 1999; *Hacker et al.*, 2003; *Yamasaki and Seno*, 2003].

Collision

Among the four places cited above, only North Izu is a collision zone where the Izu-Bonin volcanic ridge is impinging into central Honshu (fig. 19.1). The strata of the Miocene volcanic and volcaniclastic successions of the Izu-Bonin Ridge origin have been severely deformed around the northern neck of the Izu Peninsula [*Soh et al.*, 1991; *Matsuda*, 1978]. There is a marked dif-

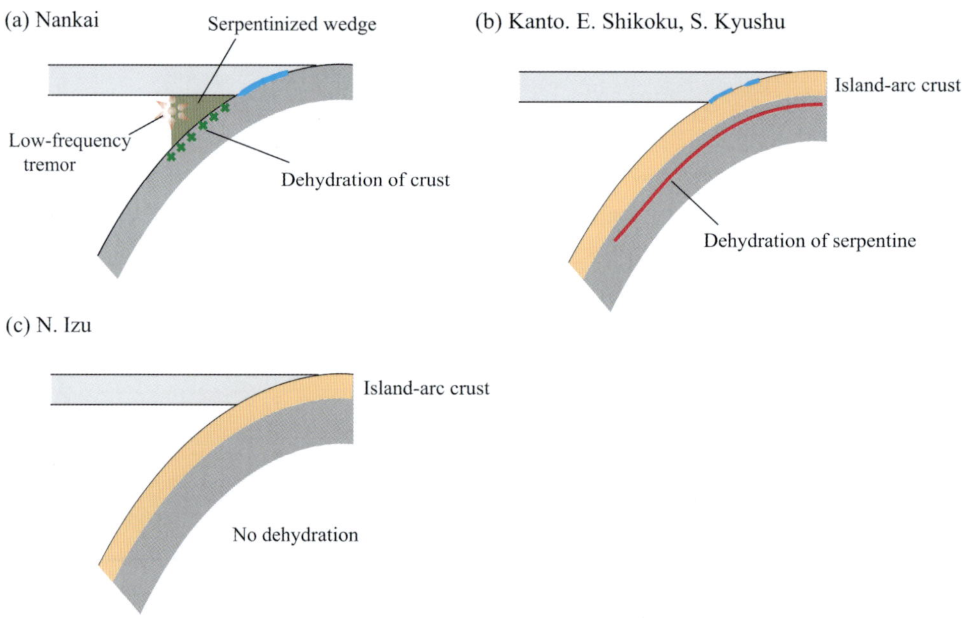

Figure 19.2 Schematic illustration of different modes of slab dehydration in a) Nankai Trough; b) Kanto, E. Shikoku and S. Kyushu; and c) N. Izu. In (a) slab with normal oceanic crust subducts with dehydration of altered crustal basalt. The water so derived migrates to thrust zone and raises pore fluid pressure (blue). It also hydrates and serpentinizes fore-arc mantle wedge and causes low frequency tremors (asterisks). In (b) slab with island-arc crust over serpentinized mantle subducts. Crustal dehydration does not occur, hence low frequency tremor does not occur. However, dehydration of slab mantle results in intraslab seismicity, lubricating the thrust zone to some extent (blue). In (c) slab with island arc crust over non-serpentinized mantle subducts. There is no slab dehydration, and hence no low frequency tremor or intraslab seismicity. The thrust zone adheres completely, resulting in collision.

ference between North Izu and the other three places. Not only LFT but also intraslab seismicity is completely missing in North Izu; it is evident that both LFT and intraslab seismicity suddenly disappear northeast of the Tokai region (fig. 19.1). *Seno and Yamasaki* [2003] suggested that this is not due to the lack of a subducted slab itself but because of the lack of dehydration of the slab. However, in Kanto, east Shikoku, and south Kyushu, there is intraslab seismicity in the mantle part. On the basis of the dehydration embrittlement hypothesis for intraslab seismicity, this implies that the mantle portion of the slab in these regions is serpentinized for some reason and dehydration of the slab mantle occurs [*Seno et al.*, 2001; *Kodaira et al.*, 2002; *Seno and Yamasaki*, 2003]. As was noted by *Seno and Yamasaki* [2003], large interplate earthquakes occur in north Izu. In contrast, large or great earthquakes occur in Kanto, east Shikoku, and south Kyushu, where dehydration from the serpentinized mantle occurs. Figure 19.2 shows schematically the difference in slab dehydration and related phenomena between these regions.

This contrast with north Izu suggests that the absence of dehydration inferred from the absence of seismicity in both the crust and mantle parts of the slab may affect the rheological nature of the plate interface and its surrounding material and characterize collision. For earthquakes to occur at the subduction zone thrust, pore-fluid pressure has to rise at least in the barrier portions of the rupture zone [*Seno*, 2003]. The source of the fluid supplied to the fault zone is likely dehydrating altered basalts in the subducted crust and/or serpentinized slab mantle [*Hacker et al.*, 2003; *Yamasaki and Seno*, 2003], because dehydration from sediments occurs shallower than 10 km [*Hyndman et al.*, 1995]. It is then expected that the absence of slab dehydration in the collision zone makes the thrust zone very strong.

The complete lack of intraslab seismicity is also seen in many other collision zones. Figure 19.3a shows a map view and a cross-section view of seismicity in Zagros (ISC data from 1980 to 1998, $M \geq 4.5$); most of the hypocenters are shallower than 50 km, and there is no indication of inclined seismic activity in association with the underthrusting Arabian plate (see also *Ni and Barazangi* [1986]). Figure 19.3b shows seismicity in Nepal and farther north (ISC data from 1980 to 1998, $M \geq 4.5$). Similarly, there is no indication of intraslab seismicity associated with the subducted Indian plate (see also *Baranowski et al.* [1984] and *Ni and Barazangi* [1984]).

Figure 19.4 shows the resistivity model in central Nepal with microseismic activity [*Lemonnier et al.*, 1999; see also *Pandey et al.*, 1995]. The microseismicity is concentrated only in the midcrustal ramp of the low angle décollement (the main Himalayan thrust, MHT), where stress concentration is expected as discussed later. The low resistivity in the shallower portion of the MHT beneath the Lesser Himalaya may indicate that no dehydration occurs in the subducting Indian continental lithosphere, except for the shallowest sediments subducted beneath the sub-Himalaya. *Lemonnier et al.* [1999] suggested that the low resistivity in the midcrustal ramp was caused by the release of metamorphic fluid, but the lack of significant intraslab seismicity (note that the microseismic activ-

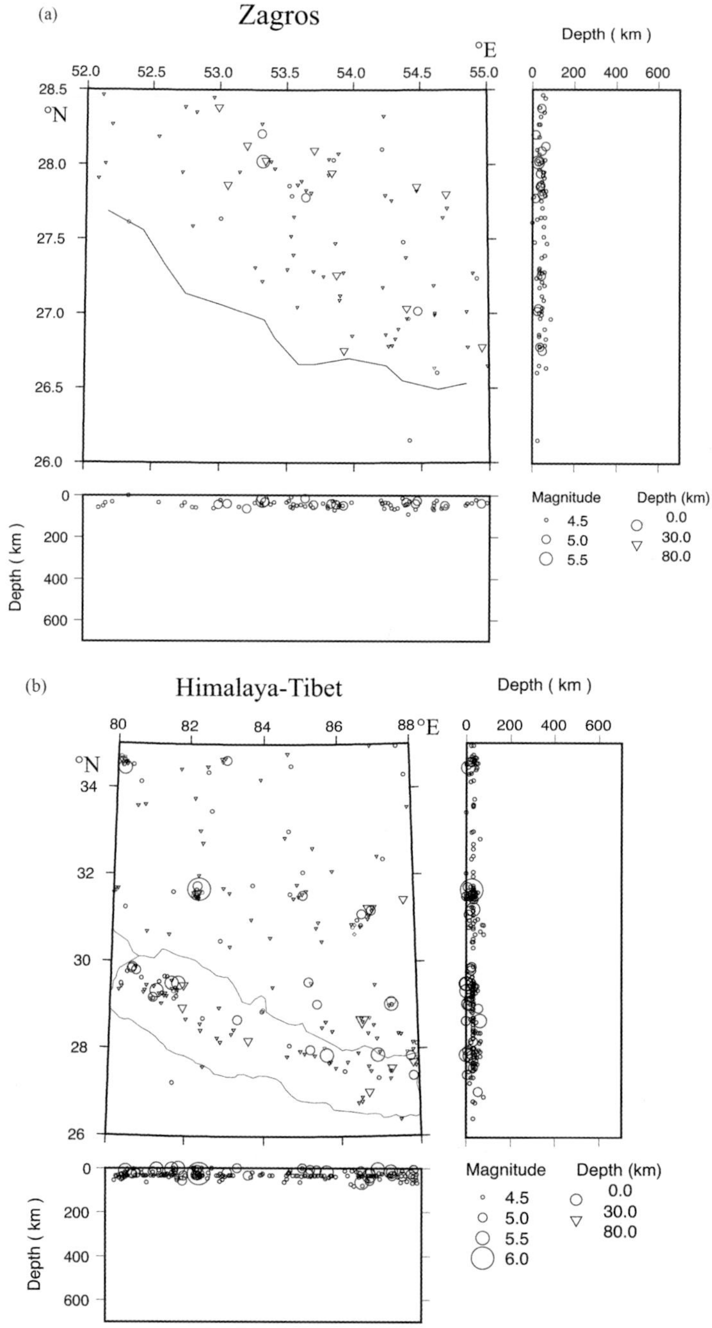

Figure 19.3 (a) Seismicity in a map view and N-S and E-W cross sections in the Nepal Himalaya and its north. The earthquakes are from International Seismological Centre Bulletins during the period between 1980 and 1998 having magnitudes >4.5. (b) Seismicity in a map view and N-S and E-W cross sections in Zagros. The earthquakes are from International Seismological Centre Bulletins during the period between 1980 and 1998 having magnitudes >4.5.

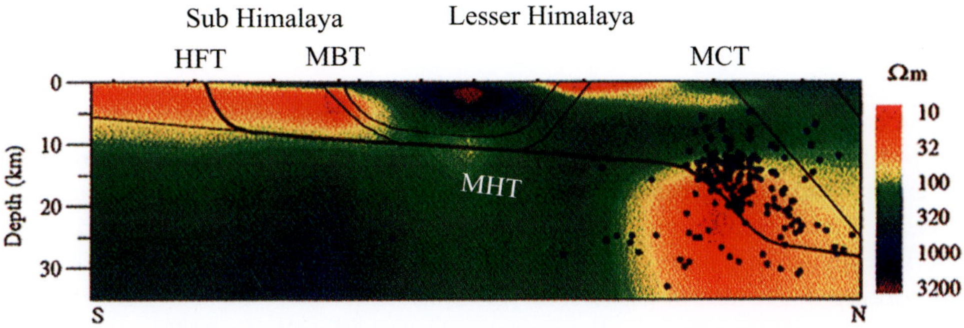

Figure 19.4 N20°E cross section of resistivity model derived from the two-dimensional inversion of main thrust (MT) data [*Lemonnier et al.*, 1999]. The Himalayan frontal thrust (HFT), main boundary thrust (MBT), main central thrust (MCT) and main Himalayan thrust (MHT) are shown.

ity is concentrated in the upper plate, fig. 19.4) suggests that the low resistivity is more consistent with high temperatures owing to shear heating.

The upper continental crust is composed of granitic-granodioritic rocks [*Taylor and McLennan*, 1995]. The lower crust is generally believed to be composed of granulite facies rocks [*Rudnick and Fountain*, 1995], although they may not represent residues left behind the intracrustal melting producing granitic rocks [*Taylor and McLennan*, 1995]. For these compositions it is inferred that the water content contained in the continental crust would be much smaller than that in the basaltic layers of the oceanic crust, which can contain water up to several wt % [*Poli and Schmidt*, 1995]. This is consistent with the no-slab dehydration hypothesis for collision presented above.

Collision Zones with Intermediate-Depth Seismic Activities

The above examples suggest that the lack of slab dehydration inferred from lack of intraslab seismicity might characterize collision. However, some collision zones are known to have intermediate-depth seismic activities. Subduction of an oceanic plate generally occurs prior to collision. Here I examine those collision zones with intermediate-depth seismic activities to see whether a subducting oceanic plate is leading a trailing continental plate. In Pamir-Hindu Kush, there are two groups of intermediate-depth seismic activity distributed down to the depth of 300 km; the western group is dipping to the north and the northeastern group is dipping to the southeast [*Chatelain et al.*, 1980]. There is a seismicity gap in the depth of ~160 km in the western group; *Roecker* [1982] interpreted this gap as representing the boundary between the oceanic and continental plates. As for the seismicity dipping to the southeast, *Mattauer* [1986] suggested intracontinental subduction in Pamir from the north. The seismic tomography study by *Bijwaard et al.* [1998] showed that a high seismic velocity is associated with the activity. The high velocity may represent a subducted back-arc basin that once existed between continental blocks.

Off the east coast of central Taiwan, there is an intermediate-depth seismicity dipping to the east down to the depth of 80–100 km [*Lin and Roecker,* 1993]. According to the tomographic study of *Rau and Wu* [1995], the Asian continental crust is subducted beneath the Luzon arc to the depth of 40 km in central Taiwan. The intermediate-depth activity is then mainly associated with the subducting oceanic plate, i.e., the former South China Sea that existed east of the continent (see also *Bijwaard et al.* [1998]). The oceanic slab seen beneath east of Taiwan would result from the recent initiation of collision in Taiwan around 3–6.5 Ma [*Chi et al.*, 1981; *Huang et al.*, 1997].

The southeasternmost part of the Carpathian arc in Romania is associated with the intermediate-depth seismicity down to 220-km depth in the continental environment. Seismic tomography studies [*Fan et al.*, 1998; *Lorenz et al.*, 1997; *Bijwaard and Spakman*, 2000] showed that high seismic velocity is associated with the seismicity, and *Fan et al.* [1998] and *Wortel and Spakman* [2000] suggested that oceanic lithosphere represents the high seismic velocity.

The Australian continent has been colliding with the eastern Sunda arc near Timor. The intermediate-depth seismicity is dipping to the north down to the 300-km depth [*McCaffrey et al.*, 1985]. *McCaffrey et al.* [1985] estimated that the Australian continent has been subducted down to the 150-km depth from a plate kinematic reconstruction. If so, the activity farther down has to be associated with the subducted Indian Ocean lithosphere.

Therefore most of the deeper portions of the intermediate-depth seismicities in these regions are likely to be associated with the subduction of oceanic plates leading trailing continents. It is often difficult to discriminate exactly the continental part from the oceanic part. However, the intermediate-depth activities imply that dehydration of the subducted oceanic lithosphere occurs in these regions, if we accept the dehydration embrittlement hypothesis for intermediate-depth earthquakes.

Comparison of Crustal Deformation Between Collision Zones and Subduction Zones

It is well known that interseismic crustal deformation in subduction zones can be described by a back-slip model (fig. 19.5) [*Savage,* 1983]. I point out that interseismic crustal deformation in collision zones is, in contrast, likely to be better described by a detachment model, in which the shallow part of the thrust zone is locked and the deeper one is stable sliding. Both models produce a similar crustal deformation pattern having a constant offset to each other (K. Wang, personal communication, 2004) because the detachment model is equivalent to the back-slip model plus a steady slip at the plate boundary (fig. 19.5). However, there is a clear distinction between the two. In the back-slip model the upper plate above the locked thrust zone sinks with respect to

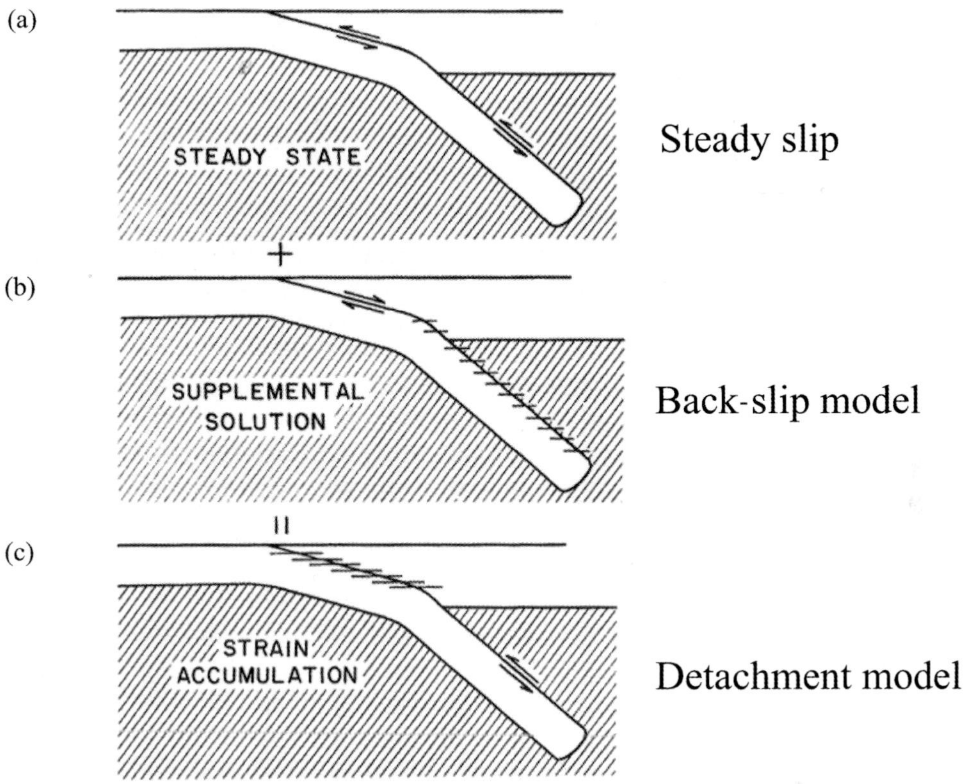

Figure 19.5 Schematic cross section of plate subduction showing the steady slip, back-slip model, and detachment model [*Savage*, 1983]. In the back-slip model the deformation caused by the steady slip is neglected, and only the normal fault-type dislocation at the thrust zone contributes to the deformation. In the detachment model the contribution from the steady slip is added to that from the back slip, resulting in the contribution from the steady slip in the detachment deeper than the thrust zone. This model is equivalent to the model by *Matsu'ura and Sato* [1989].

the lower plate seaward of the trench axis during the interseismic period. In contrast, in the detachment model, that portion of the upper plate is uplifted with respect to the lower plate seaward of the trench axis during this period. Figure 19.6 shows the crustal deformation in central Taiwan prior to the 1999 ChiChi earthquake [*Yu et al.*, 2001]. An island offshore west of Taiwan, i.e., on the subducting plate far west of the deformation front, is taken as the reference point of these movements. No subsidence of the upper plate above the fault plane of the 1999 ChiChi earthquake was observed, and both of the horizontal and vertical displacements are fitted by the slip on the detachment extending deeper than the fault of the ChiChi earthquake with a rate of 70 mm/yr (fig. 19.6). This slip rate is comparable to the relative motion between the Philippine Sea and Eurasian plates in this area [*Seno et al.*, 1993; *Yu et al.*, 1999].

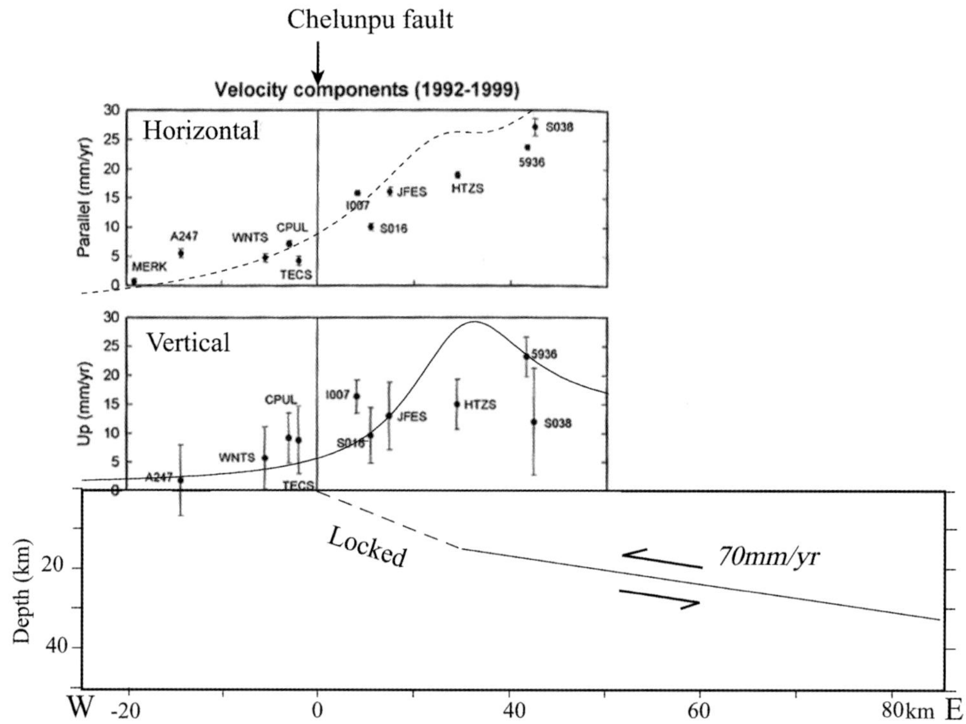

Figure 19.6 Horizontal and vertical velocities of the GPS stations in and near the rupture zone of the 1999 ChiChi earthquake before the occurrence of the event [*Yu et al.,* 2001]. These velocities are fitted by the slip at the detachment.

The crustal deformation across the Himalayas in Nepal revealed by GPS and precise leveling has been fitted by the slip on the detachment with a rate of ~20 mm/yr (fig. 19.7) [*Jackson and Bilham,* 1994; *Freymueller et al.,* 1996; *Bilham et al.,* 1997; *Larson et al.,* 1999; *Jouanne et al.,* 1999]. In this case also, no subsidence of the upper plate (Sub and Lesser Himalayas) with respect to the stations south of the Himalayan frontal thrust (HFT) on the approaching Indian plate has been seen (see *Jackson and Bilham* [1994, figure 1] and *Bilham et al.* [1997, figure 3]). *Takada and Matsu'ura* [2004] also succeeded to explain the Quaternary uplift of the Himalayas based on the steady slip model of *Matsu'ura and Sato* [1989], which is equivalent to the detachment model (see also *Cattin and Avouac* [2000]). These show a marked contrast with the deformation seen in subduction zones like the Nankai Trough and Japan Trench that shows subsidence of the coast with respect to the sea level and is fitted by the back-slip model [*Savage and Thatcher,* 1992; *Hyndman et al.,* 1995; *Ito et al.,* 1999; *Aoki and Scholz,* 2003].

The difference in interseismic crustal deformation between collision and subduction zones can be explained in terms of the slab dehydration (fig. 19.8). In the collision zone, slab dehydration does not occur, hence pore fluid in the plate interface at 10–40 km depth is minimal. The thrust zone is then com-

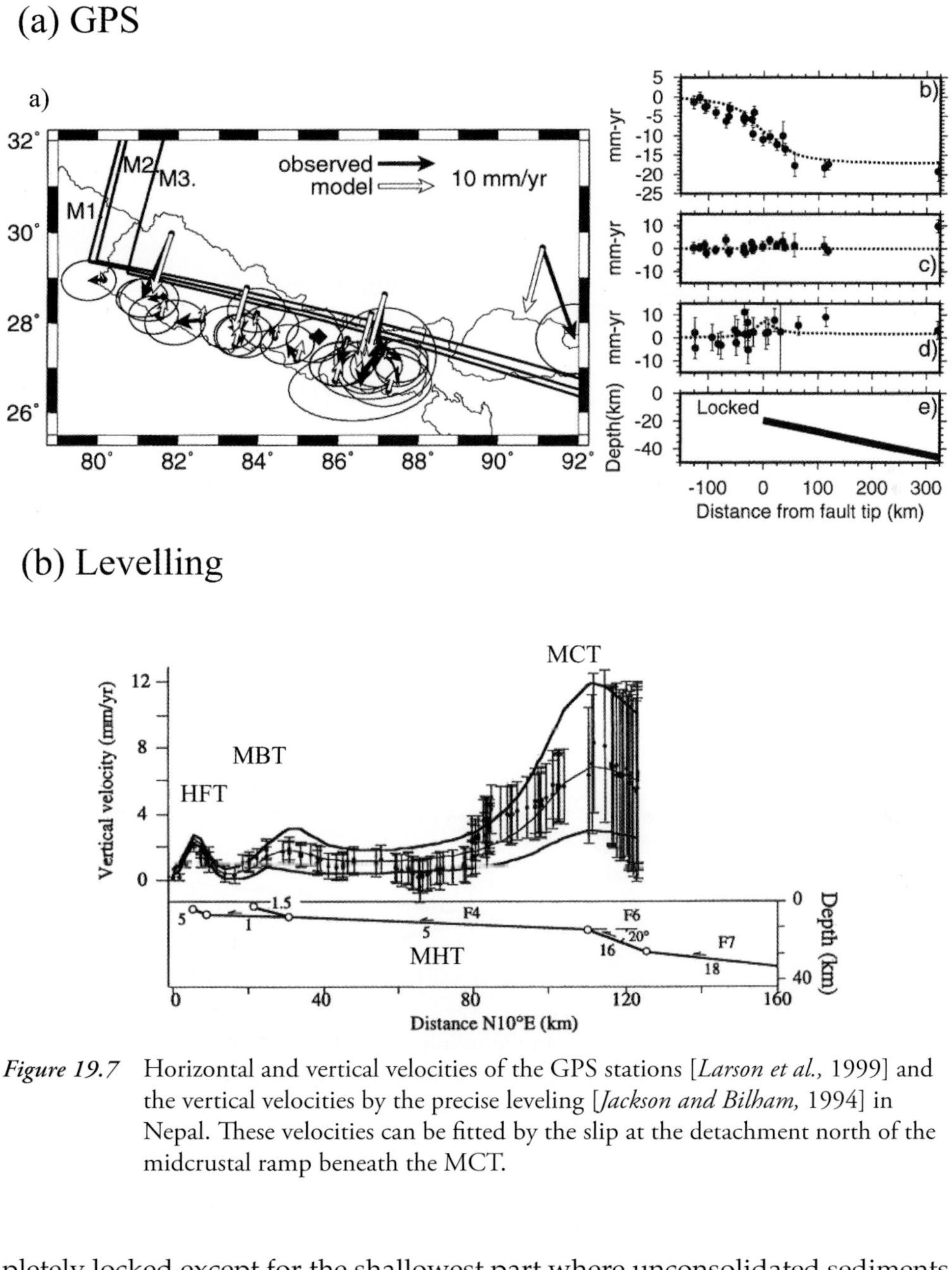

Figure 19.7 Horizontal and vertical velocities of the GPS stations [*Larson et al.,* 1999] and the vertical velocities by the precise leveling [*Jackson and Bilham,* 1994] in Nepal. These velocities can be fitted by the slip at the detachment north of the midcrustal ramp beneath the MCT.

pletely locked except for the shallowest part where unconsolidated sediments are underthrusting and dewatering (fig. 19.8a). In this case, the upper and lower plates cannot be discriminated mechanically across the thrust zone and can be regarded as continuous. In other words, the thrust zone cannot be recognized as a plate boundary until it breaks. Further downdip of the coupled zone, stable sliding occurs because of elevated temperatures (>350°C). It is then expected that the deformation is described by the detachment model, in which the shallower portion of the thrust zone is locked and the deeper por-

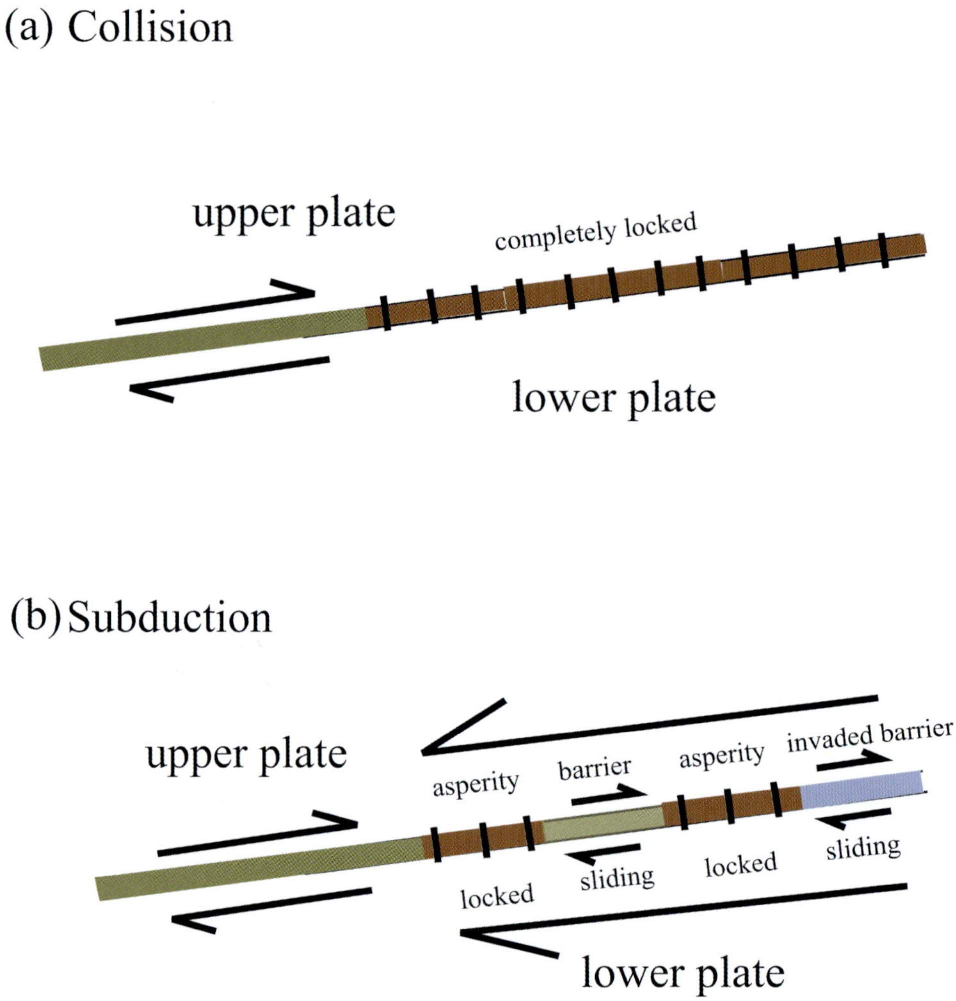

Figure 19.8 Schematic cross sections of the plate boundary in (a) the collision zone and in (b) the subduction zone. In the collision zone the shallow thrust zone is not provided with fluid, and the zone is completely coupled. In the subduction zone the thrust zone is provided with fluid, and the lower plate moves with respect to the upper plate beneath the barriers. The upper plate is dragged through asperities by the lower plate.

tion is sliding in a stable manner. High stress concentration occurs at the updip end of the stable sliding zone. The concentration of microseismic activities in the upper plate and the low resistivity near the midcrustal ramp of the MHT (fig. 19.4) may be caused by this stress concentration.

In contrast, in the subduction zone, the thrust zone is provided with the fluid released from the dehydrating slab (fig. 19.8b). Large interplate earthquakes often occur there. The fault zone consists of barriers and asperities, characterized by the negative and positive a-b of the rate and state dependent friction law, respectively [e.g., *Seno,* 2003]. The barrier portion is sliding in a stable manner. The pore-fluid pressure is likely to be raised nearly to the lithostatic in barriers to reduce the strength of the fault (see literature cited in the work of *Seno* [2003]). This further makes barriers to slide aseismically. Because a major portion of the fault zone is occupied by barriers [*Seno,* 2003], the lower plate can move with a convergence velocity with respect to the upper plate. Since the asperity portions are locked, the upper plate is dragged along with the underthrusting lower plate at the thrust zone. It is then expected that the deformation is described by the back-slip model.

In Kanto, east Shikoku, and south Kyushu, island-arc type crust is subducting. However, the mantle part of the slab has seismicity, and dehydration of serpentine would occur in the slab mantle. The thrust zone is provided with fluid released from the slab, and the pore pressure would be raised there; the situation becomes as shown in figure 19.8b. However, because the pore-fluid pressure is less efficient to be raised at the thrust zone than subduction zones where normal oceanic crust is subducting, the occurrence of great earthquakes in these regions then becomes infrequent [*Seno and Yamasaki,* 2003].

A situation similar to Kanto, east Shikoku, and south Kyushu might occur in the collision zones with oceanic lithosphere attached in the deeper part. The fluid dehydrating from the oceanic crust and/or mantle migrates to the shallower depth. The thrust zone might be lubricated by this fluid to some extent. This would make the continental subduction easier in these regions.

Discussion

Arc-Arc Collision

Oblique subduction of an oceanic plate beneath a continental or island arc often induces along-arc migration of the fore arc [*Fitch,* 1972; *Beck,* 1983; *McCaffrey,* 1992; *Kimura,* 1996]. The southwestward migration of the Kuril fore arc due to the oblique subduction of the Pacific plate is such an example [*Kimura,* 1996]. The collision has been taking place in the Hidaka mountains since the late Miocene exposing the lower continental crust (Figure 19.9) [*Komatsu et al.,* 1983; *Tsumura et al.,* 1999]. It is interesting to compare this with the case of Kanto, where the Izu-Bonin fore arc is subducting. *Seno* [2005] inferred the extent of serpentinization of the fore-arc mantle wedges near the Japanese islands on the assumption that the serpentinization prevents occurrence of interplate earthquakes [*Hyndman et al.,* 1997]. In the Bonin fore arc the mantle deeper than 20 km is likely to be serpentinized, judging from distribution of interplate events down to the 20-km depth. The *P*-wave velocity structure [*Suyehiro et al.,* 1996] also indicates that the uppermost mantle east of the volcanic front is likely to be serpentinized as shown in figure 19.10a. On the contrary, the fore-arc mantle of north Honshu east of the aseismic front is presumably not serpentinized on the basis of the distribution of interplate seismic activity

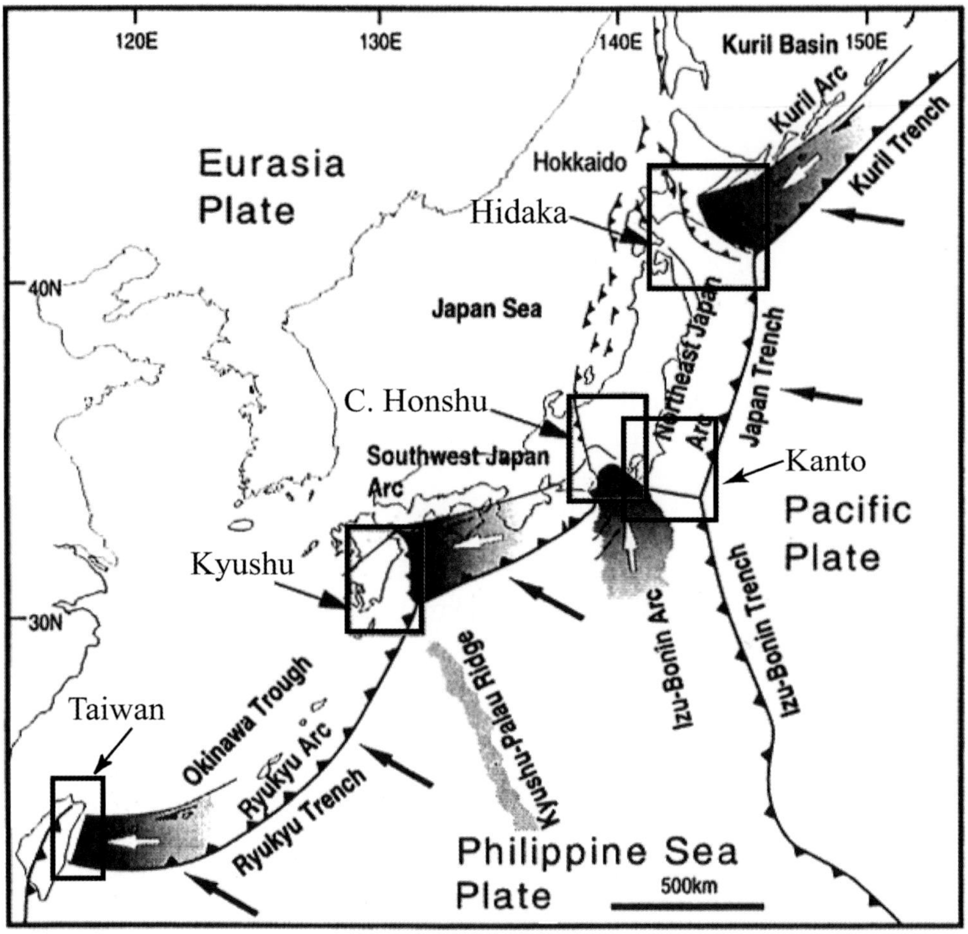

Figure 19.9 Arc-arc collision zones near Japan [*Kimura,* 1996]. The Kuril and north Honshu fore arcs are collided with each other in Hidaka. The Izu-Bonin Ridge is colliding with central Honshu; the Bonin fore arc is, however, subducting beneath Kanto. The southwest Japan fore arc is colliding with Kyushu. The Ryukyu fore arc is colliding with Taiwan.

down to the 60-km depth (fig. 19.10b). Serpentine is not expected to exist east of the volcanic front in this case. The northernmost Honshu fore arc is underthrusting the Kuril fore arc in the Hidaka collision zone (fig. 19.9). On the one hand, no dehydration from the subducted north Honshu fore arc is expected beneath Hidaka. This would have facilitated collision there. On the other hand, in Kanto, dehydration of the serpentinized Bonin fore arc has been occurring [*Seno et al.,* 2001; *Seno and Yamasaki,* 2003], and collision did not occur.

I then infer generally that, if the laterally migrating fore-arc mantle wedge at an arc-arc junction is not serpentinized, collision will happen as in Hidaka; if it is serpentinized, collision will not happen as in Kanto. If the Bonin arc

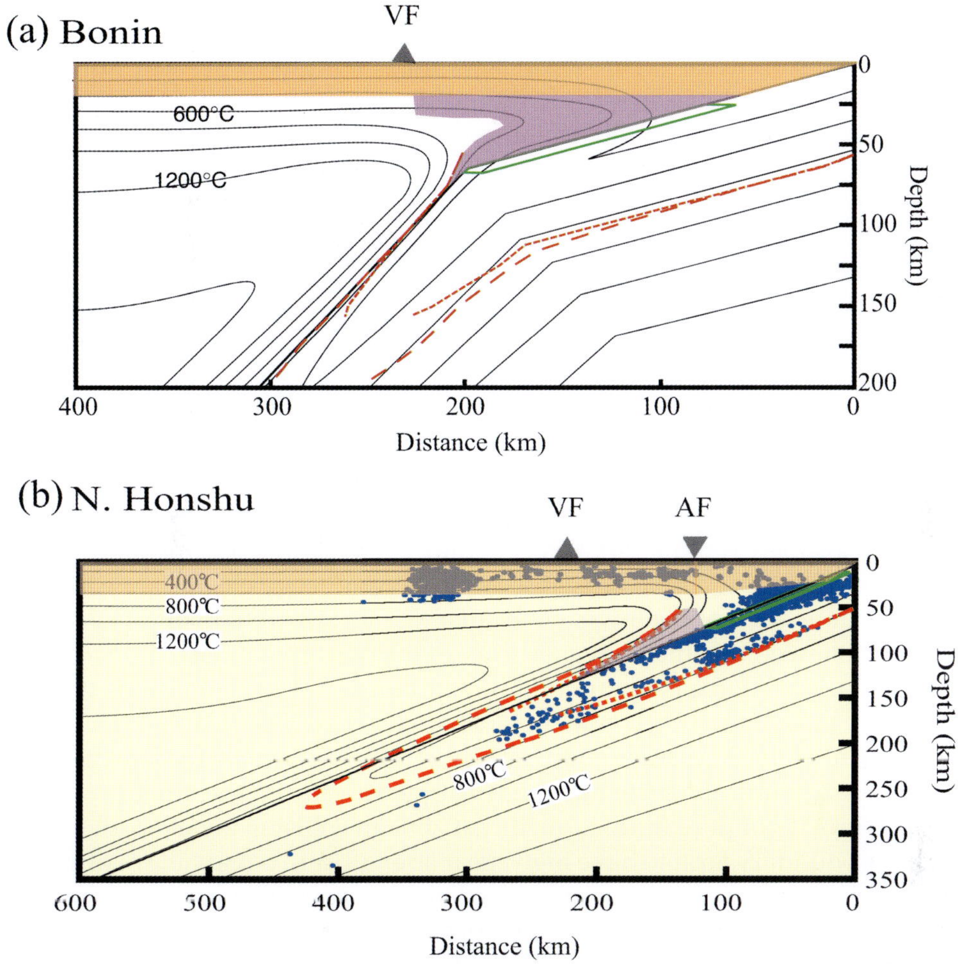

Figure 19.10 Extent of serpentinization of the mantle wedge of (a) the Bonin fore arc and (b) the north Honshu fore arc. Serpentinization of the fore-arc wedge is inferred from the distribution of interplate earthquakes [*Seno*, 2004]. Temperatures calculated by the finite element method and stability field of serpentinite (purple) are shown. Temperatures in the Bonin section are from Yamasaki and Seno (unpublished manuscript) and those in the north Honshu section are from *Yamasaki and Seno* [2003]. The dehydration loci of serpentine [*Ulmer and Trommsdorff*, 1995; *Wunder and Schreyer*, 1997] are indicated by the red lines in the depth >50 km. In the depth <50 km the 650°C isotherm is regarded as the dehydration loci. The wedge mantle seaward of the aseismic front (AF) in north Honshu is not serpentinized [*Seno*, 2004]. VF is volcanic front.

generates large interplate earthquakes down to the 60-km depth as in north Honshu, that is, if the fore-arc mantle wedge were not serpentinized, Kanto would have turned into a collision zone and the landscape of Metropolitan Tokyo would be quite different from seen at present. It would be interesting to examine other arc-arc junctions, such as the Mariana-Palau, west Aleutian-Kamchatka, southwest Japan-south Kyushu and Ryukyu-Taiwan junctions, from the viewpoint of serpentinization of the fore-arc mantle. For example, *Kimura* [1996] showed that deformation at the Ryukyu-southwest Japan junction is much less intense as compared with that of Hidaka. This might be due to the serpentinization of the mantle wedge beneath the southwest Japan fore arc (S. *Kamiya and Y. Kobayashi,* manuscript in preparation, 2004; see also *Seno and Yamasaki* [2003]).

Delamination and Intracontinental Deformation

In a collision zone the crust is stacked and thickened mainly by low-angle thrusting along the crust-mantle or upper crust-lower crust décollement, forming crustal duplex [*Mattauer,* 1986]. The lithospheric mantle and possibly part of the lower crust underthrust the upper plate and may be delaminated into the asthenosphere [*Bird,* 1978]. It is generally believed that the buoyancy of the underthrusting continental or island-arc crust produces extra resistive force to cause delamination of the lower crust/mantle. The calculation by *van den Beukel* [1992] showed that the buoyancy force of the crust is at most ~10^{13} N/m and an extra resistive force is required for delamination if the heat flow at the surface of the underthrusting continental plate is less than ~60 mW/m^2. However, as I have noted in the introduction, not only the net buoyancy force is reduced, but also it is doubtful for the buoyancy to operate only on the crust.

The scarcity of pore fluid at the thrust zone in collision zones would cause the plate boundary fault to have high shear strength and be capable of offscraping of the crust. The fault strength would become comparable to $m\sigma_z$, where μ is the coefficient of friction; this amounts to the order of 100–1000 MPa for the 6–60 km depth range for $\mu = 0.6$. Even if μ is as low as 0.3 that estimated by *Cattin and Avouac* [2000] for the flat portion of the MHT, the strength is reduced only by a factor 2. The driving forces of the underthrusting continental plate have to overcome this large resistive force to cause relative plate motion. The driving forces are ultimately provided by convection caused by heat within the Earth's interior.

The large resistive force at the thrust also explains why the deformation of the Asian continent collided by the Indian plate spreads over thousands of kilometer distance [e.g., *Molnar and Tapponnier,* 1977; *Tapponnier et al.,* 1982]. The high shear stress at the thrust of the order of 100–1000 MPa can be transmitted to the hinterland of the collision zone, resulting in intense deformation. This provides a reason why only a fraction of the relative plate motion is consumed at the thrust beneath the Himalayas [*Molnar,* 1984; *Lyon-Caen and Molnar,* 1985; *Armijo et al.,* 1986; *Schelling and Arita,* 1991; *Holt and Haines,* 1993; *Jackson and Bilham,* 1994; *Freymueller et al.,* 1996; *Mugnier et al.,* 2003], and the rest is used for the intracontinental deformation of Asia. If the shear stress at the thrust is of the order of 1–10 MPa as in subduction zones [*Seno,* 1999; *Wang and He,*

1999], it could not afford mechanical power large enough for such extensive intracontinental deformation.

The Miocene anatexis that generated granites above the CMT [*Le Fort,* 1986] might be explained by heating due to this high shear stress [*Bird,* 1978; *Molnar et al.,* 1983; *England et al.,* 1992]. However, recent modeling including erosion [*Royden,* 1993; *Henry et al.,* 1997] showed that the high shear stress is rather contradictory to the temperature distribution in the deeper part of the upper wedge favorable to the anatexis. If it is true, it may imply that during the Miocene (~20 Ma), the shear stress at the thrust might not be high, possibly in association with the subducted oceanic plate at depth.

In this context, it is noted that the deformation of the upper plate in the hinterland of collision zones with intermediate-depth seismic activity is much less intense compared with that seen in Tibet. North of Hindu Kush, the intense deformation zone extends only within the distance of ~500 km [e.g., *Tapponnier et al.,* 1982]. The deformation of the hinterland of the Carpathian arc is small. These might be because the pore fluid has been supplied to some extent to the thrust zone from the subducted slab during the recent past, resulting in the lower level of the shear stress. The modest deformation in the hinterland of the Alps (N. Italy) [see *Butler,* 1986] implies that the collision that occurred in the Alps may have been such that associated with intermediate-depth seismic activities until the recent past, like the many arcs in the present Mediterranean region, as seen by the seismic tomography studies [*Wortel and Spakman,* 2000].

Interplate Large Earthquakes in Collision Zones

I have noted that large interplate earthquakes are lacking in north Izu. It is also noted that interplate events are lacking in Zagros [*Ni and Barazangi,* 1986]. In the Himalayas, small interplate earthquakes during recent years have been occurring at the low-angle MHT [*Baranowski et al.,* 1984; *Ni and Barazangi,* 1984]. Historically, some great earthquakes, such as the 1934 Bihar (M_s 8.3) and 1905 Kangra (M_s 8.0) earthquakes likely ruptured the MHT extending close to the HFT [*Seeber et al.,* 1981]. Great earthquakes such as the 1803 and 1833 events are also known to have occurred during the 19th century but in other segments [*Bilham and England,* 2001]. I note that the occurrence of these great earthquakes is much more sporadic than, for example, in the circum-Pacific subduction zones. In addition, no repeating events are known to have ruptured the same fault segment in the Himalayas. This suggests that recurrence intervals of great events, if any, in collision zones are very long. In fact, in central Taiwan a trenching study showed the recurrence of the 1999 ChiChi earthquake to be around 1200–2000 years [*Shishikura et al.,* 2002].

In collision zones a considerable fraction of the relative plate motion is consumed in deformation of the hinterland, and this may provide one reason for the long recurrence intervals of great interplate earthquakes. However, this might not be enough because there remains a ~20 mm/yr slip rate on the thrust in the Himalayas and larger in central Taiwan. Because there is no indication of abnormally high stress drops for historical earthquakes in collision zones [*Chen and Molnar,* 1977; *Kikuchi et al.,* 2000], a considerable portion of the slip might be consumed as stable slow slip under a high shear stress.

Conclusions

On the basis of the observation that there is no intraslab seismicity or low-frequency tremor north of the Izu collision zone in central Honshu, I suggest that the lack of dehydration of the subducted slab may characterize collision. The plate-boundary thrust is devoid of pore fluid and hence is strong. I show that crustal deformation in the collision zone is described by the detachment model, whereas in the subduction zone by the back-slip model. In the subduction zone the fluid hydrated from the slab is supplied to the thrust zone and lubricates the barrier portions of the fault zone of large earthquakes. The lower plate can move with respect to the upper plate under the barriers. Because the asperity portions are locked, the upper plate above the thrust zone is dragged down with the lower plate. The crustal deformation is then described by the back-slip model in the subduction zone. In the collision zone, lack of dehydration of the subducted slab means that the thrust zone is fluid-poor, making the lower plate adhere completely to the upper plate. There is no distinction between the lower and upper plates across the thrust zone mechanically. With the stable sliding in the deeper part the crustal deformation is described by the detachment model.

The strength of the thrust zone in collision zones would be of the order of 100–1000 MPa. This explains the offscraping of the whole or upper crust (i.e., delamination of the mantle lithosphere and/or lower crust) and the extensive intracontinental deformation in the hinterland.

Collision zones with intermediate-depth seismic activities may be a type between collision zones and subduction zones. The fluid dehydrated from the slab will lubricate the thrust zone to some extent. This reduces the shear stress at the thrust zone and causes less deformation in the hinterland and less offscraping of the crust in the foreland. The apparent small amount of offscraping and deformation in the hinterland of the Alps [*Butler,* 1986] suggests that the collision in the Alps may have been associated with the oceanic slab until the recent past.

The collision at the arc-arc junction due to lateral migration of the fore arc caused by oblique subduction may be expected when the underthrusting fore-arc mantle wedge is not serpentinized, as that of the north Honshu fore arc in the Hidaka collision zone. If the underthrusting fore-arc mantle wedge is serpentinized as in the Bonin fore arc, subduction occurs as in Kanto, central Honshu.

Acknowledgments

I thank Kelin Wang for critical review of the manuscript. I also thank Makoto Omura for valuable information on the crustal deformation in the Himalayas, Yukitoshi Fukahata for discussions, and Youichiro Takada for sending his Ph.D. thesis and reprint.

References

Ando, M. (1975), Source mechanisms and tectonic significance of historical earthquakes along the Nankai Trough, Japan, *Tectonophysics*, *27*, 119–140.

Aoki, Y., and C. H. Scholz (2003), Interseismic deformation at the Nankai subduction zone and the Median Tectonic Line, southwest Japan, *J. Geophys. Res.*, *108*(B10), 2470, doi:10.1029/2003JB002441.

Armijo, R., P. Tapponnier, J. L. Mercier, and T.-L. Han (1986), Quaternary extension in southern Tibet: Field observations and tectonic implications, *J. Geophys. Res.*, *91*, 13,803–13,872.

Baranowski, J., J. Armbruster, L. Seeber, and P. Molnar (1984), Focal depths and fault plane solution of earthquake and active tectonics of the Himalaya, *J. Geophys. Res.*, *89*, 6918–6928.

Beck, M. E., Jr. (1983), On the mechanism of tectonic transport in zones of oblique subduction, *Tectonophysics*, *93*, 1–11.

Bijwaard, H., and W. Spakman (2000), Non-linear global *P*-wave tomography by iterated linearized inversion, *Geophys. J. Int.*, *141*, 71–82.

Bijwaard, H., W. Spakman, and E. R. Engdahl (1998), Closing the gap between regional and global travel time tomography, *J. Geophys. Res.*, *103*, 30,055–30,078.

Bilham, R., and P. England (2001), Plateau 'pop-up' in the great 1897 Assam earthquake, *Nature*, *410*, 806–809.

Bilham, R., K. Larson, J. Freymueller, and Project Idylhim Members (1997), GPS measurements of present-day convergence across the Nepal Himalaya, *Nature*, *386*, 61–64.

Bird, P. (1978), Initiation of intracontinental subduction in the Himalaya, *J. Geophys. Res.*, *83*, 4975–4987.

Butler, R. W. H. (1986), Thrust tectonics, deep structure and crustal subduction in the Alps and Himalaya, *J. Geol. Soc. London*, *143*, 875–873.

Cattin, R., and J. P. Avouac (2000), Modeling mountain building and the seismic cycle in the Himalaya of Nepal, *J. Geophys. Res.*, *105*, 13,389–13,407.

Chatelain, J. L., S. W. Roecker, D. Hatzfeld, and P. Molnar (1980), Microearthquake seismicity and fault plane solutions in the Hindu Kush region and their tectonic implications, *J. Geophys. Res.*, *85*, 1365–1387.

Chen, W. -P., and P. Molnar (1977), Seismic moments of major earthquakes and the average rate of slip in central Asia, *J. Geophys. Res.*, *82*, 2945–2969.

Chi, W-R., J. Namson, and J. Suppe (1981), Stratigraphic record of plate interactions in the Coastal Range of eastern Taiwan, *Mem. Geol. Soc. China*, *4*, 155–194.

Chopin, C. (1987), Very-high-pressure metamorphism in the western Alps: Implications for subduction of continental crust, *Philos. Trans. R. Soc. London, Ser. A*, *321*, 183–197.

Cloos, M. (1993), Lithospheric buoyancy and collisional orogenesis: subduction of oceanic plateaus, continental margins, island arcs, spreading ridges, and seamounts, *Geol. Soc. Am. Bull.*, *105*, 715–737.

Dewey, J., and J. M. Bird (1970), Mountain belts and the new global tectonics, *J. Geophys. Res.*, *75*, 2625–2647.

England, P., P. Le Fort, P. Molnar, and A. Pecher (1992), Heat sources for Tertiary metamorphism and anatexis in the Annapuruna-Manaslu region Central Nepal, *J. Geophys. Res., 97*, 2107–2128.

Fan, G., T. C. Wallace, and D. Zhao (1998), Tomographic imaging of deep velocity structure beneath the eastern and southern Carpathians, Romania: Implications for continental collision, *J. Geophys. Res., 103*, 2705–2724.

Fitch, T. J. (1972), Plate convergence, transcurrent faults, and internal deformation adjacent to southeast Asia and the western Pacific, *J. Geophys. Res., 77*, 4432–4457.

Freymueller, J., R. Bilham, R. Burgmann, K. M. Larson, J. Paul, S. Jade, and V. Gaur (1996), Global positioning system measurements of Indian plate motion and convergence across the Lesser Himalaya, *Geophys. Res. Lett., 23*, 3107–3110.

Goto, K. (2001), Intermediate-depth earthquakes and seismotectonics, in *Abstracts of Research Project, Grant-in-Aid for Scientific Research*, pp. 1–139, Minist. of Educ., Cult., Sports, Sci. and Technol., Kyushu, Jpn.

Hacker, B. R., S. M. Peacock, G. A. Abers, and S. D. Holloway (2003). Subduction factory: 2. Are intermediate-depth earthquakes in subducting slabs linked to metamorphic dehydration reactions? *J. Geophys. Res., 108*(B1), 2030, doi:10.1029/2001JB001129.

Henry, P., X. Le Pichon, and B. Goffe (1997). Kinematic, thermal and petrological model of the Himalayas: constraints related to metamorphism within the underthrust Indian crust and topographic elevation, *Tectonophysics, 273*, 31–56.

Holt, W. E., and A. J. Haines (1993), Velocity fields in deforming Asia from the inversion of earthquake-relative strains, *Tectonics, 12*, 1–20.

Honda, S., T. Nakakuki, Y. Tatsumi, and T. Eguchi (2000), A simple model of mantle convection including a past history of yielding, *Geophys. Res. Lett., 27*, 1559–1562.

Huang, C.-Y., W.-Y. Wu, C.-P. Chang, S. Tsao, P. B. Yuan, C.-W. Lin, and X. Kuan-Yuan (1997), Tectonic evolution of accretionary prism in the arc-continent collision terrane of Taiwan, *Tectonophysics, 281*, 31–51.

Hyndman, R. D., K. Wang, and M. Yamano (1995), Thermal constraints on the seismogenic portion of the southwestern Japan subduction thrust, *J. Geophys. Res., 100*, 15,373–15,392.

Hyndman, R. D., M. Yamano, and D. A. Oleskevich (1997), The seismogenic zone of subduction thrust faults, *Island Arc, 6*, 244–260.

Ishibashi, K. (1981), Specification of a soon-to-occur seismic faulting in the Tokai district, central Japan, based upon seismotectonics, in *Earthquake Prediction: An International Review, Maurice Ewing Ser.*, vol. 4, edited by D. W. Simpson and P. G. Richards, pp. 297–332, AGU, Washington, D. C.

Ito, T., S. Yoshioka, and S. Miyazaki (1999), Interplate coupling in southwest Japan deduced from inversion analysis of GPS data, *Earth Planet. Sci. Lett., 115*, 17–34.

Iwase, Y., and S. Honda (1993), Deformation of the crust and upper mantle beneath the back-arc caused by the subducting plate, *J. Phys. Earth, 41*, 347–363.

Jackson, M., and R. Bilham (1994), Constraints on Himalaya deformation inferred from vertical velocity field in Nepal and Tibet, *J. Geophys. Res., 99*, 13,897–13,912.

Jouanne, F., J. L. Mugnier, M. R. Pandey, J. F. Gamond, P. Le Fort, L. Serrurier, C. Vigny, J.P. Avouac, and Idylhim Members (1999), Oblique convergence in the Himalayas of western Nepal deduced from preliminary results of GPS measurements, *Geophys. Res. Lett., 26*, 1933–1936.

Kasahara, K., J. Yamada, and M. Ando (1973), Crustal movements in the southern Kanto district, and a related working hypothesis, in *Publications for the 50th Anniversary of the Great Kanto Earthquake, 1923*, pp. 103–116, Earthquake Res. Inst., Univ. of Tokyo, Tokyo.

Katsumata, A., and N. Kamaya (2003), Low-frequency continuous tremor around the Moho discontinuity away from volcanoes in the southwest Japan, *Geophys. Res. Lett., 30*(1), 1020, doi:10.1029/2002GL015981.

Kimura, G. (1996), Collision orogeny at arc-arc junctions in the Japanese islands, *Island Arc*, *5*, 262–275.
Kirby, S., E. R. Engdhal, and R. Denlinger (1996), Intermediate-depth intraslab earthquakes and arc volcanism as physical expressions of crustal and uppermost mantle metamorphism in subducting slabs (Overview), in *Subduction Top to Bottom, Geophys. Monogr. Ser.*, vol. 94, edited by G. E. Bebout et al., pp. 195–214, AGU, Washington, D. C.
Kodaira, S., E. Kurashimo, N. Takahashi, A. Nakanishi, S. Miura, J.-O. Park, T. Iwasaki, N. Hirata, K. Ito, and Y. Kaneda (2002), Structural factors in controlling a rupture process of a megathrust earthquake at the Nankai trough seismogenic zone, *Geophys. J. Int.*, *149*, 815–835.
Komatsu, M., S. Miyashita, J. Maeda, Y. Osanai, and T. Toyoshima (1983), Disclosing of a deepest section of continental-type crust up-thrust as the final event of collision of arcs in Hokkaido, north Japan, in *Accretion Tectonics in the Circum-Pacific Region*, edited by M. Hashimoto and S. Uyeda, pp. 149–165, Terra Sci., Tokyo.
Kikuchi, M., Y. Yagi, and Y. Yamanaka (2000), Source process of the Chi-Chi, Taiwan earthquake of Sept. 21, 1999 inferred from teleseismic body waves, *Bull. Earthquake Res. Inst., Univ. Tokyo*, *75*, 1–13.
Larson, K.M., R. Burgmann, R. Bilham, and J. T. Freymueller (1999), Kinematics of the India-Eurasia collision zone from GPS measurements, *J. Geophys. Res.*, *104*, 1077–1093.
Le Fort, P. (1986), Metamorphism and magmatism during the Himalayan collision, in *Collision Tectonics*, edited by M. O. Coward and A. C. Ries, *Spec. Publ. Geol. Soc. Am.*, *19*, 159–172.
Lemonnier, C., G. Marquis, F. Perrier, J.-P. Avouac, G. Chitrakar, B. Kafle, S. Sapkota, U. Gautam, D. Tiwari, and M. Bano (1999), Electrical structure of the Himalaya of central Nepal: high conductivity around the mid-crustal ramp along the MHT, *Geophys. Res. Lett.*, *26*, 3261–3264.
Lin, C. H., and S. W. Roecker (1993), Deep earthquakes beneath central Taiwan: mantle shearing in an arc-continent collision, *Tectonics*, *12*, 745–755.
Lorenz, F. P., F. Wenzel, and M. Popa (1997), Teleseismic travel-time tomography of the compressional-wave velocity structure in the Vrancia zone, Romania, *Eos Trans. AGU*, *78*, Fall Meet. Suppl., F497.
Lyon-Caen, H., and P. Molnar (1983), Constraints on the structure of the Himalaya from an analysis of gravity anomalies and a flexural model of the lithosphere, *J. Geophys. Res. 88*, 8171–8191.
Lyon-Caen, H., and P. Molnar (1985), Gravity anomalies, flexure of the Indian plate, and the structure, support and evolution of the Himalaya and Ganga basin, *Tectonics*, *4*, 513–538.
Matsu'ura, M., and T. Sato (1989), A dislocation model for the earthquake cycle at convergent plate boundaries, *Geophys. J.*, *96*, 23–32.
Matsuda, T. (1978), Collision of the Izu-Bonin arc with central Honshu: Cenozoic tectonics of the Fossa Magna, Japan, *J. Phys. Earth*, *26*, supplement, S409–S421.
Mattauer, M. (1986), Intracontinental subduction, crust-mantle decollement and crustal-stacking wedge in the Himalayas and other collision belts, in *Collision Tectonics*, edited by M. P. Coward and A. C. Ries, *Spec. Publ. Geol. Soc. Am.*, *19*, 37–50.
McCaffrey, R. (1992), Oblique plate convergence, slip vectors, and forearc deformation, *J. Geophys. Res.*, *97*, 8905–8915.
McCaffrey, R., P. Molnar, and S. W. Roecker. (1985), Microearthquake seismicity and fault plane solutions related to arc-continent collision in the eastern Sunda arc, Indonesia, *J. Geophys. Res.*, *90*, 4511–4528.
McKenzie, D. P. (1969), Speculations on the consequences and causes of plate motions, *Geophys. J. R. Astron Soc.*, *18*, 1–32.
Molnar, P. (1984), Structure and tectonics of the Himalaya: Constraints and implications of geophysical data, *Annu. Rev. Earth Planet Sci.*, *12*, 489–518.

Molnar, P., and P. Tapponnier (1977), Relation of the tectonics of eastern China to the India-Eurasia collision: Application of slip-line field theory to large-scale continental tectonics, *Geology*, *5*(4), 212–216.

Mugnier, J.-L., P, Huyghe, P. Leturmy, and F. Jouanne (2003), Episodicity and rates of thrust sheet motion in the Himalayas, in *Thrust Tectonics and Hydrocarbon Systems*, edited by K. McKay, *AAPG Mem.*, *82*, 91–114.

Nakamura, M., H, Watanabe, T. Konomi, S. Kimura, K. and K. Miura (1997), Characteristics activities of subcrustal earthquakes along the outer zone of southwestern Japan, *Ann. Disaster Prev. Res. Inst., Kyoto Univ.*, *40*, 1–20.

Ni, J., and M. Barazangi (1984), Seismotectonics of the Himalayan collision zone: Geometry of the underthrusting Indian plate beneath the Himalaya, *J. Geophys. Res.*, *89*, 1147–1163.

Ni, J., and M. Barazangi (1986), Seismotectonics of the Zagros continental collision zone and a comparison with Himalaya, *J. Geophys. Res.*, *91*, 8205–8218.

Noguchi, S., and S. Sekiguchi (2001), Subduction and deformation of the Philippine Sea and Pacific plates beneath the Kanto region (in Japanese), *Earth Mon.*, *23*, 733–741.

Obara, K. (2002), Nonvolcanic deep tremor associated with subduction in southwest Japan, *Science*, *296*, 1679–1681.

Pandey, M. R., R. P. Tandukar, J. P. Avouac, J. Lave, and J. P. Massot (1995), Interseismic strain accumulation on the Himalayan crustal ramp (Nepal), *Geophys. Res. Lett.*, *22*, 751–754.

Peacock, S. M., and K. Wang (1999), Seismic consequences of warm versus cool subduction metamorphism: Examples from southwest and northeast Japan, *Science*, *286*, 937–939.

Poli, S., and M. W. Schmidt (1995), H_2O transport and release in subduction zones: Experimental constraints on basaltic and andesitic systems, *J. Geophys. Res.*, *100*, 22,299–22,314.

Rau, R. J., and F. T. Wu (1995), Tomographic imaging of lithospheric structures under Taiwan, *Earth Planet. Sci. Lett.*, *133*, 517–532.

Roecker, S. W. (1982), Velocity structure of the Pamir-Hindu Kush region: Possible evidence of subducted crust, *J. Geophys. Res.*, *87*, 945–959.

Royden, L. (1993), The steady state thermal structure of eroding orogenic belts and accretionary prism, *J. Geophys. Res.*, *98*, 4487–4507.

Rudnick, R. L., and D. M. Fountain (1995), Nature and composition of the continental crust: A lower crustal perspective, *Rev. Geophys.*, *33*, 267–309.

Sacks, P. E., and D. T. Secor Jr. (1990), Delamination in collisional orogens, *Geology*, *18*(10), 999–1002.

Savage, J. C. (1983), A dislocation model of strain accumulation and release at a subduction zone, *J. Geophys. Res.*, *88*, 4984–4996.

Savage, J. C., and W. Thatcher (1992), Interseismic deformation at the Nankai Trough, Japan, subduction zone, *J. Geophys. Res.*, *97*, 11,117–11,135.

Schelling, D., and K. Arita (1991), Thrust tectonics, crustal shortening, and the structure of the far-eastern Nepal Himalaya, *Tectonics*, *10*, 851–862.

Schreyer, W. (1988), Subduction of continental crust to mantle depth: Petrological evidence, *Episodes*, *11*, 97–104.

Seeber, L., J. G. Armbruster, and R. C. Quittmeyer. (1981), Seismicity and continental subduction in the Himalayan arc, in *Zagros, Hindu Kush Himalaya: Geodynamic Evolution*, *Geodyn. Ser.*, vol. 3, edited by H. K. Gupta and F. M. Delany, pp. 215–242, AGU, Washington, D. C.

Seno, T. (1999), Syntheses of the regional stress fields of the Japanese islands, *Island Arc*, *8*, 66–79.

Seno, T. (2003), Fractal asperities, invasion of barriers, and interplate earthquakes, *Earth Planets Space*, *55*, 649–665.

Seno, T. (2005), Variation of downdip limit of the seismogenic zone near the Japanese islands: Implications for the serpentinization mechanism of the forearc mantle wedge, *Earth Planet. Sci. Lett.*, *231*(3-4), 249–262.

Seno, T., S. Stein, and A. E. Gripp. (1993), A model for the motion of the Philippine Sea plate consistent with NUVEL-1 and geological data, *J. Geophys. Res.*, *98*, 17,941–17,948.

Seno, T., and Y. Yamanaka. (1996), Double seismic zones, deep compressional trench: Outer rise events and superplumes, in *Subduction Top to Bottom*, *Geophys. Monogr. Ser.*, vol. 96, edited by G. E. Bebout et al., pp. 347–355, AGU, Washington, D. C.

Seno, T., and T. Yamasaki (2003), Low-frequency tremors, intraslab and interplate earthquakes in southwest Japan-from a viewpoint of slab dehydration, *Geophys. Res. Lett.*, *30*(22), 2171, doi:10.1029/2003GL018349.

Seno, T., D. Zhao, Y. Kobayashi, and M. Nakamura (2001), Dehydration in serpentinized slab mantle: Seismic evidence from southwest Japan, *Earth Planet Space*, *53*, 861–871.

Shishikura, M., et al. (2002), Paleoseismological study at Fengyuan site on the northern part of the Chelungpu fault, central Taiwan (in Japanese), *Rep. Active Faults Paleoseismol.*, *2*, 279–289

Soh, W., K. T. Pickering, A. Taira, and H. Tokuyama (1991), Basin evolution in the arc-arc Izu Collision Zone, Mio-Pliocene Miura Group, central Japan, *J. Geol. Soc. London*, *148*, 317–330.

Suyehiro, K., N. Takahashi, Y. Ariie, Y. Yokoi, R. Hino, M. Shinohara, T. Kanazawa, N. Hirata, H. Tokuyama, and A. Taira (1996), Continental crust, crustal underplating, and low-Q upper mantle beneath an oceanic island arc, *Science*, *272*, 390–392.

Taira, A., et al. (1998), Nature and growth rate of the Northern Izu-Bonin (Ogasawara) arc crust and their implications for continental crust formation, *Island Arc*, *7*, 395–407.

Takada, Y., and M. Matsu'ura (2004), A unified interpretation of vertical movement in Himalaya and horizontal deformation in Tibet on the basis of elastic and viscoelastic dislocation theory, *Tectonophysics*, *383*, 105–131.

Tapponnier, P., G., Peltzer, A. Y. Le Dain, and R. Armijo (1982), Propagating extrusion tectonics in Asia: New insights from simple experiments with plasticine, *Geology*, *10*(12), 611–616.

Taylor, S. R., and S. M. McLennan (1995), The geochemical evolution of the continental crust, *Rev. Geophys.*, *33*, 241–265.

Tsumura, N., H. Ikawa, T. Ikawa, M. Shinohara, T. Ito, K. Arita, T. Moriya, G. Kimura, and T. Ikawa (1999), Delamination-wedge structure beneath the Hidaka collision zone, central Hokkaido, Japan inferred from seismic reflection profiling, *Geophys. Res. Lett.*, *26*, 1057–1060.

Ulmer, P., and V. Trommsdorff (1995), Serpentine stability to mantle depths and subduction-related magmatism, *Science*, *268*, 858–859.

Utsu, T. (1974), Space-time pattern of large earthquakes occurring off the Pacific coast of the Japanese islands, *J. Phys. Earth*, *22*, 325–342.

van den Beukel, J. (1992), Some thermomechanical aspects of the subduction of continental lithosphere, *Tectonics*, *11*, 316–329.

Wang, K., and J. He. (1999), Mechanics of low-stress forearcs: Nankai and Cascadia, *J. Geophys. Res.*, *104*, 15,191–15,205.

Wortel, M. J. R., and W. Spakman (2000), Subduction and slab detachment in the Mediterranean-Carpathian region, *Science*, *290*, 1910–1917.

Wunder, B., and W. Schreyer (1997), Antigorite: High-pressure stability in the system MgO-SiO_2-H_2O (MSH), *Lithos*, *41*, 213–227.

Yamasaki, T., and T. Seno (2003), Double seismic zones and dehydration reactions in the subducting slab, *J. Geophys. Res.*, *108*(B4), 2214, doi:10.1029/2002JB001918.

Yu, S.-B., L.-C. Kuo, R. S. Punnngbayan, and E. G. Ramos (1999), GPS observation of crustal deformation in the Taiwan-Luzon region, *Geophys. Res. Lett.*, *27*, 923–926.

Yu, S.-B., et al. (2001), Preseismic deformation and coseismic displacements associated with the 1999 Chi-Chi, Taiwan, earthquake, *Bull. Seismol. Soc. Am.*, *91*, 995–1021.

20 Subduction and Mountain Building in the Central Andes

Jonas Kley and Tim Vietor

Abstract

Subduction-related, noncollisional orogeny in the central Andes was probably triggered and is maintained by a combination of favorable plate kinematic parameters with a strong subducting plate and a weakened overriding South American plate. The Nazca-South America plate convergence rate appears to control continental strain rates in the early evolution of the Andes, up to ~25 Ma. Later, the decreasing convergence rate is not mimicked by Andean strain rates that remain constant or even increase. Increased trenchward absolute motion of South America and subduction of younger lithosphere may play a role here. However, the pattern of displacement and strain in the Andes is strongly controlled by the plate convergence direction throughout the Neogene. Like the convergence direction, the overall strain pattern seems to have remained fairly constant for the last 30 Ma. The continentward convex shape of the subducting Nazca plate is likely to increase its resistance to shape changes, allowing it to indent the South American plate. The mechanical weakness of the South American margin in the central Andes apparently has two causes: a thin mantle lithosphere and a thick conformable sedimentary succession that reduced the strength of the upper crust on the foreland side of the mountain belt. It is still not clear which process thinned the lithospheric mantle and how long it operated before the onset of major contractional tectonics. Maximum shortening values of 250–300 km are sufficient to thicken the crust of the central Andes to its present state when the arc shape of the mountain belt is taken into account. Numerical models successfully simulate the two-stage deformation history of the Central Andean plateau with an early phase of internal shortening and a later phase of thrusting concentrated on the foreland side. They fail to predict the uplift history deduced from geologic data where internal deformation of the plateau ceases long before it has reached its present-day elevation.

Introduction

The Andes are the active type example of a subduction-related, or Cordilleran-type, mountain belt [*Dewey and Bird*, 1970]. At a total length of some 7500 km and maximum width of ~700 km, with elevations peaking over 6000 m and maximum crustal thickness exceeding 70 km, the Andes are the world's second largest mountain belt. At present, no other active continental margin has developed an orogen of similar dimensions. However, also the South American margin, despite a geologic history of more than 200 Myr of continuous subduction, did not begin to grow high topography until ~35 Ma ago, and even then it did not do so along its entire length. Contrary to earlier held views that attributed a substantial portion of crustal growth to addition of subduction-related magmas [*James*, 1971; *Kono et al.*, 1989], there is now a consensus that the thick crust and high topography in the Andes essentially reflect margin-normal tectonic shortening, quite similar to collisional orogens [*Allmendinger et al.*, 1990; *Baby et al.*, 1997; *Roeder*, 1988; *Schmitz*, 1994; *Sheffels*, 1990]. The spatially limited and episodic nature of subduction-related mountain building invites the question how and under which circumstances subduction is able to induce strong, protracted shortening of a continental margin and to create a high mountain range. Unusually strong and long-lasting deformation can either be attributed to increased forcing by the subducting plate or to reduced strength of the overriding plate. There is a large body of literature investigating the correlations between various plate tectonic parameters and the stress state of the overriding plate in modern subduction zones [e.g., *Jarrard*, 1986; *Pacheco et al.*, 1993; *Scholz and Campos*, 1995; *Uyeda and Kanamori*, 1979]. However, while a compressive regime in the overriding plate is clearly necessary for subduction-related orogeny, it is apparently not sufficient. For example, the southernmost central Andes near 40°S have a plate tectonic history very similar to the high central Andes between 12° and 27°S [*Pardo-Casas and Molnar*, 1987] but are topographically low and have no major thrust belt. In the present paper we review current concepts of how the Andes developed and discuss what may have been the main factors controlling their evolution. We report on results of numerical modeling by the second author and by other workers but do not intend to present and discuss a new numerical model in detail here. We focus our discussion on the central Andes (fig. 20.1) because this is the highest and widest part of the mountain range, and the only segment that has been exclusively built during the subduction of oceanic lithosphere. The northern and southern Andes have more complicated tectonic histories involving the collision and accretion of oceanic and continental terranes and island arcs

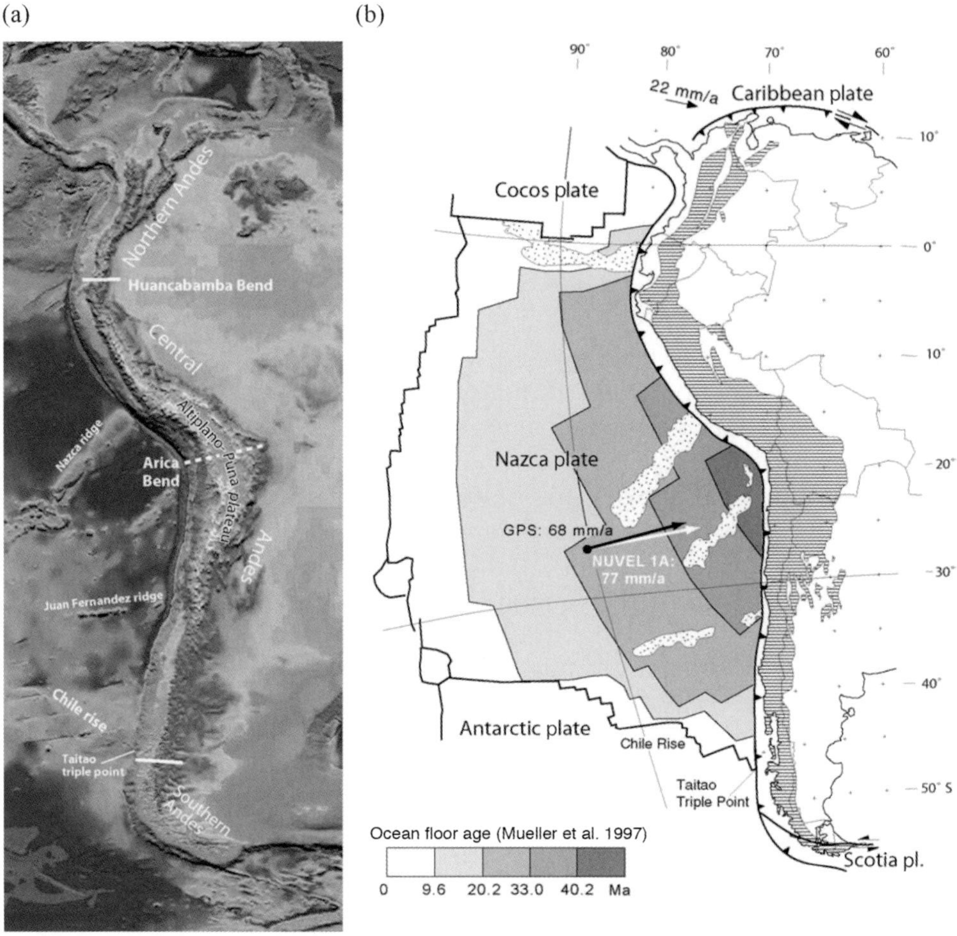

Figure 20.1 (a) Topography and geologic subdivisions and (b) plate tectonic setting of the Andes. This paper deals with the middle segment of the central Andes that contains the Altiplano/Puna or central Andean high plateau, approximately between the Nazca and Juan Fernández aseismic ridges.

[*Case et al.*, 1990] or the opening and closure of marginal basins floored by oceanic crust [*Dalziel et al.*, 1974].

Geologic Framework and Tectonics of the Central Andes

Tectonic Evolution Before the Cenozoic

From the Peru-Chile Trench in the west to the Brazilian shield in the east, the central Andes comprise the offshore fore arc, the onshore fore arc or "western monocline" [*Isacks*, 1988], the active magmatic arc whose volcanoes build the

Western Cordillera, the Altiplano-Puna high plateau and the thrust-faulted terrains that underlie the eastern slope and part of the foreland (Eastern Cordillera, Subandean Ranges, Sierras Pampeanas; fig. 20.2). A first-order feature

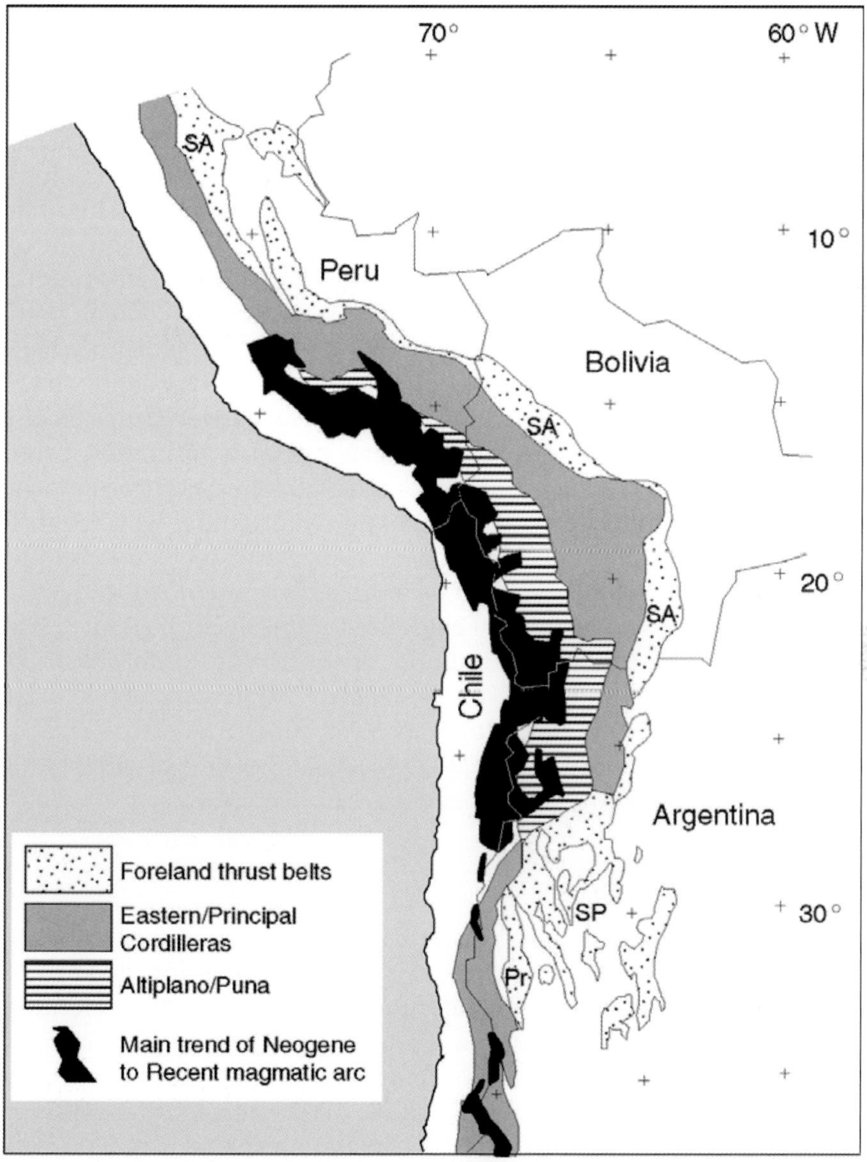

Figure 20.2 Main geologic and structural units of the central Andes between 5° and 35°S latitude. The onshore fore-arc region (unornamented) exhibits little Cenozoic deformation. The central Andean plateau comprises the Altiplano/Puna and part of the Eastern Cordillera. Foreland thrust belts comprise the Subandean Ranges (SA) in Peru, Bolivia and northern Argentina, the Sierras Pampeanas (SP) basement thrusts, and the Argentinian Precordillera (Pr).

in map view is the arc shape of the entire orogen (the "Bolivian orocline"), with a rather sharp change in structural trend at the Arica bend near 18°S.

In the central Andes the "Andean" tectonic cycle of uninterrupted eastward subduction is considered to begin with the establishment of a magmatic arc in Jurassic time [*Coira et al.*, 1982]. The remnants of this arc today extend from the leading edge of the South American plate to the coastal mountains. The complete absence of a Jurassic fore-arc assemblage (and possibly of the Jurassic arc in Peru) points to considerable subduction erosion of the South American margin [*Rutland*, 1971]. Possibly caused by tectonic erosion of the plate edge, the magmatic arc later shifted three times landward: in early late Cretaceous, late Cretaceous and Oligocene time, when it reached its present-day location at the western edge of the central Andean high plateau, some 200 km inboard of the Jurassic arc [*Scheuber and Reutter*, 1992]. In the middle segment of the central Andes, between ~14° and 24°S, Oligocene to Miocene arc magmatism initially extended right across the area later occupied by the plateau and narrowed to its present width and location along the western plateau edge only in late Miocene to Pliocene time [*Allmendinger et al.*, 1997; *Sébrier and Soler*, 1991]. The early stages of subduction in Jurassic to Lower Cretaceous time were associated with the formation of extensional back-arc basins, probably over a steeply dipping slab. Contraction of the overriding plate began in early late Cretaceous time but, compared to Cenozoic shortening, was subordinate in magnitude and localized. Early Tertiary shortening (Late Eocene "Incaic phase") is traditionally considered to be widespread in the Andes [e.g., *Mon and Salfity*, 1995], but clear structural evidence for this event is so far restricted to Peru and northern Chile, with amounts of shortening estimated at 30 km [*Mégard*, 1987] and <10 km [*Haschke and Günther*, 2003], respectively. A phase of early to middle Cenozoic extension may have punctuated the contractional tectonics of the overriding plate [*Jordan et al.*, 2001; *Rochat et al.*, 1999], but the spatial extent and importance of this event are not yet clear.

Cenozoic Shortening

The onset of strong shortening in the central Andes is not very well constrained in time. There is stratigraphic and structural evidence that it started at ~25 Ma [*Sempere et al.*, 1990], but apatite fission track data indicate that accelerated denudation began earlier, at ~35–30 Ma [*Ege et al.*, 2001]. These data in conjunction with constraints on the timing of fault activity and landscape evolution [*Elger*, 2003; *Gubbels et al.*, 1993; *Müller et al.*, 2002] suggest that folding and thrusting progressed in two steps: First, deformation spread rapidly across the entire region that was to become the high plateau, with no systematic propagation direction recognizable. Thrust belts on the eastern plateau border began to develop slowly since ~20 Ma. At ~10 Ma, deformation on the plateau (Altiplano and Eastern Cordillera) ceased and shifted to the Subandean thrust belt, which rapidly and systematically built out eastward toward the craton

[*Echavarria et al.*, 2003]. Recently, this view of the deformation history has been challenged, and a model of the Andes as a continuously evolving critically tapered thrust wedge has been proposed instead [*McQuarrie*, 2002b]. Estimates of maximum Cenozoic shortening from balanced cross sections range from 230 to 340 km, [*Baby et al.*, 1997; *Kley et al.*, 1997; *McQuarrie and DeCelles*, 2001; *Müller et al.*, 2002; *Roeder*, 1988; *Sheffels*, 1990]. Much higher estimates in excess of 500 km [*DeCelles and Horton*, 2003; *McQuarrie*, 2002a] derive the additional shortening not from any structural evidence but from inferences about foreland basin migration. There is a consensus that much of the arc shape of the central Andes is due to stronger shortening in the center than on the wings [*Hindle and Kley*, 2002; *Isacks*, 1988; *Kley and Monaldi*, 1998]. These shortening gradients are also a main cause of the regional pattern of paleomagnetically determined vertical axis rotations (see reviews in the works of *Randall* [1998] and *Lamb* [2001]).

It has been proposed that the central Andean high plateau has grown along strike, beginning near the present-day Arica bend and propagating outward to the northwest and south [*Jordan et al.*, 1997; *Müller et al.*, 2002; *Riller and Oncken*, 2003]. Estimates of lateral propagation rates are essentially based on interpretations of synorogenic basins and very vaguely constrained. Whereas *Riller and Oncken* [2003] proposed a linear southward growth of the Altiplano/Puna plateau at a rate of 26 km/Ma between 18° and 28°S, other authors concluded that synorogenic basins developed more or less simultaneously over the entire Puna [*Adelmann*, 2001; *Jordan and Alonso*, 1987]. The data of *Jordan et al.* [1997] allow for lateral propagation rates of foreland thrusting between ~50 and 200 km/Ma in the 18°–28°S segment. The onset of shortening and uplift in the thick-skinned Sierras Pampeanas foreland province (27° to 33°S) has propagated southward at a rate of ~150 km/Ma since 7 Ma as calculated from data in the work of *Ramos et al.* [2002]. In summary, the available evidence suggests that Neogene shortening of the overriding plate has started in the center of the orocline where it remains most intense today and that the region affected by shortening expanded not only across strike but also along strike.

The late Cenozoic to Present stress state in the high plateau areas is quite variable in space and time as indicated by earthquake focal mechanisms and fault kinematic analyses [*Cladouhos et al.*, 1994; *Dalmayrac and Molnar*, 1981; *Marrett et al.*, 1994]. In particular, there is evidence for episodic extension in varying directions, from nearly orogen-parallel to nearly orogen-normal [*Liu et al.*, 2002; *Mercier et al.*, 1992].

Present-Day Deformation

GPS data now permit to directly monitor present-day short-term deformation in the Andes [*Kendrick et al.*, 2001; *Norabuena et al.*, 1998]. Velocities relative to "stable" South America decrease systematically from coastal stations to those farther inland and drop to near zero at the deformation front as defined from geologic structures. Much of this signal is attributed to elastic deformation that will eventually recover and not be preserved in the geologic record [*Bevis et al.*, 2001; *Leffler et al.*, 1997; *Liu et al.*, 2000]. Close to the deformation front, however, space geodetically determined strains are probably permanent and strain rates can be compared to those estimated from geologic reconstructions. Although the strain rates estimated from techniques such as cross-section balancing are long-term averages, they are similar to the GPS rates [*Hindle et al.*, 2002]. Displacement or velocity fields derived from geologic data or a combination of geologic and paleomagnetic data [*Kley*, 1999; *Lamb*, 2000] are also closely similar to the GPS velocity fields [*Hindle et al.*, 2002] (fig. 20.3). The vectors in both

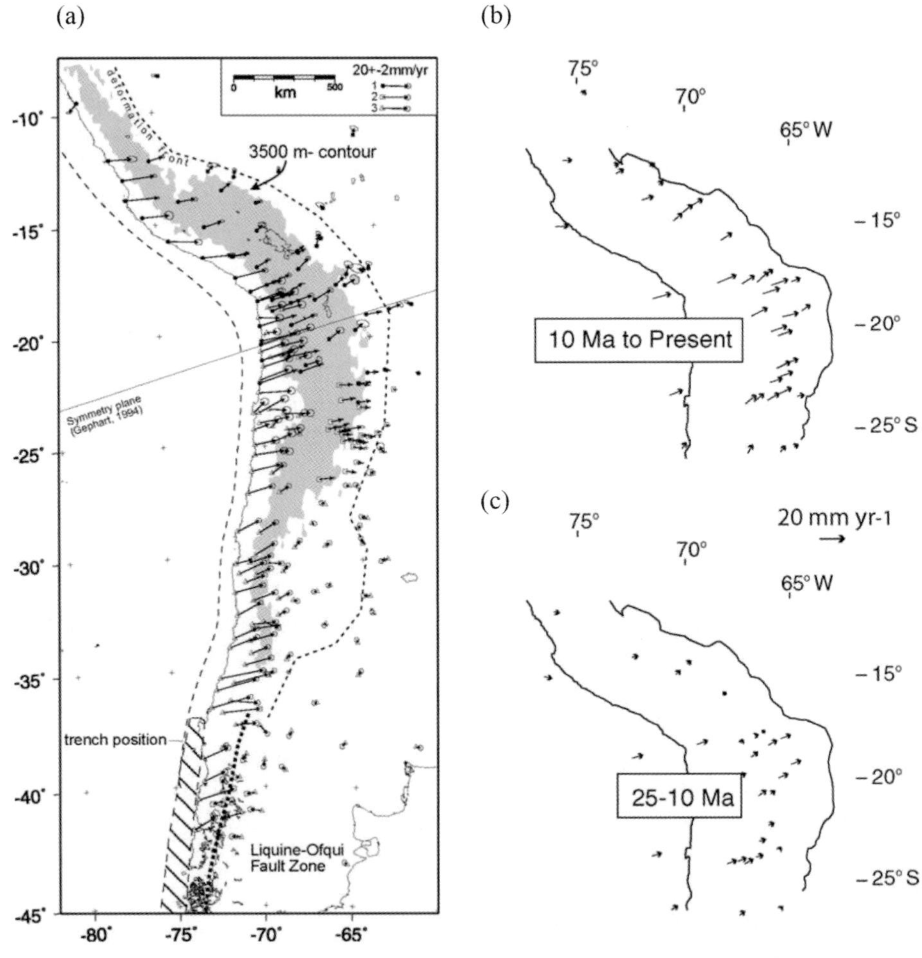

Figure 20.3 (a) Present-day displacement vector field of the central Andes as derived from GPS data (compiled by J. Klotz from the SNAPP [*Norabuena et al.*, 1998] plus CAP [*Kendrick et al.*, 2001] and SAGA networks). (b) Displacement vector fields for two time slices [*Hindle et al.*, 2002] from a map-view balanced block mosaic based on geologic data (see fig. 20.13). Notice the similarity of the geodetic and geologic vector fields.

tend to be subparallel to the Nazca-South America plate convergence direction that has changed little over the past 50 Ma [*Pardo-Casas and Molnar*, 1987]. This suggests that the central Andes have been built by a relatively constant long-term displacement field strongly controlled by the convergence direction. As topography in the little eroded central Andes roughly reflects bulk strain [*Isacks*, 1988; *Kley and Monaldi*, 1998], the high symmetry of the arc-shaped orogen about the plate convergence vector [*Gephart*, 1994] supports this notion. The extension of small magnitude observed over much of the Altiplano-Puna plateau (see above) is probably due to superimposed gravitational spreading induced by the high topography [*Liu et al.*, 2002].

Factors Controlling the Deformation of the Central Andes

In this chapter we review and discuss various factors that were proposed as causes of mountain building in the central Andes. From a physical point of view, there are only two fundamental controls: the horizontal stress applied and the mechanical strength of South America's margin. The applied stress may vary as a function of global plate dynamics, of the mechanical strength of the subducting plate, or of the level of mechanical coupling across the plate interface (fig. 20.4). The rheology of the overriding plate varies mainly as a function of composition, thermal state, and of the time scale of deformation. All these parameters are themselves influenced by various factors and are likely to change through time. We have grouped the different factors according to their assumed effect on either the stress state or the rheology of the overriding plate as far as possible, but note that there are some empirical observations of which the underlying processes are not well understood.

Factors Increasing the Stress in the Overriding Plate

Eastward Subduction

A long-term correlation between specific subduction settings and mountain building is suggested by the global asymmetry of subduction zones and associated orogens. The two most recent episodes of major subduction-related mountain building are the Late Mesozoic to Early Cenozoic Laramide orogeny of North America and the Cenozoic to Recent Andean orogeny [*Burchfiel and Davis*, 1975; *Dickinson and Snyder*, 1978; *Jordan et al.*, 1983]. Both occurred on the eastern Pacific margins above gently (0°–30°) dipping slabs [*Bird*, 1984] and with trenchward absolute motion of the overriding plate [*O'Connor and Duncan*, 1990; *Silver et al.*, 1998]. There is a fundamental contrast with the western Pacific margins that have extensional marginal basins above more steeply dipping slabs [e.g., *Uyeda and Kanamori*, 1979]. It has been proposed that the asymmetry of east- and west-dipping subduction zones is due to a global, roughly eastward directed flow of the deeper mantle relative to the litho-

(a)

(b)

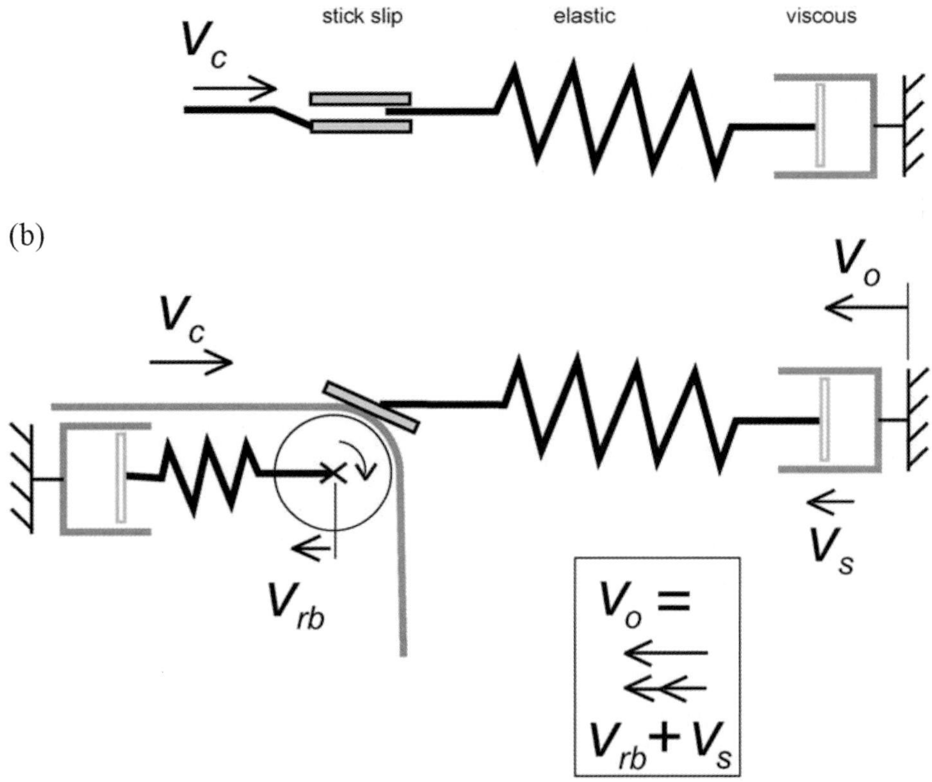

Figure 20.4 (a) Mechanical analog of the South American margin consisting of a slider (stick slip motion on plate interface), spring (elastic deformation in the Andes), and dashpot (permanent aseismic deformation in the marginal thrust belt), after *Liu et al.* [2000]. (b) Modified analog model including the effect of subduction rollback. In a nonaccreting subduction zone, even fast convergence will not induce shortening of the overriding plate if coupled with fast rollback.

sphere, which tends to steepen slabs dipping west and to flatten slabs dipping east [*Doglioni et al.*, 1999; *Uyeda and Kanamori*, 1979]. A flattening slab would increase the normal stress on the plate interface (see below), explaining why subduction-related orogens preferentially grow over east-dipping subduction zones. However, unless the global mantle flow varies in intensity, this model fails to explain the switch from back-arc extension to back-arc contraction and probably from steep to lower angle subduction that occurred in the Andes in mid-Cretaceous time, some 100 Ma after the onset of subduction [*Mpodozis and Ramos*, 1989].

Subhorizontal Subduction Present-day slab dips along the Andes are generally low, <30°, but there are two "flat slab" segments where the subducted Nazca plate becomes subhorizontal for several hundred km before returning to a steeper dip [*Barazangi and Isacks*, 1976; *Cahill and Isacks*, 1992] (fig. 20.5). The southern boundaries of the two Nazca flat slab segments are sharp and occur where the aseismic Nazca and Juan Fernandez ridges intersect the plate boundary. Thus buoyancy and slab dip can be only weakly influenced by the lateral variation of plate age, which is smooth, and hence hardly able to ex-

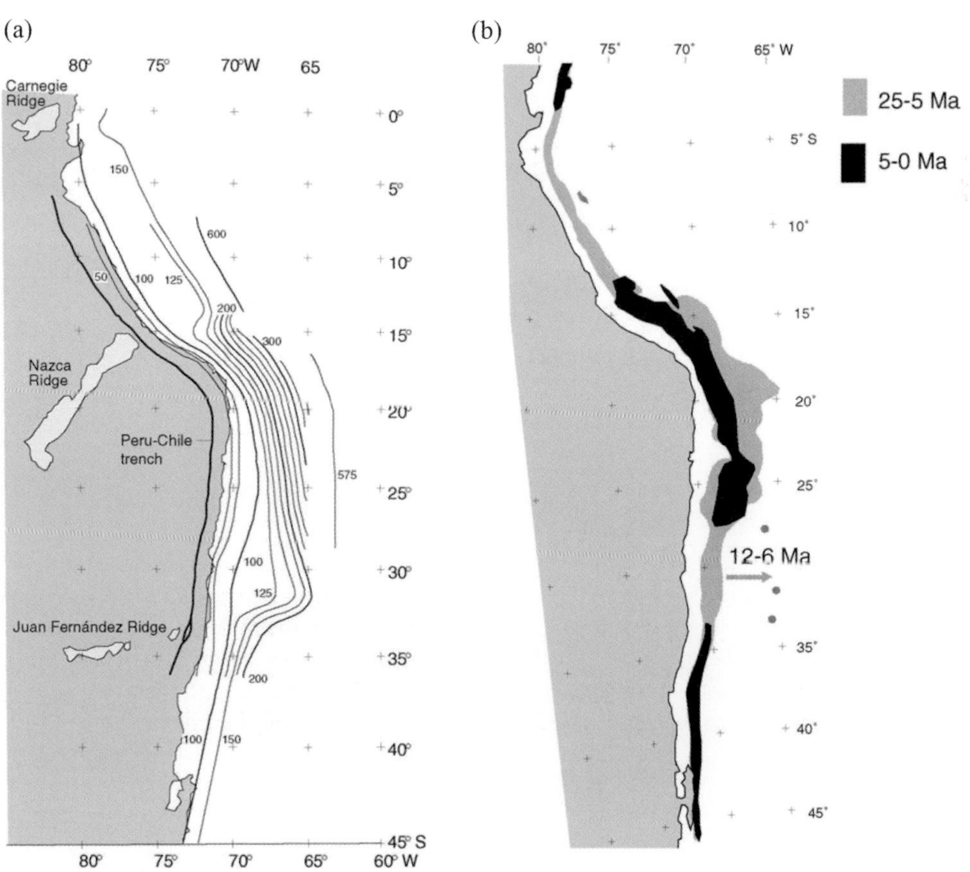

Figure 20.5 (a) Depth contours of the subducting Nazca plate, from *Cahill and Isacks* [1992]. Notice "flat slab" segments between 2° and 15°S and between ~27° and 33°S. The southern limits of these segments approximately coincide with the aseismic Nazca and Juan Fernández ridges, respectively. Active volcanism is restricted to the more steeply dipping segments. (b) transition from the broad Miocene to the narrower present-day magmatic arc in the center of the Andean orocline, interpreted as the result of a steepening slab, and evolution of the volcanic gaps over flattening slabs. Data from *Kay and Abbruzzi* [1996], S. M. Kay in the work of *Allmendinger et al.* [1997], *Sébrier and Soler* [1991].

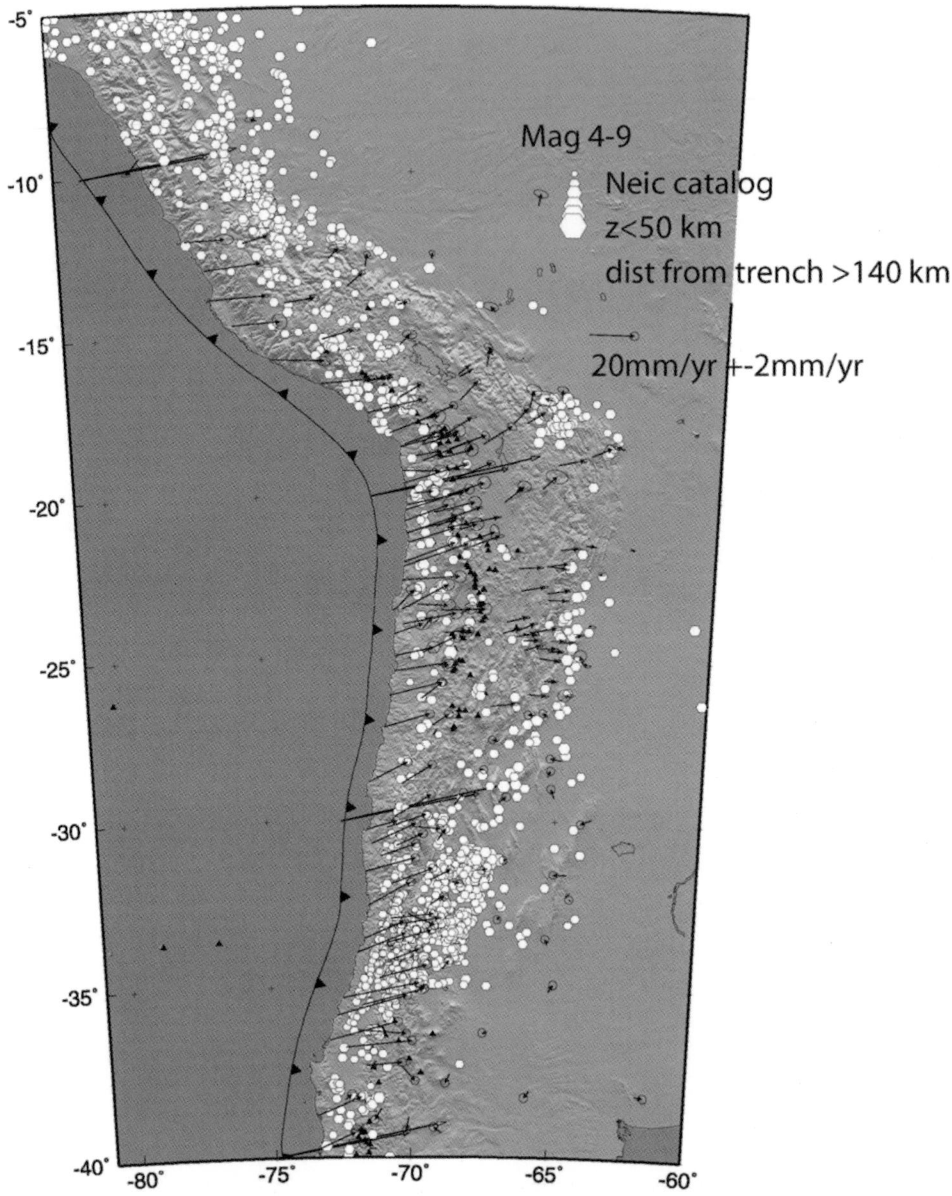

Figure 20.6 Comparison of present-day displacement vector field from GPS data (sources as in fig. 20.3) with seismicity in the South American plate. High strain rates in the middle segment of the central Andes correlate with low levels of seismicity, indicating a high proportion of aseismic strain.

plain the abrupt changes in plate dip. They rather appear to be dependent on density anomalies of the lower plate. It has been proposed that increased slab buoyancy and the enlarged plate interface in the flat slab segments increase the stress transmitted to the overriding plate, resulting in stronger upper plate seismicity and Laramide-style basement thrusting in the foreland [*Gutscher et al.*, 2000; *Jordan et al.*, 1983; *Pilger*, 1981]. However, present-day strain rates neither correlate with the flat slab segments nor with overriding plate seismicity. The GPS data demonstrate that present-day strain rates are highest in the middle segment of the central Andes that overlies a normally dipping slab (fig. 20.6). The low level of seismicity in this segment thus indicates a high proportion of aseismic deformation [*Klosko et al.*, 2002]. If stress in the overriding plate is higher above the flat slabs, it must be counteracted there by higher strength of the South American plate in order to keep strain rates low. A low correlation of slab dip and interplate force is further supported by the analysis of *Meijer et al.* [1997]. They tried to fit the present-day stress state of the South American plate as compiled and interpreted in the world stress map project [*Zoback*, 1992] by varying the contributions of the interplate forces under the constraint of dynamic equilibrium. They found that a uniform interplate force along the active western margin of South America fits the actual stress distribution better than higher stresses in the flat slab regions combined with lower stresses in the more steeply dipping sections.

Upward Convex Slab Oceanic plates are generally strong, but in the case of the Nazca plate the rigidity of the plate is probably further increased by the convex-up shape of the downgoing slab. We speculate that the "wrong" bend with respect to the normal curvature of slabs in subduction zones can also provide the necessary stability that enables large distance stress transfer along the strike of the orogen that is required for the high symmetry of the Andean orocline.

High Convergence Rate and Trenchward Absolute Plate Motion Analysis of Nazca-South America relative motion from Cretaceous time to Present suggests that phases of high convergence rates coincide with phases of tectonic shortening in the overriding plate [*Pardo-Casas and Molnar*, 1987]. In particular, the late Eocene "Incaic" tectonic phase, the following interval of quiescence and the onset of the "Quechua" tectonic phases in late Oligocene time appear to coincide with high and low convergences rates, respectively. However, as pointed out above, the Incaic phase is not recorded throughout the central Andes, and Quechua deformation may have begun earlier than previously assumed, thus somewhat weakening the correlation. Since convergence rate is only moderately correlated with the state of stress in the overriding plate [*Jarrard*, 1986] and uncorrelated with seismic coupling in modern subduction zones [*Pacheco et al.*, 1993], the mechanism linking periods of fast convergence with shortening pulses in the overriding plate is not clear. From 25 Ma to the Present, plate convergence and Andean strain rates appear decoupled (fig. 20.7): while plate convergence rates decrease, orogenic strain rates remain high or even accelerate [*Hindle et al.*, 2002], indicating that the orogen becomes increasingly independent of convergence rate with time. Absolute motion of the

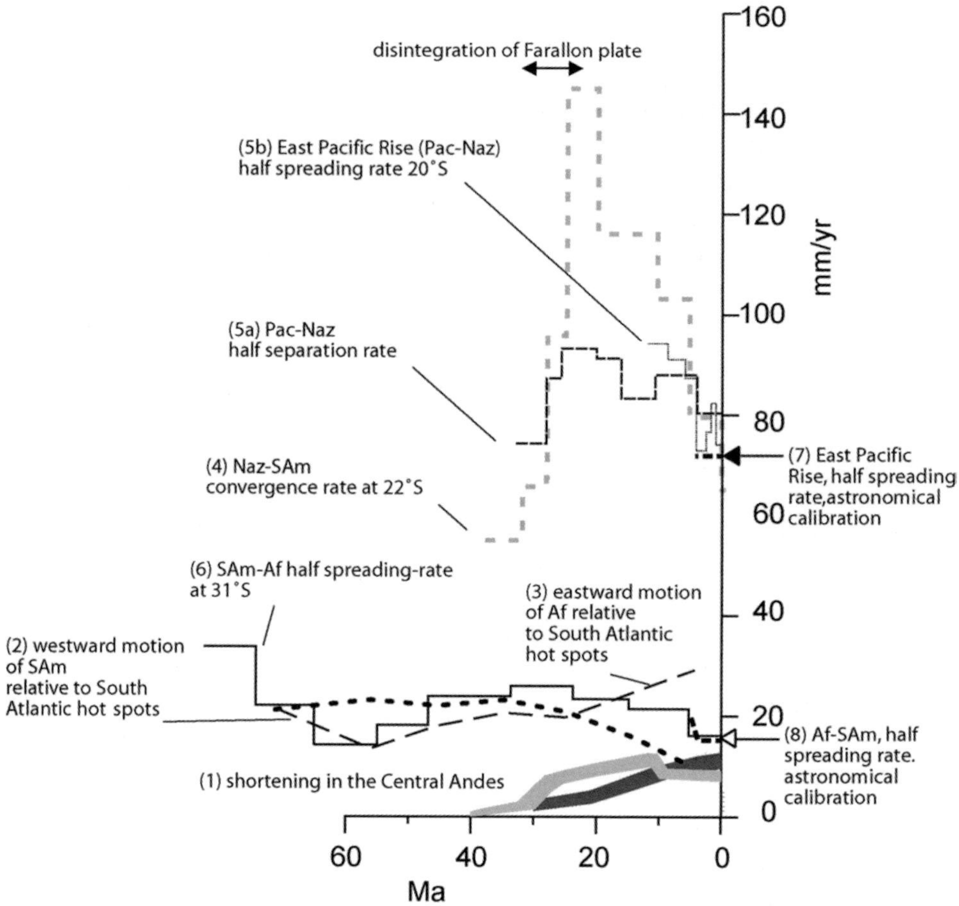

Figure 20.7 Variations of absolute and relative plate motions during the Andean orogeny. (1) Strain rate in the central Andes (light shading: finite shortening of 265 km, dark shading: 205 km) is positively correlated with the eastward velocity of South America relative to South Atlantic hot spots (2), which in turn closely follows or coincides with slowing eastward drift of Africa (3). The early shortening also correlates with the increase in relative convergence between Nazca and South America (4), the spreading rates at the East Pacific Rise (5a,b) and in the South Atlantic (6). However, in the later stage of the orogeny, after 20 Ma, all these rates decrease significantly while the shortening rate in the central Andes either increases or remains high. Data sources: (1) *Hindle and Kley* [2002]; (2,3) *Silver et al.* [1998], (4) *Somoza* [1998]. (5a) Pacific-Nazca half separation rate calculated for the Euler equator from stage rotations of *Tebbens and Cande* [1997]. (5b) *Cande and Kent* [1992] recalculated for timescale revision in the work of *Cande and Kent* [1995], (6) *Cande and Kent* [1995], (7,8) Astronomical calibrations of Plio-Pleistocene half spreading rates at the East Pacific Rise and the Mid Atlantic Ridge from *Wilson* [1993].

plates with respect to the mantle creates viscous forces that resist the horizontal translation of the slab through the mantle (the "slab anchor" or "sea anchor" force [*Uyeda and Kanamori*, 1979]). This effect has been proposed as responsible for the Neogene Andean orogeny [*Russo and Silver*, 1996]. *Silver et al.* [1998] argued that at 25 Ma the closure of the Tethys Ocean slowed the northeastward motion of Africa while the opening of the South Atlantic continued at constant speed. They calculated the resulting westward acceleration of South America and concluded that it strongly correlates with the onset of strong shortening within the South American plate. Contrary to convergence rates, the westward component of South America's absolute motion appears to have increased during the Cenozoic as have Andean strain rates (fig. 20.7). However, the overriding plate contraction of 120 km calculated from the westward acceleration of South America since 25 Ma [*Silver et al.*, 1998] falls short of the ~250–300 km estimated from geological data. *Marrett and Strecker* [2000] proposed that a change in the direction of South America's absolute motion that occurred around 3 Ma is reflected in the latest Cenozoic kinematics of Andean deformation. However, the close link of the long-term deformation pattern and of the orogen's symmetry [*Gephart*, 1994] with the convergence direction indicates that the deformation of the overriding plate is not controlled in a simple way by its absolute motion direction. South America's accelerated trenchward absolute motion must have induced accelerated slab rollback, thus increasing the resisting slab anchor force and the strength of the indenting Nazca plate while the kinematics were determined by the plate convergence direction.

Strong Interplate Coupling Increased friction or shear strength of the plate interface should be able to increase the transmitted stress and induce high strain rates in the overriding plate. In order to demonstrate this some indicator of the stress level on the plate interface is required. Many authors have investigated the relations between the current stress or strain regime ranging from back-arc spreading to back-arc thrusting in overriding plates, and modern subduction zone parameters like slab dip, slab age, convergence rate, absolute motion of the overriding plate, or the seismic coupling factor [e.g., *Jarrard*, 1986; *Pacheco et al.*, 1993; *Scholz and Campos*, 1995; *Uyeda and Kanamori*, 1979]. Modern subduction zones show only moderate correlation of compression in the overriding plate with single parameters like low slab dip, young slabs, and trenchward absolute motion [*Jarrard*, 1986]. *Scholz and Campos* [1995] argued that the seismic coupling factor (the ratio of the slip occurring in earthquakes to the total slip along the plate interface) is directly dependent on the normal stress acting on the plate interface, which in turn depends on the magnitude of the slab pull force and the direction and rate at which the slab moves through the mantle horizontally. Seismic coupling would then reflect a combination of slab buoyancy, absolute fore-arc motion, and possibly variations in the frictional properties of the plate interface. Mountain building could itself increase seismic coupling by increasing the topographic load on the plate interface (fig. 20.8). Either the seismic coupling factor or the highly correlated cumulative earthquake moment magnitude [*Pacheco et al.*, 1993] should be good indicators of the loading of the overriding plate. If this is true, then the South American plate is presently strongly loaded all along the central Andes, though not as strongly as the overriding plates in Nankai or Guam [*Scholz and Campos*, 1995]. Uncertainties on the seismic coupling estimates are probably too high to confidently identify variations along the central Andes. Also, the sampling time of 40 years may be too short to reliably infer the seismic coupling value, especially in those areas with large earthquakes (>8.5) that have recurrence times of several hundreds of years.

Another method that has been used to determine the degree of kinematic locking across the Central Andean plate interface is elastic modeling of GPS displacements. Estimates vary from the interface being ~50% locked to almost 100% locked (no interseismic displacement) [*Bevis et al.*, 2001; *Leffler et al.*, 1997; *Norabuena et al.*, 1998]. These indications of substantial to almost complete locking apparently support qualitative arguments about factors thought to favor strong mechanical coupling in southern Peru and northern Chile: the incoming Nazca plate has a pronounced horst-graben structure and topography [*von Huene et al.*, 1999], the trench is practically devoid of sediment due to hyperarid conditions on land [*Bangs and Cande*, 1997], and no accretionary prism is developed. If this combination of factors would actually increase forcing of the upper plate due to increased friction, it would also provide a mechanism linking deformation in the Andes to climate change [*Hindle et al.*, 2002; *Lamb and Davis*, 2003]. However, on the basis of high-quality bathymetric

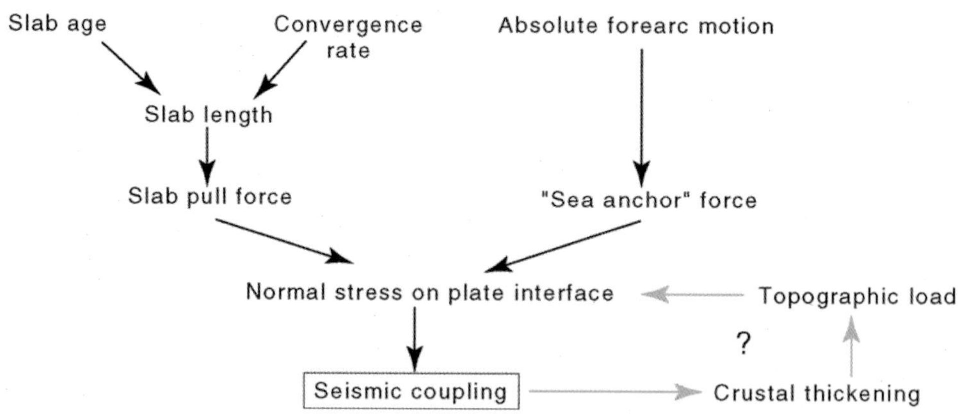

Figure 20.8 Causal links between seismic coupling and a variety of subduction zone parameters as proposed by *Scholz and Campos* [1995]. Within this concept, mountain building could enhance seismic coupling through an increased topographic load on the plate interface.

and seismic data from the offshore fore arc in northern Chile, *von Huene and Ranero* [2003] concluded that submarine mass-wasting debris shed from the steep inner trench wall smothers the ocean bottom relief before it is subducted and that the friction coefficient at the base of the fore-arc wedge is in the same range as for margins with accretionary prisms. *Lamb and Davis* [2003] proposed that the shear strength of the plate interface everywhere just balances the lithostatic head of the Andes and thus can be easily calculated from the elevation difference between trench and high mountains on the basis of a static force balance. The shear strength of the plate interface should therefore be higher in the high central Andes than in adjacent areas, an effect the authors ascribe to the sediment-starved trench. This analysis relies on two implicit assumptions: first, that all forces balancing the lithostatic head are frictional, static, and act across the plate interface and second, that the downdip length of the plate interface area is constant along strike. As to the first point, it has been argued that the deeper parts of the Andean fore-arc wedge are hot enough to deform by ductile flow [e.g., *Adam and Reuther*, 2000]. *Yañez and Cembrano* [2004] presented a numerical model of viscous plate coupling in the Andes where viscosity and coupling increase with the age of the subducted plate. Since trench fill thickness and slab age in the central Andes vary in about the same way, both the frictional and the viscous model fit the observed variations in elevation and dynamic trench topography [*Yañez and Cembrano*, 2004]. *Liu et al.* [2002] found from three-dimensional (3-D) viscoelastic modeling that stress is probably transmitted not only across the plate interface but that basal drag on the South American lithosphere may be necessary to explain the actual stress state. Regarding the assumption of constant interface area, *Wang and Suyehiro* [1999] concluded that upper plate compression in NE Japan is due to an increased plate interface area caused by a cooler slab and thus an increased total force rather than higher shear strength of the plate interface. Given the age variations of the Nazca slab (fig. 20.1), this argument would also hold for the central Andes. In summary, while it is likely that the horizontal force acting on South America's margin is higher in the central Andes than to the north and south, it is difficult to separate whether this is due to increased friction, a larger frictional plate interface area, higher viscosity in the deeper plate coupling zone, or some combination of all three.

Factors Decreasing Upper Plate Strength

Two main factors have been invoked to explain why the South American plate is particularly weak in the central Andes. First, thermal anomalies in the upper mantle and crust, and second, inherited weak zones related to old fault zones or thick, low-strength sedimentary basin fills.

Thermal Anomalies *Isacks* [1988] suggested a thermal anomaly in the upper mantle and lower crust as a prerequisite for strong Cenozoic shortening and plateau formation in the Andes. The mechanical feasibility of this model was later demonstrated by numerical modeling (see below). Refraction seismic data [*Wigger et al.*, 1994] and receiver function tomography [*Yuan et al.*, 2000] show an extensive low-velocity body at midcrustal depths beneath the plateau, interpreted to reflect partial melting of the crust. Different mechanisms were envisaged to provide the necessary heat source. *Isacks* [1988] took the widening of the late Oligocene to early Miocene magmatic arc as evidence of nearly flat subduction that produced a wide asthenospheric wedge and advected heat and fluids to a broad swath of the upper plate. An Oligocene flat slab episode was also inferred from genetic studies of Andean metal deposits and attributed to the subduction of the mirror image of the Ontong-Java Plateau [*Sillitoe and McKee*, 1996]. Other authors have proposed that delamination of the mantle lithosphere was the cause for thermally induced weakening of the crust [*Lamb and Hoke*, 1997]. A problem with both assumptions is that the onset of the volcanism indicating mid-Cenozoic low-angle subduction is approximately coeval with the onset of strong upper plate shortening. Numerical modeling [*Babeyko et al.*, 2002] indicates that heat transport through the crust is very slow unless convection in the lower and middle crust is induced by deformation. Even convection needs at least 5 Myr to transport the heat to the surface. So there is a dilemma: deformation is needed to drive convection and heat transfer, but heating is needed to weaken the upper plate and start shortening. Thus there should be some other process that provides the initial heat input, one that is capable of strongly thinning or removing the lithospheric mantle in the absence of strong shortening. *Kay and Kay* [1993] found that dense cumulates added to the lower crust by long-lasting magmatism could eventually induce sinking of the lower crust and lithospheric mantle. Since there is no time lapse between the onset of magmatism and of shortening in the central Andes, this mechanism does not seem viable here. *Kay and Abbruzzi* [1996] suggested that as the slab flattens, the lithosphere above the thinning wedge cools, inducing spinel to garnet transition and piecewise sinking of lithospheric mantle. *Pope and Willett* [1998] proposed mechanical removal of the mantle lithosphere as it gets in contact with the slab. This process can only work with no asthenospheric wedge separating the slab from the lithospheric mantle of the overriding plate, that is in periods with no arc magmatism like the early Oligocene. It is possible that at this time one of the proposed mechanisms has strongly thinned the lithosphere, setting the stage for the wide Miocene magmatic arc and weakening the South American plate sufficiently to initiate rapid shortening. Later continuous removal of mantle lithosphere may have further weakened the upper plate in order to keep the locus of shortening here [*Babeyko et al.*, 2002].

Inherited Zones of Mechanical Weakness Before the onset of the Andean orogeny, South America's western margin had a long and complex geologic history resulting in strong and often sharp stratigraphic and structural variations across and along strike. This inherited segmentation coincides in places with pronounced changes in Cenozoic structural style and with the along-strike segmentation of the subducting Nazca plate (fig. 20.9). The most striking correlation is of the thin-skinned Subandean foreland thrust belt with a thick conformable succession of Paleozoic and Mesozoic strata [*Allmendinger et al.*, 1983] that hosts two important shaly décollement levels in its lower parts [*Belotti et al.*, 1995]. The thin-skinned thrust belt terminates in the south where

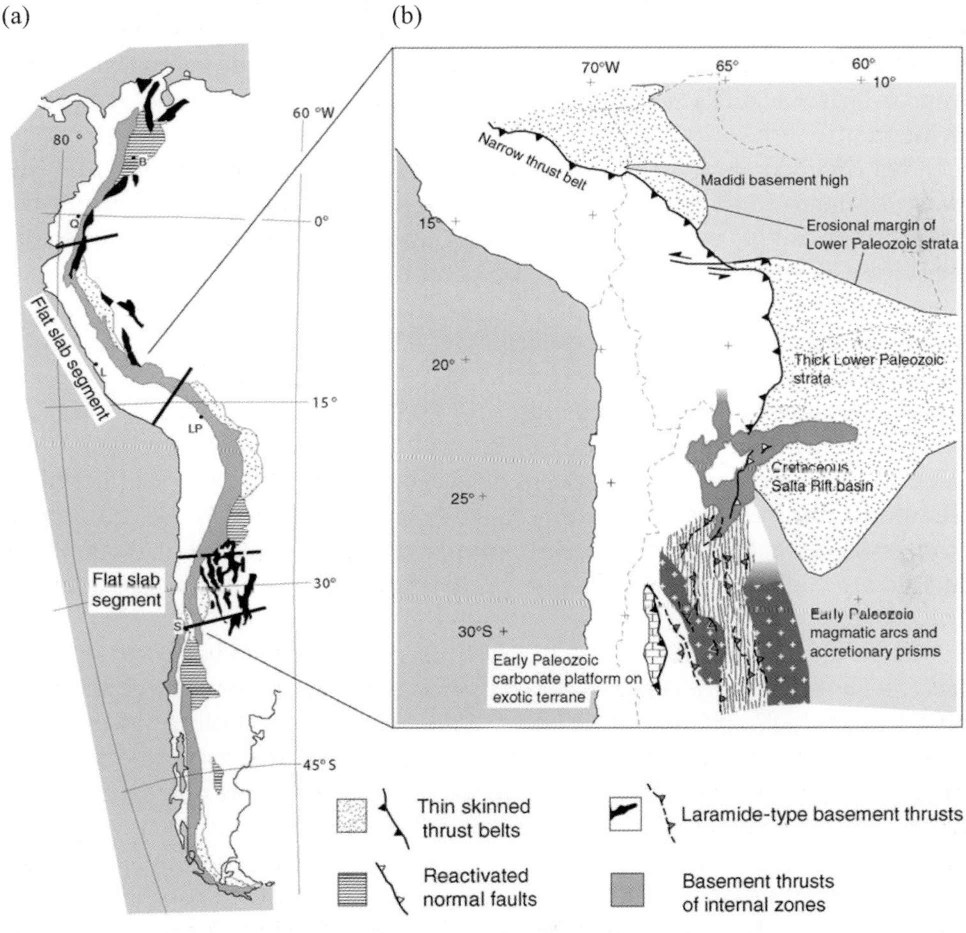

Figure 20.9 (a) Along-strike segmentation of foreland thrust belts according to structural style compared to the location of the "flat slab" segments of the Nazca plate. Styles of foreland deformation are indicated by map patterns. Notice the coincidence of the flat slabs with Laramide-type basement thrusts. (b) Thrust belt segmentation and pre-Cenozoic stratigraphic and structural features. Styles of foreland deformation are indicated by symbols. Notice the coincidence of the Subandean thin-skinned thrust belt with thick Paleozoic strata and the Cochabamba strike slip zone decoupling the freely propagating southern Subandean belt from its buttressed northern counterpart.

these décollements have been eroded over a basement arch. In the north, an abrupt narrowing of the foreland belt coincides with thinning of the sedimentary cover over a less pronounced basement arch. At the center of the bend, the foreland side of the Arica bend comprises important strike-slip faults. These faults decouple the southern part of the Subandean thrust belt that is still well within thick sedimentary strata from the northern part where the thrust front has reached the continentward edge of the cover succession and can no longer propagate freely [*Sheffels*, 1995; *Welsink et al.*, 1995]. The thin-skinned foreland thrust belts in the center of the Andean arc have accommodated a substantially higher magnitude of contraction than the adjacent segments that exhibit different structural styles [*Kley and Monaldi*, 1998]. Thus stratigraphic variations appear to have controlled the strain pattern and accentuated the central Andean arc.

Not only along-strike but also across-strike structural segmentation is apparently determined by inherited inhomogeneities. The NE boundary of the Central Andean plateau coincides with the cratonward limit of pre-Andean deformation as demonstrated by the extent of different unconformities (fig. 20.10). The effect of pre-Andean deformation here seems essentially to define an internal (western) limit to the thick conformable sedimentary succession and thus to thin-skinned deformation.

Reactivation of preexisting faults is demonstrated in several cases for oblique transpressive faults striking NE and NW on the southern limb of the Andean arc (e.g., the El Toro Lineament [*Allmendinger et al.*, 1983; *Salfity*, 1985]). These faults allow for steep along-strike gradients in bulk orogenic contraction, and they have accommodated some orogen-parallel extension, thereby facilitating the evolution of the arc [*Kley*, 1999; *Müller et al.*, 2002; *Riller and Oncken*, 2003].

Normal faults from Mesozoic rifts within the Cordillera and in the foreland have been reactivated as basement-involved thrust faults in several instances [*Bianucci et al.*, 1982; *Cristallini et al.*, 1997; *Kley and Monaldi*, 2002; *Ramos*, 1988; *Uliana et al.*, 1995]. However, the former rift areas have shortened less and hence were presumably still stronger than the unfaulted regions with a thick sedimentary cover prone to thin-skinned thrusting.

Controls on Slab Geometry from Deformation of the Overriding Plate

The along-strike shortening gradients in the finite deformation field of the South American plate induce deformation of the plate boundary in map view by increasing its curvature and accentuating the Arica bend. This requires a progressive shape change of the subducting plate and the displacement of asthenospheric material. Owing to its flexural rigidity and viscous forces the subducting Nazca plate will transfer stresses laterally and load the strongest parts of the plate margin at any time. This implies that there should only be

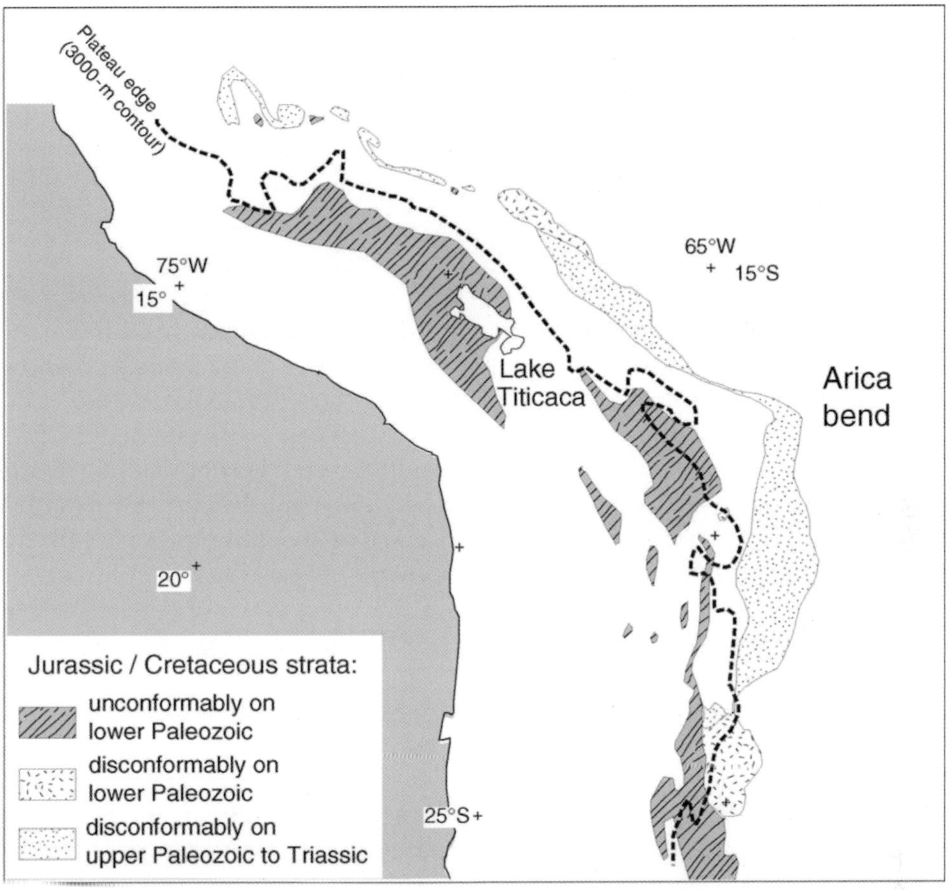

Figure 20.10 The eastern/northeastern margin of the Altiplano-Puna plateau as outlined by the 3000-m contour coincides with the area affected by pre-Cretaceous deformation events.

subdued gradients in total shortening and that tectonic events such as shortening pulses should be orogen-wide and nearly synchronous. This is in line with the acceleration of foreland shortening at around 10 Ma that can apparently be observed in the entire orogen.

A special case is the flat slab segments and their correlation with the tectonic segmentation of the upper plate [*Jordan et al.*, 1983]. Since the tectonic segmentation also correlates with inhomogeneities much older than the flat slabs [*Allmendinger et al.*, 1983; *Kley et al.*, 1999], it is unlikely to merely reflect an influence of slab dip on the structural style of the overriding plate. Unless the coincidence is accidental, upper plate segmentation must have conditioned different slab dips. *Isacks* [1988] and *Marrett and Strecker* [2000] proposed that the flat slabs were "flattened out" by stiff segments of the South American plate advancing over them. By contrast, where the upper plate was weak, it shortened and left the lower plate undeformed (fig. 20.11a) However, there

Figure 20.11 Different models of how overriding plate deformation might influence slab dip: (a) subducting plate "flattened out" where rigid overriding plate does not deform, (b) slab dip controlled by curvature of the plate boundary, and (c) slab interacting with thickening mantle lithosphere of the shortened overriding plate. Figures 20.11a and 20.11b are sketches after ideas outlined by *Isacks* [1988] and *Cahill and Isacks* [1992]; figure 20.11c is modified from *Kley et al.* [1999].

is no clear correlation of the flat slabs with shortening minima and the maximum differences in shortening are insufficient to account for the full downdip length of the flat slabs.

An alternative model was introduced by *Cahill and Isacks* [1992], who proposed that the flat slabs form where the trench is seaward convex in order to maintain a constant, concave upward slab curvature (fig. 20.11b). Since neither the "flattening out" nor the constant curvature model can easily explain the coincidence of flat slab boundaries with aseismic ridges, these would have to be fortuitous.

The problem of the slab interfering with the overriding plate also concerns the lithospheric mantle. Shortening of the overriding plate would not only thicken the crust but also the mantle lithosphere. However, evidence for a ~ 150-km-thick lithosphere was only found for the Altiplano from weak attenuation of seismic waves [*Whitman et al.*, 1992]. The lithospheric mantle appears to be thin under the Puna part of the plateau and very thin over the flat slabs [*Whitman et al.*, 1996]. A similar pattern was inferred from studies of lithospheric flexural rigidity [*Watts et al.*, 1995]. This raises the need to remove large volumes of lithospheric mantle during the Andean orogeny [*Kay and Abbruzzi*, 1996] or to have a thin lithosphere from the start. The latter scenario is probably at odds with the absence of any indicative magmatism. *Whitman et al.* [1992] noted that the lithosphere beneath the Altiplano could be tectonically thickened, an idea supported by the coincidence of thick lithosphere with the most strongly shortened segment of the central Andes [*Kley et al.*, 1999]. *Kley et al.* [1999] proposed that in this segment the mantle lithosphere may have interacted with the gently dipping Miocene slab and pushed it back to its present steeper dip (fig. 20.11c).

Open Questions

Material Balance and the "Shortening Deficit"

A long-standing problem in the central Andes is material balance. Many authors concluded that the Andean crust was mostly thickened by tectonic shortening, but there was also a consensus that in the plateau region the documented

(a)

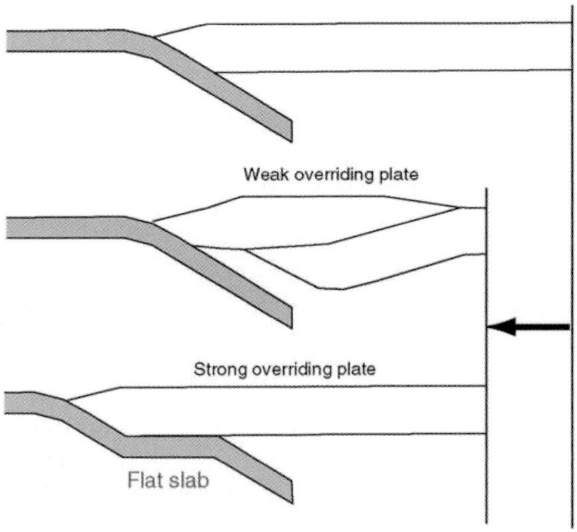
Weak overriding plate
Strong overriding plate
Flat slab

(b)

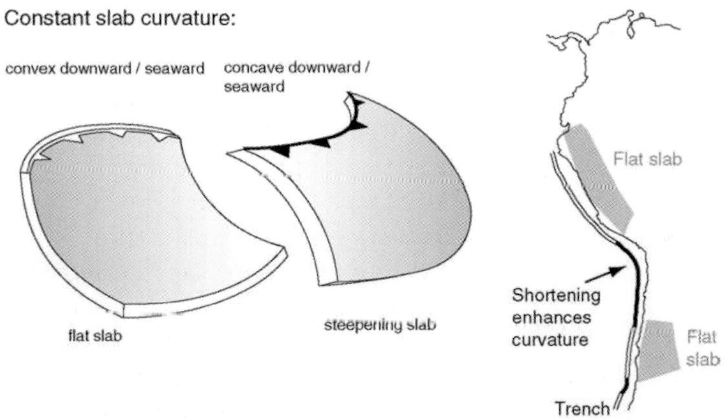
Constant slab curvature:
convex downward / seaward
concave downward / seaward
flat slab
steepening slab
Flat slab
Shortening enhances curvature
Flat slab
Trench

(c)

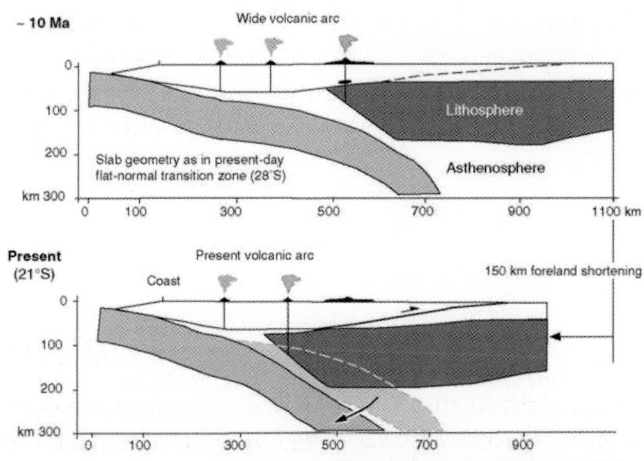
~ 10 Ma
Wide volcanic arc
Lithosphere
Slab geometry as in present-day flat-normal transition zone (28°S)
Asthenosphere
0
100
200
km 300
0 100 300 500 700 900 1100 km
Present (21°S)
Present volcanic arc
Coast
150 km foreland shortening
0
100
200
km 300
0 100 300 500 700 900
Lithosphere thickness from Whitman et al. (1992)

shortening could produce no more than ~70% of the observed crustal volume, even if material loss through erosion was neglected [*Baby et al.*, 1997; *Kley and Monaldi*, 1998; *Müller et al.*, 2002; *Roeder*, 1988; *Schmitz*, 1994; *Sheffels*, 1990]. Geochemical arguments indicate that the missing volume cannot be filled in by intruding magmas [*Francis and Hawkesworth*, 1994]. Claims to substantially higher shortening amounts have been made recently [*McQuarrie*, 2002b] but are not substantiated by any structural data.

The material balance problem could be largely an artifact of analyzing shortening from cross sections constructed and balanced in two dimensions. As in any arc-shaped orogen [*Hindle et al.*, 2000], shortening the central Andes requires either substantial along-strike extension of thrust sheets or along-strike displacement of material, most probably accommodated by strike-slip components on thrust faults (fig. 20.12). Map-view block mosaic models constructed from serial balanced cross sections [*Laubscher*, 1965] have been used to estimate the magnitudes of these strains in the central Andes [*Baby et al.*, 1993; *Hindle et al.*, 2002; *Kley*, 1999]. A simple comparison of the deformed and undeformed states of our block model (fig. 20.13) suggests that some 30–40% of the material that has gone into the Andean orogen was added along strike

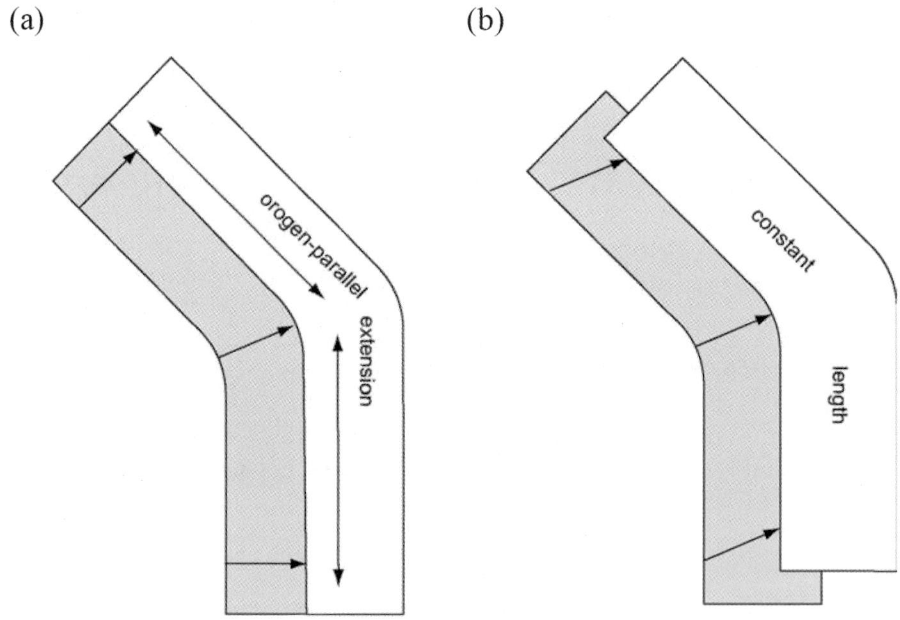

Figure 20.12 Basic geometric problem of a forelandward convex thrust sheet or thrust belt: (a) divergent displacements cause an increase in length and along-strike extension of thrust sheets or (b) parallel displacements cause strike-slip components on the thrust faults. There is little evidence for major orogen-parallel extension in the central Andes, suggesting that material is added to the mountain belt along strike.

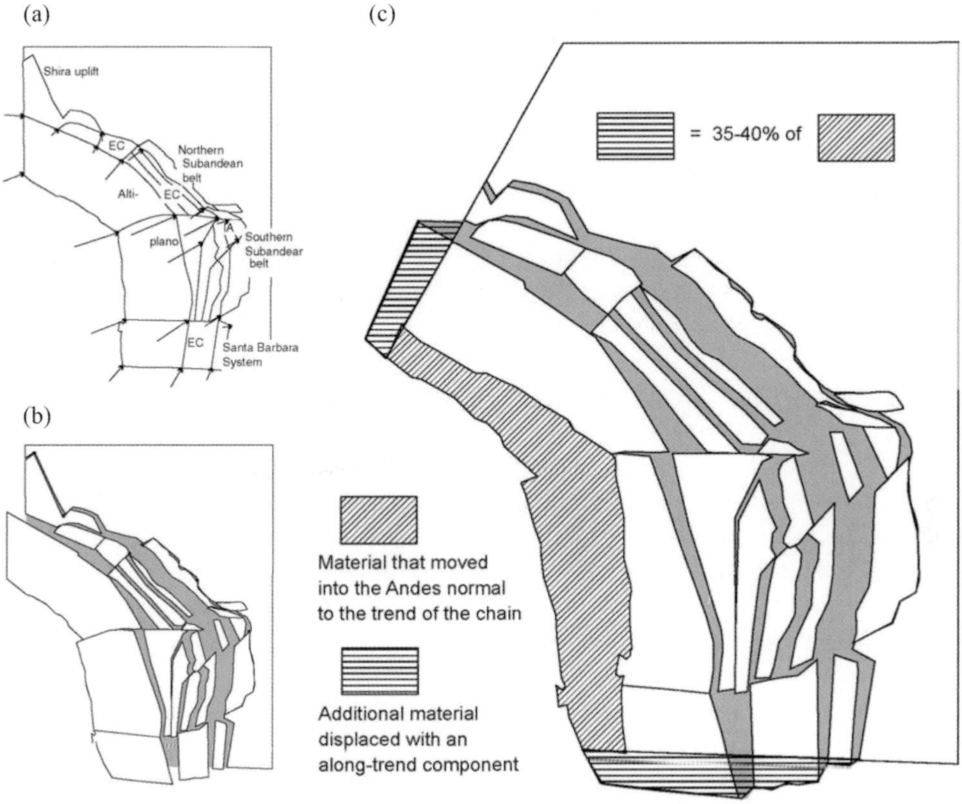

Figure 20.13 Material balance of the central Andes estimated from a restorable block mosaic [*Kley*, 1999, *Hindle et al.*, 2002]: (a) deformed (present) and (b) restored state. Arrows in figure 20.13a are displacement vectors connecting identical points in the deformed and restored states. Northwestern model boundary is still arbitrarily oriented. (c) Model clipped along boundaries perpendicular to fold trends, deformed and restored states overlain. Total area difference between the deformed and restored states is marked by ruling. Only the diagonally ruled area is accounted for in estimates from balanced cross sections drawn perpendicular to the fold trends. The areas shown with horizontal ruling are material added along strike, corresponding to 30–40% of the total added area. This is about the percentage "missing" in two-dimensional cross-section area balance calculations.

[*Hindle et al.*, 2005]. This is enough to account for the total crustal volume with no need to invoke more shortening than is actually documented in structures. A similar result was obtained by *Yang et al.* [2003] from 3-D finite element modeling. However, these authors concluded that the initial crustal thickness was too poorly constrained to rule out the high shortening estimates. Both the 3-D model and the 2-D block mosaic predict that crustal material must be redistributed within the orogen. Along-strike variations in crustal thickness and topography are only moderately correlated with shortening variations [*Kley and Monaldi*, 1998]. This suggests that hot ductile middle and lower crust may flow from strongly shortened to less deformed areas as modeled by *Yang et al.* [2003], resulting in smoother topographic than shortening variations. The details of such a process depend on the composition and rheology of the deeper crust that are not yet fully understood. Seismologic [*Zandt et al.*, 1994] and localized xenolith data [*Lucassen et al.*, 1999] suggest that the lower crust of the central Andes has a felsic composition while receiver function tomography has revealed a fast layer above the Moho interpreted as "dry, refractory lower crust" [*Yuan et al.*, 2000].

Plateau Evolution: Numerical Models and Geologic Data

Even if we consider the material balance problem solved, major questions concerning the timing and the mechanics of plateau evolution remain. Numerical models of the Andean orogeny started in the 1990s, after shortening had become widely accepted as the major agent in plateau formation. A weak base was rapidly identified to be responsible for the flat surface of the plateau. *Williams et al.* [1994] showed that the morphology of the plateau edges could be explained with a transition from frictional deformation in the foothills and within the flank to ductile deformation under the plateau. Here, the detachment becomes so weak that almost no topography can be supported. In order to obtain such a flat plateau region, *Wdowinski and Bock* [1994a, 1994b] used a viscoplastic model with a prescribed zone of weakness resulting from high mantle heat flow. They modeled crustal deformation assuming a constant motion of the subduction zone hinge with respect to the interior of South America. In these models, shortening of the lithosphere initially leads to thickening and the uplift of a flat plateau in the region that has the weaker base. Depending on the strength parameters assigned to the crust and its base, in a later stage the models display a change of their shortening mode. Upper crustal shortening becomes progressively localized at the continentward side of the thickening region. Lower crustal shortening becomes localized within the plateau region itself. This change in deformation mode is due to the increasing body forces during plateau uplift. The experiments of *Vietor* [2001] have shown that the lateral expansion of the orogen can be substantially modified by the expansion of the weak zone at the base of the plateau. In particular, deformation within the plateau can be shut down entirely if the weak base is allowed to expand. Figure 20.14 presents a model plateau evolution in which the weak zone at the base expands at convergence rate after 125 km of total contraction. After initial distributed shortening within the plateau section, the expansion of the subcrustal detachment leads to relocalization of the deformation. Shortening and uplift within the plateau region stops, while it proceeds in a narrow belt at the margin of the orogen. Hence a widening detachment can completely switch the growth mode of the plateau from vertical thickening to lateral expansion.

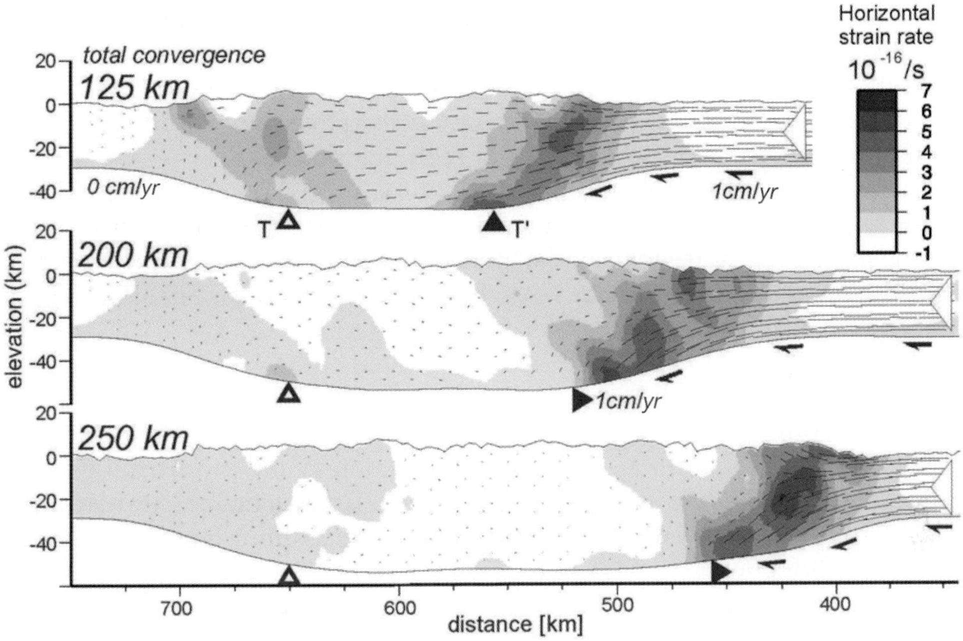

Figure 20.14 Two-stage deformation of the central Andean plateau in a numerical model [*Vietor*, 2001]. This numerical experiment uses the distinct element technique to handle the large strains involved in orogeny. The starting configuration representing a vertical crustal profile is a two-dimensional rectangular box, 900 × 30 km in size, containing some 5500 disk-shaped elements (linear radius distribution between r_{max} = 1500 m and r_{min} = 750 m, E = 30 GPa, surface friction 0.58). The bulk rheology approximates a noncohesive Navier-Coulomb behavior with an angle of internal friction of 0.25 and weak strain softening. The lower boundary of the box is split to allow convergence of the model boundaries. Deformation of the model material is driven by kinematic boundary conditions applied as fixed velocities of the right side wall and lower boundary (1 cm/yr). The friction coefficient of the driving walls is 0.58. The properties of the disks are homogeneous and the lower boundary contains a decoupled (zero shear stress) section in its center. During convergence of the model the width of the decoupling zone (T-T′) may be changed arbitrarily in order to test the impact of a widening or narrowing detachment. In the experiment shown, an initially 100-km-wide decoupling zone at the base of the crust starts to spread toward the right (foreland) side after 125 km of convergence. It expands at convergence velocity. Coeval with the start of the expansion, crustal shortening becomes confined to the right-hand flank of the orogen. Shortening and uplift of the flat plateau region dies down after ~150 km of convergence and the orogen subsequently grows by accretion toward the foreland.

As these model evolutions closely mimic the two-stage evolution of the Altiplano, they suggest that initial weakening of the crustal base plays an important role for the kinematic evolution of the Altiplano. The anomalously high mantle heat flow assumed for the plateau region is supported by the surface heat flow data compiled by *Springer and Förster* [1998]. Subsequent thermal modeling [*Springer*, 1999] showed that conductive heat flow alone is unable to fit these high values and the large variability of the surface heat flow in the plateau. In a thermomechanical model of the plateau region, *Babeyko et al.* [2002] further explored the influence of high mantle heat flow on the process of crustal shortening. Assuming a felsic composition for both the upper and lower crust, they showed that convective overturn of the lower crust following the removal of large parts of the lithospheric mantle is required to produce both the observed surface heat flow distribution and crustal melting within the time frame given for plateau formation in the Andes.

Pope and Willett [1998] tried to avoid the initial heat flow anomaly that controlled the evolution in the models of *Wdowinski and Bock* [1994a, 1994b]. Using kinematic boundary conditions, mantle deformation was prescribed in that study in order to obtain the crustal deformation field and the crustal thermal evolution. The mantle kinematics assumed the entrainment of the overriding plate's lithospheric mantle in the subduction process and its continuous removal at subduction velocity. The experiment resulted in an outward growing orogenic plateau with low viscosity flow in the lower crust due to thickening of the high-heat production upper crust. However, the assumed heat production of the entire crust of 2 mW/m^3, which were necessary to explain the observed surface heat flow, is twice as high as the values inferred from geochemical studies of the basement rocks [*Lucassen et al.*, 2001]. Furthermore the assumed upper plate shortening rate of 24 mm/yr is 2–6 times higher than the observed values [*Hindle and Kley*, 2002], and finally the two-stage kinematic evolution of the Central Andean plateau [*Allmendinger and Gubbels*, 1996; *Isacks*, 1988] is not reproduced by the model. Instead it predicts a continuous migration of the deformation front toward the continent. Recent 3-D models [*Liu et al.*, 2002; *Yang et al.*, 2003] have explored the stress state, material flow, and volume balance in the arc-shaped Andean orogen.

As shown above, several numerical models have successfully simulated the two-stage deformation history of the Central Andean plateau. However, there is an important difference in the uplift history: while in the numerical simulations the plateau reaches some limiting elevation and then begins to grow laterally at constant height [e.g., *Beaumont et al.*, 2004; *Wdowinski and Bock*, 1994b] (fig. 20.15 a), a balance of paleoelevation data from the central Andes [*Gregory-Wodzicki*, 2000] (fig. 20.15b) suggests that the plateau was at only half its present elevation 10 Ma ago but nearly as wide as today when thrusting shifted from the plateau interior to the foreland thrust belt. Substantial plateau uplift after 10 Ma makes sense in terms of material balance because a large volume of material was incorporated into the mountain belt after the onset of rapid foreland thrusting, and much of this material must have underthrust the plateau rather than contributing to the subdued topography of the foreland thrust belt. The 3-D models of *Yang et al.* [2003] do predict continuous uplift as shortening proceeds but not the concentration of strain on the eastern plateau flank. Thus it seems that some rheologic and dynamic

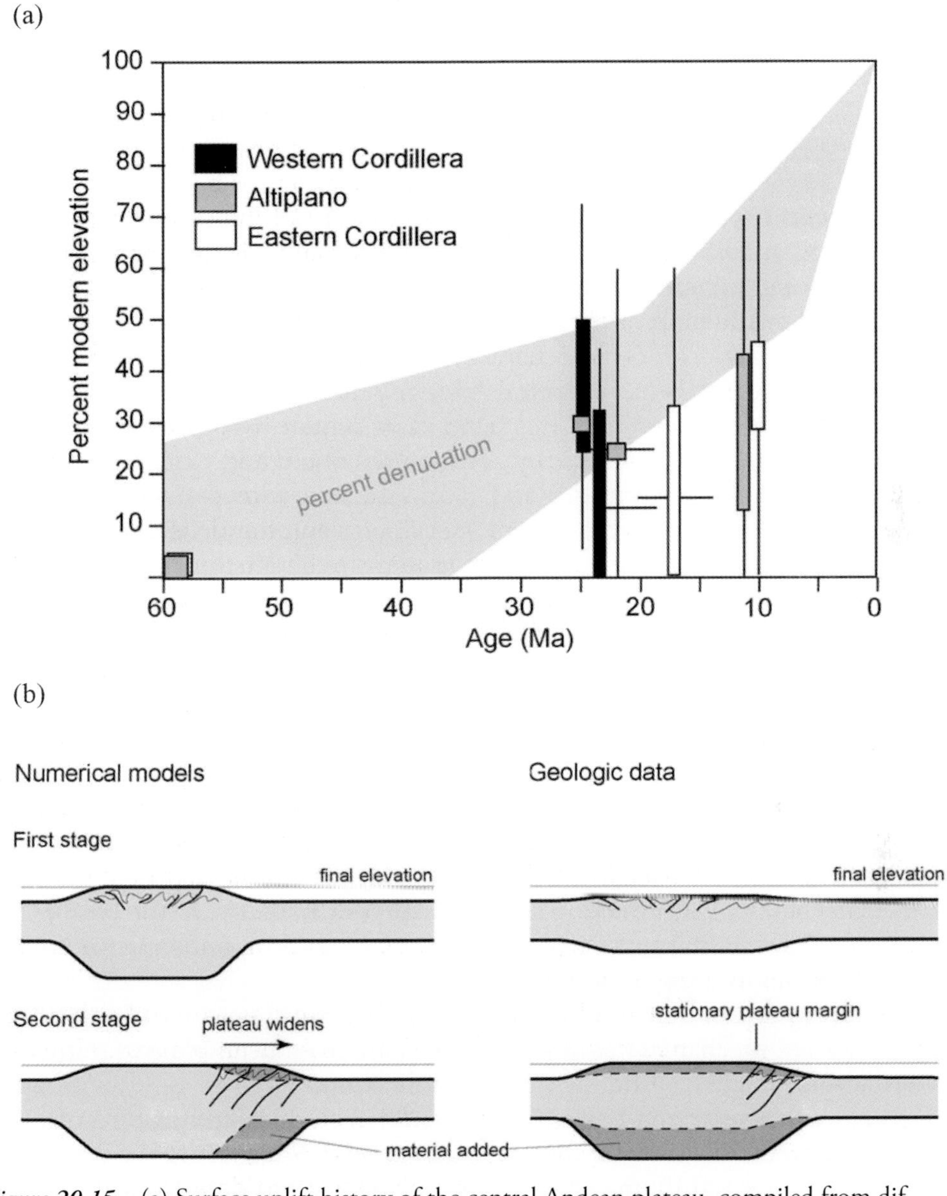

Figure 20.15 (a) Surface uplift history of the central Andean plateau, compiled from different sets of data (modified from *Gregory-Wodzicki* [2000]). Grey shading indicates exhumation history of the Eastern Cordillera defined by pairs of apatite and zircon fission track ages (data from *Benjamin et al.* [1987] and *Masek et al.* [1994]). By 10 Ma the plateau had probably attained about half its present elevation and uplift has since accelerated. (b) Sketches comparing model predictions of plateau deformation and uplift (left) with the uplift and deformation history of the central Andean plateau as suggested by geologic data (right). Models predict early growth of narrow plateau to final elevation followed by lateral accretion, geologic data suggest early growth of wide plateau to intermediate elevation followed by further uplift and very limited lateral growth.

conditions for simultaneous plateau uplift and rapid lateral accretion remain to be explored.

Conclusions

Subduction-related orogeny in the central Andes is probably due to the combined effects of favorable subduction and plate parameters, like rapid convergence of young oceanic lithosphere and South America's trenchward absolute motion, and weakening of the South American plate through high heat flow and inherited zones of low mechanical strength. Isolating the most relevant factors is made difficult in the central Andes by the high symmetry of the entire subduction system about the Arica bend in its center. Topographic elevations and trench depth, crustal seismicity, slab geometry and age, climate and trench sediment fill, strain magnitudes and rates and structural style, and foreland flexural rigidity are all to a higher or lesser degree symmetric about the central bend and the plate convergence vector. It seems likely that only some of these correlations reflect cause-and-effect relationships, whereas others may be fortuitous. The factors controlling the evolution of the Andes have also probably changed with time: during the early stages of mountain building, strain rate variations in the orogen reflect changes in plate convergence rate. Later, from ~20 Ma on, the two rates are decoupled although the plate convergence direction still controls the strain field of the central Andes.

The accumulation of permanent deformation away from the plate interface requires some mechanism weakening the overriding plate. Geophysical and geological evidence for partial melting in the thick crust of the central Andes and numerical modeling suggest that heat advected to the base of the crust by a strongly thinned lithospheric mantle was a prerequisite for the onset of strong shortening. It is still not clear which process has initially thinned the South American lithosphere, although several mechanisms of lithospheric delamination, mechanical ablation, or thermal erosion have been proposed. Thermal-mechanical modeling furthermore suggests that even with very high temperatures at the base of the crust, deformation-induced convection is required to efficiently heat the continental crust. Subsequent weakening of South America's margin may have been entirely internally controlled. A thick conformable sedimentary succession was important as an upper crustal low strength zone on the continentward side of the mountain belt.

Differential shortening of the arc-shaped central Andes, high in the center, decreasing toward the wings, induces deformation of the plate boundary and the subducting Nazca plate. The slab is made strong by its flexural rigidity, probably by its continentward convex shape, and by forces resisting its translation through the mantle, enabling it to transfer stresses laterally and to always load the strongest parts of the overriding plate. This may provide an explanation for the apparently orogen-wide correlation of some deformation events.

The long-standing problem that the observed tectonic shortening does not account for the entire crustal volume is mostly an artifact of two-dimensional analyses. New estimates that take the mountain belt's arc shape into account suggest that published shortening estimates of 250–300 km are sufficient to create the total crustal volume if the initial crustal thickness was near the high end of the range considered plausible. Nevertheless, deep crustal material has to be redistributed by ductile flow to explain the observed crustal thickness variations.

Numerical models of plateau formation are only partly successful. Many of them simulate the major shift in deformation mode from distributed strain throughout the plateau to focused deformation on one plateau margin but typically predict this shift to occur when the plateau has reached some limiting elevation. By contrast, most geologic data suggest that the Central Andean plateau was only at half its present height when the marginal thrust belt began to grow. This fundamental contradiction either indicates a problem with the current interpretation of the geologic data or a major flaw in our understanding of the dynamic and rheologic evolution of the Central Andean plateau.

Acknowledgments

Jonas Kley would like to thank Tim Dixon and Seth Stein for inviting him to the very stimulating SEIZE theoretical institute at Snowbird and all participants of that meeting for giving him a glance at the oceanic side of subduction-related orogeny. We both thank Jürgen Klotz (GFZ Potsdam) for the permission to use his compilation of GPS data. Constructive reviews by Mian Liu, Seth Stein, and Youqing Yang helped us to substantially improve this paper.

References

Adam, J., and C. D. Reuther (2000), Crustal dynamics and active fault mechanics during subduction erosion. Application of frictional wedge analysis on to the North Chilean Forearc, *Tectonophysics, 321,* 297–325.

Adelmann, D. (2001), Känozoische Beckenentwicklung des zentralandinen Puna-Plateaus (NW-Argentinien) - Das Gebiet um den Salar de Antofalla und ein Vergleich zur nördlichen Puna, *Berl. Geowissensch. Abhandl., Ser. A, 210,* 1–142.

Allmendinger, R. W., and T. Gubbels (1996), Pure and simple shear plateau uplift, Altiplano-Puna, Argentina and Bolivia, *Tectonophysics, 259,* 1–13.

Allmendinger, R. W., V. A. Ramos, T. E. Jordan, M. Palma, and B. L. Isacks (1983), Paleogeography and Andean structural geometry, northwest Argentina, *Tectonics, 2,* 1–16.

Allmendinger, R. W., D. Figueroa, D. B. Snyder, J. Beer, C. Mpodozis, and B. L. Isacks (1990), Foreland shortening and crustal balancing in the Andes at 30°S latitude, *Tectonics, 9,* 789–809.

Allmendinger, R. W., T. E. Jordan, S. M. Kay, and B. L. Isacks (1997), The evolution of the Altiplano-Puna plateau of the central Andes, *Annu. Rev. Earth Planet. Sci., 25,* 139–174.

Babeyko, A. Y., S. V. Sobolev, R. B. Trumbull, O. Oncken, and L. L. Lavier (2002), Numerical models of crustal scale convection and partial melting beneath the Altiplano-Puna Plateau, *Earth Planet. Sci. Lett., 199*, 373–388.

Baby, P., B. Guillier, J. Oller, and G. Montemurro (1993), Modèle cinématique de la Zone subandine du coude de Santa Cruz (entre 16°S et 19°S, Bolivie) déduit de la construction de cartes equilibrées, *C. R. Acad. Sci. Paris, Ser. 2, 317*, 1477–1483.

Baby, P., P. Rochat, G. Mascle, and G. Hérail (1997), Neogene shortening contribution to crustal thickening in the back arc of the central Andes, *Geology, 25*(10), 883–886.

Bangs, N. L., and S. C. Cande (1997), Episodic development of a convergent margin inferred from structures and processes along the southern Chile margin, *Tectonics, 16*, 489–503.

Barazangi, M., and B. L. Isacks (1976), Spatial distribution of earthquakes and subduction of the Nazca plate beneath South America, *Geology, 4*, 686–692.

Beaumont, C., R. A. Jamieson, M. H. Nguyen, and S. Medvedev (2004), Crustal channel flows: 1. Numerical models with applications to the tectonics of the Himalayan-Tibetan orogen, *J. Geophys. Research, 109*, B06406, doi:10.1029/2003JB002809.

Belotti, H. J., L. L. Saccavino, and G. A. Schachner (1995), Structural styles and petroleum occurrence in the Sub-Andean fold and thrust belt of northern Argentina, in *Petroleum Basins of South America*, edited by H. J. Welsink, *AAPG Mem., 62*, 545–555.

Benjamin, M. T., N. M. Johnson, and C. W. Naeser (1987), Recent rapid uplift in the Bolivian Andes: Evidence from fission-track dating, *Geology, 15*(7), 680–683.

Bevis, M., E. Kendrick, R. Smalley, B. Brooks, R. Allmendinger, and B. Isacks (2001), On the strength of interplate coupling and the rate of back arc convergence in the central Andes: An analysis of the interseismic velocity field, *Geochem. Geophys. Geosyst., 2*(11), doi:10.1029/2001GC000198.

Bianucci, H., J. F. Homovc, and O. M. Acevedo (1982), Inversión tectónica y plegamientos resultantes en la comarca Puesto Guardian - Dos Puntitas, Dpto. Orán, Provincia de Salta. Paper presented at Primer Congreso Nacional de Hidrocarburos, *Inst. Argentina del Pet. y del. Gas*, pp. 23–30, Buenos Aires.

Bird, P. (1984), Laramide crustal thickening event in the Rocky Mountain foreland and Great Plains, *Tectonics, 3*, 741–758.

Burchfiel, B. C., and G. A. Davis (1975), Nature and controls of Cordilleran orogenesis, western United States, extension of an earlier synthesis, *Am. J. Sci., 275A*, 363–396.

Cahill, T., and B. L. Isacks (1992), Seismicity and shape of the subducted Nazca plate, *J.Geophys. Res., 97*, 17,503–17,529.

Cande, S. C., and D. V. Kent (1992), A new geomagnetic polarity time scale for the Late Cretaceous and Cenozoic, *J. Geophys. Res., 97*, 13,917–13,951.

Cande, S. C., and D. V. Kent (1995), Revised calibration of the geomagnetic polarity timescale for the Late Cretaceous and Cenozoic, *J. Geophys. Res., 100*, 6093–6095.

Case, J. E., R. Shagam, and R. F. Giegengack (1990), Geology of the northern Andes; an overview, in *The Geology of North America*, edited by J. E. Case, pp. 177–200, Geol. Soc. of Am., Boulder, Colo.

Cladouhos, T. T., R. W. Allmendinger, B. Coira, and E. Farrar (1994), Late Cenozoic deformation in the central Andes: Fault kinematics from the northern Puna, northwest Argentina and southern Bolivia, *J. S. Am. Earth Sci., 7*(2), 209–228.

Coira, B., J. Davidson, C. Mpodozis, and V. Ramos (1982), Tectonic and magmatic evolution of the Andes of Northern Argentina and Chile., *Earth Sci. Rev., 18*, 303–332.

Cristallini, E., A. H. Cominguez, and V. A. Ramos (1997), Deep structure of the Metán-Guachipas Region: tectonic inversion in Northwestern Argentina, *J. S. Am. Earth Sci., 10*(5–6), 403–421.

Dalmayrac, B., and P. Molnar (1981), Parallel thrust and normal faulting in Peru and constraints on the state of stress, *Earth Planet. Sci. Lett., 55*, 473–481.

Dalziel, I. W. D., M. J. De Wit, and K. F. Palmer (1974), A fossil marginal basin in the southern Andes, *Nature, 250*, 291–294.
DeCelles, P. G., and B. K. Horton (2003), Early to middle Tertiary foreland basement development and the history of Andean crustal shortening in Bolivia, *Geol. Soc. Am. Bull., 115*(1), 58–77.
Dewey, J. F., and J. M. Bird (1970), Mountain belts and the new global tectonics, *J. Geophys. Res., 75*, 2625–2647.
Dickinson, W. R., and W. S. Snyder (1978), Plate tectonics of the Laramide orogeny, in *Laramide Folding Associated With Basement Block Faulting in the Western United States*, edited by V. Matthews, pp. 355–366, Geol. Soc. of Am., Boulder, Colo.
Doglioni, C., P. Harabaglia, S. Merlini, F. Mongelli, A. Peccerillo, and C. Piromallo 1999), Orogens and slabs vs. their direction of subduction, *Earth Sci. Rev., 45*(3–4), 167–208.
Echavarria, L., R. Hernández, R. Allmendinger, and J. Reynolds (2003), Subandean thrust and fold belt of northwestern Argentina: Geometry and timing of the Andean evolution, *AAPG Bull., 87*, 965–985.
Ege, H., V. Jacobshagen, E. Scheuber, E. Sobel, and T. Vietor (2001), Thrust-related exhumation revealed by apatite fission track dating, central Andes (southern Bolivia), *Geophys. Res. Abstr., 3*, 624.
Elger, K. (2003), *Analysis of Deformation and Tectonic History of the Southern Altiplano Plateau (Bolivia) and Their Importance for Plateau Formation*, 152 pp., GeoforschungsZentrum, Potsdam, Germany.
Francis, P. W., and C. J. Hawkesworth (1994), Late cenozoic rates of magmatic activity in the central Andes and their relationships to continental crust formation and thickening, *J. Geol. Soc., 151*, 845–854.
Gephart, J. W. (1994), Topography and subduction geometry in the central Andes; clues to the mechanics of a non-collisional orogen, *J. Geophys. Res., 99*, 12,279–12,288.
Gregory-Wodzicki, K. (2000), Uplift history of the central and northern Andes: A review, *Geol. Soc. Am. Bull., 112*, 1091–1105.
Gubbels, T. L., B. L. Isacks, and E. Farrar (1993), High-level surfaces, plateau uplift, and foreland development, Bolivian central Andes, *Geology, 21*(8), 695–698.
Gutscher, M.-A., W. Spakman, H. Bijwaard, and E. R. Engdahl (2000), Geodynamics of flat subduction: seismicity and tomographic constraints from the Andean margin, *Tectonics, 19*, 814–833.
Haschke, M., and A. Günther (2003), Balancing crustal thickening in arcs by tectonic vs. magmatic means, *Geology, 31*(11), 933–936.
Hindle, D., and J. Kley (2002), Displacements, strains and rotations in the Central Andean plate boundary zone, in *Plate Boundary Zones, Geodyn. Ser.*, vol. 30, edited by S. Stein and J. T. Freymueller, pp. 135–144, AGU, Washington, D. C.
Hindle, D., O. Besson, and M. Burkhard (2000), A model of displacement and strain for arc-shaped mountain belts applied to the Jura, *J. Struct. Geol., 22*, 1285–1296.
Hindle, D., J. Kley, E. Klosko, S. Stein, T. Dixon, and E. Norabuena (2002), Consistency of geologic and geodetic displacements during Andean orogenesis, *Geophys. Res. Lett., 29*(8), 1188, doi:10.1029/2001GL013757.
Hindle, D., J. Kley, O. Oncken, and S. Sobolev (2005), Crustal balance and crustal flux from shortening estimates in the central Andes, *Earth Planet. Sci. Lett., 230*, 113–124.
Isacks, B. L. (1988), Uplift of the Central Andean Plateau and bending of the Bolivian orocline, *J. Geophys. Res., 93*, 3211–3231.
James, D. E. (1971), Plate tectonic model for the evolution of the central Andes, *Geol. Soc. Am. Bull., 82*, 3325–3346.
Jarrard, R. D. (1986), Relations among subduction parameters, *Rev. Geophys., 24*, 217–284.

Jordan, T., B. Isacks, R. W. Allmendinger, J. A. Brewer, V. A. Ramos, and C. J. Ando (1983), Andean tectonics related to geometry of subducted Nazca plate, *Geol. Soc. Am. Bull., 94,* 341–361.

Jordan, T. E., and R. N. Alonso (1987), Cenozoic stratigraphy and basin tectonics of the Andes mountains, 20°-28° south latitude, *AAPG Bull., 71,* 49–64.

Jordan, T. E., J. H. Reynolds, and J. P. Erikson (1997), Variability in age of initial shortening and uplift in the central Andes, 16-33°30′S, in *Tectonic Uplift and Climate Change,* edited by W. F. Ruddiman, pp. 41–61, Springer, New York.

Jordan, T. E., W. M. Burns, R. Veiga, F. Pángaro, P. Copeland, S. Kelley, and C. Mpodozis (2001), Extension and basin formation in the southern Andes caused by increased convergence rates: A mid-Cenozoic trigger for the Andes, *Tectonics, 20,* 308–324.

Kay, R. W., and S. M. Kay (1993), Delamination and delamination magmatism, *Tectonophysics, 219,* 177–189.

Kay, S. M., and J. M. Abbruzzi (1996), Magmatic evidence for Neogene lithospheric evolution of the central Andean "flat-slab" between 30°S and 32°S, *Tectonophysics, 259,* 15–28.

Kendrick, E., M. Bevis, R. Smalley, and B. Brooks (2001), An integrated crustal velocity field for the central Andes, *Geochem. Geophys. Geosyst., 2*(11), doi:10.1029/2001GC000191.

Kley, J. (1999), Geologic and geometric constraints on a kinematic model of the Bolivian orocline, *J. S. Am. Earth Sci., 12,* 221–235.

Kley, J., and C. R. Monaldi (1998), Tectonic shortening and crustal thickness in the central Andes: How good is the correlation?, *Geology, 26*(8), 723–726.

Kley, J., and C. R. Monaldi (2002), Tectonic inversion in the Santa Barbara System of the central Andean foreland thrust belt, northwestern Argentina, *Tectonics, 21*(6), 1061, doi:10.1029/2002TC902003.

Kley, J., J. Müller, S. Tawackoli, V. Jacobshagen, and E. Manutsoglu (1997), Pre-Andean and Andean-age deformation in the eastern Cordillera of southern Bolivia, *J. S. Am. Earth Sci., 10*(1), 1–19.

Kley, J., C. R. Monaldi, and J. A. Salfity (1999), Along-strike segmentation of the Andean foreland: Causes and consequences, *Tectonophysics, 301,* 75–94.

Klosko, E. R., S. Stein, D. Hindle, J. Kley, E. Norabuena, T. Dixon, and M. Liu (2002), Comparison of GPS, seismological, and geological observations of Andean mountain building, in *Plate boundary zones, Geodyn. Ser.*, vol. 30, edited by S. Stein and J. T. Freymuller, pp. 123–133, AGU, Washington, D. C.

Kono, M., Y. Fukao, and A. Yamamoto (1989), Mountain building in the central Andes, *J. Geophys. Res., 94,* 3891–3905.

Lamb, S. (2000), Active deformation in the Bolivian Andes, South America, *J. Geophys. Res., 105,* 25,627–25,653.

Lamb, S. (2001), Vertical axis rotation in the Bolivian orocline, South America 1. Paleomagnetic analysis of Cretaceous and Cenozoic rocks, *J. Geophys. Res., 106* 26,605–26,632.

Lamb, S., and P. Davis (2003), Cenozoic climate change as a possible cause for the rise of the Andes, *Nature, 425,* 792–797.

Lamb, S., and L. Hoke (1997), Origin of the high plateau in the central Andes, Bolivia, South America, *Tectonics, 16,* 623–649.

Laubscher, H. P. (1965), Ein kinematisches Modell der Jurafaltung., *Eclogae Geol. Helv., 58,* 232–318.

Leffler, L., S. Stein, A. Mao, T. Dixon, M. A. Ellis, L. Ocola, and I. S. Sacks (1997), Constraints on present-day shortening rate across the central eastern Andes from GPS data, *Geophys. Res. Lett., 24,* 1031–1034.

Liu, M., Y. Yang, S. Stein, Y. Zhu, and J. Engeln (2000), Crustal shortening in the Andes: Why do GPS rates differ from geological rates?, *Geophys. Res. Lett., 27,* 3005–3008.

Liu, M., Y. Yang, S. Stein, and E. R. Klosko (2002), Crustal shortening and extension in the central Andes: Insights from a viscoelastic model, in *Plate Boundary Zones, Geodyn. Ser.*, vol. 30, edited by S. Stein and J. T. Freymueller, pp. 325–339, Washington, D. C.

Lucassen, F., S. Lewerenz, G. Franz, J. Viramonte, and K. Mezger (1999), Metamorphism, isotopic ages and composition of lower crustal granulite xenoliths from the Cretaceous Salta Rift, Argentina, *Contrib. Mineral. Petrol., 134*, 325–341.

Lucassen, F., R. Becchio, R. Harmon, S. Kasemann, G. Franz, R. Trumbull, H. G. Wilke, R. L. Romer, and P. Dulski (2001), Composition and density model of the continental crust at an active continental margin; the central Andes between 21 degrees and 27 degrees S, *Tectonophysics, 341*, 195–223.

Marrett, R., and M. R. Strecker (2000), Response of intracontinental deformation in the central Andes to late Cenozoic reorganization of South American Plate motions, *Tectonics, 19*, 452–467.

Marrett, R. A., R. W. Allmendinger, R. N. Alonso, and R. E. Drake (1994), Late Cenozoic tectonic evolution of the Puna Plateau and adjacent foreland, northwestern Argentine Andes, *J. S. Am. Earth Sci., 7*(2), 179–207.

Masek, J. G., B. L. Isacks, T. L. Gubbels, and E. J. Fielding (1994), Erosion and tectonics at the margins of continental plateaus, *J. Geophys. Res., 99*, 13,941–13,956.

McQuarrie, N. (2002a), Initial plate geometry, shortening variations, and evolution of the Bolivian orocline, *Geology, 30*(10), 867–870.

McQuarrie, N. (2002b), The kinematic history of the central Andean fold-thrust belt, Bolivia: Implications for building a high plateau, *Geol. Soc. Am. Bull., 114*, 950–963.

McQuarrie, N., and P. DeCelles (2001), Geometry and structural evolution of the central Andean backthrust belt, Bolivia, *Tectonics, 20*, 669–692.

Mégard, F. (1987), Structure and evolution of the Peruvian Andes, in *The Anatomy of Mountain Ranges*, edited by J. P. Schaer and S. Rodgers, pp. 179–210, Princeton Univ., Princeton, N. J.

Meijer, P. T., R. Govers, and M. J. R. Wortel (1997), Forces controlling the present-day state of stress in the Andes, *Earth and Planetary Science Letters, 148*, 157–170.

Mercier, J. L., M. Sebrier, A. Lavenu, J. Cabrera, O. Bellier, J.-F. Dumont, and J. Machare (1992), Changes in the tectonic regime above a subduction zone of Andean type: The Andes of Peru and Bolivia during the Pliocene-Pleistocene, *J. Geophys. Res., 97*, 11,945–11,982.

Mon, R., and J. A. Salfity (1995), Tectonic evolution of the Andes of northern Argentina, in *Petroleum Basins of South America*, edited by H. J. Welsink, *AAPG Mem., 62*, 269–283.

Mpodozis, C., and V. A. Ramos (1989), The Andes of Chile and Argentina, in *Geology of the Andes and Its Relation to Hydrocarbon and Mineral Resources*, edited by J. A. Reinemund, pp. 59–90, Circum-Pac. Counc. for Energy and Miner. Resour., Houston, Tex.

Mueller, R. D., W. R. Roest, J.-Y. Roger, L. M. Gahagan, and J. G. Sclater (1997), Digital isochrons of the world's ocean floor, *J. Geophys. Res., 102*, 3211–3214.

Müller, J., J. Kley, and V. Jacobshagen (2002), Structure and Cenozoic kinematics of the Eastern Cordillera, southern Bolivia (21°S), *Tectonics, 21*(5), 1037, doi:10.1029/2001TC001340.

Norabuena, E., L. Leffler-Griffin, A. Mao, T. Dixon, S. Stein, I. S. Sacks, L. Ocala, and M. Ellis (1998), Space geodetic observations of Nazca-South America convergence along the central Andes, *Science, 279*, 358–362.

O'Connor, J. M., and R. A. Duncan (1990), Evolution of the Walvis Ridge-Rio Grande Rise hot spot system: Implications for African and South American plate motions over plumes, *J. Geophys. Res., 95*, 17,475–17,502.

Pacheco, J. F., L. R. Sykes, and C. H. Scholz (1993), Nature of seismic coupling along simple plate boundaries of the subduction type, *J. Geophys. Res., 98*, 14,133–14,159.

Pardo-Casas, F., and P. Molnar (1987), Relative motion of the Nazca (Farallon) and South American plates since late Cretaceous time, *Tectonics, 6*, 233–248.

Pilger, R. H., Plate reconstructions, aseismic ridges, and low-angle subduction beneath the Andes (1981), *GSA Bulletin, 92*, 448–456.

Pope, D. C., and S. D. Willett (1998), Thermal-mechanical model for crustal thickening in the central Andes driven by ablative subduction, *Geology, 26*(6), 511–514.

Ramos, V. A. (1988), The tectonics of the central Andes; 30° to 33°S latitude, in *Processes in Continental Lithospheric Deformation*, edited by J. Suppe, *Spec. Pap. Geol. Soc. Am., 218*, 31–54.

Ramos, V. A., E. O. Cristallini, and D. J. Pérez (2002), The Pampean flat-slab of the central Andes, *J. S. Am. Earth Sci., 15*, 59–78.

Randall, D. E. (1998), A new Jurassic-Recent apparent polar wander path for South America and a review of central Andean tectonic models, *Tectonophysics, 299*, 49–74.

Riller, U., and O. Oncken (2003), Growth of the Central Andean plateau by tectonic segmentation is controlled by the gradient in crustal shortening, *J. Geol., 11*, 367–384.

Rochat, P., G. Hérail, P. Baby, and G. Mascle (1999), Bilan crustal et contrôle de la dynamique érosive et sédimentaire sur les mécanismes de formation de l'Altiplano, *C. R. Acad. Sci. Paris, 328*, 189–195.

Roeder, D. (1988), Andean-age structure of Eastern Cordillera (Province of La Paz, Bolivia), *Tectonics, 7*, 23–39.

Russo, R. M., and P. G. Silver (1996), Cordillera formation, mantle dynamics, and the Wilson cycle, *Geology, 24*(6), 511–514.

Rutland, R. W. R. (1971), Andean orogeny and ocean floor spreading, *Nature, 233*, 252–255.

Salfity, J. A. (1985), Lineamientos transversales al rumbo Andino en el noroeste Argentino, in *4 Congreso Geologico Chileno*, edited by S. Espinoza, pp. 2119–2137, Univ. del Norte Chile, Antofagasta, Chile.

Scheuber, E., and K.-J. Reutter (1992), Magmatic arc tectonics in the central Andes between 21° and 25°S, *Tectonophysics, 205*, 127–140.

Schmitz, M. (1994), A balanced model of the southern central Andes, *Tectonics, 13*, 484–492.

Scholz, C. H., and J. Campos (1995), On the mechanism of seismic decoupling and back arc spreading at subduction zones, *J. Geophys. Res., 100*, 22,103–22,115.

Sébrier, M., and P. Soler (1991), Tectonics and magmatism in the Peruvian Andes from late Oligocene time to the Present, in *Andean Magmatism and Its Tectonic Setting*, edited by C. W. Rapela, pp. 259–278, Geol. Soc. of Am., Boulder, Colo..

Sempere, T., G. Herail, J. Oller, and M. Bonhomme (1990), Late Oligocene - early Miocene major tectonic crisis and related basins in Bolivia, *Geology, 18*(10), 946–949.

Sheffels, B. (1995), Is the bend in the Bolivian Andes an Orocline?, in *Petroleum Basins of South America*, edited by H. J. Welsink, *AAPG Mem., 62*, 511–522.

Sheffels, B. M. (1990), Lower bound on the amount of crustal shortening in the central Bolivian Andes, *Geology, 18*(9), 812–815.

Sillitoe, R. H., and E. H. McKee (1996), Age of supergene oxidation and enrichment in the Chilean porphyry copper province, in *A Group of Papers Devoted to the Metallogeny of Black Shales*, edited by J. Pasava, pp. 164–179, *Czech Geol. Surv., Prague, Czech Republic.*

Silver, P. G., R. M. Russo, and C. Lithgow-Bertelloni (1998), Coupling of South American and African plate motion and plate deformation, *Science, 279*, 60–63.

Somoza, R. (1998), Updated Nazca (Farallon)-South America relative motions during the last 40 My: implications for mountain building in the central Andean region, *J. S. Am. Earth Sci., 11*, 211–215.

Springer, M. (1999), Interpretation of heat-flow density in the central Andes, in *Thermal Regimes in the Continental and Oceanic Lithosphere*, edited by L. Rybach, pp. 377–395, Elsevier, New York.

Springer, M., and A. Förster (1998), Heat-flow density across the central Andean subduction zone, *Tectonophysics, 291*, 123–139.

Tebbens, S. F., and S. C. Cande (1997), Southeast Pacific tectonic evolution from early Oligocene to present, *J. Geophys. Res., 102*, 12,061–12,084.

Uliana, M. A., M. E. Arteaga, L. Legarreta, J. J. Cerdán, and G. O. Peroni (1995), Inversion structures and hydrocarbon occurrence in Argentina, in *Basin Inversion*, edited by J. G. Buchanan and P. G. Buchanan, *Geol. Soc. Spec. Publ.*, *88*, 211–233.
Uyeda, S., and H. Kanamori (1979), Back-arc opening and the mode of subduction, *J. Geophys. Res.*, *84*, 1049–1061.
Vietor, T. (2001), Numerical modeling of plateau growth including some similarities to the central Andes, paper presented at *2001 MARGINS Meeting*, Kiel, Germany.
von Huene, R., and C. Ranero (2003), Subduction erosion and basal friction along the sediment-starved convergent margin off Antofagasta, Chile, *J. Geophys. Res.*, *108*, 2079, doi:10.1029/2001JB001569.
von Huene, R., W. Weinrebe, and F. Heeren (1999), Subduction erosion along the north Chile margin, *J. Geodyn.*, *27*(3), 345–358.
Wang, K., and K. Suyehiro (1999), How does plate coupling affect crustal stresses in Northeast and Southwest Japan?, *Geophys. Res. Lett.*, *26*, 2307–2310.
Watts, A. B., S. H. Lamb, J. D. Fairhead, and J. F. Dewey (1995), Lithospheric flexure and bending of the central Andes, *Earth Planet. Sci. Lett.*, *134*, 9–21.
Wdowinski, S., and Y. Bock (1994a), The evolution of deformation and topography of high elevated plateaus: 1. Model, numerical analysis and general results, *J. Geophys. Res.*, *99*, 7103–7119.
Wdowinski, S., and Y. Bock (1994b), The evolution of deformation and topography of high elevated plateaus: 2. Application to the central Andes, *J. Geophys. Res.*, *99*, 7121–7130.
Welsink, H. J., A. Franco M., and C. Oviedo (1995), Andean and pre-Andean deformation, Boomerang Hills area, Bolivia, in *Petroleum Basins of South America*, edited by H. J. Welsink, *AAPG Mem.*, *62*, 481–499.
Whitman, D., B. L. Isacks, J.-L. Chatelain, J.-M. Chiu, and A. Perez (1992), Attenuation of high-frequency seismic waves beneath the Central Andean Plateau, *J. Geophys. Res.*, *97*, 19,929–19,947.
Whitman, D., B. L. Isacks, and S. M. Kay (1996), Lithospheric structure and along-strike segmentation of the central Andean Plateau: Topography, tectonics, and timing, *Tectonophysics*, *259*, 29–40.
Wigger, P. J., et al. (1994), Variation in the crustal structure of the southern central Andes deduced from seismic refraction investigations., in *Tectonics of the Southern Central Andes*, edited by P. J. Wigger, pp. 23–48, Springer, New York.
Williams, C. A., C. Conners, F. A. Dahlen, E. J. Price, and J. Suppe (1994), Effect of the brittle-ductile transition on the topography of compressive mountain belts on Earth and Venus, *J. Geophys. Res.*, *99*, 19,947–19,974.
Wilson, D. S. (1993), Confirmation of the astronomical calibration of the magnetic polarity timescale from sea-floor spreading rates, *Nature*, *364*, 788–790.
Yáñez, G., and J. Cembrano (2004), Role of viscous plate coupling in the late Tertiary Andean tectonics, *J. Geophys. Res.*, *109*, B02407, doi:10.1029/2003JB002494.
Yang, Y., M. Liu, and S. Stein (2003), A 3-D geodynamic model of lateral crustal flow during Andean mountain building, *Geophys. Res. Lett.*, *30*(21), 2093, doi:10.1029/2003GL018308.
Yuan, X., et al. (2000), Subduction and collision processes in the central Andes constrained by converted seismic phases, *Nature*, *408*, 958–961.
Zandt, G., A. A. Velasco, and S. L. Beck (1994), Composition and thickness of the southern Altiplano crust, Bolivia, *Geology*, *22*(11), 1003–1006.
Zoback, M. L. (1992), First- and second-order patterns of stress in the lithosphere; the World Stress Map Project, *J. Geophys. Res.*, *97*, 11,703–11,728.

Contributors

Imtiaz Ahmed
University of Texas Institute for Geophysics
Austin, TX 78759

Maria V. S. Ask
Division of Mining and Geotechnical Engineering
Luleå University of Technology
SE-971 87 Luleå
Sweden

N. M. Beeler
U. S. Geological Survey
Menlo Park, CA
and
Brown University
Providence, RI

Barbara A. Bekins
U.S. Geological Survey
Menlo Park, CA

Arnim Berhorst
IFM-GEOMAR
Kiel
Germany

Susan L. Bilek
Department of Earth and Environmental Science
New Mexico Institute of Mining and Technology
Socorro, NM 87801

Heather R. DeShon
Earth Sciences and IGPP
University of California Santa Cruz
Santa Cruz, CA 95064
Now at
Department of Geology and Geophysics
University of Wisconsin–Madison
1215 West Dayton St.
Madison, WI 53706

C. Diamente
School of Earth and Ocean Sciences
University of Victoria
Victoria, BC
Canada

Timothy H. Dixon
Rosenstiel School of Marine and Atmospheric Science
University of Miami
4600 Rickenbacker Causeway
Miami, FL 33149

A. T. Fisher
Department of Earth Sciences
University of California Santa Cruz
Santa Cruz, CA 95064

Ernst R. Flueh
IFM-GEOMAR
Kiel
Germany

R. Harris
Department of Geology and Geophysics
University of Utah
Salt Lake City, UT 84112

Akira Hasegawa
Research Center for Prediction of Earthquakes and Volcanic Eruptions
Graduate School of Science
Tohoku University
Aramaki, Aoba-ku, Sendai, Miyagi 980-8578
Japan

Kosuke Heki
Division of Earth Rotation
National Astronomical Observatory,
2-12 Hoshigaoka, Mizusawa, Iwate 023-0861
Japan

Now at
Division of Earth and Planetary Sciences
Hokkaido University
Kita-ku, Sapporo-city, Hokkaido 060-0810
Japan

M. Hutnak
Department of Earth Sciences
University of California Santa Cruz
Santa Cruz, CA 95064

Roy D. Hyndman
Pacific Geoscience Centre
Geological Survey of Canada & SEOS
University of Victoria
9860 W. Saanich Road
Sidney, BC V8L4B2
Canada

Toshihiro Igarashi
Earthquake Research Institute
University of Tokyo
Yayoi 1-1-1, Bunkyo-ku
Tokyo 113-003
Japan

Robyn K. Kelly
Woods Hole Oceanographic Institution
Woods Hole, MA 02543

Jonas Kley
Institut für Geowissenschaften
Universität Jena
Burgweg 11
D-07749 Jena
Germany

Thorne Lay
Institute of Geophysics and Planetary Physics
Earth Sciences Department
University of California Santa Cruz
Santa Cruz, CA 95064

David A. Lockner
U. S. Geological Survey
Menlo Park, CA 94025

R. MacKnight
Department of Earth Sciences
University of California Santa Cruz
Santa Cruz, CA 95064

Chris Marone
Department of Geosciences
Pennsylvania State University
University Park, PA 16802

Toru Matsuzawa
Earthquake Research Institute
University of Tokyo
Yayoi 1-1-1, Bunkyo-ku
Tokyo 113-0032
Japan

Kirk D. McIntosh
Institute for Geophysics
Jackson School of Geosciences
University of Texas at Austin
4412 Spicewood Springs Rd., Bldg. 600
Austin, TX 78759-8500

Francesca Meneghini
Dipartimento di Sienze della Terra
University of Pisa
Pisa
Italy

Satoshi Miura
Research Center for Prediction of Earthquakes and Volcanic Eruptions
Graduate School of Science
Tohoku University,
Aramaki, Aoba-ku, Sendai, Miyagi 980-8578
Japan

Diane E. Moore
U. S. Geological Survey
Menlo Park, CA 94025

J. Casey Moore
Department of Earth Sciences
University of California Santa Cruz
Santa Cruz, CA 95064

Julia K. Morgan
Department of Earth Science
Rice University
Houston, TX 77005

Tomomi Okada
Research Center for Prediction of Earthquakes and Volcanic Eruptions
Graduate School of Science
Tohoku University
Aramaki, Aoba-ku, Sendai, Miyagi 980-8578
Japan

M. Pfender
FB Geowissenschaften
Universität Bremen
Klagenfurter Strasse
28359 Bremen
Germany

P. Costa Pisani
Department of Earth Sciences
University of California Santa Cruz
Santa Cruz, CA 95064
and
Geology and Geophysics Department
University of Hawaii
Honolulu, HI 96822

E. Blanche Ramsey, Geologist
Southwestern Energy
2350 N. Sam Houston Parkway East
Houston, TX 77032

né
E. Blanche Sunderland
Department of Earth Science
Rice University
Houston, TX 77005

Cesar R. Ranero
ICREA at Instituto de Ciencias del Mar, CSIC
Pg. Marítim de la Barceloneta 37-49
08003 Barcelona
Spain

Christie Rowe
Department of Earth Sciences
University of California Santa Cruz
Santa Cruz, California 95064

Demian M. Saffer
Department of Geosciences
Pennsylvania State University
University Park, PA 16802

Susan Y. Schwartz
Earth Sciences and IGPP
University of California Santa Cruz
Santa Cruz, CA 95064

Elizabeth J. Screaton
Geological Sciences Department
University of Florida
Gainesville, FL 32611

Tetsuzo Seno
Earthquake Research Institute
University of Tokyo
Yayoi 1-1-1, Bunkyo-ku
Tokyo 113-003
Japan

Eli A. Silver
Department of Earth Sciences
University of California Santa Cruz
Santa Cruz, CA 95064

G. Spinelli
Department of Earth and Environmental Science
New Mexico Tech
Socorro, NM 87801

C. A. Stein
Department of Earth & Environmental Sciences
University of Illinois at Chicago
Chicago, IL 60607

Yoko Suwa
Research Center for Prediction of Earthquakes and Volcanic Eruptions
Graduate School of Science
Tohoku University
Aramaki, Aoba-ku, Sendai, Miyagi 980-8578
Japan

Naoki Uchida
Research Center for Prediction of Earthquakes and Volcanic Eruptions
Graduate School of Science
Tohoku University
Aramaki, Aoba-ku, Sendai, Miyagi 980-8578
Japan

Michael B. Underwood
Department of Geological Sciences
University of Missouri
Columbia, MO 65203

Tim Vietor
GeoForshungZentrum
Potsdam
Germany

H. Villinger
FB Geowissenschaften
Universität Bremen
Klagenfurter Strasse
28359 Bremen
Germany

Kelin Wang
Pacific Geoscience Centre
Geological Survey of Canada
9860 West Saanich Road
Sidney, BC
Canada

and
School of Earth and Ocean Sciences
University of Victoria
Victoria, BC
Canada

Index

Page numbers in *italics* indicate that the information will be found in illustrations or tables.